DYNAMICS OF POLYMERIC LIQUIDS

Volume 1 Fluid Mechanics

Volume 2 Kinetic Theory

Board of Advisors, Engineering

DYNAMICS OF POLYMERIC LIQUIDS

VOLUME 1
FLUID MECHANICS

R. BYRON BIRD
Chemical Engineering Department
and Rheology Research Center
University of Wisconsin
Madison, Wisconsin

ROBERT C. ARMSTRONG
Chemical Engineering Department
Massachusetts Institute of Technology
Cambridge, Massachusetts

OLE HASSAGER
Instituttet for Kemiteknik
Danmarks tekniske Højskole
Lyngby, Danmark

John Wiley & Sons
New York Santa Barbara London Sydney Toronto

Library of Congress Cataloging in Publication Data

Bird, Robert Byron, 1924–
Dynamics of polymeric liquids.

Includes bibliographical references and indexes.
CONTENTS: v. 1. Fluid mechanics.
1. Fluid dynamics. 2. Polymers and polymerization.
I. Armstrong, Robert Calvin, 1948– joint author.
II. Hassager, Ole, joint author. III. Title.
TA357.B57 620.1'06 76-15408
ISBN 0-471-07375-X

Printed in the United States of America

10 9 8 7 6 5 4 3 2 1

PREFACE

Liquid motion has fascinated many generations of scientists and engineers. Although untold man-years of research have been devoted to the study of fluids of low molecular weight, which are well described by the Navier-Stokes equations, there still remain many challenging problems in both theory and applications. But even more challenging are polymeric liquids whose motions cannot be described at all by the Navier-Stokes equations. In Volume 1 ("Fluid Mechanics") we summarize the key experiments that show how polymeric fluids differ from structurally simple fluids, and we then present, roughly in historical order, various methods for solving polymer fluid dynamics problems using continuum mechanics; in Volume 2 ("Kinetic Theory") we utilize molecular models and the methods of statistical mechanics to try to elucidate bulk flow behavior in terms of polymer structure. Table I shows how we have chosen to organize the material of the two volumes.

Understanding polymer fluid dynamics is important in connection with plastics manufacture, performance of lubricants, application of paints, processing of foodstuffs, and movement of biological fluids. Rapid advances in all these areas, as well as in the associated fundamental sciences, have made it desirable to have an introductory textbook for students and research workers. Examples, questions, and problems have been included to make the book teachable, and almost all the material in both volumes has been classroom tested. These volumes should, in addition, be useful as an introduction to the many advanced monographs and to the research literature on continuum physics, rheometry, molecular theory, and polymer processing. Literature citations have been supplied throughout the text to assist the reader who wishes to pursue some topics in greater depth. Although these volumes have been prepared with fourth or fifth year chemical engineering students in mind, they may also be of interest to students or research personnel in physical chemistry, molecular biology, polymer engineering and science, mechanical engineering, materials science, and applied mechanics.

Both experiment and theory in this field have undergone rapid development in recent years. Except for a few papers of historical interest, all of the literature citations are from the period after 1950, and many are from the 1960s and 1970s. The present volumes could not have been prepared ten years ago: the key experimental facts, the continuum theories, and the molecular theories needed for organizing our presentation are to a large extent products of the decade just past. Unfortunately the following subjects have had to be omitted from the text: optical and electrical phenomena, stability analysis, computer simulation, drag reduction, boundary-layer theory, two-phase systems, thermodynamics, and liquid crystals. Research is, of course, going on in all these areas, but we have considered them outside the scope of an elementary textbook, either for reasons of difficulty or for lack of adequate understanding.

In these two volumes there is enough material for two one-semester courses: either an elementary survey course (Chapters 1 to 6 and 10 to 11) followed by an advanced topics course (Chapters 7 to 9, 12 to 15); or a fluid dynamics course (Chapters 1 to 9) followed by a kinetic theory course (Chapters 10 to 15). Needless to say, problem solving is an important activity in mastering the material, and for this purpose many illustrative examples have been included in the text. Other aids to study are the "Questions for Discussion," which

TABLE I. ORGANIZATION OF THE SUBJECT MATERIAL OF *DYNAMICS OF POLYMERIC LIQUIDS*

Volume		Chapter
VOLUME 1	I. *Background Information*	
	A. Newtonian Liquids	
	Review of Newtonian Fluid Dynamics	1
	B. Polymeric Liquids	
	The Structure of Polymeric Fluids	2
	Flow Phenomena of Polymeric Fluids	3
	Material Functions for Polymeric Fluids	4
	II. *Fluid Mechanics*	
	A. Two Elementary Models for Special Flows	
	Generalized Newtonian Models (Steady Shear Flows)	5
	Linear Viscoelastic Models (Small Deformation Flows)	6
	B. Models for General Flows	
	Quasilinear Corotational Models	7
	Nonlinear Corotational Models	8
	Codeformational Models	9
	Appendix A (Vector and Tensor Notation)	
	Appendix B (Curvilinear Coordinates)	
VOLUME 2	III. *Kinetic Theory*	
	A. Two Elementary Models for Dilute Solutions	
	Elastic Dumbbell Models for Flexible Polymers	10
	Rigid Dumbbell Model for Rodlike Polymers	11
	B. More Realistic Models for Dilute Solutions	
	Chainlike Models for Flexible Polymers	12
	General Bead-Rod-Spring Models for Polymers	13
	C. General Theory for Dilute and Concentrated Solutions	
	Phase-Space Kinetic Theory of Polymer Solutions	14
	D. Concentrated Solutions and Melts	
	Network Theories	15
	Appendix C (Continuum Mechanics)	

are intended to be useful for review and occasionally provocative, and the "Problems" at the end of the chapter, which are divided into four classes:

CLASS A. Problems involving elementary numerical calculations using equations from the chapter.

CLASS B. Problems involving elementary analyses similar to those in the text.

CLASS C. Problems requiring theoretical developments more difficult or lengthy than those encountered in Class B problems.

CLASS D. Problems requiring advanced mathematical topics or primarily emphasizing the mathematical aspect of a problem.

Material that can be omitted on first reading or for an elementary course is indicated by an asterisk in the contents. No effort has been spared to provide generous cross-referencing within the book, and several indexes and appendices are included to assist the reader in finding material quickly. Everywhere in the text and problems we have used SI units: we have somewhat reluctantly abandoned the "poise" for the SI-recommended N·s/m^2 or Pa·s (1 poise = 10^{-1} N·s/m^2) and the "Ångström unit" for the SI-recommended nm (1 Å = 10^{-1} nm).

Volumes 1 and 2 encompass material from fluid dynamics, rheology, continuum physics, and statistical mechanics, and it is impossible to settle on notation that will please everyone. Only recently the Society of Rheology published a list of recommended symbols, and we have followed their recommendations, using $\eta^* = \eta' - i\eta''$ for complex viscosity, η for the shear-rate-dependent viscosity, η_0 for zero-shear-rate viscosity, Ψ_1 and Ψ_2 for the normal-stress coefficients, and $\dot{\gamma}$ for shear rate. Like many workers in kinetic theory and transport phenomena we write the stress tensor as $\boldsymbol{\pi} = p\boldsymbol{\delta} + \boldsymbol{\tau}$, where p is the pressure, $\boldsymbol{\delta}$ the unit tensor, and $\boldsymbol{\tau}$ that part of the stress tensor which vanishes at equilibrium; that is, both p and τ_{ii} are positive in compression, and furthermore $\boldsymbol{\tau}$ has the same sign convention that is used for heat and mass fluxes. Usually continuum mechanics and rheology authors use a sign convention opposite to ours for $\boldsymbol{\pi}$ and $\boldsymbol{\tau}$, but the reader, properly forewarned, should find this at most only a minor annoyance. Bear in mind that more serious misunderstandings can result because of the differences in the research literature in the definitions of tensor operations such as $[\nabla \cdot \boldsymbol{\tau}]$, a factor of two in defining the rate-of-strain tensor $\dot{\boldsymbol{\gamma}} = \nabla \boldsymbol{v} + (\nabla \boldsymbol{v})^\dagger$, and factors of -1, 2, or -2 in defining the vorticity tensor $\boldsymbol{\omega} = \nabla \boldsymbol{v} - (\nabla \boldsymbol{v})^\dagger$. In these volumes we have attempted to use notation that has wide currency, and every effort has been made to insure that the notation is consistent throughout the entire fifteen chapters. Lists of notation are appended at the end of each volume; in addition, the appendices dealing with vectors, tensors, and continuum mechanics should be helpful in interpreting notation.

In preparing this book we have clearly drawn extensively on the research literature of the past fifty years; we acknowledge with deep appreciation our many colleagues who have been developing the frontiers and exploring hitherto uncharted terrain. Several students have been particularly helpful in reading portions of the manuscript and offering constructive criticisms: Michael J. Riddle, Alberto Co, Roger J. Grimm, Robert K. Prud'homme, Moshe Gottlieb, Carlos Narvaez, James W. Miller, Alan L. Soli, Tan Hung Hou, Li-Yar Lee Kao, Robin W. Rosser, Jacques Noordermeer, Timothy M. Lohman, Shingo Ishikawa, and Cary Wasserman.

Acknowledgment of our colleagues in the Rheology Research Center is a particular pleasure: Professors J. D. Ferry, M. W. Johnson, Jr., A. S. Lodge, and J. L. Schrag. Every chapter has been influenced in some way by the work of their research groups. Very special thanks go to Professor A. S. Lodge, Professor J. D. Goddard (University of Michigan), Professor M. C. Williams (University of California), and Dr. S. Bhumiratana (University of Wisconsin) for detailed critical comments; Dr. Bhumiratana checked the solutions to many of the problems in Volume 1 and John Torkelson assisted with the preparations of the author indexes. Undoubtedly this book would not have been possible without the financial support of the National Science Foundation and the Petroleum Research Fund of the American Chemical Society for our program on polymer fluid dynamics and kinetic theory; we are most appreciative of this generous support. Manuscript preparation was made easier by financial assistance from the Vilas Trust Fund at the University of Wisconsin (to R.B.B.), from the DuPont Young Faculty Grant at the Massachusetts Institute of Technology (to R.C.A.), and from Danmarks tekniske Højskole (The Technical University of Denmark) (to O.H.).

R. Byron Bird
Robert C. Armstrong
Ole Hassager
Charles F. Curtiss

July 4, 1976

Comments on Volume 1

Undoubtedly this book will be used by people trained in a variety of fields; therefore we have included several introductory chapters. Some readers will find Chapter 1 necessary for familiarizing themselves with classical fluid dynamics; others may find Chapter 2 useful for obtaining a minimal background in polymer synthesis, structure, and kinetic theory. Almost all readers will find the next two chapters on the experimental facts of polymer fluid mechanics essential for providing qualitative ideas about flow phenomena as well as quantitative estimates of the rheological "material functions." By the time one has seen the photographs of the funny flow behavior in Chapter 3 and the extensive data tabulations of Chapter 4, it should be evident that the task of describing the flow of polymeric liquids is an immense challenge.

In Chapters 5 and 6 we summarize two older methods of analyzing limited classes of flows; the theories discussed in these chapters are not particularly difficult mathematically and, despite their limitations, they have proven very useful. Certainly for engineers the most important property of polymeric liquids is their shear-rate-dependent viscosity, and the "generalized Newtonian fluid" models in Chapter 5 have been formulated to incorporate this salient feature. Even if the models discussed in this chapter are not sophisticated, they have proven to give good results for steady-state shear flow and heat-transfer calculations. Next, in Chapter 6, we discuss the "linear viscoelastic models," which can describe time-dependent phenomena in flows involving only very small deformations. These models have proven to be invaluable to polymer chemists in elucidating polymer structure by studying the mechanical response in a wide variety of carefully controlled flow situations; moreover it is advantageous to master linear viscoelasticity prior to embarking on the study of nonlinear viscoelastic flow problems in the remainder of this volume.

Even though the models of Chapters 5 and 6 are unquestionably useful, they leave much to be desired; one would like to have rheological models (or "rheological equations of state") of greater generality that can be used to solve not only the problems of Chapters 5 and 6, but also wider classes of flow problems. Nearly all the flow phenomena of Chapters 3 and 4 can be described semiquantitatively by using the "quasilinear corotational models" of Chapter 7; these are just the linear viscoelastic models of Chapter 6, but reformulated in a reference frame that follows a fluid particle and rotates with it. None of the currently available continuum mechanics or rheology textbooks have used the corotational models for introducing the study of nonlinear viscoelasticity and, hence, our treatment represents a distinct departure from the standard mode of presentation; there are pedagogical advantages to this treatment and somewhat less mathematical background is required. In Chapter 8 we proceed to "nonlinear corotational models," including several specific models (e.g., the Oldroyd models) and several general expansions (the memory-integral expansion and the retarded-motion expansion); the models of this chapter, together with the equations of motion and continuity, provide the tools for solving a wide variety of flow problems, although extensive numerical analysis may be required for their application. Although we choose to stress the corotational formalism in this book, most research papers and other books use a codeformational formalism; hence in Chapter 9 we present this alternative viewpoint and its relation to the methods of Chapters 7 and 8. Learning polymer fluid dynamics from both viewpoints leads to an enhanced appreciation of a very difficult subject, and researchers in the field ought to be familiar with both.

R. B. Bird
R. C. Armstrong
O. Hassager

CONTENTS

VOLUME 1

* An asterisk designates material that can be omitted on first reading or for an elementary course.

CHAPTER 1

REVIEW OF NEWTONIAN FLUID DYNAMICS

In this introductory chapter we begin by giving the equations that describe the flow of fluids. These are the "equations of change" which indicate how the mass, momentum, and energy of the fluid change with position and time. In the first section we present these equations in terms of the fluxes so that they are valid for any kind of fluid. In the second section we specialize the equations for the "Newtonian fluid," so that we can review briefly some of the results of classical hydrodynamics to which we will want to refer in subsequent chapters. It is presumed that the content of the first two sections is familiar to the reader; this material is included just as a review and to have several standard results in a form useful for later reference. Following this review we outline for the reader what the motivation is for a separate textbook on the fluid dynamics of macromolecular liquids.

At the end of this chapter there is a rather long set of problems. Many of these problems involve the solution of the hydrodynamic equations for Newtonian fluids. Almost all of these flow problems are encountered again in later chapters, where they are solved for polymeric fluids using various "non-Newtonian" models. Then it will be very convenient to have the Newtonian fluid solutions available.

§1.1 THE EQUATIONS OF FLUID DYNAMICS

The motion of any fluid is described by the equations of conservation of mass, momentum, and energy. These equations are shown in two forms in Table 1.1–1. Since these equations form the basis for all of the subject material of the book, it is reasonable to discuss their derivation briefly. A more elementary derivation can be found elsewhere.[1]

In the derivation we consider an arbitrary *fixed* region in space of volume V and surface S, as shown in Fig. 1.1–1; sometimes such a mathematical region is referred to as a "control volume." On every surface element dS there is an outwardly directed normal unit vector $\boldsymbol{n}$. We imagine that this fixed region is in the midst of a fluid flow field and that the fluid moves across the boundaries of the region. We now want to apply the laws of conservation of mass, momentum, and energy to the fluid contained within this fixed region. It is presumed that the reader is familiar with the material on vector and tensor notation in §§A.1–A.7.

a. Conservation of Mass

Suppose that at the infinitesimal surface element dS the fluid is crossing the surface of V with a velocity $\boldsymbol{v}$. Then the local volume rate of flow of a fluid across dS is

[1] See, for example, R. B. Bird, W. E. Stewart, and E. N. Lightfoot, *Transport Phenomena*, Wiley, New York (1960), Chapters 3 and 10.

TABLE 1.1–1

The Equations of Change in Terms of the Fluxes

	In terms of $\partial/\partial t$	
Mass:	$\frac{\partial}{\partial t}\rho = -(\nabla \cdot \rho \boldsymbol{v})$	(A)
Momentum:	$\frac{\partial}{\partial t}\rho \boldsymbol{v} = -[\nabla \cdot \rho \boldsymbol{v}\boldsymbol{v}] - [\nabla \cdot \boldsymbol{\pi}] + \rho \boldsymbol{g}$	(B)
Energy:	$\frac{\partial}{\partial t}\rho \hat{U} = -(\nabla \cdot \rho \hat{U}\boldsymbol{v}) - (\nabla \cdot \boldsymbol{q}) - (\boldsymbol{\pi} : \nabla \boldsymbol{v})$	(C)
	In terms of $D/Dt = \partial/\partial t + (\boldsymbol{v} \cdot \nabla)$	
Mass:	$\frac{D\rho}{Dt} = -\rho(\nabla \cdot \boldsymbol{v})$	(D)
Momentum:	$\rho \frac{D\boldsymbol{v}}{Dt} = -[\nabla \cdot \boldsymbol{\pi}] + \rho \boldsymbol{g}$	(E)
Energy:	$\rho \frac{D\hat{U}}{Dt} = -(\nabla \cdot \boldsymbol{q}) - (\boldsymbol{\pi} : \nabla \boldsymbol{v})$	(F)

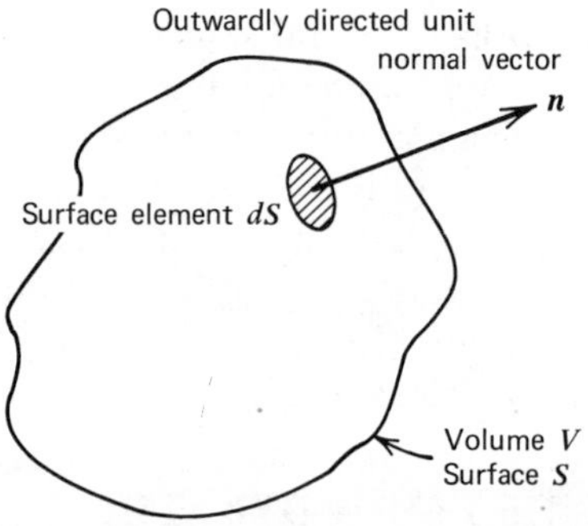

FIGURE 1.1–1. Arbitrary "control volume," fixed in space, over which mass, momentum, and energy balances are made.

$(\boldsymbol{n} \cdot \boldsymbol{v})\, dS$. If the flow is outward, then $(\boldsymbol{n} \cdot \boldsymbol{v})\, dS$ is positive, whereas if the flow is inward $(\boldsymbol{n} \cdot \boldsymbol{v})\, dS$ is negative.[2] The local mass rate of flow is then $(\boldsymbol{n} \cdot \rho\boldsymbol{v})\, dS$. Note that $\rho\boldsymbol{v}$ is the mass flux (i.e., mass per unit area per unit time).

According to the law of conservation of mass, the total mass of fluid within V will increase only because of a net influx of fluid across the bounding surface S. Mathematically this is stated as:

$$\frac{d}{dt}\int_V \rho\, dV = -\int_S (\boldsymbol{n} \cdot \rho\boldsymbol{v})\, dS \tag{1.1–1}$$

Rate of increase of mass of fluid within V	Rate of addition of mass across the surface S

When Gauss's divergence theorem (§A.5) is used, the surface integral can be transformed

[2] Recall that $(\boldsymbol{n} \cdot \boldsymbol{v})$ can be interpreted as the component of $\boldsymbol{v}$ in the direction of $\boldsymbol{n}$.

into a volume integral:

$$\frac{d}{dt}\int_V \rho \, dV = -\int_V (\nabla \cdot \rho \boldsymbol{v}) \, dV \tag{1.1–2}$$

The equation may be rearranged by bringing the time derivative inside the integral. This is permissible since the volume V is fixed. This gives:

$$\int_V \left[\frac{\partial \rho}{\partial t} + (\nabla \cdot \rho \boldsymbol{v})\right] dV = 0 \tag{1.1–3}$$

We now have an integral over an arbitrary volume and this integral is equated to zero. Because of the arbitrariness of the volume, the integrand may now be set equal to zero. This gives:

$$\boxed{\frac{\partial \rho}{\partial t} = -(\nabla \cdot \rho \boldsymbol{v})} \tag{1.1–4}$$

which is usually called the *equation of continuity*. If the fluid has constant density, then Eq. 1.1–4 simplifies to:

$$(\nabla \cdot \boldsymbol{v}) = 0 \tag{1.1–5}$$

Use is often made of this relation for "incompressible fluids."

b. Conservation of Momentum

As pointed out above, the local volume rate of flow of a fluid across the surface element dS is $(\boldsymbol{n} \cdot \boldsymbol{v})\, dS$. If this is multiplied by the momentum per unit volume of the fluid, we then get $(\boldsymbol{n} \cdot \boldsymbol{v})\rho \boldsymbol{v}\, dS$; this is the rate at which momentum is carried across the element of surface dS just because the fluid itself flows across dS. This latter expression can be rearranged as $[\boldsymbol{n} \cdot \rho \boldsymbol{v}\boldsymbol{v}]\, dS$, and the quantity $\rho \boldsymbol{v}\boldsymbol{v}$ is the momentum flux (i.e., momentum per unit area per unit time) associated with the bulk flow of fluid. Sometimes this transport associated with bulk flow is referred to as "convective transport."

Note that there is a parallelism between the first paragraph of this subsection and the first paragraph of the preceding subsection, but that the tensorial order of the quantities involved is different. In the preceding discussion the entity being transported is mass (a scalar), and the mass flux is a vector ($\rho \boldsymbol{v}$). Here the entity being transported is momentum (a vector), and the momentum flux is a tensor ($\rho \boldsymbol{v}\boldsymbol{v}$).

In addition to momentum transport by flow, there will also be momentum transferred by virtue of the molecular motions and interactions within the fluid. This additional momentum flux will be designated by the symbol $\boldsymbol{\pi}$, again a second-order tensor. We use the convention that the *ij*-component of this tensor π_{ij} represents the flux of positive *j*-momentum in the positive *i*-direction, associated with molecular processes. The rate of flow of momentum, resulting from molecular motions, across the element of surface dS with orientation $\boldsymbol{n}$ is then $[\boldsymbol{n} \cdot \boldsymbol{\pi}]\, dS$. It will be assumed throughout the entire book that $\boldsymbol{\pi}$ is a symmetric tensor (i.e., $\pi_{ij} = \pi_{ji}$). Thus far all of the kinetic theories for simple and macromolecular fluids yield symmetric momentum-flux tensors, and no experiments have been performed that enable one to measure any nonsymmetrical contributions.

We are now ready to write down the law of conservation of momentum. According to this law, the total momentum of the fluid within V will increase because of a net influx of momentum across the bounding surface—both by bulk flow and by molecular motions—and because of the external force of gravity acting on the fluid. When translated into mathematical terms this becomes:

$$\frac{d}{dt}\int_V \rho \boldsymbol{v}\, dV = -\int_S [\boldsymbol{n} \cdot \rho \boldsymbol{v}\boldsymbol{v}]\, dS - \int_S [\boldsymbol{n} \cdot \boldsymbol{\pi}]\, dS + \int_V \rho \boldsymbol{g}\, dV \tag{1.1-6}$$

Rate of increase of momentum of fluid within V	Rate of addition of momentum across S by bulk flow	Rate of addition of momentum across S by molecular motions and interactions	Force on fluid within V by gravity

where $\boldsymbol{g}$ is the force per unit mass due to gravity. Application of the Gauss divergence theorem enables us to rewrite the surface integrals as volume integrals:

$$\int_V \frac{\partial}{\partial t} \rho \boldsymbol{v}\, dV = -\int_V [\nabla \cdot \rho \boldsymbol{v}\boldsymbol{v}]\, dV - \int_V [\nabla \cdot \boldsymbol{\pi}]\, dV + \int_V \rho \boldsymbol{g}\, dV \tag{1.1-7}$$

Then since the volume V is arbitrary, the integral signs may be removed to obtain:

$$\boxed{\frac{\partial}{\partial t} \rho \boldsymbol{v} = -[\nabla \cdot \rho \boldsymbol{v}\boldsymbol{v}] - [\nabla \cdot \boldsymbol{\pi}] + \rho \boldsymbol{g}} \tag{1.1-8}$$

and this is the *equation of motion.*

Before continuing it is appropriate to give an alternative interpretation of the tensor $\boldsymbol{\pi}$ and its components. In the derivation of Eq. 1.1–8 we could have used a somewhat different physical statement leading up to Eq. 1.1–6: the total momentum of the fluid within V will increase because of a net influx of momentum across the bounding surfaces by bulk flow, and because of the external forces acting on the fluid—both the surface forces exerted by the surrounding fluid and the body forces exerted by gravity. The surface force term in Eq. 1.1–6 would then have the form $-\int \boldsymbol{\pi}_n\, dS$, where $\boldsymbol{\pi}_n\, dS$ is a vector describing the force exerted by the fluid on the negative side of dS on the fluid on the positive side of dS (Fig. 1.1–2).

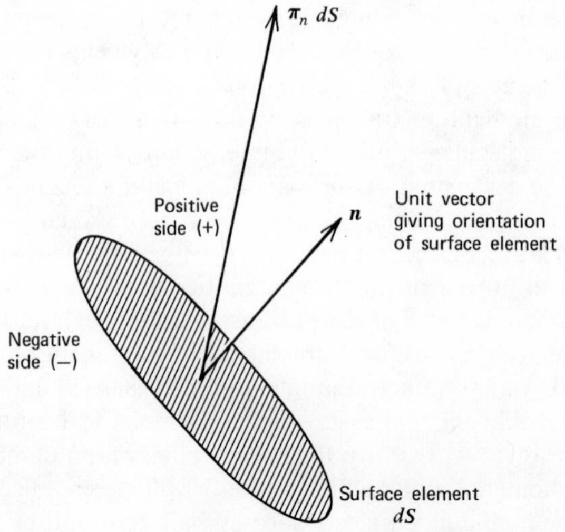

FIGURE 1.1–2. Element of surface dS across which a force $\boldsymbol{\pi}_n\, dS$ is transmitted.

Comparison of the above integral with the corresponding term in Eq. 1.1–6 shows that $\boldsymbol{\pi}_n = [\boldsymbol{n} \cdot \boldsymbol{\pi}]$. That is, the force $\boldsymbol{\pi}_n \, dS$ corresponding to any orientation $\boldsymbol{n}$ of dS can be obtained from the tensor $\boldsymbol{\pi}$. When this interpretation is used, it is more natural to refer to $\boldsymbol{\pi}$ as the "stress tensor" (the term "pressure tensor" is also used). The component π_{ij} is the stress acting in the positive j-direction on a surface perpendicular to the i-direction, the stress being exerted by the negative material on the positive material. (For a further exercise on the interpretation of $[\boldsymbol{n} \cdot \boldsymbol{\pi}]$ see Problem 1B.13.)

If one uses this viewpoint, then the integral $-\int [\boldsymbol{n} \cdot \boldsymbol{\pi}] \, dS$ in Eq. 1.1–6 can be reinterpreted as "force of the fluid outside V acting on the fluid inside V across S." For some purposes it is useful to think of $\boldsymbol{\pi}$ as a momentum flux, whereas in some situations the concept of stress is more natural. We shall feel free to use both interpretations and the terms "momentum flux tensor" and "stress tensor" will be used interchangeably.[3]

We conclude this subsection by giving two additional equations that may be obtained from Eq. 1.1–8. By forming the dot product of the local fluid velocity vector $\boldsymbol{v}$ with the entire equation of motion, we obtain an equation of change for kinetic energy:

$$\frac{\partial}{\partial t}(\tfrac{1}{2}\rho v^2) = -(\nabla \cdot \tfrac{1}{2}\rho v^2 \boldsymbol{v}) - (\boldsymbol{v} \cdot [\nabla \cdot \boldsymbol{\pi}]) + \rho(\boldsymbol{v} \cdot \boldsymbol{g}) \tag{1.1–9}$$

Furthermore, by forming the cross product of the position vector $\boldsymbol{r}$ with the equation of motion, we obtain an equation of change for angular momentum:

$$\frac{\partial}{\partial t}(\rho[\boldsymbol{r} \times \boldsymbol{v}]) = -[\nabla \cdot \rho \boldsymbol{v}[\boldsymbol{r} \times \boldsymbol{v}]] - [\nabla \cdot \{\boldsymbol{r} \times \boldsymbol{\pi}\}^{\dagger}] + [\boldsymbol{r} \times \rho \boldsymbol{g}] \tag{1.1–10}$$

in which the symbol † stands for the transpose of a tensor. These last two equations will be referred to only infrequently, but it is convenient to have them for the sake of completeness.

c. Conservation of Energy

Once again we return to the point made earlier that the local volume rate of flow of a fluid across the surface element dS is $(\boldsymbol{n} \cdot \boldsymbol{v}) \, dS$. If this is multiplied by $\frac{1}{2}\rho v^2$, which is the kinetic energy per unit volume, then we get $(\boldsymbol{n} \cdot \frac{1}{2}\rho v^2 \boldsymbol{v}) \, dS$, which is the rate of convective flow of kinetic energy across the surface. Similarly, if $\hat{U}$ is the internal energy per unit mass, then

[3] In most treatises on applied mechanics and mechanical engineering a different convention is used for the stress tensor. A stress tensor $\boldsymbol{\sigma}$ is defined by $\boldsymbol{\pi}_{-n} \, dS = -\boldsymbol{\pi}_n \, dS = [\boldsymbol{\sigma} \cdot \boldsymbol{n}] \, dS$; that is, one thinks about the fluid on the "positive side" exerting a force on the fluid on the "negative side." The tensors $\boldsymbol{\pi}$ and $\boldsymbol{\sigma}$ are related by $\boldsymbol{\pi} = -\boldsymbol{\sigma}^{\dagger}$, where † stands for "transpose"; thus they differ in sign and in the order of the indices. Since the stress tensor is usually assumed to be symmetric, the change in the order of the indices is not particularly worrisome, but the difference in sign convention is sometimes quite confusing. We have two reasons for preferring the convention adopted here: (i) In one-dimensional heat conduction described by Fourier's law $q_y = -k(dT/dy)$, it is customary to define q_y so that it is positive when heat is moving in the $+y$-direction, that is, when the temperature decreases with increasing y. Similarly in one-dimensional diffusion, described by Fick's law, $j_{Ay} = -\mathscr{D}_{AB} \, dc_A/dy$, the mass flux is defined as positive when species A is moving in the $+y$-direction, in the direction of decreasing concentration. Therefore in a shear flow described by Newton's law $\pi_{yx} = -\mu \, dv_x/dy$ it seems natural to define π_{yx} so that it is positive when x-momentum is moving in the $+y$-direction, that is, in the direction of decreasing velocity. Thus the linear laws for all three transport phenomena are formulated with the same sign convention. (ii) When the total stress tensor is broken down into two parts, a pressure contribution and a viscous contribution, as shown in Eq. 1.2–1, both parts have the same sign—that is, compression is positive in both terms.

$(\boldsymbol{n} \cdot \rho \hat{U} \boldsymbol{v})\, dS$ is the rate of convective flow of internal energy across the surface element.

In addition to the flow of kinetic and internal energy across dS because of the fluid flowing across dS, there may be energy transferred by molecular motions. This additional mode of transport is associated with heat conduction, and we designate this heat flux by the symbol $\boldsymbol{q}$. Then the heat flow across dS by heat conduction will be $(\boldsymbol{n} \cdot \boldsymbol{q})\, dS$.

As the fluid moves outwardly across the surface of V, it will do work on the fluid outside of V. It was mentioned earlier that $[\boldsymbol{n} \cdot \boldsymbol{\pi}]\, dS$ is the force exerted by the fluid *inside* V on the fluid *outside* V across dS. The rate of doing work[4] by the fluid inside V on the fluid outside V will be given by the scalar product of the force and the velocity. Hence the rate of doing work across dS is $([\boldsymbol{n} \cdot \boldsymbol{\pi}] \cdot \boldsymbol{v})\, dS$, which may also be written $(\boldsymbol{n} \cdot [\boldsymbol{\pi} \cdot \boldsymbol{v}])\, dS$.

We are now in a position to write down the law of conservation of energy (or the first law of thermodynamics) in mathematical terms. According to this law, the rate of increase of the sum of the kinetic and internal energies will equal the rate of energy addition (by flow and by heat conduction) minus the rate at which the fluid inside V is doing work (this includes work against the surrounding fluid through dS, and work against the external force of gravity). When translated into mathematical symbols this becomes:

$$\frac{d}{dt}\int_V (\tfrac{1}{2}\rho v^2 + \rho \hat{U})\, dV = -\int_S (\boldsymbol{n} \cdot (\tfrac{1}{2}\rho v^2 + \rho \hat{U})\boldsymbol{v})\, dS$$

Rate of increase of kinetic and internal energy within V	Rate of addition of kinetic and internal energy across S by bulk flow

$$-\int_S (\boldsymbol{n} \cdot \boldsymbol{q})\, dS - \int_S (\boldsymbol{n} \cdot [\boldsymbol{\pi} \cdot \boldsymbol{v}])\, dS + \int_V (\boldsymbol{v} \cdot \rho \boldsymbol{g})\, dV \qquad (1.1\text{–}11)$$

Rate of addition of energy across S by molecular motions	Rate at which fluid outside V does work against fluid inside V	Rate at which gravity does work on the fluid

Applying the divergence theorem and then removing the integral signs as before, we get.

$$\frac{\partial}{\partial t}(\tfrac{1}{2}\rho v^2 + \rho \hat{U}) = -(\nabla \cdot (\tfrac{1}{2}\rho v^2 + \rho \hat{U})\boldsymbol{v}) - (\nabla \cdot \boldsymbol{q}) - (\nabla \cdot [\boldsymbol{\pi} \cdot \boldsymbol{v}]) + (\rho \boldsymbol{v} \cdot \boldsymbol{g}) \qquad (1.1\text{–}12)$$

From this equation, which is an equation of change for the sum of the kinetic and internal energies, we can subtract the equation of change for kinetic energy alone, given in Eq. 1.1–9. This gives, if we make use of the assumption that $\boldsymbol{\pi}$ is symmetrical

$$\boxed{\frac{\partial}{\partial t}\rho \hat{U} = -(\nabla \cdot \rho \hat{U} \boldsymbol{v}) - (\nabla \cdot \boldsymbol{q}) - (\boldsymbol{\pi} : \nabla \boldsymbol{v})} \qquad (1.1\text{–}13)$$

which is the *internal energy equation.*

This completes the derivation of the three equations of change. In Table 1.1–1, where these equations are summarized, they are given in two forms: one in terms of $\partial/\partial t$, and the other in terms of D/Dt, which gives the time rate of change of a quantity as seen by an observer

[4] See Exercise 5 at the end of §A.1.

who is moving with the fluid, but not rotating with it. The operator D/Dt is called the "substantial derivative" or the "material derivative," since it describes time changes taking place at a particular element of the "substance" or "material". It should be emphasized that there are *no* assumptions involved in going from Eqs. A, B, and C to Eqs. D, E, and F in Table 1.1–1. (See Problem 1D.3.)

The analytical solution to most of the problems in this book will begin with one or more of the three "boxed" equations above. Usually one will want them written out in component form in one of the standard orthogonal coordinate systems. For the reader's convenience, in Appendix B we give the equation of motion in rectangular, cylindrical, and spherical coordinates; in Appendix A many ∇-operations are tabulated in the same three coordinate systems and also in bipolar coordinates.

§1.2 NEWTONIAN FLUID DYNAMICS

The equations in §1.1 are valid for any fluid. In this section we specialize these results for "Newtonian fluids" to obtain the equations of classical hydrodynamics. Then we give several examples of solutions of classical hydrodynamics problems. In doing so we select those problems that pertain to viscometry and to which we shall wish to refer in subsequent chapters. Additional examples may be found in textbooks on transport phenomena.[1]

For structurally simple fluids such as gases, gaseous mixtures, and low-molecular-weight liquids and their mixtures, it has been established experimentally that in a simple shearing motion $v_x = v_x(y)$ the flux of x-momentum in the positive y-direction is given by "Newton's law of viscosity," $\pi_{yx} = -\mu \, dv_x/dy$, where μ is the fluid *viscosity*. The appropriate generalization for arbitrary, time-dependent flows is:[2,3]

$$\begin{aligned} \boldsymbol{\pi} &= p\boldsymbol{\delta} + \boldsymbol{\tau} \\ &= p\boldsymbol{\delta} - \mu[\nabla \boldsymbol{v} + (\nabla \boldsymbol{v})^{\dagger}] + (\tfrac{2}{3}\mu - \kappa)(\nabla \cdot \boldsymbol{v})\boldsymbol{\delta} \end{aligned} \tag{1.2–1}$$

where $(\nabla \boldsymbol{v})^{\dagger}$ is the transpose of the dyadic $\nabla \boldsymbol{v}$, and $\boldsymbol{\delta}$ is the unit tensor. This expression reduces to the hydrostatic pressure when there are no velocity gradients; it contains all possible combinations of first derivatives of velocity components that are allowed if one assumes that the fluid is isotropic and that the momentum flux tensor is symmetric.[2,3] The symbol p represents the thermodynamic pressure,[2] which is related to the density ρ, temperature T, and mole fractions x_i through a "thermal equation of state," $p = p(\rho, T, x_i)$; that is, this is taken to be the same function that one uses in thermal equilibrium.

The tensor $\boldsymbol{\tau}$ is the part of the momentum flux tensor or stress tensor that is associated with the viscosity of the fluid. We shall usually refer to it simply as the "momentum flux tensor" or "stress tensor," and use the terms "total momentum flux tensor" or "total stress tensor" for $\boldsymbol{\pi}$ when a distinction seems necessary.

Note that in generalizing Newton's law of viscosity to arbitrary flows an additional transport property κ, the *dilatational viscosity*, arises. The dilatational viscosity is identically zero for ideal, monatomic gases; for incompressible liquids $(\nabla \cdot \boldsymbol{v}) = 0$, and the term containing κ vanishes. Consequently the dilatational viscosity is of no importance in two key limiting cases, and no further mention will be made of it in this book.

[1] See, for example, R. B. Bird, W. E. Stewart, and E. N. Lightfoot, *Transport Phenomena*, Wiley, New York (1960), Chapters 3, 4, 10, and 11.

[2] L. Landau and E. M. Lifshitz, *Fluid Mechanics*, Addison-Wesley, Reading, Mass. (1959), pp. 47–48, pp. 187–188.

[3] G. K. Batchelor, *An Introduction to Fluid Dynamics*, Cambridge University Press (1970), Sections 3.3 and 3.4.

For all fluids the density ρ depends on the local thermodynamic state variables, such as pressure, temperature, and composition. However for liquids it is often a very good assumption to take the density to be constant. Such an idealized fluid is often called an "incompressible fluid," and the *momentum flux tensor* simplified to:

$$\begin{aligned}\boldsymbol{\pi} &= p\boldsymbol{\delta} + \boldsymbol{\tau}\\ &= p\boldsymbol{\delta} - \mu[\nabla \boldsymbol{v} + (\nabla \boldsymbol{v})^{\dagger}]\\ &= p\boldsymbol{\delta} - \mu\dot{\boldsymbol{\gamma}}\end{aligned} \tag{1.2-2}$$

in which $\dot{\boldsymbol{\gamma}}$ is the *rate-of-deformation tensor*. When the incompressibility assumption is made, a problem arises as to the meaning of p. For example, for a pure incompressible fluid at constant temperature a plot of p *vs.* ρ is a vertical straight line; that is, the function $p(\rho)$ is many-valued and not specified uniquely. This poses no difficulty in solving hydrodynamic problems since only the gradient of p needs to be known. However, in connection with determining pressures at surfaces, an incompressible fluid theory can predict only pressure differences and not absolute values. For all discussions of Newtonian fluids in this book, Eq. 1.2–2 will be used for the momentum flux tensor.

The *heat flux* $\boldsymbol{q}$ for pure fluids and nondiffusing mixtures is given by "Fourier's law of heat conduction":

$$\boldsymbol{q} = -k\nabla T \tag{1.2-3}$$

in which k is the *thermal conductivity* and T is the temperature. For diffusing mixtures there are additional contributions to $\boldsymbol{q}$, but we do not discuss them here.[1,2]

Now that we have given the expressions in Eqs. 1.2–2 and 1.2–3 for the fluxes, let us turn to the equations of change, and particularly the *equations of change for incompressible Newtonian fluids*. These are listed in Table 1.2–1 and given in Appendix B in various coordinate systems. The equation of continuity was given earlier in Eq. 1.1–5. The equation of motion is obtained by substituting Eq. 1.2–2 into Eq. 1.1–8 and simplifying. The energy equation is obtained by first transforming Eq. 1.1–13 into an equation for temperature

TABLE 1.2–1

Equations of Change for Newtonian Fluids with Constant ρ, μ, and k

Continuity:	$(\nabla \cdot \boldsymbol{v}) = 0 \quad \text{or} \quad \dfrac{D}{Dt}\rho = 0$	(A)
Motion:	$\rho \dfrac{D\boldsymbol{v}}{Dt} = -\nabla p + \mu \nabla^2 \boldsymbol{v} + \rho \boldsymbol{g}$	(B)[a]
Energy:	$\rho \hat{C}_p \dfrac{DT}{Dt} = k\nabla^2 T + \frac{1}{2}\mu(\dot{\boldsymbol{\gamma}} : \dot{\boldsymbol{\gamma}})$	(C)

[a] It is often convenient to combine the pressure and gravity terms as $\nabla \mathscr{P} = \nabla p - \rho \boldsymbol{g}$, where $\mathscr{P}$ is called the "modified pressure." This nomenclature was suggested by G. K. Batchelor, *An Introduction to Fluid Dynamics*, Cambridge University Press (1967), p. 176.

(by using standard thermodynamic transformations)[4] and then inserting Fourier's law (Eq. 1.2–3) for $\boldsymbol{q}$. This process is outlined in Problem 1B.15; the final equation contains $\hat{C}_p$, which is the heat capacity at constant pressure per unit mass.

The equations of change in Table 1.2–1 are easy to interpret physically:

A. The *equation of continuity* (in the second form given there) states that as one follows along with the fluid, the density does not change with time.

B. The *equation of motion* states that the mass-times-acceleration of a fluid element equals the sum of the pressure, viscous, and gravitational forces acting on the element.

C. The *energy equation* states that the temperature of a fluid element changes as it moves along with the fluid because of heat conduction (the k-term) and heat production by viscous heating (the μ-term).

These are the basic equations for the study of the classical hydrodynamics of constant-density fluids. In the following examples their use is illustrated for several laminar flow problems. We put the emphasis here on several widely used approximations. In Example 1.2–2 we illustrate the use of *perturbation theory*, in which the deviations from some standard simplified problem are obtained approximately. In Examples 1.2–3 and 5 we use a *lubrication approximation*, in which flow between nonparallel surfaces is approximated locally as flow between equivalent parallel surfaces. In Example 1.2–4 we use the *creeping-flow assumption* for a very low Reynolds number flow around a submerged object. And in Example 1.2–6 we introduce the *quasi-steady-state approximation*. All of these methods are encountered later in connection with non-Newtonian fluids.

EXAMPLE 1.2–1 Laminar Newtonian Flow between Parallel Plane Surfaces

An incompressible Newtonian fluid is located in the space between two parallel plates that are separated by a distance B (see Fig. 1.2–1). The upper plate is moving in the $+x$-direction with a velocity v_0, thus contributing to the motion of the fluid. An additional contribution to the fluid motion is that due to a constant applied pressure gradient $\partial p/\partial x$. Find the velocity profile and the volume rate of flow. Assume that the flow is sufficiently slow that viscous heating is not important.

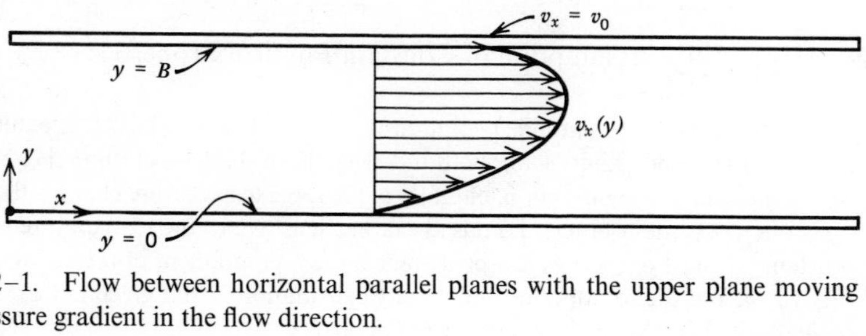

FIGURE 1.2–1. Flow between horizontal parallel planes with the upper plane moving and with an imposed pressure gradient in the flow direction.

SOLUTION

We postulate that in this system $v_x = v_x(y)$, $v_y = 0$, $v_z = 0$, $p = p(x,y)$, and $T =$ constant. We now apply these postulates to the equations of change in order to get the differential equations to describe

[4] See Problem 1B.15. For a rather complete summary of the many forms the energy equation can take, see R. B. Bird, W. E. Stewart, and E. N. Lightfoot, *op. cit.*, §§10.1–4 and §18.3.

the system. The equations of continuity and energy are clearly unimportant. The y-component of the equation of motion just gives the vertical pressure gradient that is of no interest here. The x-component of the equation of motion becomes:

$$0 = -\frac{\partial p}{\partial x} + \mu \frac{d^2 v_x}{dy^2} \tag{1.2–4}$$

in which $\partial p/\partial x$ was stated to be a constant. This equation has to be integrated with respect to y with the boundary conditions:

$$\text{At } y = 0 \qquad v_x = 0 \tag{1.2–5}$$

$$\text{At } y = B \qquad v_x = v_0 \tag{1.2–6}$$

The result is the velocity distribution:

$$v_x = v_0 \left(\frac{y}{B}\right) - \frac{B^2}{2\mu}\frac{\partial p}{\partial x}\left[\left(\frac{y}{B}\right) - \left(\frac{y}{B}\right)^2\right] \tag{1.2–7}$$

The volume rate of flow Q for plates of width W is:

$$\begin{aligned} Q = WB\langle v_x \rangle &= \int_0^W \int_0^B v_x \, dy \, dz \\ &= WB \int_0^1 v_x \, d\left(\frac{y}{B}\right) \\ &= \tfrac{1}{2} WBv_0 - \frac{WB^3}{12\mu}\frac{\partial p}{\partial x} \end{aligned} \tag{1.2–8}$$

Here the angular brackets $\langle\ \rangle$ indicate an average over the cross section. This result contains the solution for the problem where the pressure gradient and the wall motion both tend to drive the fluid in the same direction and also the problem where the pressure gradient and wall motion oppose one another.

EXAMPLE 1.2–2 Viscous Heating in Flow between Plane Parallel Surfaces

Consider the system discussed in Example 1.2–1 with $\partial p/\partial x = 0$ and with flows sufficiently fast that viscous heating is important. Find the temperature profile in the system if both the upper and lower plates are maintained at a temperature T_0.

SOLUTION (a) With Assumption of Constant Physical Properties

We begin by considering the simplified solution to the problem in which it is assumed that the viscosity μ and the thermal conductivity k are both independent of the temperature. For small viscous heating effects this is a reasonable assumption, but if there are large temperature changes the assumption is a poor one, since the viscosities of many liquids are rather sensitive to the temperature.

When constant physical properties can be assumed, the x-component of the equation of motion is the same as Eq. 1.2–4, and the solution for the velocity distribution is that given in Eq. 1.2–7 (in both equations we set $\partial p/\partial x = 0$ for the problem considered here).

The energy equation appropriate for the postulate that $T = T(y)$ is obtained by simplifying Eq. C of Table 1.2–1. This gives:

$$0 = k\frac{d^2 T}{dy^2} + \mu\left(\frac{dv_x}{dy}\right)^2 \tag{1.2–9}$$

or, using v_x from Eq. 1.2–7 (with $\partial p/\partial x = 0$):

$$0 = k\frac{d^2 T}{dy^2} + \mu\left(\frac{v_0}{B}\right)^2 \tag{1.2–10}$$

This is to be solved with the boundary conditions:

$$\text{At } y = 0 \qquad T = T_0 \tag{1.2–11}$$

$$\text{At } y = B \qquad T = T_0 \tag{1.2–12}$$

The solution is easily found to be:

$$T - T_0 = \frac{\mu v_0^2}{2k}\left[\left(\frac{y}{B}\right) - \left(\frac{y}{B}\right)^2\right] \tag{1.2–13}$$

This result is useful for estimating the magnitude of viscous heating in lubrication systems and in other situations where a fluid is located between two rapidly moving surfaces, both at the same temperature T_0.

(b) Without Assumption of Constant Physical Properties

Deviations from the above results can be estimated by using a perturbation theory analysis.[5] Here we allow both the viscosity and thermal conductivity to vary with the temperature by using simple Taylor series expansions about $T = T_0$:

$$\frac{k}{k_0} = 1 + \alpha_1\Theta + \alpha_2\Theta^2 + \cdots \tag{1.2–14}$$

$$\frac{\mu_0}{\mu} = 1 + \beta_1\Theta + \beta_2\Theta^2 + \cdots \tag{1.2–15}$$

where $\Theta = (T - T_0)/T_0$ is a dimensionless temperature rise, and the subscript "0" on μ and k means "evaluated at the wall temperature T_0". The α's and β's are dimensionless quantities that can be determined from experimental data on the temperature variation of the physical properties. See Problem 1C.6 for the solution to the same problem with k = constant and μ decreasing exponentially with temperature.

The equations of motion and energy are now (in dimensionless form):

$$\frac{d}{d\eta}\left(\frac{\mu}{\mu_0}\frac{d\phi}{d\eta}\right) = 0 \tag{1.2–16}$$

$$\frac{d}{d\eta}\left(\frac{k}{k_0}\frac{d\Theta}{d\eta}\right) + \text{Br}\,\frac{\mu}{\mu_0}\left(\frac{d\phi}{d\eta}\right)^2 = 0 \tag{1.2–17}$$

in which $\eta = y/B$, $\phi = v_x/v_0$, and $\text{Br} = \mu_0 v_0^2/k_0 T_0$ (the Brinkman number).

The equation for the dimensionless velocity distribution may be integrated once to give $d\phi/d\eta = C_1(\mu_0/\mu)$ where C_1 is a constant of integration. When this is substituted into the energy equation we get:

$$\frac{d}{d\eta}\left[(1 + \alpha_1\Theta + \alpha_2\Theta^2 + \cdots)\frac{d\Theta}{d\eta}\right] + \text{Br}\,C_1^2(1 + \beta_1\Theta + \beta_2\Theta^2 + \cdots) = 0 \tag{1.2–18}$$

This is a second-order differential equation for Θ. We postulate that Θ and C_1 can be expanded in the form:

$$\Theta = \text{Br}\,\Theta_1(\eta) + \text{Br}^2\,\Theta_2(\eta) + \cdots \tag{1.2–19}$$

$$C_1 = C_{10} + \text{Br}\,C_{11} + \text{Br}^2\,C_{12} + \cdots \tag{1.2–20}$$

[5] R. B. Bird, W. E. Stewart, and E. N. Lightfoot, *op. cit.*, pp. 306–307 (there were several errors in the solution to Problem 9.O in the first printing that were corrected in later printings). See also R. M. Turian and R. B. Bird, *Chem. Eng. Sci.*, *18*, 689–696 (1963) for application of the results cited here to the cone-and-plate viscometer. For an extensive literature review and experimental data, see P. C. Sukanek and R. L. Laurence, *A.I.Ch.E. Journal*, *20*, 474–484(1974). A survey of perturbation techniques can be found in A. Nayfeh, *Perturbation Methods*, Wiley, New York (1973).

These expressions are now substituted into Eq. 1.2–18 and coefficients of like powers of Br are equated to zero.

(i) The terms in the first power of Br give:

$$\frac{d^2\Theta_1}{d\eta^2} + C_{10}^2 = 0 \qquad (1.2\text{–}21)$$

When this is solved and the integration constants determined by the fact that $\Theta_1 = 0$ at $\eta = 0$ and $\eta = 1$, the result is:

$$\Theta_1 = \tfrac{1}{2}C_{10}^2(\eta - \eta^2) \qquad (1.2\text{–}22)$$

The function Θ_1 is then substituted into the equation $d\phi/d\eta = C_1(\mu_0/\mu)$, in which the ϕ is expanded thus:

$$\phi = \phi_0(\eta) + \text{Br}\,\phi_1(\eta) + \cdots \qquad (1.2\text{–}23)$$

This gives:

$$\frac{d}{d\eta}(\phi_0 + \text{Br}\,\phi_1 + \cdots)$$
$$= (C_{10} + \text{Br}\,C_{11} + \text{Br}^2\,C_{12} + \cdots)(1 + \beta_1(\text{Br}\,\Theta_1 + \text{Br}^2\,\Theta_2 + \cdots) + \cdots) \qquad (1.2\text{–}24)$$

We get for the coefficients of the zeroth power of Br:

$$\frac{d}{d\eta}\phi_0 = C_{10} \qquad (1.2\text{–}25)$$

Then integration gives:

$$\phi_0 = C_{10}\eta + C_2 \qquad (1.2\text{–}26)$$

The boundary conditions that $\phi = 0$ at $\eta = 0$ and $\phi = 1$ at $\eta = 1$ enable one to find that $C_{10} = 1$ and $C_2 = 0$. This gives then

$$\phi_0 = \eta \qquad (1.2\text{–}27)$$

and

$$\Theta_1 = \tfrac{1}{2}(\eta - \eta^2) \qquad (1.2\text{–}28)$$

(ii) The next higher-order terms are obtained in analogous fashion:

$$\phi_1 = -\tfrac{1}{12}\beta_1(\eta - 3\eta^2 + 2\eta^3) \qquad (1.2\text{–}29)$$

$$\Theta_2 = -\tfrac{1}{8}\alpha_1(\eta^2 - 2\eta^3 + \eta^4) - \tfrac{1}{24}\beta_1(\eta - 2\eta^2 + 2\eta^3 - \eta^4) \qquad (1.2\text{–}30)$$

The final results for the temperature and velocity profiles are then obtained by substituting these last four functions into Eqs. 1.2–19 and 23. The latter equation enables one to estimate the perturbation in the velocity distribution due to viscous heating.

EXAMPLE 1.2–3 Laminar Flow in a Circular Tube

A fluid flows through a circular tube of radius R and length L. The tube makes an angle χ with the vertical direction. The pressures at the tube axis at $z = 0$ and $z = L$ (see Fig. 1.2–2) are p_0 and p_L, respectively. Find the steady-state velocity profile and the volume rate of flow, neglecting entrance and exit effects and assuming negligible viscous heating.

$\dfrac{1}{r}\dfrac{d}{dr}\left(r\tau_{rz}\right)$

$\mu = \tau\dfrac{\partial u_z}{\partial r}$

$\mu\dfrac{1}{r}\dfrac{\partial}{\partial r}\left(\dfrac{\partial u_z}{\partial r}\right)$

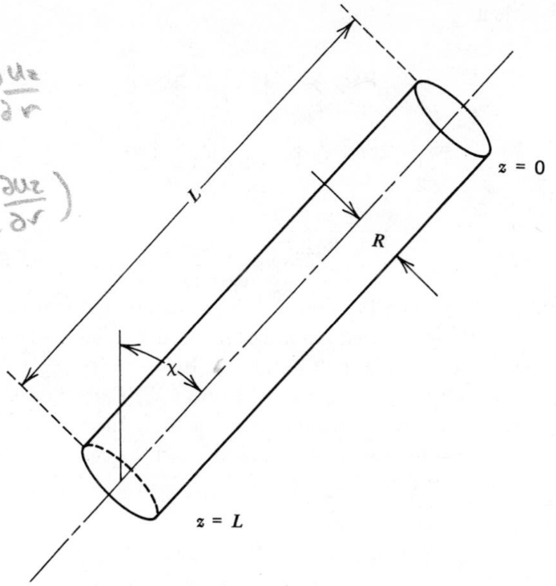

FIGURE 1.2–2.   Flow through a circular tube that is inclined at an angle χ to the vertical.

SOLUTION

We postulate a solution of the form $v_z = v_z(r)$, $v_\theta = 0$, $v_r = 0$, and $p = p(r, \theta, z)$. The equation of continuity is satisfied identically, and only the z-component of the equation of motion is of interest:

$$0 = -\frac{\partial p}{\partial z} + \mu\frac{1}{r}\frac{d}{dr}\left(r\frac{dv_z}{dr}\right) + \rho g \cos\chi \tag{1.2–31}$$

This is to be solved with $v_z = 0$ at $r = R$ and v_z finite at $r = 0$.

Next we introduce[6] the "modified pressure" $\mathcal{P}(z) = p - \rho g z \cos\chi$; then:

$$\frac{d\mathcal{P}}{dz} = \mu\frac{1}{r}\frac{d}{dr}\left(r\frac{dv_z}{dr}\right) \tag{1.2–32}$$

The left side is a function of z alone, whereas the right side is a function of r alone. Hence, both sides must equal a constant K_0, from which it follows that:

$$\mathcal{P} = K_0 z + K_1 \tag{1.2–33}$$

The constants are determined by the boundary conditions on p, and the final result is

$$\mathcal{P} = -\frac{(\mathcal{P}_0 - \mathcal{P}_L)}{L}z + \mathcal{P}_0 \tag{1.2–34}$$

where $\mathcal{P}_0 = p_0$ and $\mathcal{P}_L = p_L - \rho g L \cos\chi$. Then Eq. 1.2–32 can be integrated using the boundary conditions on v_z; this gives the well-known parabolic velocity profile as:

$$v_z = \frac{(\mathcal{P}_0 - \mathcal{P}_L)R^2}{4\mu L}\left[1 - \left(\frac{r}{R}\right)^2\right] \tag{1.2–35}$$

[6] In general, $\mathcal{P}$ is defined as $\mathcal{P} = p + \rho g h$, where h is the distance *upward* (i.e., in the direction opposed to gravity) from some arbitrarily chosen reference plane. See also Table 1.2–1.

The volume rate of flow is then:

$$Q = \pi R^2 \langle v_z \rangle = \int_0^{2\pi} \int_0^R v_z \, r \, dr \, d\theta$$
$$= 2\pi R^2 \int_0^1 v_z \cdot \left(\frac{r}{R}\right) d\left(\frac{r}{R}\right)$$
$$= \frac{\pi(\mathscr{P}_0 - \mathscr{P}_L)R^4}{8\mu L} \tag{1.2–36}$$

which is the famous result of Hagen and Poiseuille.[7,8] This relation (accompanied by additional information about end corrections) is the basic equation needed to determine viscosity from tube flow data. It is valid for Re < 2100, where the Reynolds number $\text{Re} = 2R\langle v_z \rangle \rho/\mu = 2Q\rho/\pi\mu R$, the angular brackets indicating an average over the cross section. For Re > 2100, the flow will usually be turbulent.

Equation 1.2–36 can be used to determine approximately the relation between flow rate and pressure drop in a slightly tapered tube which has a radius R_0 at the entrance and radius R_L at the exit. Then the tube radius at any distance z from the inlet is:

$$R(z) = R_0 + \left(\frac{R_L - R_0}{L}\right) z \tag{1.2–37}$$

If the taper is very slight, one feels intuitively that Eq. 1.2–36 can probably be applied locally, by replacing R, a constant, by the $R(z)$ of Eq. 1.2–37 and by replacing $(\mathscr{P}_0 - \mathscr{P}_L)/L$ by $(-d\mathscr{P}/dz)$ (this kind of approximation is called a "lubrication approximation"):

$$Q = \frac{\pi[R(z)]^4}{8\mu}\left(-\frac{d\mathscr{P}}{dz}\right) \tag{1.2–38}$$

Rather than using z as the independent variable, we can use R:

$$Q = \frac{\pi R^4}{8\mu}\left(-\frac{d\mathscr{P}}{dR}\right)\left(\frac{R_L - R_0}{L}\right) \tag{1.2–39}$$

But Q is constant for all z (and hence all R). Therefore the differential equation for $\mathscr{P}$ as a function of R can be integrated to give:

$$Q = \frac{3\pi}{8\mu}\frac{\mathscr{P}_0 - \mathscr{P}_L}{L}\frac{R_0 - R_L}{R_L^{-3} - R_0^{-3}}$$
$$= \frac{\pi(\mathscr{P}_0 - \mathscr{P}_L)R_0^4}{8\mu L}\left[1 - \frac{1 + (R_L/R_0) + (R_L/R_0)^2 - 3(R_L/R_0)^3}{1 + (R_L/R_0) + (R_L/R_0)^2}\right] \tag{1.2–40}$$

Hence the final result may be expressed as the Hagen-Poiseuille result multiplied by a correction factor.

EXAMPLE 1.2–4 Creeping Flow around a Sphere

The low Reynolds number flow around a nonrotating sphere is an important problem in classical fluid dynamics. For such a "creeping flow" one can omit the $[\boldsymbol{v} \cdot \nabla \boldsymbol{v}]$ term in the equation of motion. We consider here the situation where the center of the sphere is located at the origin of coordinates. A fluid of viscosity μ flows toward the sphere with velocity $\boldsymbol{v}_\infty$; that is, at infinite distance from the sphere, in any direction, the fluid velocity is $\boldsymbol{v}_\infty$ (see Fig. 1.2–3).

[7] G. Hagen, *Ann. Phys. Chem.*, *46*, 423–442 (1839).
[8] J. L. Poiseuille, *Comptes Rendus*, *11*, 961 and 1041 (1840); *12*, 112 (1841).

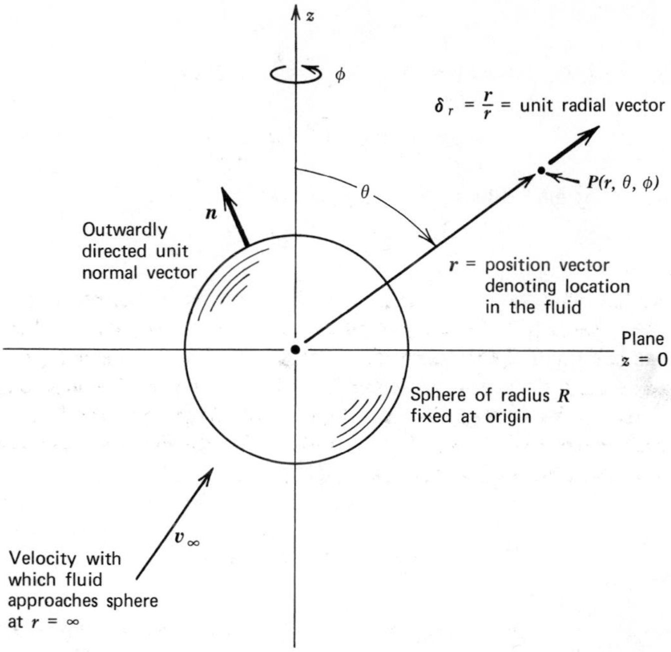

FIGURE 1.2–3. Flow around a sphere with fluid velocity $\boldsymbol{v}_\infty$ far from the sphere. Note that $\boldsymbol{n}$ is an outward unit normal on the surface of the sphere, whereas $\boldsymbol{\delta}_r$ is a radially directed unit vector anywhere in the fluid.

The equations of change for incompressible, low-Reynolds-number flow are

Continuity:
$$(\nabla \cdot \boldsymbol{v}) = 0 \tag{1.2–41}$$

Motion:
$$0 = -\nabla\mathscr{P} + \mu\nabla^2\boldsymbol{v} \tag{1.2–42}$$

To obtain the velocity and pressure distributions for flow around a sphere, several methods are available. The best discussions of this subject can be found in the excellent hydrodynamics texts of Landau and Lifshitz[9] and of Batchelor.[10] The final expressions are, for a sphere of radius R,

$$\boldsymbol{v} = \boldsymbol{v}_\infty\left[1 - \tfrac{3}{4}\left(\frac{R}{r}\right) - \tfrac{1}{4}\left(\frac{R}{r}\right)^3\right] - \tfrac{3}{4}[\boldsymbol{v}_\infty \cdot \boldsymbol{\delta}_r\boldsymbol{\delta}_r]\left[\left(\frac{R}{r}\right) - \left(\frac{R}{r}\right)^3\right] \tag{1.2–43}$$

$$\mathscr{P} = \mathscr{P}_0 - \tfrac{3}{2}\mu\frac{R}{r^2}(\boldsymbol{v}_\infty \cdot \boldsymbol{\delta}_r) \tag{1.2–44}$$

Here $\mathscr{P}_0$ is the value of the "modified pressure" at the plane $z = 0$ going through the center of the sphere. The unit vector $\boldsymbol{\delta}_r$ points in the r-direction.

(a) Use the above expressions to obtain $\dot{\boldsymbol{\gamma}}$ and $\boldsymbol{\pi}$ at all points in the fluid (i.e., for $r \geq R$).
(b) From the results in (a) obtain the expression for the drag force on the sphere (i.e., Stokes' law).
(c) Find the disturbance in a quiescent fluid due to the movement of a sphere through the fluid, in terms of the force the sphere exerts on the fluid.

[9] L. Landau and E. M. Lifshitz, *op. cit.*, pp. 63–71, 95–98.
[10] G. K. Batchelor, *op. cit.*, pp. 229–244.

SOLUTION

(a) If we abbreviate Eq. 1.2–43 by

$$\boldsymbol{v} = \boldsymbol{v}_\infty f(r) + [\boldsymbol{v}_\infty \cdot \boldsymbol{\delta}_r \boldsymbol{\delta}_r] g(r) \tag{1.2–45}$$

then $\nabla \boldsymbol{v}$ can be obtained by differentiation (see Problem 1D.1):

$$\nabla \boldsymbol{v} = f'(r)\boldsymbol{\delta}_r \boldsymbol{v}_\infty + \frac{g(r)}{r} \boldsymbol{v}_\infty \boldsymbol{\delta}_r + \frac{g}{r}(\boldsymbol{v}_\infty \cdot \boldsymbol{\delta}_r)\boldsymbol{\delta} + \left(g' - 2\frac{g}{r}\right)(\boldsymbol{v}_\infty \cdot \boldsymbol{\delta}_r)\boldsymbol{\delta}_r \boldsymbol{\delta}_r \tag{1.2–46}$$

and hence the rate-of-deformation tensor is:

$$\begin{aligned} \dot{\boldsymbol{\gamma}} &= \nabla \boldsymbol{v} + (\nabla \boldsymbol{v})^\dagger \\ &= \tfrac{3}{2}\frac{\boldsymbol{\delta}_r \boldsymbol{v}_\infty + \boldsymbol{v}_\infty \boldsymbol{\delta}_r}{r}\left(\frac{R}{r}\right)^3 - \tfrac{3}{2}\frac{(\boldsymbol{v}_\infty \cdot \boldsymbol{\delta}_r)\boldsymbol{\delta}}{r}\left[\left(\frac{R}{r}\right) - \left(\frac{R}{r}\right)^3\right] \\ &\quad + \tfrac{3}{2}\frac{(\boldsymbol{v}_\infty \cdot \boldsymbol{\delta}_r)\boldsymbol{\delta}_r \boldsymbol{\delta}_r}{r}\left[3\left(\frac{R}{r}\right) - 5\left(\frac{R}{r}\right)^3\right] \end{aligned} \tag{1.2–47}$$

The total momentum flux tensor is then obtained from Eq. 1.2–2:

$$\begin{aligned} \boldsymbol{\pi} &= p\boldsymbol{\delta} - \mu\dot{\boldsymbol{\gamma}} \\ &= \mathscr{P}\boldsymbol{\delta} - \rho g z \boldsymbol{\delta} - \mu\dot{\boldsymbol{\gamma}} \\ &= (\mathscr{P}_0 - \rho g z)\boldsymbol{\delta} - \tfrac{3}{2}\frac{R\mu}{r^2}(\boldsymbol{v}_\infty \cdot \boldsymbol{\delta}_r)\boldsymbol{\delta} - \mu\dot{\boldsymbol{\gamma}} \end{aligned} \tag{1.2–48}$$

(b) Next we want to get the force acting at each point on the surface of the sphere. We know that the force per unit area on a surface with normal unit vector $\boldsymbol{n}$ is $-[\boldsymbol{n} \cdot \boldsymbol{\pi}]$. Hence at each point on the surface of the sphere:

$$\begin{aligned} [\boldsymbol{n} \cdot \boldsymbol{\pi}]_{\text{surf}} &= [\boldsymbol{\delta}_r \cdot \boldsymbol{\pi}]_{r=R} \\ &= (\mathscr{P}_0 - \rho g R \cos\theta)\boldsymbol{\delta}_r - \frac{3\mu}{2R}(\boldsymbol{v}_\infty \cdot \boldsymbol{\delta}_r)\boldsymbol{\delta}_r - \frac{3\mu}{2R}(\boldsymbol{v}_\infty + (\boldsymbol{v}_\infty \cdot \boldsymbol{\delta}_r)\boldsymbol{\delta}_r) + \frac{3\mu}{R}(\boldsymbol{v}_\infty \cdot \boldsymbol{\delta}_r)\boldsymbol{\delta}_r \\ &= (\mathscr{P}_0 - \rho g R \cos\theta)\boldsymbol{\delta}_r - \frac{3\mu \boldsymbol{v}_\infty}{2R} \end{aligned} \tag{1.2–49}$$

Note that the term containing μ is constant over the entire spherical surface. The total force of the fluid on the sphere is then obtained by integration. The integral of $\mathscr{P}_0$ is exactly zero. The next term gives the *buoyancy force* $\boldsymbol{F}_b$, and the last term gives the *kinetic force* $\boldsymbol{F}_k$:

$$\begin{aligned} \boldsymbol{F} &= -\int_0^{2\pi}\int_0^{\pi} [\boldsymbol{n} \cdot \boldsymbol{\pi}]\Big|_{r=R} R^2 \sin\theta \, d\theta \, d\phi \\ &= \rho g R^3 \int_0^{2\pi}\int_0^{\pi} \cos\theta \sin\theta (\boldsymbol{\delta}_x \sin\theta\cos\phi + \boldsymbol{\delta}_y \sin\theta \sin\phi + \boldsymbol{\delta}_z \cos\theta)\, d\theta\, d\phi + 4\pi R^2 \cdot \frac{3\mu \boldsymbol{v}_\infty}{2R} \\ &= \tfrac{4}{3}\pi R^3 \rho g \boldsymbol{\delta}_z + 6\pi\mu R \boldsymbol{v}_\infty \end{aligned} \tag{1.2–50}$$

Buoyancy force   Kinetic force

This final result, that $\boldsymbol{F}_k = 6\pi\mu R\boldsymbol{v}_\infty$, is known as *Stokes' law*[11] for the force of a fluid on a sphere in the "creeping flow region," for which $\text{Re} = 2Rv_\infty\rho/\mu < 0.1$.

(c) We now turn the problem around and let the fluid be at rest at infinity but let the sphere move linearly with a constant velocity $\boldsymbol{v}_s = -\boldsymbol{v}_\infty$, with its center instantaneously at the origin (see Fig. 1.2–4).

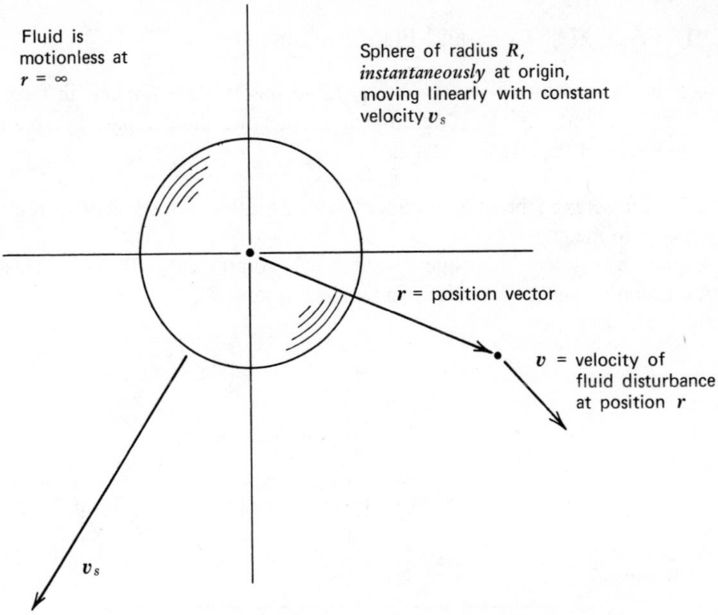

FIGURE 1.2–4. Velocity disturbance caused by a sphere moving linearly through a fluid with constant velocity $\boldsymbol{v}_s$ at that instant when the center of the sphere is at the origin.

Then we inquire as to the velocity $\boldsymbol{v}$ of the fluid at some position $\boldsymbol{r}$. From Eq. 1.2–43 this must be:

$$\boldsymbol{v} = [-\boldsymbol{v}_s \cdot \boldsymbol{\delta}]\left[-\tfrac{3}{4}\left(\frac{R}{r}\right) - \tfrac{1}{4}\left(\frac{R}{r}\right)^3\right] - \tfrac{3}{4}[(-\boldsymbol{v}_s)\cdot\boldsymbol{\delta}_r\boldsymbol{\delta}_r]\left[\left(\frac{R}{r}\right) - \left(\frac{R}{r}\right)^3\right] \tag{1.2–51}$$

But the force $\boldsymbol{F}_s = -\boldsymbol{F}_k$ exerted by the sphere on the fluid by virtue of its motion is $\boldsymbol{F}_s = 6\pi\mu R\boldsymbol{v}_s$ so that

$$\boldsymbol{v}_s = (6\pi\mu R)^{-1}\boldsymbol{F}_s \tag{1.2–52}$$

Substitution of this expression for $\boldsymbol{v}_s$ into the preceding equation gives:

$$\begin{aligned}\boldsymbol{v}(\boldsymbol{r}) &= \frac{1}{8\pi\mu r}\left[[(\boldsymbol{\delta} + \boldsymbol{\delta}_r\boldsymbol{\delta}_r) + \left(\frac{R}{r}\right)^2(\tfrac{1}{3}\boldsymbol{\delta} - \boldsymbol{\delta}_r\boldsymbol{\delta}_r)]\cdot\boldsymbol{F}_s\right] \\ &= [\boldsymbol{\Omega}\cdot\boldsymbol{F}_s] + O\left(\frac{1}{r^3}\right)\end{aligned} \tag{1.2–53}$$

in which $O(\ )$ means "of the order of ( )" and

$$\boldsymbol{\Omega} = \frac{1}{8\pi\mu r}(\boldsymbol{\delta} + \boldsymbol{\delta}_r\boldsymbol{\delta}_r) \tag{1.2–54}$$

[11] R. B. Bird, W. E. Stewart, and E. N. Lightfoot, *op. cit.*, §§2.6 and 4.2; G. G. Stokes, *Mathematical and Physical Papers*, Vol. 1, pp. 38–43, Cambridge University Press (1880).

is called the *Oseen tensor*[12] or *hydrodynamic interaction tensor.* This is used in estimating how much the solvent velocity field near one part of a macromolecule is perturbed because of the motion of another part of the macromolecule (see §§10.6, 11.6 and 12.4 for a discussion of "hydrodynamic interaction"). The relation $\boldsymbol{v}(\boldsymbol{r}) = [\boldsymbol{\Omega} \cdot \boldsymbol{F}_s]$ is exact at large distances where the $1/r^3$ term is not important; it is also exact when a force is applied to a point particle (i.e., $R = 0$) at the origin.

EXAMPLE 1.2–5 The Cone-and-Plate Viscometer

The cone-and-plate geometry shown in Fig. 1.2–5 has become popular in recent years for the measurement of viscosity (and other rheological properties, as we shall see later). Obtain the analytical relations needed to interpret the instrumental data:

(a) A relation between the angular velocity W and the $\theta\phi$-component of the rate-of-deformation tensor $\dot{\boldsymbol{\gamma}}$ in the gap

(b) A relation between the torque $\mathcal{T}$ and the $\theta\phi$-component of the stress tensor $\boldsymbol{\tau}$ in the gap.

(c) A relation giving the viscosity μ in terms of W and $\mathcal{T}$.

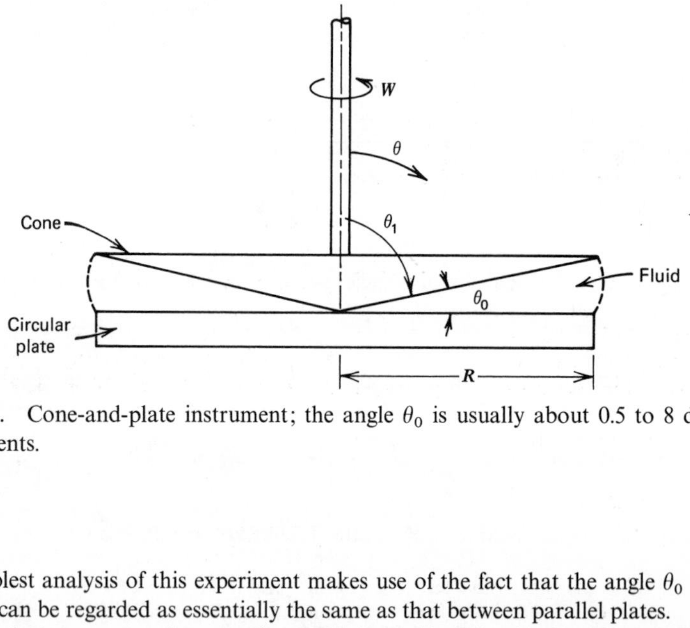

FIGURE 1.2–5. Cone-and-plate instrument; the angle θ_0 is usually about 0.5 to 8 degrees in commercial instruments.

SOLUTION

The simplest analysis of this experiment makes use of the fact that the angle θ_0 is so small that locally the flow can be regarded as essentially the same as that between parallel plates.

(a) The velocity v_ϕ at a radius r can be approximated by adapting Eq. 1.2–7 with $\partial p/\partial x = 0$ and with v_x replaced by v_ϕ. For the cone-and-plate system, at a distance r from the cone apex, the velocity of the cone will be Wr (this corresponds to v_0 in Eq. 1.2–7), and the plate-cone separation will be $r \sin\theta_0 \approx r\theta_0$ (this corresponds to B in Eq. 1.2–7). Hence the velocity profile will, to a good approximation, be:

$$v_\phi = Wr\left(\frac{\dfrac{\pi}{2} - \theta}{\dfrac{\pi}{2} - \theta_1}\right) \tag{1.2–55}$$

The $\theta\phi$-component of the $\dot{\boldsymbol{\gamma}}$-tensor is then (cf. Eqs. X and Z in Table A.7–3):

$$\dot{\gamma}_{\theta\phi} = \frac{\sin\theta}{r}\frac{\partial}{\partial\theta}\left(\frac{v_\phi}{\sin\theta}\right) \approx \frac{1}{r}\frac{\partial}{\partial\theta}v_\phi = -\frac{W}{\theta_0} \tag{1.2–56}$$

[12] C. W. Oseen, *Neuere Methoden und Ergebnisse in der Hydrodynamik*, Akad. Verlag, Leipzig (1927); J. M. Burgers, *Second Report on Viscosity and Plasticity*, Amsterdam Academy of Sciences, Amsterdam (1938), Chapter 3.

The approximation made here is that θ is so close to $\pi/2$ that sin θ can be taken to be unity; this should be an excellent approximation. We see from Eq. 1.2–56 that $\dot{\gamma}_{\theta\phi}$ is constant throughout the cone-plate gap. This is one reason why this kind of instrument is useful for macromolecular fluids where the viscosity depends on the shear rate.

(b) The torque required to maintain the motion will be obtained by integrating the product of the lever arm r and the force $\tau_{\theta\phi}|_{\theta=\pi/2} \cdot r\,dr\,d\phi$ over the surface of the plate:

$$\mathscr{T} = \int_0^{2\pi} \int_0^R \tau_{\theta\phi}|_{\theta=\pi/2}\, r^2\, dr\, d\phi \tag{1.2–57}$$

Since $\dot{\gamma}_{\theta\phi}$ is constant throughout the gap, $\tau_{\theta\phi}$ will also be constant. Hence the integration is easily performed and one gets:

$$\tau_{\theta\phi} = \frac{3\mathscr{T}}{2\pi R^3} \tag{1.2–58}$$

(c) Since $\boldsymbol{\tau} = -\mu\dot{\boldsymbol{\gamma}}$ for a Newtonian fluid, the $\theta\phi$-component of this equation, combined with Eqs. 1.2–56 and 58, gives:

$$\mu = \frac{\tau_{\theta\phi}}{-\dot{\gamma}_{\theta\phi}} = \frac{3\mathscr{T}\theta_0}{2\pi R^3 W} \tag{1.2–59}$$

This gives the viscosity of the fluid in the gap in terms of the geometrical quantities R and θ_0 and the measurements of torque $\mathscr{T}$ and angular velocity W.

More complete analyses can be found elsewhere.[13]

EXAMPLE 1.2–6 Squeezing Flow between Two Parallel Disks[14]

A fluid is placed in the gap between two parallel disks separated by a distance $2h_0$. The fluid completely fills the gap. A constant force F is applied to each disk as shown in Fig. 1.2–6. It is desired to obtain an expression for the change in gap separation with time. A "quasi-steady state" solution will be

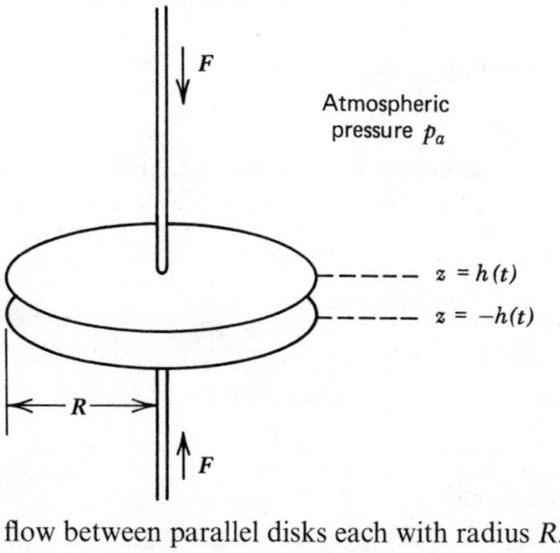

FIGURE 1.2–6. Squeezing flow between parallel disks each with radius R. The initial disk separation is $2h_0$.

[13] R. B. Bird, W. E. Stewart, and E. N. Lightfoot, *op. cit.*, p. 119, Problem 3.T; S. Oka, in *Rheology*, F. R. Eirich (Ed.), Academic Press, New York (1960), Vol. 3, Chapter 2, pp. 61–62; K. Walters, *Rheometry*, Chapman and Hall, London (1975), Chapter 4.

[14] J. Stefan, *Sitzber. K. Akad. Wiss. Math. Natur.*, Wien, *69*, Part 2, 713–735 (1874); see also L. Landau and E. M. Lifshitz, *op. cit.*, pp. 70–71.

used; that is, at any time t the radial flow problem will be treated as a steady-state hydrodynamic problem, but the time rate of change of the mass of material between the two disks will be accounted for properly.

SOLUTION

We postulate here that $v_r = v_r(r,z)$, $v_z = v_z(z)$, and $p = p(r)$. We assume initially that inertial terms are not important. The equations of change are:

Continuity:
$$\frac{1}{r}\frac{\partial}{\partial r}(rv_r) + \frac{dv_z}{dz} = 0 \tag{1.2–60}$$

Motion:
$$0 = -\frac{dp}{dr} + \mu\left[\underline{\frac{\partial}{\partial r}\left(\frac{1}{r}\frac{\partial}{\partial r}(rv_r)\right)} + \frac{\partial^2 v_r}{\partial z^2}\right] \tag{1.2–61}$$

From Eq. 1.2–60 and the postulate that v_z does not depend on r we see that v_r must have the form $v_r = rf(z)$. Hence the dashed-underlined term in Eq. 1.2–61 vanishes. Integration of the equation of continuity over the closed region from $z = 0$ to $z = h$ and from $r = 0$ to $r = r$ gives:

$$2\pi r \int_0^h v_r\, dz + \pi r^2 \dot{h} = 0 \tag{1.2–62}$$

where $\dot{h} = dh/dt = v_z(z = h)$.

Then we integrate the equation of motion with respect to z, using the fact that $f = 0$ at $z = h$ and $f'(z) = 0$ at $z = 0$:

$$v_r = rf = \frac{h^2 r}{2}\left(-\frac{1}{\mu r}\frac{dp}{dr}\right)\left[1 - \left(\frac{z}{h}\right)^2\right] \tag{1.2–63}$$

Substitution of this into Eq. 1.2–62 gives:

$$-\frac{1}{\mu r}\frac{dp}{dr} = \frac{-3\dot{h}}{2h^3} \tag{1.2–64}$$

Combination of the last two equations gives:

$$v_r = \tfrac{3}{4}\frac{(-\dot{h})}{h} r\left[1 - \left(\frac{z}{h}\right)^2\right] \tag{1.2–65}$$

This is the radial velocity of the fluid in terms of h, which describes the plate motion. Equation 1.2–64 for the pressure gradient may be integrated to give:

$$p - p_a = \frac{3(-\dot{h})\mu R^2}{4h^3}\left[1 - \left(\frac{r}{R}\right)^2\right] \tag{1.2–66}$$

which is the radial pressure distribution. We also must calculate τ_{zz}:

$$\begin{aligned}\tau_{zz} &= -2\mu\frac{\partial v_z}{\partial z}\\ &= +2\mu\left(\frac{1}{r}\frac{\partial}{\partial r}(rv_r)\right)\\ &= 3\mu\left(\frac{-\dot{h}}{h}\right)\left[1 - \left(\frac{z}{h}\right)^2\right]\end{aligned} \tag{1.2–67}$$

Now the force F on one plate may be calculated:

$$F = \int_0^{2\pi} \int_0^R (p - p_a + \tau_{zz})|_{z=h}\, r\, dr\, d\theta$$

$$= 2\pi R^2 \int_0^1 (p - p_a) \left(\frac{r}{R}\right) d\left(\frac{r}{R}\right)$$

$$= \frac{3\pi R^4 \mu(-\dot{h})}{8h^3} \tag{1.2–68}$$

This is the *Stefan equation*, which shows how much force $F(t)$ must be applied in order to maintain the disk motion $h(t)$.

If we now ask what the disk motion will be for a *constant applied force* F, we have to solve the differential equation for $h(t)$ in Eq. 1.2–68 to give:

$$\frac{1}{h^2} - \frac{1}{h_0^{\,2}} = \frac{16Ft}{3\pi R^4 \mu} \tag{1.2–69}$$

This gives the disk separation as a function of the elapsed time.

It is possible to correct Eq. 1.2–68 for inertial effects by means of a perturbation approach.[15] This is done by using the result in Eq. 1.2–65 to compute the inertial terms $\rho[\partial v_r/\partial t + v_r(\partial v_r/\partial r) + v_z(\partial v_r/\partial z)]$ which are then put into the left side of Eq. 1.2–61. The steps described in Eqs. 1.2–63 and 68 can then be repeated and one finally obtains:

$$F = \frac{3\pi R^4 \mu(-\dot{h})}{8h^3}\left[1 + \frac{5\rho h(-\dot{h})}{7\mu} + \frac{2\rho h^2 \ddot{h}}{5\mu \dot{h}}\right] \tag{1.2–70}$$

This result can be used to estimate the importance of inertia effects in lubrication squeeze films and in parallel plate plastometers.

§1.3 POLYMER FLUID DYNAMICS

From the foregoing section it is evident that many kinds of fluid dynamics problems can be solved by starting with the equations of change and making appropriate simplifying assumptions. Of course, there are also many problems so complex that, even though we can write down the starting equations, they are virtually impossible to solve. But at least in classical fluid dynamics the fundamental equations for the subject are well known and thoroughly tested experimentally. Through the years many useful compilations of solutions to the equations of change have been published, and hence we have available many reference works to assist us in hydrodynamic problem solving.[1]

We now turn to the main subject of this book, namely, the motion of fluids composed entirely or in part of macromolecules, that is, molecules of very large molecular weight—over 10^4, say. These fluids are structurally complex, and because of their complicated nature, they behave quite differently from fluids like water, ethanol, and benzene. The differences

[15] S. Ishizawa, *JSME Bull.*, *9*, 533–550 (1966); D. C. Kuzma, *Appl. Sci. Res.*, *A18*, 15–20 (1967); A. F. Jones and S. D. R. Wilson, *J. Lubr. Technol.*, *97*, 101–104 (1975). R. J. Grimm, *Appl. Sci. Res.*, *32*, 000–000 (1976).

[1] For example, H. Lamb, *Hydrodynamics*, Dover, New York (1945); L. M. Milne-Thompson, *Theoretical Hydrodynamics*, Macmillan, New York, (1967) Fifth Edition; H. L. Dryden, F. D. Murnaghan, and M. Bateman, *Hydrodynamics*, Dover, New York (1956); R. Berker, "Intégrations des équations du mouvement d'un fluide visqueux incompressible," *Handbuch der Physik*, Springer, Heidelberg (1963), Volume VIII/2, pp. 1–384.

in behavior are not just small quantitative differences or slight nuances; the differences are, in fact, major qualitative differences, with a wide variety of new phenomena appearing. Hence we have to go back and reexamine the basic equations of hydrodynamics.

Certainly the equations of change in terms of the fluxes are not open to question, for they are the mathematical statements of well-known conservation laws. However, the expression for the momentum flux, given in Eq. 1.2–1, must now be replaced by some more complicated expression. It will turn out that, in general, the expression (or expressions) appropriate for macromolecular fluids will have to be nonlinear in the velocity gradients and will have to contain information about the time-dependent mechanical response of the fluids. The replacement of Eq. 1.2–1 by these much more complicated "rheological equations of state" requires then an entirely new field of study, and in fact all of the various aspects of classical hydrodynamics—analytical solutions, numerical analysis, stability theory, turbulence, boundary-layer theory, nonisothermal flows, reacting flows, surface effects—have to be reformulated and reexamined for macromolecular fluids. This is a colossal and challenging task!

We have mentioned structural *complexity* of the macromolecular fluids as the main factor in requiring that the momentum flux in Eq. 1.2–1 be replaced by some new kinds of expressions. A second important factor in dealing with macromolecular fluids is their *diversity*. Whereas all structurally simple fluids obey the momentum flux expression in Eq. 1.2–1, we cannot hope that a single expression for $\boldsymbol{\pi}$ will be capable of describing the many kinds of macromolecular fluids that exist in nature or can be synthesized by the organic chemist. Yet in spite of this diversity, we shall find that the situation is not as confusing as one might expect. The research in continuum mechanics and kinetic theory during the past quarter of a century has shown that these complex fluids of great diversity share many of the same "anomalous" effects and can be described by the same theoretical techniques. The next three chapters are devoted to the structure, flow phenomena, and material function measurements for polymeric fluids.

QUESTIONS FOR DISCUSSION

1. In Table 1.1–1 is it assumed that ρ is a constant in going from Eq. B to Eq. E?
2. Give the physical meaning of each of the terms in Eqs. 1.1–4, 1.1–8, and 1.1–13.
3. By means of a carefully labeled sketch, illustrate the meaning of the components of the quantity $\rho\mathbf{v}\mathbf{v}$.
4. In writing down the statement of conservation of energy in Eq. 1.1–11, why is no mention made of potential energy?
5. Discuss the physical meaning of each of the terms in the equations in Table 1.2–1. Verify that the equations are dimensionally consistent.
6. How would the velocity distribution in Fig. 1.2–1 appear if the sign of the pressure gradient were reversed?
7. Sketch the temperature and velocity profiles for the two parts of Example 1.2–2.
8. If you want to use Eq. 1.2–36 to find viscosity from measurements of flow rate vs. pressure drop, how would you estimate the error inherent in the final measurement? What approximations are implied in Eq. 1.2–36?
9. Explain carefully how one goes from Eq. 1.2–43 to Eq. 1.2–51.
10. Is the shear rate constant throughout the region occupied by the fluid in Fig. 1.2–2? In Fig. 1.2–5?
11. In Eq. 1.2–68, does the term τ_{zz} contribute to F? Explain in words the physical meaning of τ_{zz}.
12. What is the range of applicability of Stokes' law? Of the Hagen-Poiseuille law? Of the Stefan equation for squeezing flow?
13. What kinds of experiments can be used to measure viscosity?
14. How does viscosity depend on temperature and pressure?
15. Note that Eq. 1.2–13 predicts that the maximum temperature in the region between the plates does *not* depend on the spacing between the plates. Does this seem strange?
16. In Eq. 1.2–15, why is μ_0/μ expanded in a Taylor series rather than μ/μ_0?

PROBLEMS

1A.1 Volume Rate of Flow through a Circular Tube

The chlorinated biphenyl used by Sukanek and Laurence[1] in their research on viscous heating has the following properties at 313 K:

$$\mu = 400 \text{ poise} = 40 \text{ N·s/m}^2$$

$$\rho = 1.6 \text{ g/cm}^3 = 1.6 \times 10^3 \text{ kg/m}^3$$

$$k = 9.4 \times 10^3 \text{ g·cm/s}^3\text{·K} = 9.4 \times 10^{-2} \text{ W/m·K}$$

$$\hat{C}_p = 0.235 \text{ cal/g·C} = 983 \text{ J/kg·K}$$

a. What will the volume rate of flow be for the flow of this fluid through a horizontal capillary tube of radius 3.2 mm and length 10.3 cm when the pressure drop is 1.67×10^5 N/m^2?
b. Compute the Reynolds number for the flow in order to verify that the flow is laminar.
c. What will happen if the radius of the tube is doubled?

Answers: a. 1.67×10^{-6} m^3/s
b. 1.33×10^{-2}

1A.2 Terminal Velocity of a Falling Sphere

a. By making a force balance on a sphere falling in a liquid and using the result in Eq. 1.2–50, show that the terminal velocity (the velocity attained at steady state) is

$$v_t = 2R^2(\rho_s - \rho)g/9\mu \qquad (1\text{A}.2\text{–}1)$$

where ρ and ρ_s are the densities of the liquid and the sphere, respectively.
b. What will be the terminal velocity of a steel sphere of radius 0.25 cm with density 7850 kg/m^3 falling in the liquid in Problem 1A.1?
c. Verify that it was permissible to use Eq. 1A.2–1 by computing the Reynolds number for the flow.
d. What is the maximum radius of steel spheres that can be used and still use Stokes' law for describing the viscous drag on the sphere falling in the liquid of Problem 1A.1?

Answers: b. 2.13×10^{-3} m/s
c. 4.26×10^{-4}
d. 1.54 cm

1A.3 Measurement of Viscosity in a Cone-and-Plate Viscometer

The system in Fig. 1.2–5 has the following geometrical measurements:

$$R = 5.2 \text{ cm} = 0.052 \text{ m}$$

$$\theta_0 = 0.35° = 0.00611 \text{ rad}$$

[1] P. C. Sukanek and R. L. Laurence, *A.I.Ch.E. Journal*, *20*, 474–484 (1974).

With a fluid completely filling the gap, a torque of 2.47 N·m is required to maintain an angular velocity of 1.28 rad/s.

a. What is the shear stress $\tau_{\theta\phi}$ in the gap?
b. What is $\dot{\gamma}_{\theta\phi}$ in the gap?
c. What is the viscosity of the fluid?

Answers: a. 8400 N/m^2
b. -209 s^{-1}
c. 40 N·s/m^2

1A.4 Squeezing Flow Experiment

a. From Eq. 1.2–69 obtain a formula for the time required to squeeze out half of the material that is initially in the gap; call this quantity $t_{1/2}$.

b. For a silicone oil with a viscosity of 106 N·s/m^2 Leider[2] found experimentally that $t_{1/2}$ was 499 s for a squeezing flow system in which the disk radius was 2.54 cm, the initial disk separation (i.e., $2h_0$) was 0.01209 cm, and a mass of 4.07 kg was placed on the upper disk. What value of $t_{1/2}$ is calculated from the formula in (a)?

Answer: b. 535 s

1A.5 Viscous Heating in a Shear Flow

a. For the fluid in Problem 1A.1, Sukanek and Laurence have given a plot of $\log_{10} \mu$ vs. $1/T$. At 313 K this curve has a slope of 1.05×10^4 K. From this fact deduce the value of β_1 that should be used in Eq. 1.2–15.

b. What is the maximum speed that the upper surface can have if the temperature rise in the gap is not to exceed 0.2 K? The gap between the two plates is 3.142 mm.

Answers: a. $\beta_1 = 77$
b. $v_{0,\max} = 0.062$ m/s

1A.6 Calibration Constants for Rolling-Ball Viscometer[3]

The rolling-ball viscometer used by Šesták and Ambros[4] in their studies on non-Newtonian fluids had the following characteristics:

Tube:	Diameter, D (mm)	15.954		
	Length, L (mm)	100		
Balls:	Diameter, d (mm)	15.627	15.565	15.000
	Density, ρ_s (g/cm^3)	2.408	8.118	7.70
	Calibration constant, K (cm^2/s^2) (for $\beta = 80°$)	0.0007880	0.001282	0.01229

The calibration constants K are those recommended by the instrument manufacturer to be used in the equation:

$$\mu = K(\rho_s - \rho)t \tag{1A.6–1}$$

where t is the time required for the ball to fall a distance L.

[2] P. J. Leider, *Ind. Eng. Chem. Fundamentals*, *13*, 342–346 (1974).
[3] See Problem 1D.7.
[4] J. Šesták and F. Ambros, *Rheol. Acta*, *12*, 70–76 (1973).

a. Obtain an expression for the calibration constant K from the Lewis equation, Eq. 1D.7–1.

b. Compare the theoretical K values computed from (a) with those recommended by the manufacturer for Newtonian liquids.

1B.1 Flow between Parallel Plane Surfaces

a. Obtain expressions for the maximum velocity in the slit flow problem of Example 1.2–1 for various ranges of v_0 and $\partial p/\partial x$.

b. Rework Example 1.2–1 with the fixed surface at $y = B$ and the surface at $y = 0$ moving with velocity v_0.

c. Rewrite the velocity profile in Eq. 1.2–7 as $v_x(\bar{y})$ where $\bar{y}$ is a variable that goes from $-B/2$ at the lower plate to $+B/2$ at the upper plate.

d. How would Eq. 1.2–4 have to be modified if the viscosity were a function of position because of thermal gradients in the system?

e. A fluid is flowing axially in the annular space between cylinders of radii κR and R, respectively, where κ is just slightly less than unity—that is, in a very thin annulus. The inner cylinder moves axially with velocity v_0, and there is a pressure gradient $\partial p/\partial z$ in the axial direction in the annulus. Show how Eqs. 1.2–7 and 8 can be adapted to describe the velocity distribution and volume rate of flow through the thin annular region; it is convenient to use $\epsilon = 1 - \kappa$ in describing the geometry of the annulus.

1B.2 Axial Flow in an Annulus

a. An incompressible Newtonian fluid is in laminar flow in an annulus formed by two fixed coaxial cylinders of radii κR and R, where κ is less than unity. By following the procedure of Example 1.2–3, obtain the following expressions for the velocity distribution and the volume rate of flow:

$$v_z = \frac{(\mathscr{P}_0 - \mathscr{P}_L)R^2}{4\mu L}\left[1 - \left(\frac{r}{R}\right)^2 + \left(\frac{1-\kappa^2}{\ln 1/\kappa}\right)\ln\frac{r}{R}\right] \tag{1B.2–1}$$

$$Q = \frac{\pi(\mathscr{P}_0 - \mathscr{P}_L)R^4}{8\mu L}\left[(1-\kappa^4) - \frac{(1-\kappa^2)^2}{\ln(1/\kappa)}\right] \tag{1B.2–2}$$

in which $\mathscr{P}_0$ and $\mathscr{P}_L$ are the values of the modified pressure at the entrance ($z = 0$) and the exit ($z = L$) to the annulus.

b. Show how the result in Eq. 1B.2–2 simplifies for the case of a very thin annulus, where $\epsilon = 1 - \kappa$ is very small. Do this by expanding Q in powers of ϵ. This operation involves using the first four terms of the Taylor series

$$\ln(1-\epsilon) = -\epsilon - \tfrac{1}{2}\epsilon^2 - \tfrac{1}{3}\epsilon^3 - \tfrac{1}{4}\epsilon^4 - \cdots \tag{1B.2–3}$$

and then performing a long division. Show that one obtains the same result as in Problem 1B.1(e).

c. Obtain the velocity profile and volume rate of flow for the axial annular flow in which there is no axial pressure gradient, the annulus is arranged horizontally, and the inner tube is moving axially with a velocity v_0.

1B.3 Effect of Compressibility on Volume Rate of Flow in Tubes

In Eqs. 1.2–37 to 40 it is shown how the Hagen-Poiseuille result may be used to obtain the volume rate of flow through a circular tube of varying cross section. The same method may be used for approximately taking into account the effect of fluid compressibility.

a. Assume a simple linear relationship between the density and pressure:

$$\rho = \rho_0\left[1 - a\left(\frac{p_0 - p}{p_0 - p_L}\right)\right] \tag{1B.3–1}$$

in which ρ_0 is the density at $p = p_0$, and a is a constant. Then show that when the Hagen-Poiseuille law is applied locally one obtains the following differential equation for the density:

$$w = \frac{\pi R^4 \rho}{8\mu}\left[-\frac{p_0 - p_L}{a\rho_0}\frac{d\rho}{dz} + \rho g \cos\chi\right] \tag{1B.3-2}$$

where w is the mass rate of flow through the tube. Show that this equation can be solved by using ρ^2 as the dependent variable, and that the mass rate of flow through the tube is:

$$w = \frac{\pi R^4 \rho_0{}^2 g \cos\chi}{8\mu}\left[\frac{(1 - a)^2 - b}{1 - b}\right] \tag{1B.3-3}$$

where $b = \exp\left[2a\rho_0 gL \cos\chi/(p_0 - p_L)\right]$.

b. How does this result have to be modified for $\chi = 90°$?

1B.4 Flow between Coaxial Cylinders and Spheres

a. The space between two coaxial cylinders is filled with an incompressible, Newtonian fluid at constant temperature. The radii of the inner and outer wetted surfaces are κR and R, respectively. The angular velocities of rotation of the inner and outer cylinders are W_i and W_o, respectively. Determine the velocity distribution of the fluid, and the torques on the two cylinders needed to maintain the motion.

b. Repeat part (a) for two concentric spheres. Assume that the flow is sufficiently slow that no secondary flows occur.

c. Obtain the temperature profiles for the system in (a) resulting from viscous dissipation when both cylinders are maintained at temperature T_0; assume constant physical properties.

Answers: a. $$v_\theta = \frac{\kappa R}{1 - \kappa^2}\left[(W_o - W_i\kappa^2)\left(\frac{r}{\kappa R}\right) + (W_i - W_o)\left(\frac{\kappa R}{r}\right)\right]$$

b. $$v_\phi = \frac{\kappa R}{1 - \kappa^3}\left[(W_o - W_i\kappa^3)\left(\frac{r}{\kappa R}\right) + (W_i - W_o)\left(\frac{\kappa R}{r}\right)^2\right]\sin\theta$$

1B.5 Laminar Newtonian Flow in a Triangular Duct[1]

A straight duct has a length L and a cross section of triangular shape, bounded by the plane surfaces $y = H$, $y = \sqrt{3}x$, and $y = -\sqrt{3}x$. Verify that the velocity profile for the laminar flow of a Newtonian fluid in a duct of this type is:

$$v_z = \frac{(\mathscr{P}_0 - \mathscr{P}_L)}{4\mu LH}(y - H)(3x^2 - y^2) \tag{1B.5-1}$$

Then obtain the volume rate of flow:

$$Q = \frac{\sqrt{3}(\mathscr{P}_0 - \mathscr{P}_L)H^4}{180\mu L} \tag{1B.5-2}$$

[1] L. D. Landau and E. M. Lifshitz, *Fluid Mechanics*, Addison-Wesley, Reading, Mass. (1959), p. 58. See also R. Berker, *Handbuch der Physik*, Springer, Berlin (1963), Vol. VIII/2, pp. 67–77, for a summary of formulas for flow in conduits of various cross sections. See Problem 1D.4 for the use of calculus of variations for solving problems of this type.

1B.6 Laminar Flow in a Square Duct[2]

A straight duct has a length L and a cross section of square shape, bounded by the plane surfaces $x = \pm B$, $y = \pm B$. Is the correct velocity distribution for this flow given by the following expression?

$$v_z = \frac{(\mathscr{P}_0 - \mathscr{P}_L)B^2}{4\mu L}\left[1 - \left(\frac{x}{B}\right)^2\right]\left[1 - \left(\frac{y}{B}\right)^2\right] \tag{1B.6–1}$$

1B.7 Simplification of Expression for the Momentum Flux Tensor

a. Show that the general expression in Eq. 1.2–1 simplifies to $\pi_{yx} = -\mu\, dv_x/dy$ for the simple shearing flow $v_x = v_x(y)$, $v_y = 0$, $v_z = 0$.

b. Show how Eq. 1.2–1 simplifies for the flow in an incompressible liquid surrounding an expanding gas bubble; in the liquid $v_r = v_r(r,t)$, $v_\theta = v_\phi = 0$.

1B.8 Velocity Distribution for Flow around a Sphere

Use Eq. 1.2–43 to write out the expressions for v_r, v_θ, v_ϕ as functions of r and θ for the special case that $\mathbf{v}_\infty = \boldsymbol{\delta}_z v_\infty$. Show also how the pressure distribution in Eq. 1.2–44 can be simplified for this case.

1B.9 Parallel-Disk Viscometer

A schematic diagram is given in Fig. 1B.9 of a parallel-disk viscometer. The fluid is placed in the gap of thickness B between the two circular disks. Develop a formula for deducing the fluid viscosity from the measurement of the torque $\mathscr{T}$ required to turn the upper disk at an angular velocity W.

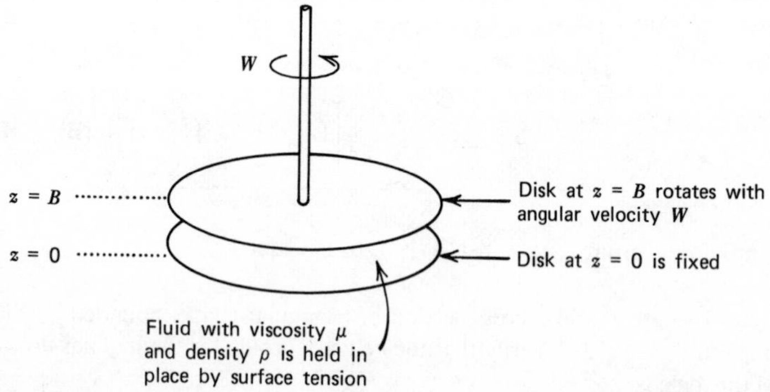

FIGURE 1B.9. Parallel-disk viscometer with two disks of radius R separated by a distance B, with dimensions such that $R \gg B$.

a. Postulate that for small values of W the velocity and pressure profiles have the form: $v_r = 0$, $v_z = 0$, and $v_\theta = rf(z)$; and $p = p(r,z)$. Write down the resulting simplified equations of continuity and motion.

b. From the θ-component of the equation of motion, obtain a differential equation for $f(z)$. Solve this equation and evaluate the constants of integration. Finally obtain the velocity distribution $v_\theta = Wrz/B$.

c. Show that the torque required to turn the upper disk is $\mathscr{T} = \pi\mu WR^4/2B$.

[2] M. J. Boussinesq, *J. Math. Pures et Appliquées*, Série 2, *13*, 377–424 (1868).

1B.10 Radial Flow between Parallel Disks[3]

Consider the flow system in Fig. 1B.10. Fluid enters the system through the vertical tube and then flows radially between the two circular disks. It is desired to obtain the volume rate of flow in terms of the geometrical variables, the fluid viscosity, and the difference in the pressures at r_1 and r_2.

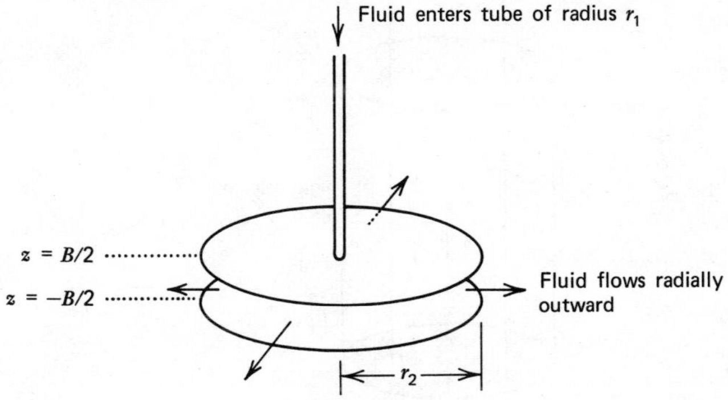

FIGURE 1B.10. Radial flow between two disks of radius r_2 separated by a distance B, with $B \ll r_2$.

a. First apply Eq. 1.2–8 locally by recognizing that at all points between the disks the flow will be similar to flow between parallel plates, provided that the radial velocity is not too large. This leads directly to an equation for p as a function of r, which is easily integrated. The final result is:

$$Q = \frac{\pi B^3(p_1 - p_2)}{6\mu \ln r_2/r_1} \tag{1B.10–1}$$

b. Obtain the same result by solving the equations of continuity and motion, neglecting the inertial terms.[4] This leads to the velocity distribution:

$$v_r(r,z) = \frac{B^2(p_1 - p_2)}{8\mu r \ln r_2/r_1}\left[1 - \left(\frac{z}{B/2}\right)^2\right] \tag{1B.10–2}$$

which can then be integrated to give the volume rate of flow in Eq. 1B.10–1.

1B.11 Steady Simple Elongational Flow and Elongational Viscosity

Here we study the type of flow that occurs when a cylindrical filament of fluid is stretched slowly in the absence of external body forces (see Fig. 1B.11). For steady-state simple elongational flow:

$$v_z = \dot{\epsilon} z \qquad v_r = -\tfrac{1}{2}\dot{\epsilon} r \qquad v_\theta = 0 \tag{1B.11–1}$$

in which $\dot{\epsilon}$ is the (constant) elongation rate.

a. Verify that this velocity distribution satisfies the equation of continuity identically. Show further that for an incompressible Newtonian fluid in steady simple elongational flow:

$$\tau_{rr} = \tau_{\theta\theta} = \mu\dot{\epsilon} \qquad \tau_{zz} = -2\mu\dot{\epsilon} \tag{1B.11–2}$$

[3] See R. B. Bird, W. E. Stewart, and E. N. Lightfoot, *Transport Phenomena*, Wiley, New York (1960), p. 114.
[4] This problem has been studied exhaustively by J. -L. Peube [*J. de Mécanique*, *2*, 337–395 (1963)] theoretically, and by C. Che-Pen [*ibid.*, *5*, 245–259 (1966)] experimentally.

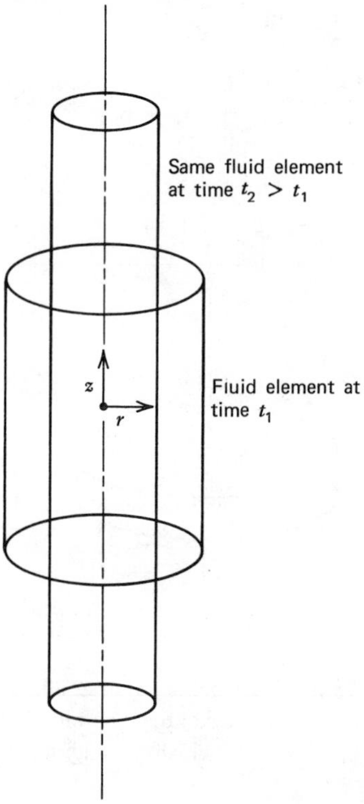

FIGURE 1B.11. Steady simple elongational flow, showing how a cylindrical portion of fluid deforms with time. The velocity components are $v_z = \dot{\epsilon}z$, $v_r = -\frac{1}{2}\dot{\epsilon}r$, $v_\theta = 0$ where $\dot{\epsilon}$ is the (constant) elongation rate.

b. Next use the equation of motion to gain further information about the components of the stress tensor. For steady-state flow with negligible inertial terms ($\nabla \cdot \rho \boldsymbol{v}\boldsymbol{v} = \mathbf{0}$) and no external forces ($\boldsymbol{g} = \mathbf{0}$), show that the components of $\boldsymbol{\pi}$ are constant throughout the fluid, and that specifically $p + \tau_{rr} = p_0$, where p_0 is the ambient pressure outside of the fluid fiber.

c. Show that the results of (b) lead to the conclusion that $\pi_{zz} - p_0 = \tau_{zz} - \tau_{rr} = -3\mu(dv_z/dz)$. The coefficient of proportionality between the normal stress difference, $\tau_{zz} - \tau_{rr}$, and the negative of the elongation rate is called $\bar{\eta}$, the *elongational viscosity*[5] or Trouton viscosity.[6] Thus for Newtonian fluids we have the well-known result that:

$$\bar{\eta} = 3\mu \tag{1B.11-3}$$

This result has been used for calculating the shape of a freely falling jet of liquid.[7]

1B.12 Flow in a Duct with a Converging Section[8]

A viscous fluid flows slowly in the system in Fig. 1B.12. In the limit of very slow flow it is a good approximation to assume that there is a parabolic profile at each position along the tube, with the maximum velocity so adjusted that the amount of fluid flowing through each cross section is a constant.

[5] A. S. Lodge, *Elastic Liquids*, Academic Press, New York (1964), pp. 97–98, 114–118. See also §4.6.
[6] F. T. Trouton, *Proc. Roy. Soc.*, *A77*, 426 (1906).
[7] W. R. Marshall, Jr., and R. L. Pigford, *The Application of Differential Equations to Chemical Engineering Problems*, University of Delaware Press, Newark, Delaware (1947). See also Problem 1C.10.
[8] A study of this system has been made by J. L. Sutterby, Ph.D. Thesis, University of Wisconsin, Madison (1964).

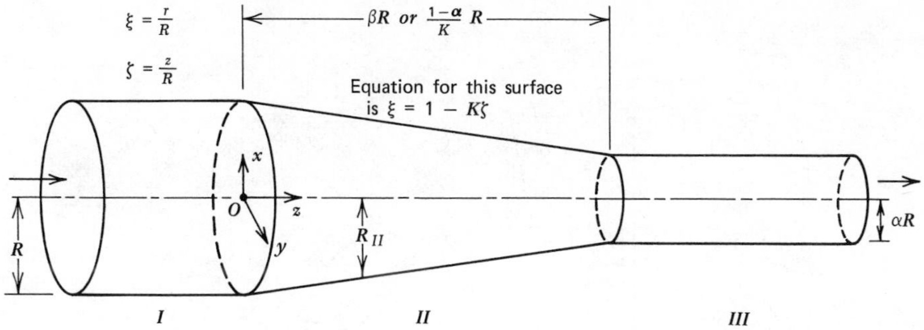

FIGURE 1B.12. Flow in a converging section; the *maximum* velocity in section I is v_I, and the pressure at O is p_0.

a. Show that the velocity profiles in the three sections of the tube are:

Region I:
$$\frac{v_z}{v_I} = 1 - \xi^2 \tag{1B.12–1}$$

Region II:
$$\frac{v_z}{v_I} = \frac{1}{(1 - K\zeta)^2} - \frac{\xi^2}{(1 - K\zeta)^4} \tag{1B.12–2}$$

Region III:
$$\frac{v_z}{v_I} = \frac{1}{\alpha^2} - \frac{\xi^2}{\alpha^4} \tag{1B.12–3}$$

b. Show that the pressure profiles are:

Region I:
$$\frac{p_0 - p}{4\mu v_I/R} = \zeta \tag{1B.12–4}$$

Region II:
$$\frac{p_0 - p}{4\mu v_I/R} = \frac{1}{3K}\left[\frac{1}{(1 - K\zeta)^3} - 1\right] \tag{1B.12–5}$$

Region III:
$$\frac{p_0 - p}{4\mu v_I/R} = \frac{\zeta}{\alpha^4} - \frac{(1 - \alpha)^2(\alpha^2 + 2\alpha + 3)}{3K\alpha^4} \tag{1B.12–6}$$

Draw a sketch of the pressure profiles in the system.

1B.13 Stress on a Plane of Arbitrary Orientation

a. The force per unit area acting across a surface perpendicular to the x-direction is given the symbol $\boldsymbol{\pi}_x$; this force has components π_{xx}, π_{xy}, and π_{xz}. Thus π_{xz} is the stress in the z-direction on a surface perpendicular to the x-direction. Keep in mind that $\boldsymbol{\pi}_x$ is the force per unit area exerted *by* the material on the negative-x side *on* the material on the positive-x side. Verify that $\boldsymbol{\pi}_x$ is the same as $[\boldsymbol{\delta}_x \cdot \boldsymbol{\pi}]$ where $\boldsymbol{\delta}_x$ is the unit vector in the x-direction.

b. Denote by $\boldsymbol{\pi}_n$ the force per unit area acting across a surface with orientation given by the unit vector $\boldsymbol{n}$; here again we mean the force exerted from the negative side toward the positive side. Show that $\boldsymbol{\pi}_n = [\boldsymbol{n} \cdot \boldsymbol{\pi}]$. Do this by considering an infinitesimal element of volume as shown in Fig. 1B.13, which is at equilibrium under the forces acting on all four faces. It is necessary to show first that the area of the face OBC is $(\boldsymbol{n} \cdot \boldsymbol{\delta}_x)S$, where S is the area of face ABC.

c. Is $[\boldsymbol{n} \cdot \boldsymbol{\pi}]$ the same as $[\boldsymbol{\pi} \cdot \boldsymbol{n}]$?

d. Is $([\boldsymbol{n} \cdot \boldsymbol{\pi}] \cdot \boldsymbol{v})$ the same as $(\boldsymbol{n} \cdot [\boldsymbol{\pi} \cdot \boldsymbol{v}])$, where $\boldsymbol{v}$ is an arbitrary vector?

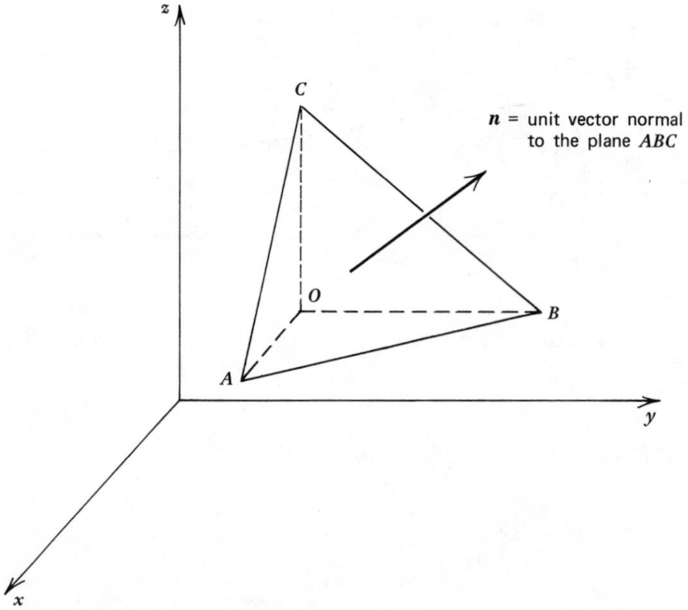

FIGURE 1B.13. Infinitesimal volume element with slanting face ABC of area S and orientation given by $\boldsymbol{n}$.

1B.14 Flow in a Falling Film

a. A film of liquid is flowing down an inclined plate, which makes an angle β with the vertical (i.e., $\beta = 0$ would be a vertical plate). If the z-direction is the direction of flow, and x is the distance into the film from the liquid-air interface, show that the liquid velocity distribution is:

$$v_z = \frac{\rho g \delta^2 \cos \beta}{2\mu}\left[1 - \left(\frac{x}{\delta}\right)^2\right] \tag{1B.14–1}$$

and that the volume rate of flow in a film of width W is:

$$Q = \frac{\rho g W \delta^3 \cos \beta}{3\mu} \tag{1B.14–2}$$

In these equations μ is the liquid viscosity, δ is the thickness of the film, and g is the acceleration due to gravity.

b. Adapt the above result to describe the flow of the film that results when a cylindrical jet of liquid impinges on the vertex of a right circular cone, as shown in Fig. 1B.14. Show that the film thickness depends on the distance s from the cone apex in the following way:

$$\delta = \sqrt[3]{\frac{3\mu Q}{\pi \rho g L \sin 2\beta}}\left(\frac{L}{s}\right)^{1/3} \tag{1B.14–3}$$

1B.15 Alternative Form of the Energy Equation

It is desired to show how the energy equation in terms of the internal energy (Eq. C of Table 1.1–1) can be transformed into the energy equation in terms of the temperature (Eq. C of Table 1.2–1).

a. First show that $(\boldsymbol{\pi} : \nabla \boldsymbol{v}) = p(\nabla \cdot \boldsymbol{v}) + (\boldsymbol{\tau} : \nabla \boldsymbol{v})$ by using Eq. 1.2–1 and the definitions of the ∇-operations.

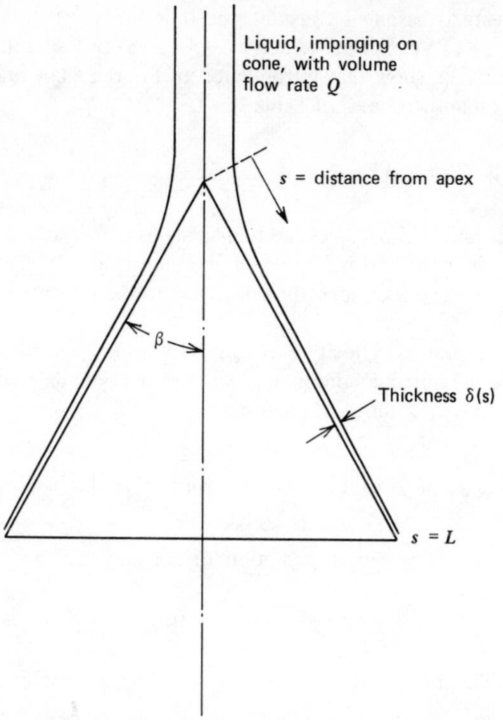

FIGURE 1B.14. Flow of a film down a cone.

b. Next replace $\hat{U}$ by $\hat{H} - p\hat{V}$, where $\hat{H}$ is the enthalpy per unit mass. Show that

$$\frac{D}{Dt} p\hat{V} = \frac{p}{\rho}(\nabla \cdot v) + \frac{1}{\rho}\frac{Dp}{Dt} \tag{1B.15–1}$$

by using the equation of continuity (Eq. D of Table 1.1–1). This enables us to rewrite the energy equation as an equation of change for enthalpy:

$$\rho \frac{D\hat{H}}{Dt} = -(\nabla \cdot q) - (\tau : \nabla v) + \frac{Dp}{Dt} \tag{1B.15–2}$$

c. Next we assume that $\hat{H}$ can be expressed in terms of two state variables p and T, so that we can use the thermodynamic relation

$$\begin{aligned} d\hat{H} &= \left(\frac{\partial \hat{H}}{\partial T}\right)_p dT + \left(\frac{\partial \hat{H}}{\partial p}\right)_T dp \\ &= \hat{C}_p\, dT + \left[\hat{V} - T\left(\frac{\partial \hat{V}}{\partial T}\right)_p\right] dp \end{aligned} \tag{1B.15–3}$$

Use of this relation enables us to rewrite Eq. 1B.15–2 as:

$$\rho \hat{C}_p \frac{DT}{Dt} = -(\nabla \cdot q) + \left(\frac{\partial \ln \hat{V}}{\partial \ln T}\right)_p \frac{Dp}{Dt} - (\tau : \nabla v) \tag{1B.15–4}$$

Note that up to this point the only assumption that has been made is that $\hat{H}$ depends only on two state variables, and not on the state of strain in the system.

d. When the density is assumed constant, the $(\partial \ln \hat{V}/\partial \ln T)_p$ term in Eq. 1B.15–4 can be omitted. Next show that $(-\boldsymbol{\tau} : \nabla \boldsymbol{v})$ can be written as $+\frac{1}{2}\mu(\dot{\boldsymbol{\gamma}} : \dot{\boldsymbol{\gamma}})$ by using the expression in Eq. 1.2–2 for incompressible fluids. Then, show that introduction of Fourier's law and assumption of constant thermal conductivity leads finally to Eq. C of Table 1.2–1.

1B.16 The Rayleigh Problem[9]

An incompressible Newtonian fluid occupies the half-space from $y = 0$ to $y = \infty$, and is bounded at $y = 0$ by a solid plane surface. At and before $t = 0$, the fluid is at rest. Then at time $t = 0$ the solid surface at $y = 0$ is made to move with constant velocity v_0 in the positive x-direction. It is desired to find the velocity distribution $v_x = v_x(y,t)$.

a. Introduce a dimensionless velocity $\phi = v_x/v_0$. Inasmuch as there is no natural unit of length for the problem, and since the initial condition at $t = 0$ and the boundary condition at $y = \infty$ are the same, it seems reasonable to try a solution of the form

$$\phi = \phi(\eta) \qquad \text{where} \qquad \eta = \left(\frac{\rho y^2}{4\mu t}\right)^{1/2} \tag{1B.16–1}$$

Show that with these variables the equation of motion for this flow becomes:

$$\phi'' + 2\eta\phi' = 0 \tag{1B.16–2}$$

in which primes denote differentiation with respect to η. What are the boundary conditions that go with this equation?

b. Show that this equation can be solved to give:

$$\phi = 1 - \frac{2}{\sqrt{\pi}} \int_0^\eta e^{-\eta^2}\, d\eta = 1 - \operatorname{erf} \sqrt{\frac{\rho y^2}{4\mu t}} \tag{1B.16–3}$$

in which $\operatorname{erf} \eta$ is the error function of η.

1C.1 Circulating Flow in an Annulus

A rod (radius κR) moves upward with constant velocity v_0 through a cylindrical container (radius R) containing an incompressible viscous fluid. The fluid circulates in the cylinder, moving upward along the moving central core and moving back downward along the fixed wall. It is desired to find the velocity distribution in the annular region, away from the end disturbances (see Fig. 1C.1).

a. First consider the problem when the annular region is quite narrow—that is, κ is just slightly less than 1. In that case the annulus may be approximated by a thin plane slit, and its curvature can be neglected. Show that in this limit, the velocity distribution is given by:

$$\frac{v_z}{v_0} = 3\zeta^2 - 4\zeta + 1 \tag{1C.1–1}$$

where $\zeta = (\xi - \kappa)/(1 - \kappa)$, and $\xi = r/R$.

b. Next, work the problem without the thin-slit assumption. Show that the velocity distribution is given by:

$$\frac{v_z}{v_0} = \frac{(1 - \xi^2)\left(1 - \dfrac{2\kappa^2}{1 - \kappa^2} \ln \dfrac{1}{\kappa}\right) - (1 - \kappa^2) \ln \dfrac{1}{\xi}}{(1 - \kappa^2) - (1 + \kappa^2) \ln \dfrac{1}{\kappa}} \tag{1C.1–2}$$

[9] Lord Rayleigh (John William Strutt), *Phil. Mag.* (6), *21*, 697–711 (1911). See also R. B. Bird, W. E. Stewart, and E. N. Lightfoot, *op. cit.* pp. 124–126.

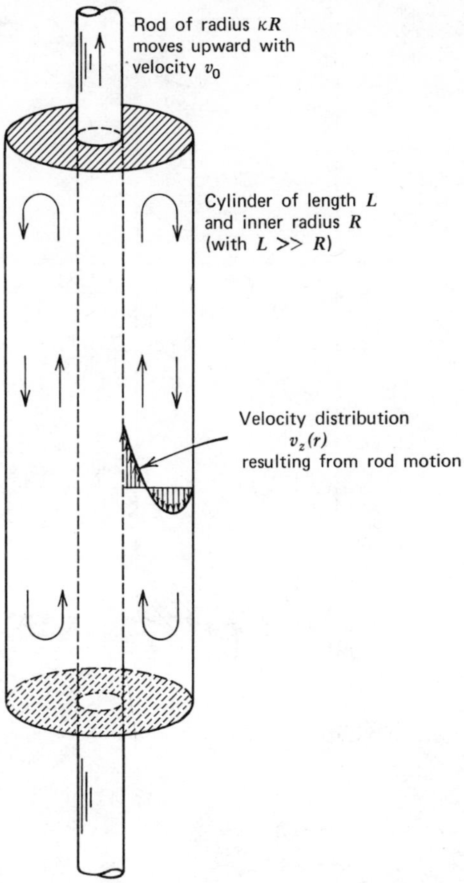

FIGURE 1C.1. Steady-state circulating flow in an annulus with the interior surface moving axially.

1C.2 Analysis of a Falling Cylinder Viscometer[1]

A falling cylinder viscometer consists of a long vertical cylinder (radius R) and a cylindrical slug (radius κR and height H) as shown in Fig. 1C.2. The slug is equipped with very thin fins so that its axis is maintained coincident with the axis of the tube.

One can observe the rate of descent of the cylindrical slug in the cylinder when the latter is filled with an incompressible fluid. It is desired to obtain an equation that gives the viscosity of the fluid in terms of the speed of fall, v_0, and the various geometric quantities. It is, of course, assumed that the measurement is made when a constant velocity of descent has been attained.

a. First show that the velocity distribution in the annular slit is given by:

$$\frac{v_z}{v_0} = -\frac{(1 - \xi^2) - (1 + \kappa^2)\ln(1/\xi)}{(1 - \kappa^2) - (1 + \kappa^2)\ln(1/\kappa)} \tag{1C.2–1}$$

in which $\xi = r/R$ is a dimensionless radial coordinate.

b. Make a force balance on the cylindrical slug and obtain finally:

$$\mu = \frac{(\rho_0 - \rho)g(\kappa R)^2}{2v_0}\left[\left(\ln\frac{1}{\kappa}\right) - \left(\frac{1 - \kappa^2}{1 + \kappa^2}\right)\right] \tag{1C.2–2}$$

in which ρ is the density of the fluid and ρ_0 is the density of the slug.

[1] J. Lohrenz, G. W. Swift, and F. Kurata, *A.I.Ch.E. Journal*, *6*, 547–550 (1960); *ibid.*, *7*, 6S (1961); E. Ashare, R. B. Bird, and J. A. Lescarboura, *A.I.Ch.E. Journal*, *11*, 910–916 (1965).

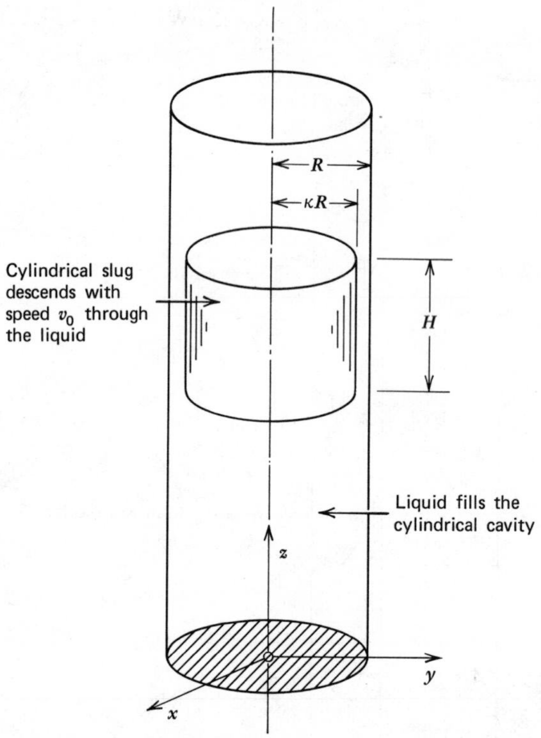

FIGURE 1C.2. Falling-cylinder viscometer, consisting of a tightly fitting cylindrical slug that descends through the liquid, which is forced upwards through the annular space.

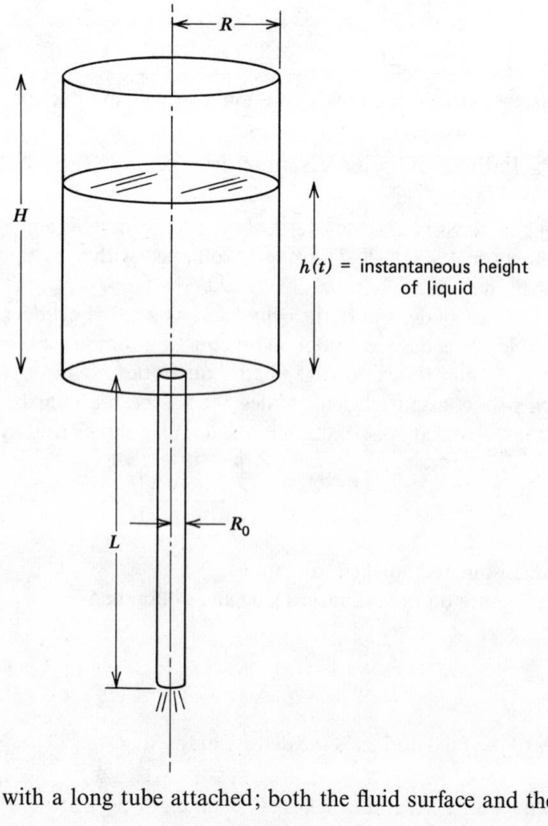

FIGURE 1C.3. Tank with a long tube attached; both the fluid surface and the tube exit are open to the atmosphere.

c. Verify that for small slit widths the result in (b) may be expanded in powers of $\epsilon = 1 - \kappa$ to give:

$$\mu = \frac{(\rho_0 - \rho)gR^2\epsilon^3}{6v_0}\left(1 - \tfrac{1}{2}\epsilon - \tfrac{13}{20}\epsilon^2 + \cdots\right) \tag{1C.2–3}$$

d. Draw a sketch showing the pressure distribution in the system.

e. Which of the two forces—pressure or viscous drag—is primarily responsible for balancing the gravitational force on the slug?

1C.3 Draining of a Tank with an Exit Tube[2]

a. The tank-and-tube assembly in Fig. 1C.3 is initially filled with an incompressible Newtonian liquid of viscosity μ and density ρ. During the draining process it is assumed that the flow in the tube is laminar. Show that the time required to drain the tank (but not the pipe), t_{efflux}, is given by:

$$t_{\text{efflux}} = \frac{8\mu LR^2}{\rho g R_0^{\,4}} \ln\left(1 + \frac{H}{L}\right) \tag{1C.3–1}$$

What assumptions were made in the derivation?

b. Rework the problem in which the tank is conical instead of cylindrical.

1C.4 Inertial Effects in a Parallel-Plate Plastometer

Derive Eq. 1.2–70 by following the procedure outlined in the text, using a perturbation analysis.

1C.5 Residence-Time Distribution in Tube Flow[3]

The residence-time distribution function $F(t)$ for a steady-state continuous flow system is defined as the volume fraction of the fluid leaving the system that has resided in the system during a time less than t. The average residence time $\bar{t}$ for the system is then given by:

$$\bar{t} = \int_0^1 t \, dF(t) \tag{1C.5–1}$$

Show that for the Newtonian flow through a circular tube, discussed in Example 1.2–3:

$$F(t) = 0 \qquad \text{for} \qquad 0 < t/\bar{t} < \tfrac{1}{2}$$

$$F(t) = 1 - \left(\frac{\bar{t}}{2t}\right)^2 \qquad \text{for} \qquad t/\bar{t} \geq \tfrac{1}{2} \tag{1C.5–2}$$

1C.6 Viscous Heating Effects in Shear Flow[4]

In Example 1.2–2(b) it was shown how viscous heating problems with variable properties can be solved by means of perturbation methods. If one assumes that the thermal conductivity does not vary with

[2] R. B. Bird, W. E. Stewart, and E. N. Lightfoot, *Transport Phenomena*, Wiley, New York (1960), p. 237.

[3] See, for example, H. Kramers and K. R. Westerterp, *Elements of Chemical Reactor Design and Operation*, Netherlands University Press, Amsterdam (1963), Chapter III, and O. Levenspiel, *Chemical Reaction Engineering*, Wiley, New York (1972), Second Edition, Chapter 9.

[4] R. M. Turian and R. B. Bird, *Chem. Eng. Sci.*, *18*, 689–696 (1963); J. Gavis and R. L. Laurence, *Ind. Eng. Chem. Fundamentals*, 7, 232–239 (1968). In both of these references other boundary conditions are considered as well.

temperature and that the viscosity varies in an exponential fashion:

$$\frac{k}{k_0} = 1 \qquad (\text{or } \alpha_j = 0) \tag{1C.6–1}$$

$$\frac{\mu}{\mu_0} = \exp(-\beta\Theta) \qquad \left(\text{or } \beta_j = \frac{\beta^j}{j!}\right) \tag{1C.6–2}$$

then an analytical solution can be obtained. This means that the problem can be reformulated thus (after performing one integration of the equation of motion):

$$e^{-\beta\Theta}\frac{d\phi}{d\eta} = S \tag{1C.6–3}$$

$$\frac{d^2\Theta}{d\eta^2} + \text{Br}\, e^{-\beta\Theta}\left(\frac{d\phi}{d\eta}\right)^2 = 0 \tag{1C.6–4}$$

in which S is a constant of integration; note that $S = \tau_0 B/\mu_0 v_0$ where τ_0 is the shear stress applied to the upper plate, so that although S is an unknown constant at this point it can be interpreted as a dimensionless wall shear stress.

When these last two equations are combined, show that the temperature distribution is found to be:

$$e^{\beta\Theta} = \left(\frac{8\beta C_1{}^2}{\text{Br}\, S^2}\right)\text{sech}^2\,[2\beta C_1(\eta + C_2)] \tag{1C.6–5}$$

in which C_1 and C_2 are additional constants of integration. Show that application of the boundary conditions on the temperature then gives:

$$C_2 = -\tfrac{1}{2} \tag{1C.6–6}$$

$$\frac{8\beta C_1{}^2}{\text{Br}\, S^2}\,\text{sech}^2\,\beta C_1 = 1 \tag{1C.6–7}$$

This last relation gives C_1 in terms of the still unknown constant S. Show that no solution can be obtained for C_1 for $\beta\,\text{Br}\,S^2 > 3.5138$. What happens for $\beta\,\text{Br}\,S^2 \leq 3.5138$? What is the physical interpretation of this situation? Verify that the temperature distribution can be rewritten as:

$$e^{\beta\Theta} = \frac{\text{sech}^2\,[2\beta C_1(\eta - \frac{1}{2})]}{\text{sech}^2\,\beta C_1} \tag{1C.6–8}$$

Next show that the velocity equation can be integrated to give:

$$\phi = \frac{S}{2\beta C_1}\,\frac{\tanh\,[2\beta C_1(\eta - \frac{1}{2})]}{\text{sech}^2\,\beta C_1} + C_3 \tag{1C.6–9}$$

in which C_3 is an additional constant of integration. Then apply the boundary conditions on the velocity to determine C_3 and S:

$$C_3 = +\tfrac{1}{2} \tag{1C.6–10}$$

$$S = \frac{2\beta C_1}{\sinh 2\beta C_1} \tag{1C.6–11}$$

Equations 1C.6–7 and 11 can be combined to give C_1 and S:

$$\beta C_1 = \text{arcsinh}\,\sqrt{\beta\,\text{Br}/8} \tag{1C.6–12}$$

$$S = \frac{\text{arcsinh}\,\sqrt{\beta\,\text{Br}/8}}{\sqrt{(\beta\,\text{Br}/8) + (\beta\,\text{Br}/8)^2}} \tag{1C.6–13}$$

This last expression can be used to prepare a graph of the wall shear stress versus the wall velocity, in suitable dimensionless quantities (see Fig. 1C.6). It can be seen that S cannot exceed the value 0.6255. Can it be concluded from the figure that the wall shear stress first increases and then decreases with increasing wall velocity?

Show that the foregoing results can be combined to give for the temperature and velocity profiles:

$$e^{\beta\Theta} = \left(1 + \frac{\beta\,\text{Br}}{8}\right)\text{sech}^2\left[(2\eta - 1)\,\text{arcsinh}\,\sqrt{\frac{\beta\,\text{Br}}{8}}\right] \tag{1C.6–14}$$

$$\phi = \tfrac{1}{2}\left\{\sqrt{1 + \frac{8}{\beta\,\text{Br}}}\tanh\left[(2\eta - 1)\,\text{arcsinh}\,\sqrt{\frac{\beta\,\text{Br}}{8}}\right] + 1\right\} \tag{1C.6–15}$$

Plots of the profiles for both isothermal and adiabatic wall conditions have been given by Gavis and Laurence.[4]

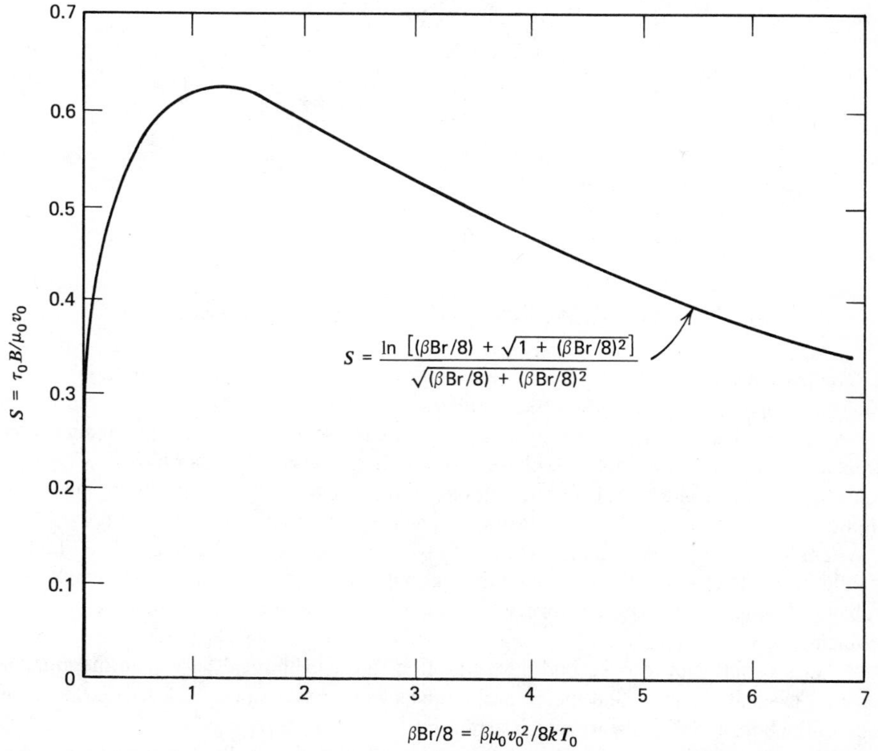

FIGURE 1C.6. Shear stress at the wall τ_0 as a function of wall speed v_0, presented in dimensionless form.

1C.7 Rotating Sphere in Viscous Liquid

A sphere of radius R is made to rotate with an angular velocity W in a Newtonian liquid of infinite extent, which is quiescent far from the sphere.

a. Find the velocity distribution in the surrounding liquid when the rotation is so slow that the centrifugal forces can be neglected. Show that

$$v_\phi = \frac{R^3 W \sin\theta}{r^2} \tag{1C.7–1}$$

b. Show that the torque required to maintain the motion is

$$\mathcal{T} = 8\pi\mu R^3 W \tag{1C.7–2}$$

Obtain this result by two methods: (1) compute the torque by integrating the product of the surface shear stress and the lever arm over the surface of the sphere, and (2) by equating the rate of doing work, $\mathcal{T}W$, to the rate of energy dissipation in the liquid outside the sphere.

c. Draw a sketch to show what the flow field will look like when inertial effects are included.

d. What is the fluid angular velocity field for this flow, where the fluid angular velocity is defined as one half the "vorticity," $\boldsymbol{w} = \frac{1}{2}[\nabla \times \boldsymbol{v}]$?

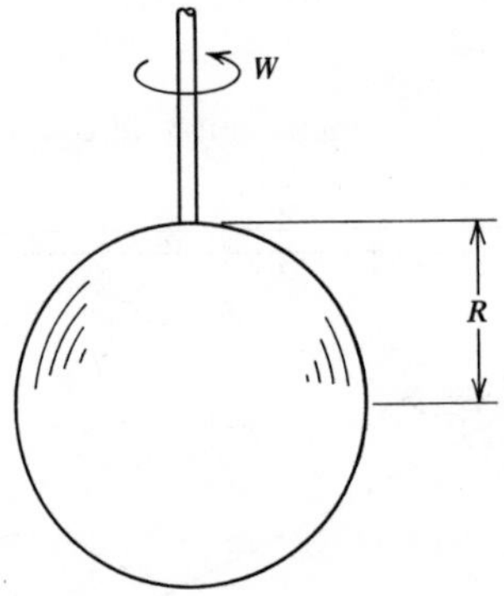

FIGURE 1C.7. Rotating sphere in a viscous liquid; radius of sphere is R.

1C.8 Analysis of Torsional Oscillatory Viscometer for Small Slit Width[5]

In the torsional oscillatory viscometer, the fluid is placed between a "cup" and a "bob" as shown in Fig. 1C.8. The cup is made to undergo small oscillations of a sinusoidal nature in the tangential direction. This motion causes the bob, suspended by a torsion wire, to oscillate with the same frequency but with a different amplitude and with a different phase. The amplitude ratio (ratio of amplitude of input function to output function) and phase shift both depend upon the viscosity of the fluid; hence measurement of the amplitude ratio and phase shift lead to measurement of the viscosity. Here we study the behavior of this system for the Newtonian fluid, inasmuch as later we shall treat the analogous case for the viscoelastic fluid in Example 4.5–4. It is assumed throughout that oscillations are of *small* amplitude. Since the problem is a linear one, it can be worked by using Laplace transforms, or by the method outlined in this problem.

a. First, apply Newton's second law of motion to the cylindrical bob when the annular space is completely evacuated. Show that the natural frequency of the system is $\sqrt{K/I}$, where I is the moment of inertia of the bob, and K is the torsion constant for the torsion wire.

b. Next, apply Newton's second law of motion when there is a fluid in the annular space with viscosity μ. Let θ_R equal the angular displacement of the bob at time t, and let v_θ be the tangential velocity of the fluid as a function of r. Show that the equation of motion of the bob is:

$$I\frac{d^2\theta_R}{dt^2} = -K\theta_R + (2\pi RL)(R)\left(\mu r \frac{\partial}{\partial r}\frac{v_\theta}{r}\right)\bigg|_{r=R} \tag{1C.8–1}$$

[5] H. Markovitz, *J. Appl. Phys.*, *23*, 1070–1077 (1952); S. Oka, in *Rheology*, F. R. Eirich (Ed.), Academic Press, New York (1960), Vol. 3, Chapter 2.

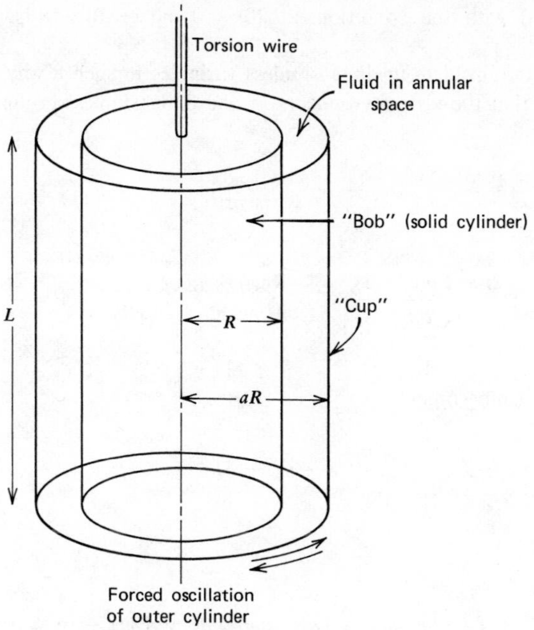

FIGURE 1C.8. Torsional oscillatory viscometer.

If the system starts from rest, we also have the initial condition:

$$I.C.: \quad \text{at } t = 0 \quad \theta_R = 0 \tag{1C.8–2}$$

c. Now write the equation of motion for the fluid. Show that the equation for v_θ is:

$$\rho \frac{\partial v_\theta}{\partial t} = \mu \frac{\partial}{\partial r}\left(\frac{1}{r}\frac{\partial}{\partial r}(r v_\theta)\right) \tag{1C.8–3}$$

with boundary and initial conditions as follows:

$$B.C.: \quad \text{at } r = R \quad v_\theta = R\frac{d\theta_R}{dt} \tag{1C.8–4}$$

$$B.C.: \quad \text{at } r = aR \quad v_\theta = aR\frac{d\theta_{aR}}{dt} \tag{1C.8–5}$$

$$I.C.: \quad \text{at } t = 0 \quad v_\theta = 0 \tag{1C.8–6}$$

where θ_{aR} is a sinusoidal driving function, which is given. Draw a sketch showing θ_{aR} and θ_R as functions of time; in terms of the sketch, define "amplitude ratio" and "phase shift."

d. Simplify Eqs. 1C.8–1 to 6 by assuming that $a \approx 1$ so that curvature may be neglected. (Markovitz[5] does *not* make this assumption.) Introduce a dimensionless length variable x:

$$x = \frac{r - R}{(a - 1)R} \tag{1C.8–7}$$

and show that:

Bob:
$$I\ddot{\theta}_R = -K\theta_R + \frac{2\pi R^2 L\mu}{(a-1)R}\left.\frac{\partial v_\theta}{\partial x}\right|_{x=0} \tag{1C.8–8}$$

Fluid:
$$\rho\frac{\partial v_\theta}{\partial t} = \mu\frac{1}{(a-1)^2R^2}\frac{\partial^2 v_\theta}{\partial x^2} \tag{1C.8–9}$$

which are to be solved with the conditions: $\theta_R(0) = 0$, $v_\theta(x = 0) = R\dot{\theta}_R$, $v_\theta(x = 1) = R\dot{\theta}_{aR}$, and $v_\theta(t = 0) = 0$.

e. Next, recast the problem in dimensionless variables in such a way that $\sqrt{I/K}$ is used as a characteristic time, and that the viscosity appears in one dimensionless group only. The only choice turns out to be:

$$\textit{Time:}\quad \tau = \sqrt{\frac{K}{I}}\,t \qquad\qquad \textit{Viscosity:}\quad M = \frac{\mu/\rho}{(a-1)^2R^2}\sqrt{\frac{I}{K}}$$

$$\textit{Velocity:}\quad \phi = \frac{2\pi R^3 L\rho(a-1)}{\sqrt{KI}}\,v_\theta \qquad\qquad \textit{Reciprocal of Moment of Inertia:}\quad A = \frac{2\pi R^4 L\rho(a-1)}{I} \tag{1C.8–10}$$

Show that the problem then becomes:

$$\textit{Bob:}\qquad \frac{d^2\theta_R}{d\tau^2} = -\theta_R + M\left.\frac{\partial\phi}{\partial x}\right|_{x=0} \tag{1C.8–11}$$

$$\textit{Fluid:}\qquad \frac{\partial\phi}{\partial\tau} = M\frac{\partial^2\phi}{\partial x^2} \tag{1C.8–12}$$

with $\theta_R(0) = 0$, $\phi(x = 0) = A\,d\theta_R/d\tau$, $\phi(x = 1) = A\,d\theta_{aR}/d\tau$, and $\phi(\tau = 0) = 0$. From this we can get θ_R and ϕ as functions of x and τ, with M and A as parameters.

f. We consider only the sinusoidal solution—that is, the solution after the initial transient has damped out. Hence, we take the driving function to be

$$\theta_{aR} = \theta_{aR}{}^0\,\mathcal{Re}\{e^{i\omega\tau}\} \tag{1C.8–13}$$

where $\theta_{aR}{}^0$ is real, and postulate solutions of the form

$$\theta_R = \mathcal{Re}\{\theta_R{}^0 e^{i\omega\tau}\}$$

$$\phi = \mathcal{Re}\{\phi^0 e^{i\omega\tau}\} \tag{1C.8–14}$$

where $\theta_R{}^0$ and ϕ^0 are complex. Verify that the amplitude ratio is given by $|\theta_R{}^0|/\theta_{aR}{}^0$ where $|\;|$ indicates the absolute magnitude of a complex quantity; and show that the phase angle is given by $\tan\alpha = \mathcal{Im}\{\theta_R{}^0\}/\mathcal{Re}\{\theta_R{}^0\}$ where $\mathcal{Im}$ and $\mathcal{Re}$ stand for imaginary and real parts, respectively.

g. Substitute the trial solutions of (f) into the equations in (e) and obtain equations for $\theta_R{}^0$ and ϕ^0.

h. Solve the equation for ϕ^0 and then find $d\phi^0/dx|_{x=0}$:

$$\left.\frac{d\phi^0}{dx}\right|_{x=0} = \frac{A(i\omega)^{3/2}}{\sqrt{M}}\left[\frac{\theta_{aR}{}^0 - \theta_R{}^0\cosh\sqrt{i\omega/M}}{\sinh\sqrt{i\omega/M}}\right] \tag{1C.8–15}$$

i. Next, solve the $\theta_R{}^0$ equation and obtain:

$$\frac{\theta_R{}^0}{\theta_{aR}{}^0} = \frac{AM\omega i}{(1-\omega^2)\dfrac{\sinh\sqrt{i\omega/M}}{\sqrt{i\omega/M}} + AM\omega i\cosh\sqrt{i\omega/M}} \tag{1C.8–16}$$

from which $|\theta_R{}^0|/\theta_{aR}{}^0$ and α can be found.

j. Expand the sinh and cosh functions in Eq. 1C.8–16 and obtain $\theta_{aR}{}^0/\theta_R{}^0$ as a power series in $1/M$. Keep the first three terms and show that:

$$\frac{\theta_{aR}{}^0}{\theta_R{}^0} = 1 + \frac{i}{M}\left(\frac{\omega^2-1}{A\omega} + \frac{\omega}{2}\right) - \frac{1}{M^2}\left(\frac{\omega^2-1}{6A} + \frac{\omega^2}{24}\right) + O\left(\frac{1}{M^3}\right) \tag{1C.8–17}$$

From this find the amplitude ratio

$$\frac{|\theta_R{}^0|}{\theta_{aR}{}^0} = \left[\left[1 + \frac{1}{M^2}\left\{-\frac{\omega^2}{24} + \frac{1-\omega^2}{6A}\right\}\right]^2 + \left[\frac{1}{M}\left\{\frac{\omega}{2} - \frac{1-\omega^2}{A\omega}\right\}\right]^2\right]^{-1/2} \tag{1C.8–18}$$

and the phase angle.

k. Let us take some sample values:

$$\frac{\mu}{\rho} = 1.0 \times 10^{-6}\ \mathrm{m^2/s}$$

$$(a-1)R = 1.0 \times 10^{-4}\ \mathrm{m}$$

$$I = 2.5 \times 10^{-4}\ \mathrm{kg{\cdot}m^2}$$

$$K = 0.40\ \mathrm{J}$$

$$L = 0.25\ \mathrm{m}$$

$$R = 0.055\ \mathrm{m}$$

$$\rho = 10^3\ \mathrm{kg/m^3}$$

Then $M = 2.5$ and $A = 0.1$. Show that Eq. 1C.8–18 then gives

ω	$\frac{1}{4}$	$\frac{1}{2}$	1	2	4
$\frac{\lvert\theta_R{}^0\rvert}{\theta_{aR}{}^0}$	0.07	0.17	0.99	0.16	0.07

Note that $\omega = 1$ corresponds to the natural frequency of the bob in a vacuum. Sketch the curve of $|\theta_R{}^0|/\theta_{aR}{}^0$ vs. ω. Interpret the result.

1C.9 Journal-Bearing Problem[6]

A journal-bearing system is shown in Fig. 1C.9. The journal of radius r_1 rotates with angular velocity W_1 in a stationary outer bearing of radius r_2. The journal and bearing are of length L in the z-direction, and the eccentric annular gap is completely filled with a Newtonian liquid. The origins of the rectangular and cylindrical coordinate systems are taken to be on the journal axis. The axes of the journal and bearing are separated by a distance a, and the difference in radii, $r_2 - r_1$, is called c, with $c \ll r_1$. It is desired to find the velocity distribution and the pressure distribution in the system. In addition, we wish to know the torque and the force exerted by the fluid on the rotating journal.

a. We assume that the tangential flow in the annular gap can be approximated locally as that in a plane slit with one wall moving and with an imposed pressure gradient (cf. Example 1.2–1). Then at each θ we introduce a local XYZ-coordinate system with X pointing in the flow direction, Y pointing radially, and Z coinciding with z. Thus $Y = r - r_1$, and the surface at $Y = 0$ is moving with the linear

[6] This problem is patterned very closely after the discussion by A. Sommerfeld, *Vorlesungen über Theoretische Physik*, Vol. 2, Dietrich, Wiesbaden (1947), Sec. 36. For a comprehensive theoretical treatment on lubrication, see N. Tipei, *Theory of Lubrication*, Stanford University Press (1962). For an analytical solution in bipolar coordinates, see G. J. Farns cited by R. Ehrlich and J. C. Slattery, *Ind. Eng. Chem. Fundamentals*, 7, 239–246 (1968). Appendix B.

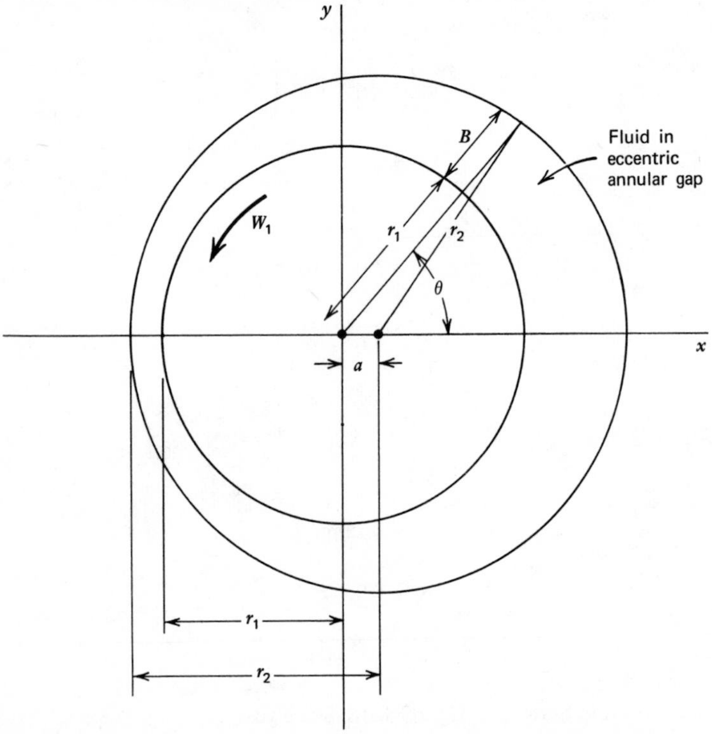

FIGURE 1C.9. Journal-bearing arrangement: cylinder of radius r_1 rotating with angular velocity W_1 in a cylindrical cavity of radius r_2. The gap width $B(\theta)$ is given approximately as $B \approx c + a\cos\theta$, where $c = r_2 - r_1$.

velocity $W_1 r_1$, whereas the surface at $Y = B$ is fixed. Show that the local velocity profile will be:

$$v_X = W_1 r_1 \left(1 - \frac{Y}{B}\right) - \frac{B^2}{2\mu}\frac{dp}{dX}\frac{Y}{B}\left(1 - \frac{Y}{B}\right) \tag{1C.9–1}$$

and that the volume rate of flow will be:

$$Q = \tfrac{1}{2}BLW_1 r_1 - \frac{B^3 L}{12\mu}\frac{dp}{dX} \tag{1C.9–2}$$

In the journal-bearing problem we do not know what Q is; it is convenient to replace the unknown constant Q by an unknown constant B_0 defined by $Q = \tfrac{1}{2}B_0 L W_1 r_1$. Then show that:

$$\frac{dp}{dX} = 6\mu W_1 r_1 \left(\frac{1}{B^2} - \frac{B_0}{B^3}\right) \tag{1C.9–3}$$

$$\tau_{YX}|_{Y=0} = -\mu \left.\frac{dv_X}{dY}\right|_{Y=0} = \mu W_1 r_1 \left(\frac{4}{B} - \frac{3B_0}{B^2}\right) \tag{1C.9–4}$$

b. Next apply these last two equations to the system in Fig. 1C.9 to obtain:

$$\frac{1}{r_1}\frac{dp}{d\theta} = 6\mu W_1 r_1 \left(\frac{1}{B^2} - \frac{B_0}{B^3}\right) \equiv \mu f(\theta) \tag{1C.9–5}$$

$$\tau_{r\theta}|_{r=r_1} = \mu W_1 r_1 \left(\frac{4}{B} - \frac{3B_0}{B^2}\right) \tag{1C.9–6}$$

in which it is understood that $B = B(\theta)$ as described in the figure caption for Fig. 1C.9. In addition, verify that Eq. 1C.9–1 can be rewritten as:

$$v_\theta = \tfrac{1}{2}Y^2 f(\theta) + Yg(\theta) + W_1 r_1 \tag{1C.9–7}$$

where $f(\theta)$ is given in Eq. 1C.9–5 and $g(\theta)$ is:

$$g(\theta) = -\frac{W_1 r_1}{B} - \tfrac{1}{2}Bf(\theta) \tag{1C.9–8}$$

Then show that from the equation of continuity:

$$v_r = -\frac{1}{r_1}\left[\tfrac{1}{6}Y^3 f'(\theta) + \tfrac{1}{2}Y^2 g'(\theta)\right] \tag{1C.9–9}$$

c. Next, in preparation for (d), we need to know how to evaluate the following integrals:

$$J_n = \int_0^{2\pi} \frac{d\theta}{(c + a\cos\theta)^n} \tag{1C.9–10}$$

$$K_n = \int_0^{2\pi} \frac{\cos\theta\, d\theta}{(c + a\cos\theta)^n} \tag{1C.9–11}$$

Show that:

$$J_1 = 2\pi(c^2 - a^2)^{-1/2} \tag{1C.9–12}$$

$$J_2 = -\frac{\partial J_1}{\partial c} = 2\pi c(c^2 - a^2)^{-3/2} \tag{1C.9–13}$$

$$J_3 = -\tfrac{1}{2}\frac{\partial J_2}{\partial c} = 2\pi(c^2 + \tfrac{1}{2}a^2)(c^2 - a^2)^{-5/2} \tag{1C.9–14}$$

$$K_2 = \left(\frac{1}{a}\right)(J_1 - cJ_2) \tag{1C.9–15}$$

$$K_3 = \left(\frac{1}{a}\right)(J_2 - cJ_3) \tag{1C.9–16}$$

d. Integrate Eq. 1C.9–5 between $\theta = 0$ and $\theta = 2\pi$. Then use the fact that p has to be the same at $\theta = 0$ and $\theta = 2\pi$ in Fig. 1C.9, to obtain:

$$B_0 = \frac{J_2}{J_3} = c\,\frac{c^2 - a^2}{c^2 + \tfrac{1}{2}a^2} \tag{1C.9–17}$$

e. Now obtain the torque that the fluid exerts on the rotating journal:

$$\begin{aligned}\mathscr{T} &= L\int_0^{2\pi} [-\tau_{r\theta}]_{r=r_1}\, r_1^2\, d\theta \\ &= -\mu L W_1 r_1^3 (4J_1 - 3B_0 J_2) \\ &= -\frac{2\pi\mu L W_1 r_1^3}{\sqrt{c^2 - a^2}}\,\frac{c^2 + 2a^2}{c^2 + \tfrac{1}{2}a^2}\end{aligned} \tag{1C.9–18}$$

f. Next we want to get the components of the force that the fluid exerts on the rotating journal. We first consider the force in the positive y-direction:

$$\begin{aligned} F_y &= L\int_0^{2\pi} [-(p+\tau_{rr})\sin\theta - \tau_{r\theta}\cos\theta]_{r=r_1} r_1\,d\theta \\ &= L\int_0^{2\pi}\left[-\frac{dp}{d\theta} - \tau_{r\theta}\right]_{r=r_1} \cos\theta\, r_1\,d\theta \\ &= -6\mu L W_1 r_1^{\,3}(K_2 - B_0 K_3) + \text{higher-order terms from } \tau_{r\theta} \\ &\cong -\frac{6\pi\mu L W_1 r_1^{\,3} a}{\sqrt{c^2 - a^2}\,(c^2 + \frac{1}{2}a^2)} \end{aligned} \tag{1C.9–19}$$

In going from the first to the second line we omit the τ_{rr} contribution, since in the lubrication approximation we assert that locally the flow is the same as between parallel plates; we also perform an integration by parts. In going from the second to the third line, we omit the $\tau_{r\theta}$ contribution, since from Eqs. 1C.9–5, 6, and 17 we see that $dp/d\theta$ is of order $(r_1/c)^2$, whereas $\tau_{r\theta}$ is of order (r_1/c).

Next show that for the force in the positive x-direction:

$$F_x = L\int_0^{2\pi} [-(p+\tau_{rr})\cos\theta + \tau_{r\theta}\sin\theta]_{r=r_1} r_1\,d\theta = 0 \tag{1C.9–20}$$

Interpret these results.

g. Finally, integrate Eq. 1C.9–5 to obtain the pressure distribution in the system:

$$\begin{aligned} p &= p_0 + 6\mu W_1 r_1^{\,2}\left[\int \frac{d\theta}{(c + a\cos\theta)^2} - B_0\int\frac{d\theta}{(c+a\cos\theta)^3}\right] \\ &= p_0 + \frac{6\mu W_1 r_1^{\,2} a\sin\theta\cdot(c + \frac{1}{2}a\cos\theta)}{(c^2+\frac{1}{2}a^2)(c+a\cos\theta)^2} \end{aligned} \tag{1C.9–21}$$

where p_0 is an arbitrary constant pressure. Both integrals can be written as $\int(c + a\cos\theta)^{-1}\,d\theta$ + additional terms, and when the indicated subtraction in [] is performed, it is found that the coefficient of $\int(c + a\cos\theta)^{-1}\,d\theta$ is zero so that this integral does not have to be evaluated.

1C.10 Shape of a Vertically Falling Stream of a Newtonian Liquid

A liquid flowing axially in a vertical circular tube with a volume rate of flow Q emerges from the tube and falls downward. Its radius is $R(z)$, where z is the distance downward from the end of the tube. As an approximation we assume that v_z is a function of z alone, so that the velocity v_z and radius of the jet $R(z)$ are related by $Q = \pi[R(z)]^2 v_z(z)$.

Assume that the flow in the jet is locally simple elongational flow with elongational viscosity 3μ (see Eq. 1B.11–3). Then show that for steady state the equation for the velocity $v_z(z)$ is given by:[7]

$$\rho\frac{dv_z}{dz} = 3\mu\frac{d}{dz}\left(\frac{1}{v_z}\frac{dv_z}{dz}\right) + 2\sigma\sqrt{\frac{\pi}{Q}}\frac{d}{dz}\left(\frac{1}{\sqrt{v_z}}\right) + \frac{1}{v_z}\rho g \tag{1C.10–1}$$

[7] A. Kaye and D. G. Vale, *Rheol. Acta*, *8*, 1–5 (1969); the same equation, without surface tension, was given and solved numerically by W. R. Marshall, Jr., and R. L. Pigford, *The Application of Differential Equations to Chemical Engineering Problems*, University of Delaware Press (1947), pp. 73–76. See also N. S. Clarke, *Mathematika*, *13*, 51–53 (1966) and *J. Fluid Mech.*, *31*, 481–500 (1968); A. Ziabicki, *Bull. Acad. Polon. Sci., Sér. Sci. Techn.*, *12*, 1[717]–8[724], 9[725]–20[736], 21[821]–28[828], 29[925]–35[931] (1964).

in which σ is the surface tension of the liquid. Kaye and Vale[7] report that the numerical solution to this equation agrees well with experimental observations on $R(z)$ for a high-viscosity silicone fluid ($\mu = 30$ N·s/m^2), but not for a low-viscosity fluid ($\mu = 1$ N·s/m^2).

1C.11 Tangential Flow between Two Coaxial Cones

The space between two coaxial cones is filled with a Newtonian fluid (see Fig. 1C.11). The angular velocities of the inner and outer cones are W_i and W_o respectively. Find the velocity distribution and the

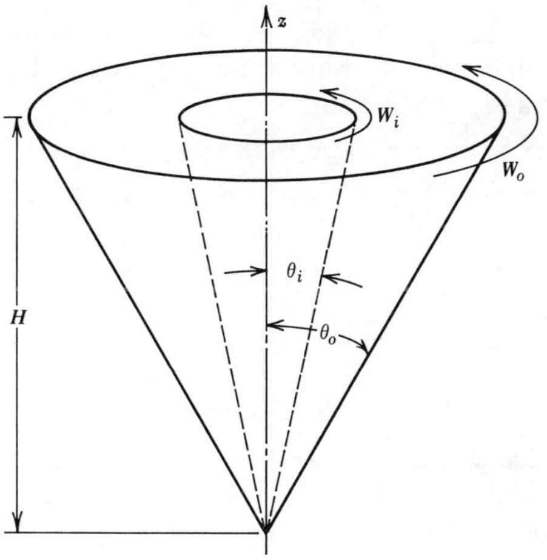

FIGURE 1C.11. Tangential flow between coaxial rotating cones.

torques on the cones. (*Hint*: Postulate a solution of the form $v_\phi(r,\theta) = w(\theta)r\sin\theta$, and obtain the differential equation for $w(\theta)$. What is the physical meaning of $w(\theta)$?)

$$\text{Answer:}\quad \frac{w - W_i}{W_o - W_i} = \frac{\left[\dfrac{\cos\theta_i}{\sin^2\theta_i} - \dfrac{\cos\theta}{\sin^2\theta} + \ln\left(\dfrac{\tan\frac{1}{2}\theta}{\tan\frac{1}{2}\theta_i}\right)\right]}{\left[\dfrac{\cos\theta_i}{\sin^2\theta_i} - \dfrac{\cos\theta_o}{\sin^2\theta_o} + \ln\left(\dfrac{\tan\frac{1}{2}\theta_o}{\tan\frac{1}{2}\theta_i}\right)\right]}$$

1D.1 Creeping Flow around a Sphere

a. Verify that Eqs. 1.2–43 and 1.2–44 satisfy the equations of change for incompressible, creeping flow of a Newtonian liquid.

b. Verify Eqs. 1.2–47 and 48.

c. Show that Stokes' law can be obtained by equating the rate of doing work on the sphere, $(\boldsymbol{F}_k \cdot \boldsymbol{v}_\infty)$, to the total rate of viscous dissipation of energy throughout the fluid outside the sphere. (*Hint*: In obtaining the expression for $\nabla \boldsymbol{v}$ in Eq. 1.2–46, use the expression for ∇ given in Eq. A.7–9, and the expressions for differentiating unit vectors in Eqs. A.7–6, 7, and 8. This will give:

$$\begin{aligned} \nabla \boldsymbol{v}_\infty f(r) &= (\nabla \boldsymbol{v}_\infty) f + (\nabla f)\boldsymbol{v}_\infty \\ &= 0 + \left(\boldsymbol{\delta}_r \frac{\partial}{\partial r} f\right)\boldsymbol{v}_\infty \\ &= \boldsymbol{\delta}_r \boldsymbol{v}_\infty f' \end{aligned} \qquad (1D.1\text{–}1)$$

$$\nabla[\boldsymbol{v}_\infty \cdot \boldsymbol{\delta}_r\boldsymbol{\delta}_r]g(r) = \{\boldsymbol{\delta}_r\boldsymbol{v}_\infty \cdot \boldsymbol{\delta}_r\boldsymbol{\delta}_r\}g' + \left\{\boldsymbol{\delta}_\theta\boldsymbol{v}_\infty \cdot \frac{1}{r}(\boldsymbol{\delta}_\theta\boldsymbol{\delta}_r + \boldsymbol{\delta}_r\boldsymbol{\delta}_\theta)\right\}g$$

$$+ \left\{\boldsymbol{\delta}_\phi\boldsymbol{v}_\infty \cdot \frac{1}{r\sin\theta}(\boldsymbol{\delta}_\phi \sin\theta\, \boldsymbol{\delta}_r + \boldsymbol{\delta}_r\boldsymbol{\delta}_\phi \sin\theta)\right\}g \qquad \text{(1D.1–2)}$$

Then to get Eq. 1.2–46 it is necessary to use $\boldsymbol{\delta} = \boldsymbol{\delta}_r\boldsymbol{\delta}_r + \boldsymbol{\delta}_\theta\boldsymbol{\delta}_\theta + \boldsymbol{\delta}_\phi\boldsymbol{\delta}_\phi$ (cf. Eq. A.3–11).)

1D.2 Oscillatory Flow of a Viscous Fluid

A viscous fluid occupies the region $y > 0$, and is bounded at the plane $y = 0$ by a solid surface that executes an oscillatory motion in the x-direction with a frequency ω. Find the velocity distribution in the system after the initial transients fade away.

It is thus desired to solve the differential equation

$$\frac{\partial v_x}{\partial t} = \nu \frac{\partial^2 v_x}{\partial y^2} \qquad \left(\nu = \frac{\mu}{\rho}\right) \qquad \text{(1D.2–1)}$$

with the boundary conditions:

$$\text{At } y = 0: \qquad v_x = v_0 \cos \omega t = v_0 \mathcal{R}e\{e^{i\omega t}\} \qquad \text{(1D.2–2)}$$

$$\text{At } y = \infty: \qquad v_x = 0 \qquad \text{(1D.2–3)}$$

a. Postulate a solution of the form:

$$v_x(y, t) = \mathcal{R}e\{v^0(y)e^{i\omega t}\} \qquad \text{(1D.2–4)}$$

in which $v^0(y)$ may be complex. Show that the substitution of this function into the differential equation leads to a differential equation for $v^0(y)$

$$\frac{d^2v^0}{dy^2} - \left(\frac{i\omega}{\nu}\right)v^0 = 0 \qquad \text{(1D.2–5)}$$

b. Show that this equation can be solved with the appropriate boundary conditions to give:

$$v^0(y) = v_0 e^{-(1+i)\sqrt{\omega/2\nu}\, y} \qquad \text{(1D.2–6)}$$

c. Show that substitution of this result into Eq. 1D.2–4 gives for the velocity distribution:

$$v_x(y, t) = v_0 e^{-\sqrt{\omega/2\nu}\, y} \cos\left(\omega t - \sqrt{\omega/2\nu}\, y\right) \qquad \text{(1D.2–7)}$$

Sketch the result.

d. Rework the problem if the fluid, instead of extending to $y = \infty$, is bounded[1] by a fixed plane at $y = B$. Show that the velocity profiles will be very nearly linear if $\sqrt{\rho\omega/2\mu}\, B \ll 1$.

1D.3 Manipulations Involving the Equations of Change

a. Show how to go from Eqs. A, B, and C of Table 1.1–1 to Eqs. D, E, and F.
b. Derive Eqs. 1.1–9 and 10 from Eq. 1.1–8.

[1] See, for example, H. Schlichting, *Boundary-Layer Theory*, McGraw-Hill, New York (1960), Fourth Edition, p. 76.

c. Show how to get the equations in Table 1.2–1 from those in Table 1.1–1.

d. Show that for an incompressible fluid with constant μ, the equation of motion can be written in the form:

$$\frac{\partial}{\partial t}\boldsymbol{v} = [\boldsymbol{v} \times [\nabla \times \boldsymbol{v}]] - \nabla\tfrac{1}{2}v^2 - \nabla(p/\rho) + \nu\nabla^2\boldsymbol{v} + \boldsymbol{g} \tag{1D.3–1}$$

in which $\nu = \mu/\rho$ is the kinematic viscosity.

e. Show that for a fluid with constant ρ and μ, and with $\boldsymbol{g} = -\nabla\hat{\Phi}$ (i.e., the external force is derivable from a potential), one may take the curl of the equation of motion to get the following equation:

$$\frac{\partial}{\partial t}\boldsymbol{w} = [\nabla \times [\boldsymbol{v} \times \boldsymbol{w}]] + \nu\nabla^2\boldsymbol{w} \tag{1D.3–2}$$

which is sometimes called the *equation of change for the fluid angular velocity*, $\boldsymbol{w} = \frac{1}{2}[\nabla \times \boldsymbol{v}]$.

1D.4 Variational Principle for Viscous Flow

Consider the flow of an incompressible, Newtonian fluid in a volume V bounded by a surface S. The latter consists of two nonoverlapping regions: S_π is that portion on which $\boldsymbol{\pi} = p\boldsymbol{\delta} + \boldsymbol{\tau}$ is specified, and S_v is that on which the velocity $\boldsymbol{v}$ is specified. We consider only steady flows for which $[\boldsymbol{v} \cdot \nabla\boldsymbol{v}]$ in the equation of motion is either identically zero or else negligible. For such problems there exists the following variational principle due to von Helmholtz:[2] Of all the possible fluid motions within V that are compatible with the equation of continuity ($(\nabla \cdot \boldsymbol{v}) = 0$) and the prescribed boundary conditions, the motion that minimizes

$$J = \int_V [\tfrac{1}{2}\mu(\dot{\boldsymbol{\gamma}} : \dot{\boldsymbol{\gamma}}) - 2\rho(\boldsymbol{v} \cdot \boldsymbol{g})]\, dV + 2\int_{S_\pi} ([\boldsymbol{\pi} \cdot \boldsymbol{v}] \cdot \boldsymbol{n})\, dS \tag{1D.4–1}$$

will be the steady-state motion. Here $\boldsymbol{n}$ is the outwardly directed unit normal. For flow in conduits of constant cross section S, the second integral in J becomes $-2\Delta p \int_S v\, dS$, where v is the axial velocity and Δp is the pressure drop through the conduit. Let us now apply this variational principle to the flow in a horizontal pipe of triangular cross section.

a. Let the cross section of the pipe be bounded by the straight lines $y = \pm mx$, and $x = a$, with the fluid flowing axially in the z-direction. Verify that the function to be minimized is:

$$J = \mu L \int_0^a \int_{-mx}^{+mx} \left[\left(\frac{\partial v_z}{\partial x}\right)^2 + \left(\frac{\partial v_z}{\partial y}\right)^2\right] dy\, dx - 2\,\Delta p \int_0^a \int_{-mx}^{+mx} v_z\, dy\, dx \tag{1D.4–2}$$

b. To apply the variational principle we now choose a trial velocity distribution that satisfies the equation of continuity and the boundary condition that the fluid velocity is zero at the walls of the pipe. A reasonable function would be:

$$v_z = (y^2 - m^2x^2)(x - a)(b + cx + dx^2 + ex^3 + \cdots) \tag{1D.4–3}$$

where $b, c, d, e, \ldots$ are variational parameters. For purposes of illustration we take $d, e, \ldots$ to be zero and find the optimal values of b and c by minimizing J. To do this substitute Eq. 1D.4–3 into Eq. 1D.4–2 and then set $\partial J/\partial b$ and $\partial J/\partial c$ both equal to zero. Show that this leads to:

$$b = \tfrac{7}{2}(\Delta p/\mu La)(17m^2 - 1)/(55m^4 + 38m^2 + 3) \tag{1D.4–4}$$

[2] See, for example, H. Lamb, *Hydrodynamics*, Dover, New York (1945), pp. 617–619.

$$c = -14(\Delta p/\mu La^2)(3m^2 - 1)/(55m^4 + 38m^2 + 3) \tag{1D.4–5}$$

as the best values for b and c.

c. Next obtain the volume rate of flow through the pipe using the velocity distribution found in (b):

$$Q = \frac{7}{90}\frac{\Delta p}{\mu L} m^3 a^4(27m^2 + 5)/(55m^4 + 38m^2 + 3) \tag{1D.4–6}$$

This result was found by Itō;[3] it gives the exact result[4] for an equilateral triangle, that is, with $m = 1/\sqrt{3}$. (See Problem 1B.5).

1D.5 Viscous Heating in Oscillatory Flow[5]

A Newtonian fluid of viscosity μ and thermal conductivity k is located in the region between two parallel plates separated by a distance b. Both plates are maintained at a temperature T_0. The lower plate (at $x = 0$) is made to oscillate sinusoidally in the z-direction with an amplitude v_0 and a circular frequency ω. The upper plate (at $x = b$) is held fixed. Estimate the temperature rise that results from viscous heating. Consider the high-frequency limit only.

Answer: $$T - T_0 = \left(\frac{\mu v_0{}^2}{4k}\right)[(1 - e^{-2a\xi}) - (1 - e^{-2a})\xi]$$

where $\xi = \dfrac{x}{b}$ and $a = b\sqrt{\dfrac{\rho\omega}{2\mu}}$

1D.6 Contraction of a Newtonian Jet

The behavior of Newtonian jets issuing from circular tubes has been studied experimentally by Middleman and Gavis.[6] They found that the jets swell for Re < 16 and contract for Re > 16, where the Reynolds number Re is defined as $\text{Re} = 2R\langle v_z\rangle\rho/\mu$ (the angular brackets indicate an average over the tube cross section).

We assume that the flow is isothermal, and we consider the equation of continuity (Eq. 1.1–4), the equation of motion (Eq. 1.1–8), and the equation for mechanical energy (Eq. 1.1–9) written for steady-state, incompressible flow.

a. First show that these three equations can be integrated over a volume V enclosed by a surface S to obtain:

$$\int_S (\boldsymbol{n} \cdot \boldsymbol{v})\, dS = 0 \tag{1D.6–1}$$

$$\int_S [\boldsymbol{n} \cdot \rho\boldsymbol{v}\boldsymbol{v}]\, dS + \int_S [\boldsymbol{n} \cdot \boldsymbol{\pi}]\, dS - \int_V \rho\boldsymbol{g}\, dV = 0 \tag{1D.6–2}$$

$$\int_S (\boldsymbol{n} \cdot \tfrac{1}{2}\rho v^2\boldsymbol{v})\, dS + \int_S (\boldsymbol{n} \cdot [\boldsymbol{\pi} \cdot \boldsymbol{v}])\, dS - \int_V (\boldsymbol{\tau} : \nabla\boldsymbol{v})\, dV - \int_V \rho(\boldsymbol{v} \cdot \boldsymbol{g})\, dV = 0 \tag{1D.6–3}$$

When these equations are applied to the fluid contained in the region between planes "1" and "2" in

[3] H. Itō *Trans. Japan. Soc. Mech. Eng.*, *17*, 99–102 (1951)

[4] L. D. Landau and E. M. Lifshitz, *Fluid Mechanics*, Addison-Wesley, Reading, Mass. (1959), p. 58, Problem 3.

[5] R. B. Bird, *Chem. Eng. Prog.*, Symposium Series Number 58, *61*, 1–15 (1965); see Illustrative Example 1.

[6] S. Middleman and J. Gavis, *Phys. Fluids*, *4*, 355–359 (1961).

Fig. 1D.6, they become:

$$\langle v_z \rangle_1 S_1 - v_2 S_2 = 0 \tag{1D.6–4}$$

$$\rho\langle v_z^2 \rangle_1 S_1 - \rho v_2^2 S_2 + \langle \pi_{zz} \rangle_1 S_1 - p_a S_2 + p_a(S_2 - S_1) = 0 \tag{1D.6–5}$$

$$\tfrac{1}{2}\rho\langle v_z^3 \rangle_1 S_1 - \tfrac{1}{2}\rho v_2^3 S_2 + \langle \pi_{zz} v_z \rangle_1 S_1 - p_a v_2 S_2 + \int_V (\boldsymbol{\tau} : \nabla \mathbf{v})\, dV = 0 \tag{1D.6–6}$$

b. It is desired to use the equations developed in (a) to estimate S_2/S_1 for the jet, that is, the extent of expansion or contraction. To do this requires that various quantities in these equations be estimated, and this is not easy. The very simplest theory is to assume that

1. The flow remains parabolic up to plane "1," so that $\langle v_z^2 \rangle_1 / \langle v_z \rangle_1^2 = \frac{4}{3}$. (Verify this numerical value.)
2. $\langle \pi_{zz} \rangle_1$ can be approximated as p_a.

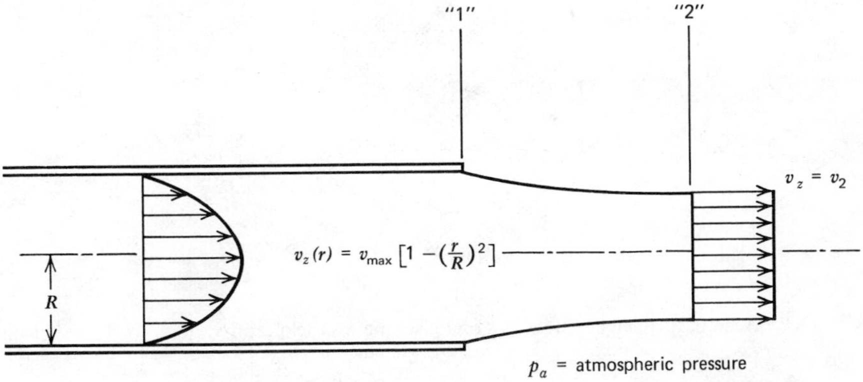

FIGURE 1D.6. Contraction of a Newtonian jet emerging from a circular tube. Within the tube the velocity profile is parabolic. At plane "2" the velocity is uniform. The cross sections at "1" and "2" are S_1 and S_2, respectively. Newtonian jets have been found to contract for Re $= 2R\langle v_z \rangle \rho/\mu > 16$ and to swell for Re < 16. For Re > 100, it is found experimentally that $S_2/S_1 = \frac{3}{4}$.

Neither of these approximations has been justified. If now one solves Eqs. 1D.6–4 and 5 simultaneously (and ignores Eq. 1D.6–6), then one obtains $S_2/S_1 = \frac{3}{4}$; verify this. This result was obtained by Harmon.[7] The value of $\frac{3}{4}$ agrees with the experimental results of Middleman and Gavis for Reynolds numbers of more than 100. However at Re $= 16$, it was found that $S_2/S_1 = 1$, and as the Reynolds number approaches zero, S_2/S_1 becomes about $\frac{5}{4}$. By taking other approximations for the velocity profile and the momentum flux at the tube exit, and by estimating the value of the integral in Eq. 1D.6–6, it may be possible to describe the entire range of fluid behavior.

1D.7 The Rolling-Ball Viscometer[8,9]

A rolling-ball viscometer consists of an inclined tube containing a sphere whose diameter is but slightly smaller than the internal diameter of the tube (see Fig. 1D.7). The viscosity of a Newtonian fluid is obtained by observing the speed v_0 with which the ball rolls down the tube when the latter is filled with liquid of viscosity μ.

[7] D. B. Harmon, *J. Franklin Inst.*, *259*, 519–522 (1955).
[8] H. W. Lewis, *Anal. Chem.*, *25*, 507–508 (1953).
[9] J. R. Van Wazer, J. W. Lyons, K. Y. Kim, and R. E. Colwell, *Viscosity and Flow Measurement*, Wiley, New York (1963), pp. 276–281.

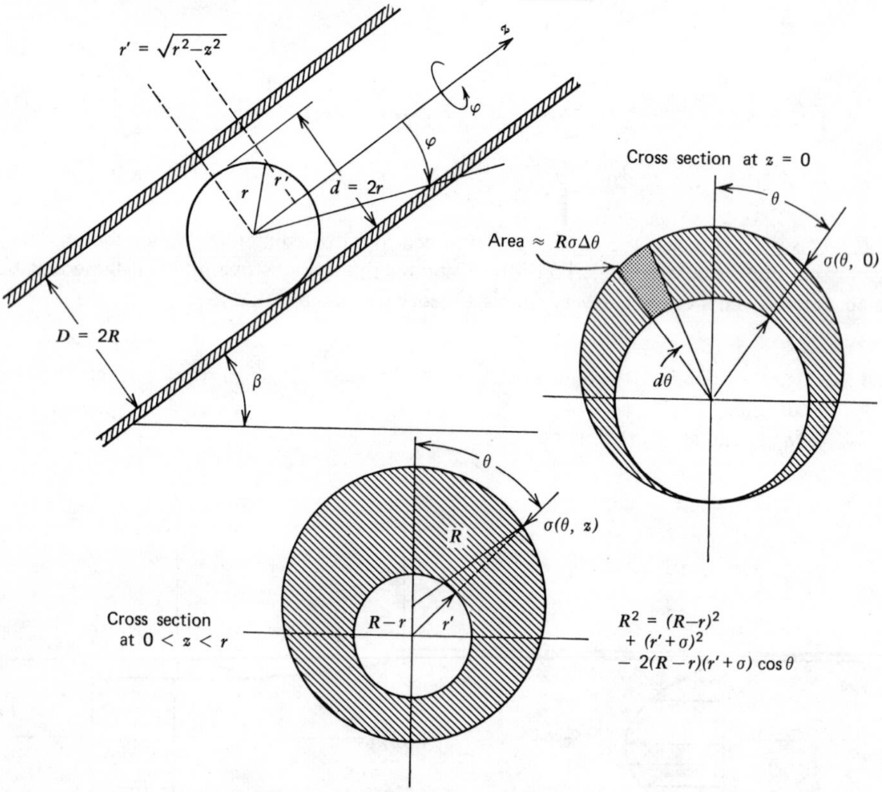

FIGURE 1D.7. Schematic drawing of rolling ball viscometer; the quantity $\sigma(\theta,z)$ is given approximately by

$$\sigma = 2(R - r)\left[\cos^2 \tfrac{1}{2}\theta + \frac{R - \sqrt{R^2 - z^2}}{2(R - r)}\right]$$

Work through the development given by Lewis[8] and obtain the final result:

$$\mu = \frac{4}{9\pi J}\frac{R^2(\rho_s - \rho)g \sin \beta}{v_0}\left(\frac{R - r}{R}\right)^{5/2} \qquad \text{(1D.7–1)}$$

where R is the tube radius, r the ball radius, g the gravitational acceleration, and β the angle made by the tube with the horizontal; J is a constant:

$$J = \tfrac{4}{3}\left[\sqrt{2} - \frac{1}{\sqrt{5}}[\sqrt{10} + 2]^{1/2}\right] = 0.531 \qquad \text{(1D.7–2)}$$

CHAPTER 2

THE STRUCTURE OF POLYMERIC FLUIDS

At the molecular level there are several important differences between polymeric (or "macromolecular") fluids and "small-molecule" fluids. Because of these differences the flow behavior of polymeric fluids is quite unlike that of the small-molecule fluids, which are satisfactorily described by Newtonian fluid dynamics. There are several salient features of macromolecular architecture that influence the flow behavior. First, the molecular weights of the constituent molecules are very high, usually in the range 10^4–10^9. Second, we seldom deal with polymers in which all molecules have exactly the same molecular weight but rather with mixtures in which there is a distribution of molecular weights, and it is found that the rheological properties are very sensitive to the molecular weight distribution. Third, the polymer molecules can assume a tremendous number of configurations, even at equilibrium; then, in flow, the distribution of configurations can be greatly altered with the result that stretching and alignment of the molecules can cause the flow properties to change. Fourth, in concentrated solutions or undiluted polymers (the latter often being called "melts") the molecules can form a temporary entanglement network with entanglement junctions whose number can change with time in various flow situations. These and other features of polymeric fluids are discussed in this chapter.

In §2.1 we discuss briefly the various kinds of chemical reactions that are used to make polymers and the connection between reaction mechanism and polymer structure. Then §2.2 is devoted to molecular weight distributions and the two molecular weight averages in common use. In §2.3 we discuss the configurations of polymer molecules at equilibrium and the calculation of average quantities; this requires a short discussion of equilibrium statistical mechanics which is included to make the explanation self-contained. The following section, §2.4, deals with several points associated with polymer-solvent and polymer-polymer interactions. The final section, §2.5, is concerned with the configurations of polymer molecules in flow as well as with the calculation of the stress tensor by kinetic theory; no derivations are included, but instead the key results from Volume 2 are cited for reference.

The treatment is an elementary introduction intended for the physicist or engineer with little background in the physical chemistry of macromolecules. The reader wishing a more comprehensive treatment should consult the many standard references on the subject.[1]

[1] See, for example, F. W. Billmeyer, Jr., *Textbook of Polymer Science*, Wiley, New York (1962), Second Edition; P. J. Flory, *Principles of Polymer Chemistry*, Cornell University Press, Ithaca (1953); C. Tanford, *Physical Chemistry of Macromolecules*, Wiley, New York (1961); P. Meares, *Polymers: Structure and Bulk Properties*, Van Nostrand, London (1965).

§2.1 STRUCTURE AND SYNTHESIS

Chemically a *macromolecule* (or *polymer*)[2] is a large molecule composed of many small simple chemical units, generally called *structural units.* In some polymers each structural unit is connected to precisely two other structural units, and the resulting chain structure is called a *linear* macromolecule. In other polymers most structural units are connected to two other units, although some structural units connect three or more units, and we talk of *branched* molecules. Where the chains terminate, special units called *end groups* are found. Figure 2.1–1 shows symbolic representations of linear and branched macromolecules.

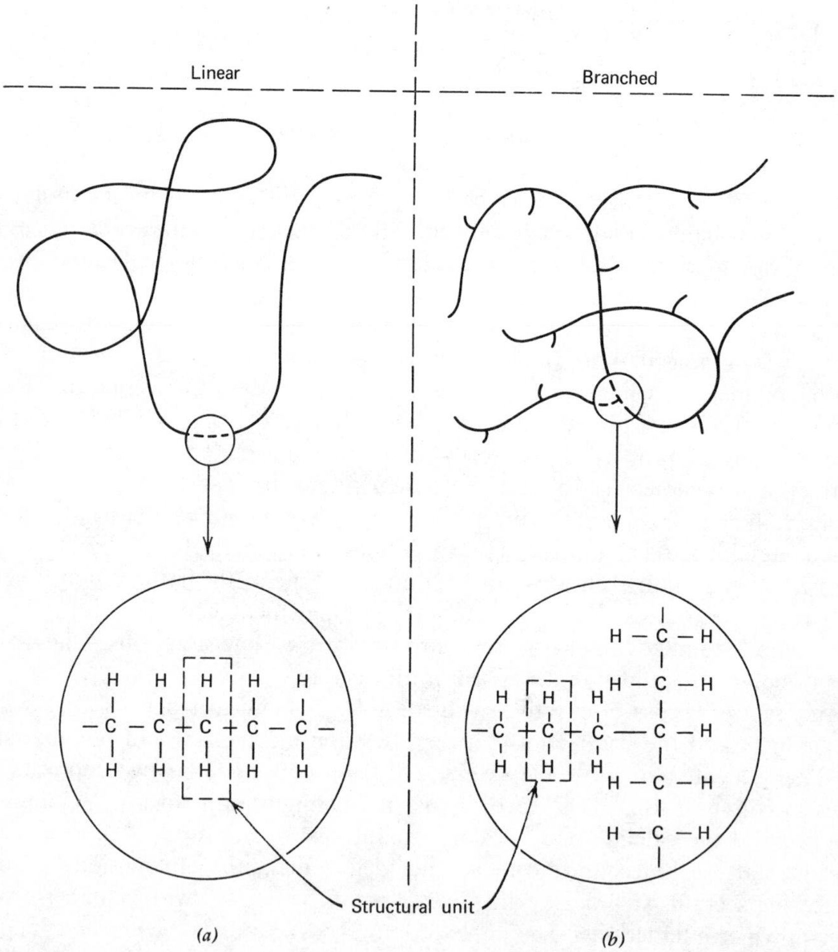

FIGURE 2.1–1. Symbolic representations of linear and branched macromolecules (high-density polyethylene and low-density polyethylene, respectively). (*a*) Linear; (*b*) branched.

cules. For the sake of completeness we mention also that in some macromolecular materials all structural units are interconnected resulting in a three-dimensional *crosslinked* or *network structure* rather than in separate molecules. Such materials, however, generally possess no fluid phase and are therefore outside the scope of this book.

[2] Some scientists are careful to distinguish between *macromolecules* (very large molecules) and *polymers* (macromolecules made up of repeating structural units). This fine distinction seems to be generally ignored, and the terms are used interchangeably. We shall use the two terms synonymously in this book.

It is sometimes useful to distinguish between synthetic and natural (biological) macro-molecules. Many synthetic polymers are built from a single structural unit, and the polymer is then referred to as a *homopolymer*. Typical examples of synthetic homopolymers are polyethylene, polystyrene, and polyvinylchloride (see Table 2.1–1). By contrast *copoly-mers* are built from two or more different structural units. According to the manner in which the structural units combine, copolymers are further classified as random copolymers, block

TABLE 2.1–1

Some Synthetic Polymers, Their Monomers and Their Structural Units

Polymer	Monomer(s)	Structural Unit
Polyethylene	$CH_2{=}CH_2$	$-CH_2-$
Poly(vinyl chloride)	$CH_2{=}CHCl$	$-CH_2-CHCl-$
Polystyrene	$CH_2{=}CH$ (phenyl ring)	$-CH_2-CH-$ (phenyl ring)
Polyisobutylene	$CH_2{=}C$ with CH_3 above and CH_3 below	$-CH_2-C-$ with CH_3 above and CH_3 below
Polyisoprene (natural rubber)	$CH_2{=}C-CH{=}CH_2$ with CH_3 above	$-CH_2-C{=}CH-CH_2-$ with CH_3 above
Poly(dimethylsiloxane)	$HO-Si-OH$ with CH_3 above and CH_3 below	$-Si-O-$ with CH_3 above and CH_3 below
Poly(ethylene oxide) (Polyox)	CH_2-CH_2 (epoxide, O above)	$-O-CH_2-CH_2-$
Poly(hexamethylene adipamide) (nylon 66)	$NH_2-(CH_2)_6-NH_2$ and $HO-\overset{O}{\overset{\|}{C}}-(CH_2)_4-\overset{O}{\overset{\|}{C}}-OH$	$-NH-(CH_2)_6-NH-\overset{O}{\overset{\|}{C}}-(CH_2)_4-\overset{O}{\overset{\|}{C}}-$
Poly(ethylene terephthalate) (polyester)	$HO-CH_2-CH_2-OH$ and $HO-\overset{O}{\overset{\|}{C}}$—(phenyl ring)—$\overset{O}{\overset{\|}{C}}-OH$	$-O-CH_2-CH_2-O-\overset{O}{\overset{\|}{C}}$—(phenyl ring)—$\overset{O}{\overset{\|}{C}}-$

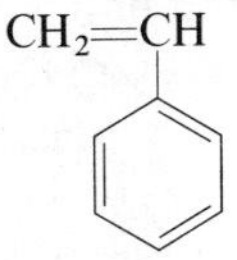

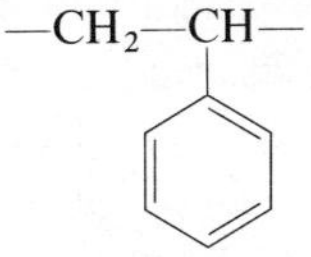

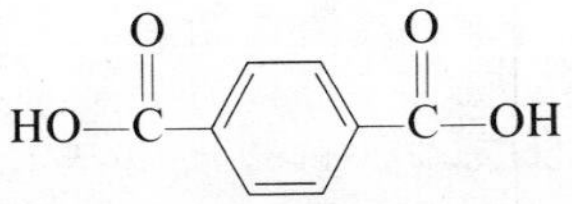

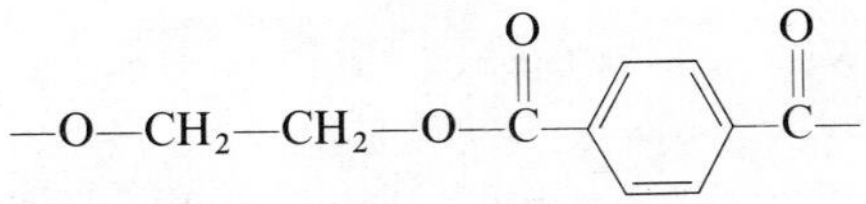

```
(a) •••AAABABBAABABBAB •••

(b) •••BBAAAAAAAABBBBBBAA •••

(c) •••AAAAAAAAAAAAAAAAA •••
        B        B     B
        B        B     B
        B        B     B
        :        :     :
```

FIGURE 2.1–2. Schematic representations of (*a*) random, (*b*) block, and (*c*) graft copolymers. A and B represent two different kinds of structural units.

copolymers, or graft copolymers, as illustrated in Fig. 2.1–2. The motivation for producing copolymers is to obtain materials with a wider range of mechanical properties than is possible with the homopolymers alone. For example, we mention the random copolymer of styrene and butadiene from which SBR rubber is made. This rubber has better characteristics for skid-resistance of tires than the rubber from pure butadiene.

Biological macromolecules, in contrast with synthetic macromolecules, generally contain a large number of different structural units. The polypeptide chains that make up proteins, for instance, consist of about 20 different structural units. Among other examples of biological macromolecules are the viruses and the interesting DNA molecules that carry in their structure the key to the inherited characteristics of organisms. More on biological polymers can be found elsewhere.[3]

Polymers are formed from *monomers* in a chemical reaction called *polymerization.* The monomer is generally of a fairly simple chemical composition and resembles the structural units. During the polymerization reaction, the monomer molecules are joined together to form the polymer chains. Polymerization reactions are usually classified according to their reaction mechanism: *step reactions* and *chain reactions.*

Monomers that polymerize by step reactions generally contain a number of reactive functional groups (such as carboxyl groups, alcohol groups, or amino groups). The nature of the polymer formed in step reactions is determined by the number of reactive functional groups per monomer. Bifunctional monomers, that is, monomers containing two functional groups, react to form linear polymers. On the other hand, if the reaction mixture contains polyfunctional monomers, that is, monomers with more than two functional groups, branched or crosslinked polymers are formed. As a typical example of a step reaction polymerization let us consider the reactions involved in the formation of linear nylon-6 from ω-aminocaproic acid:

$$NH_2(CH_2)_5COOH + NH_2(CH_2)_5COOH \rightarrow H[NH(CH_2)_5CO]_2OH + H_2O \qquad (2.1\text{–}1)$$

$$NH_2(CH_2)_5COOH + H[NH(CH_2)_5CO]_2OH \rightarrow H[NH(CH_2)_5CO]_3OH + H_2O \qquad (2.1\text{–}2)$$

and, in general,

$$H[NH(CH_2)_5CO]_nOH + H[NH(CH_2)_5CO]_mOH \rightarrow H[NH(CH_2)_5CO]_{n+m}OH + H_2O \qquad (2.1\text{–}3)$$

We see that the reaction involves the elimination of water. The elimination of a small molecule is typical of step reactions, and these are often alternatively called "condensation reactions." The important point to notice, however, is that any monomer or polymer in the reaction mixture is free to react at any time with any other monomer or polymer in the

[3] See, for example, C. Tanford, *Physical Chemistry of Macromolecules*, Wiley, New York (1961).

mixture. Consequently if the polymerization is performed in a well-mixed batch reactor, the weight fraction of monomer decreases rapidly at an early stage of the reaction. The average molecular weight of the polymers then increases steadily during the reaction, but polymers of very high molecular weight are not formed in substantial numbers before the final stages of the polymerization.

Chain reactions, in contrast with step reactions, generally require the presence of an initiator to take place. The initiation of one monomer molecule typically leads to the rapid polymerization of thousands of monomers; when the chain reaction is finally terminated, the result is a stable macromolecule that will not usually react again with either monomers or other polymers. Chain reactions are further classified according to their detailed reaction mechanism. By way of illustration we consider the important radical chain polymerization. The initiator E_2 here decomposes to form free radicals E·,

$$E_2 \rightarrow 2E\cdot \tag{2.1-4}$$

The free radicals then initiate the chain reaction. For example, the chain polymerization of ethylene proceeds as follows:

$$E\cdot + CH_2{=}CH_2 \rightarrow E{-}CH_2{-}CH_2\cdot \tag{2.1-5}$$

$$E{-}CH_2{-}CH_2\cdot + CH_2{=}CH_2 \rightarrow E{-}(CH_2)_3{-}CH_2\cdot \tag{2.1-6}$$

In general,

$$E{-}(CH_2)_{2n+1}{-}CH_2\cdot + CH_2{=}CH_2 \rightarrow E{-}(CH_2)_{2n+3}{-}CH_2\cdot \tag{2.1-7}$$

The chain reaction may then be terminated by the combination of two growing chains,

$$E{-}(CH_2)_n{-}CH_2\cdot + E{-}(CH_2)_m{-}CH_2\cdot \rightarrow E{-}(CH_2)_{n+m+2}{-}E \tag{2.1-8}$$

resulting in a stable polyethylene molecule. The entire growth process of the chain is very rapid. As a result the reaction mixture contains at any time both monomers and stable high molecular weight polymers but only very few growing chains. Thus in a flow reactor one may polymerize only 20% of the monomer, then separate polymer from monomer and recycle the monomer.

A listing of some synthetic polymers, their monomers, and their structural units is given in Table 2.1-1.

§2.2 MOLECULAR WEIGHT AND MOLECULAR WEIGHT DISTRIBUTIONS

In the previous section we have considered the chemical composition of single macromolecules. Aside from unimportant corrections from end groups and branch points, the molecular weight of a macromolecule is the product of the molecular weight of a structural unit and the number of structural units in the molecule. Typical synthetic polymer molecules may have molecular weights between 10,000 and 1,000,000 g/mol. Biological macromolecules may have even larger molecular weights; for example, the molecular weight of tobacco mosaic virus is about 40,000,000 g/mol.

Naturally one can conceive of a polymer sample in which the molecular weight of all macromolecules is the same. Such a sample is called *monodisperse*. Indeed, some biological polymers are monodisperse. Synthetic monodisperse or "almost monodisperse" polymers may be prepared by special techniques, but are seldom used commercially. Most commercial polymers by contrast are *polydisperse*, that is, they contain molecules of many different molecular weights. Thus we may talk of a distribution of molecular weights.

In order to describe molecular weight distributions in simple quantitative terms, we introduce various molecular weight averages. Let us assume that we are given a polydisperse macromolecular sample, composed of a number of monodisperse fractions. Specifically let us say that fraction "1" contains N_1 moles of molecular weight M_1, fraction "2" contains N_2 moles of molecular weight M_2 and so forth. We may introduce an average molecular weight by multiplying the molecular weight of each fraction by the number of moles in that fraction, summing and dividing by the total number of moles,

$$\boxed{\bar{M}_n = \frac{\sum_i N_i M_i}{\sum_i N_i}} \tag{2.2-1}$$

This is called the *number average molecular weight*; it is particularly sensitive to additions of small amounts of low molecular weight fractions. The mass of the ith fraction w_i is:

$$w_i = N_i M_i \tag{2.2-2}$$

We may therefore alternatively form an average by weighting the M_i with respect to the mass of the fractions:

$$\boxed{\bar{M}_w = \frac{\sum_i w_i M_i}{\sum_i w_i} = \frac{\sum_i N_i M_i^2}{\sum_i N_i M_i}} \tag{2.2-3}$$

This average is called the *weight average molecular weight*; it is particularly sensitive to the high molecular weight fractions. For a monodisperse sample $\bar{M}_w/\bar{M}_n = 1$, but for polydisperse samples $\bar{M}_w/\bar{M}_n > 1$. This ratio, the "heterogeneity index," is often taken as a measure of the polydispersity of the sample (Problem 2B.3).

Methods for molecular weight determination based on colligative properties, such as osmotic pressure and boiling point elevation, yield the number average molecular weight. On the other hand, light-scattering measurements may be used to determine the weight average molecular weight of a macromolecular sample. Viscosity measurements yield a third molecular weight average, the viscosity average molecular weight, which we discuss in §2.4. Detailed descriptions of these measurements may be found in standard references.[1]

EXAMPLE 2.2-1 Calculation of Average Molecular Weights of a Blend of Two Polydisperse Samples

A polymer blend is being prepared by mixing 80 kg of one polymer sample A (number average molecular weight $\bar{M}_{n,A} = 10{,}000$ g/mol and weight average molecular weight $\bar{M}_{w,A} = 15{,}000$ g/mol) with 20 kg of another sample B ($\bar{M}_{n,B} = 20{,}000$ g/mol and $\bar{M}_{w,B} = 50{,}000$ g/mol). It is desired to calculate the number average and weight average molecular weights of the mixture, $\bar{M}_{n,\text{mix}}$ and $\bar{M}_{w,\text{mix}}$, respectively.

[1] See, for example, E. A. Collins, J. Bareš, and F. W. Billmeyer, Jr., *Experiments in Polymer Science*, Wiley, New York (1973), or P. W. Allen, *Techniques of Polymer Characterization*, Butterworths, London (1959).

SOLUTION

Let us denote the total mass of samples A and B by W_A and W_B, respectively. Thus we know for sample A:

$$\sum_i N_{i,A} M_i = W_A \tag{2.2-4}$$

Here $N_{i,A}$ denotes the number of moles of sample A with molecular weight M_i. The definitions Eqs. 2.2–1 and 3 for sample A are then:

$$\bar{M}_{n,A} = \frac{\sum_i N_{i,A} M_i}{\sum_i N_{i,A}} = \frac{W_A}{\sum_i N_{i,A}} \tag{2.2-5}$$

$$\bar{M}_{w,A} = \frac{\sum_i N_{i,A} M_i^2}{\sum_i N_{i,A} M_i} = \frac{\sum_i N_{i,A} M_i^2}{W_A} \tag{2.2-6}$$

Similar expressions hold for sample B.

The desired averages may now be calculated directly from the definitions:

$$\bar{M}_{n,\text{mix}} = \frac{\sum_i (N_{i,A} + N_{i,B}) M_i}{\sum_i (N_{i,A} + N_{i,B})} \tag{2.2-7}$$

After substituting Eqs. 2.2–4 and 5 in the right side of Eq. 2.2–7 we find:

$$\bar{M}_{n,\text{mix}} = \frac{(W_A + W_B)\bar{M}_{n,A}\bar{M}_{n,B}}{W_A \bar{M}_{n,B} + W_B \bar{M}_{n,A}} \tag{2.2-8}$$

Similarly it follows from the definition of $\bar{M}_w$ and Eqs. 2.2–4 and 6 that

$$\bar{M}_{w,\text{mix}} = \frac{W_A \bar{M}_{w,A} + W_B \bar{M}_{w,B}}{W_A + W_B} \tag{2.2-9}$$

For the numerical example under consideration we have

$$\bar{M}_{n,\text{mix}} = \frac{(100\text{ kg})\left(10^4 \frac{\text{g}}{\text{mol}}\right)\left(2 \times 10^4 \frac{\text{g}}{\text{mol}}\right)}{(80\text{ kg})\left(2 \times 10^4 \frac{\text{g}}{\text{mol}}\right) + (20\text{ kg})\left(10^4 \frac{\text{g}}{\text{mol}}\right)}$$

$$= 1.11 \times 10^4 \frac{\text{g}}{\text{mol}} \tag{2.2-10}$$

$$\bar{M}_{w,\text{mix}} = \frac{(80\text{ kg})\left(1.5 \times 10^4 \frac{\text{g}}{\text{mol}}\right) + (20\text{ kg})\left(5 \times 10^4 \frac{\text{g}}{\text{mol}}\right)}{80\text{ kg} + 20\text{ kg}}$$

$$= 2.2 \times 10^4 \frac{\text{g}}{\text{mol}} \tag{2.2-11}$$

Notice that in the present example $\bar{M}_w$ is greater than $\bar{M}_n$, as it must always be (see Problem 2B.3).

It is often useful to have information about the general form of the molecular weight distribution of a given polymer. If we know the reaction mechanism of the polymerization process, kinetic or statistical arguments can be used to find expressions for the resulting molecular weight distributions in terms of a few parameters.[2] Particularly well-known is the so-called *most probable distribution.* This distribution is obtained in a step reaction polymerization of a linear polymer. Suppose that we start with N_0 monomer molecules in a batch reactor. After the polymerization reaction has started, the total number of monomer plus polymer molecules in the mixture N will decrease with time. The extent of reaction p is defined in terms of N and N_0 by

$$N = (1 - p)N_0 \tag{2.2–12}$$

Notice that p is also the fraction of functional groups that have reacted. At the onset of the polymerization p is zero; p then increases with time and approaches unity for large conversions. Let us denote by N_n the number of molecules in the sample that are *n-mers*, that is, molecules composed of n structural units. According to the most probable distribution (Example 2.2–2) N_n is given in terms of N and p by:

$$N_n = Np^{n-1}(1 - p) \tag{2.2–13}$$

Denoting the molecular weight of the structural units M_0 we find for the total mass w_n of the n-mers in the sample:

$$w_n = nM_0Np^{n-1}(1 - p)/\tilde{N} \tag{2.2–14}$$

where $\tilde{N}$ is Avogadro's number. From Eqs. 2.2–12 to 2.2–14 it follows (Problem 2B.1) that the number average and weight average molecular weights for the most probable distribution are:

$$\bar{M}_n = \frac{M_0}{(1 - p)} \tag{2.2–15}$$

$$\bar{M}_w = \frac{M_0(1 + p)}{(1 - p)} \tag{2.2–16}$$

From these expressions it is seen that large extents of reaction are mandatory for obtaining high molecular weight averages in step reaction polymerizations. Notice also that the ratio $\bar{M}_w/\bar{M}_n = (1 + p)$ approaches 2 for large extents of reaction.

EXAMPLE 2.2–2 The Most Probable Distribution of Molecular Weights for a Linear Step-Reaction Polymerization[3]

A monomer AB, containing two functional groups A and B, undergoes a step reaction polymerization. The functional groups B react with the functional groups A such that a typical reaction is:

$$AB + AB\text{—}AB\text{—}AB \rightarrow AB\text{—}AB\text{—}AB\text{—}AB \tag{2.2–17}$$

Assuming that the chemical reactivities of the groups A and B are independent of the size of the molecule to which they are attached, show that the number distribution of n-mers is given by Eq. 2.2–13.

[2] P. J. Flory, *Principles of Polymer Chemistry*, Cornell University Press, Ithaca (1953).
[3] P. J. Flory, *Chem. Revs.*, *39*, 137–197 (1946).

SOLUTION

Since p is the fraction of the functional groups that have reacted, it follows that p is also the probability that a randomly selected functional group has reacted, and that $(1 - p)$ is the probability that the functional group has not reacted. Moreover, these probabilities are independent of the size of the molecule to which the functional group is attached because of the assumption that the reactivity of the functional group is independent of the size of the molecule.

We now select at random one molecule in the mixture, and calculate the probability that the molecule is an n-mer. To do this let us focus our attention on the functional groups B in the chain starting from one end. The probability that the molecule is an n-mer is precisely the probability that we encounter first $(n - 1)$ consecutive reacted functional groups and then one unreacted group. The probability that the first group has reacted is p, and the probability that the first $(n - 1)$ groups have reacted is p^{n-1}. The probability that the last group is unreacted is $(1 - p)$; hence the probability that the molecule is an n-mer is $p^{n-1}(1 - p)$. Equation 2.2–13 now follows by noting that the probability that a molecule selected at random is an n-mer $(p^{n-1}(1 - p))$ precisely equals the fraction of the molecules in the sample that are n-mers (N_n/N).

The molecular weight distribution is not a permanent characteristic of a macromolecular fluid, as the macromolecules are subject to degradation. There are several kinds of degradation: thermal, radiative, chemical, biological, and mechanical. Of particular importance in macromolecular hydrodynamics is the rupture of polymer chains when polymeric fluids are subjected to high shear rates.[4] This changes the molecular weight distribution of the polymer and causes a permanent decrease in viscosity. The reduction in chain length can have a profound effect on the utility of macromolecules in many applications. Presently no reliable quantitative theory of mechanical stability of macromolecular fluids is available, and individual cases must be investigated experimentally. Higashitani[5] investigated the degradation effects of a gear pump on 2% aqueous solutions of polyacrylamide and poly(ethylene oxide). The shear stress and the "primary normal stress difference" (to be defined in Chapter 4) are both functions of the molecular weight. Measurements of these functions are compared in Fig. 2.2–1. It is seen that the poly(ethylene oxide) is degraded by the pumping whereas the polyacrylamide is not. In addition we see that the primary normal stress difference is a more sensitive measure of the polymer degradation, that is, of changes in the molecular weight distribution than the shear stress. This illustration also serves as an incentive for the reader to familiarize himself with the quantitative measurements of flow properties of macromolecular fluids to be discussed in Chapter 4.

§2.3 MACROMOLECULAR CONFIGURATIONS IN FLUIDS AT REST

Here we consider in more detail the actual shapes and configurations of macromolecules.[1] Whereas the chemical formulas in Table 2.1–1 serve as a very useful symbolic notation for the chemist, they do not convey any information about the relative sizes of the constituent atoms, or about the extent to which the molecules are able to change their

[4] A. Casale, R. S. Porter, J. F. Johnson, *Rubber Chem. Technol.*, *44*, 534–577 (1971).

[5] K. Higashitani, Ph.D. Thesis, University of Wisconsin, Madison (1973).

[1] The term configuration is used in two ways: (1) to indicate the *stereochemical configuration* or molecular architecture and (2) to describe the *spatial configuration* or instantaneous geometrical arrangement. Here we use the word configuration exclusively in the second sense above. The most exhaustive references on macromolecular configurations are P. J. Flory, *Statistical Mechanics of Chain Molecules*, Wiley, New York (1969) and M. Volkenstein, *Configurational Statistics of Polymeric Chains*, Interscience, New York (1963).

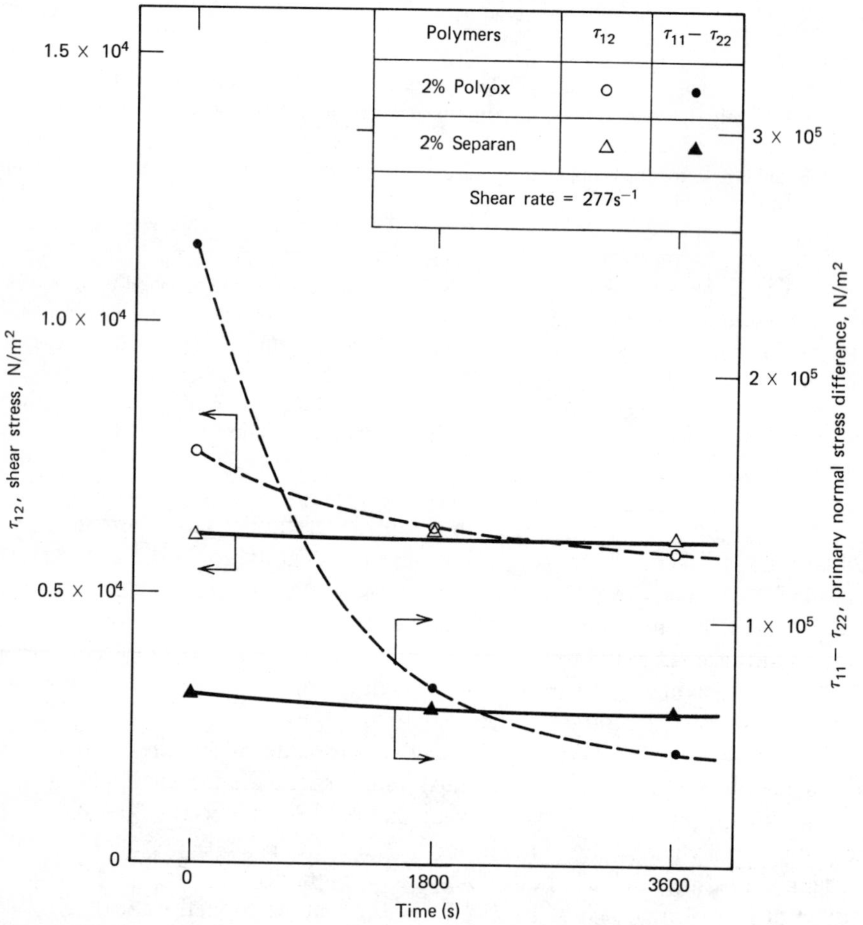

FIGURE 2.2–1. Shear degradation of solutions of 2% Polyox (poly(ethylene oxide)) in water and 2% Separan (polyacrylamide) in water by a gear pump. A volume of 300 cm^3 of the solutions was circulated through the pump with a flow rate of 15 cm^3/s for an hour. The shear stress and primary normal stress difference at a shear rate of 277 s^{-1} were measured on small test samples using a cone-and-plate viscometer (Chapter 4). [K. Higashitani, Ph.D. Thesis, University of Wisconsin, Madison (1973).]

configurations by rotations around chemical bonds. For this purpose the so-called *Fisher-Hirschfelder-Taylor* structural models prove very useful. Figure 2.3–1*a* shows Fisher-Hirschfelder-Taylor models of eight monomer units of a poly(ethylene oxide) chain in two possible configurations. Note that the chain is capable of assuming an enormous variety of different configurations by rotations around the chemical bonds. For this reason poly(ethylene oxide) is called a *flexible macromolecule*. A commonly used model of flexible macromolecules is the freely jointed chain model shown in Fig. 2.3–1*b*. This model consists of N identical mass points or "beads" connected by $(N - 1)$ rigid "rods" of length L with universal joints (i.e., an ideal ball-and-socket joint that permits two adjoining rods to have complete freedom of movement). The model is considerably simpler than the Fisher-Hirschfelder-Taylor model, but it still retains the essential feature of a flexible macromolecule, the long flexible backbone capable of assuming a huge variety of configurations. The quantitative description of the configurations of flexible macromolecules, the object of this section, is particularly important since practically all synthetic polymers are flexible. It is worthwhile to note, however, that not all macromolecules are flexible. For instance

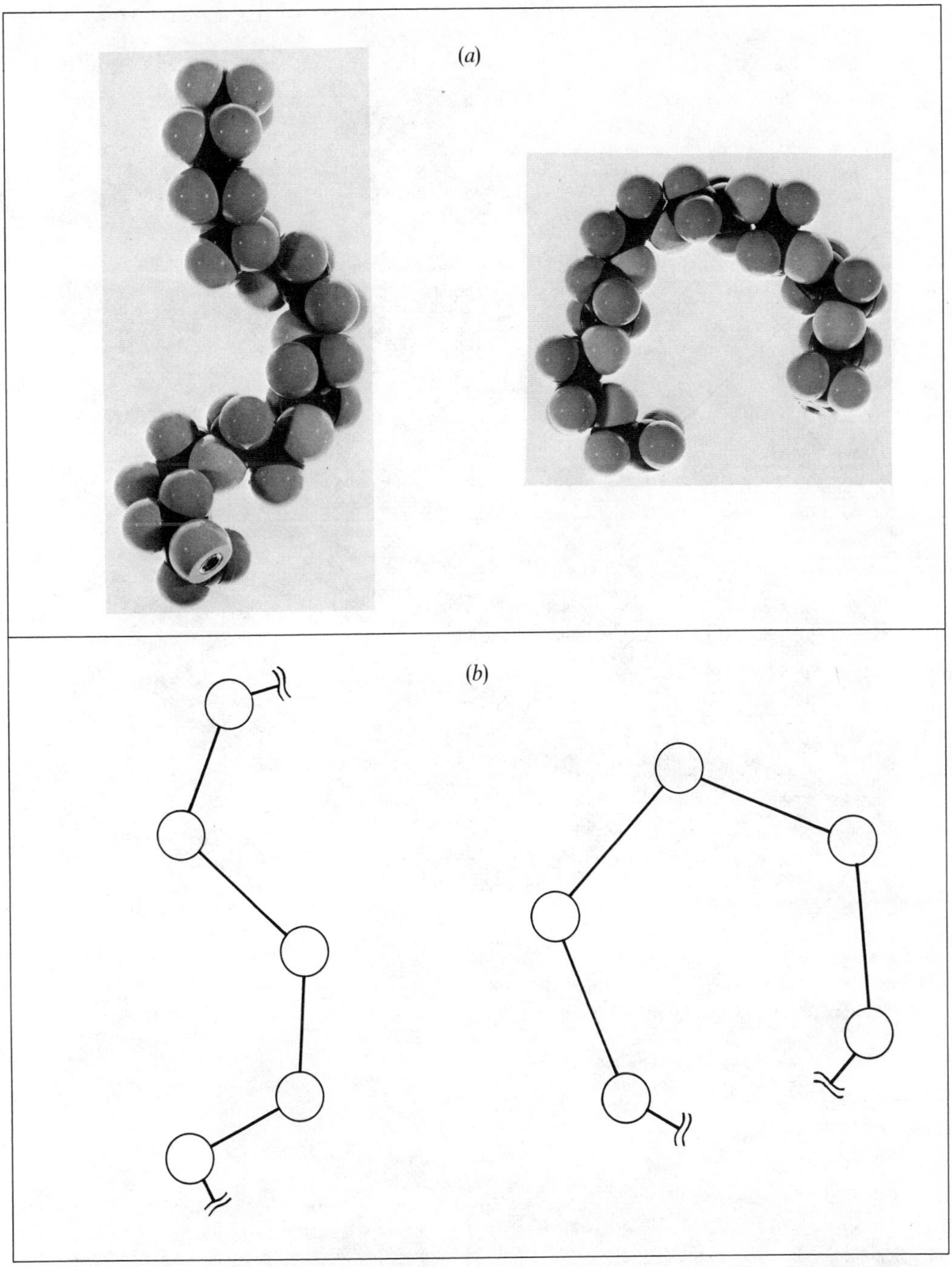

FIGURE 2.3–1. (*a*) Fisher-Hirschfelder-Taylor models of eight monomer units of a poly(ethylene oxide) chain in two of many possible configurations. (*b*) Freely jointed chain models duplicating the same configurations as the models above. Normally, however, in the idealization of a macromolecule by a freely jointed chain, each bead-rod unit corresponds to about 10 to 20 monomer units.

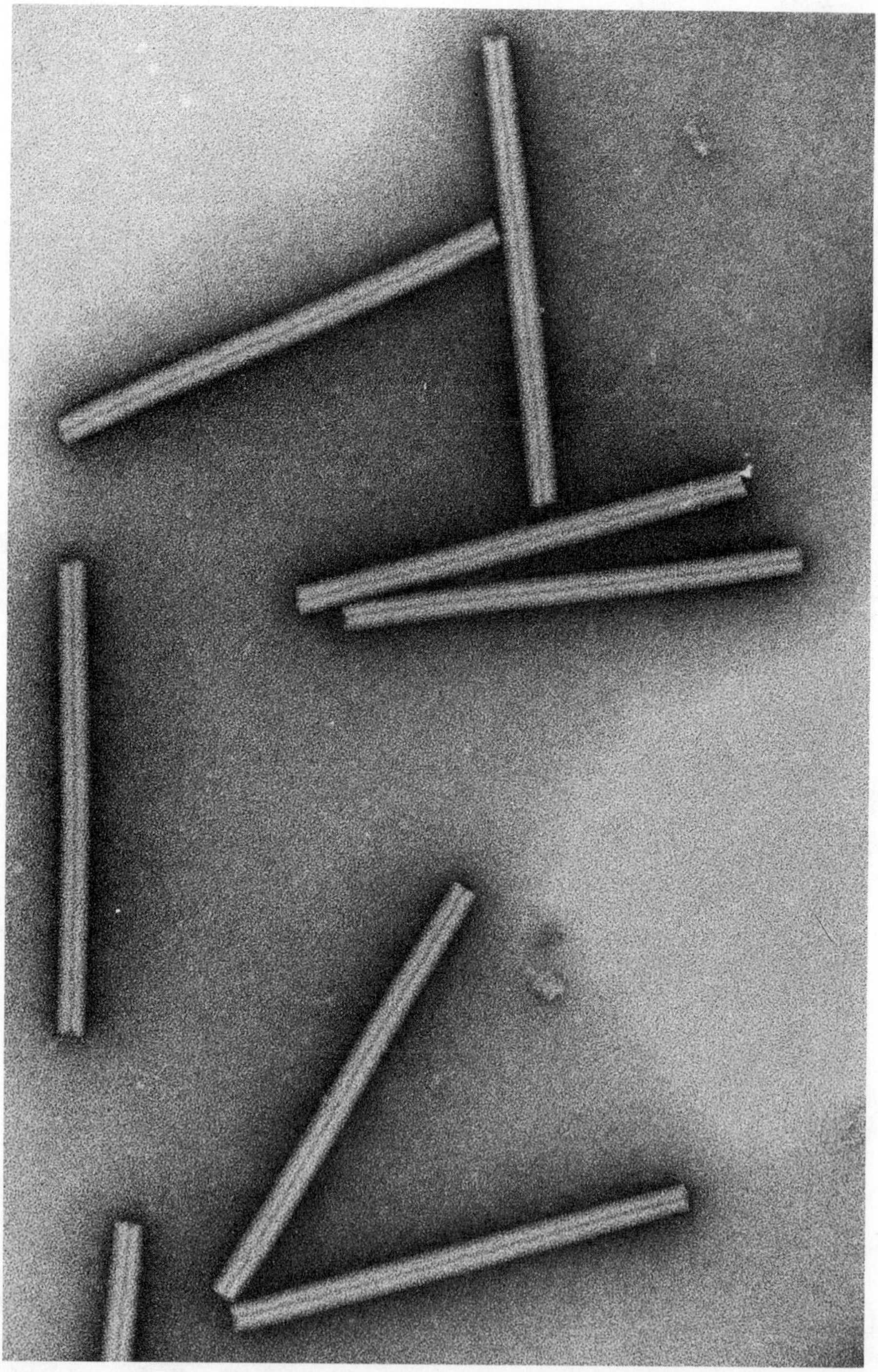

FIGURE 2.3–2. Electron micrograph of tobacco mosaic virus embedded in negative stain. The magnification is ×300,000. [Courtesy of the Virus Laboratory, University of California, Berkeley.]

Fig. 2.3–2 shows an electron micrograph of tobacco mosaic virus. We see that the molecule resembles a long, rigid rod. Such molecules are often referred to as *rigid macromolecules.*

Let us now focus our attention on a single flexible macromolecule. First we idealize the molecule as a freely jointed chain. The positions of the beads with respect to the center of mass are specified by the vectors $\boldsymbol{R}_\nu$, $\nu = 1, 2, \ldots, N$ (see Fig. 2.3–3). The *end-to-end distance r* and the *radius of gyration s* are defined by:

$$r^2 = \langle(\boldsymbol{R}_N - \boldsymbol{R}_1)\cdot(\boldsymbol{R}_N - \boldsymbol{R}_1)\rangle \qquad (2.3\text{–}1)$$

$$s^2 = \frac{1}{N}\sum_{\nu=1}^{N}\langle\boldsymbol{R}_\nu \cdot \boldsymbol{R}_\nu\rangle \qquad (2.3\text{–}2)$$

Due to random thermal motions (i.e., Brownian motion) the macromolecule will be continually changing its configuration in a highly erratic fashion. As a result, the end-to-end distance and the radius of gyration will be fluctuating wildly with time. Since we are not able to perform instantaneous measurements, the quantities of interest are therefore not the instantaneous values of r and s, but rather their time average values. To calculate these time averages we digress briefly to define distribution functions in configuration space and phase space and to explain how these may be obtained from equilibrium statistical mechanics.

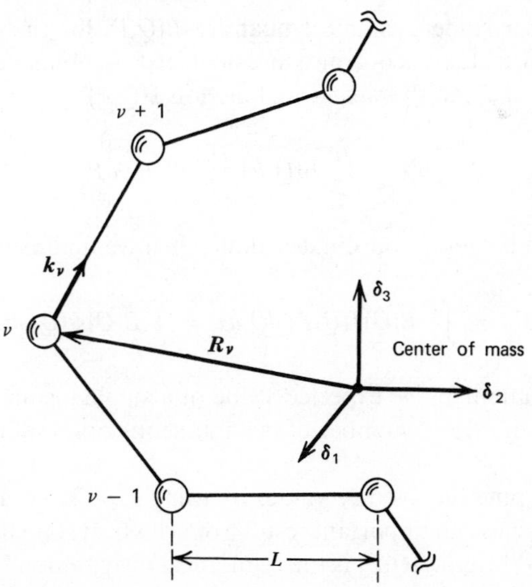

FIGURE 2.3–3. Section of the freely jointed chain of N beads connected by $(N - 1)$ rigid rods of length L. The figure shows the position vector $\boldsymbol{R}_\nu$ of the νth bead from the center of mass, the triad of unit vectors $\boldsymbol{\delta}_1$, $\boldsymbol{\delta}_2$, $\boldsymbol{\delta}_3$ with orientation fixed in space, and the unit vector $\boldsymbol{k}_\nu$ that points in the direction from bead ν toward bead $\nu + 1$.

Consider a single particle with Cartesian coordinates x_1, x_2, x_3 and momenta p_1, p_2, p_3. The configuration space for this particle is the three-dimensional space formed from the coordinates x_1, x_2, x_3, and the phase space for the particle is the six-dimensional space formed from the coordinates and momenta $x_1, x_2, x_3, p_1, p_2, p_3$. A point in the configuration space gives the location of the particle, whereas a point in the phase space gives the instantaneous location and momentum of the particle.

For a complicated many-particle system, such as the freely jointed chain under consideration, we use a set of generalized coordinates[2] $Q \equiv \{Q_1, Q_2, \ldots, Q_s\}$ and generalized momenta $P \equiv \{P_1, P_2, \ldots, P_s\}$ to specify the locations and momenta of all the particles. If there are no constraints in the system, then $s = 3N$, but when there are constraints (such as the constant bond lengths in the freely jointed chain), then $s < 3N$, where N is the number of particles in the system. The configuration space is then an s-dimensional hyperspace and the phase space has $2s$ dimensions.

Let us denote by $f(Q,P)\,dQ\,dP$ the probability that at any instant the system has coordinates and momenta in the range $dQ\,dP$ around the point Q, P in the phase space. In order for $f(Q,P)$ to have the meaning of a probability density, the integral over the entire phase space of $f(Q,P)$ must equal unity:

$$\iint f(Q,P)\,dQ\,dP = 1 \tag{2.3–3}$$

We will refer to $f(Q,P)$ as the phase-space distribution function. When it is integrated over all momenta, we obtain the configuration-space distribution function $\psi(Q)$:

$$\psi(Q) = \int f(Q,P)\,dP \tag{2.3–4}$$

Then $\psi(Q)\,dQ$ represents the probability that at any instant the system has coordinates in the range dQ about Q.

We now consider some dynamical quantity $B(Q,P)$ for the system. The expected value of a hypothetical instantaneous measurement of B is obtained by averaging B over the phase space with respect to the weighting function $f(Q,P)$:

$$\langle B \rangle = \iint B(Q,P) f(Q,P)\,dQ\,dP \tag{2.3–5}$$

If the quantity B depends on the coordinates alone, then we find using Eq. 2.3–4 that:

$$\langle B \rangle = \iint B(Q) f(Q,P)\,dQ\,dP = \int B(Q)\psi(Q)\,dQ \tag{2.3–6}$$

It is reasonable to assume that the expected value of a single instantaneous observation is also the average value of a large number of instantaneous observations—that is, the time-average value.

In order to compute the average values by using Eqs. 2.3–5 and 6 it is necessary to have $f(Q,P)$. For this we use an important result from classical statistical mechanics, namely that for systems at equilibrium $f(Q,P)$ is the "canonical distribution":[3]

$$f(Q,P) = Z^{-1} e^{-\mathscr{H}/kT} \tag{2.3–7}$$

in which k is the Boltzmann constant, T the absolute temperature, and $\mathscr{H}$ the Hamiltonian; the latter is just the sum of the kinetic and potential energies of the system expressed in terms

[2] For a short and very satisfactory treatment of generalized coordinates and statistical mechanics as applied to molecular systems, see H. Margenau and G. M. Murphy, *The Mathematics of Physics and Chemistry*, Van Nostrand, Princeton (1956), Second Edition, Chapters 9 and 12.

[3] J. O. Hirschfelder, C. F. Curtiss, and R. B. Bird, *Molecular Theory of Gases and Liquids*, Wiley, New York (1964), second corrected printing, Chapter 2.

of the generalized coordinates and momenta. The normalization constant Z in Eq. 2.3–7 can be found from Eq. 2.3–3 to be:

$$Z = \iint e^{-\mathscr{H}/kT}\, dQ\, dP \tag{2.3–8}$$

The quantity Z is called the "partition function" (the German name "Zustandsintegral"—indicating an integral over all states of the system—is perhaps more meaningful) and is related to the Helmholtz free energy A by:

$$A = -kT \ln Z \tag{2.3–9}$$

We see thus that statistical mechanics provides us with a specific recipe for obtaining $f(Q,P)$ for systems at equilibrium.

We are now in a position to return to the problem of obtaining expressions for the time-average values of the square of the end-to-end distance and the square of the radius of gyration:

$$\langle r^2 \rangle = \int ((\boldsymbol{R}_N - \boldsymbol{R}_1) \cdot (\boldsymbol{R}_N - \boldsymbol{R}_1))\psi(Q)\, dQ \tag{2.3–10}$$

$$\langle s^2 \rangle = \int \frac{1}{N} \sum_{\nu=1}^{N} (\boldsymbol{R}_\nu \cdot \boldsymbol{R}_\nu)\psi(Q)\, dQ \tag{2.3–11}$$

For the generalized coordinates Q describing the internal configuration of the freely jointed bead-rod chain we use the angles θ_i, ϕ_i ($i = 1, 2, \ldots, N - 1$) which are the spherical coordinates designating the bond directions. We also use a set of unit vectors $\boldsymbol{k}_i$ in the direction from the ith to the $(i + 1)$th bead (see Fig. 2.3–3). The $\boldsymbol{k}_i$ may be expressed in terms of the generalized coordinates as follows:

$$\boldsymbol{k}_i = \boldsymbol{\delta}_1(\sin\theta_i \cos\phi_i) + \boldsymbol{\delta}_2(\sin\theta_i \sin\phi_i) + \boldsymbol{\delta}_3(\cos\theta_i) \tag{2.3–12}$$

where $\boldsymbol{\delta}_1$, $\boldsymbol{\delta}_2$, $\boldsymbol{\delta}_3$ are the unit vectors for a Cartesian coordinate system. Strictly speaking we also need three additional coordinates to specify the location of the center of mass of the molecule, but these additional coordinates can be ignored in this discussion. In the kinetic-theory calculations of Chapters 10 to 14 we discuss carefully the separation of the translational and internal motions.

In order to evaluate the integrals in Eqs. 2.3–10 and 11 we now need an explicit expression for $\psi(\theta_1,\phi_1, \ldots, \theta_{N-1},\phi_{N-1})$. It is straightforward to write down the expression for the Hamiltonian for a freely jointed chain and to use Eqs. 2.3–4, 7, and 8 to obtain a formal expression for ψ, but obtaining a closed-form analytic expression proves to be rather difficult.[4] The exact result does show that the orientation of any link in the chain is only very weakly correlated with the orientations of neighboring links. In this chapter we shall neglect these weak correlations entirely; that is, we assume here that the orientation of the ith link is random, so that the probability that it has an orientation in a range $d\theta_i\, d\phi_i$ around θ_i, ϕ_i is:

$$\psi_i\, d\theta_i\, d\phi_i = \frac{1}{4\pi} \sin\theta_i\, d\theta_i\, d\phi_i \tag{2.3–13}$$

[4] H. A. Kramers, *J. Chem. Phys.*, *14*, 415–424 (1946); M. Fixman, *Proc. Nat. Acad. Sci. U.S.A.*, *71*, 3050–3053 (1974); see also §13.5.

independent of the orientation of neighboring links. In this approximation it follows that the distribution function for the entire chain is the product of the distribution functions for the individual links,

$$\psi = \prod_{i=1}^{N-1} \psi_i = \left(\frac{1}{4\pi}\right)^{N-1} \prod_{i=1}^{N-1} \sin\theta_i \tag{2.3–14}$$

This distribution will be referred to as the *random flight distribution.* Note that the random flight distribution has the property that (see Problem 2B.5):

$$\begin{aligned}\langle \boldsymbol{k}_i \cdot \boldsymbol{k}_j \rangle &= \int (\boldsymbol{k}_i \cdot \boldsymbol{k}_j)\psi(Q)\, dQ \\ &= \int_{\phi_{N-1}=0}^{2\pi} \cdots \int_{\theta_1=0}^{\pi} (\boldsymbol{k}_i \cdot \boldsymbol{k}_j)\left(\left(\frac{1}{4\pi}\right)^{N-1} \prod_{k=1}^{N-1} \sin\theta_k\right) \prod_{l=1}^{N-1} d\theta_l\, d\phi_l \\ &= \delta_{ij}\end{aligned} \tag{2.3–15}$$

To calculate the mean square end-to-end distance of the chain note that the end-to-end distance may be obtained by summing over all the bond vectors:

$$\boldsymbol{R}_N - \boldsymbol{R}_1 = \sum_{i=1}^{N-1} L\boldsymbol{k}_i \tag{2.3–16}$$

It now follows directly from the definition, Eq. 2.3–10, that the mean square end-to-end distance of a freely jointed chain of N beads with $(N - 1)$ links of length L is:

$$\begin{aligned}\langle r^2 \rangle &= \left\langle \left(\sum_{i=1}^{N-1} L\boldsymbol{k}_i \cdot \sum_{j=1}^{N-1} L\boldsymbol{k}_j \right) \right\rangle \\ &= L^2 \sum_{i=1}^{N-1} \sum_{j=1}^{N-1} \langle \boldsymbol{k}_i \cdot \boldsymbol{k}_j \rangle \\ &= L^2 \sum_{i=1}^{N-1} \sum_{j=1}^{N-1} \delta_{ij} \\ &= L^2 \sum_{i=1}^{N-1} 1 \\ &= L^2(N-1)\end{aligned} \tag{2.3–17}$$

Note that the root-mean-square end-to-end distance is $\sqrt{\langle r^2 \rangle} = \sqrt{N-1}L$. On the other hand the fully extended length of the chain is $(N - 1)L$. Therefore, on the average, a polymer chain may be extended by a factor $\sqrt{N-1}$ from its equilibrium configuration to its fully stretched state. Since $\sqrt{N-1}$ can be of the order of magnitude 100 for reasonable models of macromolecules, it can be expected that this enormous change in shape will profoundly affect the optical and rheological properties of macromolecular fluids.

By analogy with Eq. 2.3–17 for $\langle r^2 \rangle$ it may be shown (Example 2.3–1) that the mean square radius of gyration of the freely jointed chain of N beads with $(N - 1)$ links of length L is:

$$\langle s^2 \rangle = \tfrac{1}{6}L^2 \frac{(N-1)(N+1)}{N} \tag{2.3–18}$$

Note that $6\langle s^2\rangle = \langle r^2\rangle$ for large N. This simple relation, due to Debye,[5] indicates that either $\langle s^2\rangle$ or $\langle r^2\rangle$ may be used as a measure of the volume swept out by polymer chains.

EXAMPLE 2.3–1 Mean Square Radius of Gyration

The configuration space distribution function for a freely jointed chain is approximated by the random walk distribution, Eq. 2.3–14. Show that the mean square radius of gyration for the chain is given by Eq. 2.3–18.

SOLUTION

We need to rewrite the expression for s^2 in Eq. 2.3–2 in terms of the bond vectors. From purely geometrical considerations it may be shown (see §12.1) that:

$$\sum_{\nu=1}^{N} \boldsymbol{R}_\nu \boldsymbol{R}_\nu = L^2 \sum_{i=1}^{N-1} \sum_{j=1}^{N-1} C_{ij} \boldsymbol{k}_i \boldsymbol{k}_j \tag{2.3–19}$$

in which the matrix elements C_{ij} are (see Eq. 12.1–7):

$$C_{ij} = \begin{cases} \dfrac{i(N-j)}{N} & \text{if} \quad i \leq j \\[2ex] \dfrac{j(N-i)}{N} & \text{if} \quad j \leq i \end{cases} \tag{2.3–20}$$

The mean square radius of gyration now follows from Eqs. 2.3–11, 15, and 20:

$$\begin{aligned} \langle s^2\rangle &= \left\langle \frac{L^2}{N} \sum_{i=1}^{N-1} \sum_{j=1}^{N-1} C_{ij}(\boldsymbol{k}_i \cdot \boldsymbol{k}_j) \right\rangle \\ &= \frac{L^2}{N} \sum_{i=1}^{N-1} \sum_{j=1}^{N-1} C_{ij}\delta_{ij} \\ &= \frac{L^2}{N} \sum_{i=1}^{N-1} C_{ii} \\ &= \frac{L^2}{N} \left[\sum_{i=1}^{N-1} i - \frac{1}{N} \sum_{i=1}^{N-1} i^2 \right] \\ &= \tfrac{1}{6} L^2 \frac{(N-1)(N+1)}{N} \end{aligned} \tag{2.3–21}$$

The summations involved in the last step may be found in standard tables.[6]

All observable quantities for the chain model may be obtained as averages over the phase space distribution function or, if the quantity of interest is independent of the momenta, from the configuration space distribution function alone. We now introduce a different distribution, namely that for the end-to-end vector of the freely jointed chain. We do so by way of an illustration in which we rewrite the expression for the mean square end-to-end

[5] P. Debye, *J. Chem. Phys.*, *14*, 636–639 (1946).

[6] H. B. Dwight, *Table of Integrals and Other Mathematical Data*, Macmillan, New York (1961), p. 7; see also Eq. 12.1–10.

distance in terms of the configuration space distribution function:

$$\begin{aligned}\langle r^2 \rangle &= \int ((\boldsymbol{R}_N - \boldsymbol{R}_1) \cdot (\boldsymbol{R}_N - \boldsymbol{R}_1))\psi(Q)\, dQ \\ &= \int_{-\infty}^{\infty} \int (\boldsymbol{r} \cdot \boldsymbol{r})\delta(\boldsymbol{r} - (\boldsymbol{R}_N - \boldsymbol{R}_1))\psi(Q)\, dQ\, d\boldsymbol{r} \\ &= \int_{-\infty}^{\infty} r^2 P_{N,L}(\boldsymbol{r})\, d\boldsymbol{r} \end{aligned} \tag{2.3–22}$$

In Eq. 2.3–22 the distribution function $P_{N,L}(\boldsymbol{r})$ for the end-to-end vector of the freely jointed chain made up of $N - 1$ links of length L has been defined by:

$$P_{N,L}(\boldsymbol{r}) = \int \delta(\boldsymbol{r} - (\boldsymbol{R}_N - \boldsymbol{R}_1))\psi(Q)\, dQ \tag{2.3–23}$$

The "Dirac delta function" $\delta(\boldsymbol{x} - \boldsymbol{a})$ has the property that $\int_{-\infty}^{\infty} \delta(\boldsymbol{x} - \boldsymbol{a}) f(\boldsymbol{x})\, d\boldsymbol{x} = f(\boldsymbol{a})$. This property of the delta function establishes the connection between the first and second line of Eq. 2.3–22. The distribution $P_{N,L}(\boldsymbol{r})$ contains less information than the complete configuration space distribution function. To obtain $P_{N,L}(\boldsymbol{r})$ in the random flight approximation we have to introduce $\psi(Q)$ from Eq. 2.3–14 in the right side of Eq. 2.3–23. The mathematical details in the rigorous treatment of the resulting multiple integral are omitted here. Provided that N is large, and that the end-to-end distance is smaller than about $0.5NL$, it may be shown[7] that $P_{N,L}(\boldsymbol{r})$ is approximated by

$$\boxed{P_{N,L}(\boldsymbol{r}) \doteq \left(\frac{3}{2\pi(N - 1)L^2}\right)^{3/2} e^{-3r^2/2(N-1)L^2}} \tag{2.3–24}$$

Equation 2.3–24 is known as the *Gaussian distribution* for the end-to-end vector of the freely jointed chain with $(N - 1)$ links of length L. We see that $P_{N,L}(\boldsymbol{r})$ has a maximum at $\boldsymbol{r} = \boldsymbol{0}$; that is, if one end of the molecule is located at the point A, then the most probable position of the other end is at the same point A. This does not mean, however, that the most likely value of the end-to-end distance $r = |\boldsymbol{r}|$ is zero. Indeed we already know from Eq. 2.3–17 that the root mean square end-to-end distance is $\sqrt{\langle r^2 \rangle} = \sqrt{(N - 1)}L$. It is interesting that although the Gaussian distribution for the end-to-end vector is only approximately valid, it still gives the correct value of the mean square end-to-end distance,

$$\begin{aligned}\langle r^2 \rangle &= \int r^2 P_{N,L}(\boldsymbol{r})\, d\boldsymbol{r} \\ &= \int_0^{2\pi} \int_0^{\pi} \int_0^{\infty} r^2 \left(\frac{3}{2\pi(N - 1)L^2}\right)^{3/2} e^{-3r^2/2(N-1)L^2} r^2 \sin\theta\, dr\, d\theta\, d\phi \\ &= L^2(N - 1) \end{aligned} \tag{2.3–25}$$

There is an interesting analogy between the problem of determining $P_{N,L}(\boldsymbol{r})$ in the approximation that the correlations between the orientations of neighboring links are neglected, and the classical problem of random flights.[8] In this latter problem we imagine a particle undergoing a sequence of $(N - 1)$ displacements. Each displacement is assumed

[7] H. Yamakawa, *Modern Theory of Polymer Solutions*, Harper and Row, New York (1971), p. 15; P. J. Flory, *Statistical Mechanics of Chain Molecules*, Wiley, New York (1969), Chapter VIII.
[8] S. Chandrasekhar, *Revs. of Modern Phys.*, *15*, 1–89 (1943).

to be of length L, but the direction is random and independent of the previous displacements. The distribution function for the vector $\boldsymbol{r}$ from the initial position of the particle to the final position of the particle equals the distribution function $P_{N,L}(\boldsymbol{r})$ for the end-to-end vector of the freely jointed chain. It is from this analogy that the distribution in Eq. 2.3–14 derives its name, the "random flight distribution." Furthermore the analogy makes possible some interesting methods for estimating $P_{N,L}(\boldsymbol{r})$ as illustrated in the following example.

EXAMPLE 2.3–2 The One-Dimensional Random Flight

A particle moves on the x-axis, undergoing displacements of equal length L at time intervals Δt. Each displacement occurs with equal probability (1/2) either in the positive or negative direction. Assume that the particle started at $x = 0$ at time $t = 0$, and find the probability that it has moved a distance x from the origin after $N - 1$ random steps, that is, at time $t = (N - 1)\,\Delta t$.

SOLUTION

Consider that at time $t + \Delta t$ the particle is at location x. For this to be the case it must at time t *either* have been at location $x - L$ and have been displaced in the positive direction, *or* have been at location $x + L$ and have been displaced in the negative direction. The probability that it was at $x - L$ at time t is $P(x - L,t)$ and the probability that it was displaced in the positive direction is $\frac{1}{2}$. Thus the probability of arriving at x at time $t + \Delta t$ from $x - L$ is $\frac{1}{2}P(x - L,t)$. Similarly, the probability of arriving at x at time $t + \Delta t$ from $x + L$ is $\frac{1}{2}P(x + L,t)$. But the sum of these two probabilities must exactly equal the probability that the particle is at x at time $t + \Delta t$, that is, $P(x,t + \Delta t)$. We then arrive at the following difference equation:

$$P(x,t + \Delta t) = \tfrac{1}{2}P(x - L,t) + \tfrac{1}{2}P(x + L,t) \tag{2.3–26}$$

The appropriate initial and boundary conditions are:

$$\text{I.C.:} \quad P(x,0) = 0 \quad \text{for} \quad x \neq 0 \tag{2.3–27}$$

$$\text{B.C.:} \quad \sum_{x=0,\pm L,\pm 2L,\ldots} P(x,t) = 1 \quad \text{for all} \quad t \geq 0 \tag{2.3–28}$$

Equation 2.3–28 expresses the fact that the particle must be somewhere on the x-axis for all $t \geq 0$, and Eq. 2.3–27 expresses the fact that the particle started at the origin at $t = 0$. Rather than solving Eqs. 2.3–26 to 28 let us consider the situation where the individual steps are so small and the number of steps so large that the particle can occupy any position on the x-axis. We can pass to this continuous limit in the following fashion. First subtract $P(x,t)$ from both sides and divide by Δt to get:

$$\frac{P(x,t + \Delta t) - P(x,t)}{\Delta t} = \frac{L^2}{\Delta t}\,\frac{1}{2}\,\frac{P(x - L,t) - 2P(x,t) + P(x + L,t)}{L^2} \tag{2.3–29}$$

We now take the limit $\Delta t \to 0$ holding the ratio $a \equiv (L^2/\Delta t)$ constant. Equation 2.3–29 then becomes:

$$\frac{\partial}{\partial t} P(x,t) = \tfrac{1}{2}a \frac{\partial^2}{\partial x^2} P(x,t) \tag{2.3–30}$$

Also the boundary and initial conditions become:

$$\text{I.C.:} \quad P(x,0) = 0 \quad \text{for} \quad x \neq 0 \tag{2.3–31}$$

$$\text{B.C.:} \quad \int_{-\infty}^{\infty} P(x,t)\,dx = 1 \quad \text{for all} \quad t \geq 0 \tag{2.3–32}$$

Equation 2.3–30 is of the same form as the one-dimensional heat conduction equation. The solution with the conditions given in Eqs. 2.3–31 and 32 is known to be:[9]

$$P(x,t) = \left(\frac{1}{2\pi at}\right)^{1/2} e^{-x^2/2at} \tag{2.3–33}$$

Recall that $t = (N - 1)\,\Delta t = (N - 1)L^2/a$, so that Eq. 2.3–33 becomes:

$$P_{N,L}(x) = \left(\frac{1}{2\pi(N - 1)L^2}\right)^{1/2} e^{-x^2/2(N-1)L^2} \tag{2.3–34}$$

In this example we have gone from a discrete problem (in which the particle can be only at the points $x = kL$ where k is an integer) to a continuous problem (in which it can be located at any value of x). This was accomplished by taking the limit $L \to 0$. In this continuous description the probability of finding the particle in an interval dx around the distance x from the origin after $(N - 1)$ random displacements of length L is approximated by $P_{N,L}(x)\,dx$ with $P_{N,L}(x)$ given by Eq. 2.3–34. The function $P_{N,L}(x)$ is called a *Gaussian* or *normal density*. It is clear, however, that the Gaussian density is only an approximation to the correct distribution since it indicates incorrectly that there is a finite probability of finding the particle at distances $x > (N - 1)L$ after it has suffered $(N - 1)$ random displacements of length L. As N becomes large, however, the Gaussian density becomes a very good approximation for the end-to-end distribution of a random-flight chain for distances x up to about $0.5\ NL$.

In Problem 2C.1 we extend this analysis to the three-dimensional problem where the particle is free to move randomly in space with displacements of length L along the coordinate directions. This leads to the density in Eq. 2.3–24 for the vector $\boldsymbol{r}$ from the initial to the final position of the particle. By the analogy between the freely jointed chain problem and the random flight problem, this then represents an alternative derivation of Eq. 2.3–24 for the distribution function for the end-to-end vector of the freely jointed chain.

So far we have considered the situation in which the freely jointed chain is free to move unrestricted. Let us now consider the situation in which the two ends of the chain are held at fixed positions. In order to hold the ends of the chain fixed we must apply forces at the ends. These forces are needed because of the thermal motions of the beads in the chain and will be wildly fluctuating with time. The time average values of the forces, however, may be calculated from statistical mechanics as illustrated in the following example.

EXAMPLE 2.3–3 Time Average Tension in a Polymer Chain[10]

The two ends of a freely jointed chain are held at fixed positions $\boldsymbol{R}_1$ and $\boldsymbol{R}_N$. Consider the chain as a thermodynamic system with the vector $\boldsymbol{r}$ from $\boldsymbol{R}_1$ to $\boldsymbol{R}_N$ as an external parameter. It is desired to find the time average force $\boldsymbol{F}(\boldsymbol{r})$ exerted by the system on the point $\boldsymbol{R}_N$.

SOLUTION

We begin by assuming that the phase space distribution function for the constrained chain f_c may be written as a product containing the phase space distribution for the unconstrained chain f and a

[9] H. S. Carslaw and J. C. Jaeger, *Conduction of Heat in Solids*, Oxford University Press (1959), Second Edition, §§2.1 and 2.2; H. Margenau and G. M. Murphy, *The Mathematics of Physics and Chemistry*, Van Nostrand, Princeton (1956), Second Edition, p. 239.

[10] L. R. G. Treloar, *The Physics of Rubber Elasticity*, Oxford University Press, London (1967), Second Edition, pp. 62, 104.

delta function expressing the constraints on the end points. That is, we *assume*:

$$f_c = Z_c^{-1}\delta(\boldsymbol{r} - (\boldsymbol{R}_N - \boldsymbol{R}_1))f \tag{2.3–35}$$

where Z_c is the partition function for the constrained chain. This assumption seems to be inherent in all derivations of $\boldsymbol{F}(\boldsymbol{r})$ to our knowledge, but does not appear to have been critically analyzed. With this assumption, however, we may calculate the partition function from the normalization condition on f_c, using also Eqs. 2.3–4 and 2.3–23:

$$\begin{aligned} Z_c &= \iint \delta(\boldsymbol{r} - (\boldsymbol{R}_N - \boldsymbol{R}_1))f(Q,P)\,dQ\,dP \\ &= \int \delta(\boldsymbol{r} - (\boldsymbol{R}_N - \boldsymbol{R}_1))\psi(Q)\,dQ \\ &= P_{N,L}(\boldsymbol{r}) \end{aligned} \tag{2.3–36}$$

In Eq. 2.3–36 $\psi(Q)$ is the configuration space distribution function for the unconstrained chain, and $P_{N,L}(\boldsymbol{r})$ is the distribution function for the end-to-end vector of the unconstrained chain. By Eqs. 2.3–9 and 36 it follows that the Helmholtz free energy of the chain with fixed end points is:

$$A = -kT \ln P_{N,L}(\boldsymbol{r}) \tag{2.3–37}$$

We now allow the point $\boldsymbol{R}_N$ to move so that $\boldsymbol{r}$ changes by an amount $d\boldsymbol{r}$. Then the work performed by the system on the surroundings is $(\boldsymbol{F} \cdot d\boldsymbol{r})$. But under isothermal conditions this also equals the decrease in the Helmholtz free energy of the system. Therefore, using Eqs. 2.3–24 and 37 we find,

$$\begin{aligned} (\boldsymbol{F} \cdot d\boldsymbol{r}) &= -dA \\ &= kT\,d \ln P_{N,L}(\boldsymbol{r}) \\ &= kT\,d\left(\frac{-3(\boldsymbol{r} \cdot \boldsymbol{r})}{2(N-1)L^2}\right) \\ &= -\frac{3kT}{(N-1)L^2}(\boldsymbol{r} \cdot d\boldsymbol{r}) \end{aligned} \tag{2.3–38}$$

Equation 2.3–38 may be rearranged to give

$$\left(\left[\boldsymbol{F} + \frac{3kT}{(N-1)L^2}\boldsymbol{r}\right] \cdot d\boldsymbol{r}\right) = 0 \tag{2.3–39}$$

This scalar product must equal zero for all infinitesimal increments $d\boldsymbol{r}$, but that can be true only if the vector in the square bracket is identically zero, that is, if

$$\boxed{\boldsymbol{F}(\boldsymbol{r}) = -\frac{3kT}{(N-1)L^2}\boldsymbol{r}} \tag{2.3–40}$$

Thus the freely jointed chain with ends at specified points acts like a Hookean spring with force constant $H = 3kT/(N-1)L^2$ and length zero in the absence of an external force, that is, its natural length is zero.[11] Since this calculation is based on the Gaussian distribution, Eq. 2.3–24, one also often

[11] Some polymer scientists have suggested that instead of a force law $\boldsymbol{F} = -H\boldsymbol{r}$, we should write $\boldsymbol{F} = -H\boldsymbol{r} - K(\boldsymbol{r}/r)\,dr/dt$, the last term accounting for something called "internal viscosity." The origin of this extra contribution is not well understood and the concept of internal viscosity has engendered some controversy in polymer circles.

talks about a *Gaussian spring.* As mentioned previously the Gaussian approximation is good only for large N and extensions r smaller than about $0.5NL$. For this reason the Hookean (or Gaussian) spring law also ceases to be a valid approximation for $r > 0.5NL$. In particular, we see that Eq. 2.3–40 permits extensions beyond the fully stretched length of the chain, whereas in actuality the spring stiffens up as the chain becomes fully extended. To take account of this fact various modified force laws have been proposed, of which we note the inverse Langevin force law[10] and the Warner force law[12] given by Eqs. 2.3–41 and 42, respectively:

(i) *Inverse Langevin* (derivable from molecular arguments[10]):

$$F(r) = \frac{kT}{L}\,\mathscr{L}^{-1}\left(\frac{r}{(N-1)L}\right) \tag{2.3–41}$$

where $\mathscr{L}^{-1}$ is the function inverse to the Langevin function $\mathscr{L}$ given by $\mathscr{L}(x) = (\coth x) - x^{-1}$.

(ii) *Warner* (a useful empiricism[12]):

$$F(r) = \frac{3kTr}{(N-1)L^2\left[1 - \left(\dfrac{r}{(N-1)L}\right)^2\right]} \tag{2.3–42}$$

In Eqs. 2.3–41 and 42, $F = |\boldsymbol{F}|$ denotes the tension in the spring. These functions are shown in Fig. 2.3–4. Note that the Warner force law has all of the desired physical characteristics, linear behavior at small extensions and a finite length $(N-1)L$, without the mathematical complexity of the inverse Langevin function. For this reason we prefer its use over Eq. 2.3–41. The results given in this example will be particularly useful in connection with the molecular theories treated in Volume 2.

Evidently the freely jointed chain is a rough but useful model of a macromolecule. One particularly important feature ignored in the freely jointed chain model as presented here is the effect of *excluded volume.* In the calculations so far we have assumed that the beads of the chain are mass points, that is, two beads may occupy positions arbitrarily close to each other. This, of course, is not realistic since the space occupied by one part of a polymer chain is not available to another part of the chain; hence the name excluded volume effect. As a result of this effect the average dimensions of the chain will be somewhat larger than expected from Eqs. 2.3–17 and 18, and one generally introduces a correction factor α called the *molecular expansion factor*:

$$\langle r^2\rangle_e = \alpha^2\langle r^2\rangle \tag{2.3–43}$$

$$\langle s^2\rangle_e = \alpha^2\langle s^2\rangle \tag{2.3–44}$$

In Eqs. 2.3–43 and 44, $\langle r^2\rangle_e$ and $\langle s^2\rangle_e$ are the mean square values of the end-to-end distance and the radius of gyration corrected for molecular expansion, whereas $\langle r^2\rangle$ and $\langle s^2\rangle$ are the values predicted in the absence of excluded volume effects. It is generally assumed that the same correction factor applies for the end-to-end distance and the radius of gyration although this assumption does not seem to have been critically tested. Neutron scattering experiments[13] indicate that α is close to unity for concentrated solutions and melts. For dilute solutions, however, α may differ significantly from unity as discussed further in the next section.

[12] H. R. Warner, Jr., *Ind. Eng. Chem. Fundamentals, 11*, 379–387 (1972). A spring obeying Eq. 2.3–42 is sometimes called a FENE connector (*f*initely *e*xtendable *n*onlinear *e*lastic).

[13] R. G. Kirste, W. A. Kruse, and J. Schelten, *Makromoleculare Chemie, 162*, 299–303 (1972); P. J. Flory, *Statistical Mechanics of Chain Molecules*, Wiley, New York (1969).

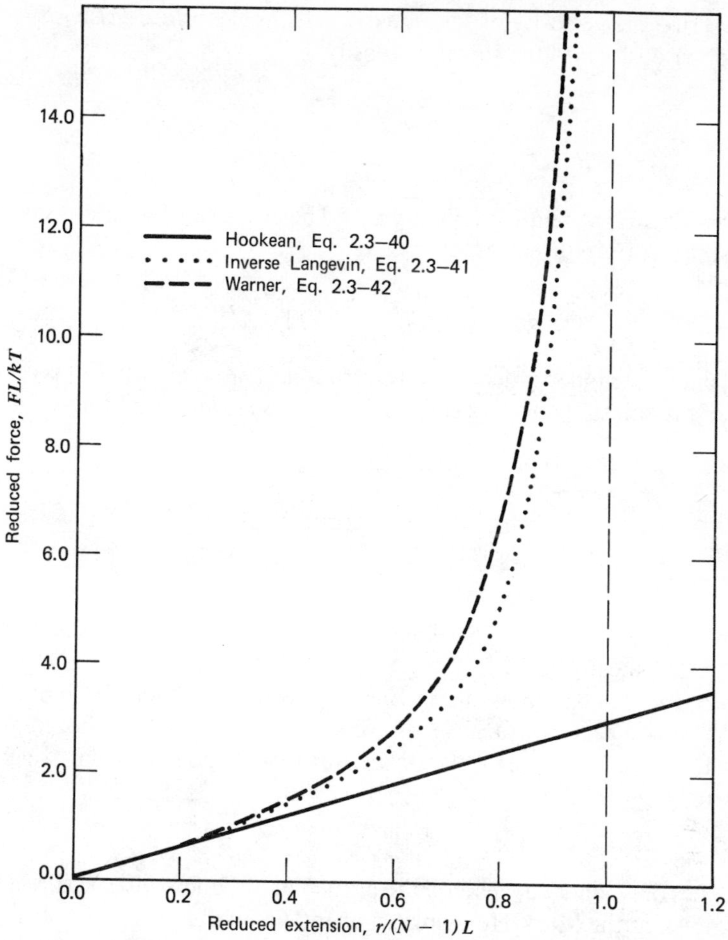

FIGURE 2.3–4. Plots of three approximate expressions for the time average tension F in a freely jointed chain of $(N - 1)$ links of length L with end-to-end distance r.

In this section we have considered the equilibrium configurations of macromolecules. The statistical mechanical methods for calculating equilibrium properties are well established, and have been illustrated using a simple molecular model, namely the freely jointed chain model. More realistic models may be treated using the same general methods; and results for equilibrium properties of models including features such as fixed valence cone angles and hindered rotations around chemical bonds are available.[1] Our main concern in this book, however, is with fluids undergoing flow rather than with fluids at rest. Chapters 10 to 15 are devoted to the prediction of the stress tensor of polymeric fluids; a brief summary of the main results of these chapters is given in §2.5.

§2.4 POLYMER-SOLVENT AND POLYMER-POLYMER INTERACTIONS

Frequently macromolecules are encountered in solution. The process of dissolving a bulk polymer generally takes place in two stages. First, the polymer slowly absorbs solvent to form a swollen gel; second, this gel goes into solution. If the polymer is chemically cross-linked or if for other reasons the solution is thermodynamically unfavorable, only the swelling may take place.

To describe the viscosity of solutions[1] we define the *relative viscosity* η_r as the ratio of the solution viscosity η to the solvent viscosity, η_s:

$$\eta_r = \frac{\eta}{\eta_s} \tag{2.4-1}$$

Thus for pure solvents the relative viscosity is 1. The viscosity of solutions may be expanded in a Taylor series in the concentration c:

$$\eta_r = 1 + [\eta]c + k'[\eta]^2c^2 + \cdots \tag{2.4-2}$$

in which $[\eta]$ and k' are independent of concentration. The coefficient $[\eta]$ is called the *intrinsic viscosity* of the solution and k' is called the *Huggins coefficient*; it follows from Eq. 2.4–2 that:

$$\boxed{[\eta] = \lim_{c \to 0} \left(\frac{\eta - \eta_s}{c\eta_s} \right)} \tag{2.4-3}$$

Note that the intrinsic viscosity has dimensions of reciprocal concentration; in general it will depend on the shear rate $\dot{\gamma}$.

At very small shear rates the intrinsic viscosity approaches a limiting value $[\eta]_0$ known as the zero-shear-rate intrinsic viscosity. It is found that for homologous series of fractionated linear polymers, the relation between $[\eta]_0$ and molecular weight can be expressed as:[2]

$$[\eta]_0 = K'M^a \tag{2.4-4}$$

where the values of K' and a depend on the particular polymer-solvent pair. The parameter a, which is known as the Mark-Houwink exponent, lies in the range 0.5 to 0.8. Originally, it was thought that $[\eta]_0$ was proportional to M, this linear relation being known as *Staudinger's rule.*[3]

Equation 2.4–4 provides a convenient method for determining the molecular weight of a polydisperse sample of a linear polymer.[2] For this purpose we define a *viscosity average molecular weight* $\bar{M}_v$ by (cf. Eqs. 2.2–1 and 3):

$$\bar{M}_v = \left(\frac{\sum_i N_i M_i^{1+a}}{\sum_i N_i M_i} \right)^{1/a} \tag{2.4-5}$$

It can then be shown that for polydisperse samples:

$$[\eta]_0 = K'\bar{M}_v^{\,a} \tag{2.4-6}$$

where K' and a have the same values as for the monodisperse polymer. Equation 2.4–6 is obtained by assuming the contributions to $\eta_r - 1$ of each molecular weight fraction are

[1] In Chapter 1 we used the symbol μ for the viscosity of a Newtonian fluid. As we shall see in §2.5 and Chapters 3 and 4 the "viscosity" of macromolecular fluids depends strongly on the velocity gradient, and we chose to use the symbol η in this situation.

[2] P. J. Flory, *Principles of Polymer Chemistry*, Cornell University Press, Ithaca, New York (1953), pp. 24, 310–314; C. Tanford, *Physical Chemistry of Macromolecules*, Wiley, New York (1961), pp. 407–412.

[3] H. Staudinger and W. Huer, *Ber. der deutschen Chem. Gesellschaft*, *63*, 222–234 (1930).

additive. When $a = 1$, the viscosity average and weight average molecular weights are identical. For a in the range $0.5 < a < 0.8$, we have $\bar{M}_n < \bar{M}_v < \bar{M}_w$.

The zero-shear-rate intrinsic viscosity has also been related to the molecular dimension of the polymer molecule.[4] For a broad spectrum of polymer-solvent systems the relation:

$$[\eta]_0 = \Phi \frac{\langle r^2 \rangle^{3/2}}{M} \tag{2.4-7}$$

holds provided $M > 10^6$. The parameter Φ is nearly a universal constant, equal to about 2.5×10^{23} when $\langle r^2 \rangle^{1/2}$ is given in cm and $[\eta]_0$ in cm^3/g.

The dependence of $[\eta]$ on the type of solvent is due to the interactions between polymer and solvent molecules. Roughly speaking, a *good solvent* for a certain polymer is defined as a solvent such that polymer-solvent contacts are favored over polymer-polymer contacts. A good solvent therefore causes the polymer coils to unwind—that is, other conditions being equal the better the solvent the larger the molecular expansion factor α in Eqs. 2.3–43 and 44. Conversely in a *poor solvent* polymer-polymer contacts are favored over polymer-solvent contacts, thereby causing the polymer coils to contract, so that by using a poorer solvent one may decrease the molecular expansion factor α. We may now recall that due to excluded volume effects alone α is somewhat larger than unity. For a given polymer we may therefore expect to be able to find a solvent sufficiently poor that it exactly compensates for the excluded volume effect, that is, one for which $\alpha = 1$. Such a solvent is called a *θ-solvent*. For a given polymer-solvent pair, the molecular expansion factor will depend on the temperature. Thus a solvent which at a certain temperature is a good solvent may at a lower temperature be a θ-solvent. The temperature at which, for a given polymer-solvent pair, the solvent becomes a θ-solvent is called the *θ-temperature*. As we shall see in Chapter 12, molecular theories of dilute solutions predict that the zero-shear-rate viscosity increases with the mean square radius of gyration $\langle s^2 \rangle_e$. Thus the intrinsic viscosity of a polymer solution increases with the molecular expansion factor. Ideally the results from the dilute solution molecular theories of Volume 2 should be applied only to polymers in θ-solvents.

The Huggins coefficient defined in Eq. 2.4–2 is reported to have values in the range 0.3 to 0.4 for good solvents and between 0.5 and 0.8 for θ-solvents. It is found that k' is almost independent of molecular weight.

It is reasonable to call a solution *dilute* if the macromolecules are, on the average, so far apart that they have negligible influence on each other. For such solutions the viscosity contributions of the individual macromolecules are additive. We may therefore define a dilute solution as one in which the viscosity increases linearly with concentration, in other words one for which the term c^2 and the higher order terms in Eq. 2.4–2 are negligible. For concentrated solutions, on the other hand, the dependence of viscosity on concentration is highly nonlinear, as evidenced by Fig. 2.4–1. The range in which a solution may be considered dilute depends on the polymer and on the molecular weight. For instance it appears from Fig. 2.4–1 that the poly(vinyl acetate) sample[5] ($\bar{M}_n = 0.14 \times 10^6$ g/mol) forms a dilute solution up to concentrations of about 2 to 3%. On the other hand even at a weight fraction of 1% the viscosity of the polyisobutylene sample[6] ($\bar{M}_w = 2.5 \times 10^6$ g/mol) exceeds that which is obtained by linear extrapolation from zero concentration by a factor of about 4; at

[4] P. J. Flory and T. G. Fox, *J. Am. Chem. Soc.*, *73*, 1904–1908 (1951); P. J. Flory, *op. cit.*, pp. 616–620, W. W. Graessley, *Fortschritte der Hochpolymeren-Forschung*, *16*, 1–179 (1974).

[5] J. D. Ferry, E. L. Foster, G. V. Browning, and W. M. Sawyer, *J. Colloid Sci.*, *6*, 377–388 (1951).

[6] M. F. Johnson, W. W. Evans, I. Jordan, and J. D. Ferry, *J. Colloid Sci.*, *7*, 498–510 (1952). The factors quoted in the text are obtained directly from the data and may not be estimated from the graph, since the viscosity is plotted on a linear scale.

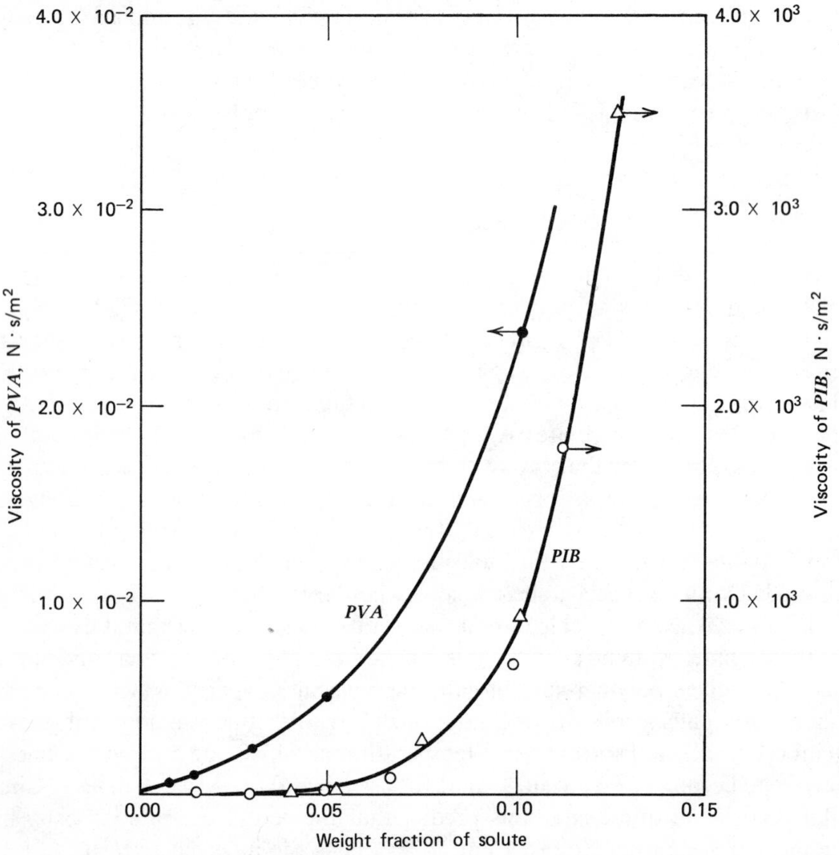

FIGURE 2.4–1. Zero-shear-rate viscosities of concentrated solutions of polyisobutylene, PIB ($\bar{M}_w = 2.5 \times 10^6$ g/mol), and poly(vinyl acetate), PVA ($\bar{M}_n = 0.14 \times 10^6$ g/mol) at 298 K. ○ = PIB in decalin. △ = PIB in a 70% decalin–30% cyclohexanol mixture. [Data of M. F. Johnson, W. W. Evans, I. Jordan, and J. D. Ferry, *J. Colloid Sci.*, 7, 498–510 (1952).] ● = PVA in methyl isobutyl ketone. [Data of J. D. Ferry, E. L. Foster, G. V. Browning, and W. M. Sawyer, *J. Colloid Sci.*, 6, 377–388 (1951).]

the same concentration the solution viscosity exceeds that of the solvent by a factor of about 25. At a weight fraction of about 10% the same factors are 5000 and 300,000, respectively. From this highly nonlinear dependence of viscosity on concentration it is concluded that interactions between polymer molecules are very important in determining the rheological properties of concentrated polymer solutions and polymer melts. These interactions may take many forms, and their exact nature is not yet completely understood. The hypothesis of so-called *entanglement couplings*, however, has become well established and deserves special mention. It is assumed that macromolecules may interact at isolated points along the chain by forming loops around each other or by forming knots, such that for short periods of time the macromolecules behave as if they were permanently crosslinked. For longer periods of time, however, the entanglements differ from the permanent chemical crosslinks in that the loops and knots may untie themselves because of thermal motions; the interaction is then lost. There is a considerable amount of experimental evidence in favor of the existence of entanglements.[7]

[7] See, for example, J. D. Ferry, *Viscoelastic Properties of Polymers*, Wiley, New York (1970), Second Edition; and W. W. Graessley, *op. cit.*

The idea of entanglements is a key concept in the molecular network theories for concentrated polymer solutions and melts treated in Chapter 15. It is assumed that the macromolecules form a three-dimensional network via entanglements that extends throughout the solution. During flow the network is stretched, but is also continually being rearranged as entanglements are lost and created due to the thermal motions and stresses generated by flow. It is worthwhile pointing out at this stage, however, that understanding the molecular mechanism for flow of concentrated solutions and melts, despite intensive research, is still one of the unanswered challenging problems in rheology.

§2.5 THE STRESS TENSOR IN FLOWING POLYMERIC FLUIDS

In this final section we present some of the predictions of kinetic theory for the stress tensor of polymeric fluids. This material will serve as a useful background, in particular, in connection with the continuum mechanics discussions of Chapters 7, 8, and 9.

The statistical mechanics given in §2.3 is valid for the prediction of equilibrium properties only. Hence it cannot provide us with any information regarding nonequilibrium material functions such as viscosity. The kinetic theory development needed for the rheological behavior is discussed in Chapters 10 to 15; in this section we summarize some of the key results of those chapters without proof.

As was pointed out in §2.3 because of thermal agitation (Brownian motion) even in a fluid "at rest," each polymer molecule goes through a succession of configurations. The distribution of configurations is given by Eqs. 2.3–4 and 7. If we approximate the mechanical behavior of the polymer molecule by a mechanical "model," such as one of those shown in Table 2.5–1, then we can obtain an analytical expression for the distribution of orientations in a solution that is not in a state of flow.

For example, let us consider first the Hookean dumbbell model, shown in Table 2.5–1, which is a crude representation of a flexible macromolecule. It consists of two "beads" connected by a Hookean spring with tension equal to $H\boldsymbol{R}$, where H is the spring constant and $\boldsymbol{R}$ is the vector from bead "1" to bead "2," that is, a vector that describes the dumbbell length and orientation. The use of a Hookean spring is suggested by the result obtained in Eq. 2.3–40. For this model it can be shown (see Problem 2C.3) from Eqs. 2.3–4, 7, and 8 that the normalized configuration-space distribution function at equilibrium is given by:

$$\begin{aligned}\psi(R,\theta,\phi)\, dR\, d\theta\, d\phi &= \frac{e^{-HR^2/2kT} R^2 \sin\theta}{\int_0^{2\pi}\int_0^{\pi}\int_0^{\infty} e^{-HR^2/2kT} R^2 \sin\theta\, dR\, d\theta\, d\phi}\, dR\, d\theta\, d\phi \\ &= \left(\frac{H}{2\pi kT}\right)^{3/2} R^2 \sin\theta\, e^{-HR^2/2kT}\, dR\, d\theta\, d\phi \end{aligned} \tag{2.5–1}$$

That is, all angular orientations (θ,ϕ) are equally likely, but there is a distribution of dumbbell lengths in the solution. A second example is that of the rigid dumbbell of length L, which is a crude representation of a rigid macromolecule. For this model the distribution function in the two-dimensional θ-ϕ configuration space is just a special case of Eq. 2.3–14:

$$\psi(\theta,\phi)\, d\theta\, d\phi = \frac{\sin\theta}{\int_0^{2\pi}\int_0^{\pi} \sin\theta\, d\theta\, d\phi}\, d\theta\, d\phi = \frac{\sin\theta}{4\pi}\, d\theta\, d\phi \tag{2.5–2}$$

That is, all dumbbell orientations are equally likely.

When the polymer solution is in a state of flow, then the configurational distributions are no longer described by Eqs. 2.5–1 and 2, because the motion of the solvent tends to

TABLE 2.5–1

Summary of Kinetic Theory Results[a]

Chapter Reference	Model \ Rheological Equation of State (Flow)	Retarded Motion Expansion (Slow, Slowly Varying Flows) (Eq. 8.4–3)	Linear Viscoelastic Model (Small Displacement and Displacement Gradient Flows) (Eq. 6.1–16)	General Rheological Equation of State (Arbitrary, Homogeneous Flows)
		$\boldsymbol{\tau} = -\eta_0\dot{\boldsymbol{\gamma}} + \frac{1}{2}\Psi_{1,0}\frac{\mathscr{D}}{\mathscr{D}t}\dot{\boldsymbol{\gamma}} - (\frac{1}{2}\Psi_{1,0} + \Psi_{2,0})\{\dot{\boldsymbol{\gamma}}\cdot\dot{\boldsymbol{\gamma}}\} + \cdots$	$\boldsymbol{\tau} = -\int_{-\infty}^{t} G(t-t')\dot{\boldsymbol{\gamma}}(t')\,dt'$	
10	Hookean dumbbell  H = Hooke's law spring constant $\lambda_H = \zeta/4H$ = time constant	$\eta_0 = \eta_s + nkT\lambda_H$ $\Psi_{1,0} = 2nkT\lambda_H{}^2$ $\Psi_{2,0} = 0$	$G(t-t') = 2\eta_s\delta(t-t') + nkTe^{-(t-t')/\lambda_H}$	$\boldsymbol{\tau} + \lambda_H\frac{\mathscr{D}}{\mathscr{D}t}\boldsymbol{\tau} - \frac{\lambda_H}{2}\{\dot{\boldsymbol{\gamma}}\cdot\boldsymbol{\tau} + \boldsymbol{\tau}\cdot\dot{\boldsymbol{\gamma}}\} = -\eta_s\dot{\boldsymbol{\gamma}} - nkT\lambda_H\dot{\boldsymbol{\gamma}} - \eta_s\lambda_H\frac{\mathscr{D}}{\mathscr{D}t}\dot{\boldsymbol{\gamma}} + \eta_s\lambda_H\{\dot{\boldsymbol{\gamma}}\cdot\dot{\boldsymbol{\gamma}}\}$
11	Rigid dumbbell 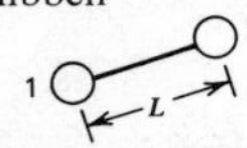 L = Length of dumbbell $\lambda = \zeta L^2/12kT$ = time constant	$\eta_0 = \eta_s + nkT\lambda$ $\Psi_{1,0} = \frac{6}{5}nkT\lambda^2$ $\Psi_{2,0} = 0$	$G(t-t') = 2\eta_s\delta(t-t') + nkT[\frac{4}{5}\lambda\delta(t-t') + \frac{3}{5}e^{-(t-t')/\lambda}]$	See Eq. 2.5–3.
12	Rouse (freely jointed bead-spring) model 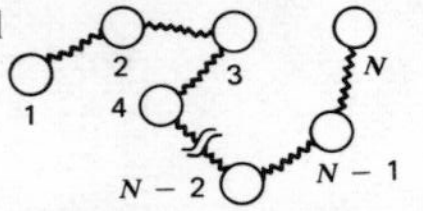 H = Hooke's law spring constant for each connector There are $N-1$ time constants $\lambda_j = \frac{\zeta/2H}{4\sin^2(j\pi/2N)}$ = jth time constant	$\eta_0 = \eta_s + nkT\left(\frac{\zeta}{4H}\right)\times\left[\frac{(N+1)(N-1)}{3}\right]$ $\Psi_{1,0} = 2nkT\left(\frac{\zeta}{4H}\right)^2\times\left[\frac{(N+1)(N-1)(2N^2+7)}{45}\right]$ $\Psi_{2,0} = 0$	$G(t-t') = 2\eta_s\delta(t-t') + nkT\sum_{j=1}^{N-1}e^{-(t-t')/\lambda_j}$	$\boldsymbol{\tau} + \eta_s\dot{\boldsymbol{\gamma}} = \sum_{j=1}^{N-1}\boldsymbol{\tau}'_{pj}$ where the $\boldsymbol{\tau}'_{pj}$ are solutions to $\boldsymbol{\tau}'_{pj} + \lambda_j\frac{\mathscr{D}}{\mathscr{D}t}\boldsymbol{\tau}'_{pj} - \frac{1}{2}\{\dot{\boldsymbol{\gamma}}\cdot\boldsymbol{\tau}'_{pj} + \boldsymbol{\tau}'_{pj}\cdot\dot{\boldsymbol{\gamma}}\} = -nkT\lambda_j\dot{\boldsymbol{\gamma}}$

solutions

Dilute	13	Kramers (freely jointed bead-rod) model 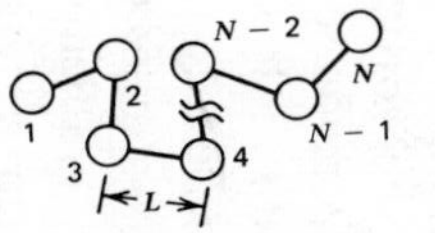 L = distance between adjacent beads	$\eta_0 \cong \eta_s + \frac{1}{36}(N-1)(N+1)n\zeta L^2$ $\Psi_{1,0} \cong \frac{(N-1)(N+1)}{16200} \times (10N^3 - 12N^2 + 35N - 12) \times \frac{n\zeta^2 L^4}{kT}$ $\Psi_{2,0} = 0$	$N = 3$: $G(t-t') = 2\eta_s\delta(t-t') + 0.1474 n\zeta L^2 \delta(t-t') + nkT \sum_{j=1}^{4} b_j e^{-\nu_j kT(t-t')/\zeta L^2}$ $b_1 = 0.46425$ $\nu_1 = 5.4376$ $b_2 = 0.75115$ $\nu_2 = 12.4378$ $b_3 = 0.00909$ $\nu_3 = 19.1832$ $b_4 = 0.07370$ $\nu_4 = 24.4595$	Not known.
	13	Shishkebob 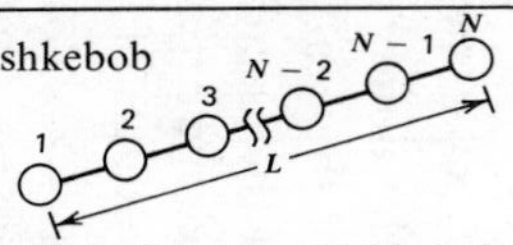 L = Length of the shishkebob N = Number of beads spaced evenly along the backbone $\lambda_N = \frac{\zeta L^2 N(N+1)}{72(N-1)kT}$ = Time constant	All results are the same as for the rigid dumbbell with λ replaced by λ_N.		
Concentrated solutions and melts	15	Network model 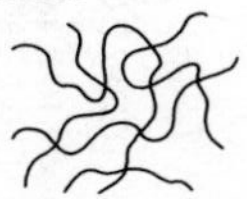 η_j, λ_j, and p are constants not determined by the model	$\eta_0 = \sum_j \eta_j$ $\Psi_{1,0} = 2\sum_j \eta_j\lambda_j$ $\Psi_{2,0} = 0$	$G(t-t') = \sum_j \frac{\eta_j}{\lambda_j} e^{-(t-t')/\lambda_j}$	$\boldsymbol{\tau} + p\boldsymbol{\delta} = \sum_j \boldsymbol{\tau}_j$ where the $\boldsymbol{\tau}_j$ are solutions to $\boldsymbol{\tau}_j + \lambda_j \frac{\mathscr{D}}{\mathscr{D}t}\boldsymbol{\tau}_j - \frac{1}{2}\{\dot{\boldsymbol{\gamma}} \cdot \boldsymbol{\tau}_j + \boldsymbol{\tau}_j \cdot \dot{\boldsymbol{\gamma}}\} = -\eta_j \dot{\boldsymbol{\gamma}}$

[a] $\mathscr{D}/\mathscr{D}t$ is defined by $(\mathscr{D}/\mathscr{D}t)\boldsymbol{\Lambda} = (D/Dt)\boldsymbol{\Lambda} + \frac{1}{2}\{\boldsymbol{\omega} \cdot \boldsymbol{\Lambda} - \boldsymbol{\Lambda} \cdot \boldsymbol{\omega}\}$.

align the molecules and, in the case of flexible molecules, stretch them. That is, the distribution of configurations now depends on the balance among the Brownian motion forces, the intramolecular forces, and the hydrodynamic forces. These hydrodynamic forces are taken into account by presuming that the beads in the model experience a Stokes' drag as they move through the solvent. That is, the bead experiences a force ζv_{rel}, where ζ is a friction coefficient and v_{rel} is the relative speed of the bead with respect to the solvent; hence $\zeta = 6\pi\eta_s r_0$, where η_s is the solvent viscosity and r_0 is the bead radius. When the balance among the Brownian, intramolecular, and hydrodynamic forces is described in mathematical terms, one obtains a partial differential equation for the configuration-space distribution function ψ in arbitrary flow fields; however, the solution of this partial differential equation is no easy matter, and analytical results have been obtained only for simple models. In Chapters 10 to 13 this question is pursued further.

Once one has found the distribution function for the molecular orientations, then a kinetic theory expression for the stress tensor can be used to obtain the effect of the molecular orientations on the stress tensor in the solution. In this way one ultimately obtains an expression for the stress tensor in terms of various kinematic tensors that describe the state of flow. The derivation of these expressions is quite lengthy; also, the mathematical difficulties encountered increase dramatically as one proceeds to more realistic models such as the freely jointed chain model of §2.3. A rather complete discussion can be found in Chapters 10 to 14 for dilute polymer solutions and a less extensive discussion on the network theories for concentrated solutions and melts is to be found in Chapter 15.

One of the few models for which the stress-tensor expression has been found is the rigid dumbbell model, and even here we have been able to obtain only the first few terms of a "memory integral expansion." For a solution of n rigid dumbbells per unit volume it has been shown that for an arbitrary, homogeneous, time-dependent flow field:

$$\begin{aligned}\boldsymbol{\tau} = &-(\eta_s + \tfrac{2}{5}nkT\lambda)\dot{\boldsymbol{\gamma}} - \tfrac{3}{5}nkT\int_{-\infty}^{t} e^{-(t-t')/\lambda}\,\dot{\boldsymbol{\gamma}}'\,dt' \\ &- \tfrac{6}{35}nkT\lambda\int_{-\infty}^{t} e^{-(t-t')/\lambda}\{\dot{\boldsymbol{\gamma}}\cdot\dot{\boldsymbol{\gamma}}' + \dot{\boldsymbol{\gamma}}'\cdot\dot{\boldsymbol{\gamma}}\}\,dt' \\ &- \tfrac{3}{10}nkT\int_{-\infty}^{t}\int_{-\infty}^{t'} e^{-(t-t'')/\lambda}\,[\tfrac{3}{7}\{\dot{\boldsymbol{\gamma}}'\cdot\dot{\boldsymbol{\gamma}}'' + \dot{\boldsymbol{\gamma}}''\cdot\dot{\boldsymbol{\gamma}}'\} \\ &- \{\boldsymbol{\omega}'\cdot\dot{\boldsymbol{\gamma}}'' - \dot{\boldsymbol{\gamma}}''\cdot\boldsymbol{\omega}'\}]\,dt''\,dt' + \cdots \\ &+ \text{isotropic terms containing integrals of } \dot{\boldsymbol{\gamma}}. \end{aligned} \tag{2.5–3}$$

Here the symbol $\dot{\boldsymbol{\gamma}}'$ is shorthand for $\dot{\boldsymbol{\gamma}}(t')$, and $\boldsymbol{\omega} = \nabla\boldsymbol{v} - (\nabla\boldsymbol{v})^\dagger$ is the vorticity tensor. The quantity $\lambda = \zeta L^2/12kT$ is a time constant for the dumbbell suspension, constructed from the dumbbell length L, the friction coefficient for the beads ζ, and kT. In Chapter 11 a more concise way of writing Eq. 2.5–3 is given, but the essential features are unchanged. It is evident that the expression is far from simple, since the expansion continues with triple, quadruple, etc. integrals and triple, quadruple, etc. products of velocity gradients.

The main conclusions we can draw from Eq. 2.5–3 and similar results for other molecular models are: (a) the stress tensor depends on multiple *products of velocity gradients*—that is, polymeric fluids will not in general exhibit linear stress vs. shear rate behavior; and (b) the stress in a fluid element at the present time t depends on the *velocity gradients for all past times* t' from $t' = -\infty$ to the present time $t' = t$; note also that, because of the exponential weighting factors, the recent kinematic events receive greater emphasis than events long past (we sometimes say that the fluids have a "fading memory"). Because of the great complexity of Eq. 2.5–3 as compared with the relatively simple stress tensor expression in Eq. 1.2–2 for Newtonian fluids, we can expect that the classical (Newtonian) hydrodynamics will be of little value for describing the flow of polymeric fluids and that a large number of new physical phenomena can be expected, phenomena that can be described only by stress-tensor expressions that are highly nonlinear and strongly time-dependent. In Chapter 3 we illustrate this by presenting the results of several decades of experimental observations.

Equation 2.5–3 is so complicated that it is far from evident what this equation would predict for the steady-state shear flow between two parallel plates, $v_x = \dot{\gamma} y$, where $\dot{\gamma}$ is the constant shear rate. As will be shown in Chapter 11, for this simple flow Eq. 2.5–3 gives expansions in $\lambda\dot{\gamma}$ that are useful up to about $\lambda\dot{\gamma} = 0.4$:

$$\tau_{yx} = -\eta_s \dot{\gamma} - nkT\lambda\dot{\gamma}\left[1 - \frac{18}{35}(\lambda\dot{\gamma})^2 + \frac{1326}{1925}(\lambda\dot{\gamma})^4 - \cdots\right] \quad (2.5\text{–}4)$$

$$\tau_{xx} - \tau_{yy} = -\tfrac{6}{5}nkT(\lambda\dot{\gamma})^2\left[1 - \frac{38}{35}(\lambda\dot{\gamma})^2 + \cdots\right] \quad (2.5\text{–}5)$$

$$\tau_{yy} - \tau_{zz} = 0 \quad \text{and} \quad \tau_{yz} = \tau_{zy} = \tau_{xz} = \tau_{zx} = 0 \quad (2.5\text{–}6)$$

This shows us that the shear stress is not proportional to the velocity gradient and, furthermore, that the normal stresses are unequal. These results are clearly quite different from those for a Newtonian fluid for which $\tau_{yx} = -\mu\dot{\gamma}$, and $\tau_{xx} = \tau_{yy} = \tau_{zz} = 0$. Steady shear flow between two parallel plates is just one of a number of rheological tests that can be performed on a polymeric fluid; in Chapter 4 we discuss this and a number of other experiments that provide further information about polymer flow behavior. Note particularly that Eq. 2.5–4 states that the "viscosity" of the fluid $\eta = -\tau_{yx}/\dot{\gamma}$ decreases with increasing $\dot{\gamma}$. This "shear-thinning" effect is quite important in industrial problems and is discussed extensively in Chapter 5.

Because of the complexity of Eq. 2.5–3, one might well ask whether it is possible to get somewhat simpler expressions by considering less general classes of flows. The answer is that for many molecular models we now know how to derive the expression for the stress tensor for two special classes of flows:

a. *Slow flows that vary slowly in time*, for which the stress tensor can be written as a "retarded motion expansion" (see §8.4), which generally describes how the fluid begins to deviate from Newtonian behavior as the rate of deformation increases; here the fluid response may be described in terms of a set of constants η_0, $\Psi_{1,0}$, $\Psi_{2,0}$, etc.

b. *Flows with small deformations and small deformation gradients*, for which the stress tensor can be written as a single integral, which describes the mechanical response to various time-dependent imposed deformations; here the fluid behavior is described by one function $G(t - t')$ (see Chapter 6).

Some of the results for these special flows are displayed in Table 2.5–1. From this table we see that the kinetic theory for specific mechanical models gives explicit expressions for the constants in the retarded motion expansion and the function $G(t - t')$ in the linear viscoelastic model, in terms of the model parameters. It is evident that the same continuum mechanics expressions for the stress tensor can describe all of the various mechanical models used to depict the polymer solutions. Thus the kinetic theory reinforces our feeling that continuum mechanics expressions provide a good framework for describing non-Newtonian behavior. Therefore in Chapters 7, 8, and 9 we devote considerable space to presenting some of the recent continuum mechanics results so that they can be used as a basis for problem solving.

QUESTIONS FOR DISCUSSION

1. Discuss the differences between step reaction polymerizations and chain reaction polymerizations.
2. How can the amount of branching be controlled in a step reaction polymerization?
3. What is the range of molecular weights for typical polymers?
4. What will happen if we stretch a piece of rubber, irradiate it with γ radiation, and let go?
5. What is the most probable distribution of molecular weights, and what are the assumptions under which this distribution is derived?
6. What are the configuration space and the phase space for a pendulum?
7. In Eq. 2.3–13 explain the occurrence of the factor $\sin \theta_i$. Why does the random flight distribution Eq. 2.3–14 indicate that the orientation of each individual link is random?
8. How is the radius of gyration of a set of mass points defined? How is it related to the moment of inertia in classical mechanics?
9. Discuss the difference between a probability and a probability density.
10. The Gaussian density in Eq. 2.3–34 in Example 2.3–2 indicates incorrectly that there is a finite probability of finding a moving particle at distances $x > (N - 1)L$ from the origin after it has suffered $(N - 1)$ random displacements of length L. Where in the derivation of Eq. 2.3–34 has the error been introduced?
11. What is the Hookean force law for a freely jointed chain with fixed end points? Why is it not valid for extensions r larger than about $0.5NL$?
12. How can the molecular expansion factor α be determined?
13. What is meant by the θ-temperature?
14. What happens to a solution of rigid macromolecules when it is placed in a viscometer as regards (a) molecular orientations and (b) viscosity?
15. Compare and contrast Eqs. 1.2–2 and 2.5–3.
16. What kinds of degradation can polymers undergo, and how does degradation affect the molecular weight distribution?
17. What important conclusions are to be drawn from Table 2.5–1?

PROBLEMS

2A.1 Intrinsic Viscosity

We are given the following measurements[1] of relative viscosities of solutions of polyisobutylene ($\bar{M}_w = 2.5 \times 10^6$) in decalin and in a 70% decalin and 30% cyclohexanol mixture at 298 K.

Decalin		Decalin-Cyclohexanol	
c g/cm^3	η_r	c g/cm^3	η_r
0.000408	1.227	0.000409	1.155
0.000543	1.311	0.000546	1.210
0.000815	1.485	0.000818	1.326
0.001304	1.857	0.001169	1.488
0.001630	2.125	0.001637	1.740

a. Calculate the intrinsic viscosity, $[\eta]$, for the polyisobutylene sample in the two solvents.
b. Which is the better solvent?

Answer: a. Decalin: 5.1×10^2 cm^3/g
Decalin-cyclohexanol: 3.1×10^2 cm^3/g

2A.2 Average Dimension of a Freely Rotating Chain

A somewhat more realistic model than the freely jointed chain is the *freely rotating chain* shown in Fig. 2A.2. For the mean square radius of gyration Kramers[2] gives the expression,

$$\langle s^2 \rangle = \tfrac{1}{6} N L^2 \cot^2 \frac{\alpha}{2} \qquad N \gg 1 \qquad \text{(2A.2–1)}$$

where $(N - 1)$ is the number of links all of length L and α is the valence angle.

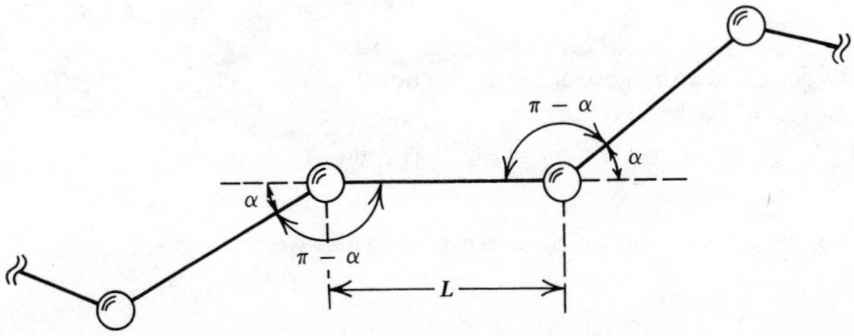

FIGURE 2A.2. Section of a freely rotating chain. The links have length L and the angle α between successive links is fixed, but no other angles are constrained and no internal potentials are included.

[1] M. F. Johnson, W. W. Evans, I. Jordan, and J. D. Ferry, *J. Colloid Sci.*, *7*, 498–510 (1952).
[2] H. A. Kramers, *Physica*, *11*, 1–19 (1944).

a. Calculate the root-mean-square radius of gyration $\sqrt{\langle s^2 \rangle}$ for a polyethylene chain of molecular weight 10^6 g/mol. Replace the C—C backbone in the polymer by a freely rotating chain such that each C atom is replaced by a bead. Assume a bond length of 0.15 nm and a valence angle of 71°.

b. How much does this result differ from the prediction obtained by using a freely jointed chain with the same values of N and L?

Answer: a. 23 nm

2B.1 Molecular Weight Averages for the Most Probable Distribution

Derive $\bar{M}_n$ and $\bar{M}_w$ (Eqs. 2.2–15 and 16) from the most probable distribution, Eq. 2.2–13. You will need the infinite sum,[1]

$$\sum_{n=1}^{\infty} P^n = \frac{P}{1 - P} \qquad P^2 < 1 \tag{2B.1–1}$$

and two other sums that may be obtained by differentiation of the above sum.

2B.2 Continuous Molecular Weight Distribution

The definitions of $\bar{M}_n$ and $\bar{M}_w$ by Eqs. 2.2–1 and 3 are based on a discrete description of molecular weight distributions. The object of this problem is to consider the corresponding continuous description. The latter description is an approximation, but will be a good one provided the discrete distribution is sufficiently broad and smooth.

Let us consider a range of molecular weights ΔM around M; the increment ΔM should be sufficiently large to contain many discrete molecular weights. Also let ΔN be the number of moles in the range ΔM. We may define a molecular weight distribution function $f(M)$ by

$$f(M) = \lim_{\Delta M \to 0} \frac{\Delta N}{\Delta M} \tag{2B.2–1}$$

The limit in Eq. 2B.2–1 is strictly speaking not defined; it may, however, be very well approximated by not allowing ΔM to become too small.

In terms of f, the differential increment in the number of moles dN given an increment of molecular weight dM is then:

$$dN = f(M)\, dM \tag{2B.2–2}$$

a. Show that the total number of moles in a sample is approximated by:

$$N = \int_0^{\infty} f(M)\, dM \tag{2B.2–3}$$

b. Let us define the nth moment of the function f by

$$f^{(n)} = \int_0^{\infty} M^n f(M)\, dM \tag{2B.2–4}$$

Show that the number average and weight average molecular weights are approximated by:

$$\bar{M}_n = \frac{f^{(1)}}{f^{(0)}} \tag{2B.2–5}$$

$$\bar{M}_w = \frac{f^{(2)}}{f^{(1)}} \tag{2B.2–6}$$

[1] L. B. W. Jolley, *Summation of Series*, Dover, New York (1961), p. 8, Eq. 39.

c. Finally rederive the mixing laws, Eqs. 2.2–8 and 9, for $\bar{M}_{n,\text{mix}}$ and $\bar{M}_{w,\text{mix}}$ using the continuous viewpoint.

2B.3 The Heterogeneity Index

a. From the definitions, Eqs. 2.2–1 and 3 show that:

$$\frac{\bar{M}_w}{\bar{M}_n} - 1 = \frac{\sum_i \sum_j N_i N_j (M_i - M_j)^2}{2(\sum_i N_i M_i)^2} \tag{2B.3–1}$$

b. What does Eq. 2B.3–1 imply about the relative sizes of $\bar{M}_n$ and $\bar{M}_w$?

2B.4 Steady-State Shear Flow of Dumbbell Solutions

a. Show that for the steady-state shear flow $v_x = \dot{\gamma} y$ (with $\dot{\gamma}$ = constant) the linear terms in Eq. 2.5–3 give for the shear stress:

$$\tau_{yx} = -(\eta_s + nkT\lambda)\dot{\gamma} \tag{2B.4–1}$$

in agreement with Eq. 2.5–4 for rigid dumbbell suspensions.

b. Repeat the derivation for elastic Hookean dumbbells using Table 2.5–1.

2B.5 A Property of the Random-Flight Distribution

a. In Eq. 2.3–15 consider the case where $i = 1, j = 2$, and show that:

$$\langle \boldsymbol{k}_1 \cdot \boldsymbol{k}_2 \rangle = \left(\frac{1}{4\pi}\right)^2 \int_0^{2\pi} \int_0^{2\pi} \int_0^{\pi} \int_0^{\pi} [\sin\theta_1 \cos\phi_1 \sin\theta_2 \cos\phi_2 + \sin\theta_1 \sin\phi_1 \sin\theta_2 \sin\phi_2$$
$$+ \cos\theta_1 \cos\theta_2] \sin\theta_1 \sin\theta_2 \, d\theta_1 \, d\theta_2 \, d\phi_1 \, d\phi_2 \tag{2B.5–1}$$

and verify that this is zero.

b. Then for $i = 1, j = 1$ show that:

$$\langle \boldsymbol{k}_1 \cdot \boldsymbol{k}_1 \rangle = \frac{1}{4\pi} \int_0^{2\pi} \int_0^{\pi} [\sin^2\theta_1 \cos^2\phi_1 + \sin^2\theta_1 \sin^2\phi_1 + \cos^2\theta_1] \sin\theta_1 \, d\theta_1 \, d\phi_1 \tag{2B.5–2}$$

and verify that this quantity is unity.

2C.1 Three-Dimensional Random Flight

Generalize Example 2.3–2 to three dimensions by now allowing the particle to "jump" in a three-dimensional lattice such that it can occupy the positions $(x,y,z) = (iL,jL,kL)$, where i, j, and k are integers. Assume that it jumps in a random manner in the lattice by changing its values of either i, j, or k by either $+1$ or -1 with all such changes equally likely. Let the particle start with $i = j = k = 0$ at time $t = 0$ and perform the random jumps at time intervals Δt.

a. Construct a difference equation for $P(x,y,z,t)$, the probability that the particle is at (x,y,z) at time t, analogous to Eq. 2.3–26 for $P(x,t)$. Formulate the appropriate initial and boundary conditions for $P(x,y,z,t)$.

b. Show that in the limit $\Delta t \to 0$ with $a = L^2/\Delta t$ held constant, P must satisfy

$$\frac{\partial}{\partial t} P = \tfrac{1}{6} a \left(\frac{\partial^2 P}{\partial x^2} + \frac{\partial^2 P}{\partial y^2} + \frac{\partial^2 P}{\partial z^2}\right) = \tfrac{1}{6} a \nabla^2 P \tag{2C.1–1}$$

$$\text{I.C.:} \qquad P(\boldsymbol{r},0) = 0 \qquad \text{for} \qquad \boldsymbol{r} \neq \boldsymbol{0} \tag{2C.1–2}$$

$$\text{B.C.:} \qquad \int_{-\infty}^{\infty} P(\boldsymbol{r},t)\, d\boldsymbol{r} = 1 \qquad \text{for all} \qquad t \geq 0 \tag{2C.1–3}$$

where $\boldsymbol{r} = (x,y,z)$ is the position vector, and $d\boldsymbol{r} = dx\, dy\, dz$.

c. Show that Eqs. 2C.1–1 to 3 have the solution:

$$P(\boldsymbol{r},t) = \left(\frac{3}{2\pi a t}\right)^{3/2} e^{-3r^2/2at} \tag{2C.1–4}$$

where $r = |\boldsymbol{r}|$.

d. Show how Eq. 2.3–24 follows from Eq. 2C.1–4 above.

Note that we obtain precisely the correct Gaussian distribution although we have used the artificial lattice in the derivation.

2C.2 Oscillatory Shear Flow of Dumbbell Suspensions

a. Show that for the small-amplitude oscillatory shearing motion $v_x = (\dot{\gamma}^0 \cos \omega t)y$, with $\dot{\gamma}^0$ sufficiently small that quadratic terms in $\dot{\gamma}^0$ can be neglected, Eq. 2.5–3 gives for the shear stress:

$$\tau_{yx} = -nkT\lambda\dot{\gamma}^0\left[1 - \tfrac{3}{5}\frac{(\lambda\omega)^2}{1 + (\lambda\omega)^2}\right]\cos \omega t - nkT\lambda\dot{\gamma}^0\left[\tfrac{3}{5}\frac{\lambda\omega}{1 + (\lambda\omega)^2}\right]\sin \omega t \tag{2C.2–1}$$

Interpret the result. How is this result related to Eq. 2B.4–1?

b. Obtain the analogous result for Hookean dumbbell suspensions using Table 2.5–1.

2C.3 Configuration-Space Distribution Function for Elastic Dumbbells

a. First verify[1] that the Hamiltonian for the internal motion of an elastic (Hookean) dumbbell is:

$$\mathscr{H} = \frac{1}{m}\left(p_R^{\,2} + \frac{p_\theta^{\,2}}{R^2} + \frac{p_\phi^{\,2}}{R^2 \sin^2 \theta}\right) + \tfrac{1}{2}HR^2 \tag{2C.3–1}$$

in which H is the spring constant and R is the interbead distance; the quantities p_R, p_θ, and p_ϕ are the (generalized) momenta conjugate to the coordinates R, θ, ϕ.

b. Then show that Eqs. 2.3–4, 7, and 8 lead to:

$$\psi(R,\theta,\phi) = \frac{\displaystyle\iiint_{-\infty}^{+\infty} e^{-\mathscr{H}/kT}\, dp_R\, dp_\theta\, dp_\phi}{\displaystyle\int_0^{2\pi}\int_0^{\pi}\int_0^{\infty}\iiint_{-\infty}^{+\infty} e^{-\mathscr{H}/kT}\, dp_R\, dp_\theta\, dp_\phi\, dR\, d\theta\, d\phi} \tag{2C.3–2}$$

and that, when the integrations are performed, we get Eq. 2.5–1.

[1] For a brief and very readable introduction to generalized coordinates see H. Margenau and G. M. Murphy, *The Mathematics of Physics and Chemistry*, Van Nostrand, Princeton, N.J., Second Edition (1956).

CHAPTER 3
FLOW PHENOMENA IN POLYMERIC FLUIDS

A fluid that's macromolecular
Is really quite weird—in particular
The abnormal stresses
The fluid possesses
Give rise to effects quite spectacular.

It is often helpful in beginning any new field of study to spend a little time familiarizing oneself with the experimental facts in that area. In the last chapter we briefly discussed the chemical structure of polymer molecules, and their flow and nonflow configurations. In this chapter we want to help the reader start developing an intuition for the ways in which macromolecular fluids flow, as opposed to the flow of liquids composed of small molecules.[1] In order to do this, we shall present a number of experiments with side by side comparisons of the behavior of Newtonian and macromolecular fluids. The reader may already have a good feeling for the way in which Newtonian fluids will respond to the simple experiments we present as a result of daily experience with these fluids or from a study of classical hydrodynamics. This background will make the response of the macromolecular fluids all the more striking. It should be emphasized that the differences that will be brought out in the ensuing sections are not small. One cannot describe the behavior of macromolecular fluids by making small quantitative corrections to Newtonian behavior; the fluid responses are qualitatively different as we have been led to suspect from Eq. 2.5–3. In addition to supplying the reader with an intuition that will serve him well in solving later problems, this chapter is also intended to provide motivation for the study of macromolecular hydrodynamics. The strange behavior that is shown by non-Newtonian fluids should serve as a challenge to the scientist who wishes to understand polymeric fluids as well as to the engineer who needs to apply that understanding to industrial processes.

It is worth emphasizing two additional points about the "fun experiments" we present in this chapter. First, they do not represent anomalous behavior, but are rather

[1] Other discussions of illustrations of these differences may be found in A. S. Lodge, *Elastic Liquids*, Academic Press, New York (1964), Chapter 10; A. G. Fredrickson, *Principles and Applications of Rheology*, Prentice-Hall, Englewood Cliffs (1964), Chapter 6; B. D. Coleman, H. Markovitz, and W. Noll, *Viscometric Flows of Non-Newtonian Fluids*, Springer, New York (1966), Chapter 5; and C. Truesdell, *Ann Rev. Fluid Mech.*, *6*, 111–146 (1974). An interesting movie about polymer fluid mechanics has been prepared: H. Markovitz, *Rheological Behavior of Fluids*, Education Services, Inc., Watertown, Mass. (1965).

typical of fluids containing very large molecules. There seem to be many polymeric materials that behave to greater or lesser degree as shown in this chapter. Indeed, many of these simple experiments can form the basis for class demonstrations on non-Newtonian fluids. Instructions are included for preparing some fluids that are particularly well suited for demonstration. Second, many of the flows can be found in industrial processing of non-Newtonian fluids. We shall try to indicate where these flows occur and how the dramatic differences between the qualitative responses of Newtonian and structurally complex fluids affect industrial and practical applications of them.

§3.1 TUBE FLOW AND "SHEAR THINNING"

In this first experiment, we consider two identical, vertical tubes, the bottoms of which are covered by a flat plate. The two tubes are filled with fluids, one Newtonian and the other polymeric, chosen in such a way as to have the same viscosity in an experiment involving very low shear rates. This criterion is satisfied, for example, if two identical small spheres fall through each sample at the same rate[1] (Fig. 3.1–1*a*). A particularly good Newtonian liquid to use in an experiment in which it is desired to match viscosities is an aqueous glycerin solution. By varying the glycerin concentration, the viscosity of the mixture can be varied from 0.001002 to 1.490 N·s/m^2 at 293 K. In Fig. 3.1–1*b* we see what happens when the plate is removed from the bottoms of the tubes and the fluids are allowed to flow out by gravity. The polymeric fluid drains much more quickly than the Newtonian fluid.

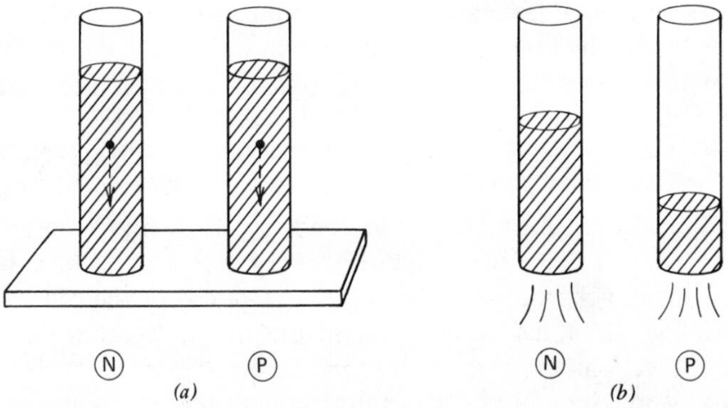

FIGURE 3.1–1. Tube flow and "shear thinning." In each part, the Newtonian behavior is shown on the left Ⓝ; the behavior of a polymer on the right Ⓟ. (*a*) A tiny sphere falls at the same rate through each; (*b*) the polymer flows out faster than the Newtonian fluid.

Recall that the law of Hagen and Poiseuille (Eq. 1.2–36) from classical hydrodynamics says that for a given pressure drop, the volumetric flow rate in a tube is inversely proportional to the (constant) fluid viscosity. We might, then, explain the results in Fig. 3.1–1 by saying that the viscosity of the macromolecular fluid appears to be lower in the higher shear rate part of the experiment (b). The decrease in viscosity with increasing shear rate is sometimes referred to as *shear thinning*, and the fluid is said to be *pseudoplastic*. This effect can be quite dramatic, with the viscosity decreasing by a factor of as much as 10^3 or 10^4. Examples

[1] The falling sphere problem has been discussed for Newtonian fluids in Example 1.2–4. The non-Newtonian fluid analog has been studied by J. C. Slattery and R. B. Bird, *Chem. Eng. Sci.*, *16*, 231–241 (1961) and by R. M. Turian, *A.I.Ch.E. Journal*, *13*, 999–1006 (1967).

of pseudoplastic fluids are molten polyethylene and polypropylene, and solutions of carboxymethylcellulose (CMC) in water, polyacrylamide in water and glycerin, and aluminum laurate in decalin and m-cresol. The variety of macromolecular fluids that show this effect is truly amazing. Almost all polymer solutions and melts that exhibit a shear-rate dependent viscosity are pseudoplastic.

A few fluids behave oppositely to what we have shown here, that is, they show *shear thickening* and flow out of the tube in Fig. 3.1–1*b* more slowly than the corresponding Newtonian liquid. A fluid whose viscosity increases with shear rate is sometimes called *dilatant*.[2] This behavior is exhibited by fairly concentrated suspensions of very small particles; two reported examples are suspensions of titanium oxide in a sucrose solution[3] and corn starch in an ethylene glycol-water mixture.[4] As far as we know, relatively few polymer solutions have been shown to be dilatant. These include moderately dilute solutions of poly(methyl methacrylate) in amyl alcohol,[5] poly(methyl methacrylate) in Aroclor, and polyisobutylene in polybutene.[6]

Still different behavior is shown by fluids that will not flow at all unless acted on by at least some critical shear stress, called the yield stress. We call these *viscoplastic fluids*. In the flow out of a tube by gravity, the shear stress is given by:

$$\tau_{rz} = \frac{\rho g r}{2} \tag{3.1-1}$$

where ρ is the fluid density, g is the acceleration due to gravity, and r is the radial coordinate. If the shear stress at some position r_0 is equal to the yield stress, then the material inside the cylindrical surface $r = r_0$ will move as a solid plug and the velocity of the fluid between r_0 and the wall $r = R$ will vary with r. If the yield stress is greater than $\rho g R/2$, then the material in the tube will not flow. Certain types of paints, greases, and pastes are examples of viscoplastic fluids. Carboxypolymethylene (pH 7) in water has also been reported to be viscoplastic.[7] Parenthetically, we note that it is the yield stress that accounts for much of the usefulness of these substances. For instance, paint is easily brushed onto a vertical wall (a high shear stress acting on the paint), but it will not run off the wall when acted on by gravity alone (a low shear stress environment). More will be said about the details of the dependence of viscosity on shear rate in the next chapter, where we discuss its measurement in polymeric fluids.

This first illustration makes an initial case for the need for intelligent application of macromolecular hydrodynamics to industrial design problems. One certainly would not want to use standard pipe-flow correlations for Newtonian fluids to predict the volumetric flow rate at a given pressure drop for the polymeric fluid shown in Fig. 3.1–1. The dramatic dependence of viscosity on shear rate has to be taken into account in making quantitative calculations of flow patterns in mold filling, in viscous heating in high speed extrusion, and in the design of distributor systems in injection molding.

[2] A review of dilatant behavior is given by W. H. Bauer and E. A. Collins, in *Rheology*, F. R. Eirich (Ed.), Academic Press, New York (1967), Vol. 4, Chapter 8, pp. 423–459. See especially their Appendix II, p. 459, which contains a listing of dilatant systems.

[3] A. B. Metzner and M. Whitlock, *Trans. Soc. Rheol.*, *2*, 239–254 (1958).

[4] R. G. Green and R. G. Griskey, *Trans. Soc. Rheol.*, *12*, 13–25 (1968).

[5] S. D. Jha, *Kolloid-Z.*, *143*, 174–175 (1955).

[6] The latter two examples were studied by S. Burow, A. Peterlin, and D. T. Turner, *Polym. Lett.*, *2*, 67–70 (1964) who suggest that dilatancy may be a general property of long macromolecules dissolved in very viscous solvents. See also S. P. Burow, A. Peterlin, and D. T. Turner, *Polymer*, *6*, 35–47 (1965).

[7] A. G. Fredrickson, *Principles and Applications of Rheology*, Prentice-Hall, Englewood Cliffs (1964), p. 178.

§3.2 ROD-CLIMBING

In Experiment 2 we consider what happens when rotating rods are inserted into two beakers, one containing a Newtonian liquid and the other a polymer solution. In Fig. 3.2–1 we see that the Newtonian liquid behaves exactly as expected. The liquid near the rotating rod is pushed outward by centrifugal force, and the characteristic dip in the liquid surface near the center of the beaker results. This is typical of the flow near the rotating shaft of a stirrer.

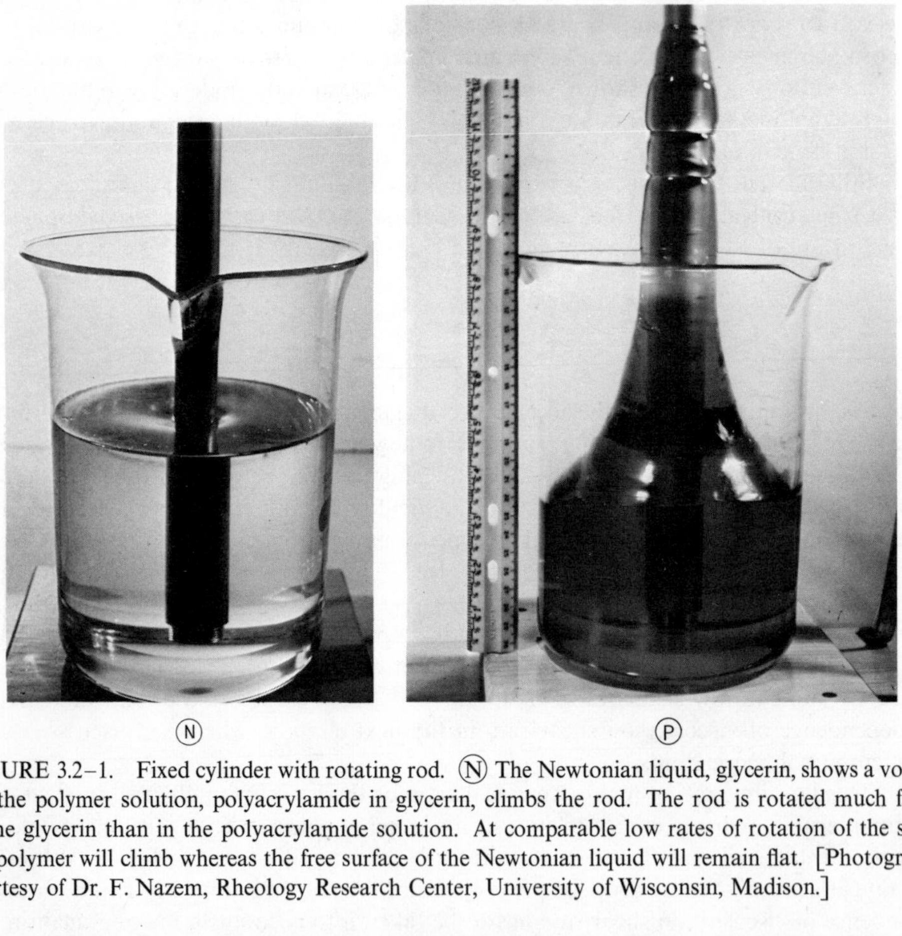

Ⓝ Ⓟ

FIGURE 3.2–1. Fixed cylinder with rotating rod. Ⓝ The Newtonian liquid, glycerin, shows a vortex; Ⓟ the polymer solution, polyacrylamide in glycerin, climbs the rod. The rod is rotated much faster in the glycerin than in the polyacrylamide solution. At comparable low rates of rotation of the shaft, the polymer will climb whereas the free surface of the Newtonian liquid will remain flat. [Photographs courtesy of Dr. F. Nazem, Rheology Research Center, University of Wisconsin, Madison.]

The contrasting behavior of the polymer solution is striking. The polymer solution moves in the opposite direction, toward the center of the beaker and climbs up the rod. Moreover, for comparable rotational speeds, the response of the polymer solution is far more dramatic than that of the Newtonian liquid, as seen in Fig. 3.2–1. This phenomenon is sometimes called the *Weissenberg effect*, as Weissenberg was the first to explain such an effect in terms of the stresses in fluids undergoing a steady shearing flow.[1] The effect had been observed earlier by Garner and Nissan.[2] This phenomenon should be kept in mind in connection with the design of mixing equipment.

[1] R. J. Russell, Ph.D. Thesis, Imperial College, University of London (1946), p. 58; K. Weissenberg, *Rep. Gen. Conf. Brit. Rheol. Club*, Nelson, Edinburgh (1946), p. 36; and K. Weissenberg, *Nature*, *159*, 310 (1947) cited in A. S. Lodge, *Elastic Liquids*, Academic Press, New York (1964), p. 194.

[2] F. H. Garner and A. H. Nissan, *Nature*, *158*, 634–635 (1946).

Because this effect is so large even at very low rates of rotation, the rod-climbing experiment makes a particularly impressive class demonstration. For this purpose the liquids shown in Fig. 3.2–1 are very suitable. The polymer solution is prepared as follows:[3] heat 1000 cm^3 of glycerin to 310 K. While stirring the heated glycerin fast enough to form a vortex, add 10 gm of Separan AP–30 (at 290 K); continue stirring the mixture for two hours. The solution should be allowed to stand overnight before use. To make the fluid more visible, 10 drops of Carmine Red are added during the mixing.

We can try to improve our understanding of this phenomenon by making another measurement. Pressure taps are mounted at points A and B, which are located the same height above the bottom of the beaker, that height being chosen such that the effects of the free surface and bottom are not important in determining the readings at A and B (see Fig. 3.2–2). For the Newtonian fluid the pressure at B exceeds that at A as a result of the centrifugal force. On the other hand, the pressure at A exceeds that at B in the polymer solution. These pressure differences can give us a way of measuring the forces in the polymer that oppose the centrifugal force.

Calculation of the shape of the free surface in this experiment is an extremely difficult task.[4] However, it is possible to interpret the pressure difference between A and B by making a hydrodynamic analysis of tangential annular flow. This we do in the following example.

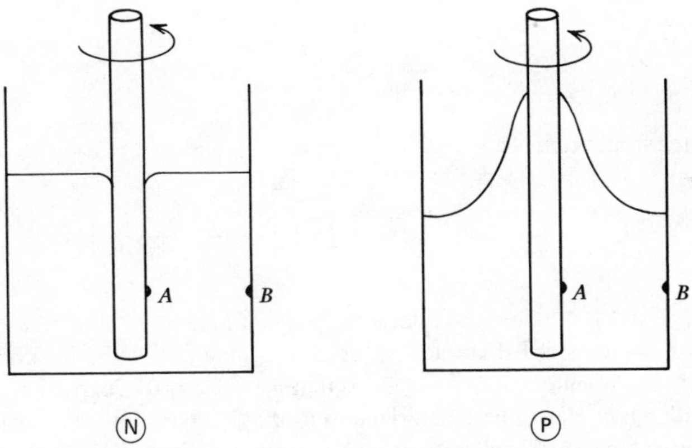

FIGURE 3.2–2. Pressure measurements in the rod-climbing experiment. Pressure taps at A and B are located so that their readings are not affected by the free surface or bottom of the cylinder. Ⓝ For the Newtonian liquid $p_B > p_A$; Ⓟ the reading at A exceeds that at B in the polymer.

EXAMPLE 3.2–1 Interpretation of a Tangential Annular Flow Experiment

Consider a fluid that is contained in the annular region between two long, vertical coaxial cylinders where the inner cylinder is caused to rotate with a constant angular velocity. This idealized system approximates the flow near the pressure taps A and B in Fig. 3.2–2. Using the equations of change in cylindrical coordinates, explain why the pressure reading at B is greater than that at A for a Newtonian fluid, whereas the opposite behavior is found for non-Newtonian fluids.

[3] This recipe was given to us by Dr. F. Nazem, Rheology Research Center, University of Wisconsin, Madison.

[4] The form of the free surface has been computed using a special rheological equation of state, the second order fluid (see Eq. 8.4–3), by D. D. Joseph and R. L. Fosdick, *Arch. Rat. Mech. Anal.*, *49*, 321–380 (1973) and A. Kaye, *Rheol. Acta*, *12*, 206–211 (1973). See also D. D. Joseph, G. S. Beavers, and R. L. Fosdick, *Arch. Rat. Mech. Anal.*, *49*, 381–401 (1973), and G. S. Beavers and D. D. Joseph, *J. Fluid Mech.*, *69*, 475–511 (1975).

SOLUTION

We begin by noting that in steady laminar flow, a fluid particle will follow a circular trajectory centered on the axis of the cylinders and lying in a horizontal plane.[5] In cylindrical coordinates, then, the only nonzero component of velocity will be v_θ. We further postulate that v_θ as well as the components τ_{ij} of the stress tensor will depend only on the radial position r. With these assumptions, the equation of continuity $(\nabla \cdot \boldsymbol{v} = 0)$ is satisfied identically. The equations of motion (Eqs. B.1–4 to 6) reduce to

r-component:
$$-\frac{\rho v_\theta^2}{r} = -\frac{\partial p}{\partial r} - \left(\frac{1}{r}\frac{d}{dr}(r\tau_{rr}) - \frac{\tau_{\theta\theta}}{r}\right) \tag{3.2–1}$$

θ-component:
$$0 = -\frac{1}{r^2}\frac{d}{dr}(r^2\tau_{r\theta}) \tag{3.2–2}$$

z-component:
$$0 = -\frac{\partial p}{\partial z} - \frac{1}{r}\frac{d}{dr}(r\tau_{rz}) + \rho g_z \tag{3.2–3}$$

For a Newtonian fluid, it is easy to show that the velocity postulates require that all of the "normal stresses" be zero: $\tau_{rr} = \tau_{\theta\theta} = \tau_{zz} = 0$. For the non-Newtonian fluids we do not know anything about the components of the stress tensor and are consequently forced to retain the normal stresses in Eq. 3.2–1.

Now we rearrange Eq. 3.2–1 to give:

$$\frac{d}{dr}(p + \tau_{rr}) = \rho\frac{v_\theta^2}{r} + \frac{\tau_{\theta\theta} - \tau_{rr}}{r} \tag{3.2–4}$$

which when integrated from point A to point B results in:

$$(p + \tau_{rr})_B - (p + \tau_{rr})_A = \int_{r_A}^{r_B}\left(\rho\frac{v_\theta^2}{r} + \frac{\tau_{\theta\theta} - \tau_{rr}}{r}\right)dr \tag{3.2–5}$$

The pressure taps at A and B measure the total force per unit area $\pi_{rr} = p + \tau_{rr}$ exerted by the fluid in a direction normal to the cylindrical surfaces. For a Newtonian fluid the integrand above is simply the centrifugal force; consequently, Eq. 3.2–5 tells us that $p_B - p_A > 0$ as expected. However, the result of the experiment with a non-Newtonian material tells us that $\tau_{\theta\theta} - \tau_{rr}$ must be negative and sufficiently large that it dominates the centrifugal force term. Thus we reach the important conclusion that for many polymeric fluids the normal stresses are nonzero and that the normal stress difference $\tau_{\theta\theta} - \tau_{rr}$ is negative.

Since normal stress differences will appear in many of the subsequent illustrations in this chapter, it will be helpful to establish some labeling conventions for referring to them. If the fluid moves along one coordinate direction only and its velocity varies only in one other coordinate direction, then we call the direction of fluid motion the "1" direction, the direction of velocity variation, the "2" direction, and the remaining neutral direction, the

[5] If the speed of rotation of the inner cylinder becomes too large, then instabilities appear in the form of Taylor vortices. These are toroidal cells that are roughly square in cross section and that circulate perpendicularly to the main flow. Photographs contrasting these cells for Newtonian and polymeric fluids are given by G. S. Beavers and D. D. Joseph, *Phys. Fluids*, *17*, 650–651 (1974). The earliest experiments on Taylor instabilities in polymer solutions were done by E. W. Merrill, H. S. Mickley, and A. Ram, *J. Fluid Mech.*, *13*, 86–90 (1962). Other papers in which data and photographs on this problem can be found are H. Rubin and C. Elata, *Phys. Fluids*, *9*, 1929–1933 (1966); H. Giesekus, *Rheol. Acta*, *5*, 239–252 (1966); M. M. Denn and J. Roisman, *A.I.Ch.E. Journal*, *15*, 454–459 (1969); H. Giesekus, in *Progress in Heat and Mass Transfer*, W. R. Schowalter (Ed.), Pergamon Press, New York (1972), Vol. 5, pp. 187–193; and J. W. Hayes and J. F. Hutton, *op. cit.*, pp. 195–209.

"3" direction. Then we call $\tau_{11} - \tau_{22}$ the *primary* (or first) *normal stress difference.* Likewise, we call $\tau_{22} - \tau_{33}$ the *secondary* (or second) *normal stress difference.* In the preceding example, θ, r, and z correspond to 1, 2, and 3, respectively. Thus the primary normal stress difference is $\tau_{\theta\theta} - \tau_{rr}$, and that is the important normal stress difference in tangential annular flow. According to the above convention, the secondary normal stress difference here would be $\tau_{rr} - \tau_{zz}$. However, this normal stress difference has not appeared in our analysis. We shall discuss the normal stress differences in greater detail in Chapter 4.

It can be shown (see Problem 3B.1) that the negative $\tau_{\theta\theta} - \tau_{rr}$ found in the preceding example corresponds to an extra tension in a fluid filament along a streamline. It is sometimes helpful to think of this tension as a "hoop stress," as the streamlines are circular. One can then imagine that the "hoop stresses" are trying to pull the fluid in toward the rod in much the same way as a rubber band stretched around a rolled up magazine tries to squeeze it into a tighter roll. The result is that the pressure taps read as previously discussed. Rod climbing also involves "hoop stresses," but it depends also on other stresses in the fluid (see Problem 3C.1). Similar "hoop stresses" will appear in analyses of the cone-and-plate and parallel disk instruments that we discuss in the next chapter.

§3.3 AXIAL ANNULAR FLOW

Here we consider axial laminar flow in the annular region between two concentric cylinders; the measurement of interest for this discussion is the difference between the readings at two pressure taps located at the same elevation on either side of the gap (*A* and *B* in Fig. 3.3–1). For Newtonian fluids, the pressure difference across the annular space

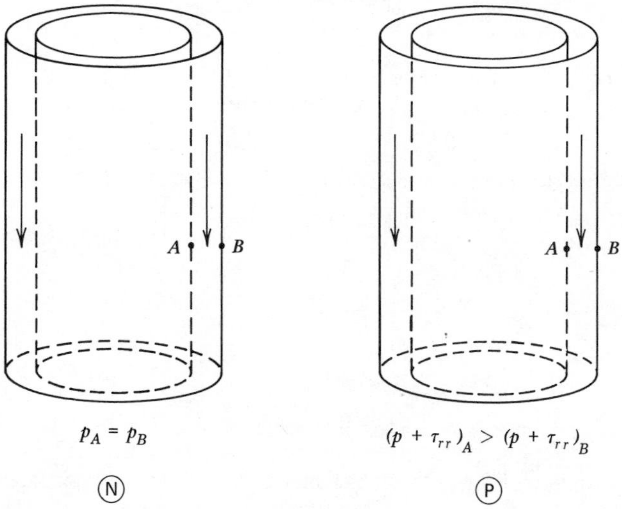

FIGURE 3.3–1. Pressure measurements in axial annular flow. For a Newtonian fluid $p_A = p_B$, whereas for a polymeric fluid, $(p + \tau_{rr})_A > (p + \tau_{rr})_B$.

is zero, as expected. For polymeric fluids, experiments have indicated a small difference in the readings at *A* and *B* with the pressure on the outer wall greater than that on the inner wall.[1] However, these measurements were made using "pressure holes," which are now known to introduce errors into the type of pressure reading taken in these studies (see §3.4).

[1] J. W. Hayes and R. I. Tanner, in *Proceedings of the Fourth International Congress on Rheology*, E. H. Lee (Ed.), Interscience, New York (1965), Part 3, pp. 389–399; J. D. Huppler, *Trans. Soc. Rheol.*, *9*, 273–286 (1965); and R. I. Tanner, *Trans. Soc. Rheol.*, *11*, 347–360 (1967).

The size of the error introduced by the presence of the hole is of the same order of magnitude as the pressure difference measured; and when the results are corrected for the "hole pressure error," the pressure at the inner wall is found to exceed that at the outer wall.[2] As we show in Example 3.3–1, this sign for the pressure difference between A and B can be identified with a positive secondary normal stress difference. The corrected result from axial annular flow then agrees with the sign for the secondary normal stress difference obtained by other methods.[3]

It would be nice to see the axial annular flow experiments repeated so as to give a positive value for the secondary normal stress difference without the necessity of applying a correction larger than the measured pressure difference itself. This is not easy to do in practice. Until more refined measurements are made, however, we cannot be absolutely certain of the sign of the pressure difference $(p + \tau_{rr})_B - (p + \tau_{rr})_A$. Although the present evidence indicates that it is negative, it may change signs with increasing shear rate and may be positive for some polymers. The effect of "pressure holes" on investigations of the pressure difference in axial annular flow emphasizes the need for careful study of measuring techniques.

EXAMPLE 3.3–1 Interpretation of an Axial Annular Flow Experiment

Consider the axial annular flow experiment shown in Fig. 3.3–1. For Newtonian liquids, it is not hard to show that there should be no difference in the pressures at A and B, or $p_A = p_B$.

Suggest why the sensor at A can have a higher reading than at B when a macromolecular fluid is used.

SOLUTION

We begin by postulating that $v_r = 0$, $v_\theta = 0$, $v_z = v_z(r)$ and that $p + \rho g z = \mathscr{P}(r,z)$. We further postulate that the τ_{ij} depend only on r. The equation of continuity is then satisfied identically. The r, θ, and z components of the equation of motion are:

$$0 = -\frac{\partial \mathscr{P}}{\partial r} - \left[\frac{1}{r}\frac{d}{dr}(r\tau_{rr}) - \frac{\tau_{\theta\theta}}{r}\right] \tag{3.3–1}$$

$$0 = -\frac{1}{r^2}\frac{d}{dr}(r^2\tau_{r\theta}) \tag{3.3–2}$$

$$0 = -\frac{\partial \mathscr{P}}{\partial z} - \frac{1}{r}\frac{d}{dr}(r\tau_{rz}) \tag{3.3–3}$$

The equation of interest is Eq. 3.3–1, which may be integrated in a horizontal plane to give:

$$(p + \tau_{rr})_B - (p + \tau_{rr})_A = -\int_{r_A}^{r_B}\left(\frac{\tau_{rr} - \tau_{\theta\theta}}{r}\right) dr \tag{3.3–4}$$

For the Newtonian fluid the normal stresses τ_{rr} and $\tau_{\theta\theta}$ vanish for the postulated velocity distribution; hence, the readings on the pressure taps are the same, and $p_A = p_B$. For polymeric liquids, if the reading at A exceeds that at B, the results of the experiment depicted in Fig. 3.3–1 suggest (1) that $\tau_{rr} - \tau_{\theta\theta}$ is

[2] A. C. Pipkin and R. I. Tanner, *Mechanics Today*, *1*, 262–321 (1972).

[3] Four tabular summaries are given in K. Walters, *Rheometry*, Chapman and Hall, London (1975), p. 89; A. C. Pipkin and R. I. Tanner, *loc. cit.*; O. Olabisi and M. C. Williams, *Trans. Soc. Rheol.*, *16*, 727–759 (1972); and R. F. Ginn and A. B. Metzner, *Trans. Soc. Rheol.*, *13*, 429–453 (1969).

positive or (2) that $\tau_{rr} - \tau_{\theta\theta}$ is positive over enough of the gap so that the integral in Eq. 3.3–4 is positive. According to the numerical labeling scheme established at the end of Example 3.2–1, this normal stress difference can be written as $\tau_{22} - \tau_{33}$. Whereas tangential annular flow shows a large, negative primary normal stress difference, axial annular flow indicates that the secondary normal stress difference is small and probably positive.

The most important point to bear in mind regarding the secondary normal stress difference is that it is small. The attempt to determine its value experimentally has been a challenge to rheologists for the past 25 years, and it is still a difficult quantity to measure. However, the secondary normal stress difference is probably not very important in many hydrodynamic calculations. One possible exception is wire coating.[4] In this process, the axial annular flow is driven by the axial motion of the inner cylinder—the wire. Analysis of the flow in a wire coating die indicates that if the wire is off-centered, the restoring force acting on it is proportional to $\tau_{22} - \tau_{33}$ and that the shear stress and primary normal stress difference are not important in stabilizing the position of the wire (see Example 8.5–2). Nonetheless, accurate measurements of $\tau_{22} - \tau_{33}$ will be useful as an analytical tool in evaluating molecular theories of polymer solutions and theoretically proposed rheological equations of state.

In §3.1, we saw that the viscosity of a macromolecular fluid can depend on the shear rate. Example 3.2–1 and the preceding example are important because they show that even in steady shear flow, there are forces at work in polymeric fluids that do not appear in Newtonian fluids. These first three sections, then, emphasize that expressions for the stress tensor will have to be different for polymeric and Newtonian fluids.

§3.4 HOLE PRESSURE ERROR

In the previous section, it was noted that early experimental efforts on the measurement of the secondary normal stress difference were hampered by "hole pressure errors." In fact, errors due to the "pressure hole" were the same order of magnitude as the normal stress being sought. The practice of positioning a pressure transducer or other sensing device at the bottom of a well drilled into the wall of an apparatus is very common in making pressure measurements in fluid flow. The reason for moving the measuring device away from the location at which pressure is to be measured is that typically the sensor is too large for the desired location. This might be because of the curvature of a channel wall or because the fluid pressure is varying too rapidly with position for the sensor to give a local reading. The well, then, provides the connection between the position where pressure is to be measured and the large sensor. Thus, it is important to understand the effect of pressure holes on the accuracy of the measurement.

In Fig. 3.4–1 we show the flow geometry with which we are concerned. For Newtonian fluids the pressure measured at the bottom of the pressure hole p_M is the same as the true pressure at the wall p, provided the hole used is small enough.[1] For many polymer solutions, on the other hand, $(p + \tau_{zz})_M$ is lower than $(p + \tau_{zz})$ no matter how small the hole is made. The difference we call a hole pressure error $p_H = (p + \tau_{zz})_M - (p + \tau_{zz})$. Note that, whereas the effect is well substantiated for polymer solutions, more study needs to be done on polymer melts.

These pressure discrepancies occur whenever a polymer solution flows over any depression in the conduit wall. Three geometries that have been studied are circular holes,

[4] Z. Tadmor and R. B. Bird, *Polym. Eng. Sci.*, *14*, 124–136 (1974); see also Example 8.5–2.

[1] Actually there can be a hole pressure error with Newtonian fluids associated with inertial effects in the flow over the hole. However, these can be made as small as desired by making the hole diameter small.

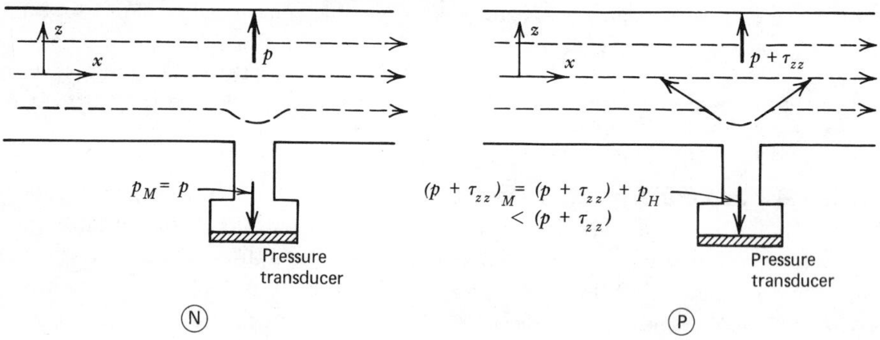

FIGURE 3.4–1. Hole pressure error. A wall pressure measurement is made by taking a reading at the bottom of a well. (N) The measured pressure gives the desired result for a Newtonian fluid but is too low for the polymer solution (P). The arrows in (P) indicate how an extra tension along a streamline lifts the fluid out of the hole resulting in a low reading. The y-direction is the vertical.

narrow slits with axes perpendicular to the direction of flow, and narrow slits aligned with the flow. For these geometries, the hole pressure error is related to the normal stress differences by rather simple looking formulas:[2]

Circular hole:

$$p_H = \tfrac{1}{3}\int_0^{\tau_w} \frac{(\tau_{11} - \tau_{22}) - (\tau_{22} - \tau_{33})}{\tau_{21}}\, d\tau_{21} \qquad (3.4\text{–}1)$$

Transverse slot:

$$p_H = \tfrac{1}{2}\int_0^{\tau_w} \frac{\tau_{11} - \tau_{22}}{\tau_{21}}\, d\tau_{21} \qquad (3.4\text{–}2)$$

Axial slot:

$$p_H = -\int_0^{\tau_w} \frac{\tau_{22} - \tau_{33}}{\tau_{21}}\, d\tau_{21} \qquad (3.4\text{–}3)$$

where τ_w is the wall shear stress, and it is assumed that the slot width or hole diameter is small relative to the channel dimensions and the slot or hole depth. In each of these results, the 1, 2, and 3 directions have the meanings given them at the end of §3.2 with the fluid velocity assumed to be constant along a streamline. Additional assumptions made in deriving Eqs. 3.4–1, 2, and 3 are that the normal force π_{11} does not change along a streamline, that the flow is slow, that there are no secondary flows, and that the fluid streamlines are symmetrical about the centerline of the hole or the center-plane of the slot. Of these assumptions, the most questionable[3] seems to be the one involving π_{11}. If the flow in Fig. 3.4–1 is generated by moving the upper wall, then it can be shown that in the absence of a hole or slot, π_{11} is constant along a streamline (i.e., $\partial\pi_{xx}/\partial x = 0$). The derivative of π_{11} with position along a streamline is probably not large in the presence of the hole or slot. On the other hand, if the flow is driven by a pressure gradient with no hole or slot in the bottom plate, $\partial\pi_{xx}/\partial x \neq 0$; presumably π_{11} also varies with position along a streamline when the hole or slot is introduced. However, experimental evidence[3] suggests that the influence of the π_{11} term on p_H is negligible.

Note how a negative primary normal stress difference and a positive secondary normal stress difference act to reduce the reading at the pressure transducer. The effect of an extra tension along a streamline in producing a low pressure reading is illustrated in Fig. 3.4–1.

[2] K. Higashitani and W. G. Pritchard, *Trans. Soc. Rheol.*, *16*, 687–696 (1972).

[3] K. Higashitani and A. S. Lodge, *Trans. Soc. Rheol.*, *19*, 307–335 (1975). Possible sources of error in measuring p_H are considered here along with a review of published data.

In Example 3.4–1 we show how to obtain the axial slot result, Eq. 3.4–3, and an analysis of flow past a circular hole is the subject of Problem 3D.1.

There are two main reasons for the great interest recently shown in the hole pressure error topic. First, many of the quantitative data in the literature on macromolecular fluids were taken before it was known that the presence of a hole affects measurements. Hence, published results of data obtained using a hole will have to be discarded or recomputed. As we saw in the preceding section, these changes in the data can be very significant when small pressure differences are involved.

Second, there is a desire to exploit the hole pressure error for the purpose of making quantitative measurements on macromolecular fluids. We have seen how the hole pressure error is related to the normal stresses. Since the primary normal stress difference is more sensitive to the molecular weight of a polymer than viscosity, accurate measurement of hole pressure error offers a new method for rheologically determining molecular weights. One application currently being developed is an on-line stress meter.[4] This device may be useful in continuously monitoring the degree of polymerization in a polymerization reaction by utilizing relations between normal stresses and molecular weight.

EXAMPLE 3.4–1 Significance of Pressure Measurements in Flow along an Axial Slit[5]

We consider here the flow between two large, parallel, horizontal, flat plates. The bottom plate has a very long, deep, narrow slit cut in it; the top plate is caused to move with velocity v_0 parallel to the axis of the slot (see Fig. 3.4–2*a*). Derive the relation between the secondary normal stress difference and the hole pressure error given in Eq. 3.4–3.

SOLUTION

To a first approximation we can assume that the fluid velocity is everywhere parallel to that of the upper plate. Neglect of secondary flows is known to be valid for slow flow of Newtonian fluids and some classes of macromolecular fluids. To facilitate the solution to this problem, we introduce an orthogonal curvilinear coordinate system[6] q_1, q_2, q_3. The coordinates are chosen such that locally $\boldsymbol{\delta}_1$ points in the direction of flow, $\boldsymbol{\delta}_3$ is perpendicular to $\boldsymbol{\delta}_1$ and tangent to a surface of constant velocity, and $\boldsymbol{\delta}_2$ is chosen so as to complete a right-handed triad (see Fig. 3.4–2*b*). Note that near the upper plate, the coordinate system will look very much like a rectangular Cartesian one.

We restrict ourselves here to creeping flows, so that nonlinear velocity terms in the equation of motion may be discarded. Furthermore, gravitational effects are assumed unimportant, and the momentum equation takes the simple form

$$[\nabla \cdot \boldsymbol{\pi}] = \mathbf{0} \qquad (3.4\text{–}4)$$

The two interesting components of Eq. 3.4–4 for the flow at hand are:

q_1-*component*:

$$\frac{1}{h_1}\frac{\partial \pi_{11}}{\partial q_1} + \frac{(\pi_{11} - \pi_{22})}{h_1 h_2}\frac{\partial h_2}{\partial q_1} + \frac{(\pi_{11} - \pi_{33})}{h_1 h_3}\frac{\partial h_3}{\partial q_1} + \frac{1}{h_2}\frac{\partial \tau_{21}}{\partial q_2} + \frac{\tau_{21}}{h_2 h_3}\frac{\partial h_3}{\partial q_2} + \frac{2\tau_{21}}{h_1 h_2}\frac{\partial h_1}{\partial q_2} = 0 \qquad (3.4\text{–}5)$$

[4] D. G. Baird and A. S. Lodge, *The Stressmeter*, Rheology Research Center Report 27, University of Wisconsin, Madison (1974).

[5] This example is patterned after K. Higashitani and W. G. Pritchard, *loc. cit.*

[6] Orthogonal curvilinear coordinate systems are discussed in Appendix A. More information regarding scale factors can be found in P. M. Morse and H. Feshbach, *Methods of Theoretical Physics*, McGraw-Hill, New York (1953), Chapter 1.

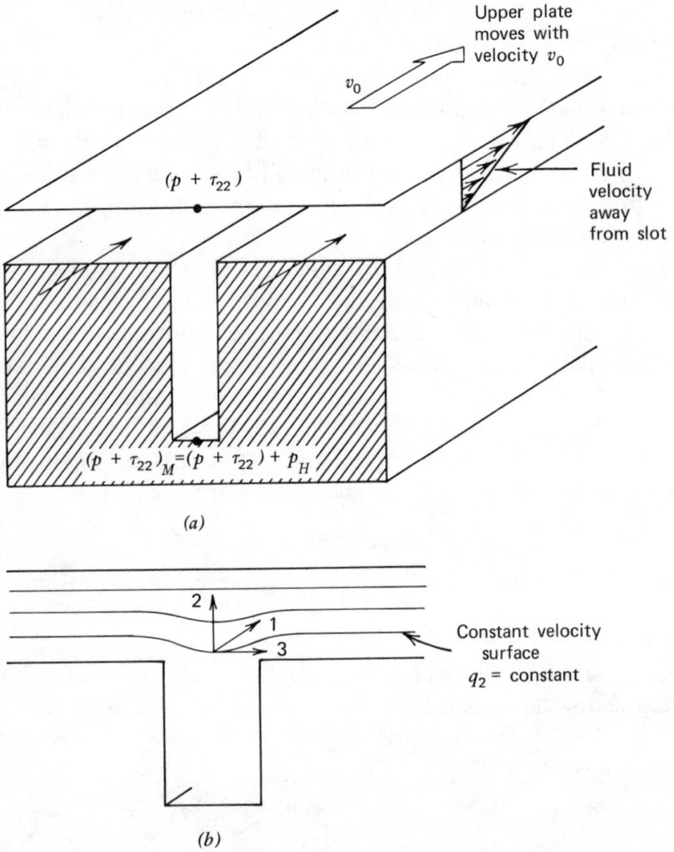

FIGURE 3.4–2. Axial flow along a slit. The flow is driven by the motion of the upper plate at velocity v_0. The locations of the pressure readings are shown in (*a*); the coordinate system used in the analysis is illustrated in (*b*).

q_2-component:

$$\frac{1}{h_2}\frac{\partial}{\partial q_2}\pi_{22} + \frac{(\pi_{22} - \pi_{33})}{h_2 h_3}\frac{\partial h_3}{\partial q_2} + \frac{(\pi_{22} - \pi_{11})}{h_1 h_2}\frac{\partial h_1}{\partial q_2} + \frac{1}{h_1}\frac{\partial \tau_{21}}{\partial q_1} + \frac{\tau_{21}}{h_1 h_3}\frac{\partial h_3}{\partial q_1} + \frac{2\tau_{21}}{h_1 h_2}\frac{\partial h_2}{\partial q_1} = 0 \quad (3.4\text{–}6)$$

where the h_α's are the scale factors (see §A.7) and we have taken[7] $\tau_{13} = \tau_{31} = \tau_{23} = \tau_{32} = 0$.

Considerable simplification of these equations can be obtained from the geometry of the flow. The easiest way to grasp this process is first to recall that a coordinate surface of constant q_i has a radius of curvature ρ_{ij} along the q_j axis given by[8]

$$\rho_{ij}^{-1} = \frac{1}{h_i h_j}\frac{\partial h_j}{\partial q_i} \quad (3.4\text{–}7)$$

Now for the axial flow we have assumed, the q_1 direction will be everywhere parallel to the axis of the slit. This means that a surface q_1 = constant will be flat and:

$$\rho_{12}^{-1} = 0$$

$$\rho_{13}^{-1} = 0 \quad (3.4\text{–}8a)$$

Similarly,

$$\rho_{21}^{-1} = 0 \quad (3.4\text{–}8b)$$

[7] For a justification of this postulate, see §4.2.
[8] P. M. Morse and H. Feshbach, *op. cit.*, pp. 26–27.

We further expect that derivatives with respect to q_1 will be zero. Equations 3.4–5 and 6 can thus be rewritten as:

$$\frac{1}{h_2}\frac{\partial \tau_{21}}{\partial q_2} + \frac{\tau_{21}}{h_2 h_3}\frac{\partial h_3}{\partial q_2} = 0 \tag{3.4–9}$$

$$\frac{1}{h_2}\frac{\partial \pi_{22}}{\partial q_2} + \frac{\pi_{22} - \pi_{33}}{h_2 h_3}\frac{\partial h_3}{\partial q_2} = 0 \tag{3.4–10}$$

The details of the coordinate system we have been using, which are contained in the scale factors, can be eliminated between these two equations to give a differential equation for π_{22}

$$\frac{\partial}{\partial q_2}\pi_{22} = \frac{\pi_{22} - \pi_{33}}{\tau_{21}}\frac{\partial \tau_{21}}{\partial q_2} \tag{3.4–11}$$

If we express π_{22} as a function of the shear stress τ_{21} instead of position, then the differential equation becomes:

$$\frac{d\pi_{22}}{d\tau_{21}} = \frac{\pi_{22} - \pi_{33}}{\tau_{21}} \tag{3.4–12}$$

Finally, we integrate this equation from a shear stress of zero to the shear stress that would exist at the wall τ_w if there were no slit cut in it. Provided the slit is deep enough, $\tau_{21} = 0$ corresponds to the shear stress at the bottom of it, and $\pi_{22}\ (\tau_{21} = 0)$ is then $\pi_{22,M}$. Similarly, $\pi_{22}(\tau_w)$ is the normal force that would act on the wall in the absence of the slit. Thus

$$\begin{aligned} p_H &= \pi_{22,M} - \pi_{22,w} \\ &= -\int_0^{\tau_w}\left(\frac{\tau_{22} - \tau_{33}}{\tau_{21}}\right) d\tau_{21} \end{aligned} \tag{3.4–13}$$

which is just Eq. 3.4–3.

§3.5 EXTRUDATE SWELL

If a macromolecular fluid is forced to flow from a large reservoir through a circular tube of diameter D, the diameter of the extrudate D_e is found to be larger than the tube diameter. This *extrudate swell* is shown in Fig. 3.5–1 where it is contrasted with the behavior of a Newtonian jet.[1] Although the Newtonian fluid (silicone) and the polymer solution (poly(methyl methacrylate) in dimethyl phthalate) shown here have comparable viscosities, the latter undergoes a 400% increase in cross-sectional area whereas the former experiences no swelling at all. Extrudate diameters of up to three or four times the tube diameter are possible with some polymers.

Actually, at high Reynolds numbers $\rho\langle v\rangle D/\mu$ the diameter of the Newtonian jet will be 13% less than the tube diameter. As seen in Problem 1D.6, this decrease in diameter can be understood from simple momentum conservation arguments. Corresponding attempts to explain extrudate swell in non-Newtonian fluids have been neither as clear nor as successful. In what follows, we present two qualitative attempts at understanding the phenomenon.

One way to explain the phenomenon concerns the ability of the fluid to remember the entrance region of the tube. The idea is to imagine a fluid element moving from the reservoir into the tube as a short, fat cylinder getting squeezed into a long, slender cylinder.

[1] Extrudate swell is known variously as jet swell, die swell, the Barus effect, and the Merrington effect. A. B. Metzner, *Trans. Soc. Rheol.*, *13*, 467–470 (1969) discusses the history of extrudate swell and argues against using the last two names.

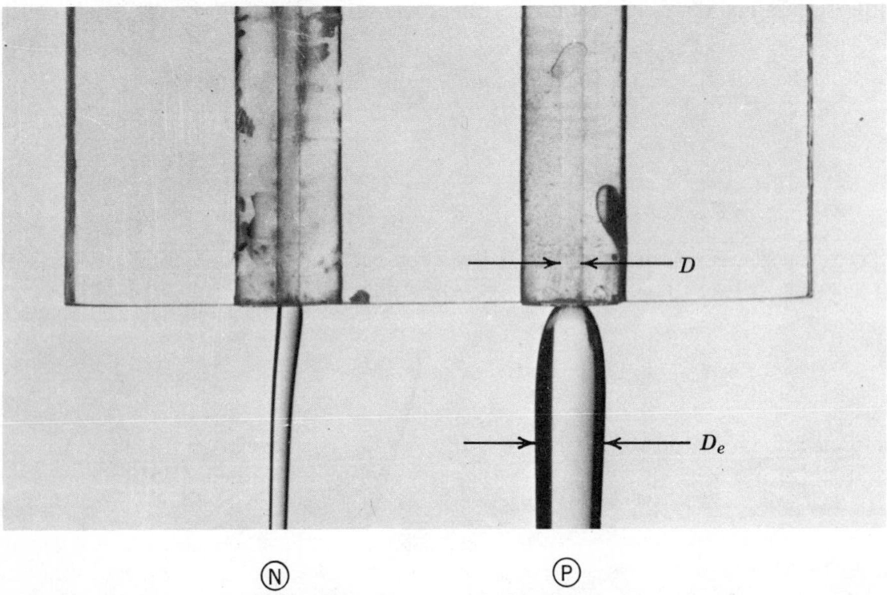

FIGURE 3.5–1. Behavior of fluids issuing from orifices. Ⓝ A stream of Newtonian fluid (silicone fluid) shows no diameter increase upon emergence from the capillary tube; Ⓟ a solution of 2.44 g of poly(methyl methacrylate) ($\bar{M}_n \approx 10^6$) in 100 cm^3 of dimethyl phthalate shows a 200% increase in diameter as it flows downward out of the tube. [A. S. Lodge, *Elastic Liquids*, Academic Press, New York (1964), p. 242.]

If it does not take too long for the fluid element to traverse the length of the tube, then upon emerging from the orifice, it will try to return to its original shape. Thus, the swelling effect results. This mechanism implies, however, that if we use longer and longer tubes, the ratio of extrudate diameter to tube diameter D_e/D should ultimately approach unity. The diameter ratio does indeed decrease with longer tubes but is found to reach a limiting value greater than unity as L/D approaches infinity.[2] Some other mechanism is needed to explain this residual swelling.

Another interesting experiment that tests the above extrudate swell explanation was done by Lodge.[3] He filled a tube with a sample of silicone "silly putty" and allowed it to remain there for a time much longer than its memory. The "silly putty" was then forced from the tube with a plunger and in the process showed a marked increase in diameter and corresponding decrease in length. The memory of an entrance flow certainly cannot explain this swelling.

Because of the long tube and long holding time results above, part of the explanation for extrudate swell must be associated with the flow in the tube. The tension along a streamline in steady-state shear flow associated with the normal stresses provides the basis for a simple, though nonrigorous, explanation. Once the sample has passed through the tube exit, the fluid will try to relax the tension along the streamline by contracting in the longitudinal direction; for incompressible materials this results in a lateral expansion. Lodge[4]

[2] P. L. Clegg in *Rheology of Elastomers*, P. Mason and N. Wookey (Eds.), Pergamon, London (1958), pp. 174–189.

[3] This experiment is described in A. S. Lodge, *Elastic Liquids*, Academic Press, New York (1964), pp. 243–244, where there are some striking photographs. The duration of the memory of "silly putty" is determined by simple recovery experiments discussed in §3.8.

[4] A. S. Lodge in *Rheology of Elastomers*, P. Mason and N. Wookey (Eds.), Pergamon, London (1958), pp. 70–85.

has given a detailed account of the "free recovery", the lateral expansion and longitudinal contraction that takes place in a viscoelastic liquid when all forces are removed from it, following a homogeneous steady shear flow. Tanner[5] has based an approximate theory of extrudate swell on Lodge's free recovery calculation. He found that for large tube length to diameter ratios the extrudate swell is given by

$$\frac{D_e}{D} = 0.1 + \left[1 + \frac{1}{2}\left(\frac{\tau_{11} - \tau_{22}}{2\tau_{21}}\right)_w^2\right]^{1/6} \tag{3.5-1}$$

where the subscript w indicates that the stresses in steady tube flow $(\tau_{11} - \tau_{22})$ and τ_{21} are to be evaluated at the wall. The simplicity of Eq. 3.5–1 and its success in describing data on extrudate swell recommend its use for estimation purposes. This result is semi-empirical, however. The 0.1 term has been added to improve the fit with data for small values of $(\tau_{11} - \tau_{22})_w/\tau_{21,w}$ and the ratio $(\tau_{11} - \tau_{22})/\tau_{21}^2$ has been taken to be constant. The reader should also bear in mind that the inhomogeneous tube flow preceding the extrudate swell is different from the homogeneous steady shear flow that preceded the free recovery that Lodge calculated. Nonetheless, free recovery following a shear flow could be responsible for the extrudate swell remaining in the long tube and long holding time experiments in which memory of the entrance region has completely faded away.

It is hoped that by this point the reader has acquired some appreciation for the possible factors that produce extrudate swell. We do not want to leave the mistaken impression that the memory of entrance effects and the release of normal stresses provide a complete explanation of extrudate swell. In addition to these effects, one must also take into account interfacial forces, compressibility, the secondary normal stress difference, viscous heating, and the complicated flow near the tube exit. Important problems also need to be solved in the flow out of noncircular dies. For example, in order to extrude a polymer with a square cross section, the die used must have a cross section that looks like a square with its sides bowed in. This is a very difficult problem to analyze. To date, understanding extrudate swell remains an active area of research.[6]

The importance of this phenomenon in the plastics and fiber industry can hardly be overlooked. In the spinning of rayon, a 10% increase in diameter is encountered;[7] much larger swelling ratios are experienced in the extrusion of molten polymers from dies. Since the diameter increase depends not only on the particular polymer but also on operating conditions such as temperature and flow rate, the industrial problems involving extrudate swell are particularly complex and challenging.

§3.6 SECONDARY FLOWS IN THE DISK AND CYLINDER SYSTEM

We now look at the flow produced by placing a rotating disk on top of a beaker filled with fluid. This system has been studied experimentally[1] and theoretically.[2] The

[5] R. I. Tanner, *J. Polym. Sci.*, A–2, *8*, 2067–2078 (1970).

[6] Extrudate swell was reviewed by E. B. Bagley and H. P. Schreiber, in *Rheology*, F. R. Eirich (Ed.), Academic Press, New York (1969), Vol. 5, Chapter 3, pp. 93–125. A more recent review of the literature is given by B. A. Whipple, Ph.D. Thesis, Washington University, St. Louis (1974).

[7] A. S. Lodge, *Elastic Liquids*, Academic Press, New York (1964), p. 243.

[1] C. T. Hill, J. D. Huppler, and R. B. Bird, *Chem. Eng. Sci.*, *21*, 815–817 (1966) and C. T. Hill, *Trans. Soc. Rheol.*, *16*, 213–245 (1972).

[2] J. M. Kramer, *Trans. Soc. Rheol.*, *16*,197–212 (1972).

primary fluid motion is in the tangential direction because of the rotation of the disk. Moreover, the fluid at the top of the container will be rotating with a larger angular velocity than that at the bottom; consequently, the fluid nearer the disk will experience a larger outward centrifugal force. For Newtonian fluids there are no forces to counter this one, and we are not surprised to find there is a weak secondary flow, everywhere perpendicular to the primary flow, which is directed radially outward near the disk, down the side of the beaker, inward along the bottom, and finally back up near the center. The magnitudes of velocities in the secondary flow are roughly 10% of those in the primary flow. One of Hill's photographs of this secondary flow is shown in Fig. 3.6–1 Ⓝ.

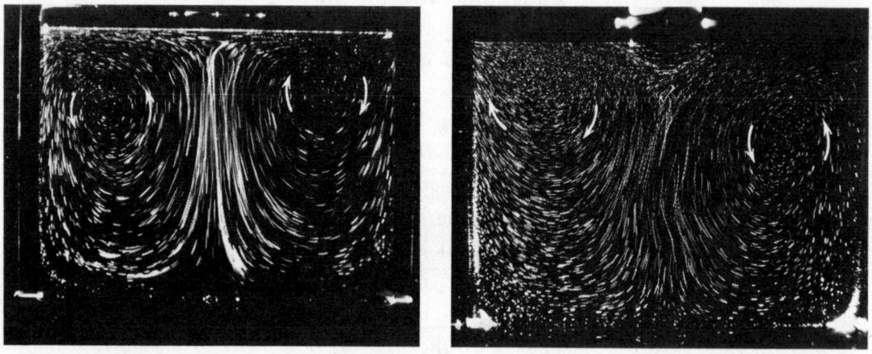

Ⓝ Ⓟ

FIGURE 3.6–1. Secondary flows in the disk-cylinder system. Ⓝ The Newtonian fluid moves up at the center whereas Ⓟ the viscoelastic fluid, polyacrylamide (Separan 30)/glycerol/water, moves down at the center. [C. T. Hill, *Trans. Soc. Rheol. 16*, 213–245 (1972).]

A polymer fluid placed in the disk and cylinder apparatus exhibits a secondary flow in the opposite direction. A typical flow pattern is shown in Fig. 3.6–1 Ⓟ for a solution of polyacrylamide in a solvent of glycerol and water. Qualitatively we expect that the "hoop stresses" associated with the primary flow that acted to produce the rod climbing are at work here. Moreover, the normal stresses do not have to be very large in an absolute sense in order to cause a reversal in the direction of secondary flow. At polymer concentrations so low that normal stresses could no longer be detected by standard techniques, the reverse secondary flow was still clearly visible at small disk rotation speeds.

Reverse secondary flows in viscoelastic liquids may be found in other geometries. For example, a rotating sphere in a large bath of fluid illustrates the same effects discussed here.[3] The addition of a small amount of polymer to water can also cause a reversal in the secondary flow generated by a cylinder oscillating normal to its axis[4] (see Fig. 3Q.13).

§3.7 CONVEX SURFACE IN A TILTED TROUGH

The secondary normal stress difference was introduced in §3.3. There we saw that very careful pressure measurements were needed in order to determine even its sign correctly. It is not necessary, though, to use exotic pressure measuring devices in order to see the secondary normal stress difference. The neatest method we know of for this purpose is

[3] H. Giesekus, *Rheol. Acta, 4*, 85–101 (1965); K. Walters and J. G. Savins, *Trans. Soc. Rheol., 9*: 1, 407–416 (1965).

[4] C. Chang and W. R. Schowalter, *Nature, 252*, 686–688 (1974); Figures 2b and 2c are unfortunately interchanged in this reference.

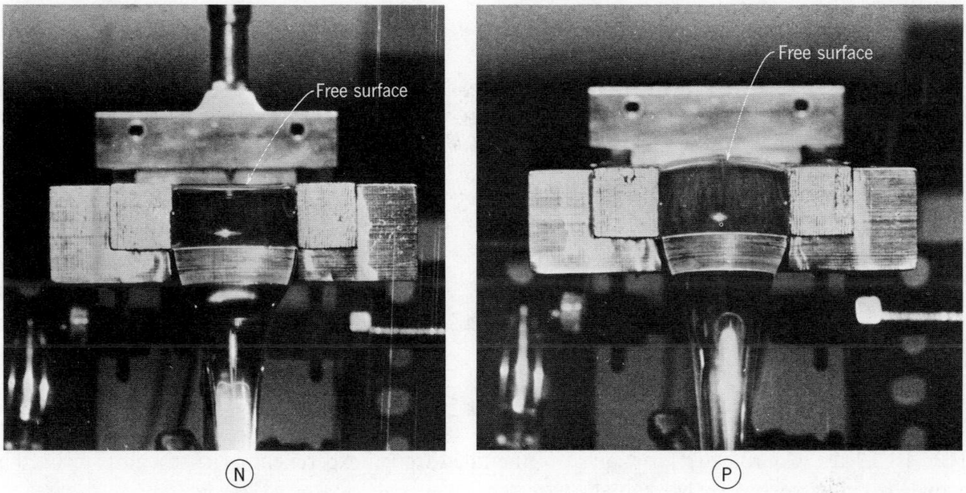

FIGURE 3.7–1. Tanner's tilted trough. A Newtonian liquid flowing down an inclined channel exhibits a flat free surface (N); the free surface of a polyisobutylene solution is convex (P). [R. I. Tanner, *Trans. Soc. Rheol.*, *14*, 483–507 (1970).]

"Tanner's tilted trough"[1] shown in Fig. 3.7–1. Here a fluid is allowed to flow by gravity down an open, inclined channel; one then observes the shape of the free surface of the fluid. A Newtonian liquid is found to have a flat surface, aside from the meniscus effect, whereas that of a typical non-Newtonian substance is slightly convex. The bulge is small but reproducible. In contrast to extrudate swell, which is a big effect, the bulge in Tanner's tilted trough is a very small effect.

EXAMPLE 3.7–1 Interpretation of Free Surface Shapes in the Tilted Trough Experiment[2]

For the flow in Fig. 3.7–1 explain why the free surface is flat for Newtonian liquids and convex for macromolecular fluids. To simplify the analysis, assume that the trough is very deep so that the presence of the bottom may be ignored. The trough is inclined at an angle β relative to the horizontal and has width W.

SOLUTION

If the trough is very deep, we anticipate a velocity field of the form $v_z = v_z(x)$, $v_x = 0$, and $v_y = 0$ where the coordinate system is chosen as shown in Fig. 3.7–2. We further expect the τ_{ij} to depend on x alone, and $p = p(x,y,z)$ since the surface is not necessarily flat. No information is gained from the continuity equation for the assumed velocity profile.

The equation of motion for this flow gives

x-component:
$$0 = -\frac{\partial p}{\partial x} - \frac{d\tau_{xx}}{dx} \tag{3.7–1}$$

y-component:
$$0 = -\frac{\partial p}{\partial y} - \frac{d\tau_{xy}}{dx} - \rho g \cos \beta \tag{3.7–2}$$

---

[1] R. I. Tanner, *Trans. Soc. Rheol.*, *14*, 483–507 (1970) was the first to perform this experiment that had been suggested earlier by A. S. Wineman and A. C. Pipkin, *Acta Mechanica*, *2*, 104–115 (1966).
[2] For a more detailed analysis see L. Sturges and D. D. Joseph, *Arch. Rat. Mech. Anal.*, *59*, 000–000 (1976).

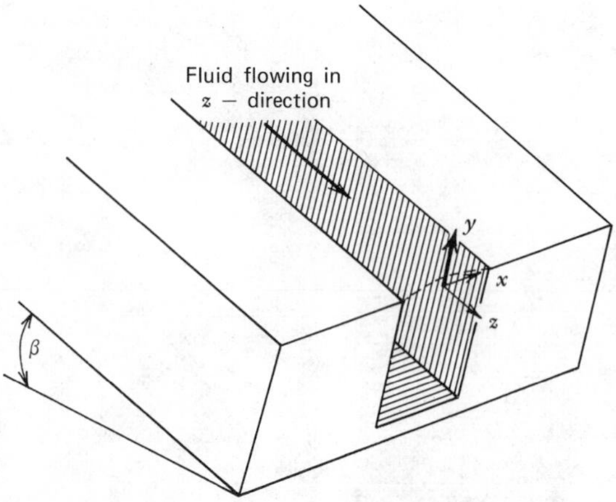

FIGURE 3.7–2. Coordinates for analysis of the tilted trough experiment. The trough is inclined at an angle β with respect to the horizontal.

z-component:
$$0 = -\frac{\partial p}{\partial z} - \frac{d\tau_{xz}}{dx} + \rho g \sin \beta \tag{3.7–3}$$

The dashed-underlined term can be neglected if we assume that the free surface is not distorted enough to alter the stress field from that in steady shear flow between two infinite parallel plates (see §4.2). From these equations we can conclude that

$$\frac{\partial p}{\partial z} = 0 \tag{3.7–4}$$

and hence that

$$\tau_{xz} = (\rho g \sin \beta)x \tag{3.7–5}$$

Equation 3.7–4 is found by taking the partial derivative with respect to z of all three components of the equation of motion. From these results it can be seen that $\partial p/\partial z$ is a constant; and since $\partial p/\partial z$ is zero along the free surface, it must be zero everywhere.

Next, since π_{yy} is an analytic function of position, we may write:

$$d\pi_{yy} = \frac{\partial \pi_{yy}}{\partial x} dx + \frac{\partial p}{\partial y} dy \tag{3.7–6}$$

By using Eqs. 3.7–1 and 2 together with Eq. 3.7–6, we find after integration that:

$$\pi_{yy}(x,y) = -(\tau_{xx} - \tau_{yy})|_x - \rho g(\cos \beta)y + C \tag{3.7–7}$$

where C is a constant of integration.

The total outward normal force per unit area exerted by the fluid at the free surface must be equal to the atmospheric pressure p_a. Provided the surface is not too curved, this requires that $\pi_{yy} = p_a$ and, thus, the free surface is given by the equation:

$$y = \frac{-(\tau_{xx} - \tau_{yy})|_x + C - p_a}{\rho g \cos \beta} \tag{3.7–8}$$

At $x = 0$ all velocity gradients vanish so that the components of the stress tensor are all zero. For $x > 0$, if $(\tau_{xx} - \tau_{yy})|_x$ is positive, then the free surface will be lower than at $x = 0$. Thus the convex free surface shown in Fig. 3.7–1 corresponds to a positive secondary normal stress difference. For Newtonian fluids $(\tau_{xx} - \tau_{yy}) = 0$ for this flow, and Eq. 3.7–8 predicts a flat liquid surface as observed.

We can regard the secondary normal stress difference to be a function of the shear stress τ_{xz} instead of x. From Eq. 3.7–8 the total height of the fluid bulge h is given by:

$$h = \frac{(\tau_{xx} - \tau_{yy})|_{\tau_w}}{\rho g \cos \beta} \tag{3.7–9}$$

where $\tau_w = \frac{1}{2}\rho g W \sin \beta$ is the wall shear stress. Thus by measuring the size of the bulge in the fluid surface for very small values of the tilt angle β, we can measure the secondary normal stress difference at vanishingly small shear stresses. Since this method is restricted to low shear stresses, any conclusions we have drawn about the secondary normal stress difference from this experiment are also limited. For larger values of β (shear stress), a more detailed analysis has been performed to give information about the complete shape of the free surface and the onset of secondary flows.[2]

Note that we have neglected surface tension in this analysis. If R denotes the radius (or half-width) of the channel, then the surface tension σ will not be important provided[1] $(\rho g R^2 \cos \beta)/2\sigma \gg 1$.

One of the nicest results of this experiment is that it clearly establishes that the secondary normal stress difference is positive, at least at low shear rates. In view of the difficulties encountered in the axial annular flow experiment, this is an important conclusion.

§3.8 RECOIL

A particularly striking contrast between the viscoelastic behavior of macromolecular fluids and the purely viscous behavior of Newtonian fluids is provided by the recoil experiment performed by Kapoor,[1] which is shown in Fig. 3.8–1. In his experiment, a streak of charcoal slurry was first introduced into a 2% (weight) solution of carboxymethylcellulose in water by means of a syringe. A pressure gradient was then applied to the solution, and the deformation of the trace line was recorded in a sequence of photographs. After a short time the pressure gradient was removed; the pictures taken after this time show the charcoal trace line recede as the fluid *recoils*. Recoil is not observed for Newtonian liquids.

The two most important points to be brought out by this experiment are the "elasticity" and the "fading memory" of the macromolecular fluid. By "elasticity" we are referring qualitatively to the ability of the fluid to snap back when the external forces are removed. By "fading memory" we mean that the viscoelastic material remembers where it came from, but that the memory of recent configurations is much better than that of configurations the polymer experienced a longer time before. Because of the fading memory, the polymeric fluid cannot recoil completely to its initial state.

Recoil can be illustrated in a host of other experiments. For example, one can see recoil in the rod climbing experiment if the torque is removed from the rod. Or, if a rotating beaker containing polymer solution is suddenly brought to a stop, then entrapped air bubbles will partially retrace their motion as the fluid recoils. Lodge[2] has used "silly putty" for further demonstrations of recoil. He rapidly stretches the sample, holds it in the elongated position for a short time, and then allows it to spring back. If the "silly putty" is held in the elongated position for a longer time, then the amount of recovery decreases dramatically, with no visible recoil occurring after waiting a few seconds. This decrease in recoil with increasing holding time illustrates the notion of fading memory, and one can speak qualitatively of a "characteristic time" for memory of the order of a few seconds for this particular substance: This should be compared to the characteristic time of 10^{-13} s for water and 10^{13} s for mountains. One reason why the behavior of viscoelastic substances often seems so peculiar is that their time constants are on the same order as our attention span.

[1] N. N. Kapoor, M.S. Thesis, University of Minnesota, Minneapolis (1964).
[2] A. S. Lodge, *Elastic Liquids*, Academic Press, New York (1964), p. 236.

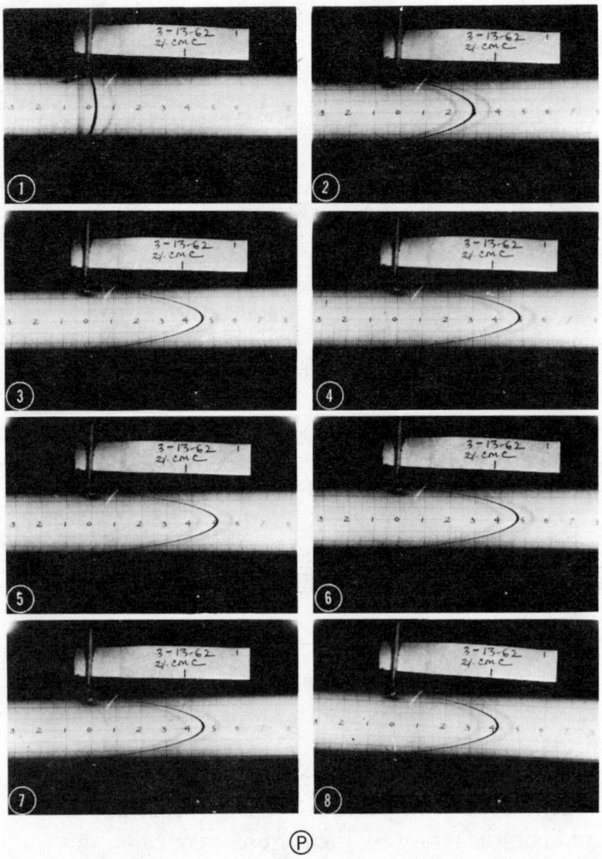

FIGURE 3.8–1. Recoil.[1] A solution of 2% carboxymethylcellulose (CMC 70H) in water is made to flow under a pressure gradient that is turned off just before frame 5. The flow and subsequent recoil are shown by the charcoal tracer line. [A. G. Fredrickson, *Principles and Applications of Rheology*, © 1964 Prentice-Hall, Inc., Englewood Cliffs, N.J., p. 120.]

In quantitative experiments, Meissner[3] found that a filament of low-density polyethylene at 423 K which is stretched rapidly from 1 to 30 cm in length, at which point it is suddenly set free, will recover to a length of just 3 cm. A recovery of a factor of 10 is enormous; a typical rubber band cannot recover this much because it breaks before it can be stretched by a factor of 10 from its natural state. Thus recoil in molten plastics can be quite large.

§3.9 THE TUBELESS SIPHON

We turn now to an experiment involving the siphoning of Newtonian and non-Newtonian fluids. Imagine two identical experiments in which fluid is being siphoned out of a container—one experiment using a Newtonian fluid, the other, a polymeric fluid (see Fig. 3.9–1). Now suppose that the tube is lifted up out of the fluid in the container. We immediately hear a slurping sound from the siphon that was in the Newtonian fluid as the

[3] J. Meissner, *Rheol. Acta*, *10*, 230–242 (1971).

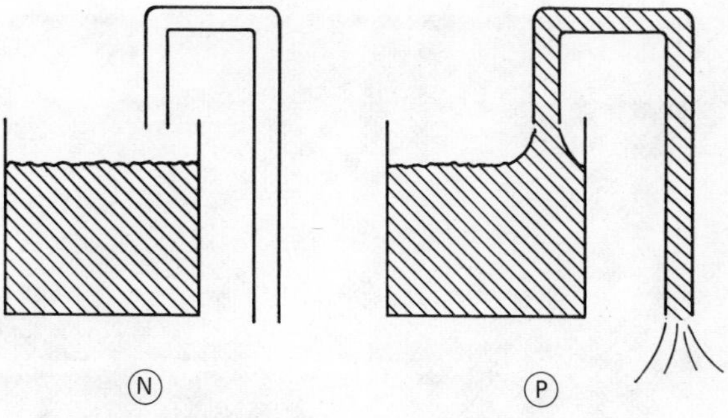

FIGURE 3.9–1. The tubeless siphon. When the siphon tube is lifted out of the fluid, the Newtonian liquid stops flowing (left); the macromolecular fluid continues to be siphoned (right).

liquid empties out of the tube and the siphoning stops. Not so with the non-Newtonian fluid; the fluid keeps right on flowing up to and through the siphon![1]

An even more amusing experiment that can be performed with a macromolecular fluid is as follows: Place a large beaker filled with a solution of 1 or 2% (by weight) poly-(ethylene oxide) (Polyox) with $\bar{M}_w \approx 4 \times 10^6$ in water on a table and an empty beaker on the floor. To start the siphoning, either dip a finger in the polymer solution and pull a filament of the fluid over the top of the beaker and down to the floor or tilt the beaker to start the liquid flowing over the lip. The poly(ethylene oxide) solution in the upper beaker follows the original filament down to the floor, thus providing a truly "tubeless" siphon. Approximately three-fourths of the upper beaker can be siphoned off in this manner. Imagine trying this experiment with a glass of water.

This experiment is closely linked to the property known as "spinnability": the stability of a stretching filament of fluid with respect to small perturbations in its cross-sectional area. This spinnability is one of the principal reasons for the commercial importance of many polymers, for example, nylons, Dacron, and polyethylene.

§3.10 THE UEBLER EFFECT AND FLOW THROUGH A SUDDEN CONTRACTION

Let us consider a polymeric fluid flowing in a tube with a sudden contraction; it flows from a section of large diameter tube into one of significantly smaller diameter (see Fig. 3.10–1). A common method for experimentally determining fluid streamlines and velocities is to introduce tiny gas bubbles (for example, of hydrogen) into the liquid well upstream of the region of interest and to observe their motion. For example, photographs of a bubble introduced into water along the center of the tube with a sudden contraction will show the forward motion and acceleration of the bubble as it moves into the smaller duct. This is observed for any Newtonian fluid and also for non-Newtonian fluids, provided the bubble is small enough.

In experiments illustrating the use of small bubbles as tracer particles, Uebler[1] found that what we have just described is not always observed when using macromolecular fluids. If the bubbles are too big, on the order of 1/6 to 1/8 of the small tube diameter, those moving along the centerline come to a sudden stop right at the entrance to the small tube.

[1] D. F. James, *Nature*, *212*, 754–756 (1966).

[1] E. A. Uebler, Ph.D. Thesis, University of Delaware, Newark (1966).

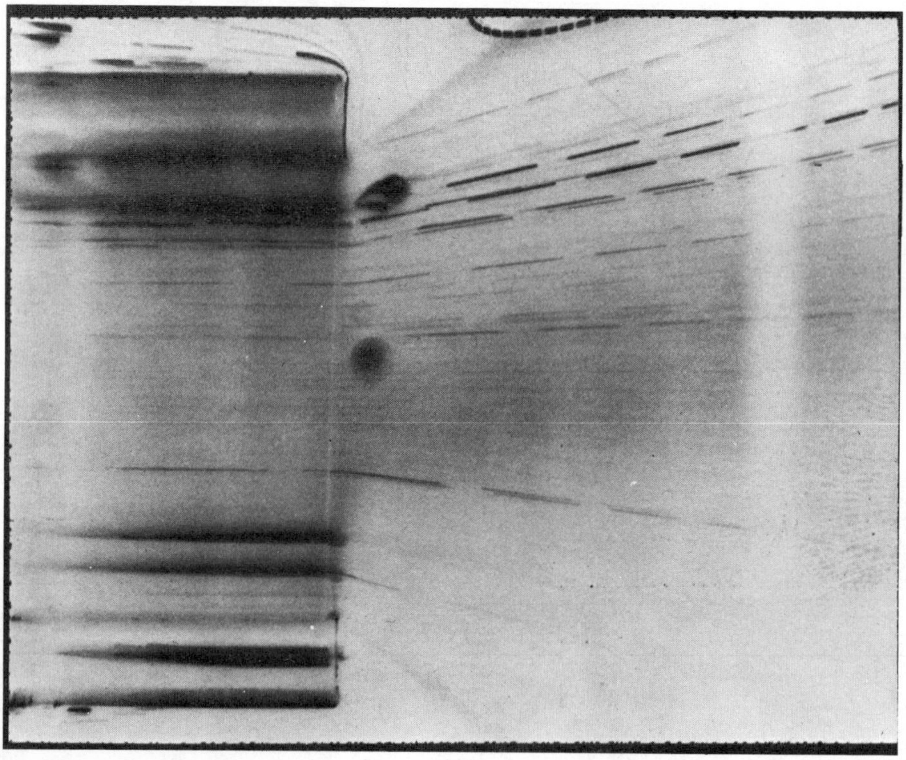

Ⓟ

FIGURE 3.10–1. The Uebler effect. A small bubble stops at the entrance to a small circular tube. The fluid is 0.45% polyacrylamide (Separan AP–30) in water flowing from right to left in the photograph at an average velocity of 1.2 m/s. [A. B. Metzner, *A.I.Ch.E. Journal*, *13*, 316–318 (1967).]

A stationary bubble can remain in that position for a minute or more before proceeding down the tube. Figure 3.10–1 illustrates this behavior, which is known as the *Uebler effect*. Away from the centerline the bubbles do not get held up and will move past a stationary bubble.

This phenomenon may be of interest to experimentalists using gas bubbles as tracers to map velocity fields in polymeric systems. In addition, several industrial processes call for the addition of a second (gas) phase to molten polymers. Nitrogen, for instance, is added to polymers such as polyethylene, polystyrene, and polypropylene in the production of foamed plastics. The Uebler effect has shown us that our intuition cannot be trusted in predicting even qualitatively what will happen in flows of two-phase systems. There may be many more undiscovered phenomena involving two-phase flows of polymeric fluids. The Uebler effect underlines the need for more study in this area.

We now focus our attention on the details of the flow pattern in the neighborhood of a sudden contraction. Figure 3.10–2 shows two streak photographs taken by Giesekus[2] contrasting the velocity fields of glycerin and aqueous polyacrylamide as they flow from a large reservoir into a small tube. The Reynolds number in both photographs is very low. The behavior of the two fluids is not even qualitatively similar. A typical streamline in the

[2] A. B. Metzner, E. A. Uebler, and C. F. Chan Man Fong, *A.I.Ch.E. Journal*, *15*, 750–758 (1969). For other photographs of converging flows of polymer solutions, see H. Giesekus, *Rheol. Acta*, *7*, 127–138 (1968).

(N)

(P)

FIGURE 3.10–2. Velocity fields near a sudden contraction. The fluid moves from top to bottom in the photographs from a large reservoir into a small circular tube. (N) The streamlines in glycerin are straight and all directed toward the exit; (P) a 1.67% aqueous polyacrylamide solution shows a large toroidal vortex. [Photographs by H. Giesekus, given in A. B. Metzner, E. A. Uebler, and C. F. Chan Man Fong, *A.I.Ch.E. Journal*, *15*, 750–758 (1969).]

Newtonian fluid is a straight line heading directly toward the entrance to the small tube; glycerin approaches the exit from a full 90 degrees in any direction about the centerline.[3] However, for a polymer solution, only the fluid in a small conical region about the centerline moves linearly toward the entrance to the small tube. A significant portion of the polymer solution is trapped in a large circulation pattern and does not enter the small tube at all. Other pictures of converging flows of polymeric fluids can be found in the literature.[2,3] The distorted velocity profiles in converging flows of non-Newtonian fluids have been partially explained theoretically.[4]

Flow through sudden contractions and flow in converging sections are common in processing operations. Our theoretical understanding of converging and diverging flows, however, is far from complete.

§3.11 DRAG REDUCTION IN TURBULENT FLOW[1]

Up to this point we have been dealing with unusual phenomena in fairly concentrated polymer solutions and melts. In this section and the following two, we are going to see some rather dramatic effects produced by adding only a minute amount of polymer to a Newtonian fluid. By far the best known of these is drag reduction, sometimes called the Toms phenomenon.[2]

In 1948 Toms discovered that if he added a small amount of poly(methyl methacrylate), approximately 10 ppm by weight, to monochlorobenzene undergoing turbulent tube flow, a rather substantial reduction in pressure drop at the given flow rate resulted. This reduction in pressure drop in the polymer solution relative to the pure solvent alone at the same flow rate is defined as *drag reduction.* Since then any number of polymer-solvent pairs have been found to show drag reduction. Three additional examples give an idea of the variety of possible systems: polyisobutylene in decalin, carboxymethylcellulose in water, and poly(ethylene oxide) in water.

Drag reduction results may be presented conveniently in terms of the Fanning friction factor f:

$$f = \frac{1}{4}\left(\frac{D}{L}\right)\frac{\mathscr{P}_0 - \mathscr{P}_L}{\frac{1}{2}\rho\langle v\rangle^2} \tag{3.11-1}$$

where $\mathscr{P}_0 - \mathscr{P}_L$ is the modified pressure drop over a length L of the tube, D is the tube diameter, and $\langle v\rangle$ is the average velocity over a cross section of the tube. The friction factor is essentially a dimensionless pressure gradient, and it is a function only of the Reynolds number $\text{Re} = D\langle v\rangle\rho/\mu$ for fully developed flow of Newtonian fluids. At the very tiny polymer concentrations of interest in drag reduction, the viscosity and density of the polymer solution differ only slightly from those of the pure solvent. Nonetheless, the effect of the polymer additive is to lower the value of the friction factor at a given Reynolds number.

[3] A. B. Metzner, E. A. Uebler, and C. F. Chan Man Fong, *A.I.Ch.E. Journal, 15*, 750–758 (1969). Also note that if the small tube extends into the interior of the large reservoir, the streamlines are straight and approach the opening from 180 degrees in any direction from the tube axis. See also A. B. Metzner, *A.I.Ch.E. Journal, 13*, 316–318 (1967).

[4] The converging flow calculations are discussed by R. S. Rivlin and K. N. Sawyers, *Ann. Rev. Fluid Mech., 3*, 117–146 (1971).

[1] Two recent review articles on drag reduction are J. L. Lumley, *Macromolecular Rev.,7*, 263–290 (1973) and J. W. Hoyt, *Trans. ASME, J. Basic Eng., 94D*, 258–285 (1972).

[2] B. A. Toms in *Proceedings of the International Congress on Rheology* (Holland, 1948), North-Holland, Amsterdam (1949), pp. II.135–II.141.

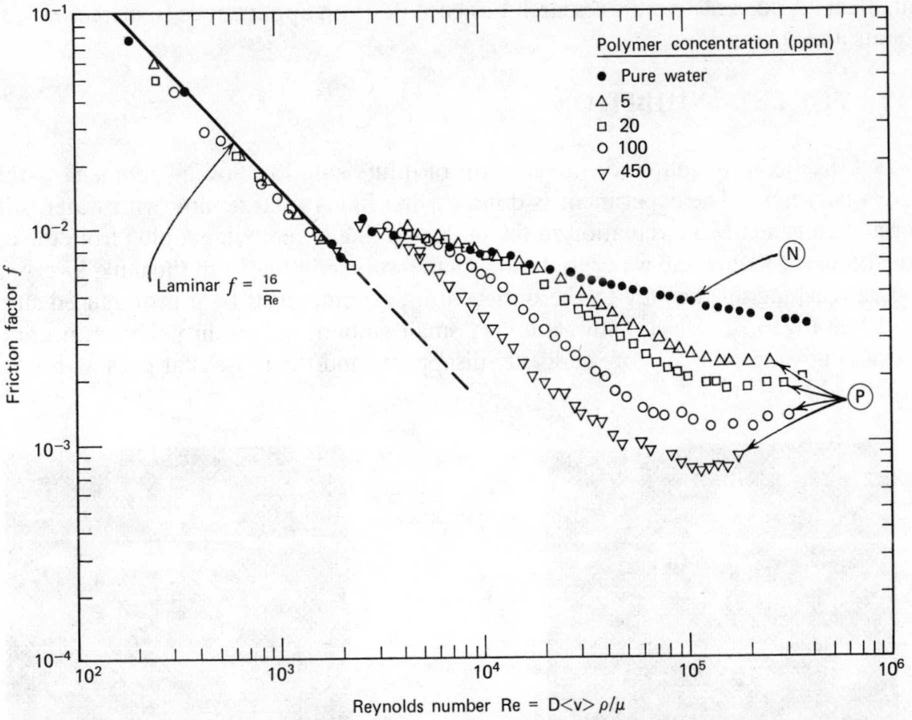

FIGURE 3.11–1. Friction factor for dilute aqueous solutions of poly(ethylene oxide), $\bar{M}_v = 6.1 \times 10^6$. In the turbulent regime, the curves for the polymer solutions lie below that of the solvent and illustrate drag reduction. [Data replotted from P. S. Virk, Sc. D. Thesis, Massachusetts Institute of Technology (1966).] See also C. B. Wang, *Ind. Eng. Chem. Fundamentals*, *11*, 546–551 (1972).

The amount by which the friction factor is lowered is a measure of the amount of drag reduction.[3]

Figure 3.11–1 shows the friction factor for water with and without a small amount of poly(ethylene oxide). It is clear from this figure that the drag reduction is strictly a turbulent flow phenomenon; it does *not* result from maintaining laminar flow past the usual transition region. Note also that whereas the addition of only 5 ppm of poly(ethylene oxide) to water gives a 40% reduction in f at Re = 1.0×10^5, the viscosity of the solution is only 1.0% greater than the viscosity of the water alone.

Although the mechanism producing drag reduction is not yet known, a number of polymer characteristics making for good drag reducers have been determined. A long-chain backbone and flexibility are important characteristics of good drag-reducing agents. For instance, of two polymers with the same molecular weight and same structural units, a linear one will be more effective than a highly branched one. Also, for two different polymers of similar configuration and the same molecular weight, the one with the lower molecular weight monomer will have the greater drag-reducing effect if both are utilized at the same weight concentration.

The potential applications of drag reduction are quite diverse; we mention just a few here. One use is in reducing pumping costs for transporting fluids. This is of interest, for example, in the petroleum industry where long pipelines for transmitting products are not uncommon. Drag-reducing agents have been used to increase the range of fire-fighting

[3] Provided the polymer additive has negligible effect on the viscosity and density of the solvent, then the drop in f at fixed Re will agree with the definition of the amount of drag reduction given before.

equipment. Also considerable research has been done on applications to increase the speed of boats and submarines.

§3.12 VORTEX INHIBITION

A particularly simple demonstration of dilute solution flow phenomena is that of *vortex inhibition.*[1] The experiment is done by first filling a large tank with water, stirring the water to generate a circulation in the tank, and finally removing a plug from the center of the bottom to allow the water to drain. As the water empties from the tank, a very stable air core reaching all the way to the outlet forms, accompanied by a pronounced slurping sound (see Fig. 3.12–1 Ⓝ). Now, if a very small amount of certain polymers is added to the draining water, the air core suddenly disappears and the noise that goes with it ceases (see Fig. 3.12–1Ⓟ).

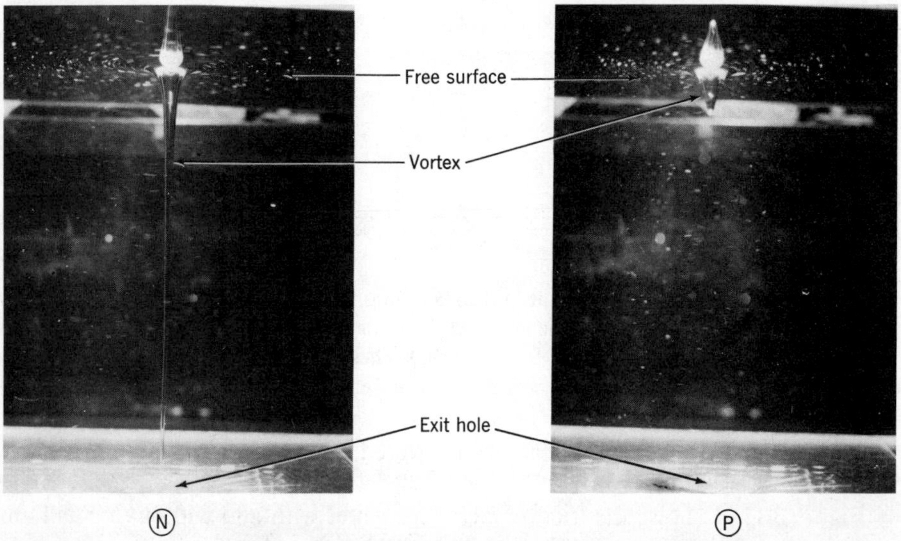

FIGURE 3.12–1. Vortex inhibition. Ⓝ shows the vortex that is formed as water drains from a tank, and Ⓟ shows the corresponding result after addition of a small amount of polymer to the water. A reflection of the vortex in the free surface of the liquid can be seen just above the vortex in either photograph. [Photographs courtesy of Professor R. J. Gordon, Department of Chemical Engineering, University of Florida.]

Among the best polymers for vortex inhibition are poly(ethylene oxide), effective at 7.5 ppm, and polyacrylamide, effective at only 3 ppm. These solutions are exceedingly dilute. It is even more interesting that the polymers that are good at vortex inhibition are generally also good in drag reduction. Furthermore, the same ordering in terms of effective concentrations seems to hold for both phenomena. For example, the two polymers mentioned above show significant drag reduction at concentrations of 9 and 5 ppm, respectively.

The mechanism for vortex inhibition is more amenable to understanding than that for drag reduction because the former involves laminar flow and the latter involves turbulent flow. Because of the tangential nature of the flow, we expect "hoop stresses" to build up after the addition of polymer to the water, leading to the demise of the air core. Dilute solution molecular theories have been applied to an analysis of vortex inhibition, and they seem to support the arguments given here[2] (see Problem 11B.8).

[1] R. J. Gordon and C. Balakrishnan, *J. Appl. Polym. Sci.*, *16*, 1629–1639 (1972).

[2] R. C. Armstrong, Ph.D. Thesis, University of Wisconsin, Madison (1973), pp. 105–112.

§3.13 EFFECT OF POLYMER ADDITIVES ON HEAT TRANSFER IN TURBULENT FLOW

As a final illustration, we look at a nonisothermal effect, namely, the influence of very small amounts of polymer additives on heat transfer. The primary concern here is with drag-reducing systems, and so we restrict ourselves to the turbulent regime. Consider, then, fluid flowing in a pipe of diameter D at a sufficiently high Reynolds number so that the flow is turbulent. The tube is jacketed by a heater so as to provide a constant heat flux q_0 to the fluid in the region of interest. As is customary, the heat transferred from the wall to the fluid may be expressed in terms of a local Stanton number St:

$$\mathrm{St} = \frac{q_0}{(T_w - T_b)\hat{C}_p G} = \frac{h}{\hat{C}_p G} \tag{3.13–1}$$

where $(T_w - T_b)$ represents the local difference between the wall temperature and the bulk temperature of the fluid, $\hat{C}_p$ is the fluid heat capacity per unit mass, G is mass velocity, and h is the local heat transfer coefficient.[1] On the order of 25 diameters down the tube from the entrance to the heated section, the local Stanton number becomes independent of position, and it is this measure of heat transfer that is of interest here.[2]

The data on this problem are scanty; however, in Fig. 3.13–1 we present some results that are sufficient for our purposes. The data were taken on a drag-reducing polymer solution consisting of 50 ppm of a high molecular weight poly(ethylene-oxide) in water.[3] Both friction factor and heat transfer results are included for comparison. The friction factor data show that drag reduction is indeed being realized. Note that varying the Prandtl number $\mathrm{Pr} = \hat{C}_p\mu/k$, where k is the thermal conductivity of the fluid, has little effect on f, as expected. The results for the Stanton number show a corresponding decrease after the addition of 50 ppm of the polymer. Even though the magnitude of the Stanton number shows a noticeable dependence on Pr, the percent decrease in St is roughly constant at a given Reynolds number. For example, at $\mathrm{Re} = 10^5$ the addition of 50 ppm poly(ethylene oxide) reduces the Stanton number approximately 82% at all three Prandtl numbers shown. The decrease in St appears to be slightly more pronounced than the drop in friction factor (74% at $\mathrm{Re} = 10^5$) for the drag-reducing system presented here. The results in Fig. 3.13–1 are for flow in a smooth tube. For a rough tube, the polymer additives produce an even larger drop in both f and St, with the Stanton number being lowered by nearly a factor of 10 by 50 ppm of polymer.

One is not really all that surprised by the lowering of heat transfer coefficients in drag-reducing systems. In pipe flow all momentum must be removed at the wall, and to reduce drag the mechanism for transporting this momentum to the wall must be impeded somehow.

[1] For more on the various ways of presenting heat transfer results see R. B. Bird, W. E. Stewart, and E. N. Lightfoot, *Transport Phenomena*, Wiley, New York (1960), Chapter 13.

[2] See M. K. Gupta, A. B. Metzner, and J. P. Hartnett, *Int. J. Heat Mass Transfer*, *10*, 1211–1224 (1967).

[3] These data are given by P. M. Debrule and R. H. Sabersky, *Int. J. Heat Mass Transfer*, *17*, 529–540 (1974). Other data on this problem are given among others by K. A. Smith, G. H. Keurohlian, P. S. Virk, and E. W. Merrill, *A.I.Ch.E. Journal*, *15*, 294–297 (1969), M. Poreh and U. Paz, *Int. J. Heat Mass Transfer*, *11*, 805–818 (1968), Y. Dimant and M. Poreh, *Advances in Heat Transfer*, *12*, 000–000 (1976), C. S. Wells, Jr., *A.I.Ch.E. Journal*, *14*, 406–410 (1968), G. Marrucci and G. Astarita, *Ind. Eng. Chem. Fundamentals*, *6*, 470–471 (1967), and M. K. Gupta, *et al.*, *loc. cit.*, along with some suggested correlations for the heat transfer data. The data in this area are not always in agreement with each other and seem inconclusive. The main point of disagreement is the relative size of the effect of polymer additives on heat transfer compared to their effect on the friction factor. Caution is therefore urged in making generalizations based on Fig. 3.13–1.

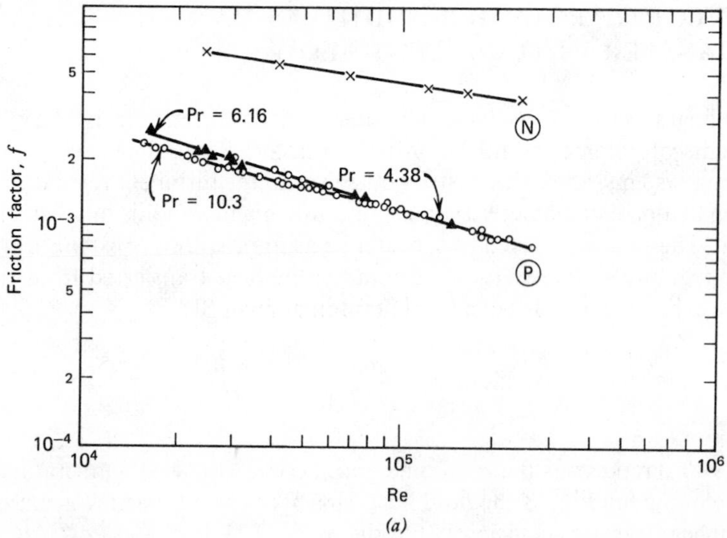

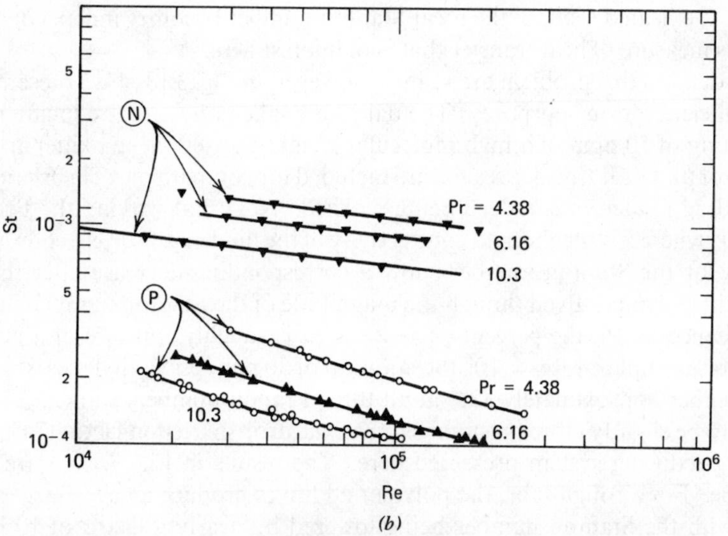

FIGURE 3.13–1. Turbulent heat transfer in water Ⓝ and in a drag-reducing polymer solution consisting of 50 ppm poly(ethylene oxide) in water Ⓟ. (*a*) The friction factor for water is lowered by the addition of a small amount of the polymer. The friction factor for the polymer solution shows a small dependence on the Prandtl number; the water curve is for all three values of Pr. (*b*) The decrease in Stanton number for the dilute polymer solution relative to the pure water is shown at three Prandtl numbers. [R. M. Debrule and R. H. Sabersky, *Int. J. Heat Mass Transfer, 17*, 529–540 (1974).]

In the heat transfer problem, the wall again serves as the source or sink. Presumably the same causes that reduce drag will also cut down the flow of heat. The exact nature of this mechanism, however, is not known; the reason why the effect of the polymer additives is greater in heat transfer is likewise a puzzle.

Whether the effect we have illustrated here is beneficial or detrimental depends on the particular application involved. Since dilute polymer solutions exhibiting drag reduction may be used in heat transfer equipment or in applications where heat transfer is important, it is necessary to take account of the large drop in heat transfer coefficient as well as the reduction in friction factor in designing processing operations.

QUESTIONS FOR DISCUSSION

Some of the questions below cannot be answered using information from Chapter 3. They are included to point out interesting phenomena, to stimulate discussion, and to provide further background for later chapters.

1. A class demonstration that Lodge uses is shown in Fig. 3Q.1. Discuss this experiment in terms of stresses in the fluid that make it behave as shown. What phenomena discussed in the chapter are illustrated here? What would happen if this experiment were tried on a Newtonian liquid like glycerin?

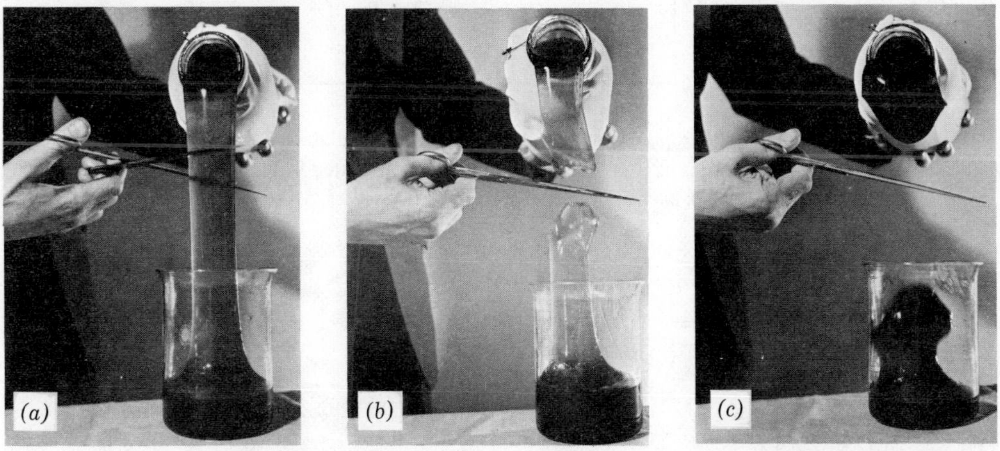

FIGURE 3Q.1. An aluminum soap solution, made of aluminum dilaurate in decalin and *m*-cresol, is (*a*) poured from a beaker and (*b*) cut in midstream. In part (*c*), note that the liquid above the cut springs back to the beaker and only the fluid below the cut falls to the container. [A. S. Lodge, *Elastic Liquids*, Academic Press, New York (1964), p. 238.] For a further discussion of aluminum soap solutions see N. Weber and W. H. Bauer, *J. Phys. Chem.*, *60*, 270–273 (1956).

2. If one tries to do the falling sphere experiment using an aluminum soap solution, a rather startling result is found. Instead of falling directly to the bottom, the sphere initially drops approximately half of the distance to the bottom of the cylinder, whereupon it bounces back almost to the top of the liquid before reversing directions again. These oscillations are slowly damped out as the sphere gradually settles to the bottom of the cylinder[1] (see Fig. 3Q.2). Explain why the motion of the sphere might be this way.
3. A device that has been under investigation recently is the conical pump,[2] which consists of a rotating conical section placed coaxially within a converging section of pipe (Fig. 3Q.3). Which way will a Newtonian fluid move when placed in this apparatus, and what is the driving force for this motion? Discuss the behavior of a polymer solution in the conical pump. What forces bear on the response of the polymer?
4. Kaye[3] has observed that when a solution of polyisobutylene in decalin is poured in a thin stream onto a shallow pool of itself, the stream of liquid will bounce off of the pool

[1] This experiment was done by W. Philippoff at the Society of Rheology Meeting, Madison (1961). The unsteady motion of a sphere, acted on by a constant force and initially at rest, in a viscoelastic liquid has been analyzed by R. H. Thomas and K. Walters, *Rheol. Acta*, *5*, 23–27 (1966) and by M. J. King and N. D. Waters, *J. Phys. D.: Appl. Phys.*, *5*, 141–150 (1972).

[2] P. A. Good, A. J. Schwartz, and C. W. Macosko, *A.I.Ch.E. Journal*, *20*, 67–73 (1974).

[3] A. Kaye, *Nature*, London, *197*, 1001–1002 (1963).

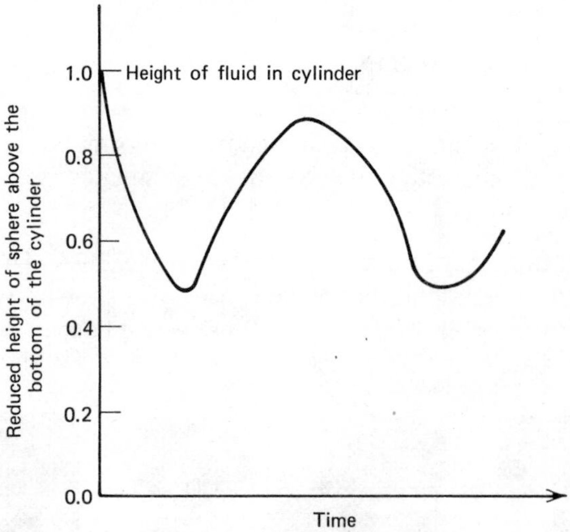

FIGURE 3Q.2. Falling sphere in an aluminum soap solution. A glass sphere dropped into a cylinder containing the solution oscillates as it proceeds to the bottom. [Plotted from a sequence of photographs given in R. S. Brodkey, *The Phenomena of Fluid Motions*, Addison-Wesley, Reading, Mass. (1967), p. 377.]

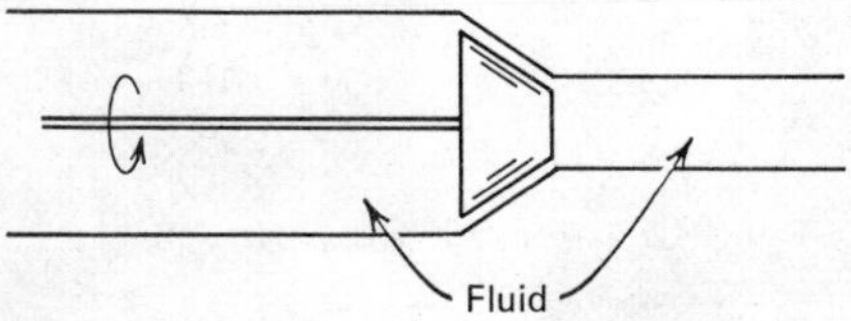

FIGURE 3Q.3. The conical pump.

as shown in Fig. 3Q.4. Furthermore, the bouncing liquid stream forms a fine, continuous element with a curiously shaped trajectory. This behavior, which lasts for a period of about one second, alternates on a regular basis with a period of similar length, during which there is no bouncing stream. In the off-period, a small hump of liquid can be seen forming where the stream being poured strikes the surface of the pool. Try to explain why the bouncing stream can form for the polymer solution.

5. It has recently been reported that if the wall of a glass beaker, which has contained a solution of a high molecular weight polyisobutylene in low molecular weight polyisobutylene, is cleaned by tearing the polymer from the wall while wrapping it onto a glass rod, an interesting phenomenon is observed. Several very thin streams of fluid are found to emanate from irregularities on the surface of the liquid (see Fig. 3Q.5). The streams increase in length at a rate of several centimeters per second.[4] It is believed that this is an electrically driven phenomenon. However, in Newtonian liquids of low conductivity similar to that of the polyisobutylene, the streaming jet breaks up into a series of droplets, instead of remaining as a continuous filament like the polymer shown here.[5] What is the reason for this difference in behavior?
6. What would be the effect of adding a small amount of polymer on mass transfer[6] to the wall in turbulent pipe flow?
7. Suppose that in pressure-driven axial tube flow a small sinusoidal oscillatory pressure gradient is superposed on the existing pressure gradient. What change is there in the

[4] B. J. S. Barnard and W. G. Pritchard, *Nature*, *250*, 215–216 (1974).
[5] G. I. Taylor, *Proc. Roy. Soc.*, London, *A313*, 453–475 (1969).
[6] G. H. Sidahmed and R. G. Griskey, *A.I.Ch.E. Journal*, *18*, 138–141 (1972).

FIGURE 3Q.4. A solution of 5.76 g of polyisobutylene ($\bar{M}_n \approx 10^6$) in 100 cm^3 of decalin is poured onto a shallow pool of itself (*a*). A filament of the liquid stream rebounds from the pool if the stream being poured has a sufficiently small diameter (*b*). [A. S. Lodge, *Elastic Liquids*, Academic Press, New York (1964), p. 251.]

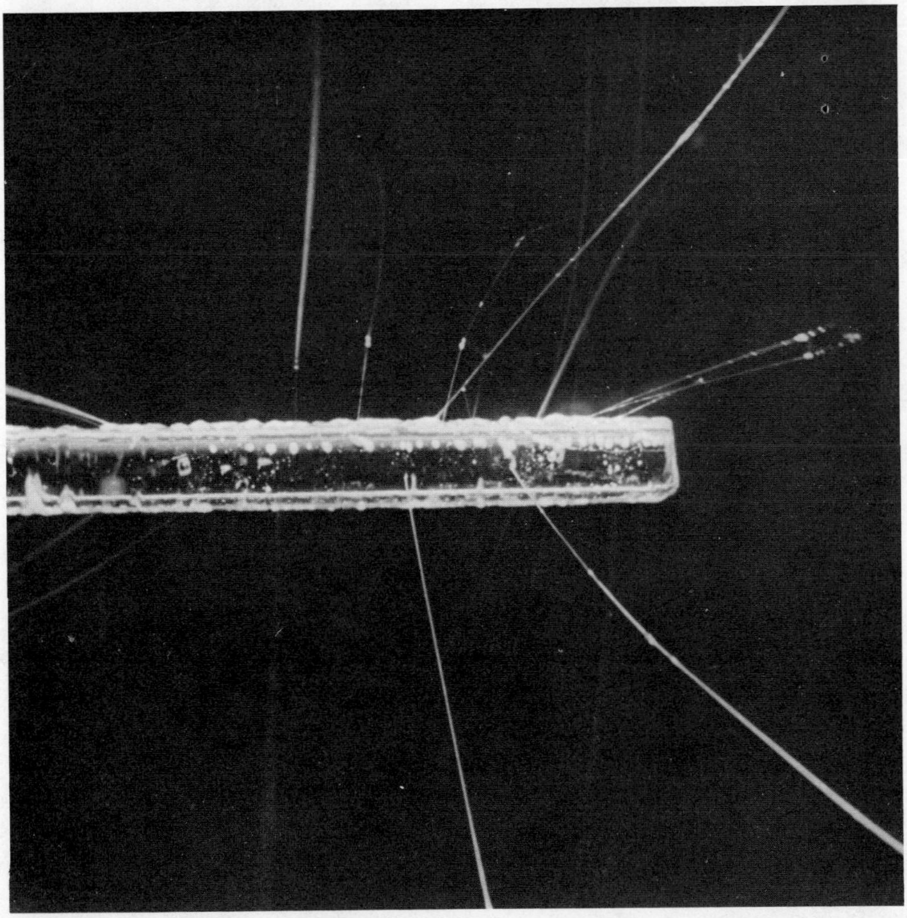

FIGURE 3Q.5. A solution of high molecular weight polyisobutylene in low molecular weight polyisobutylene forms fine, continuous streams emanating from a 6 mm diameter glass rod. [B. J. S. Barnard and W. G. Pritchard, *Nature*, *250*, 215–216 (1974).]

mean volume rate of flow for a Newtonian liquid? For moderate flow rates of an aqueous solution of polyacrylamide, it is found that a substantial increase in the mean volume flow rate results from the extra oscillatory pressure.[7] Why should this happen?

8. "Silly putty" will flow out into a flat puddle if left on a table top for an hour or so. Yet, if it is rolled up into a ball, it can be bounced off the floor like an ordinary rubber ball. Explain why silly putty can act like a liquid sometimes and like a solid at other times.
9. For the flow in the disk and cylinder apparatus, is it possible to identify the primary normal stress difference?

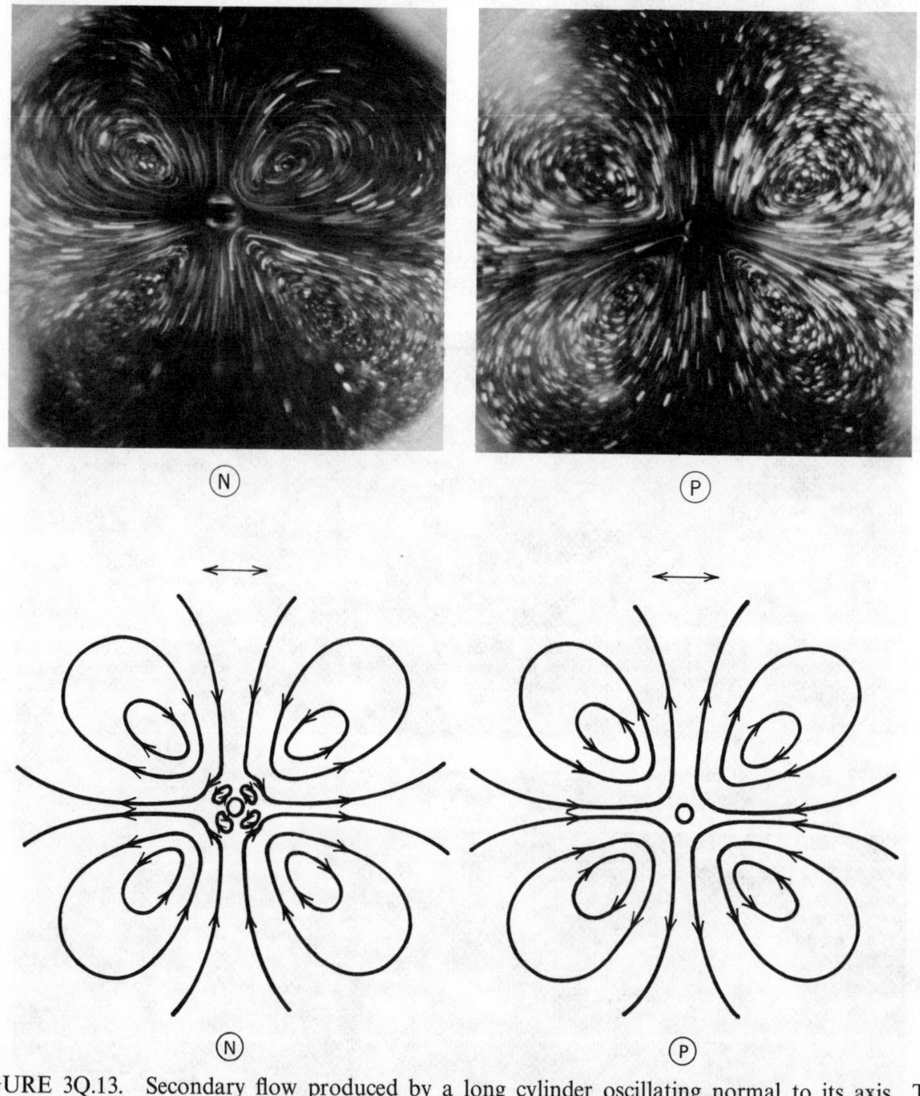

FIGURE 3Q.13. Secondary flow produced by a long cylinder oscillating normal to its axis. The cylinder is viewed on end and the direction of oscillation is shown by the double arrow. Ⓝ A water/glycerin mixture moves away from the cylinder along the axis of oscillation. Ⓟ The direction of the secondary flow is reversed when 100 ppm polyacrylamide (Separan AP 30) is added to the water/glycerin mixture. [C. Chang and W. R. Schowalter, *Nature*, *252*, 686–688 (1974).]

[7] H. A. Barnes, P. Townsend, and K. Walters, *Rheol. Acta*, *10*, 517–527 (1971).

10. The flow in the tubeless siphon is very closely approximated by an elongational flow. From Problem 1B.11, we know that in this type of flow a Newtonian fluid exhibits a normal force in the direction of stretching that is related to the elongation rate by a constant elongational viscosity $\bar{\eta}$, which is simply 3μ. Try and suggest qualitatively what form $\bar{\eta}$ might have for the polymer solution in the tubeless siphon experiment. Relate this to the spinnability of polymers.
11. In drag reduction studies, it has been found that the percentage drop in friction factor at a given Reynolds number decreases as the pipe diameter increases. What might account for this behavior?
12. Describe the primary and secondary flow patterns near a sphere rotating in a large bath of a Newtonian fluid. In what way would the velocity field be different if the fluid surrounding the sphere were like that shown in Fig. 3.6–1Ⓟ?
13. If a long, thin cylinder submerged in a Newtonian fluid is forced to oscillate normal to its axis, then a steady secondary flow (acoustic streaming) is produced. The bulk of the fluid in this secondary flow moves away from the cylinder along the axis of oscillation. If a little bit of polymer is added to the Newtonian fluid, the direction of the secondary flow is reversed; the polymer solution moves toward the cylinder along the axis of oscillation (see Fig. 3Q.13).[8] Discuss.
14. In Problem 1C.8 the oscillating cup and bob experiment (see Fig. 1C.8) was discussed for Newtonian fluids. There it was shown that the ratio of the amplitude of the induced oscillations of the inner cylinder to the amplitude of the forced oscillations of the outer cylinder has a maximum value of unity when the driving frequency is the natural frequency ω_n of the system (see Fig. 3Q.14). For polymeric materials it is found that the maximum amplitude ratio is greater than unity and occurs at frequencies higher than the natural frequency of the torsion pendulum.[9] Explain the difference.

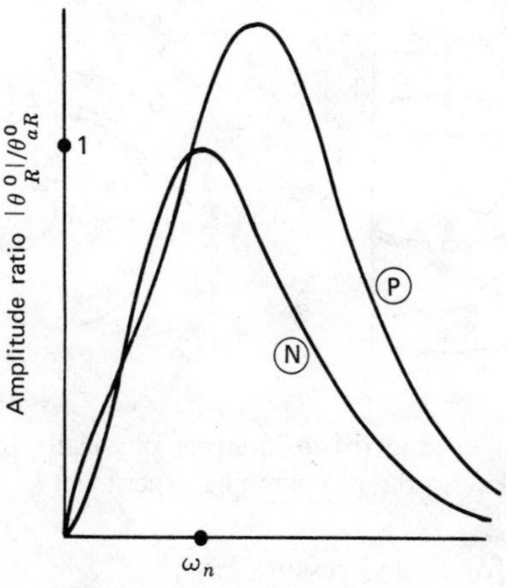

FIGURE 3Q.14. Amplitude ratio $|\theta_R{}^0|/\theta_{aR}{}^0$ in the cup-and-bob system of Problem 1C.8. The natural frequency of the system is ω_n. The cup undergoes forced oscillations with frequency ω. For Newtonian fluids Ⓝ the amplitude ratio has a maximum of unity at ω_n. For polymeric materials Ⓟ the maximum is raised and shifted to higher frequencies.

[8] C. Chang and W. R. Schowalter, *Nature*, *252*, 686–688 (1974); unfortunately Figs. 2b and 2c were interchanged in this publication.
[9] B. A. Toms, in *Rheology*, F. R. Eirich (Ed.), Academic Press, New York (1958), Vol. 2, Chapter 12, pp. 496–500.

PROBLEMS

3A.1 Comparison of Viscosity Determinations

The characteristics of two fluids, A and B, have been investigated by (1) measuring the terminal velocity of a sphere falling in a large bath of each fluid and by (2) measuring the volume flow rate of each at a specified pressure gradient. In (1) a 0.5 cm diameter sphere weighing 0.104 g fell with a terminal velocity of 0.75 cm/s in both fluids. However, in (2), fluids A and B had volumetric flow rates of 3.061×10^{-2} and 0.3455 cm^3/s, respectively, at an imposed pressure drop of 6.857×10^4 N/m^3 in a 0.210 cm diameter tube. The density of A is 1.0 g/cm^3, and the density of B is 1.014 g/cm^3. Based on these results which of the two fluids is definitely *not* Newtonian?

3B.1 Interpretation of the Primary Normal Stress Difference in Tangential Annular Flow[1]

a. In the limit that inertial forces are negligible, show that the equation of motion for tangential annular flow leads to:

$$\frac{d}{dr}(p + \tau_{rr}) = \frac{\tau_{\theta\theta} - \tau_{rr}}{r} \tag{3B.1-1}$$

b. Now consider the static forces that are present when a thin string wrapped around a cylinder of radius R is under tension T (see figure). By making a force balance on a very short isolated segment of string, show that the reaction of the cylinder is a radial force of magnitude F_r given by:

$$\left(\frac{F_r}{l}\right) = \frac{T}{R} \quad \left(\frac{l}{R} \ll 1\right) \tag{3B.1-2}$$

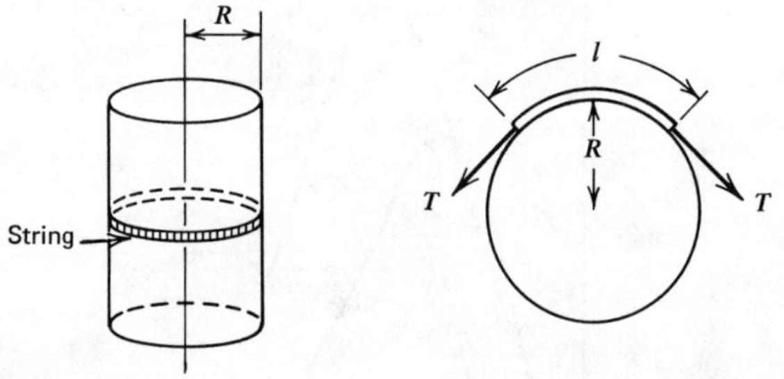

c. Use the results from (a) and (b) to interpret physically the normal stresses in tangential annular flow. How does this relate to the rod-climbing experiment?

3B.2 Simplified Results for Hole Pressure Error

For many polymeric liquids the normal stress differences can be expressed as powers of the shear stress:[2]

$$\tau_{11} - \tau_{22} = -\kappa_1 |\tau_{21}|^{n_1} \tag{3B.2-1}$$

$$\tau_{22} - \tau_{33} = \kappa_2 |\tau_{21}|^{n_2} \tag{3B.2-2}$$

in which κ_1, κ_2, n_1 and n_2 are all constants. Show that for fluids in which these relations hold, the

[1] This problem was suggested by A. S. Lodge, *Elastic Liquids*, Academic Press, New York (1964), p. 191.
[2] K. Higashitani and W. G. Pritchard, Trans. Soc. Rheol., *16*, 687–696 (1972).

hole pressure error results, Eqs. 3.4–1 to 3, simplify to

Circular hole:
$$p_H = \tfrac{1}{3}\left(\frac{\tau_{11} - \tau_{22}}{n_1} - \frac{\tau_{22} - \tau_{33}}{n_2}\right)_w \tag{3B.2–3}$$

Transverse slot:
$$p_H = \frac{1}{2n_1}(\tau_{11} - \tau_{22})_w \tag{3B.2–4}$$

Axial slot:
$$p_H = -\frac{1}{n_2}(\tau_{22} - \tau_{33})_w \tag{3B.2–5}$$

where the subscript w means "evaluated at the wall". One fluid model found in the literature, the second order fluid, gives a value of 2 for both n_1 and n_2. The results above agree with derivations of the hole pressure error made with the second order fluid model assumed from the beginning[3] (see Example 8.4–2).

3B.3 Annular Flow Experiments with Newtonian Fluids

Show that for the annular flow experiments considered in Examples 3.2–1 and 3.3–1, the normal stresses are all zero for Newtonian fluids. Explain why this leads to the pressure readings shown in Figs. 3.2–2Ⓝ and 3.3–1Ⓝ.

3B.4 Free Surface Shape in a Tilted Trough[4]

It is desired to obtain an expression for the free surface shape of a fluid flowing down a tilted trough of semicircular cross section that is inclined at an angle β to the horizontal. To do this, parallel the development in Example 3.7–1.

a. First, consider steady flow in a tube of radius R inclined at an angle β. From the symmetry of the velocity field it can be shown that $\tau_{z\theta} = \tau_{r\theta} = 0$ for this flow (see §4.2). Combine this information with the equations of motion to show that the variation of $\pi_{\theta\theta}$ within the tube is given by:

$$d\pi_{\theta\theta} = -\rho g \cos\beta \cos\theta \, r d\theta - \left(\frac{d}{dr}N_2 + \frac{N_2}{r} + \rho g \cos\beta \sin\theta\right) dr + \left(\frac{\partial p}{\partial z}\right) dz \tag{3B.4–1}$$

where $N_2 = \tau_{rr} - \tau_{\theta\theta}$ is the secondary normal stress difference and we have taken $\theta = 0$ to be horizontal.

b. Now imagine that the upper half of the tube and fluid is removed so that we are left with flow down a semicircular trough. The velocity profile in the trough will be identical to that in the lower half of the tube provided the normal stress distribution given in Eq. 3B.4–1 is maintained on the surface $\theta = 0$. Explain.

c. In order to describe the free surface it is convenient to consider a Cartesian coordinate system oriented similarly to the one in Fig. 3.7–2: the z-axis is coincident with the axis of the trough, x lies in the horizontal plane, and y is positive in the upward direction. Show that the normal stress π_{yy} is given by

$$\pi_{yy}(x,y) = -\rho g y \cos\beta + \pi_{yy}(x,0) + C$$

$$\pi_{yy}(x,0) = -\int_0^x \frac{N_2}{x}\,dx - N_2(x) \tag{3B.4–2}$$

where C is an integration constant.

[3] R. I. Tanner and A. C. Pipkin, *Trans. Soc. Rheol.*, *13*, 471–484 (1969) treat the transverse slot problem and E. A. Kearsley, *Trans. Soc. Rheol.*, *14*, 419–424 (1970) analyzes flow along an axial slot.
[4] A. C. Pipkin and R. I. Tanner, *Mechanics Today*, *1*, 262–321 (1972); R. I. Tanner, *Trans. Soc. Rheol.*, 483–507 (1970). L. Sturges and D. D. Joseph, *Arch. Rat. Mech. Anal.*, *59*, 359–387 (1976) include surface tension in their analysis and also consider the onset of secondary flow with increasing β.

d. Show that the total bulge in the free surface of the fluid is

$$h = -\frac{\pi_{yy}(R,0)}{\rho g \cos \beta} \tag{3B.4-3}$$

3C.1 Qualitative Analysis of Rod Climbing[1]

In Example 3.2–1 we analyzed tangential annular flow to determine the cause of different pressure readings at pressure taps located on the inner and outer cylinders for typical Newtonian and polymeric fluids (Fig. 3.2–2). For that purpose we examined the radial distribution of the total radial normal stress $\pi_{rr}(r)$. Here we want to look at the variation of the total axial normal stress $\pi_{zz}(r)$ with r, and from this result we wish to find whether or not the fluid under consideration will exhibit rod climbing.

a. Consider the idealized tangential annular flow discussed in Example 3.2–1; the inner cylinder is rotating, and both cylinders are so long that end effects can be ignored. The angular velocity of the rotating cylinder is so small that the centrifugal force term may be neglected in the equation of motion. Beginning with the identity

$$\pi_{zz} = \pi_{rr} - (\pi_{rr} - \pi_{zz}) \tag{3C.1-1}$$

use the r-component of the equation of motion to show:

$$\frac{d}{dr}\pi_{zz} = \frac{\pi_{\theta\theta} - \pi_{rr}}{r} - \frac{d}{dr}(\pi_{rr} - \pi_{zz}) \tag{3C.1-2}$$

b. Now use the θ-component of the equation of motion to rewrite Eq. 3C.1–2; thus

$$\frac{d}{dr}\pi_{zz} = \frac{\tau_{\theta\theta} - \tau_{rr}}{r} + \frac{2\dot{\gamma}}{r\left(\dfrac{d \ln \tau_{r\theta}}{d \ln \dot{\gamma}}\right)}\frac{d}{d\dot{\gamma}}(\tau_{rr} - \tau_{zz}) \tag{3C.1-3}$$

where the shear rate $\dot{\gamma} = -rd(v_\theta/r)/dr$ is positive for a rotating inner cylinder.

c. Equation 3C.1–3 describes the radial variation in π_{zz} necessary to have the assumed tangential annular flow. If the surface were flat, then π_{zz} would be equal to atmospheric pressure everywhere on the free surface. If according to Eq. 3C.1–3, $(d/dr)\pi_{zz} < 0$, then the fluid would be expected to climb the inner cylinder in order to provide a proper axial pressure distribution down in the bulk of the fluid. Conversely, one expects the fluid to dip at the rod if $d\pi_{zz}/dr > 0$. Note that according to this result, rod climbing will depend on both normal stress differences and the shear stress. For a fluid obeying the Weissenberg hypothesis, what sign for the primary normal stress difference corresponds to rod climbing?

d. Plots of $\tau_{11} - \tau_{22}$, $\tau_{22} - \tau_{33}$, and τ_{21} as functions of shear rate $\dot{\gamma}$ are given in Figs. 4.3–3, 6, and 8 for a polyacrylamide solution. The quantities η, Ψ_1, and Ψ_2 are defined in terms of the stresses in Eqs. 4.3–1, 2, and 3. Based on these data, use Eq. 3C.1–3 to show that the polyacrylamide will climb the rod. Recall that Fig. 3.2–1Ⓟ is a photograph of a polyacrylamide solution.

3C.2 Free Surface Shape of a Newtonian Fluid in Tangential Annular Flow

Consider the experimental geometry used in the rod-climbing experiment. Find an expression for the free surface shape of a Newtonian fluid undergoing tangential annular flow (see Fig. 3.2–2Ⓝ).

a. Postulate that for the system under consideration: $v_\theta = v_\theta(r)$, $v_r = 0$, $v_z = 0$, and $p = p(r,z)$. Then show that the free surface is described by:

$$h(R) - h(r) = \frac{1}{2g}\left(\frac{\kappa^2 R W}{1 - \kappa^2}\right)^2\left[\left(\frac{R}{r}\right)^2 - \left(\frac{r}{R}\right)^2 - 4 \ln\left(\frac{R}{r}\right)\right] \tag{3C.2-1}$$

[1] A. S. Lodge, *Elastic Liquids*, Academic Press, New York (1964), pp. 192–194.

in which h is the height of the surface, W is the angular velocity of the shaft, R is the radius of the cylindrical container, and κR is the radius of the rotating shaft.

b. What are the limitations on the result in (a)?

3D.1 Pressure Error in Flow over a Circular Hole[1]

A polymeric fluid is flowing in the gap between two large, horizontal, flat plates; in order to measure the pressure drop in the direction of flow, pressure holes have been drilled into the bottom plate, and a pressure transducer mounted at the bottom of each hole. Derive the expression for the hole pressure error p_H that is given in Eq. 3.4–1. To do this, focus on the flow near one hole (see Fig. 3D.1).

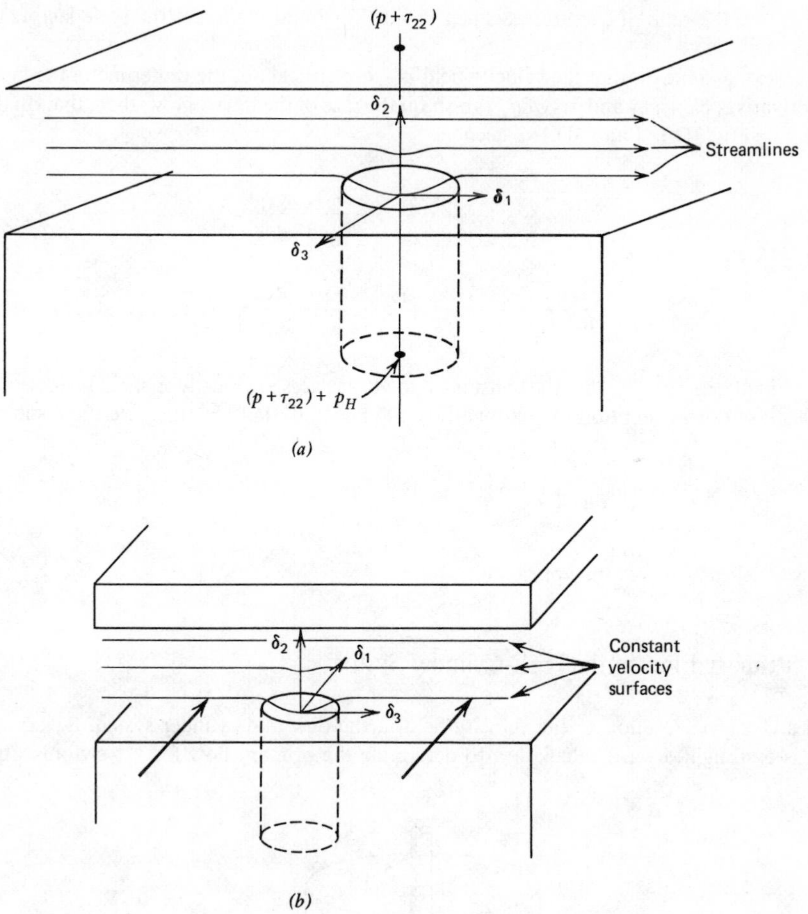

FIGURE 3D.1. Flow over a hole. Fluid streamlines are shown in (*a*), and constant velocity surfaces in (*b*).

a. Assume (i) that an orthogonal curvilinear coordinate system q_1, q_2, q_3 can be constructed, which is chosen locally as in Example 3.4–1, that is, $\boldsymbol{\delta}_1$ points in the direction of flow, $\boldsymbol{\delta}_3$ is tangent to a surface of constant velocity and perpendicular to $\boldsymbol{\delta}_1$, and $\boldsymbol{\delta}_2$ is chosen to complete the usual right-handed triad; also, we take $\tau_{13} = \tau_{31} = \tau_{23} = \tau_{32} = 0$ in this coordinate system.[2] These seem to be reasonable assumptions near the centerline of the hole. By further taking the flow to be slow and by neglecting

[1] K. Higashitani and W. R. Pritchard, *Trans. Soc. Rheol.*, *16*, 687–696 (1972).

[2] Assuming these two properties corresponds to assuming the flow is a unidirectional shearing flow as we shall see in §§4.1 and 2.

gravity, show that the equation of motion gives

$$q_1\text{-}component: \quad \frac{1}{h_1}\frac{\partial \pi_{11}}{\partial q_1} + \frac{1}{h_2}\frac{\partial \tau_{21}}{\partial q_2} + \frac{2\tau_{21}}{\rho_{21}} + \frac{\tau_{21}}{\rho_{23}} + \frac{\pi_{11} - \pi_{33}}{\rho_{13}} + \frac{\pi_{11} - \pi_{22}}{\rho_{12}} = 0 \tag{3D.1–1}$$

$$q_2\text{-}component: \quad \frac{1}{h_2}\frac{\partial \pi_{22}}{\partial q_2} + \frac{1}{h_1}\frac{\partial \tau_{21}}{\partial q_1} + \frac{2\tau_{21}}{\rho_{12}} + \frac{\tau_{21}}{\rho_{13}} + \frac{\pi_{22} - \pi_{11}}{\rho_{21}} + \frac{\pi_{22} - \pi_{33}}{\rho_{23}} = 0 \tag{3D.1–2}$$

$$q_3\text{-}component: \quad \frac{1}{h_3}\frac{\partial \pi_{33}}{\partial q_3} + \frac{\pi_{33} - \pi_{11}}{\rho_{31}} + \frac{\pi_{33} - \pi_{22}}{\rho_{32}} = 0 \tag{3D.1–3}$$

where the ρ_{ij} are the radii of curvature defined in Eq. 3.4–7 and the h_α are the scale factors defined in Eq. A.7–15.

b. Next, assume (ii) that the velocity field is symmetric about the centerline of the hole and (iii) that the derivatives $\partial\pi_{11}/\partial q_1$ and $\partial\tau_{21}/\partial q_1$ vanish on the axis of the hole. Show, then, that on the centerline of the hole Eqs. 3D.1–1 and 3D.1–2 become:

$$\frac{1}{h_2}\frac{\partial \tau_{21}}{\partial q_2} + \frac{2\tau_{21}}{\rho_{21}} + \frac{\tau_{21}}{\rho_{23}} = 0 \tag{3D.1–4}$$

$$\frac{1}{h_2}\frac{\partial \pi_{22}}{\partial q_2} + \frac{\pi_{22} - \pi_{11}}{\rho_{21}} + \frac{\pi_{22} - \pi_{33}}{\rho_{23}} = 0 \tag{3D.1–5}$$

c. Finally, assume (iv) that the constant velocity surfaces are locally spherical near the centerline of the hole. Use this assumption to combine Eqs. 3D.1–4 and 3D.1–5. Integrate the result to obtain:

$$p_H = \tfrac{1}{3}\int_0^{\tau_w} \frac{(\tau_{11} - \tau_{22}) - (\tau_{22} - \tau_{33})}{\tau_{21}}\, d\tau_{21} \tag{3D.1–6}$$

where τ_w is the shear stress at the upper wall.

3D.2 Primary Flow in the Disk-Cylinder System

Although a complete solution to the velocity field in the disk and cylinder system is very difficult to obtain, it is a straightforward calculation to determine the primary flow for a Newtonian fluid. This

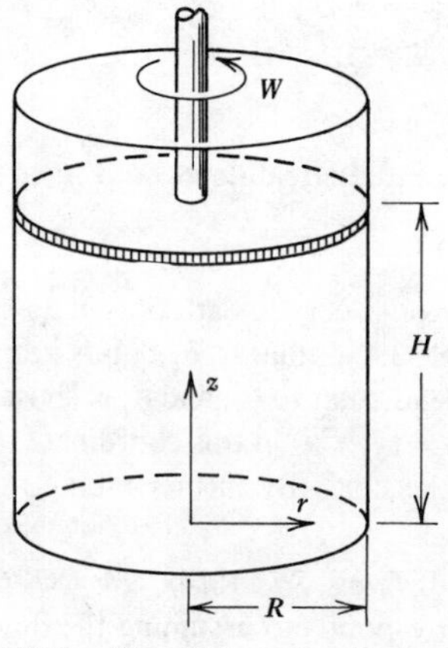

FIGURE 3D.2. The disk and cylinder apparatus.

solution provides a starting point for the analysis of the complete flow pattern for a non-Newtonian fluid.[3]

To a first approximation, the fluid contained in a cylindrical tank, the top of which is rotating with angular velocity W (Fig. 3D.2), can be assumed to be moving in the tangential direction only. Thus, postulate that the primary flow is of the form $v_\theta = v_\theta(r,z)$, $v_r = v_z = 0$. Show that for a container of height H holding a Newtonian fluid the primary flow is described by:

$$v_\theta(\xi,\zeta) = 2RW \sum_{n=1}^{\infty} \left[\frac{J_1(c_n\xi)}{c_n J_2(c_n)} \frac{\sinh(c_n\zeta\lambda)}{\sinh(c_n\lambda)} \right] \tag{3D.2–1}$$

where $\xi = r/R$, $\zeta = z/H$, and $\lambda = H/R$. The constants c_n are solutions to $J_1(c_n) = 0$ for $n = 1, 2, 3. \ldots$

[3] For details on how to proceed from the Newtonian solution see J. M. Kramer and M. W. Johnson, Jr., *Trans. Soc. Rheol.*, *16*, 197–212 (1972). Equation 3D. 2–1 was first given by W. Hort, *Z. Tech. Phys.*, *10*, 213–221 (1920).

CHAPTER 4
MATERIAL FUNCTIONS FOR POLYMERIC FLUIDS

Chapter 3 has demonstrated rather dramatically that Newton's law of viscosity is wholly inadequate for the description of macromolecular fluid flow. In this chapter we begin the task of discovering the proper way to describe these non-Newtonian fluids. The first step is the quantitative experimental characterization of their flow behavior. Recall that incompressible Newtonian fluids at constant temperature can be characterized by just two *material constants*: the density ρ and the viscosity μ. Once these quantities have been measured, the velocity distribution and the stresses in the fluid can, in principle, be found for any flow system. There are many steady-state and unsteady-state experiments from which μ can be determined.[1]

The experimental description of incompressible non-Newtonian fluids, on the other hand, is much more complicated. We can, of course, measure the density, but since we have no equation for τ analogous to the Newtonian expression Eq. 1.2–2, we do not know what other property or properties need to be measured. From the "fun experiments" that were presented in the last chapter, we know a little about what to expect. For example, if we try to measure a "viscosity" using a viscometer, this "viscosity" will not be a constant. Furthermore, we know from analyses of annular flow and hole pressure error experiments that normal stresses should also be measured. The recoil experiment in §3.8 indicates that unsteady-state experiments might also lead to additional information.

The point of the above discussion is twofold: first, whereas different isothermal experiments on a Newtonian fluid yield a single material property, namely the viscosity, a variety of experiments performed on a polymeric liquid will yield a host of material properties. Second, in contrast to the Newtonian fluid material constants, experiments on macromolecular fluids lead to *material functions* that depend on shear rate, frequency, time, and so on. These material functions serve to classify fluids, and they can be used to determine constants in specific rheological models. In this chapter, then, we shall present the standard flow patterns used in characterizing polymeric liquids and discuss the material functions that can be obtained from each. Representative fluid behavior will also be indicated by means of sample experimental data. Once the reader understands the kind of fluid responses characteristic of polymeric materials in various experiments, the rheological models developed and used in the rest of the book will be more meaningful.

The first part of this chapter is devoted to shear flows. In §4.1 we discuss the kinematics and classifications of shear flows. This allows us to define a shear flow precisely and to consider the variety of flow patterns encompassed by this definition. Next in §4.2 we look

[1] A rather complete listing of standard techniques is given in J. R. Van Wazer, J. W. Lyons, K. Y. Kim, and R. E. Colwell, *Viscosity and Flow Measurement*, Wiley, New York (1963).

at the form of the stress tensor for unidirectional shear flows. This leads naturally to the definitions of material functions for these flows that are presented along with sample experimental data in §§4.3 and 4. Methods for extracting the material functions from experimentally measurable quantities are the subject of §4.5. In §4.6 a short discussion is given of shearfree flows. And finally, in §4.7 we present some of the most important experimental findings about the concentration, temperature, and molecular weight dependence of the material functions.

The reader will notice that considerable space is devoted to shear flows. This is done for two reasons: first, there are many flows of practical importance (e.g., flow through tubes, slits, dies, and annuli) that are either shear flows or can be closely approximated as shear flows; second, shear flows have received a preponderance of attention, with a relative neglect of other types of flow. However, problems associated with fiber spinning, film blowing, foaming operations, and bubble growth have indicated that the shear flow material functions are not always the crucial properties; this realization has led in recent years to the study of shearfree flows and, in particular, elongational flows. There are clearly other flow patterns in which it might be useful to characterize the behavior of polymeric liquids. The squeezing flows between two parallel disks and flow in the rolling ball viscometer are two such examples. These were discussed in Chapter 1 for Newtonian fluids, but they are more difficult to analyze for macromolecular fluids.

This chapter is not intended to be a discussion of experimental techniques. Information on that aspect of material functions can be found elsewhere.[2]

§4.1 SHEAR FLOWS: KINEMATICS AND CLASSIFICATION

In this section we define a class of flows known as *shear flows* and catalog some of the special types of shear flows to which we will need to refer later. In Table 4.1–1 we tabulate all of the various categories of shear flows and give an example of each. This table should give some perspective to the definitions that follow. Steady simple shear flow is briefly discussed to introduce some of the kinematical concepts.

Consider the flow of a fluid contained between two parallel plates, the upper one of which moves with a constant velocity v_0 as shown in Fig. 4.1–1. The velocity profile in the gap is given by

$$v_x = \dot{\gamma} y \qquad v_y = 0 \qquad v_z = 0 \tag{4.1–1}$$

where the velocity gradient $\dot{\gamma}$ is constant. This flow is known as *steady simple shear flow*; the quantity $\dot{\gamma}$ is called the shear rate. A number of characteristics[1] of the flow described by Eq. 4.1–1 are:

(a) Fluid planes that are parallel to the bounding plates move rigidly; since these planes consist of the same set of fluid particles at all times, we call them material planes or surfaces. These form a one-parameter family of material surfaces, y = constant, and each member is denoted as a *shearing plane*. Note that the distance between any two particles in a shearing plane is constant.

(b) The volume of every material element remains constant throughout the flow. The constant volume condition is a requirement for a shear flow independent of the assumption of incompressibility used in this book.

[2] K. Walters, *Rheometry*, Chapman and Hall, London (1975).

[1] A thorough discussion of shear and other flows in terms of the response of material particles, lines, and planes is given in A. S. Lodge, *Elastic Liquids*, Academic Press, New York (1964) and A. S. Lodge, *Body Tensor Fields in Continuum Mechanics*, Academic Press, New York (1974).

TABLE 4.1–1

Classification and Examples of Shear Flows

		Rheologically Unsteady:		Rheologically Steady:
General Shear Flow	N[a]	Tangential annular flow with the inner cylinder vibrating axially.		
	H[b]	Eccentric-disk (orthogonal) rheometer flow. See Problem 4B.1.		
Unidirectional Shear Flow	N	Oscillatory tangential annular flow.	N	Helical flow. See Table 4.1–2*e*.[c]
	H	Oscillatory cone and plate flow (approximate for small cone angles and high viscosity fluids). See Table 4.1–2*d*.	H	Tangential annular flow (approximate for narrow gaps and high viscosity fluids).[c]
Rectilinear Shear Flow	N	Pulsatile pressure-driven tube flow.	N	Pressure driven slit flow.[c]
	H	Wall-driven oscillatory flow between parallel plates (approximate for narrow gaps and high viscosity fluids).[d]	H	Wall-driven flow between parallel plates.[c,d]

[a] N = Nonhomogeneous
[b] H = Homogeneous
[c] These are called "viscometric flows."
[d] These are called "simple shear flows."

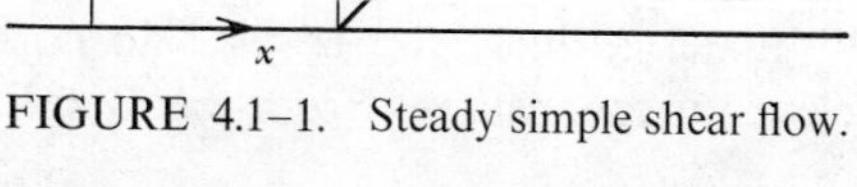

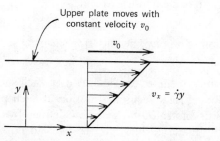

FIGURE 4.1–1. Steady simple shear flow.

(c) The direction of relative motion of the shearing planes is denoted by the unit vector along the x-coordinate axis $\boldsymbol{\delta}_x$. If at any instant we trace out a curve to which $\boldsymbol{\delta}_x$ is everywhere tangent, we generate a straight line that we call a *line of shear*. In the present flow all of the lines of shear are parallel to the x-axis, and they are coincident with fluid particle pathlines. This latter property does not hold in all shear flows as we shall see in Example 4.1–1. In simple shear flow, the lines of shear are material lines.

(d) The shear rate $\dot{\gamma}$, which is equal to the relative sliding velocity of any two shearing planes divided by the distance between them, is a constant.

(e) Finally, in Cartesian coordinates the velocity gradient tensor has the form

$$\nabla \boldsymbol{v} = \dot{\gamma}\begin{pmatrix} 0 & 0 & 0 \\ 1 & 0 & 0 \\ 0 & 0 & 0 \end{pmatrix} = \boldsymbol{\delta}_y\boldsymbol{\delta}_x\dot{\gamma} \tag{4.1–2}$$

and, hence,

$$\dot{\boldsymbol{\gamma}} = \dot{\gamma}\begin{pmatrix} 0 & 1 & 0 \\ 1 & 0 & 0 \\ 0 & 0 & 0 \end{pmatrix} = (\boldsymbol{\delta}_x\boldsymbol{\delta}_y + \boldsymbol{\delta}_y\boldsymbol{\delta}_x)\dot{\gamma} \tag{4.1–3}$$

By using Eq. 4.1–3, it is not difficult to show that the shear rate $\dot{\gamma}$ is related to the second invariant of $\dot{\boldsymbol{\gamma}}$ (see Eq. A.3–21):

$$\dot{\gamma} = \sqrt{\tfrac{1}{2}\text{II}_{\dot{\gamma}}} = \sqrt{\tfrac{1}{2}(\dot{\boldsymbol{\gamma}} : \dot{\boldsymbol{\gamma}})} \tag{4.1–4}$$

One can take as a definition for steady simple shear flow, either properties (a) to (d) or property (e).

Now that we have discussed steady simple shear flow in some detail we are ready to give the definition of general shear flow: one that is not necessarily steady-state, rectilinear, or homogeneous and that does not have plane shearing surfaces. To this general definition, we then add a sequence of further requirements that define special kinds of shear flow; this specialization process will end at our original example, steady simple shear flow.

We define a *shear flow* to be a flow in which

(i) there is a one-parameter family of material surfaces, the *shearing surfaces*, which move isometrically, that is, the distance between any two neighboring particles in the surface is constant; and

(ii) the volume of every fluid element is constant.

As an alternative to (i) and (ii) one could use (i) and

(ii′) the separation of any two neighboring shearing surfaces is constant.

At any instant, a family of curves can be drawn on each shearing surface so that they are tangent to the direction of relative sliding motion at each particle. These curves are known as *lines of shear*. For an arbitrary shear flow defined by (i) and (ii) above, the lines of shear are not necessarily material lines, that is, they do not necessarily consist of the same set of material particles at every instant. From the material point of view, then, the lines of shear are in general functions of time, since they must be drawn differently on the shearing surfaces at different times.

We now define a type of shear flow for which the lines of shear are "constant." These are *unidirectional shear flows* and they have the following property in addition to (i) and (ii):

(iii) the lines of shear are material lines.

Unidirectional shear flow is the most frequently encountered kind of shear flow. We can now think of the lines of shear as being constant in the sense that if a line of shear is drawn on a shearing surface at any particular time, then the same material particles that lie along the curve at that time also fall on the same line of shear at any other time.

In order to describe mathematically the kinematics of shear flows, it is useful to introduce a coordinate system based on the shapes of the shearing surfaces. We construct then an orthogonal curvilinear coordinate system with unit vectors $\breve{\boldsymbol{\delta}}_1$, $\breve{\boldsymbol{\delta}}_2$, and $\breve{\boldsymbol{\delta}}_3$ at every particle so that at any time $\breve{\boldsymbol{\delta}}_1$ and $\breve{\boldsymbol{\delta}}_3$ are tangent to a shearing surface and $\breve{\boldsymbol{\delta}}_2$ is normal to the shearing surface. Furthermore we take the grid of $\breve{x}_1$- and $\breve{x}_3$- coordinate curves to be imprinted on the shearing surfaces so that these curves are made up of material particles. We shall refer to the $\breve{\boldsymbol{\delta}}_i$ set of axes as "shear axes"; notice that from a spatial point of view the shear axes may be functions of time as well as position. Since the lines of shear are material lines for unidirectional shear flow, we may choose the $\breve{x}_1$-curves to coincide with the lines of shear for this kind of shearing motion. We then call the direction denoted by $\breve{\boldsymbol{\delta}}_1$ "the direction of shear". Referred to the $\breve{\boldsymbol{\delta}}_1\breve{\boldsymbol{\delta}}_2\breve{\boldsymbol{\delta}}_3$-axes, the velocity gradient tensor at every particle has the following form in unidirectional shear flow[2]:

$$\nabla\breve{\mathbf{v}} = \dot{\gamma}(t)\begin{pmatrix} 0 & 0 & 0 \\ 1 & 0 & 0 \\ 0 & 0 & 0 \end{pmatrix} = \breve{\boldsymbol{\delta}}_2\breve{\boldsymbol{\delta}}_1\dot{\gamma}(t) \tag{4.1-5}$$

and the rate-of-strain tensor can be written:

$$\dot{\boldsymbol{\gamma}} = \dot{\gamma}(t)\begin{pmatrix} 0 & 1 & 0 \\ 1 & 0 & 0 \\ 0 & 0 & 0 \end{pmatrix} = (\breve{\boldsymbol{\delta}}_1\breve{\boldsymbol{\delta}}_2 + \breve{\boldsymbol{\delta}}_2\breve{\boldsymbol{\delta}}_1)\dot{\gamma}(t) \tag{4.1-6}$$

(See Problem 4C.4 for the corresponding form for $\dot{\boldsymbol{\gamma}}$ for arbitrary shear flows.) Note that $\dot{\gamma}(t)$ is related to the second invariant of the rate-of-strain tensor just as in steady simple shear flow; it can thus be computed using the definition of $\dot{\gamma}$ in Eq. 4.1–4, provided that $\dot{\boldsymbol{\gamma}}$ is known in *any* coordinate system. By comparing Eqs. 4.1–5 and 4.1–6 with Eqs. 4.1–2 and 4.1–3 we see that at every particle unidirectional shear flow is a simple shear flow when viewed from the particle shear axes.

The motion of a typical fluid element in unidirectional shear flow is depicted in Fig. 4.1–2. Note that the same fluid particles lie along $\breve{\boldsymbol{\delta}}_1$ and $\breve{\boldsymbol{\delta}}_3$ at times t and t'; whereas different fluid particles lie along $\breve{\boldsymbol{\delta}}_2$ at the two different times. From a particle point of view, different unidirectional shear flows are obtained by changing the time dependence of $\dot{\gamma}$ in Eq. 4.1–5. The various time-dependent unidirectional shear flow experiments used in rheology are illustrated in Fig. 4.3–1 by showing $\dot{\gamma}(t)$ for simple shear flow.

[2] For the use of matrices to display the components of a tensor, see §A.9.

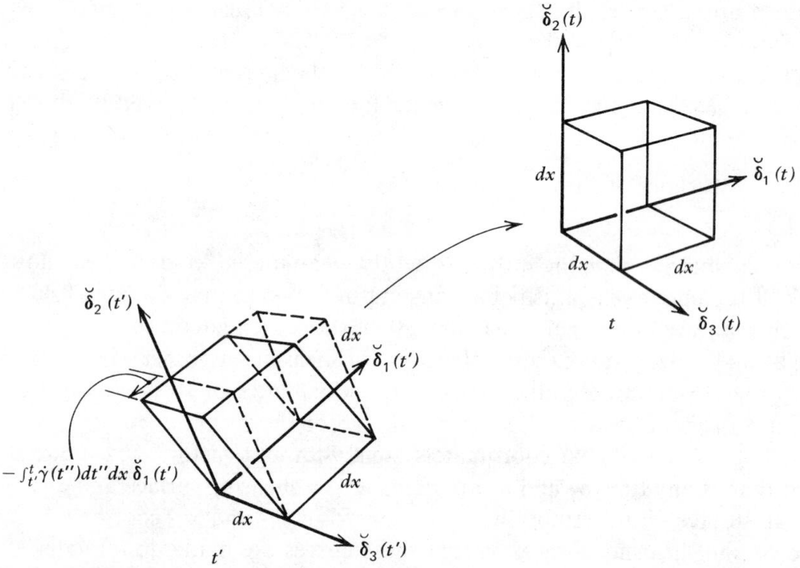

FIGURE 4.1–2. Motion of a typical fluid element of volume $(dx)^3$ undergoing unidirectional shear flow. The deformation that takes the element from its shape at some past time t' to its shape at the present time t is a simple shear flow with shear rate $\dot{\gamma}(t)$. The unit vectors $\breve{\boldsymbol{\delta}}_1$, $\breve{\boldsymbol{\delta}}_2$, $\breve{\boldsymbol{\delta}}_3$ specify the coordinate system from which the flow is seen as simple shear. The same fluid particles lie along $\breve{\boldsymbol{\delta}}_1$ and $\breve{\boldsymbol{\delta}}_3$ at times t' and t, but different fluid particles mark $\breve{\boldsymbol{\delta}}_2$ at the two times. The magnitude of shear between times t' and t is $\gamma = \int_{t'}^{t} \dot{\gamma}(t'')\, dt''$. For a viscometric flow, $\dot{\gamma}$ is a constant and $\gamma = (t - t')\dot{\gamma}$. The top and bottom faces of the cube are parts of adjacent shearing surfaces.

Next, we define *rheologically steady shear flow* or *viscometric flow*[3] as a unidirectional shear flow for which

(iv) the shear rate $\dot{\gamma}$ is independent of time at a given particle.

In view of Eq. 4.1–5, an alternative way[4] of defining viscometric flow requires that the flow history at any particle be one of steady simple shear as seen from the shear axes at the particle.[5] Note that this definition requires the flow to be steady following a particle.

[3] This definition of viscometric flow is not universal. It is the same definition used by B. D. Coleman, H. Markovitz, and W. Noll, *Viscometric Flows of Non-Newtonian Fluids*, Springer, New York (1966) and by Pipkin and Tanner (see footnote 5). Some rheologists use the word viscometric to stand for a unidirectional shear flow.

[4] For future reference the finite strain tensor $\boldsymbol{\gamma}^{[0]}(t,t')$, as defined in §9.2, has the form

$$\boldsymbol{\gamma}^{[0]} = \begin{pmatrix} 0 & -(t-t')\dot{\gamma} & 0 \\ -(t-t')\dot{\gamma} & (t-t')^2\dot{\gamma}^2 & 0 \\ 0 & 0 & 0 \end{pmatrix} \tag{4.1–6a}$$

when referred to the shear axes. Requiring $\boldsymbol{\gamma}^{[0]}$ to be given in this way is equivalent to both definitions of viscometric flow given in the text. The rate-of-strain tensor may be calculated from $\boldsymbol{\gamma}^{[0]}$ using the relation given in Table 9.1–2 that

$$\dot{\boldsymbol{\gamma}}(t) = \left(\frac{\partial}{\partial t'}\boldsymbol{\gamma}^{[0]}(t,t')\right)\bigg|_{t'=t} \tag{4.1–6b}$$

[5] This is the approach taken in the review article by A. C. Pipkin and R. I. Tanner, *Mechanics Today*, *1*, 262–321 (1972). We have drawn on this review in Example 4.1–1.

In rheologically complex fluids that exhibit memory effects (cf. §3.8), this definition of steady flow is preferable to the usual hydrodynamic one. We refer, then, to flows steady at a fluid particle as "rheologically steady." The shear rate $\dot{\gamma}$ at a particle is constant; however, $\dot{\gamma}$ is free to vary from particle to particle. If $\dot{\gamma}$ does not vary from particle to particle, then the flow is homogeneous as well. An example of a flow that is steady in the hydrodynamical sense that $\partial \boldsymbol{v}(\boldsymbol{r},t)/\partial t = 0$, but that is rheologically unsteady is eccentric-disk rheometer flow (see Problem 4B.1). Some examples of rheologically steady shear flows are steady-state tube flow, steady tangential annular flow, and steady axial annular flow.

The viscometric flows in Table 4.1–1 are specially marked. It is important to be able to identify these flows, as we shall later obtain a rheological equation of state that describes viscometric flows for a wide class of polymeric materials (see §8.5). In Example 4.1–1 we show that steady tube flow, tangential annular flow, and helical flow are viscometric flows. These and some other viscometric flows are listed in Table 4.1–2 along with some of their identifying characteristics.

Before proceeding to Example 4.1–1, we point out that *rectilinear shear flows* are defined by (i), (ii), (iii), and

(v) the fluid particle pathlines are straight lines.

TABLE 4.1–2

Examples of Viscometric Flow

Flow	Velocity Field	Shear Axes / Shear Rate	Shearing Surfaces / Lines of Shear
a. Steady tube flow 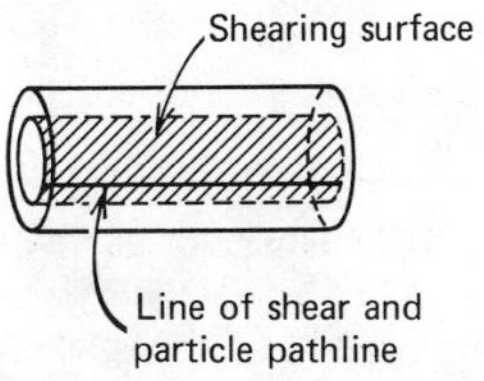 *(a)*	$v_z = v_z(r)$ $v_r = 0$ $v_\theta = 0$	$\breve{\boldsymbol{\delta}}_1 = \boldsymbol{\delta}_z$ $\breve{\boldsymbol{\delta}}_2 = \boldsymbol{\delta}_r$ $\breve{\boldsymbol{\delta}}_3 = \boldsymbol{\delta}_\theta$ $\dot{\gamma} = -\dfrac{d}{dr} v_z$	Concentric cylinders / straight lines parallel to the tube axis
b. Steady tangential annular flow 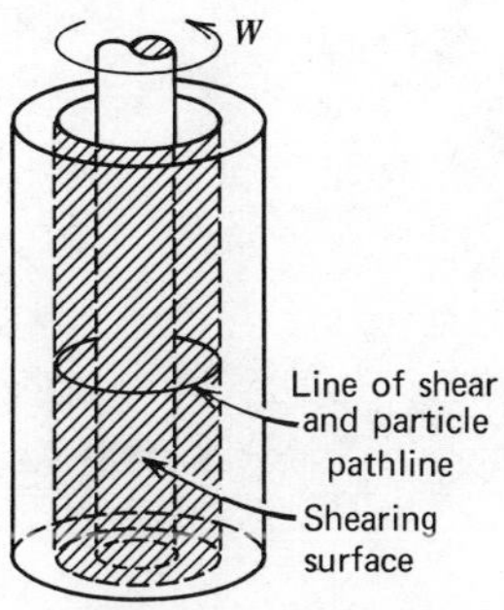 *(b)*	$v_\theta = v_\theta(r)$ $v_r = 0$ $v_z = 0$	$\breve{\boldsymbol{\delta}}_1 = \boldsymbol{\delta}_\theta$ $\breve{\boldsymbol{\delta}}_2 = \boldsymbol{\delta}_r$ $\breve{\boldsymbol{\delta}}_3 = -\boldsymbol{\delta}_z$ $\dot{\gamma} = -r\dfrac{d}{dr}\left(\dfrac{v_\theta}{r}\right)$	Concentric cylinders / circles of constant r and z

TABLE 4.1–2 (*Continued*)

Flow	Velocity Field	Shear Axes / Shear Rate	Shearing Surfaces / Lines of Shear
c. Steady torsional flow 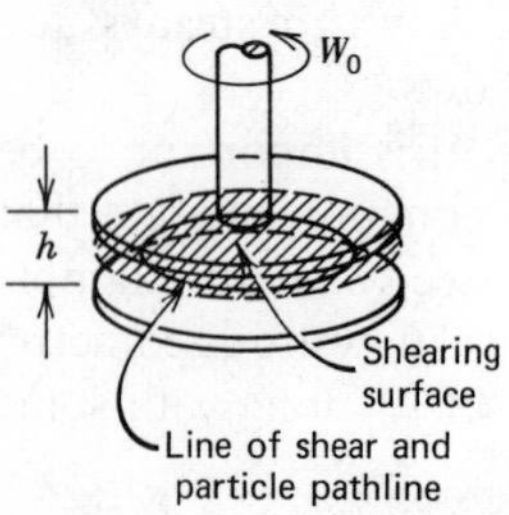 *(c)*	$v_\theta = \dfrac{rzW}{h}$ $v_r = 0$ $v_z = 0$	$\breve{\boldsymbol{\delta}}_1 = \boldsymbol{\delta}_\theta$ $\breve{\boldsymbol{\delta}}_2 = \boldsymbol{\delta}_z$ $\breve{\boldsymbol{\delta}}_3 = \boldsymbol{\delta}_r$ - - - - - $\dot{\gamma} = rW/h$	Parallel disks - - - - - circles of constant r and z
d. Steady cone and plate flow (small cone angle) 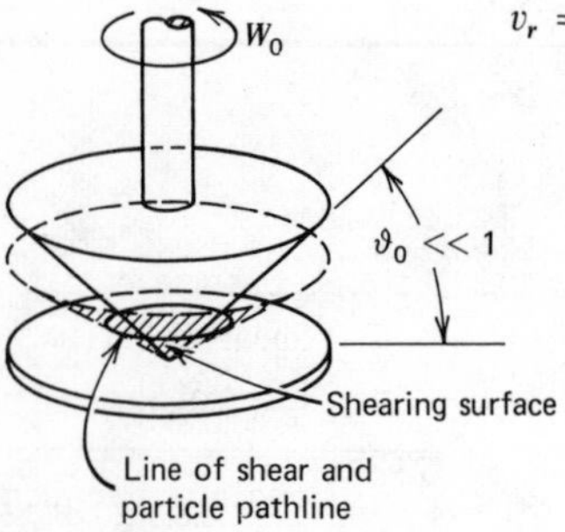 *(d)*	$v_\phi = v_\phi(r,\theta)$ $v_\theta = 0$ $v_r = 0$	$\breve{\boldsymbol{\delta}}_1 = \boldsymbol{\delta}_\phi$ $\breve{\boldsymbol{\delta}}_2 = -\boldsymbol{\delta}_\theta$ $\breve{\boldsymbol{\delta}}_3 = \boldsymbol{\delta}_r$ - - - - - $\dot{\gamma} = -\dfrac{1}{r}\dfrac{\partial v_\phi}{\partial \theta}$	Cones of constant θ - - - - - circles of constant r and z
e. Steady helical flow 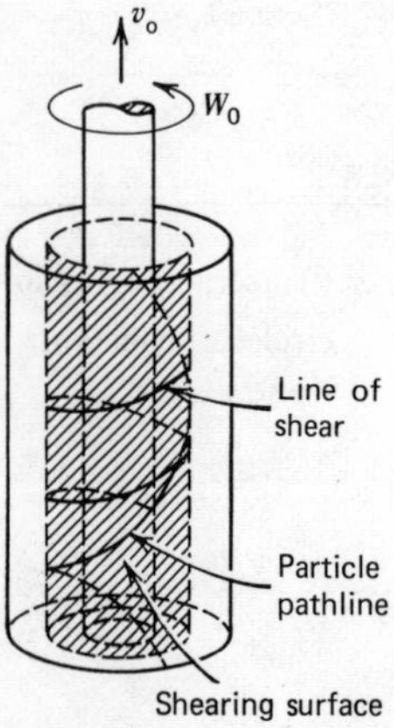 *(e)*	$v_\theta = v_\theta(r)$ $v_z = v_z(r)$ $v_r = 0$	$\breve{\boldsymbol{\delta}}_1 = \dfrac{1}{\dot{\gamma}}\left[\boldsymbol{\delta}_\theta r \dfrac{d}{dr}\left(\dfrac{v_\theta}{r}\right) + \boldsymbol{\delta}_z \dfrac{dv_z}{dr}\right]$ $\breve{\boldsymbol{\delta}}_2 = \boldsymbol{\delta}_r$ $\breve{\boldsymbol{\delta}}_3 = \dfrac{1}{\dot{\gamma}}\left[\boldsymbol{\delta}_\theta \dfrac{dv_z}{dr} - \boldsymbol{\delta}_z r \dfrac{d}{dr}\left(\dfrac{v_\theta}{r}\right)\right]$ - - - - - $\dot{\gamma} = +\sqrt{\left(r\dfrac{d}{dr}(v_\theta/r)\right)^2 + \left(\dfrac{dv_z}{dr}\right)^2}$	Concentric cylinders - - - - - helices

Finally, *steady simple shear flow* is a viscometric flow for which, in addition to (i), (ii), (iii), and (iv), we require:

(vi) The shearing surfaces are planes, that is, the shear axes $\breve{\boldsymbol{\delta}}_1$, $\breve{\boldsymbol{\delta}}_2$, $\breve{\boldsymbol{\delta}}_3$ are everywhere identical to the rectangular Cartesian axes $\boldsymbol{\delta}_x$, $\boldsymbol{\delta}_y$, $\boldsymbol{\delta}_z$.

(vii) The flow is homogeneous, that is, $\dot{\gamma}$ is independent of position.

EXAMPLE 4.1–1 Kinematics of Steady Tube Flow, Steady Tangential Annular Flow, and Steady Helical Flow

Consider the three fluid flows: (a) steady axial tube flow under a constant pressure gradient, (b) steady tangential annular flow between two concentric cylinders in relative rotation, and (c) steady helical flow between two coaxial cylinders, the inner one of which is rotating and is translating in the axial direction. In (a) the motion of the liquid will be such that $v_z = v_z(r)$, $v_r = 0$, and $v_\theta = 0$; in (b) the velocity field will have the form $v_\theta = v_\theta(r)$, $v_z = 0$, and $v_r = 0$; and in (c), $v_\theta = v_\theta(r)$, $v_z = v_z(r)$, and $v_r = 0$. Show that all of these flows are viscometric according to the definition given in the text.

SOLUTION (a) Steady Tube Flow

Since the fluid velocity is in the z-direction only and varies in the r-direction only, the flow can be pictured as the relative axial sliding motion of concentric cylinders of fluid. These rigid cylindrical material surfaces are then the shearing surfaces and perform a telescoping motion. The shear axes correspond to the cylindrical coordinate axes as follows: $\breve{\boldsymbol{\delta}}_1 = \boldsymbol{\delta}_z$, $\breve{\boldsymbol{\delta}}_2 = \boldsymbol{\delta}_r$ and $\breve{\boldsymbol{\delta}}_3 = \boldsymbol{\delta}_\theta$. It is easy to see that the lines of shear are parallel to the z-axis and are also fluid pathlines. Referred to the shear axes, the rate-of-strain tensor is

$$\dot{\boldsymbol{\gamma}} = \dot{\gamma}_{21}(r)\begin{pmatrix} 0 & 1 & 0 \\ 1 & 0 & 0 \\ 0 & 0 & 0 \end{pmatrix} \qquad (4.1\text{–}7)$$

where $\dot{\gamma}_{21}(r) = dv_z(r)/dr = -\dot{\gamma}$ is constant for each material particle since each particle moves at constant r. Thus the flow history at every particle is a steady simple shearing motion and the flow is viscometric.

(b) Steady Tangential Annular Flow

Once again the shearing surfaces are concentric cylinders. At any time t, the shear axes at a particle P are related to the cylindrical coordinate axes at the position of the particle as follows: $\breve{\boldsymbol{\delta}}_1 = \boldsymbol{\delta}_\theta$, $\breve{\boldsymbol{\delta}}_2 = \boldsymbol{\delta}_r$, and $\breve{\boldsymbol{\delta}}_3 = -\boldsymbol{\delta}_z$. In order to compute the velocity gradient tensor relative to the shear axes at P, we note that a point fixed relative to the shear axes has velocity $(v_\theta/r)_P r$ referred to the space-fixed cylindrical coordinate system. Here $(v_\theta/r)_P$ is the constant angular velocity of P as it moves in its circular trajectory. Thus when viewed from the shear axes, the velocity field is $\breve{v}_1 = v_\theta - (v_\theta/r)_P r$, $\breve{v}_2 = \breve{v}_3 = 0$. Since at any instant the shear coordinate system is coincident with the cylindrical one, we can use the relations for the components of $\nabla \boldsymbol{v}$ given in Table A.7–2 to find:

$$\nabla\breve{\boldsymbol{v}} = \begin{bmatrix} 0 & -\dfrac{\breve{v}_1}{r} & 0 \\ \dfrac{d\breve{v}_1}{dr} & 0 & 0 \\ 0 & 0 & 0 \end{bmatrix} \qquad (4.1\text{–}8)$$

(In this matrix the entries are given in the order θ, r, z.) But

$$\frac{d\breve{v}_1}{dr} = \frac{d}{dr}\left(v_\theta - \left(\frac{v_\theta}{r}\right)_P r\right) = \frac{dv_\theta}{dr} - \left(\frac{v_\theta}{r}\right)_P$$

$$= r\frac{d}{dr}\left(\frac{v_\theta}{r}\right) + \frac{v_\theta}{r} - \left(\frac{v_\theta}{r}\right)_P \tag{4.1-9}$$

When Eq. 4.1–9 is substituted into Eq. 4.1–8 and the velocity gradient tensor evaluated at the particle P, we find:

$$(\nabla\breve{\mathbf{v}})|_P = \dot{\gamma}_{21}(r)\begin{pmatrix} 0 & 0 & 0 \\ 1 & 0 & 0 \\ 0 & 0 & 0 \end{pmatrix} \tag{4.1-10}$$

where $\dot{\gamma}_{21} = r\, d(v_\theta/r)/dr$ is the constant shear rate at the particle and is equal to $-\dot{\gamma}$ if the inner cylinder is rotating and the outer one fixed. The flow at any particle P is thus steady simple shear as seen from the shear axes; according to our definition, then, tangential annular flow is a viscometric flow.

(c) Steady Helical Flow

In this flow the direction of shear is not so obvious as in the two preceding parts. Hence, in this part we show how the shear axes may be determined in the process of writing the velocity gradient tensor in the form appropriate to a steady simple shear flow. Again we compute the velocity gradient with respect to shear axes at an arbitrary particle P.

Because the shearing surfaces are concentric cylinders, at any time t the unit vector $\breve{\boldsymbol{\delta}}_2$ at P is coincident with $\boldsymbol{\delta}_r$. Furthermore, the unit vectors $\breve{\boldsymbol{\delta}}_1$ and $\breve{\boldsymbol{\delta}}_3$ are tangent to the cylindrical surfaces so that at any time $\boldsymbol{\delta}_\theta$ and $\boldsymbol{\delta}_z$ can be expressed in terms of them:

$$\boldsymbol{\delta}_\theta = \breve{\boldsymbol{\delta}}_1 a + \breve{\boldsymbol{\delta}}_3 b$$

$$\boldsymbol{\delta}_z = \breve{\boldsymbol{\delta}}_1 b - \breve{\boldsymbol{\delta}}_3 a \tag{4.1-11}$$

where a and b depend on the details of the flow field and $\sqrt{a^2 + b^2} = 1$.

We now note that a point, which is stationary as seen from the shear axes at P, has velocity $\mathbf{v} = \boldsymbol{\delta}_\theta (v_\theta/r)_P r + \boldsymbol{\delta}_z (v_z)_P$ relative to the space-fixed cylindrical coordinate system. Thus the velocity $\breve{\mathbf{v}}$ of any fluid element at time t relative to the shear axes at P may be written in terms of the cylindrical coordinate unit vectors:

$$\breve{\mathbf{v}} = \boldsymbol{\delta}_\theta\left[v_\theta - \left(\frac{v_\theta}{r}\right)_P r\right] + \boldsymbol{\delta}_z[v_z - (v_z)_P] \tag{4.1-12}$$

The velocity gradient relative to the shear axes is readily computed in the cylindrical coordinate system by using Table A.7–2:

$$\nabla\breve{\mathbf{v}} = \boldsymbol{\delta}_r\boldsymbol{\delta}_\theta \frac{d}{dr}\left[v_\theta - \left(\frac{v_\theta}{r}\right)_P r\right] + \boldsymbol{\delta}_r\boldsymbol{\delta}_z \frac{d}{dr}[v_z - (v_z)_P] + \boldsymbol{\delta}_\theta\boldsymbol{\delta}_r\left[-\frac{v_\theta}{r} + \left(\frac{v_\theta}{r}\right)_P\right] \tag{4.1-13}$$

When this is evaluated at the particle P we find

$$(\nabla\breve{\mathbf{v}})|_P = \boldsymbol{\delta}_r\boldsymbol{\delta}_\theta r\frac{d}{dr}\left(\frac{v_\theta}{r}\right) + \boldsymbol{\delta}_r\boldsymbol{\delta}_z\frac{dv_z}{dr}$$

$$= \breve{\boldsymbol{\delta}}_2\breve{\boldsymbol{\delta}}_1\dot{\gamma} \tag{4.1-14}$$

where the last line above emphasizes that when the velocity gradient is written in terms of the $\breve{\boldsymbol{\delta}}_i$, it must have the simple shear flow form. By substituting Eqs. 4.1–11 into the first line of Eq. 4.1–14 we can make the following identifications:

$$\breve{\boldsymbol{\delta}}_1 = \frac{1}{\dot{\gamma}}\left[\boldsymbol{\delta}_\theta r \frac{d}{dr}\left(\frac{v_\theta}{r}\right) + \boldsymbol{\delta}_z \frac{dv_z}{dr}\right] \tag{4.1–15}$$

$$\breve{\boldsymbol{\delta}}_3 = \frac{1}{\dot{\gamma}}\left[\boldsymbol{\delta}_\theta \frac{dv_z}{dr} - \boldsymbol{\delta}_z r \frac{d}{dr}\left(\frac{v_\theta}{r}\right)\right] \tag{4.1–16}$$

$$\dot{\gamma} = \sqrt{\left(r\frac{d}{dr}\left(\frac{v_\theta}{r}\right)\right)^2 + \left(\frac{dv_z}{dr}\right)^2} \tag{4.1–17}$$

Thus steady helical flow is a steady simple shear flow if seen from the shear axes given in Eqs. 4.1–15 and 4.1–16, and it is thus a viscometric flow. Note that Eq. 4.1–17 can also be obtained from the postulated velocity field, the definition of $\dot{\gamma}$ in Eq. 4.1–4, and Eqs. B.3–7 to 12. Equation 4.1–17 serves to emphasize that the terms "shear rate" and "velocity gradient" are not necessarily synonymous.

In parts (a) and (b) of this example, the lines of shear and the particle pathlines are identical, since $\breve{\boldsymbol{\delta}}_1$ is everywhere parallel to the local fluid velocity vector (see Table 4.1–2a and b). Steady helical flow, however, is a flow in which the particle pathlines and the lines of shear are different. The pathline is the curve traced out by a given particle as it moves about in space. Since the flow considered here is steady, the velocity vector at any position is constant, and thus the pathline is the curve that is everywhere tangent to the velocity vector $\mathbf{v} = \boldsymbol{\delta}_\theta v_\theta + \boldsymbol{\delta}_z v_z$. On the other hand the line of shear is defined as the curve that, at any instant, is everywhere tangent to $\breve{\boldsymbol{\delta}}_1$. Since the flow is steady, the line of shear is independent of time. From Eq. 4.1–15 we see that $\breve{\boldsymbol{\delta}}_1$ is not in general parallel to $\mathbf{v}$ so that the line of shear is different than the particle pathline (see Table 4.1–2e).

§4.2 UNIDIRECTIONAL SHEAR FLOWS: FORM OF THE STRESS TENSOR

Now that we have described the kinematics of shear flows, we turn to the question of determining the stresses in these flows. We begin by considering the stresses required to sustain a (time-dependent) simple shearing motion. For non-Newtonian fluids we must assume, in the absence of a rheological equation of state, that in any flow all six independent components of the stress tensor may be nonzero.[1] However, for simple shearing flows of incompressible liquids, it is possible to show that at most three independent combinations of components of the stress tensor can be measured. To do this we take advantage of the fact that simple shear flow is symmetric with respect to a 180° rotation about the 3-axis.[2] Formally we show this symmetry by making the following change of coordinates $x \to \bar{x}$:

$$\begin{aligned} \bar{x}_1 &= -x_1 \\ \bar{x}_2 &= -x_2 \\ \bar{x}_3 &= x_3 \end{aligned} \tag{4.2–1}$$

[1] We will always take the stress tensor to be symmetrical. The theoretical possibility of a nonsymmetrical stress tensor is discussed by J. S. Dahler and L. E. Scriven, *Nature, 192*, 36–37 (1961). To date no experiments have shown asymmetry in the stress tensor.

[2] A. S. Lodge, *Elastic Liquids*, Academic Press, New York (1964), pp. 62–65.

In the $\bar{x}$-coordinate system and in the x-coordinate system, the velocity field has the same form:

$$\boldsymbol{v} = \boldsymbol{\delta}_1 \dot{\gamma} x_2 = \bar{\boldsymbol{\delta}}_1 \dot{\gamma} \bar{x}_2 \tag{4.2-2}$$

so that $v_i(x_1,x_2,x_3,t) = \bar{v}_i(\bar{x}_1,\bar{x}_2,\bar{x}_3,t)$. Keep in mind that $\dot{\gamma}$ is a function of t alone.

Now if we assume that the fluid is isotropic, that is, it has no preferred direction other than one introduced by the flow itself, and if we assume that the stress depends only on the kinematics (i.e., the motion of the fluid), then the stress field must have the same symmetry as the flow field. Hence, we conclude

$$\bar{\tau}_{ij} = \tau_{ij} \tag{4.2-3}$$

We next write down the force exerted by the fluid on a plane normal to $\boldsymbol{\delta}_3$, in terms of the stress components referred to either coordinate system:

$$\begin{aligned} [\boldsymbol{\delta}_3 \cdot \boldsymbol{\tau}] &= [\boldsymbol{\delta}_3 \cdot \textstyle\sum_i \sum_j \boldsymbol{\delta}_i \boldsymbol{\delta}_j \tau_{ij}] \\ &= \boldsymbol{\delta}_1 \tau_{31} + \boldsymbol{\delta}_2 \tau_{32} + \boldsymbol{\delta}_3 \tau_{33} \end{aligned} \tag{4.2-4}$$

or

$$\begin{aligned} [\boldsymbol{\delta}_3 \cdot \boldsymbol{\tau}] &= [\boldsymbol{\delta}_3 \cdot \textstyle\sum_i \sum_j \bar{\boldsymbol{\delta}}_i \bar{\boldsymbol{\delta}}_j \bar{\tau}_{ij}] \\ &= \bar{\boldsymbol{\delta}}_1 \bar{\tau}_{31} + \bar{\boldsymbol{\delta}}_2 \bar{\tau}_{32} + \bar{\boldsymbol{\delta}}_3 \bar{\tau}_{33} \\ &= -\boldsymbol{\delta}_1 \tau_{31} - \boldsymbol{\delta}_2 \tau_{32} + \boldsymbol{\delta}_3 \tau_{33} \end{aligned} \tag{4.2-5}$$

Thus, the symmetry considerations require that:

$$\begin{aligned} \tau_{31} &= \tau_{13} = 0 \\ \tau_{32} &= \tau_{23} = 0 \end{aligned} \tag{4.2-6}$$

Notice that nothing in this argument requires that $\dot{\gamma}$ be independent of time, so that Eq. 4.2-6 holds for any unsteady simple shear flow. The most general form the total stress tensor can have for a simple shearing flow is:

$$\boxed{\boldsymbol{\pi} = p\boldsymbol{\delta} + \boldsymbol{\tau} = \begin{pmatrix} p + \tau_{11} & \tau_{21} & 0 \\ \tau_{21} & p + \tau_{22} & 0 \\ 0 & 0 & p + \tau_{33} \end{pmatrix}} \tag{4.2-7}$$

However for incompressible fluids we know that we cannot measure the pressure p, but only pressure differences. The only quantities of experimental interest, then, are the shear stress and two normal stress differences. The stresses that are customarily used in conjunction with shear flow are:

Shear stress: τ_{21}

Primary normal stress difference: $\tau_{11} - \tau_{22}$

Secondary normal stress difference: $\tau_{22} - \tau_{33}$ (4.2-8)

Thus, by some rather simple arguments, we have reduced the total number of stress measurements to be made in simple shear flow from six to only three.

In a unidirectional shear flow, either time-dependent or viscometric, the flow at every particle is a simple shear flow as viewed from the shear axes. If we assume that the stress at a fluid particle depends only on the kinematic history of that particle, and not on the histories of neighboring particles (see footnote 3 in §7.1), then the stress at every particle will be determined by the values of τ_{21}, $\tau_{11} - \tau_{22}$, and $\tau_{22} - \tau_{33}$ referred to the shear axes fixed at each particle. These in turn are just the values found for a simple shearing flow with the appropriate $\dot{\gamma}(t)$. In order to describe the stresses in any (inhomogeneous, unsteady) unidirectional shear flow, then, it is enough to know how they depend on $\dot{\gamma}$ and time in simple shear flow.[3] The shear stress and normal stress differences in a specific simple shear flow are material properties inasmuch as they depend only on the fluid being tested. In the sections that follow, we use these quantities to define material functions in simple shear flow, which can be used to characterize macromolecular fluids in any unidirectional shear flow.

§4.3 STEADY SHEAR FLOW MATERIAL FUNCTIONS

A summary of the most common unidirectional shear flow experiments used in rheology is given in Fig. 4.3–1. Here we have idealized each of the experiments as flow between parallel plates for the sake of simplicity in illustrating the dependence of the velocity gradient on time. We further take the gap between the plates to be sufficiently small that the shear rate is independent of position. As a convenient means for describing the stresses in these experiments, material functions relating the shear stress and normal stress differences to the time-dependent shear rate are customarily defined. In this section we define and discuss those material functions appropriate for steady shear flow, and in the next section we do the analogous task for the unsteady experiments shown in Fig. 4.3–1. Since these velocity fields are all variations of simple shear flow, the stresses are uniform throughout the fluid; thus the material functions we define are functions of shear rate alone. In view of the discussion at the end of the last section, they can then be used to describe the stress tensor in any unidirectional shear flow. We use this fact in §4.5 where we analyze some of the actual experiments used to measure the material functions.

In steady simple shear flow (Fig. 4.3–1*a*) all transient stresses have died out and the steady-state stresses depend only on the shear rate $\dot{\gamma}$. It is common to define a *viscosity function* η (also called the *non-Newtonian viscosity* or *shear-rate dependent viscosity*) analogously to the definition of viscosity for Newtonian fluids (cf. Eq. 1.2–1):

$$\boxed{\tau_{yx} = -\eta(\dot{\gamma})\dot{\gamma}_{yx}} \tag{4.3–1}$$

Likewise, one can define normal stress coefficients Ψ_1 and Ψ_2 as follows:

$$\boxed{\tau_{xx} - \tau_{yy} = -\Psi_1(\dot{\gamma})\dot{\gamma}_{yx}^2} \tag{4.3–2}$$

$$\boxed{\tau_{yy} - \tau_{zz} = -\Psi_2(\dot{\gamma})\dot{\gamma}_{yx}^2} \tag{4.3–3}$$

The functions Ψ_1 and Ψ_2 are known as the *primary* and *secondary normal stress coefficients*, respectively. Since η, Ψ_1, and Ψ_2 completely determine the state of stress in any rheologically steady shear flow, they are sometimes called *viscometric functions*. It is possible to show by

[3] For a proof of these statements in terms of body vectors and tensors, see A. S. Lodge, *Body Tensor Fields in Continuum Mechanics*, Academic Press, New York (1974), pp. 162–167.

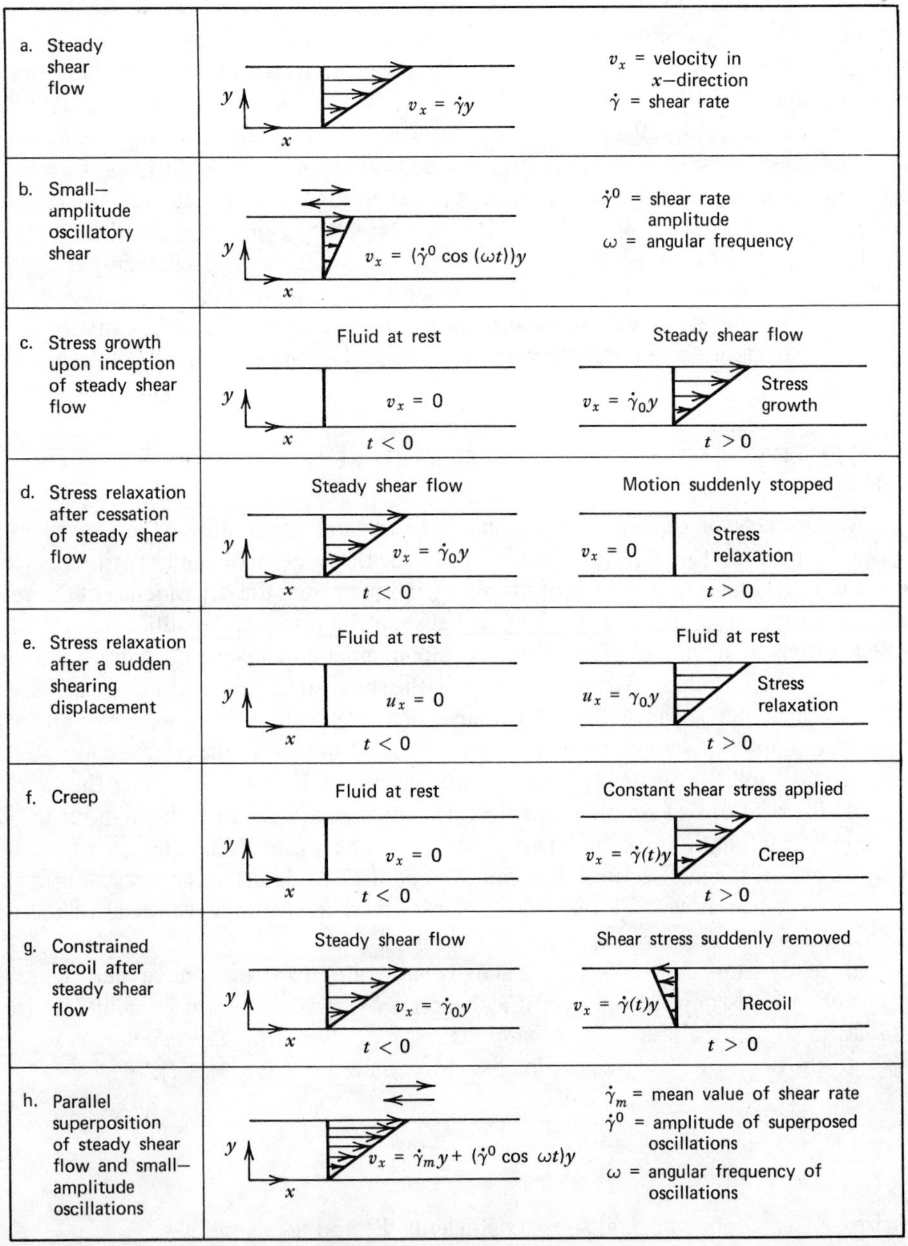

FIGURE 4.3–1. Various types of unidirectional shear flow experiments used in rheology. The shear rate $\dot{\gamma}$ (as defined by Eq. 4.1–4) is identical with $\dot{\gamma}_{yx}$ in the shear flow experiments. For the shearing displacement experiment in (*e*) u_x is the displacement of a particle in the x-direction measured from its position just before $t = 0$; also the "magnitude of shear" γ_0 in that part is the shear-displacement gradient and $\gamma_0 = \int_0^t \dot{\gamma}(t')\, dt'$.

using arguments similar to those in Eqs. 4.2–1 to 6 that all three of the viscometric functions are even functions of the shear rate $\dot{\gamma}$ (see Problem 4C.2). Sometimes it is more convenient to regard the viscometric functions as functions of the absolute value of the shear stress $\tau = |\tau_{yx}|$.

The viscosity function is the best known, experimentally, of the viscometric functions. Some typical plots of $\eta(\dot{\gamma})$ are given in Figs. 4.3–2 to 4 for a polymer melt, concentrated

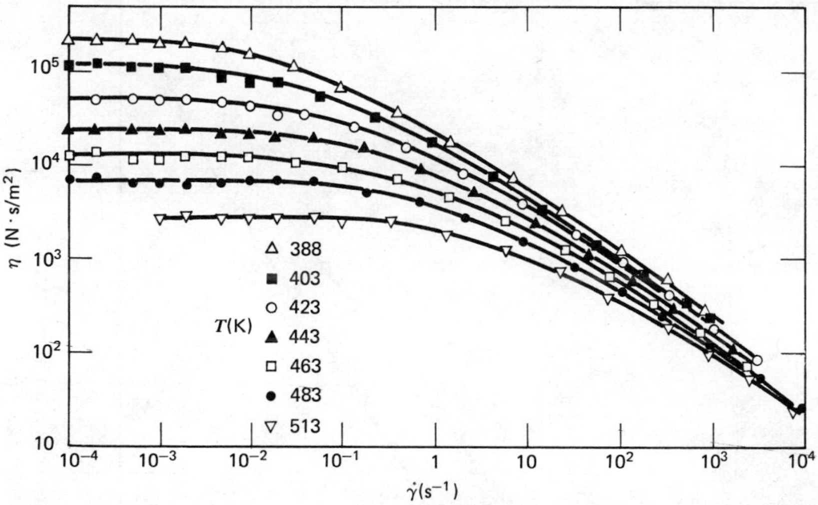

FIGURE 4.3–2. Non-Newtonian viscosity η of a low density polyethylene melt (Melt I) at several different temperatures. The data taken at shear rates below $5 \times 10^{-2}\ s^{-1}$ were obtained with a Weissenberg Rheogoniometer; the viscosity at higher shear rates was determined using a capillary viscometer. [J. Meissner, *Kunststoffe*, *61*, 576–582 (1971).]

solutions, and dilute solutions. At low shear rates, the shear stress is proportional to $\dot{\gamma}$, and the viscosity approaches a constant value η_0, the *zero-shear-rate viscosity.* At higher shear rates the viscosity decreases with increasing shear rate. For many engineering applications, this is the most important characteristic of polymeric fluids; almost all macromolecular fluids show this shear thinning or pseudoplastic behavior.[1] When plotted as log η vs. log $\dot{\gamma}$, the viscosity vs. shear rate curve exhibits a pronounced linear region at high shear rates that can persist over several decades of decreasing viscosity. The slope of this linear section is found experimentally to be between -0.4 and -0.85 for typical polymers. Since this so-called "power-law" region occurs at shear rates that are often encountered in technical applications, empirical descriptions of macromolecular fluid viscosities have been based on it (see Chapter 5). The range of shear rates over which the transition from η_0 to the power law region occurs is fairly narrow for narrow molecular weight distributions. As the molecular weight distribution of the polymer is broadened, the transition region is also broadened and shifted to lower shear rates.[2] Finally, at very high rates of shear the viscosity may again become independent of shear rate and approach η_∞, the *infinite-shear-rate viscosity.* For concentrated solutions and melts, η_∞ is not usually measurable since polymer degradation becomes a serious problem before sufficiently high shear rates can be obtained.

The primary normal stress coefficient is shown in Figs. 4.3–5 and 6. It is seen that Ψ_1 is positive and that it has a large power law region in which it decreases by as much as a factor of 10^6. Most often, the rate of decline of Ψ_1 with $\dot{\gamma}$ is greater than that of η with $\dot{\gamma}$. At low rates of shear the primary normal stress difference is proportional to $\dot{\gamma}^2$, so that Ψ_1 tends to a constant $\Psi_{1,0}$, the zero-shear-rate primary normal stress coefficient, as $\dot{\gamma}$ approaches zero. There does not seem to be a limiting value for Ψ_1 at high shear rates to correspond to η_∞. Some data do show, however, a leveling off trend at the high shear rates.

[1] Some exceptions, or dilatant fluids, are discussed by S. Burow, A. Peterlin, and D. T. Turner, *Polym. Lett.*, *2*, 67–70 (1964). See §3.1. For more recent "shear-thickening" data see M.-N. Layec-Raphalen and C. Wolff, *J. Non-Newtonian Fl. Mech.*, *1*, 159–173 (1976); they studied tube flow of dilute solutions of high molecular weight polystyrene in decalin.

[2] W. W. Graessley, *Fortschr. Hochpolym.-Forsch.*, *16*, 1–179 (1974).

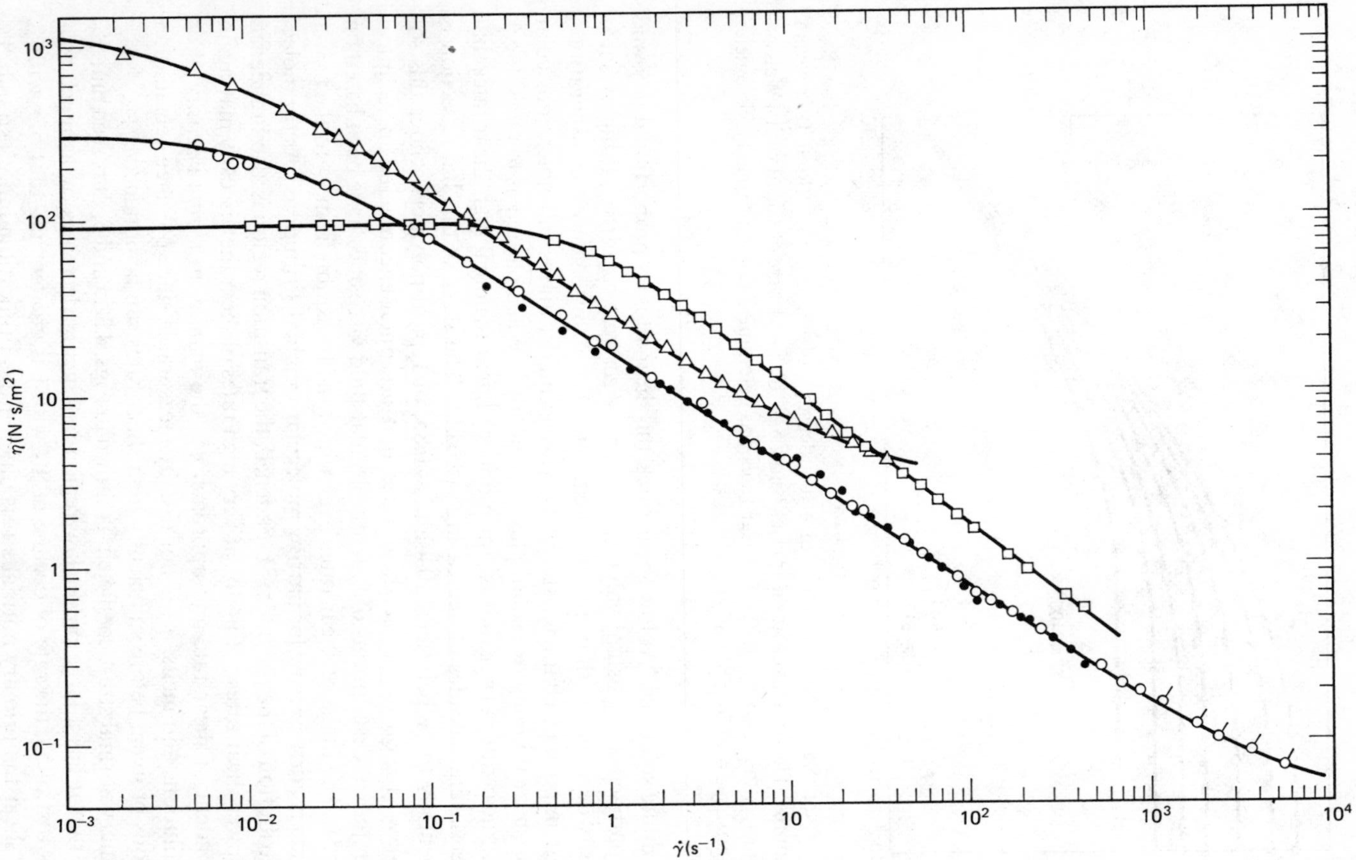

FIGURE 4.3–3. Dependence of viscosity on shear rate for two polymer solutions and an aluminum soap solution: ○ 1.5% polyacrylamide (Separan AP 30) in a 50/50 mixture by weight of water and glycerin, △ 2.0% polyisobutylene in Primol, and □ 7% aluminum laurate in a mixture of decalin and *m*-cresol. The data shown by ơ were taken on a Ferranti-Shirley viscometer; all others are from a Weissenberg Rheogoniometer. All data were taken at 298 K. [Data of J. D. Huppler, E. Ashare, and L. A. Holmes, *Trans. Soc. Rheol.*, *11*, 159–179 (1967).]

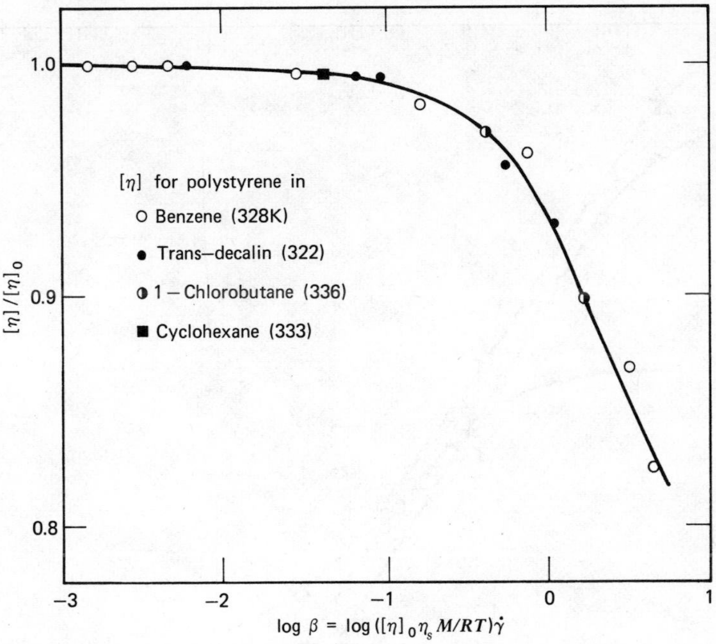

FIGURE 4.3–4. Intrinsic viscosity $[\eta]$ (see Eq. 2.4–3) of polystyrene solutions, with various solvents, as a function of reduced shear rate β. Data of T. Kotaka, H. Suzuki, and H. Inagaki, *J. Chem. Phys.*, 45, 2770–2773 (1966).

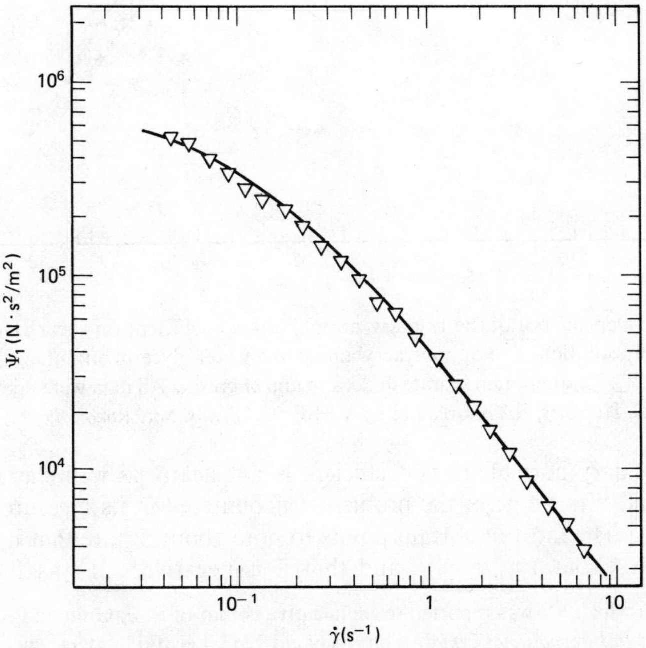

FIGURE 4.3–5. Dependence of the primary normal stress coefficient Ψ_1 on shear rate for a low density polyethylene melt. [Data replotted from I–J. Chen and D. C. Bogue, *Trans. Soc. Rheol.*, *16*, 59–78 (1972).]

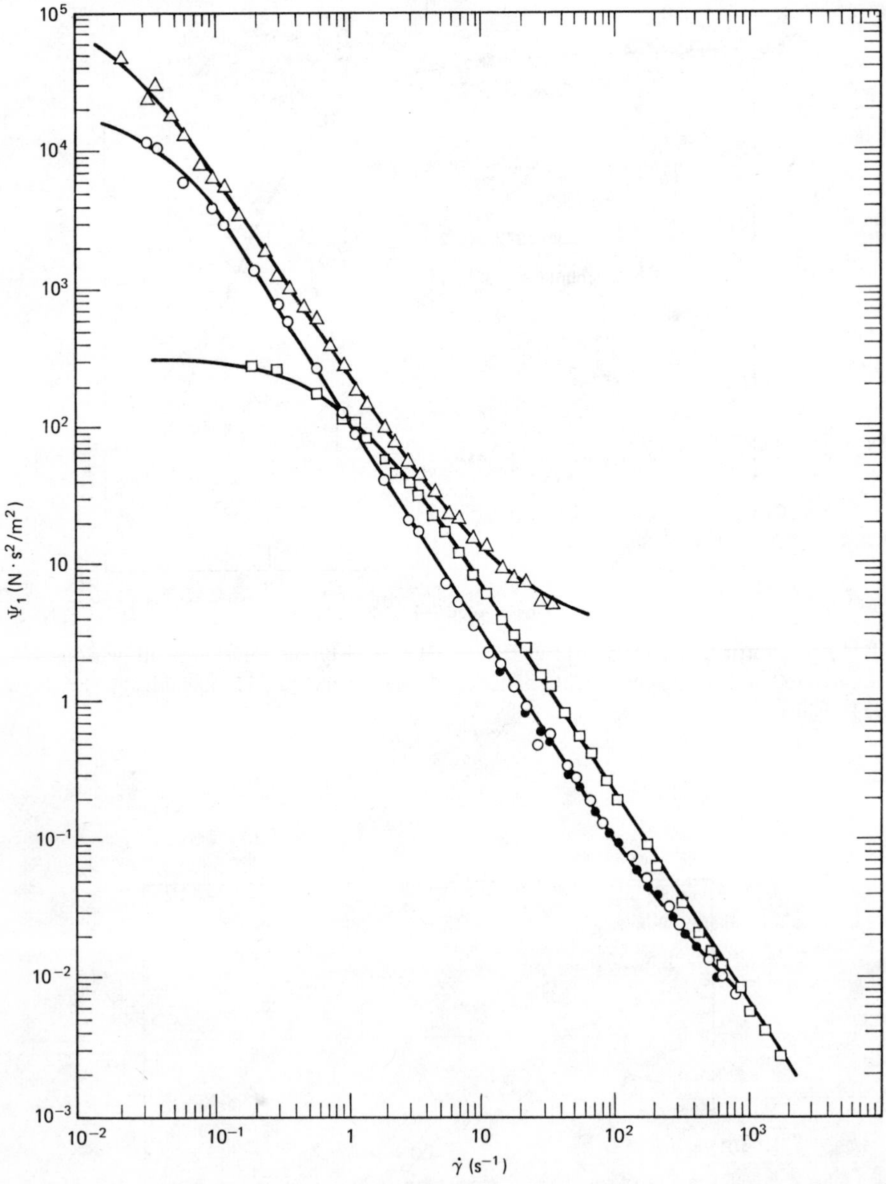

FIGURE 4.3–6. Dependence of the primary normal stress coefficient on shear rate for two polymer solutions and a soap solution: ○ 1.5% polyacrylamide in a water/glycerin mixture, △ 2.0% polyisobutylene in Primol, and □ 7% aluminum laurate in decalin and *m*-cresol. All data were taken at 298 K. [Data replotted from J. D. Huppler, E. Ashare, and L. A. Holmes, *Trans. Soc. Rheol.*, *11*, 159–179 (1967).]

The secondary normal stress coefficient is not nearly as well characterized experimentally as η and Ψ_1. Some of the problems encountered in its measurement have been discussed in §3.3. The most important points to note about Ψ_2 are that it is much smaller than Ψ_1, usually about 10% of Ψ_1, and that it is negative.[3] It was thought for some

[3] The first study in which Ψ_2 was reported to be negative is that of R. F. Ginn and A. B. Metzner, *Proceedings of the Fourth International Congress on Rheology*, E. H. Lee (Ed.), Interscience, New York (1965), Part 2, pp. 583–601; *Trans. Soc. Rheol.*, *13*, 429–453 (1969). The uncertainty in the sign of Ψ_2 was resolved by the discovery of the hole pressure error; see J. M. Broadbent, A. Kaye, A. S. Lodge, and D. G. Vale, *Nature*, *217*, 55–56 (1968); A. Kaye, A. S. Lodge, and D. G. Vale, *Rheol. Acta*, *7*, 368–379 (1968); and J. M. Broadbent and A. S. Lodge, *Rheol. Acta*, *10*, 557–573 (1972).

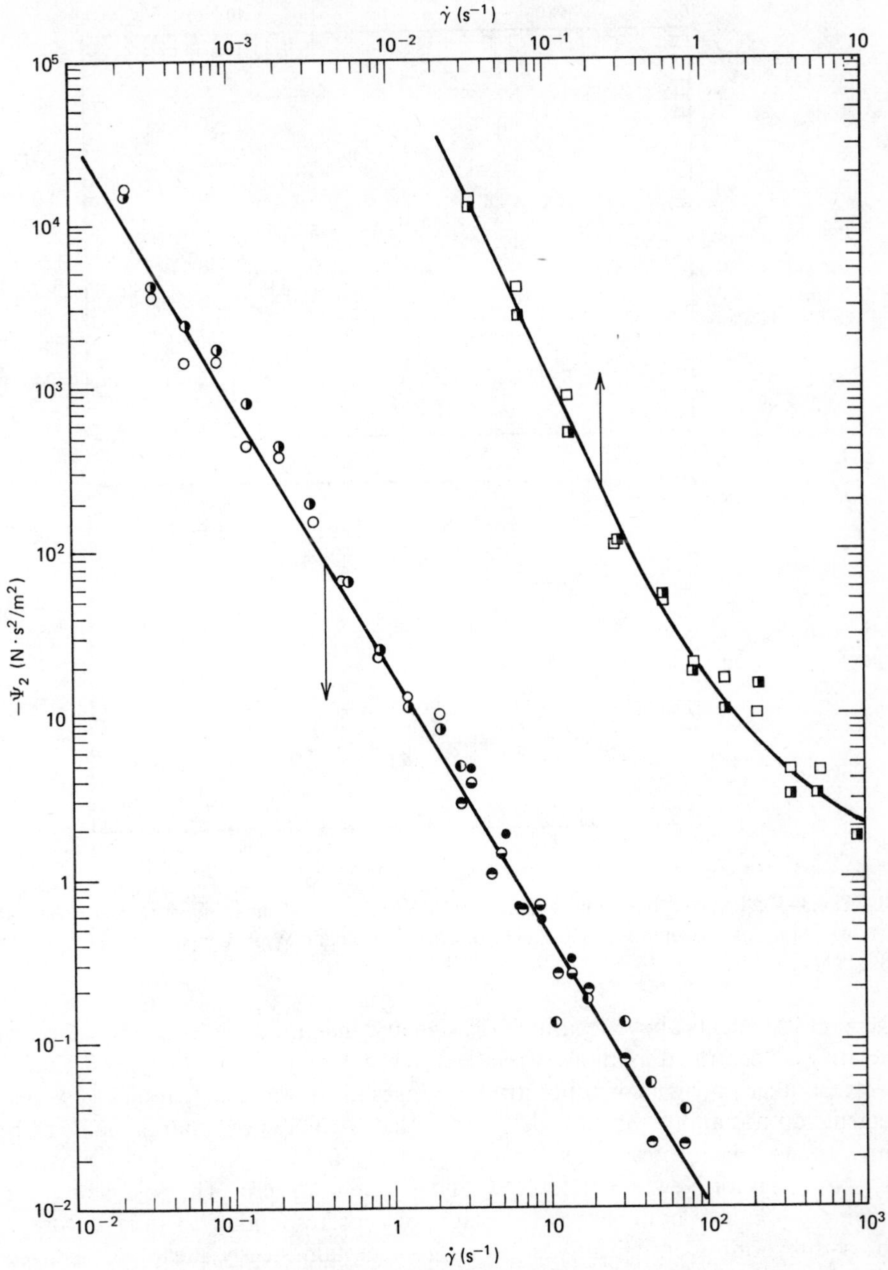

FIGURE 4.3–7. Dependence of the secondary normal stress coefficient on shear rate for two polymer solutions: the circles are for a 2.5% solution of polyacrylamide in a 50/50 mixture of water and glycerin, and the squares are for a 3% solution of poly(ethylene oxide) in a 57/38/5 mixture of water, glycerin, and isopropyl alcohol. [Data replotted from E. B. Christiansen and W. R. Leppard, *Trans. Soc. Rheol.*, *18*, 65–86 (1974).]

time that $\tau_{yy} = \tau_{zz}$ (or $\Psi_2 = 0$); this relation, called the "Weissenberg hypothesis", is no longer believed to be correct. In Fig. 4.3–7 we present data for Ψ_2. There it can be seen that Ψ_2 exhibits a large power-law region like those for $\eta(\dot{\gamma})$ and $\Psi_1(\dot{\gamma})$. No values for $\Psi_{2,0}$ or $\Psi_{2,\infty}$, the zero- and infinite-shear-rate values of Ψ_2, are observed. In Fig. 4.3–8, values of the ratio $-\Psi_2/\Psi_1$ are seen to range from 0.01 to 0.2 for a polyacrylamide solution and a poly(ethylene oxide) solution. Our knowledge about Ψ_2 is still incomplete. Furthermore,

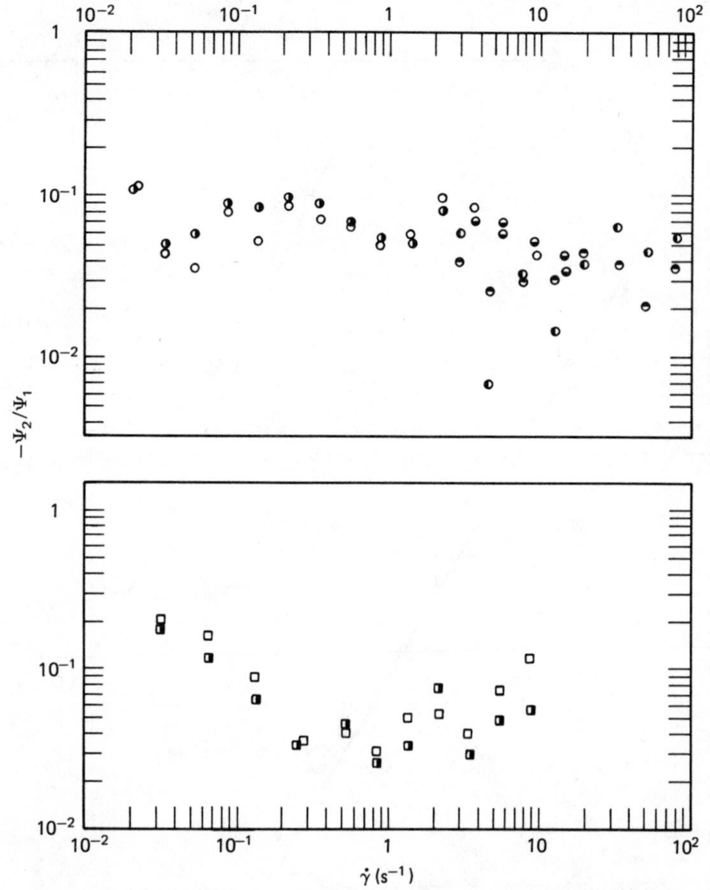

FIGURE 4.3–8. Ratio of the secondary normal stress coefficient to the primary normal stress coefficient for the two solutions shown in Fig. 4.3–7. [E. B. Christiansen and W. R. Leppard, *Trans. Soc. Rheol.*, *18*, 65–86 (1974).]

the general statements above regarding the sign and magnitude of Ψ_2 are based on data for moderately concentrated solutions of relatively few different kinds of polymers. Future work may reveal other polymers or concentration ranges for those already tested for which these statements do not apply. Some data[4] suggest that Ψ_2 may even change sign and become positive at large shear rates.

Note that for Newtonian fluids Ψ_1 and Ψ_2 are both zero. The positive $\tau_{yy} - \tau_{zz}$ and negative $\tau_{xx} - \tau_{yy}$ can both be loosely thought of as corresponding to an extra compression in the y-direction. Consequently, in order to maintain steady shear flow between two parallel plates, a normal force must be applied to the plates to prevent them from separating when the fluid is polymeric. Only a shear stress is needed to maintain steady shear flow for a Newtonian fluid.

§4.4 UNIDIRECTIONAL UNSTEADY SHEAR FLOW MATERIAL FUNCTIONS

We now consider the time-dependent stresses in unsteady unidirectional shear flows. For this purpose we examine macromolecular fluid responses to experiments b to h in

[4] W. G. Pritchard, *Phil. Trans. Roy. Soc., London, A270*, 507–556 (1971). See footnote 3 in §3.3 for references to other data on Ψ_2.

Fig. 4.3–1. As in steady shear flow, the shear stress and normal stress differences are conveniently represented in terms of material functions. Now, however, the material functions may be expected to depend on time (or frequency) as well as on the shear rate.

Experiment b: Small-Amplitude Oscillatory Shear Flow

The *small-amplitude oscillatory shear experiment* (Fig. 4.3–1*b*) involves measurement of the unsteady response of a sample that is contained between two parallel plates, the upper one of which undergoes small-amplitude oscillations in its own plane with frequency ω. The instantaneous velocity profile will be very nearly linear in y if $\omega\rho h^2/2\eta_0 \ll 1$, where h is the distance between the plates (see Problem 1D.2). For Newtonian fluids the shear stress τ_{yx} is in phase with the shear rate $\dot{\gamma}_{yx}$, and there are no normal stresses. For polymeric materials, the response is shown qualitatively in Fig. 4.4–1. The two main features to be noted from this figure are that the shear stress oscillates with frequency ω, but is out of phase with the shear rate, and that the normal stresses oscillate with a frequency 2ω about a nonzero mean value (see Problem 4C.2).

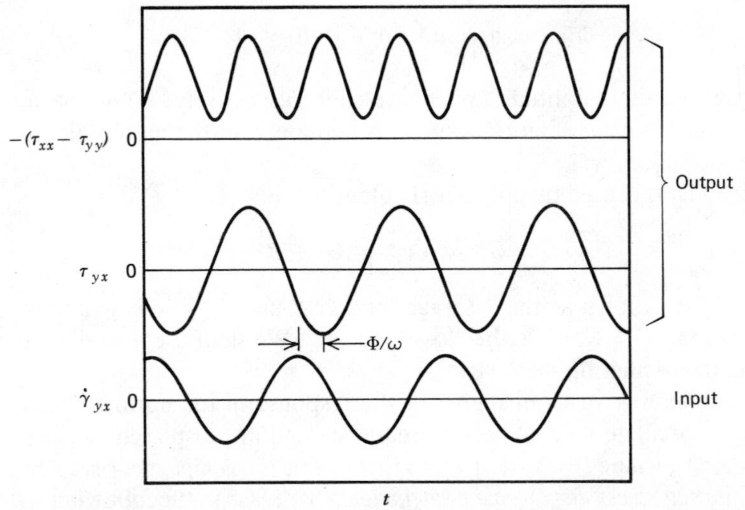

FIGURE 4.4–1. Oscillating shear rate, shear stress, and primary normal stress difference in small-amplitude oscillatory flow. [A. S. Lodge, *Elastic Liquids*, Academic Press, New York (1964), p. 113.]

For shear stresses one can measure the amplitude and phase shift as a function of the frequency ω. Often the data are presented in terms of the real and imaginary parts of the *complex viscosity* $\eta^*(\omega)$. We note that the velocity gradient is varying sinusoidally as $\dot{\gamma}_{yx} = \mathcal{Re}\{\dot{\gamma}_{yx}{}^0 e^{i\omega t}\}$, where $\dot{\gamma}_{yx}{}^0$ may in general be complex;[1] similarly the shear stress varies sinusoidally and can be represented as $\tau_{yx} = \mathcal{Re}\{\tau_{yx}{}^0 e^{i\omega t}\}$, where $\tau_{yx}{}^0$ is complex. The complex viscosity function η^* is defined by:

$$\tau_{yx}{}^0 = -\eta^* \dot{\gamma}_{yx}{}^0 \tag{4.4-1}$$

It is necessary to allow η^* to be complex in order to account for the phase difference between the shear stress and shear rate. We then break up η^* into its real and imaginary parts:

$$\eta^* = \eta' - i\eta'' \tag{4.4-2}$$

The real part η', the dynamic viscosity, may be thought of as the viscous contribution, associated with energy dissipation; the imaginary part η'' may be thought of as the elastic

[1] $\mathcal{Re}\{\ \}$ stands for the real part of the complex quantity. Recall that $e^{i\theta} = \cos\theta + i\sin\theta$ where $i = \sqrt{-1}$.

contribution, associated with energy storage. Since η^* determines the stresses that are linear in shear rate, η' and η'' are often called linear viscoelastic properties. They are important in characterizing fluids in terms of the general linear viscoelastic model discussed in Chapter 6. For more detail on these and other linear properties, standard references should be consulted.[2] For the Newtonian fluid $\eta' = \mu$ and $\eta'' = 0$.

Note that if $\dot{\gamma}_{yx}{}^0$ is taken to be real and positive, then $\dot{\gamma}_{yx} = \dot{\gamma}_{yx}{}^0 \cos \omega t$ and Eq. 4.4–1 is equivalent to:

$$\boxed{\tau_{yx} = -\eta'\dot{\gamma}_{yx}{}^0 \cos \omega t - \eta''\dot{\gamma}_{yx}{}^0 \sin \omega t} \tag{4.4–3}$$

which shows the η' term as being the in-phase component and the η'' term as being the out-of-phase component. Alternatively, one can write:

$$\tau_{yx} = -A \cos (\omega t - \Phi) \tag{4.4–4}$$

where

$$A(\omega) = \sqrt{\eta'^2 + \eta''^2}\, \dot{\gamma}_{yx}{}^0 \tag{4.4–5}$$

$$\Phi(\omega) = \arctan (\eta''/\eta') \qquad (0 \leqslant \Phi \leqslant \tfrac{1}{2}\pi) \tag{4.4–6}$$

Here A and Φ are the amplitude and phase shift, respectively. Thus the functions $\eta'(\omega)$ and $\eta''(\omega)$ contain the same information as is contained in the amplitude and phase shift as a function of frequency.

Other functions used by polymer rheologists are:

$$G^* = i\omega\eta^* = G' + iG'' \tag{4.4–7}$$

where $G' = \omega\eta''$ is known as the "storage modulus" and $G'' = \omega\eta'$ is known as the "loss modulus"; also $\tan \delta = G''/G'$ is the "loss tangent." We shall use η' and η'' in this chapter for specifying the oscillating shear stress.

In a similar way one can represent the response of the normal stresses. Since the normal stresses oscillate with twice the frequency and are displaced, we may write $\tau_{jj} = d_j + \mathcal{Re}\{\tau_{jj}{}^0 e^{2i\omega t}\}$, where $j = x, y$, or z and then define the *complex normal stress coefficients* Ψ_1^*, Ψ_2^* and *normal stress displacement coefficients* $\Psi_1{}^d$, $\Psi_2{}^d$ in the following way:

$$\tau_{xx}{}^0 - \tau_{yy}{}^0 = -\Psi_1^* (\dot{\gamma}_{yx}{}^0)^2 \tag{4.4–8}$$

$$d_x - d_y = -\Psi_1{}^d |\dot{\gamma}_{yx}{}^0|^2 \tag{4.4–9}$$

$$\tau_{yy}{}^0 - \tau_{zz}{}^0 = -\Psi_2^* (\dot{\gamma}_{yx}{}^0)^2 \tag{4.4–10}$$

$$d_y - d_z = -\Psi_2{}^d |\dot{\gamma}_{yx}{}^0|^2 \tag{4.4–11}$$

As with the complex viscosity, the complex normal stress coefficients may be split into real and imaginary parts:

$$\Psi_1^* = \Psi_1' - i\Psi_1'' \tag{4.4–12}$$

$$\Psi_2^* = \Psi_2' - i\Psi_2'' \tag{4.4–13}$$

[2] See, for example, J. D. Ferry, *Viscoelastic Properties of Polymers*, Wiley, New York (1970), Second Edition; H. Leaderman, in *Rheology*, F. R. Eirich (Ed.), Academic Press, New York (1958), Vol. 2, Chapter 1, pp. 1–61; A. V. Tobolsky, *op. cit.*, pp. 63–81; T. Alfrey, Jr., and E. F. Gurnee, in *Rheology*, F. R. Eirich (Ed.), Academic Press, New York (1956), Vol. 1, Chapter 11, pp. 387–429.

with all these functions being dependent on ω. Alternatively, if $\dot{\gamma}_{yx}{}^0$ is taken to be real

$$\tau_{xx} - \tau_{yy} = -\Psi_1{}^d (\dot{\gamma}_{yx}{}^0)^2 - \Psi_1' (\dot{\gamma}_{yx}{}^0)^2 \cos 2\omega t - \Psi_1'' (\dot{\gamma}_{yx}{}^0)^2 \sin 2\omega t \qquad (4.4\text{–}14)$$

$$\tau_{yy} - \tau_{zz} = -\Psi_2{}^d (\dot{\gamma}_{yx}{}^0)^2 - \Psi_2' (\dot{\gamma}_{yx}{}^0)^2 \cos 2\omega t - \Psi_2'' (\dot{\gamma}_{yx}{}^0)^2 \sin 2\omega t \qquad (4.4\text{–}15)$$

Sample data showing the complex viscosity functions for a polymer melt, three moderately concentrated solutions, and a dilute solution are shown in Figs. 4.4–2 to 5. The dynamic viscosity η' is found to approach the zero-shear-rate viscosity at low frequency; this is to be expected. On the other hand, the out-of-phase part of the complex viscosity associated with energy storage is found to approach zero linearly in ω. Thus there is a

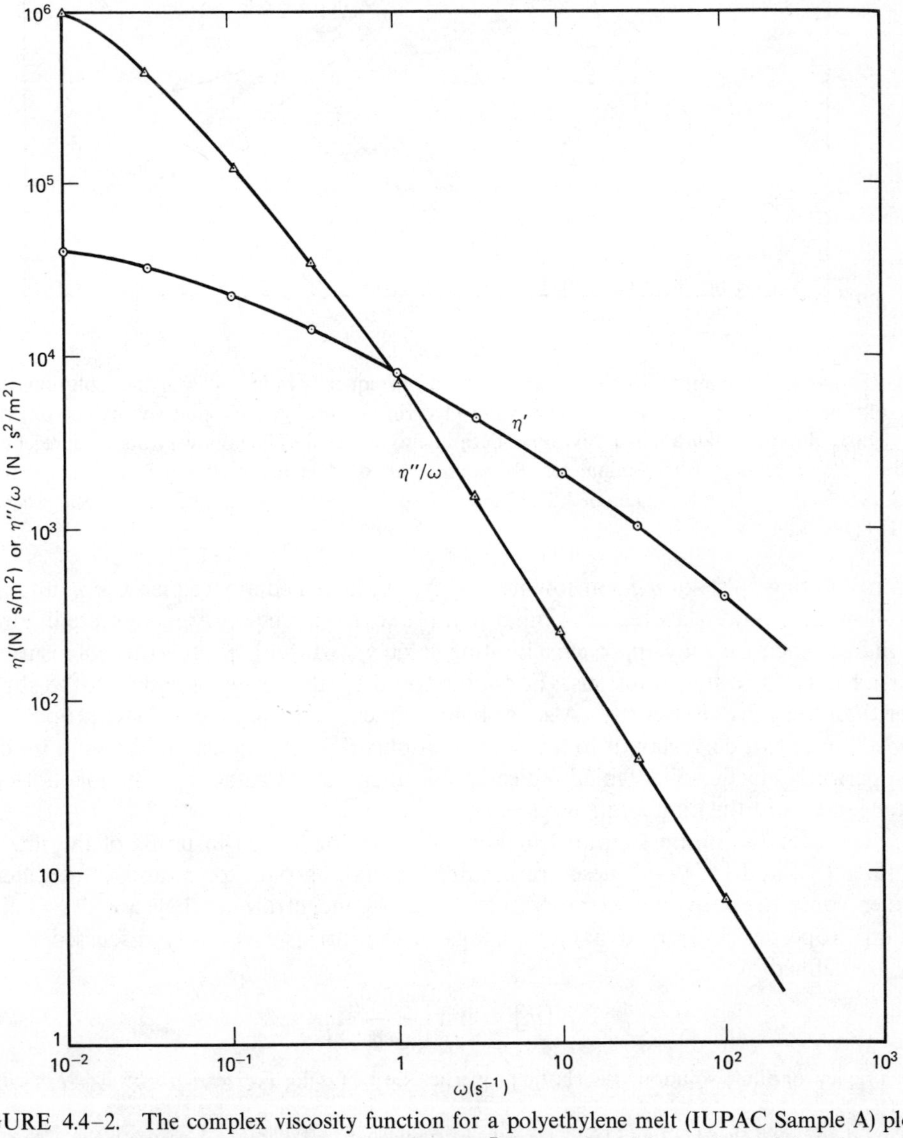

FIGURE 4.4–2. The complex viscosity function for a polyethylene melt (IUPAC Sample A) plotted as η'(○) and η''/ω(△) as functions of the frequency ω. [Data replotted from Fig. 6 of J. Meissner, *Pure and Appl. Chem.*, *42*, 551–612 (1975).]

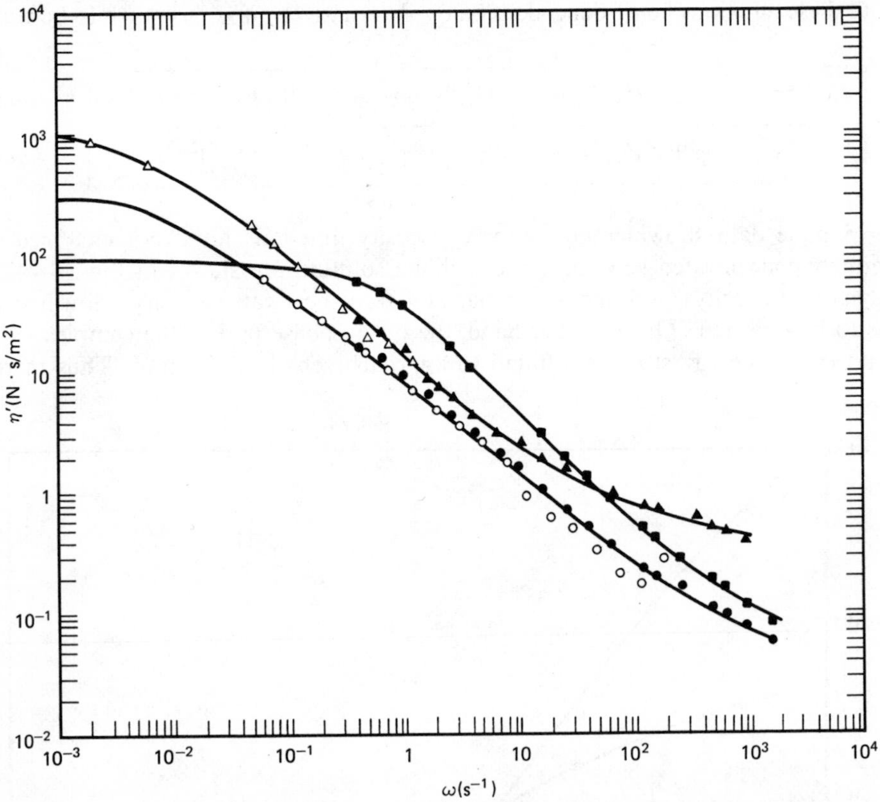

FIGURE 4.4–3. Dynamic viscosity η' as a function of frequency ω for two polymer solutions and a soap solution: ○ 1.5% polyacrylamide in a water/glycerin mixture, △ 2.0% polyisobutylene in Primol, and □ 7.0% aluminum laurate in a mixture of decalin and *m*-cresol. The hollow symbols represent data taken on a Weissenberg Rheogoniometer; the solid symbols, data taken on the Birnboim apparatus. All data were taken at 298 K. [Data of J. D. Huppler, E. Ashare, and L. A. Holmes, *Trans. Soc. Rheol*, *11*, 159–179 (1967).]

nonzero limiting value of η''/ω at low frequency. At intermediate frequencies, η' and η''/ω both show large power-law regions similar to those we have previously encountered. Finally at high frequencies, η' may approach a limiting value η'_∞. As with η_∞, this quantity is usually observed only for dilute solutions. For solutions the value of η'_∞ is found to be slightly larger than the solvent viscosity. Also, at high frequencies η''/ω may become proportional to $1/\omega^2$. Since this corresponds to a storage modulus G' that is constant, the fluid is acting like a perfectly elastic solid, which indicates that there is not sufficient time for molecular rearrangements at the high frequencies.

Data for the dilute solution[3] in Fig. 4.4–5 are presented in terms of the intrinsic quantities $[\eta']$ and $[\eta'']/\omega$. These are obtained by taking data for η' and η'' at different polymer concentrations and extrapolating to zero concentration; they are thus infinite dilution properties and are dynamic analogs of the intrinsic viscosity discussed in §2.4. They are defined by:[4]

$$[\eta'] = \lim_{c \to 0} \left(\frac{\eta' - \eta_s}{c\eta_s} \right) \tag{4.4–16}$$

[3] For a review of dilute solution viscoelastic properties see K. Osaki, *Fortschr. Hochpolym.-Forsch.*, *12*, 1–64 (1973).

[4] It should be noted that $[G']$ and $[G'']$ are defined similarly to $[\eta'']$ and $[\eta']$, respectively, except that the values to be extrapolated to infinite dilution are divided by c rather than $c\eta_s$. Thus $[G'] = \omega\eta_s[\eta'']$ and $[G''] = \omega\eta_s[\eta']$.

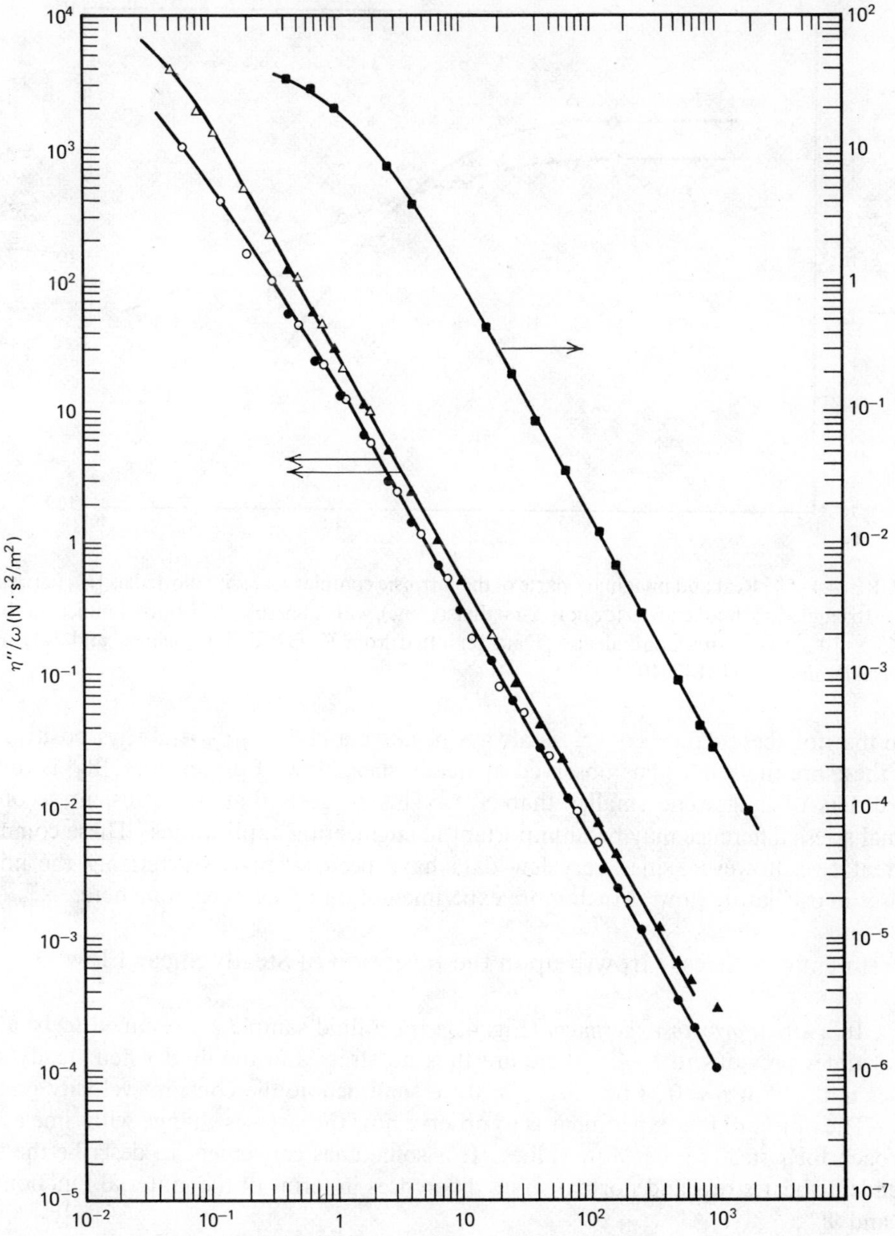

FIGURE 4.4–4. Plot of η''/ω as a function of frequency for the two polymer solutions and the aluminum soap solution shown in Fig. 4.4–3. [Data replotted from J. D. Huppler, E. Ashare, and L. A. Holmes, *Trans. Soc. Rheol.*, *11*, 159–179 (1967).]

$$[\eta'']/\omega = \lim_{c \to 0} \left(\frac{\eta''/\omega}{c\eta_s} \right) \tag{4.4–17}$$

where c is the concentration and η_s is the solvent viscosity.

In Figs. 4.4–6 and 7 we show one complete set of small-amplitude oscillatory material functions for a poly(ethylene oxide) solution. Since $\Psi_1{}^d$ and $\Psi_2{}^d$ are greater than $|\Psi_1^*|$ and $|\Psi_2^*|$ at all frequencies, each oscillating normal stress difference always has the same sign.

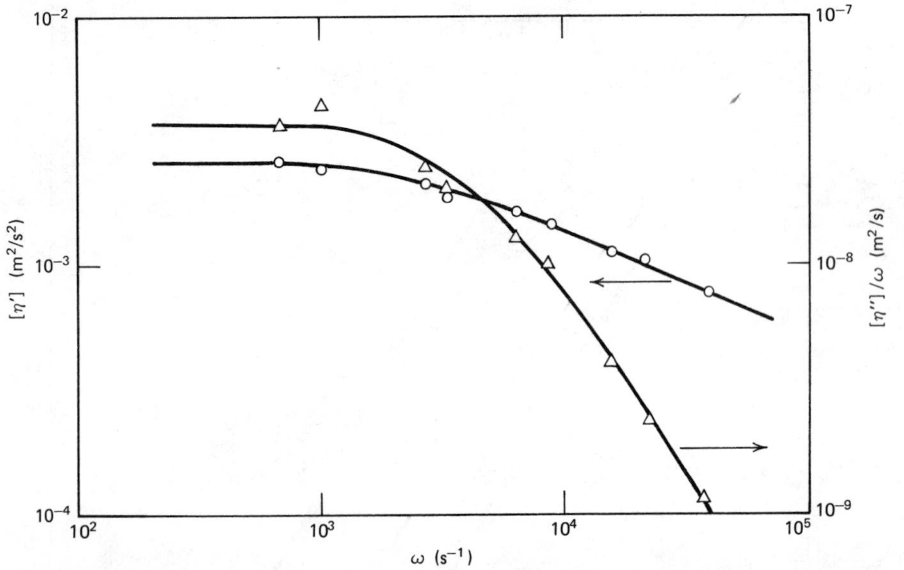

FIGURE 4.4–5. Real and imaginary parts of the intrinsic complex viscosity plotted as $[\eta']$ (circles) and $[\eta'']/\omega$ (triangles) vs. frequency ω for poly(α-methylstyrene), with a narrow-distribution molecular weight of 1.43×10^6, in α-chloronaphthalene. [Data replotted from K. Osaki, J. L. Schrag, and J. D. Ferry, *Macromolecules*, *5*, 144–147 (1972).]

Note that for these data $\tau_{xx} - \tau_{yy}$ is always negative and $\tau_{yy} - \tau_{zz}$ is always positive, and that these are the same signs observed in steady shear flow. Furthermore, $\Psi_2{}^d$ is roughly two orders of magnitude smaller than $\Psi_1{}^d$. This suggests that the transient secondary normal stress difference may be unimportant in engineering applications. These comments are tentative, however, since very few data have been taken characterizing the normal stresses in oscillatory flow. Much more experimentation needs to be done here.

Experiment c: Stress Growth upon the Inception of Steady Shear Flow

In a *stress growth experiment* (Fig. 4.3–1c), a fluid sample is presumed to be at rest for all times previous to $t = 0$; there are thus no stresses in the fluid when steady shear flow is initiated at $t = 0$. For times $t \geq 0$ we shall denote the constant velocity gradient as $\dot{\gamma}_0$. The object of this experiment is to observe how the stresses change with time as they approach their steady shear flow values. It is sometimes convenient to describe the time-dependent shear stress and normal stress differences in terms of the material functions η^+, Ψ_1^+, and Ψ_2^+:

$$\tau_{yx}(t) = -\eta^+(t;\dot{\gamma}_0)\dot{\gamma}_0 \quad (4.4\text{–}18)$$

$$\tau_{xx}(t) - \tau_{yy}(t) = -\Psi_1^+(t;\dot{\gamma}_0)\dot{\gamma}_0{}^2 \quad (4.4\text{–}19)$$

$$\tau_{yy}(t) - \tau_{zz}(t) = -\Psi_2^+(t;\dot{\gamma}_0)\dot{\gamma}_0{}^2 \quad (4.4\text{–}20)$$

where the $\dot{\gamma}_0$ appearing after the semicolon in the argument of the material functions indicates the parametric dependence of the quantities on the imposed shear rate. The plus sign on the material functions is a reminder that the steady shear flow occurs at positive times.

Qualitatively it is found that only for very small shear rates does the shear stress approach its steady-state value monotonically. For larger shear rates, η^+ goes through a

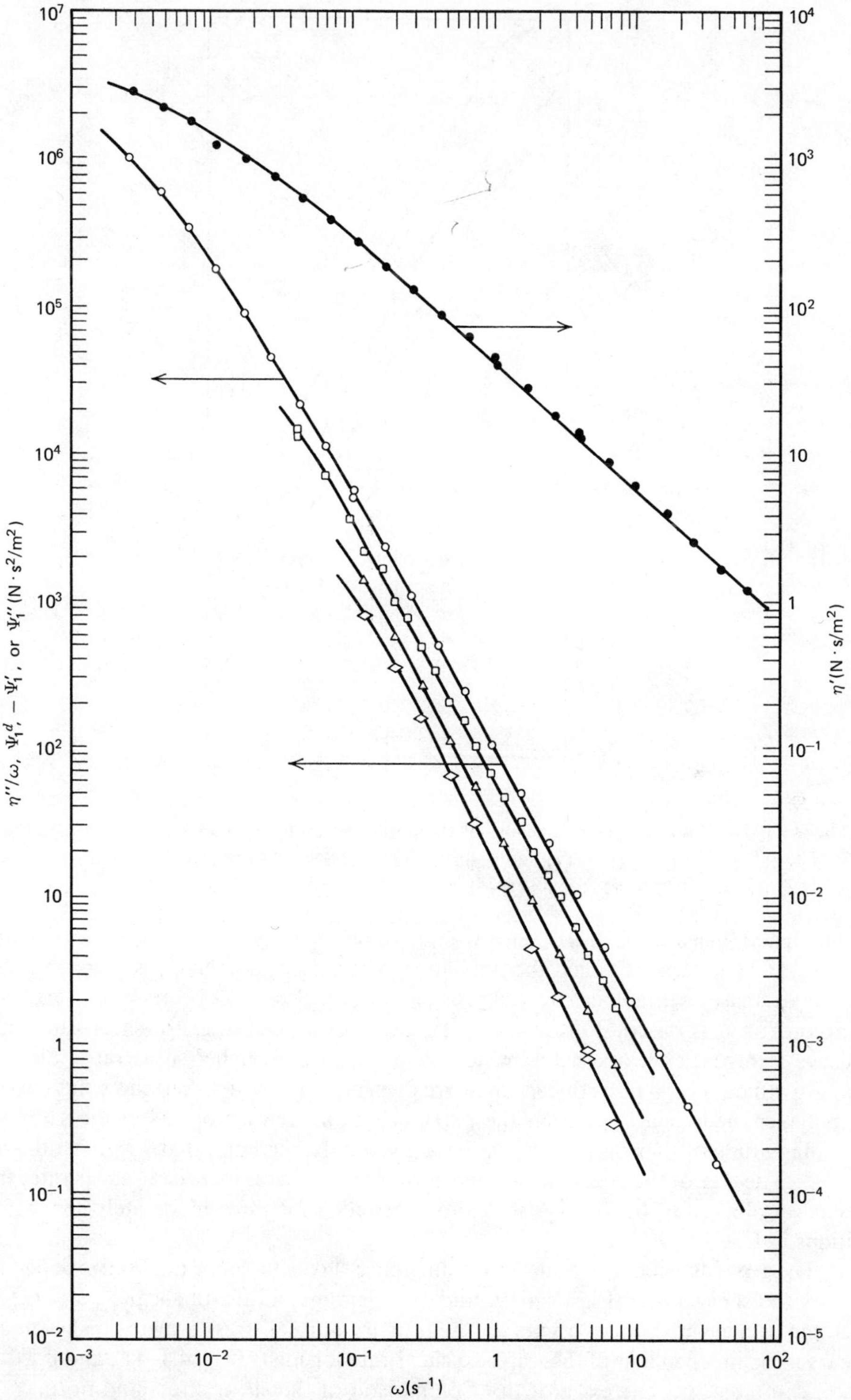

FIGURE 4.4–6. Dynamic properties of a 3% solution of poly(ethylene oxide) in a 57/38/5 mixture of water, glycerin, and isopropyl alcohol as functions of frequency: (●)η', (○)η''/ω, (□)$\Psi_1{}^d$, (△) $-\Psi_1'$, (◇)Ψ_1''. [Data replotted from E. B. Christiansen and W. R. Leppard, *Trans. Soc. Rheol.*, *18*, 65–86 (1974).]

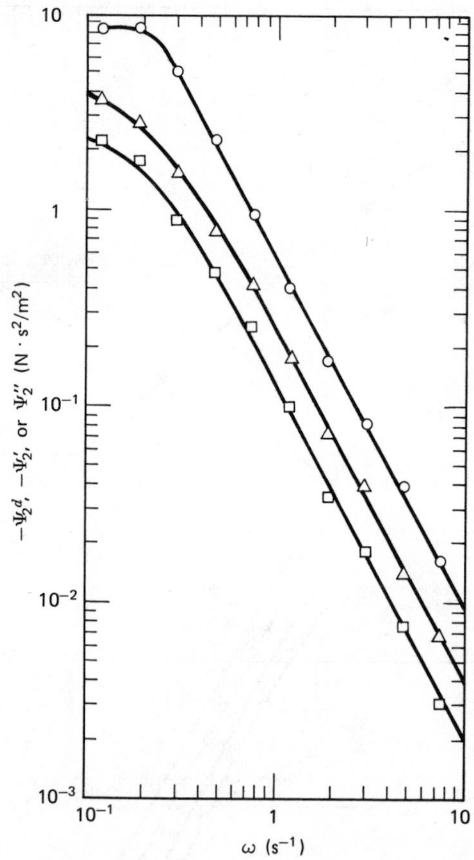

FIGURE 4.4–7. Dynamic secondary normal stress difference functions vs. frequency for the solution in Fig. 4.4–6: (○) $-\Psi_2{}^d$, (△) $-\Psi_2'$, and (□)Ψ_2''. [Data replotted from E. B. Christiansen and W. R. Leppard, *Trans. Soc. Rheol.*, *18*, 65–86 (1974).]

maximum and then approaches the steady-state value, possibly after one or more oscillations about $\eta(\dot{\gamma}_0)$. The time at which the maximum occurs decreases as $\dot{\gamma}_0$ is increased. Since viscosity decreases with increasing shear rate, η^+ approaches successively lower steady-state asymptotes as $\dot{\gamma}_0$ is raised. Furthermore, the monotone-increasing, low-shear-rate-limiting η^+ curve forms an envelope below which the η^+ curves at higher shear rates fall. Shear stress growth data for a polyethylene melt are presented in Fig. 4.4–8 to show this envelope. For polymer melts it appears that the maximum in η^+ always occurs at the same value of the magnitude of shear $\gamma_{max} = t_{max}\dot{\gamma}_0$ for a given polymer,[5] and that γ_{max} is on the order of 2 to 3. The size of the shear stress overshoot increases with increasing shear rate; this is reflected in plots of $\eta^+(t;\dot{\gamma}_0)/\eta(\dot{\gamma}_0)$, which are presented for three moderately concentrated solutions in Fig. 4.4–9.

The growing primary normal stress difference shows the same qualitative dependence on $\dot{\gamma}_0$ as the shear stress (Figs. 4.4–10 and 12). The magnitude of shear $\gamma_{max} = t_{max}\dot{\gamma}_0$ at which the primary maximum occurs for the primary normal stress growth coefficient again appears to be independent of the imposed shear rate for melts (Fig. 4.4–11), and it is larger than γ_{max} for the shear stress growth. Comparison of the shear stress and normal stress data shows that, at a given $\dot{\gamma}_0$, the quantity $\Psi_1^+(t;\dot{\gamma}_0)/\Psi_1(\dot{\gamma}_0)$ increases more slowly than $\eta^+(t;\dot{\gamma}_0)/\eta(\dot{\gamma}_0)$, reaching its primary maximum at a later time. The size of the maximum

[5] W. W. Graessley, *Fortschr. Hochpolym.-Forsch.*, *16*, 1–179 (1974).

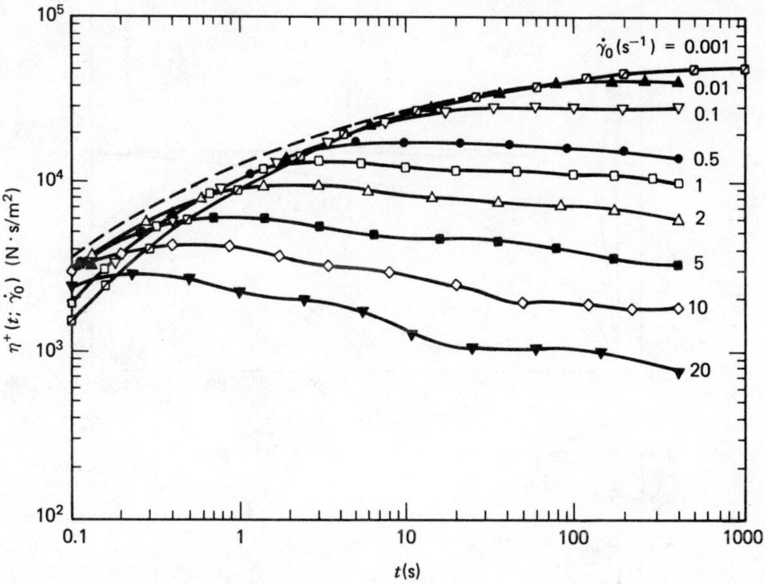

FIGURE 4.4–8. Shear stress growth function $\eta^+(t;\dot{\gamma}_0)$ data for a low density polyethylene melt (IUPAC Sample A). The maximum in η^+ occurs at smaller times as $\dot{\gamma}_0$ is increased. Note that all of the η^+ curves lie below the dashed curve envelope, which is the value of η^+ in the linear viscoelastic limit. The dashed curve was calculated from data; the experimental curves for small $\dot{\gamma}_0$ differ from it at short times because of instrumental problems involved in the start up of steady shear flow. [Data of J. Meissner, *J. Appl. Polym. Sci.*, *16*, 2877–2899 (1972).]

overshoot is less for Ψ_1^+/Ψ_1 than for η^+/η. The only data that have been reported on $\Psi_2^+(t;\dot{\gamma}_0)$ indicate[6] that $\Psi_2^+(t;\dot{\gamma}_0)/\Psi_1^+(t;\dot{\gamma}_0) = \Psi_2/\Psi_1$ for all $\dot{\gamma}_0$ and all t.

Experiment d: Stress Relaxation after Cessation of Steady Shear Flow

In the *stress relaxation experiment* shown in Fig. 4.3–1*d*, the fluid is first made to undergo steady shear flow with velocity gradient $\dot{\gamma}_0$ for $t < 0$. The constant shear rate was started sufficiently long before $t = 0$ so that all of the transients in the stress growth experiment have died out. Then at $t = 0$ the flow is stopped instantaneously so that $\dot{\gamma}_{yx} = 0$ for $t > 0$. (This can be approximated quite well for a small gap between the two plates and a very viscous fluid.) One then observes the manner in which the stresses decay. We describe the relaxing stresses by the stress relaxation functions η^-, Ψ_1^-, and Ψ_2^- defined as follows:

$$\tau_{yx}(t) = -\eta^-(t;\dot{\gamma}_0)\dot{\gamma}_0 \qquad (4.4\text{–}21)$$

$$\tau_{xx}(t) - \tau_{yy}(t) = -\Psi_1^-(t;\dot{\gamma}_0)\dot{\gamma}_0^{\,2} \qquad (4.4\text{–}22)$$

$$\tau_{yy}(t) - \tau_{zz}(t) = -\Psi_2^-(t;\dot{\gamma}_0)\dot{\gamma}_0^{\,2} \qquad (4.4\text{–}23)$$

As in the definition of the stress growth functions, the $\dot{\gamma}_0$ in the argument of each material function indicates the parametric dependence on shear rate. The minus sign on these functions is a reminder that the steady shear flow occurred at negative times.

[6] W. R. Leppard and E. B. Christiansen *A.I.Ch.E. Journal*, *21*, 999–1006 (1975).

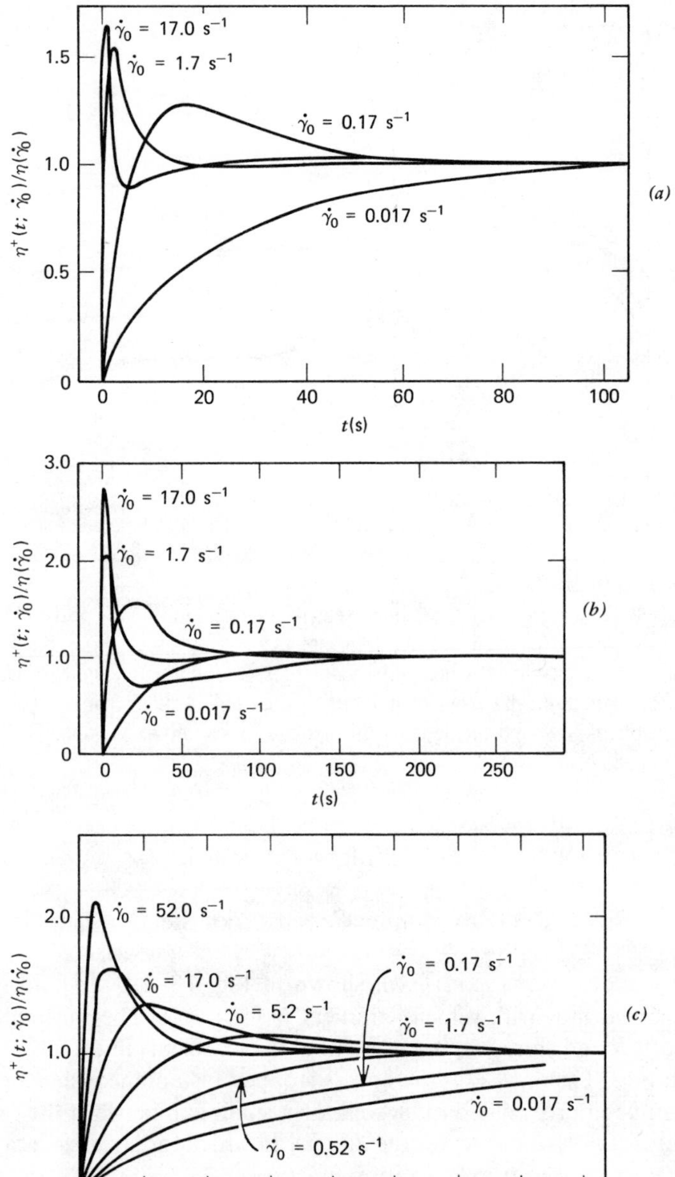

FIGURE 4.4–9. Shear stress growth function $\eta^+(t;\dot{\gamma}_0)/\eta(\dot{\gamma}_0)$ data for two polymer solutions and an aluminum soap solution: (*a*) 1.5% polyacrylamide in a 50/50 mixture of water and glycerin, (*b*) 2.0% polyisobutylene in Primol, and (*c*) 7.0% aluminum laurate in a mixture of decalin and *m*-cresol. All data were taken at 298 K. Note that the data are reduced with respect to $\eta(\dot{\gamma}_0)$ so that they all approach a common asymptote; all the data still lie within the linear viscoelastic envelope if plotted as $\eta^+(t;\dot{\gamma}_0)$. [Data of J. D. Huppler, I. F. Macdonald, E. Ashare, T. W. Spriggs, R. B. Bird, and L. A. Holmes, *Trans. Soc. Rheol.*, *11*, 181–204 (1967).]

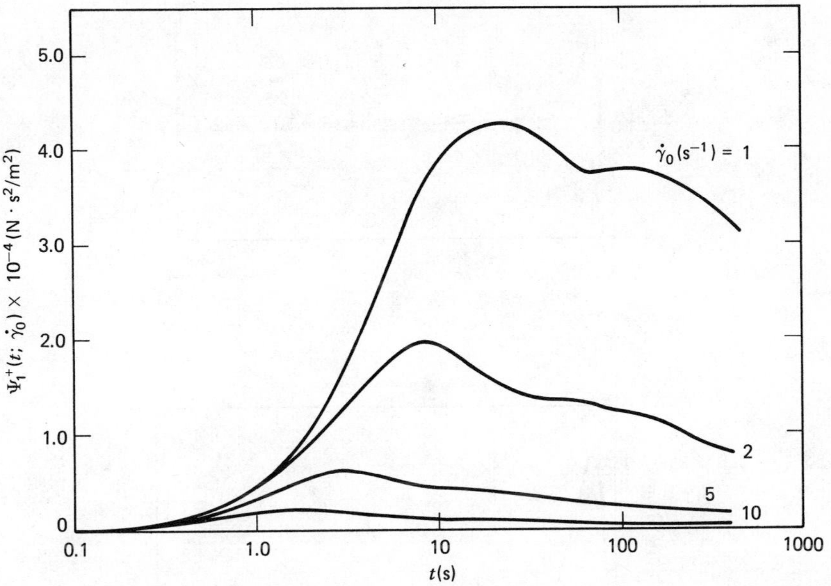

FIGURE 4.4–10. Primary normal stress growth coefficient $\Psi_1^+(t;\dot{\gamma}_0)$ data for a low density polyethylene melt (IUPAC Sample A). The data are not shown at sufficiently small shear rates to illustrate the low-shear-rate envelope. Note that the maximum in Ψ_1^+ is shifted to shorter times as the shear rate is increased. [Data replotted from J. Meissner, *J. Appl. Polym. Sci.*, *16*, 2877–2899 (1972).]

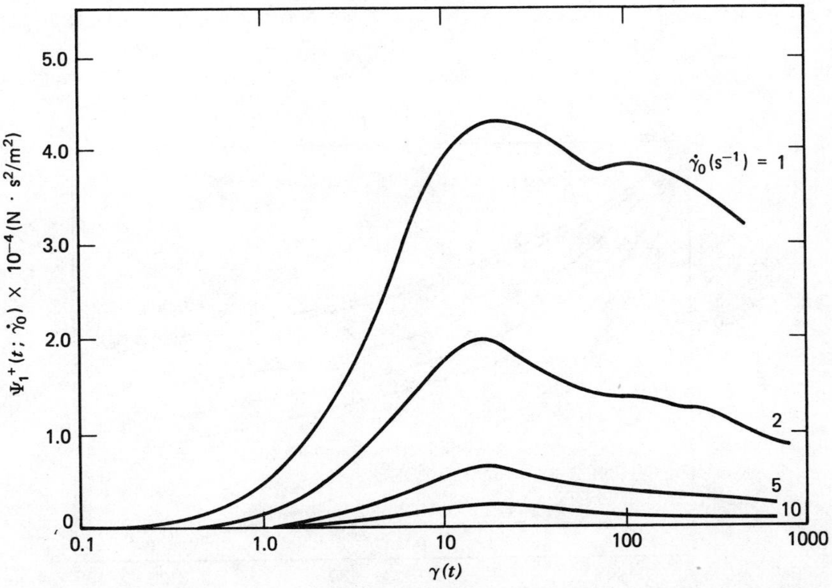

FIGURE 4.4–11. Primary normal stress growth coefficient $\Psi_1^+(t;\dot{\gamma}_0)$ data from Fig. 4.4–10 plotted as a function of the magnitude of shear $\gamma = t\dot{\gamma}_0$. The maxima occur at approximately the same value of γ independent of the imposed shear rate. [Data replotted from J. Meissner, *J. Appl. Polym. Sci.*, *16*, 2877–2899 (1972).]

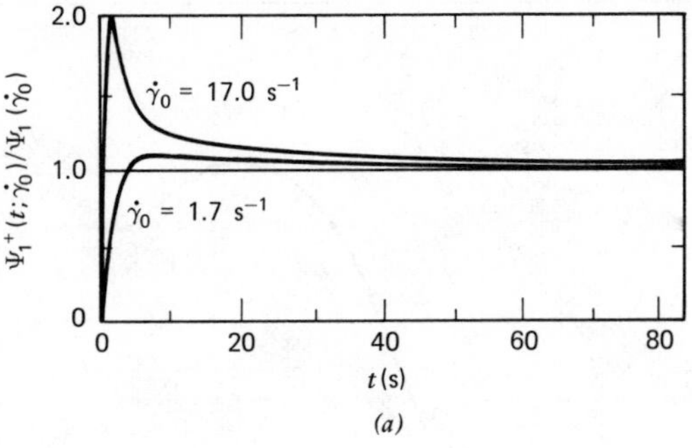

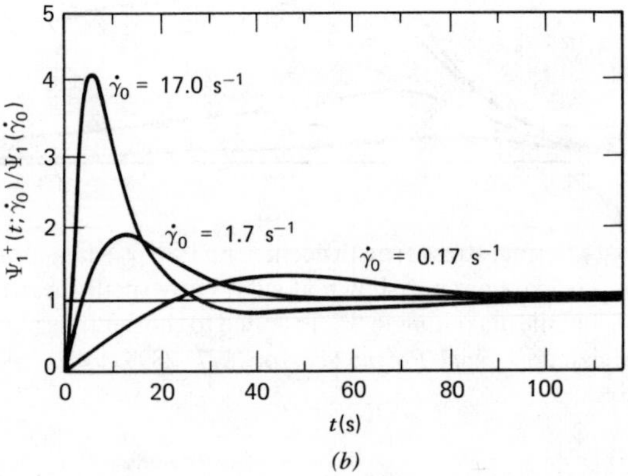

FIGURE 4.4–12. Primary normal stress growth coefficient $\Psi_1^+(t;\dot{\gamma}_0)/\Psi_1(\dot{\gamma}_0)$ data for (*a*) 1.5% polyacrylamide in a 50/50 mixture of water and glycerin and (*b*) 2.0% polyisobutylene in Primol. All data were taken at 298 K. Note that Ψ_1^+ is reduced relative to the steady shear flow value Ψ_1 so that these curves show the relative sizes of the overshoot at different shear rates. [Data of J. D. Huppler, I. F. Macdonald, E. Ashare, T. W. Spriggs, R. B. Bird, and L. A. Holmes, *Trans. Soc. Rheol.*, *11*, 181–204 (1967).]

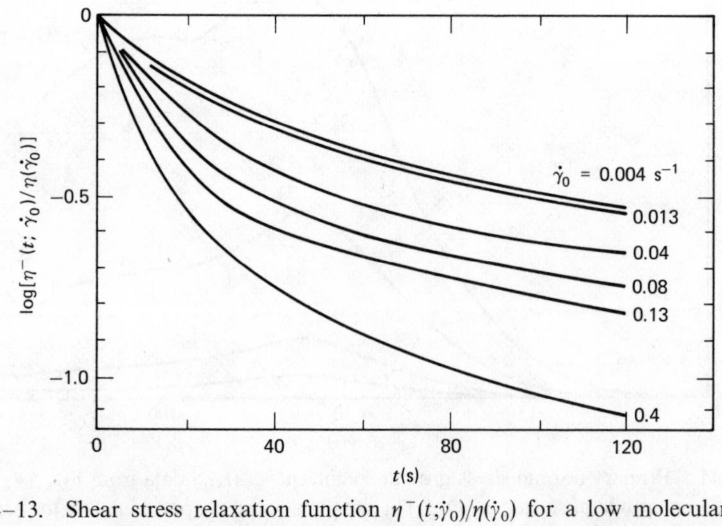

FIGURE 4.4–13. Shear stress relaxation function $\eta^-(t;\dot{\gamma}_0)/\eta(\dot{\gamma}_0)$ for a low molecular weight polyisobutylene melt ($\bar{M}_w \sim 10^4$). [E. Mustafayev, A. Ya. Malkin, Ye. P. Polotnikova, and G. V. Vinogradov, *Vysokomol. Soyed.*, *6*, 1515–1521 (1964).]

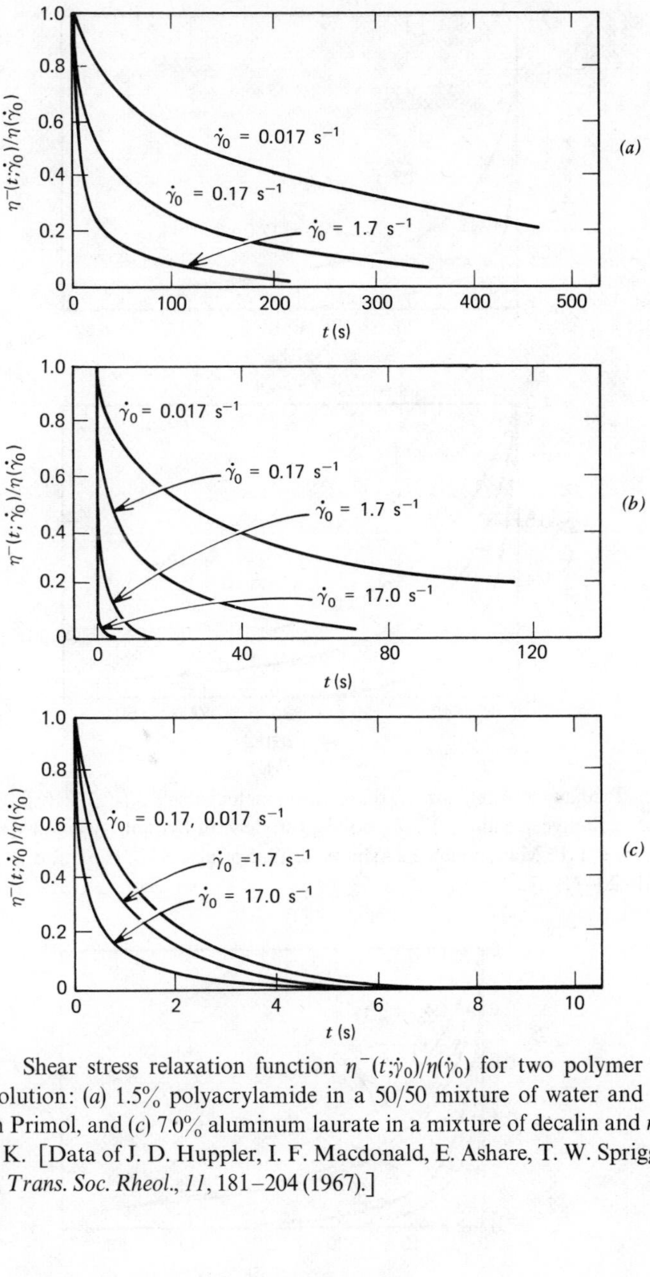

FIGURE 4.4–14. Shear stress relaxation function $\eta^-(t;\dot\gamma_0)/\eta(\dot\gamma_0)$ for two polymer solutions and an aluminum soap solution: (*a*) 1.5% polyacrylamide in a 50/50 mixture of water and glycerin, (*b*) 2.0% polyisobutylene in Primol, and (*c*) 7.0% aluminum laurate in a mixture of decalin and *m*-cresol. All data were taken at 298 K. [Data of J. D. Huppler, I. F. Macdonald, E. Ashare, T. W. Spriggs, R. B. Bird, and and L. A. Holmes, *Trans. Soc. Rheol.*, *11*, 181–204 (1967).]

Experimentally it is found that the stresses relax monotonically to zero and that they relax more rapidly as the shear rate $\dot\gamma_0$ in the preceding steady shear flow is increased. Furthermore it is found that the shear stress relaxes more rapidly than the primary normal stress difference. Sample data for η^- and Ψ_1^- are shown in Figs. 4.4–13 to 16. The only data that are available on Ψ_2^- indicate[6] that $\Psi_2^-/\Psi_1^- = \Psi_2/\Psi_1$ for all $\dot\gamma_0$ and all t.

Experiment e: Stress Relaxation after a Sudden Shearing Displacement

Another experiment in which one can observe relaxing stresses is *stress relaxation following a sudden shearing displacement* (Fig. 4.3–1*e*). A macromolecular fluid is at rest in

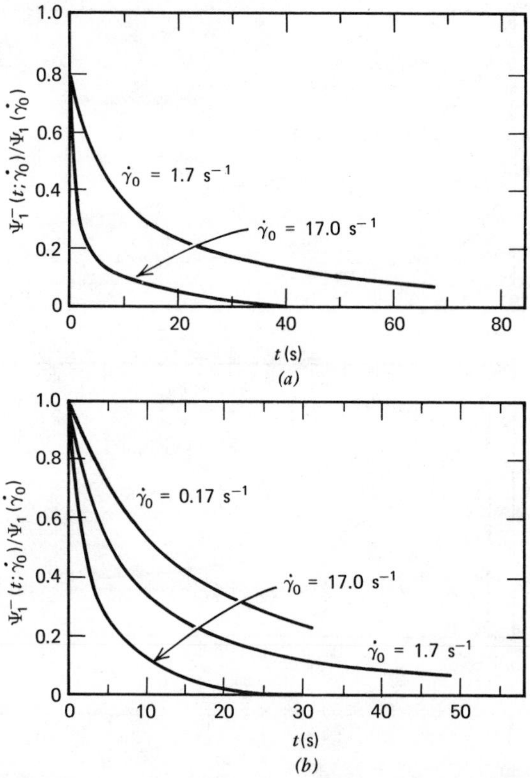

FIGURE 4.4–15. Primary normal stress relaxation coefficient $\Psi_1^-(t;\dot{\gamma}_0)/\Psi_1(\dot{\gamma}_0)$ for (*a*) 1.5% polyacrylamide in water and glycerin and (*b*) 2.0% polyisobutylene in Primol. All data were taken at 298 K. [Data of J. D. Huppler, I. F. Macdonald, E. Ashare, T. W. Spriggs, R. B. Bird, and L. A. Holmes, *Trans. Soc. Rheol.*, *11*, 181–204 (1967).]

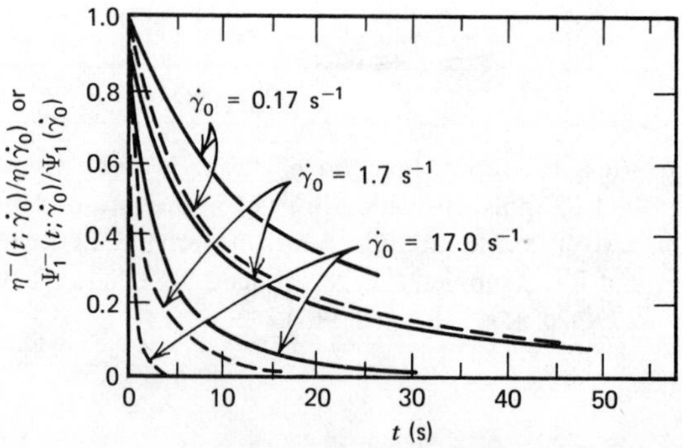

FIGURE 4.4–16. Comparison of primary normal stress relaxation coefficient $\Psi_1^-(t;\dot{\gamma}_0)/\Psi_1(\dot{\gamma}_0)$ and shear stress relaxation function $\eta^-(t;\dot{\gamma}_0)/\eta(\dot{\gamma}_0)$ for 2.0% polyisobutylene in Primol at 298 K. Note that the primary normal stress difference (———) relaxes more slowly than the shear stress (-----). [Data of J. D. Huppler, I. F. Macdonald, E. Ashare, T. W. Spriggs, R. B. Bird, and L. A. Holmes, *Trans. Soc. Rheol.*, *11*, 181–204 (1967).]

the region between two parallel plates. At time $t = t_0 - \epsilon$ a constant shear rate $\dot{\gamma}_0$ lasting only for the brief time interval ϵ is imposed on the fluid. By time $t = t_0$ the sample has experienced a total magnitude of shear $\gamma_0 = \epsilon\dot{\gamma}_0$, and the shear rate for $t > t_0$ is zero (Fig. 4.4–17). If ϵ is sufficiently small, the above can be viewed as the imposition of a shearing displacement of magnitude γ_0 on the sample at the instant t_0. The quantity of interest is the decay of the shear stress that is generated by the sudden displacement. In order to

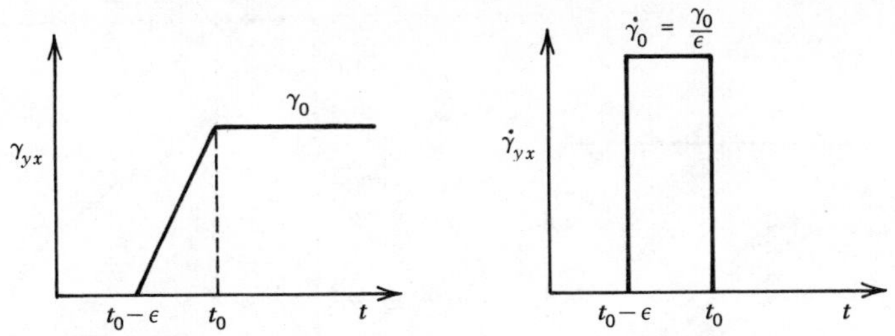

FIGURE 4.4–17. Time-dependent behavior of γ_{yx} and $\dot{\gamma}_{yx}$ in the sudden shear strain/stress relaxation experiment.

describe the relaxation, we define a relaxation modulus $G(t - t_0;\gamma_0)$ by

$$\boxed{\tau_{yx}(t) = -G(t - t_0;\gamma_0)\gamma_0} \qquad (4.4\text{–}24)$$

The material function $G(t - t_0;\gamma_0)$ is similar to the shear modulus of a perfectly elastic solid in that, for small shear strains, it is the ratio of shear stress to shear strain. Note that G depends parametrically on the magnitude of shear applied. Customarily, the time t_0 is taken to be zero.

The shear stress decreases monotonically with time. For small shear displacements the relaxation modulus is found to be independent of γ_0. In that limit, the stress is linear in strain, so that the relaxation is a linear viscoelastic property and contains the same information given by η' or η'' (see Chapter 6). As the magnitude of shear is increased from the linear limit, $G(t;\gamma_0)$ is found to decrease more rapidly at short times. However, at long times the rate of decrease of $G(t;\gamma_0)$ is again independent of γ_0, and the relaxation moduli for different shear displacements appear to be superposable by vertical shifts in this region, Data[7] illustrating the relaxation modulus for a polystyrene solution are shown in Fig. 4.4–18.

Experiment f: Creep

Instead of programing the shear rate, one can use a prescribed time-dependent shear stress and measure the resulting magnitude of shear. In a *creep experiment* (Fig. 4.3–1*f*), the fluid is at rest and stressfree for $t < 0$. At $t = 0$ a constant shear stress $\tau_{yx} = \tau_0$ is applied and held at that value for all later times. One then observes the change in the shearing displacement γ_{yx} as a function of time. This can be described in terms of the *creep compliance* $J(t;\tau_0)$:

$$\gamma_{yx}(t) = \int_0^t \dot{\gamma}_{yx}(t')\,dt' = -J(t;\tau_0)\tau_0 \qquad (4.4\text{–}25)[8]$$

The material function $J(t;\tau_0)$ is in general a nonlinear property depending on the shear stress applied for positive times. After enough time has elapsed for the flow to become steady, the magnitude of shear stress will vary linearly with time. If we then denote the intercept

[7] Earlier data on shear- and normal-stress relaxation following a sudden shearing displacement were given by N. J. Mills, *Eur. Polym. J.*, *5*, 675–695 (1969).

[8] The symbol J is sometimes used as a steady shear flow property defined by:

$$J(\dot{\gamma}) = \Psi_1/2\eta^2 \qquad (4.4\text{–}25a)$$

It can be shown from continuum mechanics (see Eq. 6.2–1 and Problem 8B.10) and from experiments that:

$$\lim_{\dot{\gamma}\to 0} J(\dot{\gamma}) = J_e^{\,0} \qquad (4.4\text{–}25b)$$

See W. W. Graessley, *Fortschr. Hochpolym.-Forsch.*, *16*, 1–179 (1974).

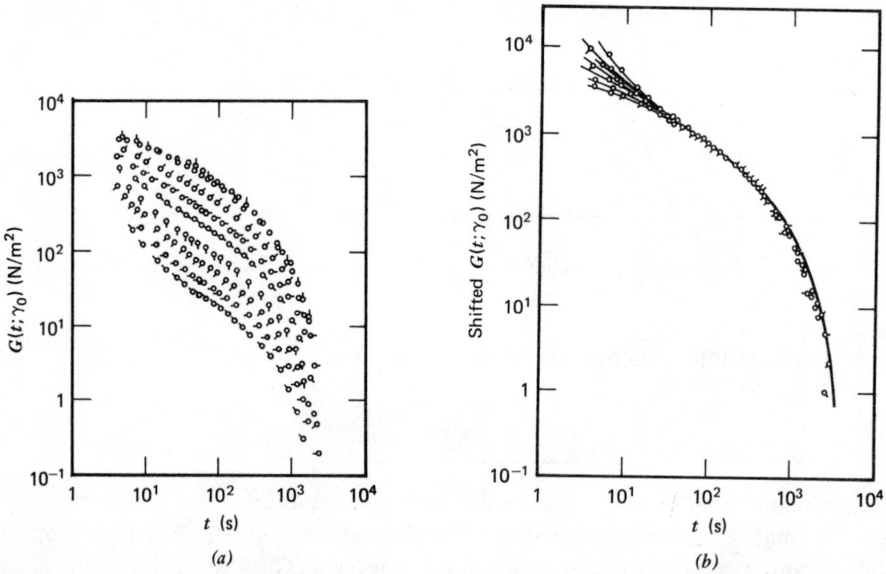

FIGURE 4.4–18. The stress relaxation modulus $G(t;\dot{\gamma}_0)$ for 20% polystyrene (narrow distribution $\bar{M}_w = 1.80 \times 10^6$) in Aroclor. The imposed magnitudes of shear are (○) 0.41, (ȯ) 1.87, (ό) 3.34, (○-) 5.22, (○̣) 6.68, (ọ) 10.0, (ọ) 13.4, (-○) 18.7, and (ˋ○) 25.4. Part (*a*) shows how $G(t;\gamma_0)$ varies with shear displacement; note that the data for γ_0 equal to 0.41 and 1.87 are in the linear regime. In (*b*) the data are superposed by vertical shifting to show the similarity in $G(t;\gamma_0)$ at large times regardless of the imposed shear displacement. [Y. Einaga, K. Osaki, M. Kurata, S. Kimura, and M. Tamura, *Polymer J.* (Japan), *2*, 550–552 (1971).]

of this steady shear flow asymptote of slope $\dot{\gamma}_\infty$ with the $t = 0$ axis as γ_0, the *steady-state compliance* $J_e{}^0(\tau_0)$ is defined by (see Fig. 4.4–19):

$$\boxed{J_e{}^0(\tau_0) = \frac{\gamma_0}{\tau_0}} \tag{4.4–26}$$

In the steady flow regime:

$$J(t;\tau_0) = J_e{}^0(\tau_0) + \frac{t}{\eta(\tau_0)} \tag{4.4–27}$$

If the driving shear stress τ_0 is small enough, $J(t;\tau_0)$ and $J_e{}^0(\tau_0)$ are found to be independent of τ_0. In this limit these are linear viscoelastic properties and are usually designated as $J(t)$ and $J_e{}^0$. Equation 4.4–27 may then be written as:

$$J(t) = J_e{}^0 + \frac{t}{\eta_0} \tag{4.4–28}$$

where η_0 is the zero-shear-rate viscosity. One may thus view the creep compliance as composed of an "elastic part," $J_e{}^0$, and a "viscous part," t/η_0. Customarily, creep data are taken in the linear viscoelastic regime, and the dependence of $J_e{}^0$ on polymer molecular weight and concentration is determined. We illustrate a typical creep compliance curve[9] in Fig. 4.4–20 for a polystyrene solution.

[9] For other examples of creep compliance data see D. J. Plazek, *Trans. Soc. Rheol.*, *9*:1, 119–134 (1965); D. J. Plazek and J. H. Magill, *J. Chem. Phys.*, *45*, 3038–3050 (1966); and D. J. Plazek and V. M. O'Rourke, *J. Polym. Sci.*, A-2, *9*, 209–243 (1971).

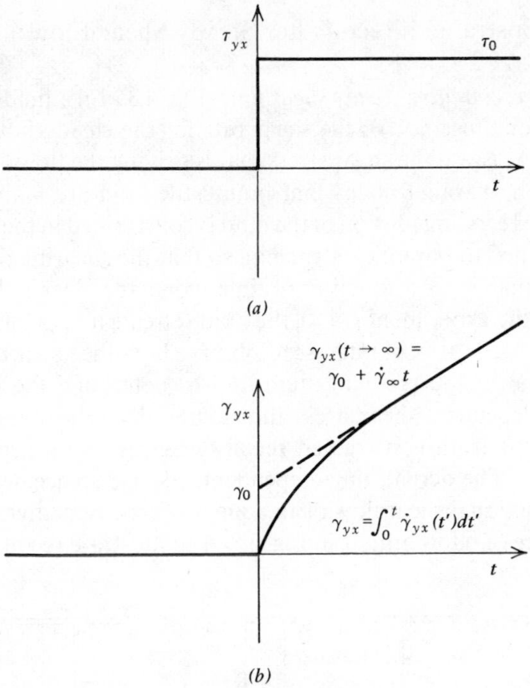

FIGURE 4.4–19. Creep experiment. (*a*) A constant shear stress τ_0 is applied at $t = 0$ and maintained for all times greater than $t = 0$. (*b*) The magnitude of shear in the creep experiment increases with time and approaches an asymptote with slope $\dot{\gamma}_\infty$ for large times.

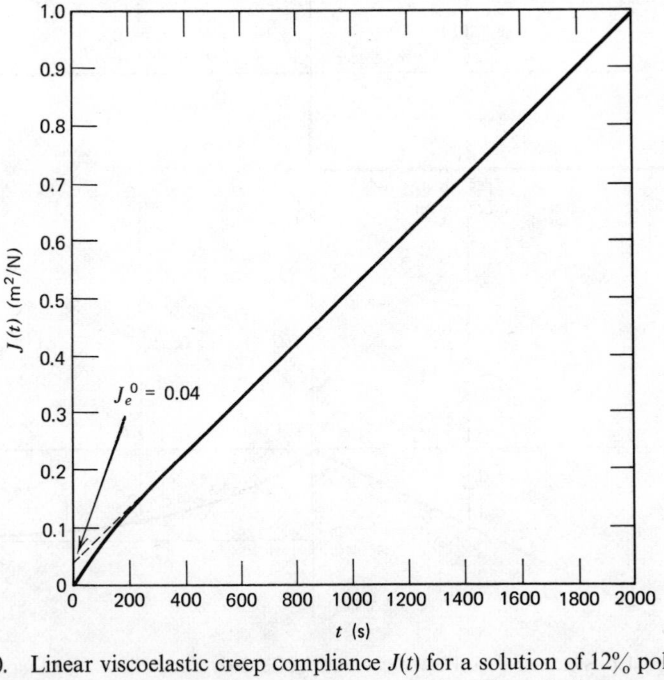

FIGURE 4.4–20. Linear viscoelastic creep compliance $J(t)$ for a solution of 12% polystyrene in chlorinated diphenyl. The curve represents data taken for several (small) applied shear stresses. Note that steady shear flow is established after approximately 400 s. [Drawn from data of K. Osaki, Y. Einaga, M. Kurata, and M. Tamura, *Macromolecules*, 4, 82–87 (1971).]

Experiment g: Constrained Recoil after Steady Shear Flow

In *constrained recoil after steady shear flow* (Fig. 4.3–1*g*) a fluid sample is undergoing a steady shear flow for times $t < 0$; the shear rate for the steady flow is $\dot{\gamma}_0$, and the shear stress is equal to τ_0. At $t = 0$ the shear stress that is driving the flow is removed and held at zero for all $t > 0$. The parallel plates that bound the fluid are maintained at a constant separation, so that the recoiling motion of the fluid is constrained in the y-direction. Furthermore, the gap is assumed to be sufficiently small so that the shearing displacement is always linear. This implies that γ_{yx} is a function of time alone and does not depend on position. During the period of the experiment $t > 0$, the fluid retreats to a position it had occupied at some previous time $t < 0$. It is of interest to observe how the shear displacement gradient $\gamma(t)$ changes with time in the recoiling fluid and to determine the ultimate recoil $\gamma_\infty = \gamma(t = \infty)$. The time-dependent shear stress and magnitude of shear are shown schematically in Fig. 4.4–21. Results from constrained recoil experiments on a polyethylene melt are shown in Fig. 4.4–22. The actual, time-dependent, shear-displacement gradient in a nonhomogeneous unidirectional shear flow of an aqueous carboxymethylcellulose solution may be seen in the sequence of photographs in Fig. 3.8–1 to illustrate recoil.

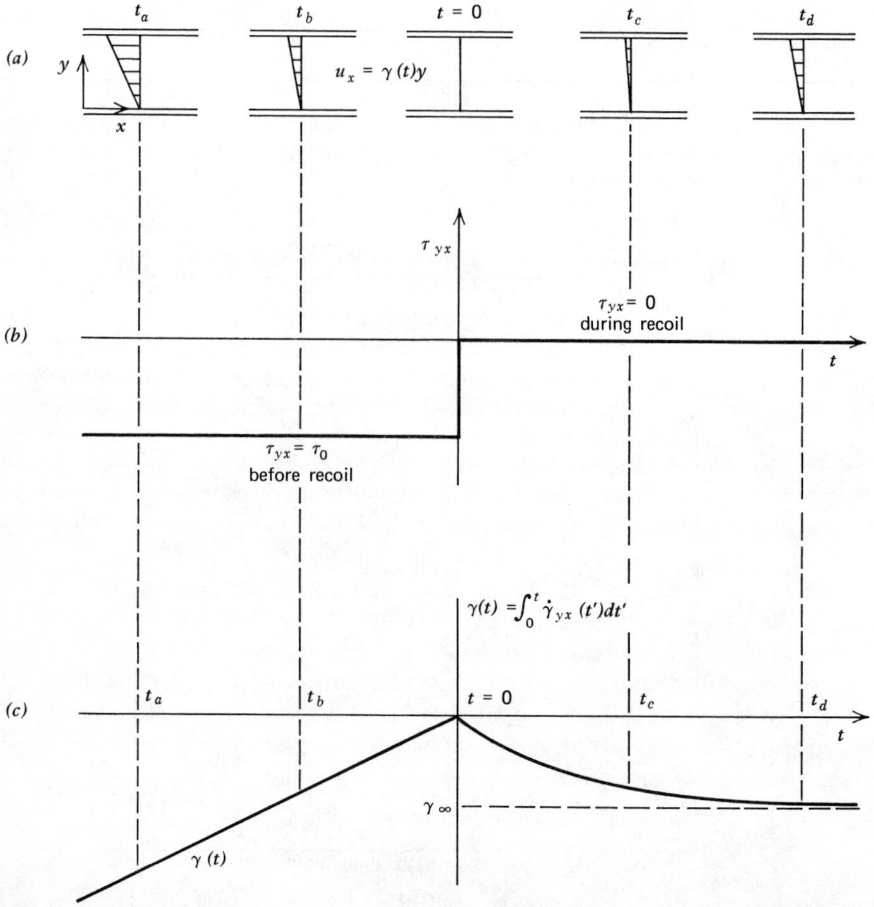

FIGURE 4.4–21. Constrained recoil after cessation of steady shear flow. (*a*) Diagram showing the shearing displacement $\gamma(t) = \int_0^t \dot{\gamma}_{yx}(t')\,dt'$ in the fluid measured with respect to the instant at which recoil begins ($t = 0$). (*b*) Sketch of stress as a function of time, showing that the shear stress is removed at $t = 0$. (*c*) Sketch of shear displacement as a function of time, showing the recoil that occurs after $t = 0$, and showing the "ultimate recoil" γ_∞.

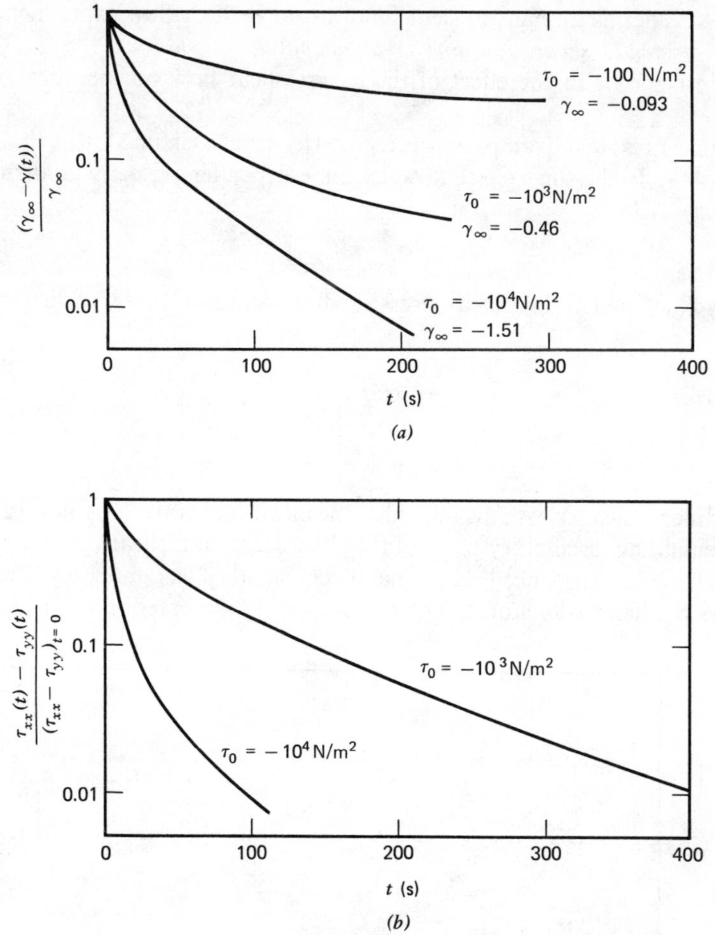

FIGURE 4.4–22. Constrained recoil after cessation of steady shear flow at the indicated shear stresses τ_0. The data are for a polyethylene melt (Melt I). (*a*) The time-dependent shear displacement $\gamma(t)$ is reduced with respect to the ultimate recoil γ_∞. Note that as the magnitude of the shear stress in the preceding steady shear flow is increased, the ultimate recoil is increased and is attained faster. (*b*) The relaxing primary normal stress difference during the recoil experiment emphasizes that $\tau_{xx} - \tau_{yy}$ does not become zero immediately when τ_{yx} is set to zero. [J. Meissner, *Rheol. Acta*, *14*, 201–218 (1975).]

A consequence of constraining the fluid recovery by maintaining the gap between the plates at a fixed size is that the normal stresses do not go to zero when the shear stress is made zero; rather, they relax over a period of time comparable to that required for the recoil (Fig. 4.4–22*b*). If all stresses were removed from the sample at $t = 0$, the fluid might expand laterally during the recoil.[10] Unconstrained or free recovery is thus not necessarily a shear flow. Since constrained recoil can be pictured as the opposite to a creep experiment, it is sometimes called "creep recovery."

Experiment h: Parallel Superposition of Steady Shear Flow and Small-Amplitude Oscillations

In addition to the experiments we have described above, we can also investigate the response of a polymer liquid to a combination of several of these. As an example of this kind

[10] A. S. Lodge, *Elastic Liquids*, Academic Press, New York (1964), pp. 131–138, 239–242.

of study, we consider as a final transient unidirectional shear flow experiment the *parallel superposition of steady shear flow and small amplitude oscillations* (Fig. 4.3–1*h*).[11] The primary interest here is in the effect of the steady shear flow on the linear viscoelelastic properties.

The shear stress in the sample is represented by a material function $\eta^*_{||}$ that is analogous to η^* in Eq. 4.4–1. In the superposed flow, the time-dependent velocity gradient is given by:

$$\dot{\gamma}_{yx} = \dot{\gamma}_m + \mathcal{R}e\{\dot{\gamma}_{yx}{}^0 e^{i\omega t}\} \tag{4.4–29}$$

where $\dot{\gamma}^0$ may be complex. The shear stress oscillates similarly and may be represented by:

$$\boxed{\begin{aligned}\tau_{yx} &= \tau_m + \mathcal{R}e\{\tau_{yx}{}^0 e^{i\omega t}\} \\ &= -\eta\dot{\gamma}_m - \mathcal{R}e\{\eta^*_{||}\dot{\gamma}_{yx}{}^0 e^{i\omega t}\}\end{aligned}} \tag{4.4–30}$$

Note that the mean shear stress τ_m is related to the mean shear rate $\dot{\gamma}_m$ by the viscosity defined in §4.3; the remaining, oscillatory parts of the shear stress and shear rate are related by the new material function $\eta^*_{||}$. Since $|\dot{\gamma}_{yx}{}^0|$ is made very small, $\eta^*_{||}$ is a function of $\dot{\gamma}_m$ and the frequency of the oscillations ω alone. The quantity $\eta^*_{||}$ is separated into real and imaginary

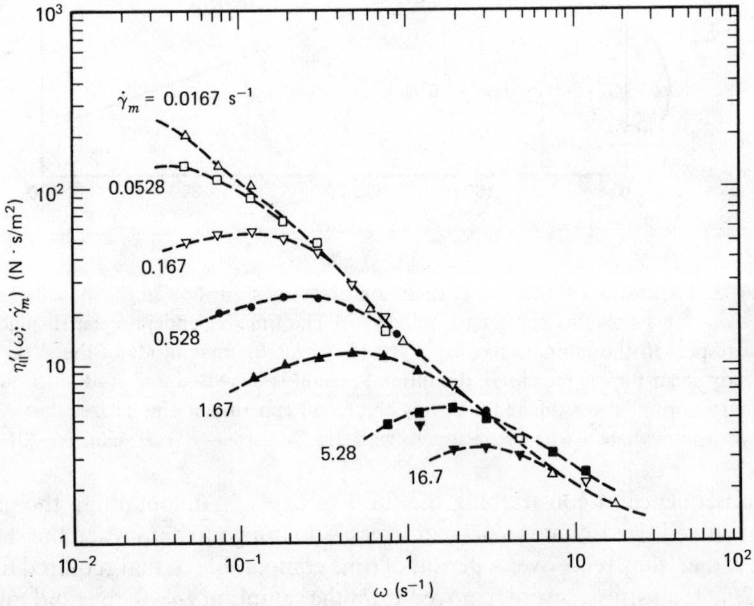

FIGURE 4.4–23. The parallel superposed flow quantity $\eta'_{||}$ as a function of frequency ω and mean shear rate $\dot{\gamma}_m$ for a 2.0% solution of polyisobutylene in Primol. All data were taken at 298 K on a Weissenberg Rheogoniometer. [I. F. Macdonald, *Trans. Soc. Rheol.*, *17*, 537–555 (1973).]

[11] I. F. Macdonald, *Trans. Soc. Rheol.*, *17*, 537–555 (1973). Data taken prior to 1970 are reviewed by H. C. Booij, Ph.D. Thesis, University of Leiden, The Netherlands (1970). His review includes K. Osaki, M. Tamura, M. Kurata, and T. Kotaka, *J. Soc. Mat. Sci. Japan*, *12*, 339–340 (1963), *J. Phys. Chem.*, *69*, 4183–4191 (1965); T. Kataoka and S. Ueda, *J. Polym. Sci.*, A–2, *7*, 475–481 (1969); K. Walters and T. E. R. Jones, in *Proceedings of the Fifth International Congress on Rheology*, S. Onogi (Ed.), University of Tokyo Press (1970), Volume 4, pp. 337–350. The data of Booij, of Kataoka and Ueda, and of Walters and Jones give $\omega_0 = 0.5\dot{\gamma}_m$, which is not supported by the data in Figs. 4.4–23 and 24.

parts as follows:

$$\eta^*_{||}(\omega; \dot{\gamma}_m) = \eta'_{||}(\omega; \dot{\gamma}_m) - i\eta''_{||}(\omega; \dot{\gamma}_m) \quad (4.4\text{–}31)$$

Data[11] illustrating the functions defined in Eq. 4.4–31 are shown in Figs. 4.4–23 and 24. At values of $\dot{\gamma}_m$ low enough so that $\eta(\dot{\gamma}) = \eta_0$, it is found that the steady shear flow does not affect the small-amplitude oscillatory shear flow properties, that is, $\eta'_{||} = \eta'$ and $\eta''_{||} = \eta''$. At higher shear rates, $\eta'_{||}$ is decreased from η' for low frequencies and is either unchanged or slightly raised compared with η' at high frequencies. As a result, a maximum can be seen in some of the $\eta'_{||}$ curves occurring at a frequency ω_0 that depends on $\dot{\gamma}_m$. The effect of the mean shear rate on $\eta''_{||}$ is even more pronounced. Again, at low frequencies, the values of $\eta''_{||}$ are reduced below their linear viscoelastic limiting values. At several of the highest shear rates, the data are even seen to become negative. The change in sign of $\eta''_{||}$ occurs approximately at ω_0.

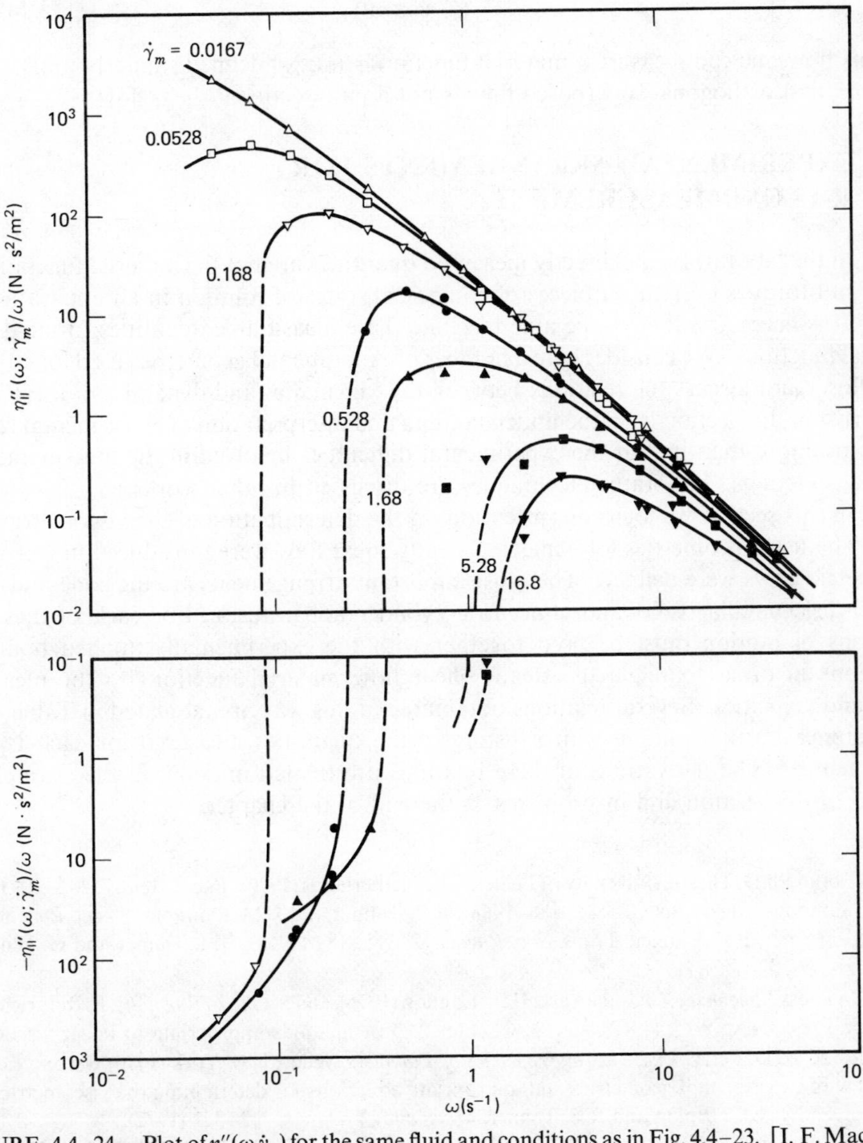

FIGURE 4.4–24. Plot of $\eta''_{||}(\omega, \dot{\gamma}_m)$ for the same fluid and conditions as in Fig. 4.4–23. [I. F. Macdonald, *Trans. Soc. Rheol., 17*, 537–555 (1973).]

The primary normal stress difference has also been studied in parallel superposed flow. As steady shear flow is added to the small-amplitude oscillatory flow, a part of the primary normal stress difference is observed that oscillates with frequency ω; this is in addition to the oscillating component with frequency 2ω that is found in the small-amplitude oscillatory shear flow experiment. At high shear rates, the term with frequency ω is the dominant one.[12]

It is also possible to superpose small-amplitude oscillations on a steady shear flow in perpendicular fashion.[13] Instead of the velocity profile given in Fig. 4.3–1*h*, the velocity pattern in a *transverse superposition of steady shear flow and small-amplitude oscillations* is:

$$v_x = \dot{\gamma}_0 y$$

$$v_y = 0$$

$$v_z = (\dot{\gamma}^0 \cos \omega t) y \tag{4.4–32}$$

For this flow one can measure a material function $\eta_{\perp}^*(\omega; \dot{\gamma}_0)$ defined similarly to $\eta_{\parallel}^*$. Note, however, that orthogonal superposed flow is not a unidirectional shear flow.

§4.5 EXPERIMENTAL ARRANGEMENTS FOR SHEAR FLOW MEASUREMENT

In the laboratory, the directly measured quantities are not the material functions but forces and torques exerted on pieces of equipment, rates of rotation in an apparatus, and so on. It is necessary, then, to be able to relate these measurable quantities to the desired material functions. We consider here a number of experimental geometries used for studying shear flows and discuss the relations between the kinematics and dynamics of these flows. The thrust of this section is on the understanding and interpretation of experimental results. We do no more than hint at the experimental difficulties involved in the measurement of material functions. Laboratory techniques are discussed in other works.[1]

In this section we focus our attention on the determination of the viscometric functions. The usual geometries for achieving steady shear flow were introduced in §4.1 where viscometric flows were defined. The most important arrangements are the cone-and-plate, parallel-disk, circular tube, and concentric cylinder apparatuses. For each of these, the equations of motion must be used together with the experimentally applied boundary conditions in order to relate the steady shear flow material functions to the measured forces and velocities. Several relations obtainable in this way are tabulated in Table 4.5–1 for reference purposes along with a listing of the quantities measured for each type of experiment.[2,3] The derivations of these relations are treated in detail in the examples at the end of this section and in problems at the end of the chapter.

[12] H. C. Booij, Ph.D. Thesis, University of Leiden, The Netherlands (1970); *Rheol. Acta*, *7*, 202–209 (1968).
[13] J. M. Simmons, *Rheol. Acta*, *7*, 184–188 (1968); R. I. Tanner and J. M. Simmons, *Chem. Eng. Sci.*, *22*, 1803–1815 (1967); R. I. Tanner, *Trans. Soc. Rheol.*, *12*, 155–182 (1968); R. I. Tanner and G. Williams, *Rheol. Acta*, *10*, 528–538 (1971).

[1] See K. Walters *Rheometry*, Chapman and Hall, London (1975), and S. Oka in *Rheology*, F. R. Eirich (Ed.), Academic Press, New York (1960), Vol. 3, Chapter 2. For methods appropriate to linear viscoelastic properties, see J. D. Ferry, *Viscoelastic Properties of Polymers*, Wiley, New York (1970), Second Edition.
[2] A review of experimental geometries and appropriate equations for determining the viscometric functions is given by A. C. Pipkin and R. I. Tanner, *Mechanics Today*, *1*, 262–321 (1972).
[3] H. Markovitz, in *Rheology*, F. R. Eirich (Ed.), Academic Press, New York (1967), Vol. 4, Chapter 6, pp. 347–410.

Table 4.5–1

Examples of Relations for Determining the Viscometric Functions (η, Ψ_1, Ψ_2) in Standard Experimental Arrangements

A. *Capillary Viscometer* (Table 4.1–2*a* and Example 4.5–3)

Q = volume rate of flow
$\Delta\mathcal{P}$ = pressure drop through tube
R = tube radius
L = tube length
$\dot{\gamma}_R$ = shear rate at tube wall
τ_R = shear stress at tube wall

$$\eta(\dot{\gamma}_R) = \frac{\tau_R}{(Q/\pi R^3)}\left[3 + \frac{d\ln(Q/\pi R^3)}{d\ln\tau_R}\right]^{-1} \qquad \text{(A–1)}$$

$$\dot{\gamma}_R = \frac{1}{\tau_R^{\,2}}\frac{d}{d\tau_R}(\tau_R^{\,3}Q/\pi R^3) \qquad \text{(A–2)}$$

$$\tau_R = \Delta\mathcal{P}R/2L \qquad \text{(A–3)}$$

B. *Cone-and-Plate Instrument* (Table 4.1–2*d* and Example 4.5–1)

R = radius of circular plate
ϑ_0 = angle between cone and plate (usually less than 4°)
W_0 = angular velocity of cone
$\mathcal{T}$ = torque on cone
$\mathcal{F}$ = force required to keep tip of cone in contact with circular plate
$\pi_{\theta\theta}(r)$ = pressure measured by flush-mounted pressure transducers located on plate
p_a = atmospheric pressure

$$\eta(\dot{\gamma}) = \frac{3\mathcal{T}}{2\pi R^3\dot{\gamma}} \qquad \text{(B–1)}$$

$$\Psi_1(\dot{\gamma}) = \frac{2\mathcal{F}}{\pi R^2\dot{\gamma}^2} \qquad \text{(B–2)}$$

$$\Psi_1(\dot{\gamma}) + 2\Psi_2(\dot{\gamma}) = -\frac{1}{\dot{\gamma}^2}\frac{\partial\pi_{\theta\theta}}{\partial\ln r} \qquad \text{(B–3)}$$

$$\Psi_2(\dot{\gamma}) = \frac{p_a - \pi_{\theta\theta}(R)}{\dot{\gamma}^2} \qquad \text{(B–4)}$$

$$\dot{\gamma} = W_0/\vartheta_0 \qquad \text{(B–5)}$$

C. *Parallel-Disk Instrument* (Table 4.1–2*c* and Example 4.5–2)

R = radius of disks
H = separation of disks
W_0 = angular velocity of upper disk
$\mathcal{T}$ = torque required to rotate upper disk
$\mathcal{F}$ = force required to keep separation of two disks constant
$\dot{\gamma}_R$ = shear rate at edge of system
$\pi_{zz}(0)$, $\pi_{zz}(R)$ = normal pressure measured on disk at center and at rim
p_a = atmospheric pressure

$$\eta(\dot{\gamma}_R) = \frac{(\mathcal{T}/2\pi R^3)}{\dot{\gamma}_R}\left[3 + \frac{d\ln(\mathcal{T}/2\pi R^3)}{d\ln\dot{\gamma}_R}\right] \qquad \text{(C–1)}$$

$$\Psi_1(\dot{\gamma}_R) - \Psi_2(\dot{\gamma}_R) = \frac{(\mathcal{F}/\pi R^2)}{\dot{\gamma}_R^{\,2}}\left[2 + \frac{d\ln(\mathcal{F}/\pi R^2)}{d\ln\dot{\gamma}_R}\right] \qquad \text{(C–2)}$$

$$\Psi_1(\dot{\gamma}_R) + \Psi_2(\dot{\gamma}_R) = \frac{1}{\dot{\gamma}_R^{\,2}}\frac{d\pi_{zz}(0)}{d\ln\dot{\gamma}_R} \qquad \text{(C–3)}$$

$$\Psi_2(\dot{\gamma}_R) = \frac{p_a - \pi_{zz}(R)}{\dot{\gamma}_R^{\,2}} \qquad \text{(C–4)}$$

$$\dot{\gamma}_R = \frac{W_0R}{H} \qquad \text{(C–5)}$$

D. *Couette Viscometer* (Table 4.1–2*b*) Narrow Gap

R_1, R_2 = radii of inner and outer cylinders
H = height of cylinders
W_1, W_2 = angular velocities of inner and outer cylinders
$\mathcal{T}$ = torque on inner cylinder

$$\eta(\dot{\gamma}) = \frac{\mathcal{T}(R_2 - R_1)}{2\pi R_1^{\,3}H|W_2 - W_1|} \qquad \text{(D–1)}$$

$$\Psi_1(\dot{\gamma}) = \frac{-[\pi_{rr}(R_2) - \pi_{rr}(R_1)]R_1}{(R_2 - R_1)\dot{\gamma}^2} + \frac{\rho R_1^{\,2}}{3\dot{\gamma}^2}(W_1^{\,2} + W_2^{\,2} + W_1W_2) \qquad \text{(D–2)}$$

Table 4.5–1 (*Continued*)

D. *Couette Viscometer* (*continued*)		
$\pi_{rr}(R_1), \pi_{rr}(R_2)$ = normal pressures measured on inner and outer cylinders	$\dot{\gamma} = \dfrac{\lvert W_2 - W_1 \rvert R_1}{R_2 - R_1}$	(D–3)
E. *Axial Annular Flow* (Fig. 3.3–1)		
R_1, R_2 = radii of inner and outer cylinders $(\pi_{rr})_i$ = reading of a flush-mounted pressure gauge at R_i	$(\pi_{rr})_1 - (\pi_{rr})_2 = -\int_{R_1}^{R_2} \dfrac{\Psi_2 \dot{\gamma}^2}{r}\, dr$	(E–1)
F. *Torsional Flow between Two Disks, the Upper One of Which is Rotating and Is Attached to a Vertical Tube.*		
h = height of rise of fluid in tube R_1 = radius of tube R_2 = radius of disks W_0 = angular velocity of tube-disk assembly H = gap between disks g = gravitational acceleration	$h = \dfrac{W_0^2}{\rho g H} \int_{R_1}^{R_2} (\Psi_1 + \Psi_2) r\, dr - \dfrac{W_0^2}{6g}(R_2^2 - R_1^2)$	(F–1)

The most widely used geometry for complete characterization of viscometric functions is the cone-and-plate configuration; the angle between the cone and plate is very small, typically less than 4°. It is available in a variety of commercial rheological instruments. At high rates of rotation in the cone-and-plate apparatus, a steady shear flow is not possible because of inertial, viscoelastic, and possibly other effects. Thus, this geometry cannot be used to obtain rheological data at very high shear rates. The cone-and-plate instrument is also used to measure unsteady unidirectional shear flow material functions.

Because of its freedom from inertial effects (see Example 4.5–3), the capillary viscometer is used to determine viscosity at very high shear rates ($\dot{\gamma} > 10^2\ \mathrm{s}^{-1}$). However, flow in the capillary viscometer yields no information on the normal stress coefficients.[4]

A parallel disk configuration can also be used to obtain all of the viscometric functions. This arrangement has the same drawback as the cone and plate at high shear rates because of inertial forces. It has the advantage, however, that the fluid motion is viscometric at all low rates of rotation, whereas the cone-and-plate flow is steady shear only in the limit of small cone angles. From Table 4.5–1, we see that to find either of the normal stress functions in torsional flow, we must know values of the pressure exerted against one of the disks at specific positions. Many early measurements were misinterpreted because of hole pressure errors. Note, on the other hand, that with a cone-and-plate instrument, the primary normal stress coefficient can be determined by measuring the total force required to keep the cone and plate together; this gives Ψ_1 directly without interference from the presence of holes.

[4] Measurements have been made of the thrust of the fluid exiting from the capillary tube on a wall, and attempts have been made to relate these measurements to the normal stress coefficients. For a summary of errors in the literature associated with the measurement of normal stresses in capillaries and slits, see J. M. Davies, J. F. Hutton, and K. Walters, *J. Phys. D.: Appl. Phys.*, *6*, 2259–2266 (1973), and K. Walters, *Rheometry*, Chapman and Hall, London (1975), pp. 96–105.

The Couette viscometer, utilizing the concentric cylinder geometry, allows a straight-forward determination of viscosity when the gap between the cylinders is small. For large gaps the analysis is complicated because the velocity field depends on the non-Newtonian viscosity function. Finally, hole pressure errors are significant in this instrument; this makes measurement of the normal stresses difficult.

In the following examples, we show how a few of the systems mentioned here are analyzed. Some of the operational difficulties encountered with each are also discussed. The final example treats an unsteady unidirectional shear flow in the oscillating cup and bob system. There we show how the complex viscosity functions are obtained from a small-amplitude, time-dependent, unidirectional shear flow. We do not give such determinations much attention in this section because they are readily available elsewhere.[5] The cup-and-bob example illustrates how an equation of motion of the equipment must often be coupled with the equation of motion for the fluid in analyzing the experiment. Equipment inertia can greatly complicate interpretation of results from time-dependent experiments.

EXAMPLE 4.5–1   Measurement of Viscosity and Normal Stress Coefficients in the Cone-and-Plate Instrument

A cone-and-plate geometry is shown in Fig. 4.5–1. The fluid to be tested is placed in the gap between the cone and plate. The cone is made to rotate with an angular velocity W. Three measure-ments can be made: the torque $\mathcal{T}$ required to turn the cone, the total normal thrust $\mathcal{F}$ on the plate, and the "pressure distribution," $(p + \tau_{\theta\theta})|_{\theta=\pi/2}$ across the plate. From these measurements we want to get the material functions, η, Ψ_1, and Ψ_2. Assume that inertial effects (i.e., the centrifugal force term in the radial equation of motion) can be neglected. Recall that in Example 1.2–5 it was shown how to find the viscosity of a Newtonian fluid from cone-and-plate viscometer data.

$$\dot{\varepsilon}_{ij} = \begin{bmatrix} \dot{\varepsilon}_{rr} & \dot{\varepsilon}_{r\theta} & \dot{\varepsilon}_{v\phi} \\ \dot{\varepsilon}_{\theta r} & \dot{\varepsilon}_{\theta\theta} & \dot{\varepsilon}_{\theta\phi} \\ \dot{\varepsilon}_{\phi r} & \dot{\varepsilon}_{\phi\theta} & \dot{\varepsilon}_{\phi\phi} \end{bmatrix}$$

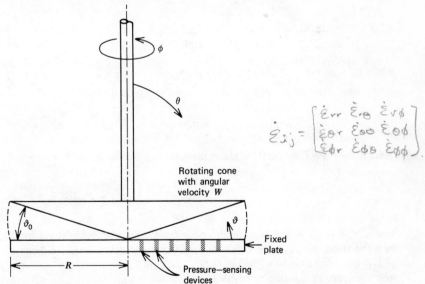

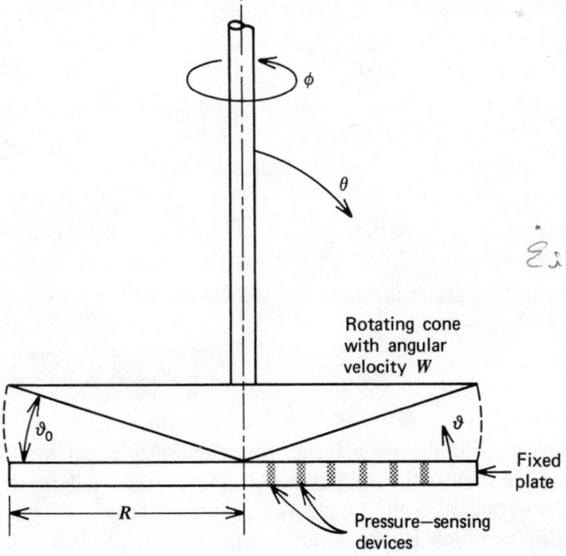

FIGURE 4.5–1.   Cone-and-plate device; ϑ_0 is quite small, of the order of several degrees.

SOLUTION   (a)   The Shear Rate

In this system, since ϑ_0 is quite small, the angle θ is very nearly equal to $\pi/2$ and consequently the $\theta\phi$-component of the rate-of-deformation tensor $\dot{\gamma}$ is very closely approximated by $(1/r)(\partial v_\phi/\partial\theta)$ (see

[5] For example, J. D. Ferry, *Viscoelastic Properties of Polymers*, Wiley, New York (1970), Second Edition; S. Oka, in *Rheology*, F. R. Eirich (Ed.), Academic Press, New York (1960), Vol. 3, Chapter 2, pp. 18–82; K. Walters, *Rheometry*, Chapman and Hall, London (1975), Chapter 6.

Table A.7–3). But since locally the flow can be regarded as simple shearing flow between two parallel plates separated by a distance $r\vartheta_0$, we have to a good approximation $v_\phi = (Wr)\vartheta/\vartheta_0$ where $\vartheta = (\pi/2) - \theta$. Then the shear rate is given by:

$$\dot{\gamma}_{\theta\phi} = -\frac{W}{\vartheta_0} \tag{4.5–1}$$

Hence $\dot{\gamma}_{\theta\phi}$ is virtually constant throughout the gap. In view of this it is known that $\tau_{\theta\phi}$, $\tau_{\phi\phi} - \tau_{\theta\theta}$, and $\tau_{\theta\theta} - \tau_{rr}$ are also virtually independent of position within the gap.

(b) The Shear Stress

The torque required to turn the cone is given by $\mathscr{T} = 2\pi \int_0^R (r\tau_{\theta\phi})r\,dr$. Since $\tau_{\theta\phi}$ is independent of r, the integration can be performed, and one obtains:

$$\tau_{\theta\phi} = \frac{\mathscr{T}}{\frac{2}{3}\pi R^3} \tag{4.5–2}$$

Hence measurement of $\mathscr{T}$ and W gives $\tau_{\theta\phi}$ and $\dot{\gamma}_{\theta\phi}$. Then the viscosity function η can be obtained from $\tau_{\theta\phi} = -\eta\dot{\gamma}_{\theta\phi}$.

(c) One Combination of Normal Stresses

From the radial equation of motion (neglecting the centrifugal force term $-\rho v_\phi{}^2/r$) we get:

$$0 = -\frac{1}{r^2}\frac{\partial}{\partial r}(r^2\pi_{rr}) + \frac{\pi_{\theta\theta} + \pi_{\phi\phi}}{r} \tag{4.5–3}$$

where $\pi_{rr} = p + \tau_{rr}$. A rearrangement gives:

$$\frac{\partial\,\pi_{rr}}{\partial \ln r} = \pi_{\theta\theta} + \pi_{\phi\phi} - 2\pi_{rr} \tag{4.5–4}$$

Since $\pi_{rr} - \pi_{\theta\theta}$ is constant throughout the gap, $\partial\pi_{rr}/\partial \ln r = \partial\pi_{\theta\theta}/\partial \ln r$ and:

$$\frac{\partial}{\partial \ln r}(p + \tau_{\theta\theta}) = (\tau_{\phi\phi} - \tau_{\theta\theta}) + 2(\tau_{\theta\theta} - \tau_{rr}) \tag{4.5–5}$$

Hence, if $p + \tau_{\theta\theta}$ is measured at the plate (where the angle θ is equal to $\pi/2$) and plotted against $\ln r$, the slope will yield the indicated combination of normal stress differences. The right-hand side of Eq. 4.5–5 can also be written as $-(\Psi_1 + 2\Psi_2)\dot{\gamma}_{\theta\phi}{}^2$. Since data in §4.3 indicate that Ψ_1 is positive and Ψ_2 is roughly $-0.1\Psi_1$, we would expect the above slope to be negative; this is illustrated is Fig. 4A.2. Note from Eq. B–4 of Table 4.5–1 or Eq. 4A.2–1 that Ψ_2 can be determined from the value $(p + \tau_{\theta\theta})$ at the "rim" (i.e., at $r = R$).

Equation 4.5–5 may be integrated along the plate from R to r to give:

$$\pi_{\theta\theta} = \pi_{\theta\theta}(R) + (\pi_{\phi\phi} + \pi_{\theta\theta} - 2\pi_{rr}) \ln\left(\frac{r}{R}\right) \tag{4.5–6}$$

Here we have used the fact that $\pi_{\theta\theta}$ at the plate is a function of r alone; this follows from the symmetry of the system.

(d) Another Combination of Normal Stresses

The total thrust of the fluid on the plate minus the thrust associated with atmospheric pressure p_a is $\mathscr{F} = 2\pi \int_0^R \pi_{\theta\theta} r\, dr - \pi R^2 p_a$. Insertion of $\pi_{\theta\theta}$ from Eq. 4.5–6 then gives after integration:

$$\mathscr{F} = \pi R^2 \pi_{\theta\theta}(R) - \tfrac{1}{2}\pi R^2(\pi_{\phi\phi} + \pi_{\theta\theta} - 2\pi_{rr}) - \pi R^2 p_a \tag{4.5–7}$$

Since the normal stress differences are independent of position, we evaluate all the π_{ii} in the second term on the right at R. Then we make use of the fact that, if the free surface is a spherical surface at $r = R$, then $p_a = \pi_{rr}(R)$. Then Eq. 4.5–7 finally yields:

$$\tau_{\phi\phi} - \tau_{\theta\theta} = -\frac{2\mathscr{F}}{\pi R^2} \tag{4.5–8}$$

Hence from a knowledge of $\mathscr{F}$ and W one can calculate $\tau_{\phi\phi} - \tau_{\theta\theta}$ and $\dot{\gamma}_{\theta\phi}$. Then from $\tau_{\phi\phi} - \tau_{\theta\theta} = -\Psi_1 \dot{\gamma}_{\theta\phi}{}^2$ one can get the primary normal stress coefficient Ψ_1.

Then from the distribution of $(p + \tau_{\theta\theta})$, one can get from Eq. 4.5–5 the difference $\tau_{\theta\theta} - \tau_{rr}$; and from the definition $\tau_{\theta\theta} - \tau_{rr} = -\Psi_2 \dot{\gamma}_{\theta\phi}{}^2$, one can obtain the secondary normal stress coefficient Ψ_2.

Note that the above considerations have made no assumptions concerning the nature of the fluid—that is, no rheological models have been used in the derivations, and all the properties η, Ψ_1, and Ψ_2 can be measured. Another method of getting Ψ_2 is by varying the vertical distance between the cone and plate.[6]

There are several assumptions that have been made in the derivations of the material functions given above that can introduce error into interpretation of experimental measurements.[2,6] First, the fluid inertia,[7] which tends to throw the fluid out of the gap, has been neglected. Consequently, the total thrust measured will be less than that given by Eq. 4.5–7 for the true primary normal stress difference, and Ψ_1 will be underestimated. In practice, the best way to correct for the centrifugal force is to calibrate the instrument using Newtonian liquids. Inertial effects can also be accounted for approximately by theoretical means (see Problem 4B.3).

Second, the rim condition assumed is that the fluid-air interface is spherical, and experimentally this is not found exactly. The nonspherical free surface implies that the flow is not viscometric all the way to the edge of the instrument. However, varying the free surface shape does not seem to affect the total thrust measurements,[8] and errors associated with the nonspherical interface appear to be no more than 5 percent. To achieve reproducibility the cone-and-plate device is sometimes operated in a sea of liquid to avoid a free surface.

Third, for large cone angles and high speeds of rotation, secondary flows are observed in the gap as a result of the competing centrifugal and normal stress effects.[9] Secondary flows may occur for cone angles as small as 10°. If secondary flows are present, the overall fluid motion will not be viscometric and the analysis presented here is invalid.

Fourth, the temperature of the fluid in the gap may not be uniform due to viscous heating. An approximate analysis of nonisothermal effects in the cone-and-plate viscometer gives a maximum tem-

[6] R. Jackson and A. Kaye, *Brit. J. Appl. Phys.*, *17*, 1355–1360 (1966); B. D. Marsh and J. R. A. Pearson, *Rheol. Acta*, *7*, 326–331 (1968); K. Walters, *Rheometry*, Chapman and Hall, London (1975), pp. 56–58.

[7] Neglect of fluid inertia can also lead to errors in the measurement of small amplitude oscillatory shear properties in the cone-and-plate geometry. The errors can be serious at high frequencies, particularly for η''. See K. Walters and R. A. Kemp, in *Polymer Systems*, R. E. Welton and R. W. Whorlow (Eds.), Macmillan, London (1968), pp. 237–250.

[8] E. Ashare, Ph.D. Thesis, University of Wisconsin, Madison (1968), pp. 50–52, 67–68, 105.

[9] H. Giesekus, *Rheol. Acta*, *4*, 85–101 (1965); W. H. Hoppmann, II, and C. E. Miller, *Trans. Soc. Rheol.*, *7*, 181–193 (1963); K. Walters and N. D. Waters, in *Polymer Systems*, R. E. Welton and R. W. Whorlow (Eds.), Macmillan, London (1968), pp. 211–235; and R. M. Turian, *Ind. Eng. Chem. Fundamentals*, *11*, 361–368 (1972).

perature rise within the gap:[10]

$$(T - T_0)_{\max} = \frac{3\mathcal{T} W \vartheta_0}{16\pi k R} \tag{4.5–9}$$

where T_0 is the temperature at which the cone-and-plate surfaces are held constant, and k is the thermal conductivity of the fluid. Equation 4.5–9 can be used to estimate the importance of viscous heating; if a temperature rise of 1 or 2 K is predicted, then serious errors can result from use of the isothermal analysis presented here. Viscous heating for some simple non-Newtonian fluid models is discussed in §5.4.

Finally, there are other problems associated with polymer degradation,[11] melt fracture, bubble inclusion, solvent evaporation, possible slip at the wall, balling of the material, and alignment of the instrument that add uncertainty to the reported values.

EXAMPLE 4.5–2 Measurement of the Viscometric Functions in the Parallel-Disk Instrument

The parallel-disk system (Fig. 4.5–2) is very similar in operation to the cone-and-plate device discussed in the preceding example. As before, when the upper disk is rotated with constant angular velocity W, the torque $\mathcal{T}$ required to achieve this rotation as well as the total force $\mathcal{F}$ required to maintain the disks at a separation H, and the pressure distribution $(p + \tau_{zz})|_{z=0}$ across one of the plates can be measured. Show how to obtain the viscometric functions in terms of these measurements when inertial effects are unimportant. (See Problem 1B.9 for Newtonian flow in this system.)

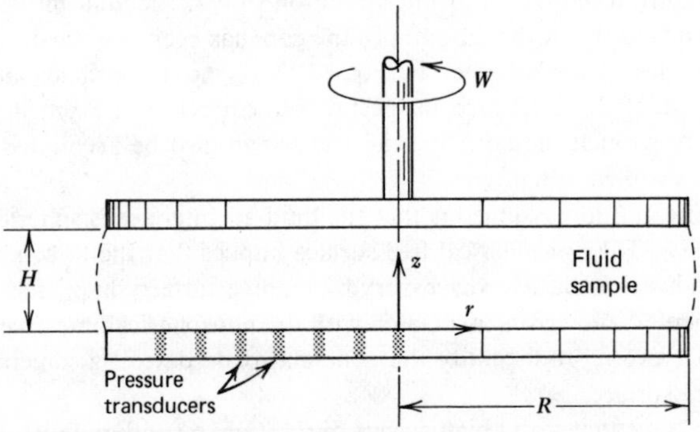

FIGURE 4.5–2. Parallel-disk instrument.

SOLUTION (a) Velocity Field and Shear Rate

We begin by identifying the shearing surfaces as fluid planes of constant z; each shearing surface is rotating with an angular velocity $w(z)$ that depends on its position between the two disks. The velocity field is then $v_\theta = rw(z)$, $v_r = 0$, and $v_z = 0$; accordingly, the shear rate is $\dot{\gamma} = r\,dw/dz$. If fluid inertia is neglected, the equation of motion in cylindrical coordinates takes the form:

r-component:
$$0 = -\frac{\partial p}{\partial r} - \left(\frac{1}{r}\frac{\partial}{\partial r}(r\tau_{rr}) - \frac{\tau_{\theta\theta}}{r}\right) \tag{4.5–10}$$

θ-component:
$$0 = \frac{\partial}{\partial z}\tau_{\theta z} \tag{4.5–11}$$

[10] This result was obtained by R. B. Bird and R. M. Turian, *Chem. Eng. Sci.*, *17*, 331–334 (1961) using a variational method.

[11] J. M. Lipson and A. S. Lodge, *Rheol. Acta*, *7*, 364–379 (1968); W. G. Pritchard, *Phil. Trans. Roy. Soc. London*, *A270*, 507–556 (1971).

z-component:
$$0 = -\frac{\partial p}{\partial z} - \frac{\partial \tau_{zz}}{\partial z} + \rho g_z \tag{4.5-12}$$

where we have set $\tau_{r\theta} = \tau_{rz} = 0$ in accordance with the symmetry requirements set forth in §4.2 (see Table 4.1–2 for the relation between the shear axes and the cylindrical coordinate system). Equation 4.5–11 says that the shear stress, and hence the shear rate, is independent of z. Thus the expression for the shear rate given above can be integrated to yield w, which in turn gives:

$$v_\theta = \frac{rWz}{H} \tag{4.5-13}$$

Thus the shear rate is a function of r alone:

$$\dot{\gamma} = \frac{rW}{H} \tag{4.5-14}$$

(b) The Viscosity

In order to find the viscosity of the sample, we consider the total torque $\mathscr{T}$ required to rotate the upper disk:

$$\begin{aligned} \mathscr{T} &= 2\pi \int_0^R (-r\tau_{z\theta}) r\, dr \\ &= 2\pi \int_0^R \eta\dot{\gamma} r^2\, dr \\ &= 2\pi \frac{R^3}{\dot{\gamma}_R{}^3} \int_0^{\dot{\gamma}_R} \eta(\dot{\gamma})\dot{\gamma}^3\, d\dot{\gamma} \end{aligned} \tag{4.5-15}$$

where $\dot{\gamma}_R = \dot{\gamma}(R)$ is the shear rate at the rim of the device. In going from the second to third line of Eq. 4.5–15 we have made the change of variable indicated by Eq. 4.5–14. Because of the inhomogeneity of the shear field in torsional flow, the viscosity does not naturally appear as an explicit function of the applied torque. However by differentiating this last result with respect to $\dot{\gamma}_R$ we find:[12]

$$\eta(\dot{\gamma}_R) = \frac{(\mathscr{T}/2\pi R^3)}{\dot{\gamma}_R}\left[3 + \frac{d \ln (\mathscr{T}/2\pi R^3)}{d \ln \dot{\gamma}_R}\right] \tag{4.5-16}$$

Thus by varying $\dot{\gamma}_R$ and computing the change in torque as indicated above, the viscosity function may be determined explicitly.

(c) One Combination of Normal Stresses[13]

A slight rearrangement of the radial component of the equation of motion gives:

$$\frac{\partial \pi_{zz}}{\partial \ln r} = (\tau_{\theta\theta} - \tau_{rr}) + \frac{\partial}{\partial \ln r}(\tau_{zz} - \tau_{rr}) \tag{4.5-17}$$

[12] Use the "Leibnitz formula" for differentiating an integral:

$$\frac{d}{dt}\int_{a_1(t)}^{a_2(t)} f(x,t)\, dx = \int_{a_1(t)}^{a_2(t)} \frac{\partial f}{\partial t}\, dx + f(a_2,t)\frac{da_2}{dt} - f(a_1,t)\frac{da_1}{dt}$$

[13] The treatment of normal stresses given here is patterned after that of T. Kotaka, M. Kurata, and M. Tamura, *J. Appl. Phys.*, *30*, 1705–1712 (1959).

This result may now be integrated from r to R to give:

$$\pi_{zz}(r) - p_a = -\int_r^R \frac{(\tau_{\theta\theta} - \tau_{rr})}{r}\, dr + (\tau_{zz}(r) - \tau_{rr}(r)) \tag{4.5–18}$$

where we have taken the radial force at the rim $\pi_{rr}(R)$ to be atmospheric pressure p_a.

Along the axis of this system, $r = 0$, the shear rate is zero. If we assume, as we have done in §4.2, that the stress in a fluid element depends only upon its deformation history, then the fluid located on the z-axis must be stressfree. This follows because the shear rate $\dot{\gamma}$ is zero for $r = 0$ and a fluid particle on the axis of the torsional flow device will remain there. Hence, all of the normal stress differences are zero there[14] and consequently $\pi_{rr}(0) = \pi_{\theta\theta}(0) = \pi_{zz}(0)$. Setting $r = 0$ in Eq. 4.5–18 then gives:

$$\pi_{zz}(0) - p_a = -\int_0^R \frac{(\tau_{\theta\theta} - \tau_{rr})}{r}\, dr \tag{4.5–19}$$

or if we change the variable of integration to $\dot{\gamma}$ and introduce the normal stress coefficients:

$$\pi_{zz}(0) - p_a = \int_0^{\dot{\gamma}_R} (\Psi_1(\dot{\gamma}) + \Psi_2(\dot{\gamma}))\dot{\gamma}\, d\dot{\gamma} \tag{4.5–20}$$

Thus we have a relation between the normal force measured at the center of the bottom disk and the sum of the normal stress coefficients.

In order to have the material functions explicitly, we differentiate Eq. 4.5–20 with the result that:

$$\frac{d}{d\dot{\gamma}_R}(\pi_{zz}(0) - p_a) = (\Psi_1(\dot{\gamma}_R) + \Psi_2(\dot{\gamma}_R))\dot{\gamma}_R \tag{4.5–21}$$

(d) The Secondary Normal Stress Coefficient

The secondary normal stress difference can easily be found by evaluating Eq. 4.5–18 at the rim:

$$\begin{aligned} \pi_{zz}(R) - p_a &= \tau_{zz}(R) - \tau_{rr}(R) \\ &= -\Psi_2(\dot{\gamma}_R)\dot{\gamma}_R^2 \end{aligned} \tag{4.5–22}$$

(e) Another Combination of Normal Stresses

The total force $\mathscr{F}$ that must be applied to keep the disks from separating is found by integrating $\pi_{zz}(r)$ over the area of one of the disks. Using π_{zz} from Eq. 4.5–18 gives:

$$\begin{aligned} \mathscr{F} &= 2\pi \int_0^R [\pi_{zz}(r) - p_a] r\, dr \\ &= -2\pi \int_0^R \int_{r'}^R \frac{(\tau_{\theta\theta}(r'') - \tau_{rr}(r''))}{r''}\, dr''\, r'\, dr' + 2\pi \int_0^R (\tau_{zz}(r') - \tau_{rr}(r'))r'\, dr' \\ &= -2\pi \int_0^R \int_0^{r''} \frac{(\tau_{\theta\theta}(r'') - \tau_{rr}(r''))}{r''}\, r'\, dr'\, dr'' + 2\pi \int_0^R (\tau_{zz}(r') - \tau_{rr}(r'))r'\, dr' \\ &= \frac{\pi R^2}{\dot{\gamma}_R^2} \int_0^{\dot{\gamma}_R} (\Psi_1(\dot{\gamma}) - \Psi_2(\dot{\gamma}))\dot{\gamma}^3\, d\dot{\gamma} \end{aligned} \tag{4.5–23}$$

[14] This is equivalent to the experimental observations that the normal stress coefficients, Ψ_1 and Ψ_2, approach finite values at zero shear rate (see §4.3).

Solving Eq. 4.5–23 for the normal stress coefficients gives

$$\Psi_1 - \Psi_2 = \frac{1}{\dot{\gamma}_R^{\,2}}\left(\frac{\mathscr{F}}{\pi R^2}\right)\left[2 + \frac{d \ln (\mathscr{F}/\pi R^2)}{d \ln \dot{\gamma}_R}\right] \tag{4.5–24}$$

By utilizing only one set of disks in a parallel disk rheometer, one can determine the non-Newtonian viscosity and both normal stress coefficients by varying the angular velocity of the upper disk and the separation of the disks. Note that as with the cone-and-plate analysis, we have made no assumption regarding the forms of the viscometric functions in deriving Eqs. 4.5–16, 21, 22, and 24.

EXAMPLE 4.5–3 Obtaining the Non-Newtonian Viscosity Function from Capillary Viscometer Data

In the cone-and-plate instrument, the shear stress and shear rate are virtually constant throughout the gap. However, in tube flow both shear stress and shear rate vary over the cross section, so that it is not obvious how to extract the viscosity function $\eta(\dot{\gamma})$ from flow rate vs. pressure drop measurements in capillary tubes. Show how these items are related.

SOLUTION (a) The Shear Stress

It is not difficult to show, using the equation of motion, that for steady tube flow the shear stress varies linearly with distance from the center of the tube (see Example 5.2–1):

$$\tau_{rz} = \tau_R \frac{r}{R} \tag{4.5–25}$$

where τ_R is the shear stress at the tube wall and is given by:

$$\tau_R = (\mathscr{P}_0 - \mathscr{P}_L)R/2L \tag{4.5–26}$$

(b) The Shear Rate[15]

The volume rate of flow is given by:

$$Q = 2\pi \int_0^R v_z r \, dr \tag{4.5–27}$$

Integration by parts gives:

$$Q = -\pi \int_0^R \left(\frac{dv_z}{dr}\right) r^2 \, dr = \pi \int_0^R \dot{\gamma} r^2 \, dr \tag{4.5–28}$$

In Eq. 4.5–28 we have introduced the shear rate $\dot{\gamma} = -dv_z/dr$. Next we change the variable of integration from r to τ_{rz} according to Eq. 4.5–25:

$$\left(\frac{Q}{\pi R^3}\right) = \frac{1}{\tau_R^{\,3}} \int_0^{\tau_R} \dot{\gamma} \tau_{rz}^{\,2} \, d\tau_{rz} \tag{4.5–29}$$

The shear rate is to be regarded as a function of shear stress in this last equation. To obtain the desired

[15] K. Weissenberg as cited by B. Rabinowitsch, *Z. Physik-Chemie*, *A145*, 1–26 (1929) and R. Eisenschitz, *Kolloid-Z.*, *64*, 184–195 (1933).

expression for the shear rate at the wall, we differentiate Eq. 4.5–29 with respect to τ_R:

$$\dot{\gamma}_R = \frac{1}{\tau_R{}^2}\frac{d}{d\tau_R}\left(\tau_R{}^3 \frac{Q}{\pi R^3}\right) \tag{4.5–30}$$

This well-known result is sometimes called the "Weissenberg-Rabinowitsch equation." It tells how the wall shear rate $\dot{\gamma}_R$ can be obtained by differentiating pressure drop-flow rate data. Once $\dot{\gamma}_R$ and τ_R are known the viscosity is easily found to be:

$$\eta(\dot{\gamma}_R) = \frac{\tau_R}{\dot{\gamma}_R} = \frac{\tau_R}{(Q/\pi R^3)}\left[3 + \frac{d \ln (Q/\pi R^3)}{d \ln \tau_R}\right]^{-1} \tag{4.5–31}$$

(c) End Effects

We now know how to calculate the viscosity, given data on the volume flow rate and pressure gradient for fully developed steady shear flow in a tube. The volume rate of flow is obtained in standard capillary viscometer measurements; however, the steady shear flow pressure gradient needed to compute the wall shear stress is not. Figure 4.5–3 illustrates a typical experimental arrangement for this instru-

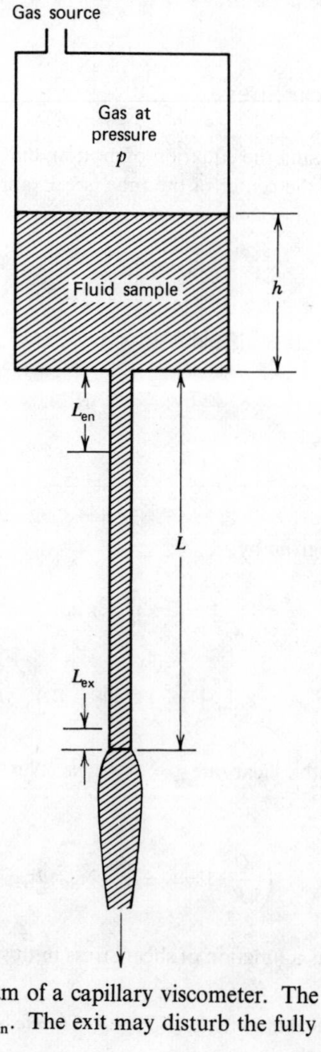

FIGURE 4.5–3. Schematic diagram of a capillary viscometer. The steady shear flow velocity profile develops over an entrance length L_{en}. The exit may disturb the fully developed profile over a distance L_{ex} from the tube end.

ment. The test fluid is pushed from a reservoir through the tube by applying a gas pressure p above the reservoir fluid. It is this pressure that is regulated and measured experimentally. Thus, in order to use the formulas just developed, we must know how to compute the steady shear flow pressure gradient (or equivalently τ_R) from p.

The desired wall shear stress can be obtained by performing two experiments,[16] one with a tube of length L_A attached to the reservoir and the other with a tube of length L_B. Both tubes have radius R and are longer than the sum of the entrance length L_{en} required for the establishment of fully developed velocity profiles and the exit length L_{ex} over which the velocity profiles may be influenced by the end of the tube. This requirement ensures that there will be steady shear flow in some section of each tube. In the first experiment, the pressure p_A required to produce some volumetric flow rate Q through the tube of length L_A is measured. A similar measurement is made in the second experiment to find the pressure p_B necessary to obtain the same flow rate Q. In each experiment, the level of fluid in the reservoir, h, is adjusted to be the same at the time of the pressure measurement. The pressure gradient that exists in the steady shear flow portions of the two tubes may then be computed by applying the macroscopic mechanical energy balance[17] to the two experimental arrangements (see Problem 4C.1). By subtracting these two results, one can show that

$$\tau_R = \left(\frac{p_B - p_A}{L_B - L_A} + \rho g \cos \beta\right)\frac{R}{2} = -\left(\frac{d\mathscr{P}}{dz}\right)\cdot\frac{R}{2} \qquad \begin{cases} L_A > L_{en} + L_{ex} \\ L_B > L_{en} + L_{ex} \end{cases} \tag{4.5–32}$$

where β is the angle between the tube axis and the vertical. This is then the corrected wall shear stress to be used with a flow rate Q through a tube of radius R. On the right-hand side of Eq. 4.5–32, $d\mathscr{P}/dz$ is the constant pressure gradient corresponding to τ_R.

We now wish to make a connection between Eq. 4.5–32 and the *Bagley end correction factor e*.[18] For simplicity we consider Eq. 4.5–32 only for horizontal tubes. According to Bagley's scheme, one can obtain the steady shear flow τ_R by measuring the driving pressure p in the reservoir for various flow rates and for tubes of different lengths and radii. Then, for a fixed value of $Q/\pi R^3$, the pressure p is plotted as a function of L/R. The result is a straight line with slope $2\tau_R$ which intersects the L/R axis at $-e$:

$$p = 2\tau_R\left(\frac{L}{R}\right) + 2e\tau_R \tag{4.5–33}$$

To compare Eq. 4.5–32 to the Bagley result, we take the measurements p_A, L_A to be reference measurements and the corresponding "B" measurements to be for any other tube length. We then drop the subscript B in Eq. 4.5–32. This equation is next rearranged to give:

$$p = 2\tau_R\frac{L}{R} + \left(p_A - 2\tau_R\frac{L_A}{R}\right) \qquad \begin{cases} L_A > L_{en} + L_{ex} \\ L > L_{en} + L_{ex} \end{cases} \tag{4.5–34}$$

where R is a constant and the flow rate Q is the same for all pairs of values of p and L. If e is given by:

$$e = \left(\frac{1}{2}\frac{p_A}{\tau_R} - \frac{L_A}{R}\right) \tag{4.5–35}$$

it is clear that Eq. 4.5–32 agrees with the Bagley equation for tubes of the same radius R. It is interesting that Bagley's data[18] down to $L/R = 0$ all fall on the same straight line. This suggests that $L_{en} + L_{ex}$ is very small.

[16] B. A. Toms, in *Rheology*, F. R. Eirich (Ed.), Academic Press (1958), Vol. 2, pp. 475–501. The correction given in Eq. 4.5–32 was first suggested by Couette.
[17] A. G. Fredrickson, *Principles and Applications of Rheology*, Prentice-Hall, Englewood Cliffs, N.J. (1964), pp. 196–200.
[18] E. B. Bagley, *J. Appl. Phys.*, *28*, 624–627 (1957). In this paper the symbol n is used for the correction factor which we call e. For additional experimental data, see W. Philippoff and F. H. Gaskins, *Trans. Soc. Rheol.*, *2*, 263–284 (1958).

Now if tubes of some different radius R' are used in the capillary viscometer, a plot of p' vs. L/R' will be parallel to the plot of p vs. L/R provided $Q/\pi R^3 = Q'/\pi R'^3$. This follows from the fact that the slope of either curve is twice the wall shear stress; and from Eq. 4.5–29, the wall shear stress depends only on the quantity $Q/\pi R^3$ (for a given fluid). However, the Bagley correction procedure assumes, in addition, that all data for different tube lengths and radii will fall on the *same* line for constant $Q/\pi R^3$. Although data[18] do seem to confirm this, it is not evident why this is so.

In addition to end effects, there are other effects which make it difficult to interpret tube flow data: viscous dissipation heating, fluid compressibility, change of viscosity with pressure, and flow instabilities. At the present, the best practice is to arrange laboratory testing so that these effects can be minimized.

EXAMPLE 4.5–4 Analysis of a Torsional Oscillatory Viscometer for Small Slit Width[19]

In Problem 1C.8, an analysis of the torsional oscillatory viscometer (see Fig. 1C.8) is presented which shows how the viscosity of a Newtonian fluid can be found from measurements of the amplitude ratio and phase shift between the driving and response functions. Repeat the procedure for a non-Newtonian fluid and show how the complex viscosity functions can be determined from the same measurements. Use the notation of Problem 1C.8, and let R be the radius of the "bob" and aR be the radius of the "cup."

SOLUTION

We begin by applying Newton's second law of motion to the bob. The rate of change in its angular momentum, $Id^2\theta_R/dt^2$, is set equal to the sum of the torques acting on it. When the bob has rotated through an angle θ_R, the torsion wire exerts a restoring torque $-K\theta_R$ on the bob; in addition the fluid exerts a torque on the bob through the shear stress at the bob surface $2\pi RL \cdot R \cdot (-\tau_{r\theta})|_R$. Combining these contributions gives an equation of motion for the bob:

$$I\frac{d^2\theta_R}{dt^2} = -K\theta_R + 2\pi R^2 L(-\tau_{r\theta})|_R \tag{4.5–36}$$

subject to the initial condition in Eq. 1C.8–2.

For the fluid in the gap, we assume that $v_\theta = v_\theta(r,t)$, $v_r = v_z = 0$ so that the θ-component of the equation of motion (Eq. B.1–5) becomes

$$\rho\frac{\partial v_\theta}{\partial t} = -\frac{1}{r^2}\frac{\partial}{\partial r}(r^2\tau_{r\theta}) \tag{4.5–37}$$

The no-slip boundary conditions on the bob and cup surfaces are

$$B.C.\ 1: \quad \text{at } r = R \qquad v_\theta = R\frac{d\theta_R}{dt} \tag{4.5–38}$$

$$B.C.\ 2: \quad \text{at } r = aR \qquad v_\theta = aR\frac{d\theta_{aR}}{dt} \tag{4.5–39}$$

The initial condition is given in Eq. 1C.8–6; the initial conditions are not needed here since we shall be interested in the bob motion after all start-up transients have died out.

[19] H. Markovitz, *J. Appl. Phys.*, *23*, 1070–1077 (1952) analyzes this problem without neglecting curvature, that is, without the narrow gap assumption. See also S. Oka, in *Rheology*, F. R. Eirich (Ed.), Academic Press, New York (1960), Vol. 3, Chapter 2.

Next we use the natural frequency of the system $\sqrt{K/I}$ to define a dimensionless time $\tau = t\sqrt{K/I}$; a dimensionless driving frequency ω is similarly defined. The driving function, that is, the forced angular oscillations of the cup, may then be written

$$\theta_{aR} = \theta_{aR}{}^0 \mathscr{Re}\{e^{i\omega\tau}\} \tag{4.5–40}$$

The amplitude of the cup oscillations, which is small, is taken here to be real, so that the cup motion is described by a cosine function. After the initial transients have died out, the shear stress in the fluid, the fluid velocity, and the bob are all expected to oscillate at the same frequency ω but with different amplitudes and phase angles than the cup. We thus postulate

$$\theta_R = \mathscr{Re}\{\theta_R{}^0 e^{i\omega\tau}\} \tag{4.5–41}$$

$$v_\theta = \mathscr{Re}\{v_\theta{}^0 e^{i\omega\tau}\} \tag{4.5–42}$$

$$\begin{aligned} \tau_{r\theta} &= \mathscr{Re}\{\tau_{r\theta}{}^0 e^{i\omega\tau}\} \\ &= \mathscr{Re}\left\{-\eta^* \frac{\partial v_\theta{}^0}{\partial r} e^{i\omega\tau}\right\} \end{aligned} \tag{4.5–43}$$

where $\theta_R{}^0$, $v_\theta{}^0$, and $\tau_{r\theta}{}^0$ are the complex amplitudes of the bob angle, fluid velocity and shear stress. In the last line of Eq. 4.5–43 we have inserted the complex viscosity using its definition in Eq. 4.4–1.

When Eqs. 4.5–41 to 43 are substituted into Eqs. 4.5–36 and 4.5–37, these latter equations become:

$$(1-\omega^2)K\mathscr{Re}\{\theta_R{}^0 e^{i\omega\tau}\} = 2\pi R^2 L \mathscr{Re}\left\{\eta^* \left.\frac{\partial v_\theta{}^0}{\partial r}\right|_{r=R} e^{i\omega\tau}\right\} \tag{4.5–44}$$

$$\omega\rho\sqrt{\frac{K}{I}}\mathscr{Re}\{i v_\theta{}^0 e^{i\omega\tau}\} = \mathscr{Re}\left\{\frac{\eta^*}{r^2}\frac{\partial}{\partial r}\left(r^2 \frac{\partial}{\partial r} v_\theta{}^0\right) e^{i\omega\tau}\right\} \tag{4.5–45}$$

Now, to remove the $\mathscr{Re}$-operators from these equations, we use the fact that if z_1 and z_2 are complex quantities, then $\mathscr{Re}\{z_1 e^{i\omega t}\} = \mathscr{Re}\{z_2 e^{i\omega t}\}$, for arbitrary frequencies ω, implies that $z_1 = z_2$; this gives:

$$\theta_R{}^0 = \frac{2\pi R^2 L \eta^*}{(1-\omega^2)K}\left.\left(\frac{\partial v_\theta{}^0}{\partial r}\right)\right|_{r=R} \tag{4.5–46}$$

$$v_\theta{}^0 = -\frac{i\eta^*}{\rho\omega}\sqrt{\frac{I}{K}}\frac{1}{r^2}\frac{\partial}{\partial r}\left(r^2 \frac{\partial}{\partial r} v_\theta{}^0\right) \tag{4.5–47}$$

Equations 4.5–46 and 47 can be simplified by assuming that $a \approx 1$, so that curvature effects may be neglected. To specify positions between the cup and bob we introduce a dimensionless length variable $x = (r - R)/(a - 1)R$. Equations 4.5–46 and 47 may thus be written as:

$$\theta_R{}^0 = \frac{M}{(1-\omega^2)} \left.\frac{\partial \phi^0}{\partial x}\right|_{x=0} \tag{4.5–48}$$

$$\frac{\partial^2 \phi^0}{\partial x^2} - \frac{i\omega}{M}\phi^0 = 0 \tag{4.5–49}$$

where Eq. 4.5–49 must be solved with the boundary conditions:

$$B.C.\ 1': \quad \text{at } x = 0 \qquad \phi^0 = iA\omega\theta_R{}^0 \tag{4.5–50}$$

$$B.C.\ 2': \quad \text{at } x = 1 \qquad \phi^0 = iA\omega\theta_{aR}{}^0 \tag{4.5–51}$$

In Equations 4.5–48 to 51 we have used the dimensionless groups introduced in Problem 1C.8:

$$\phi^0 = \frac{2\pi R^3 L\rho(a-1)}{\sqrt{KI}} v_\theta{}^0$$

$$M = \frac{\eta^*/\rho}{(a-1)^2 R^2}\sqrt{\frac{I}{K}}$$

$$A = 2\pi R^4 \frac{L\rho(a-1)}{I} \tag{4.5–52}$$

except that M is now a complex quantity involving η^* rather than μ.

One can now proceed as in Problem 1C.8 to determine $\theta_{aR}{}^0/\theta_R{}^0$ by first solving for ϕ^0 and then using that answer in Eq. 4.5–48. The result is conveniently expressed as a power series in $1/M$:

$$\frac{\theta_{aR}{}^0}{\theta_R{}^0} = 1 + \frac{i}{M}\left(\frac{\omega^2-1}{A\omega} + \frac{\omega}{2}\right) - \frac{1}{M^2}\left(\frac{\omega^2-1}{6A} + \frac{\omega^2}{24}\right) + O\left(\frac{1}{M^3}\right) \tag{4.5–53}$$

The experimentally measurable quantities are the amplitude ratio $\mathscr{A} = |\theta_R{}^0|/\theta_{aR}{}^0$ and the phase angle δ. In terms of these quantities we may write the left-hand side of Eq. 4.5–53 as (see Problem 4B.2):

$$\frac{\theta_{aR}{}^0}{\theta_R{}^0} = \frac{1}{\mathscr{A}e^{i\delta}} = \mathscr{A}^{-1}(\cos\delta - i\sin\delta) \tag{4.5–54}$$

To illustrate how η' and η'' are found from $\mathscr{A}$ and δ, we assume terms of $O(1/M^2)$ in Eq. 4.5–53 are negligible. That equation may then be easily solved for M:

$$\begin{aligned} M &= \frac{iW}{\mathscr{A}^{-1}(\cos\delta - i\sin\delta) - 1} \\ &= -\frac{W\mathscr{A}\sin\delta}{1+\mathscr{A}^2 - 2\mathscr{A}\cos\delta} - \frac{iW\mathscr{A}(\mathscr{A}-\cos\delta)}{1+\mathscr{A}^2-2\mathscr{A}\cos\delta} \end{aligned} \tag{4.5–55}$$

where W is given by

$$W = \frac{\omega^2-1}{A\omega} + \frac{\omega}{2} \tag{4.5–56}$$

Then from the definition of M in Eq. 4.5–52 and from the definition $\eta^* = \eta' - i\eta''$ we find

$$\eta' = \frac{-\rho BW\mathscr{A}\sin\delta}{1+\mathscr{A}^2-2\mathscr{A}\cos\delta} \tag{4.5–57}$$

$$\eta'' = \frac{\rho BW\mathscr{A}(\mathscr{A}-\cos\delta)}{1+\mathscr{A}^2-2\mathscr{A}\cos\delta} \tag{4.5–58}$$

in which B is an instrument parameter defined by

$$B = (a-1)^2R^2\sqrt{\frac{K}{I}} \tag{4.5–59}$$

Equations 4.5–57 and 58 are the desired expressions for the complex viscosity functions. Here W, $\mathscr{A}$, and δ are all functions of ω, which, in this example, is a *dimensionless* frequency and *not* the same as ω in Eq. 4.4–3.

§4.6 SHEARFREE FLOWS

In this section we discuss a different class of flows called *shearfree flows*. These flows have not received as much attention as shear flows until recently; this relative neglect has been at least partly due to the difficulties involved in generating these flows in controlled experiments. The study of shearfree flows is important because they occur to some extent in such diverse processing operations as vacuum forming, blow molding, foaming operations, and spinning. Data on shearfree flows are urgently needed for testing continuum models and for testing molecular theories. Inasmuch as it may be possible to characterize the complete rheological behavior of a fluid from its response in an arbitrary unsteady irrotational flow (see Problem 9D.1), data from unsteady shearfree flows would be particularly useful. Here we parallel the treatment given for shear flows: the kinematics of shearfree flows are discussed first; then the form of the stress tensor for these flows is given; finally, material functions are defined for elongational flows and some sample data are given.

A *shearfree flow*[1] is defined as a flow for which it is possible to select for every fluid element an orthogonal set of unit vectors fixed in the element so that referred to these axes the rate-of-strain tensor has a diagonal form:

$$\dot{\gamma} = \begin{pmatrix} \dot{\gamma}_{11} & 0 & 0 \\ 0 & \dot{\gamma}_{22} & 0 \\ 0 & 0 & \dot{\gamma}_{33} \end{pmatrix} \tag{4.6–1}$$

In addition we require that the volume remain constant in a shearfree flow so that $\dot{\gamma}_{11} + \dot{\gamma}_{22} + \dot{\gamma}_{33} = 0$. By "fixed in a fluid element" we mean that at every time t the same fluid particles lie along each of the three coordinate axes.[2] The particle-fixed axes used in Eq. 4.6–1 are known as the principal axes of strain. At any instant, the rate-of-strain tensor has the same form given above when referred to an orthogonal space-fixed coordinate system which has the same orientation everywhere as the principal axes of strain.

Different catagories[3] of shearfree flows are defined by specifying the $\dot{\gamma}_{ii}$. Since the constant volume condition gives one relation between the $\dot{\gamma}_{ii}$'s, only two can be chosen independently. Two special shearfree flows that have been studied are:

(i) *Elongational flow*:

$$\dot{\gamma}_{11} = 2\dot{\epsilon}; \qquad \dot{\gamma}_{22} = -\dot{\epsilon}; \qquad \dot{\gamma}_{33} = -\dot{\epsilon} \tag{4.6–2}$$

where $\dot{\epsilon} = \dot{\epsilon}(t)$ is known as the elongation rate. In later chapters we often take the principal axes of strain to be Cartesian axes and let the 1-direction

[1] See also A. S. Lodge, *Elastic Liquids*, Academic Press, New York (1964), pp. 35–38; A. S. Lodge, *Body Tensor Fields in Continuum Mechanics*, Academic Press, New York (1974) pp. 81–83, 175–176; B. D. Coleman, *Proc. Roy. Soc.*, *A 306*, 449–476 (1968). Coleman calls these flows "extensions". Thorough reviews of theoretical and experimental aspects of shearfree flows are given by J. M. Dealy, *Polym. Eng. Sci.*, *11*, 433–445 (1971) and by K. Walters, *Rheometry*, Chapman and Hall, London (1975), Chapter 7.

[2] Since $\dot{\gamma}$ is symmetric, it is always possible at any time t to select an orthogonal set of unit vectors at a given fluid element such that $\dot{\gamma}$ is diagonal when referred to these axes. However, in general, these axes will not consist of the same material particles at different times, and they are therefore not fixed in the element in the sense used above.

[3] For a method of classifying shearfree flows see J. F. Stevenson, S. C.-K. Chung, and J. T. Jenkins, *Trans. Soc. Rheol.*, *19*, 397–405 (1975).

coincide with the z-direction so that the velocity field is

$$v_x = -\frac{\dot{\epsilon}}{2}x; \qquad v_y = -\frac{\dot{\epsilon}}{2}y; \qquad v_z = \dot{\epsilon}z \tag{4.6–3}$$

Equation 4.6–3 describes *simple elongational flow.* This type of flow is closely approximated in a thin filament of fluid that is being elongated by forces exerted on the two ends; it is sometimes called "uniaxial extensional" or "uniaxial elongational flow."

If the elongation rate is negative then the flow is similar to the flow in a square sheet of material that is being stretched by equal forces on all four edges; this type of elongational flow is sometimes called "biaxial extensional flow."

(ii) *Planar elongational* (or *pure shear*) *flow*:

$$\dot{\gamma}_{11} = \dot{\epsilon}; \qquad \dot{\gamma}_{22} = -\dot{\epsilon}; \qquad \dot{\gamma}_{33} = 0 \tag{4.6–4}$$

Again $\dot{\epsilon}$ can be a function of time. This is the flow that is very nearly achieved in the Taylor four-roller apparatus.[4]

Note that both of the flows defined above are homogeneous, since $\dot{\epsilon}$ is allowed to depend on time but not on position. For steady flows $\dot{\epsilon}$ is a constant.

Next we consider the form of the stress tensor for shearfree flows. Because the rate-of-strain tensor is diagonal when referred to special axes fixed in the fluid element, the flow field is symmetric with respect to rotations of 180° about the 1-, 2-, and 3- axes. By using arguments similar to those presented in §4.2, one can use these symmetry relations to show that for shearfree flows the stress tensor, when referred to the x_1, x_2, x_3 coordinate system, must also be diagonal:

$$\boxed{\tau = \begin{pmatrix} \tau_{11} & 0 & 0 \\ 0 & \tau_{22} & 0 \\ 0 & 0 & \tau_{33} \end{pmatrix}} \tag{4.6–5}$$

Since we are dealing with incompressible fluids, we can measure only normal stress differences. Thus for any shearfree flow, there are at most two stress measurements that can be made; these can in turn be related to two material functions (see Eq. 8.5–4). For elongational flows described by Eq. 4.6–2, additional symmetry considerations lead to $\tau_{22} = \tau_{33}$. Hence, only one normal stress difference can be measured for these flows.

In steady elongational flow a material function $\bar{\eta}$ known as the *elongational viscosity* is used to describe the normal stress difference. For a steady simple elongational flow $\bar{\eta}$ is given by

$$\boxed{\tau_{zz} - \tau_{xx} = -\bar{\eta}(\dot{\epsilon})\dot{\epsilon}} \tag{4.6–6}$$

The quantity $\bar{\eta}$ is sometimes called the "Trouton viscosity." Sample data[5] for $\bar{\eta}$ are shown in

[4] G. I. Taylor, *Proc. Roy. Soc.*, *A146*, 501–523 (1934). This instrument has been used for studying the behavior of suspended particles in planar elongational flow by H. Giesekus, *Rheol. Acta*, *2*, 112–122 (1962) and W. K. Lee, Ph.D. Thesis, University of Houston (1972).

[5] J. F. Stevenson, *A.I.Ch.E. Journal*, *18*, 540–547 (1972).

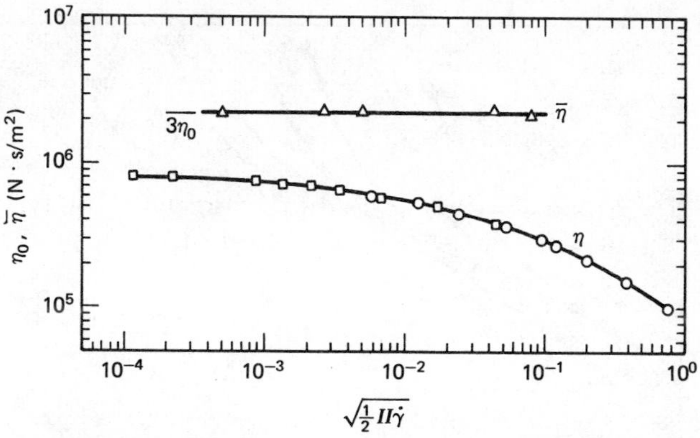

FIGURE 4.6–1. Elongational viscosity $\bar{\eta}$ and viscosity η for a polyisobutylene-isoprene copolymer (Butyl 035). Both functions are plotted vs. $\sqrt{\frac{1}{2}\mathrm{II}_{\dot{\gamma}}} = \sqrt{\frac{1}{2}(\dot{\boldsymbol{\gamma}} : \dot{\boldsymbol{\gamma}})}$, which is $\sqrt{3}\dot{\epsilon}$ for the elongational flow and $\dot{\gamma}$ for the shear flow. Note that whereas η decreases with shear rate $\bar{\eta}$ remains constant over the elongation rates tested. Also note that $\bar{\eta}$ agrees well with $3\eta_0$. Some of the viscosity data (○) were obtained from complex viscosity data using the Cox-Merz rule (see Eq. 4.7–7). [Data of J. F. Stevenson, *A.I.Ch.E. Journal*, *18*, 540–547 (1972).]

Fig. 4.6–1 for an isobutylene-isoprene copolymer. Also shown for comparison are data on the viscosity for the same material. The *zero-elongation-rate elongational viscosity* $\bar{\eta}_0$ is approximately three times the zero-shear-rate viscosity η_0. Recall a similar relation is valid for Newtonian fluids at all rates of strain (cf. Eq. 1B.11–3). As the elongation rate is increased, $\bar{\eta}$ appears to remain constant. This is contrasted in Fig. 4.6–1 with the viscosity that decreases at shear rates comparable to elongation rates shown. Similar results have been found for other polymer melts.[6] A number of continuum theories and molecular theories predict sharp increases in $\bar{\eta}$ with elongation rate.[7] Data on polymer solutions and additional data for melts are needed to evaluate these predictions.

Measuring $\bar{\eta}$ has proven to be an extremely difficult task. Probably the most straightforward arrangement for generating a simple elongational flow consists of clamping the ends of a filament of test material and using these clamps to stretch the sample (see Fig. 1B.11). We denote the direction of stretching as the z-direction. The axial normal component of stress τ_{zz} in the filament is determined by dividing the force required to pull the clamps apart by the cross-sectional area of the specimen. If p_a represents the ambient pressure, then $\pi_{zz} - p_a$ is identical to the normal stress difference $\tau_{zz} - \tau_{xx}$ in the sample (Problem 1B.11). In order for the flow to be steady, both the elongation rate and stress must be constant. The steady elongation rate is achieved by pulling the ends of the sample apart at a rate that increases exponentially with time. Then, once the force measurement reaches steady state, Eq. 4.6–6 is used to compute $\bar{\eta}$. However, even at modest elongation rates, it is difficult to achieve the steady stress measurement before either the sample breaks or the limits of the instrument are exceeded.

The start up of steady elongational flow is described by an elongational stress growth function $\bar{\eta}^+$. For $t < 0$ the fluid is assumed to be motionless and stressfree; for $t \geq 0$ the velocity field is given by Eq. 4.6–3 with $\dot{\epsilon} = \dot{\epsilon}_0$, the constant rate of elongation. Then $\bar{\eta}^+$

[6] R. L. Ballman, *Rheol. Acta*, *4*, 137–140 (1965); B. V. Radushkevich, V. D. Fikhman, and G. V. Vinogradov, *Mekh. Polimerov*, *2*, 343–348 (1968), *Dokl. Akad. Nauk, SSSR*, *180*, 404–407 (1968); G. V. Vinogradov, B. V. Radushkevich, and V. D. Fikhman, *J. Polym. Sci.*, A–2, *8*, 1–17 (1970); G. V. Vinogradov, V. D. Fikhman, B. V. Radushkevich, and A. Ya. Malkin, *J. Polym. Sci.*, A–2, *8*, 657–678 (1970); F. N. Cogswell, *Rheol. Acta*, *8*, 187–194 (1969), *Plast. Polym.*, *36*, 109–111 (1968). See reference 5 above for a tabular summary of these results.

[7] See, for instance, Examples 9.4–2 and 10.4–2.

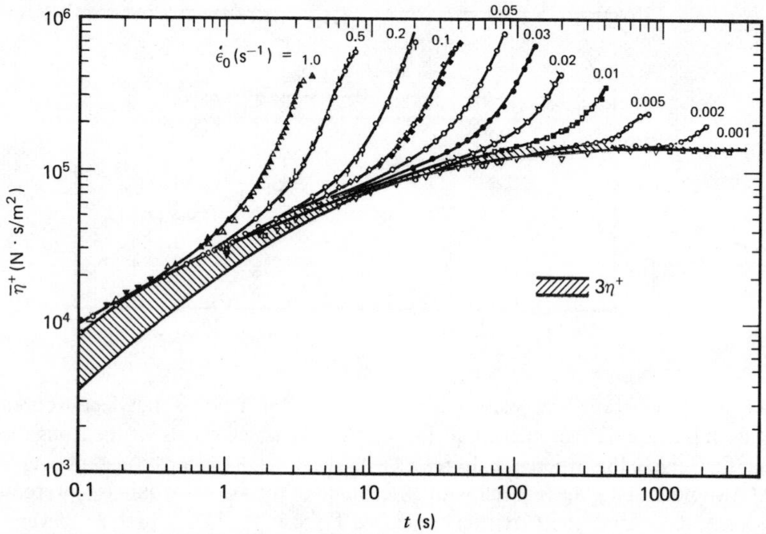

FIGURE 4.6–2. Elongational stress growth function $\bar{\eta}^+$ vs. time at various elongation rates for a polyethylene melt (Melt I). The shaded area is $3\eta^+$ where η^+ is taken from data on the same sample during the start up of steady shear flow at very small shear rates. Note the curves do not show an approach to steady state. [Data of J. Meissner, *Rheol. Acta, 10*, 230–242 (1971).]

is defined by:

$$\boxed{\tau_{zz}(t) - \tau_{xx}(t) = -\bar{\eta}^+(t;\dot{\epsilon}_0)\dot{\epsilon}_0} \tag{4.6–7}$$

for $t \geq 0$. Data for a polyethylene melt are shown in Fig. 4.6–2. The most striking feature of this figure is that apparently no steady state is attained even at the very low rates of elongation. The $\bar{\eta}^+$ curve for the lowest elongation rates follows very closely $3\eta^+$ computed from data on the start-up of steady shear flow at low shear rates. The unattainability of a steady state has been observed for other polymer melts.[5]

If a fluid sample undergoes steady elongational flow prior to $t = 0$, at which time the flow is stopped and the fluid held motionless for $t \geq 0$, then one can describe the relaxing stresses for $t \geq 0$ by the elongational stress relaxation function $\bar{\eta}^-$:

$$\boxed{\tau_{zz}(t) - \tau_{xx}(t) = -\bar{\eta}^-(t;\dot{\epsilon}_0)\dot{\epsilon}_0} \tag{4.6–8}$$

where $\dot{\epsilon}_0$ is the elongation rate in the preceding elongational flow. One can define other material functions for time-dependent elongational flows that are analogous to unidirectional shear flow material functions in §4.4.

§4.7 USEFUL CORRELATIONS FOR MATERIAL FUNCTIONS[1–4]

In §§4.3 to 4.6 we concentrated on describing how the material functions depend on the kinematics, or fluid motion. These material properties also depend on the chemical

[1] W. W. Graessley, *Fortschr. Hochpolym.-Forsch., 16*, 1–179 (1974).
[2] S. Middleman, *The Flow of High Polymers*, Interscience, New York (1968), Chapter 5.
[3] A. Peterlin, *Adv. Macromolecul. Chem., 1*, 225–281 (1968).
[4] G. C. Berry and T. G. Fox, *Fortschr. Hochpolym.-Forsch., 5*, 261–357 (1968).

constitution of the polymeric fluid—for example, solvent, type of polymer, and molecular weight—and on the physical state of the fluid as measured by variables such as polymer concentration, temperature, and pressure. In this final section we wish to illustrate how these other variables affect the material functions. To do this we focus our attention on the viscosity η and the dynamic viscosity η', since these have been the most extensively studied properties.

Temperature Effects[5,6]

The material functions are strong functions of temperature. In Fig. 4.3–2, the influence of temperature on the viscosity of a polyethylene melt is shown. There it can be seen that the zero-shear-rate viscosity decreases by two orders of magnitude as the temperature is raised from 388 to 513 K. More important, the viscosity-shear-rate curves evidently have similar shapes at the different temperatures. This similarity provides the basis for an important empirical method, known as the "method of reduced variables", for combining data taken at several different temperatures into one master curve for the sample.

In order to obtain a master curve for the viscosity function at an arbitrary reference temperature T_0 from plots of log η vs. log $\dot{\gamma}$ for a variety of temperatures T we follow a two step procedure: (1) the curve at temperature T is first shifted vertically upward by an amount $\log [\eta_0(T_0)/\eta_0(T)]$ and (2) the resulting curve is then shifted horizontally in such a way that any overlapping regions of the T_0-curve and shifted T-curve superpose. The amount by which $\eta(\dot{\gamma})_T$ must be translated to the right in order to achieve superposition is defined as $\log a_T$. Thus the method of reduced variables predicts that a single curve can be obtained by plotting viscosity data as $\log [\eta(\dot{\gamma})\eta_0(T_0)/\eta_0(T)]$ vs. $\log a_T\dot{\gamma}$ in which $a_T(T)$ is an adjustable parameter that may depend on temperature. The need for vertical shifting may be eliminated by using a reduced viscosity η/η_0. However, frequently η_0 data are not available. Then, either η_0 must be estimated by extrapolating available data[6] or else both shifts, (1) and (2), must be determined simultaneously in such a way that the best superposition is obtained. It is important to note that the reduction of data to a single curve can be achieved only when the different isotherms are of similar shape, and that the existence of a master curve implies that temperature variations influence viscosity only by altering η_0 and the shear rate (denoted by $\dot{\gamma}_0$) at which $\eta(\dot{\gamma})$ begins to decrease.

As a result of the method of reduced variables, the complete dependence of viscosity on temperature is given by the three pieces of information: a master curve, $\eta_0(T)$, and $a_T(T)$. Some useful expressions for describing a_T are given below in connection with the discussion of the dynamic viscosity. To describe the temperature variation of η_0 we recall that the viscosity of many low-molecular weight liquids varies with temperature according to:

$$\mu = Ae^{B/T} \tag{4.7–1}$$

where A and B are constants. For polymers, this type of relation may still be useful. However, a plot of log η_0 vs. $1/T$ is slightly curved so that Eq. 4.7–1 is only a rough approximation.

One of the principal uses for the method of reduced variables is that it allows us to extend the effective shear rate range of an experimental geometry. Because of the factor a_T that multiplies $\dot{\gamma}$ in the master curve, varying temperature at a fixed shear rate is equivalent to varying shear rate at a fixed temperature. Thus, by measuring viscosity over two or three decades of shear rate at several different temperatures and then combining the results from the separate tests into a composite curve, it is possible to obtain the viscosity function

[5] J. D. Ferry, *Viscoelastic Properties of Polymers*, Wiley, New York (1970), Second Edition, Chapter 11.
[6] G. V. Vinogradov and A. Ya. Malkin, *J. Polym. Sci.*, Part A, *21*, 2357–2372 (1964).

over 10 or more decades of shear rate. To obtain reliable results by this method, it is essential that substantial portions of the superposed segments overlap. Even then we must be careful in using results obtained this way, since the cumulative error in estimating several shifts may be large when a wide range of shear rates is used. Finally, when the method of reduced variables is used for dilute solutions, it is more convenient to work in terms of $\eta - \eta_s$ and $\eta_0 - \eta_s$ in place of η and η_0.

The method of reduced variables may be applied to obtain a master curve for dynamic viscosity data in exactly the same way we have described for viscosity. Actually, the method was developed in this latter context. Complete information about $\eta'(\omega)$ as a function of temperature is given by a master curve, $\eta_0(T)$, and $a_T(T)$. Since a_T is determined empirically in the process of superposing data, there is no reason to expect a priori that a_T found from dynamic viscosity data will be the same as that found from viscosity. However, since it is known experimentally that the frequency ω_0 at which η' begins to decrease is approximately the same as $\dot{\gamma}_0$, we would not be surprised to find the two shifting functions very nearly the same.

It has been found that a_T determined empirically by shifting linear viscoelastic properties may be fit by an equation of the form (see Eq. 12.3–21):

$$\log a_T = \frac{-c_1{}^0(T - T_0)}{c_2{}^0 + (T - T_0)} \tag{4.7–2}$$

for a wide variety of polymers. Equation 4.7–2 is known as the *WLF Equation*;[7] it should be used only for $T_g < T < (T_g + 100)$ where T_g is the glass transition temperature. If T_0 is taken to be T_g then $c_1{}^0 = 17.44$ and $c_2{}^0 = 51.6$ K (with T and T_0 measured in K) are found to be "universal constants" that provide a rough fit with data for an extremely broad range of polymers. Use of these values is recommended only in the absence of specific data for a polymer under consideration. A better fit of the data is obtained by using $c_1{}^0 = 8.86$ and $c_2{}^0 = 101.6$ K along with a T_0 chosen to provide a best fit with available data. For many polymers, the reference temperature chosen in this manner is found to be $T_g + 50$, within ± 5 K.

A theoretical basis for the method of reduced variables is provided by molecular theories (see Chapter 12). For dilute solutions it is predicted that

$$a_T = \frac{[(\eta_0 - \eta_s)/c]_T}{[(\eta_0 - \eta_s)/c]_{T_0}} \cdot \frac{T_0}{T} \tag{4.7–3}$$

Since $(\eta_0 - \eta_s)$ is much more sensitive to temperature than cT, the horizontal as well as vertical shifting is seen to be governed by the temperature dependence of the zero-shear-rate viscosity.

As for the temperature dependence of other shear flow material functions, it has been suggested[8] that these are governed solely by the magnitude of the shear stress. For example, when Ψ_1 is plotted as a function of $\eta\dot{\gamma}$, the result is essentially temperature independent.

Effects of Concentration and Molecular Weight[9]

We have already been introduced to the effects of concentration on the viscosity function in §2.4 where we used the intrinsic viscosity to describe the initial rate of increase

[7] M. L. Williams, R. F. Landel, and J. D. Ferry, *J. Amer. Chem. Soc.*, *77*, 3701–3707 (1955).
[8] W. W. Graessley, *op. cit.*, p. 129.
[9] W. W. Graessley, *op. cit.*, §§5 and 8. We have drawn heavily on Graessley's review article for this subsection.

of η with increasing c. In this subsection we shall extend that discussion to cover a full range of concentrations, from infinite dilution to the undiluted state. Experimentally, it is found that at low concentrations (defined roughly as $c[\eta]_0 < 1$ to 5) the material functions correlate well with $c[\eta]_0$ and at high concentrations they correlate with cM. If we recall from §2.4 that $[\eta]_0$ is proportional to M^a where a is the Mark-Houwink exponent, then we can combine the above and say that cM^x should be a good correlating parameter with x approximately equal to 0.68 at low concentration and x equal to unity at high concentrations. Because c and M occur together in both of these regimes, we shall discuss molecular weight and concentration effects together. The quantity $c[\eta]_0$, which gives a rough estimate of the degree to which the domains of different molecules are likely to overlap, is sometimes called the "coil overlap" parameter. Similarly cM gives an approximate measure of the number of intermolecular contacts per molecule. The fact that each is dominant in one concentration regime suggests two different kinds of interaction between molecules at high and low concentrations.

From the discussion in §4.3 we know that the non-Newtonian viscosity can be described to a very good approximation by giving three quantities: the zero-shear-rate viscosity η_0, the shear rate $\dot{\gamma}_0$ at which η begins to decrease from η_0, and the slope $n - 1$ (see §5.1) of the power-law region of a log η vs. log $\dot{\gamma}$ curve. A convenient way to present solution data is to plot $(\eta - \eta_s)/(\eta_0 - \eta_s)$ as a function of a reduced shear rate $\lambda\dot{\gamma}$ where λ is some characteristic time constant for the polymer solution. The form of λ suggested by molecular theory (see Chapters 10 to 12) leads to a reduced shear rate $\beta \equiv \lambda\dot{\gamma} = (\eta_0 - \eta_s)M\dot{\gamma}/cRT$. Note that in the limit as c approaches zero this definition becomes $\beta = [\eta]_0\eta_s M\dot{\gamma}/RT$. This latter quantity was used in constructing Fig. 4.3–4 in which intrinsic viscosity data were presented for a variety of polystyrene solutions at different temperatures. We shall consider, then, the influence of concentration and molecular weight on η_0, $\dot{\gamma}_0$ (or β_0), and $(n - 1)$.

The behavior of η_0 as a function of c and M is fairly well understood. At low concentrations an expression of the form $\eta_0 = \eta_0(c[\eta]_0)$ may be used to consolidate data for a given polymer-solvent system over a wide range of both concentration and molecular weight of the polymer. An example is the Martin equation:[10]

$$\eta_0 - \eta_s = \eta_s c[\eta]_0 e^{k''c[\eta]_0} \tag{4.7–4}$$

in which k'' is an arbitrary constant. Actually, a slightly better fit with data is obtained by replacing $[\eta]_0$ with $M^{a'}$ and choosing a' to fit data. It is found that a' is close to the Mark-Houwink exponent.

At high concentrations η_0 is governed by the product cM. The most striking feature of $\eta_0(cM)$ is illustrated in Fig. 4.7–1 for a variety of undiluted polymers. It is seen that η_0 goes from a linear to a 3.4 power dependence on M at some critical molecular weight M_c:

$$\eta_0 \sim M \qquad (M < M_c)$$

$$\eta_0 \sim M^{3.4} \qquad (M > M_c) \tag{4.7–5}$$

Actually there is no discontinuity in the slope of $\eta_0(cM)$; the transition is smooth and appears sharp because of the way in which the figure is plotted. The $M^{3.4}$ dependence is followed by an amazingly wide variety of linear polymers. For concentrated solutions, similar behavior is found with $(M_c)_{\text{solution}} = M_c(\rho/c)$ where ρ is the density of the solution. The $M^{3.4}$ behavior of melts and concentrated polymer solutions is thought by many polymer chemists to be due to the existence of "entanglements," or temporary, physical junctions between

[10] L. Utracki and R. Simha, *J. Polym. Sci.*, Part A, *1*, 1089–1098 (1963).

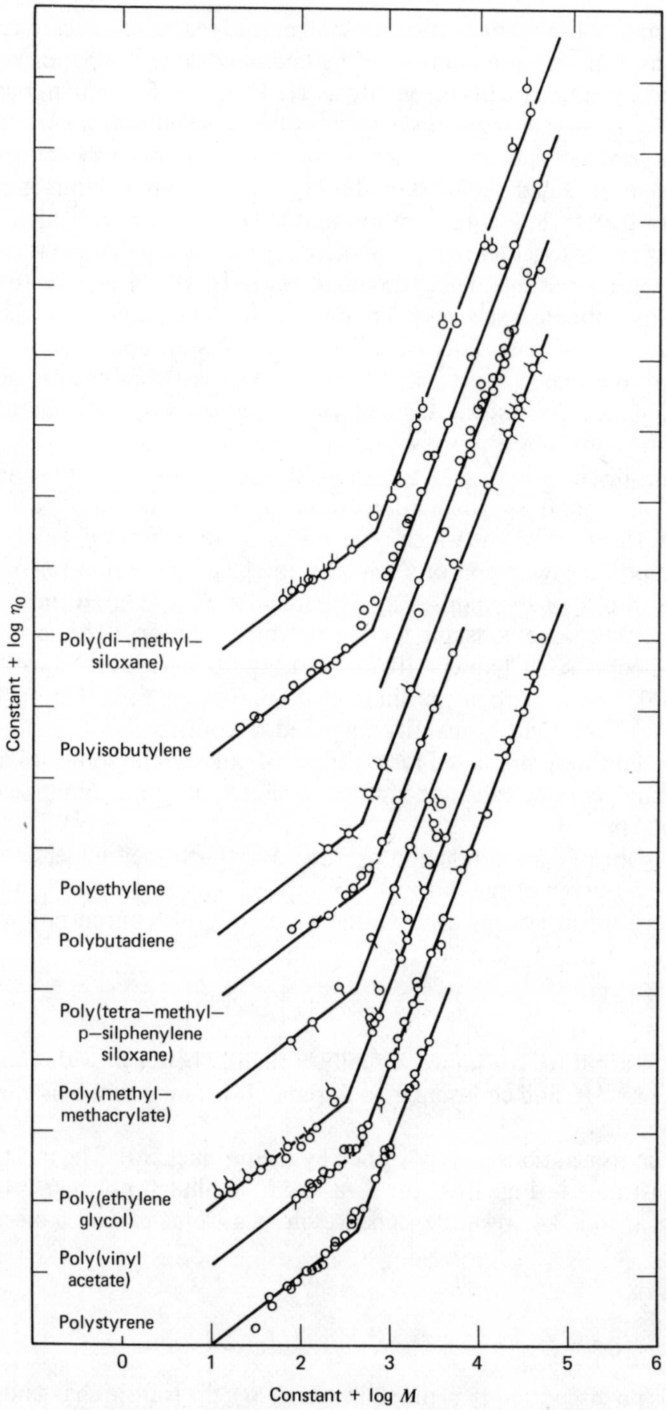

FIGURE 4.7–1. Plots of constant + log η_0 vs. constant + log M for nine different polymers. The two constants are different for each of the polymers, and the one appearing in the abscissa is proportional to concentration, which is constant for a given undiluted polymer. For each polymer the slopes of the left and right straight line regions are 1.0 and 3.4, respectively. [G. C. Berry and T. G. Fox, *Fortschr. Hochpolym.-Forsch.* (*Adv. Polym. Sci.*), *5*, 261–357 (1968).]

different polymer molecules (see Chapter 15). It is believed that M_c is a rough measure (within a factor of 2 or 3) of the molecular weight of polymer between two entanglement points.

Next we consider the variation[11] of the critical shear rate β_0 (or $\dot{\gamma}_0$) with c and M. For the sake of definiteness, β_0 is arbitrarily taken to be the value of the dimensionless shear rate at which $(\eta - \eta_s)/(\eta_0 - \eta_s) = 0.8$. At low concentrations β_0 is a constant between 1.5 and 2.0; this value persists down to infinite dilution. At high concentrations, β_0 increases with cM; β_0 begins this increase at a value of cM slightly greater than ρM_c.

Finally, the power-law slope $(n - 1)$ is approximately -0.1 at infinite dilution. It increases with $c[\eta]_0$ and approaches $(n - 1) = -0.8$ for $c[\eta]_0 > 20$. The fact that the power-law slope approaches a constant value at high concentrations and molecular weights suggests that data taken for concentrated polymer solutions (melts) at varying concentrations, molecular weights, and temperatures should all lie on a single curve when plotted as η/η_0 vs. $\dot{\gamma}/\dot{\gamma}_0$. This prediction is impressively confirmed in Fig. 4.7–2.

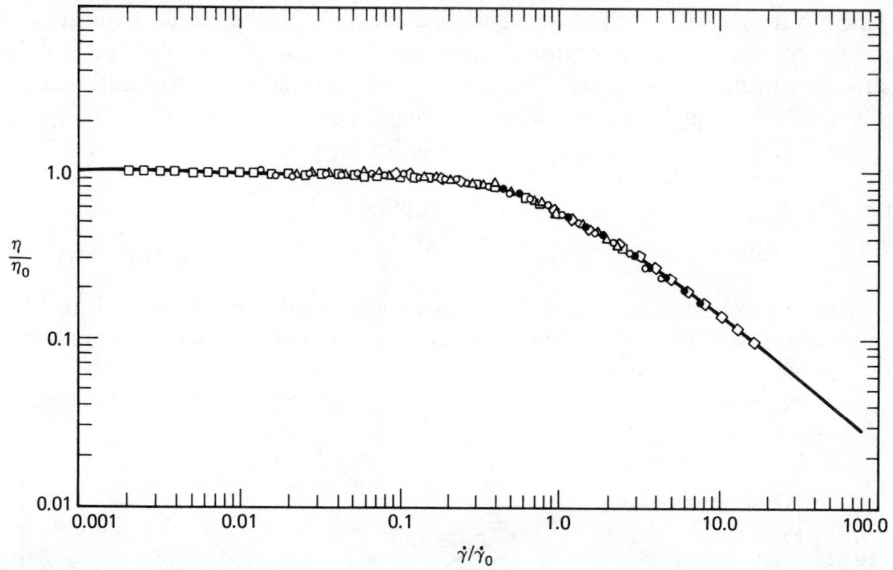

FIGURE 4.7–2. Composite plot of reduced viscosity η/η_0 vs. reduced shear rate $\dot{\gamma}/\dot{\gamma}_0$ for several different concentrated polystyrene/*n*-butyl benzene solutions. Molecular weights varied from 1.6×10^5 to 2.4×10^6; concentrations, from 0.255 to 0.55 g/cm^3; temperatures, from 303 to 333 K. [W. W. Graessley, *Fortschr. Hochpolym.-Forsch.* (*Adv. Polym. Sci.*), *16*, 1–179 (1974).]

Most of the above statements are for polymers with narrow-distribution molecular weights. Roughly speaking, broadening the distribution lowers $\dot{\gamma}_0$ and broadens the range of shear rates over which the transition from η_0 to a power-law behavior occurs. Some work has been done on constructing phenomenological blending laws from which the properties of polymers of broad molecular weight distributions can be predicted from monodisperse polymer behavior. As an example of effort along these lines we mention the study of Harris on polystyrene solutions.[12] It was found that η_0 for a two-component blend could be related to η_0 of the separate components by:

$$\log (\eta_0)_{\text{blend}} = w_1 \log (\eta_0)_1 + w_2 \log (\eta_0)_2 \tag{4.7–6}$$

[11] These results as well as those for the power-law slope are based on data for narrow molecular weight distribution polystyrene and poly(α-methyl styrene) solutions and melts. See Reference 1.

[12] E. K. Harris, Jr., *J. Appl. Polym. Sci.*, *17*, 1679–1692 (1973).

where w_1 and w_2 are the weight fractions of the components of the blend. This relation does best when the difference between molecular weights of the components is not too great. Molecular weight distribution did not appear to affect either the power-law slope or $\dot{\gamma}_0$ inasmuch as $\dot{\gamma}_0$ was the same function of c and $\bar{M}_w$ and $(n - 1)$ the same function of $c\bar{M}_n$ as for narrow distribution polymers. This study included molecular weights in the range $3.94 \times 10^5 < \bar{M}_w < 1.9 \times 10^6$ and concentrations of polystyrene in Aroclor between 0.0357 and 0.1052 g/cm^3.

The dynamic viscosity $\eta'(\omega)$ can be described by η_0, a characteristic frequency ω_0 at which $\eta'(\omega)$ begins to decrease, and a power-law slope for high frequencies. The dynamic viscosity, η' is often presented as $(\eta' - \eta_s)/(\eta_0 - \eta_s)$ as a function of $\lambda\omega$ where λ is given by molecular theory as before. Experimentally, it is found that the characteristic frequency ω_0 is practically the same as $\dot{\gamma}_0$ for *all* polymeric fluids.[13] Hence the only aspect of η' that we need to describe is the high frequency behavior.[5] For molecules that are not of too high a molecular weight, η' is proportional to $\omega^{-1/3}$ at low concentrations. At $c[\eta]_0$ approximately equal to 3, the dependence on ω begins to change until for high concentrations and undiluted polymers, η' varies as $\omega^{-1/2}$. This sequence is sometimes referred to as a transition from "Zimm-like" to "Rouse-like" behavior, as these are molecular theories that account for the respective dependencies on ω (see Chapter 12). For completeness, we note that as one proceeds from low to high concentrations, the high frequency behavior of η''/ω goes from $\omega^{-4/3}$ to $\omega^{-3/2}$.

Relations between Linear Viscoelastic Properties and Viscometric Functions

From the examples of viscosity and dynamic viscosity data presented in §§4.3 and 4.4 it is evident that $\eta(\dot{\gamma})$ and $\eta'(\omega)$ are similar functions of their arguments. In fact, we know

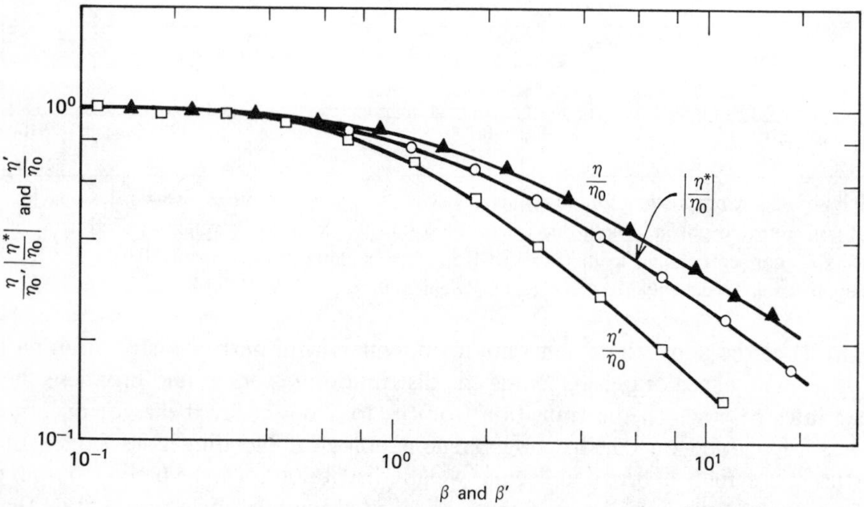

FIGURE 4.7–3. Comparison of the viscosity η, dynamic viscosity η' and the magnitude of the complex viscosity $|\eta^*|$ for a 5.0% solution of narrow-distribution molecular weight ($\bar{M}_w = 8.6 \times 10^5$) polystyrene in Aroclor. The reduced shear rate β and reduced frequency β' are defined by $(\eta_0 - \eta_s)M\dot{\gamma}/cRT$ and $(\eta_0 - \eta_s)M\omega/cRT$ respectively. All data were taken at 298 K. Note that the Cox-Merz rule Eq. 4.7–7 predicts that $|\eta^*|$ and η should be equal. Data of E. K. Harris, Jr., Ph.D. Thesis, University of Wisconsin, Madison (1968). [W. W. Graessley, *Fortschr. Hochpolym.-Forsch.*, *16*, 1–179 (1974).]

[13] W. W. Graessley, *op. cit.*, p. 126.

that both approach η_0 as their arguments go to zero and also that both begin to decrease at comparable values of $\dot{\gamma}$ and ω. The main difference between these functions is their behavior at large shear rates and frequencies; it is found that η' decreases more rapidly with ω than η does with $\dot{\gamma}$. The *Cox-Merz rule*[14] has been suggested as a way of obtaining an improved relation between the linear viscoelastic properties and the viscosity. This empiricism predicts that the magnitude of the complex viscosity should be compared with the viscosity at equal values of frequency and shear rate:

$$|\eta^*(\omega)| = \sqrt{\eta'(\omega)^2 + \eta''(\omega)^2} = \eta(\dot{\gamma})|_{\dot{\gamma}=\omega} \qquad (4.7\text{–}7)$$

The Cox-Merz rule has proven very useful in predicting η when only linear viscoelastic data are available. In Fig. 4.7–3 we show an experimental test of Eq. 4.7–7. Also shown are data for the dynamic viscosity. It is clear that $|\eta^*|$ follows η more closely than η' does and the Cox-Merz rule does not do too badly. An almost perfect match between $|\eta^*|$ and η has been found for molten polystyrene[14] and polyethylene.[15]

Other relations among shear flow material functions have been suggested by continuum theories of polymer fluids. Examples of these may be found in Tables 6.2–1 and 7.5–1 and 2.

[14] W. P. Cox and E. H. Merz, *J. Polym. Sci.*, *28*, 619–622 (1958).

[15] S. Onogi, T. Fujii, H. Katō, and S. Ogihara, *J. Phys. Chem.*, *68*, 1598–1601 (1964).

QUESTIONS FOR DISCUSSION

1. Is it necessary for a fluid undergoing a shearing flow to be incompressible in order for the flow to be a constant volume one?
2. Sketch the shapes of G' and G'' for the polymer solution shown in Fig. 4.4–5.
3. What is the Weissenberg hypothesis? How accurately does it describe macromolecular fluids?
4. Compare and contrast the definitions of shear flows and shearfree flows.
5. In what ways are the shapes of log-log plots of $\eta(\dot{\gamma})$, $\Psi_1(\dot{\gamma})$, $\eta'(\omega)$, and $\eta''(\omega)/\omega$ similar? What differences are there?
6. What is the relationship between $\tau_{11} - \tau_{22}$ in a steady shear flow and in a steady elongational flow?
7. Discuss secondary flows in the cone-and-plate viscometer in terms of centrifugal forces and the primary normal stress difference. How does this compare to the flow in the disk-cylinder apparatus in §3.6?
8. In Fig. 4.4–1 the primary normal stress difference is shown oscillating with twice the frequency of the shear rate and shear stress in the small amplitude oscillatory shear experiment. Explain.
9. Why is η_0' equal to η_0?
10. In Eq. 4.5–53 what does the expansion in terms of $(1/M)$ correspond to physically? What quantity is assumed to be small?
11. Sketch the velocity profiles in simple shear flow, simple elongational flow, and simple biaxial elongational flow.
12. Why can the power-law slope of the $\eta(\dot{\gamma})$ curve not be less than -1? Is there a similar limitation on $\Psi_1(\dot{\gamma})$? $\eta'(\omega)$? $\eta''(\omega)/\omega$?

PROBLEMS

4A.1 Analysis of Capillary Viscometer Data[1]

The volumetric flow rate through a capillary tube has been measured at a series of imposed pressure drops for a 3.5% (weight) solution of carboxymethylcellulose in water at 303 K. The data have been corrected for end effects using the procedure of Example 4.5–3:

$4Q/\pi R^3$, s^{-1}	τ_R, N/m^2
250	220
350	255
500	298
700	341
900	382
1250	441
1750	509
2500	584
3500	670
5000	751
7000	825
9000	887
12500	1000
17500	1070
25000	1200

Construct a graph of log η vs. log $\dot{\gamma}$ for this fluid. What is the slope of this curve in the power-law region?

4A.2 Normal Stress Measurements in the Cone-and-Plate Instrument

In Fig. 4A.2 data are shown on the radial variation of the total normal stress $\pi_{\theta\theta}$ exerted on the plate in the cone-and-plate device. The data are for a 2.5% polyacrylamide solution and were taken using flush-mounted pressure transducers at the positions indicated. The radius of the plate is 5 cm. From these data determine Ψ_1 and Ψ_2 for the polyacrylamide solution at the four shear rates shown. To do this the relation:

$$[\pi_{\theta\theta}(R) - p_a] = (\tau_{\theta\theta} - \tau_{rr}) \tag{4A.2–1}$$

is needed. Where did this come from?

4A.3 Viscous Heating in the Cone-and-Plate Instrument

The data in Fig. 4.3–3 were obtained using a cone-and-plate geometry; the cone angle was 1.5° and the plate diameter was 7.5 cm. The cone-and-plate surfaces were maintained at 298 K. We want to estimate the effect of viscous heating on this data. For the 1.5% polyacrylamide solution, what is the maximum rate of rotation that can be allowed to insure that the temperature rise in the gap is less than 1 K? What shear rate does this correspond to?

[1] Tabular data from A. G. Fredrickson, *Principles and Applications of Rheology*, © 1964. Reprinted by permission of Prentice-Hall, Inc., Englewood Cliffs, N.J., pp. 309–310.

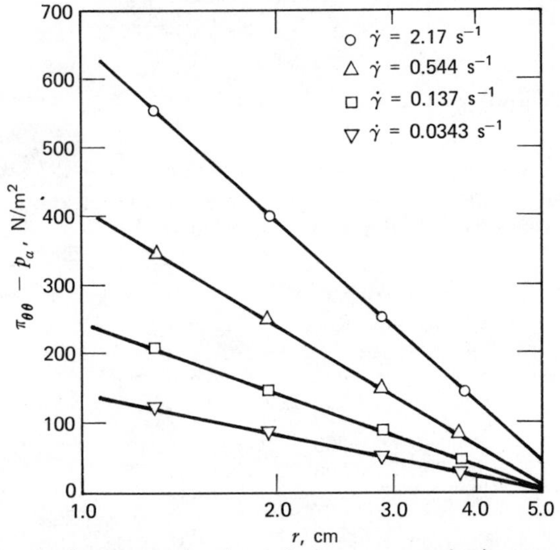

FIGURE 4A.2. Total normal stress on the plate relative to atmospheric pressure $\pi_{\theta\theta} - p_a$ vs. distance from the center of the plate for a 2.5% polyacrylamide solution. [E. B. Christiansen and W. R. Leppard, *Trans. Soc. Rheol.*, *18*, 65–86 (1974).]

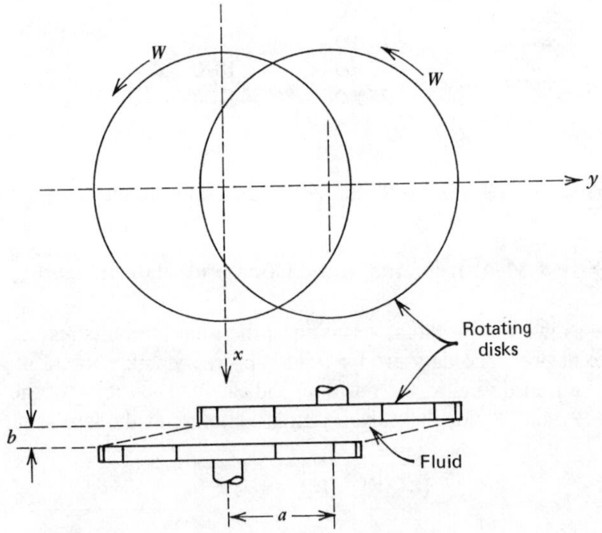

FIGURE 4B.1. Schematic diagram of the eccentric disk (Maxwell orthogonal) rheometer; both circular disks rotate with the same constant angular velocity W.

4A.4 Relation between Viscosity and Complex Viscosity

Use the Cox-Merz rule[2] to construct plots of $\eta(\dot{\gamma})$ for the polyacrylamide and aluminum soap solutions shown in Fig. 4.3–3 and Figs. 4.4–3 and 4. Also on the same plots, graph $\eta'(\omega)$ and some of the actual $\eta(\dot{\gamma})$ data. How well is the viscosity described by $|\eta^*|$? η'? In Chapter 7 we shall encounter a rheological equation of state that predicts $\eta(\dot{\gamma}) = \eta'(\omega)|_{\omega=\dot{\gamma}}$ (cf. Eqs. A and B in Table 7.5–1).

[2] W. P. Cox and E. H. Merz, *J. Polym. Sci.*, *28*, 619–622 (1958).

4A.5 Applicability of Concentric-Cylinder Instrument Results

In order to use the formulas presented in Table 4.5–1*D*, the gap between the two cylinders of this instrument must be small. How must the ratio R_1/R_2 be restricted if the shear stress is to vary no more than two percent within the gap?

4A.6 Viscosity Measurements with an Epprecht Viscometer

An Epprecht viscometer utilizes a concentric cylinder geometry in which the inner cylinder is rotated at angular velocity W_1 and the outer cylinder is held in a fixed position. The shear stress at the inner cylinder τ_1 is measured for a series of rotation rates. Sample data[3] taken on a 1% solution of carboxymethylcellulose are given below. For the particular instrument used, the inner cylinder had a radius $R_1 = 1.54$ cm, and the outer cylinder had a radius $R_2 = 1.94$ cm. Use the results of Problem 4C.3 to construct a plot of $\eta(\dot{\gamma})$ for this polymer solution.

τ_1 (N/m^2)	W_1 (rad/s)
4.1	2.63
5.2	3.53
6.5	4.64
8.1	6.19
10.1	8.17
13.6	11.8
17.0	15.9
20.7	20.9
25.4	28.0
30.9	36.9

4B.1 Eccentric Disk Rheometer

The eccentric disk (or Maxwell orthogonal) rheometer[1] consists of two parallel disks, both of which rotate with a constant angular velocity W. The axes of rotation of the disks are separated by a distance a, and the gap between the disks is b (Fig. 4B.1). Fluid disks of constant z are assumed to rotate rigidly about points on the line connecting the centers of the bounding disks.

a. Is this flow a shear flow?

b. Show that the velocity field corresponding to the motion described above is:

$$v_x = -W(y - Az)$$

$$v_y = Wx$$

$$v_z = 0 \qquad (4B.1\text{–}1)$$

where $A = a/b$.

In Problem 7C.2 it is shown that η' and η'' can be obtained from measurements on this instrument (see also Problem 11C.5). For more on the kinematics of the eccentric disk flow see Problem 4C.4.

[3] Tabular data from S. Middleman, *The Flow of High Polymers*, Interscience, New York (1968), p. 23.

[1] A. N. Gent, *Brit. J. Appl. Phys.*, *11*, 165–167 (1960); B. Maxwell and R. P. Chartoff, *Trans. Soc. Rheol.*, *9*:1, 41–52 (1965); K. Walters, *Rheometry*, Chapman and Hall, London (1975), pp. 168–200, discusses other off-center instruments as well.

4B.2 Complex Representation of Oscillatory Quantities

Let $z = a + ib$ be a complex quantity whose complex conjugate is $\bar{z} = a - ib$ (here a and b are real).

a. Verify that the magnitude of z is:

$$|z| \equiv \sqrt{a^2 + b^2} = (z\bar{z})^{1/2} \tag{4B.2-1}$$

b. Show that z may be written as:

$$z = |z|e^{i\theta} \tag{4B.2-2}$$

where $\theta = \tan^{-1}(b/a)$.

c. In some oscillatory rheological experiment, a part of the apparatus is forced to undergo sinusoidal oscillations with frequency ω which we represent as a cosine function, $\cos \omega t$. The response of some other part of the equipment is observed to oscillate with frequency ω and amplitude R, but it oscillates out of phase with the driving function. Thus we represent the response function by $r = R \cos(\omega t + \delta)$; here δ is the angle by which the response lags the input. Show that r can also be represented by:

$$r = \mathcal{Re}\{r^0 e^{i\omega t}\} \tag{4B.2-3}$$

where r^0 is a complex amplitude with:

$$\mathcal{Re}\{r^0\} = R \cos \delta \tag{4B.2-4a}$$

$$\mathcal{Im}\{r^0\} = R \sin \delta \tag{4B.2-4b}$$

Sketch the driving and response functions and label the angle δ.

4B.3 Inertial Corrections in the Cone-and-Plate Instrument

In obtaining Eq. 4.5–8 for the total normal thrust in the cone-and-plate system, inertial forces were assumed to be negligible. The size of the error introduced in this way can be estimated theoretically in the following manner by considering the equation of motion for a Newtonian fluid:

a. Show that for a Newtonian fluid, the pressure distribution between the cone and plate (for small cone angles) is

$$p - p_a = \tfrac{1}{2}\rho W^2 R^2 \left(\frac{\vartheta}{\vartheta_0}\right)^2 \left(\frac{r^2}{R^2} - 1\right) \tag{4B.3-1}$$

b. Determine the average pressure in the gap[2] at any radial position $\bar{p}(r) = (1/\vartheta_0)\int_0^{\vartheta_0} p\, d\theta$ and use this result to compute the total normal thrust (over that due to atmospheric pressure) that a Newtonian fluid exerts on the plate:

$$\mathcal{F}_{\text{Newtonian}} = -\frac{\pi}{12}\rho R^4 W^2 \tag{4B.3-2}$$

c. Interpret the result in (b). The averaged form of Eq. 4B.3–1 has been compared with experimental data by Adams and Lodge[3] who find a reasonable fit. It appears that this method slightly overestimates the inertial terms.

4B.4 End Correction for the Couette Viscometer

The assumption of tangential annular flow used in obtaining the results in Table 4.5–1D does not hold near the bottom of the viscometer. In order to compensate for this end effect, the bottom of the inner

[2] This averaging procedure was first used by H. W. Greensmith and R. S. Rivlin, *Phil. Trans. Roy. Soc.* (London), *A245*, 399–428 (1953) to fit experimental data for a torsional flow system; see also K. Walters, *Rheometry*, Chapman and Hall, London (1975), pp. 61–66.

[3] N. Adams and A. S. Lodge, *Phil. Trans. Roy. Soc.* (London), *A256*, 149–184 (1964).

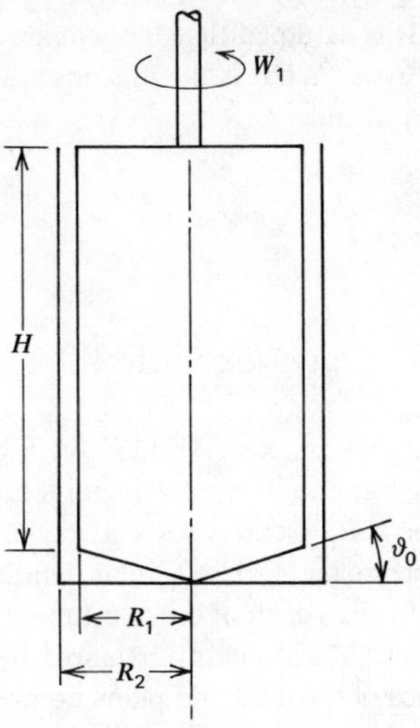

FIGURE 4B.4. Concentric cylinder viscometer with conical bottom. The inner cylinder rotates and the outer cylinder is fixed.

cylinder is actually constructed with a conical shape; the flow at the bottom is thus like a cone-and-plate flow. The gap is assumed to be narrow, that is, $(R_2 - R_1)/R_1 \ll 1$.

a. For what choice of the cone angle ϑ_0 is the shear rate uniform throughout the fluid?

b. Obtain an expression for η for this geometry that is analogous to Eq. D–1 in Table 4.5–1.

Answer (b): $$\eta = \frac{\mathscr{T}(R_2 - R_1)}{2\pi R_1{}^3 H[1 + (R_1/3H)]W_1}$$

4B.5 Form of the Stress Tensor for Shearfree Flows

In a shearfree flow, the velocity field near an arbitrary fluid particle is given relative to the principal axes of strain at that particle by (cf. Eq. 4.6–1):

$$\mathbf{v} = \boldsymbol{\delta}_1 \tfrac{1}{2}\dot{\gamma}_{11} x_1 + \boldsymbol{\delta}_2 \tfrac{1}{2}\dot{\gamma}_{22} x_2 + \boldsymbol{\delta}_3 \tfrac{1}{2}\dot{\gamma}_{33} x_3 \tag{4B.5–1}$$

where the $\dot{\gamma}_{ii}$ may be functions of time at the given particle. By considering a 180° rotation about the 1-axis show that:

$$\tau_{12} = \tau_{21} = 0$$

$$\tau_{13} = \tau_{31} = 0 \tag{4B.5–2}$$

Use the discussion in §4.2 as a guide. Similarly show that:

$$\tau_{23} = \tau_{32} = 0 \tag{4B.5–3}$$

Thus the stress tensor must have the form shown in Eq. 4.6–5 for shearfree flows.

4B.6 Viscous Heating in a Concentric Cylinder Viscometer

Estimate the temperature rise caused by viscous heating in a concentric cylinder viscometer by considering steady tangential annular flow between two cylinders. The inner cylinder has radius κR and rotates with constant angular velocity W; the outer cylinder has radius R and is held fixed. In addition, the temperature of the outer cylinder is kept at T_0. It is assumed that the temperature rise is small enough so that the viscosity is not substantially altered by it. What is the maximum temperature rise in the fluid if $\kappa \approx 1$ and no heat is transferred to the inner cylinder?

$$\text{Answer:} \qquad T_{\text{max}} - T_0 = \frac{\eta R^2 W^2}{4k}$$

4C.1 End Correction for a Capillary Viscometer[1]

Consider the operation of two capillary viscometers A and B (see Fig. 4.5–3). The reservoirs are identical and are located at the same elevation, but one is equipped with a tube of length L_A and the other, a tube of length L_B. The radius of each tube is R. Denote, respectively, by h_A and p_A the height of fluid in the reservoir and driving pressure for viscometer A. Use similar definitions for B. The pressures p_A and p_B are adjusted so that the flow rate Q through each of the two tubes is the same.

a. Apply the macroscopic mechanical energy balance[2] to the fluid in the viscometer, taking plane number 1 to be the upper surface of the fluid and plane number 2 to be at the exit of the tube. For the purposes of this balance, take h to be constant. Show that for viscometer A:

$$(\hat{\Phi}_{A_2} - \hat{\Phi}_{A_1}) + \tfrac{1}{2}\left(\left.\frac{\langle v_z^3\rangle}{\langle v_z\rangle}\right|_{A_2} - \left.\frac{\langle v_z^3\rangle}{\langle v_z\rangle}\right|_{A_1}\right) + \frac{1}{\rho Q}\iint_{A_2} p v_z\, dS - \frac{p_A}{\rho} = -(\hat{E}_v)_A \qquad (4C.1\text{–}1)$$

and that a similar relation holds for B. Here $\hat{\Phi}$ is the potential energy per unit mass, v_z is the axial velocity, $\langle v_z^3\rangle$ and $\langle v_z\rangle$ are cross-sectional averages of v_z^3 and v_z, and $\hat{E}_v$ is defined by:

$$\hat{E}_v = -\frac{1}{\rho Q}\iiint_V (\boldsymbol{\tau} : \nabla \mathbf{v})\, dV \qquad (4C.1\text{–}2)$$

b. Subtract the two mechanical energy balances to obtain:

$$(\hat{\Phi}_{B_2} - \hat{\Phi}_{B_1}) - (\hat{\Phi}_{A_2} - \hat{\Phi}_{A_1}) - \frac{1}{\rho}(p_B - p_A) = +(\hat{E}_v)_A - (\hat{E}_v)_B \qquad (4C.1\text{–}3)$$

c. Next, by splitting the total volume of each system into a volume in which there is steady shear flow plus volumes associated with the flow in the reservoir and in the entrance and exit regions, show that:

$$(\hat{E}_v)_A - (\hat{E}_v)_B = \frac{2\pi(L_B - L_A)}{\rho Q}\int_0^R \tau_{rz}\left(\frac{dv_z}{dr}\right) r\, dr = -\frac{2}{\rho}\frac{\tau_R}{R}(L_B - L_A) \qquad (4C.1\text{–}4)$$

Assume that $h_A = h_B$ in deriving this result. What other assumptions are needed?

d. Choose the reference potential to be at the bottom of the tube of length B so that $\hat{\Phi}_{B_2} = 0$. Use the resulting expressions for the potentials together with the results from (b) and (c) above to show:

$$\tau_R = \left(\frac{p_B - p_A}{L_B - L_A} + \rho g \cos\beta\right)\frac{R}{2} \qquad (4C.1\text{–}5)$$

where β is the angle between the tube axis and the vertical.

[1] A. G. Fredrickson, *Principles and Applications of Rheology*, Prentice-Hall, Englewood Cliffs, N.J. (1964), pp. 196–200.

[2] See, for example, R. B. Bird, W. E. Stewart, and E. N. Lightfoot, *Transport Phenomena*, Wiley, New York (1960), pp. 211–214.

4C.2 Functional Forms of the Viscometric Functions

Consider a steady simple shear flow with $v_x = \dot{\gamma}y$, $v_y = v_z = 0$. By utilizing a rotation of the coordinate system through 180° about the y-axis and arguments similar to those in §4.2, show that η, Ψ_1, and Ψ_2 are all even functions of the shear rate $\dot{\gamma}$. How can arguments like those proposed here be used to explain the shapes of the normal-stress and shear-stress responses in the small-amplitude oscillatory shear experiment?

4C.3 Measurements in a Wide-Gap, Couette Viscometer[3]

a. Show that for tangential annular flow between concentric cylinders with a wide gap, the appropriate generalizations of Eqs. D–1 and D–2 in Table 4.5–1 are:

$$\Delta\dot{W} = W_2 - W_1 = \tfrac{1}{2}\int_{\tau_1}^{\tau_2} \eta^{-1}(\tau)\, d\tau \tag{4C.3–1}$$

$$\Delta\pi_{rr} = \pi_{rr}(R_2) - \pi_{rr}(R_1) = \int_{R_1}^{R_2} \left(\rho\frac{v_\theta^2}{r} + \frac{\tau_{\theta\theta} - \tau_{rr}}{r}\right) dr \tag{4C.3–2}$$

In Eq. 4C.3–1, $\tau \equiv \tau_{r\theta} = \mathscr{T}/2\pi r^2 H$.

b. For wide gaps the above formulas may be conveniently inverted by differentiating with respect to the torque $\mathscr{T}$. The resulting expressions, which are valid for any value of $\mathscr{T}$, must also hold for a torque equal to $\beta\mathscr{T}$, where $\beta \equiv R_1^2/R_2^2$. Use this fact to obtain:

$$\dot{\gamma}_{r\theta}(\tau_1) = \sum_{j=0}^{\infty} \left(2\mathscr{T}'\frac{\partial\,\Delta W}{\partial\mathscr{T}'}\right)\Bigg|_{\mathscr{T}'=\mathscr{T}\beta^j} \tag{4C.3–3}$$

$$(\tau_{\theta\theta} - \tau_{rr})|_{\tau_1} = \sum_{j=0}^{\infty} \left(2\mathscr{T}'\frac{\partial\,\Delta\pi_{rr}^{(c)}}{\partial\mathscr{T}'}\right)\Bigg|_{\mathscr{T}'=\mathscr{T}\beta^j} \tag{4C.3–4}$$

where

$$\Delta\pi_{rr}^{(c)} = \Delta\pi_{rr} - \int_{R_1}^{R_2} \rho\frac{v_\theta^2}{r}\, dr \tag{4C.3–5}$$

c. Explain how Eqs. 4C.3–3 and 4 can be used to interpret experimental data. Why are they useful only for wide gaps?

4C.4 Eccentric Disk Rheometer Flow and Kinematics of General Shear Flow[4]

In Eq. 4.1–6 we give the rate-of-strain tensor referred to the shear axes for a unidirectional shear flow. Here we give the corresponding result for general shear flows and then ask how this result applies to the eccentric disk flow of Problem 4B.1.

We consider the same shear axes defined above Eq. 4.1–5 with $\breve{x}_1$- and $\breve{x}_3$-coordinate curves embedded in the shearing surfaces. It can be shown for a general shear flow that the rate-of-strain tensor

[3] B. D. Coleman, H. Markovitz, and W. Noll, *Viscometric Flows of Non-Newtonian Fluids*, Springer, New York (1966), pp. 42–44; S. Middleman, *The Flow of High Polymers*, Interscience, New York (1968), pp. 19–25.

[4] A. S. Lodge, *Body Tensor Fields in Continuum Mechanics*, Academic Press, New York (1974), pp. 60–72, 76–78.

is always of the form

$$\dot{\boldsymbol{\gamma}} = \begin{pmatrix} 0 & \dot{\gamma}_1 & 0 \\ \dot{\gamma}_1 & 0 & \dot{\gamma}_3 \\ 0 & \dot{\gamma}_3 & 0 \end{pmatrix}$$
$$= (\breve{\boldsymbol{\delta}}_1\breve{\boldsymbol{\delta}}_2 + \breve{\boldsymbol{\delta}}_2\breve{\boldsymbol{\delta}}_1)\dot{\gamma}_1 + (\breve{\boldsymbol{\delta}}_3\breve{\boldsymbol{\delta}}_2 + \breve{\boldsymbol{\delta}}_2\breve{\boldsymbol{\delta}}_3)\dot{\gamma}_3 \quad (4C.4\text{–}1)$$

when referred to this coordinate system. Here $\dot{\gamma}_1$ and $\dot{\gamma}_3$ can both be functions of time. Furthermore it is found that a line of shear makes an angle $\zeta(t)$ with the $\breve{x}_1$-curve where

$$\zeta(t) = \tan^{-1}(\dot{\gamma}_3/\dot{\gamma}_1) \quad (4C.4\text{–}2)$$

and that the shear rate is

$$\dot{\gamma} = \sqrt{\tfrac{1}{2}\text{II}_{\dot{\gamma}}} = \sqrt{\dot{\gamma}_1{}^2 + \dot{\gamma}_3{}^2} \quad (4C.4\text{–}3)$$

Note that if ζ is a function of time then the lines of shear cannot be material lines and thus the flow cannot be a unidirectional shear flow.

Now consider the eccentric disk flow described by Eq. 4B.1–1.

a. What are the shearing surfaces for this flow?

b. Next examine the flow as seen from the shear axes at an arbitrary particle P. For convenience take $\breve{\boldsymbol{\delta}}_1 = \boldsymbol{\delta}_x$ and $\breve{\boldsymbol{\delta}}_3 = -\boldsymbol{\delta}_y$ at $t = 0$. Obtain relations between $\breve{\boldsymbol{\delta}}_1$, $\breve{\boldsymbol{\delta}}_3$ and $\boldsymbol{\delta}_x$, $\boldsymbol{\delta}_y$ at any time t.

c. Determine the velocity field near P relative to the shear axes and show that

$$\nabla\breve{\mathbf{v}} = \breve{\boldsymbol{\delta}}_2\breve{\boldsymbol{\delta}}_1 AW\cos Wt - \breve{\boldsymbol{\delta}}_2\breve{\boldsymbol{\delta}}_3 AW\sin Wt \quad (4C.4\text{–}4)$$

d. Show further that the shear rate is AW (independent of position and time) and that the direction of shear is always parallel to the x-axis.

e. Is this flow a unidirectional shear flow? Explain,

CHAPTER 5

THE GENERALIZED NEWTONIAN FLUID

In Chapters 3 and 4 we have discussed the many kinds of flow phenomena that are encountered in the study of macromolecular fluids. In the remainder of the book we seek to find ways of describing these phenomena in mathematical language and to understand the interrelationships among them. This is a long and arduous task, because of the diversity of the phenomena themselves and because of the diversity in the structure of the fluids with which we are dealing. It is also difficult because we must understand the experimental observations from two viewpoints: the continuum theory and the molecular theory.

This part of the book deals with the continuum point of view, in which the molecular structure is not brought into consideration. We could begin by presenting some very general continuum theories capable of describing all of the phenomena described in Chapters 3 and 4, and then—by specializing the theory—descend to the description of relatively simple problems. From a logical point of view this might be preferable and, indeed, for advanced students this is desirable. For the beginner, and for the person who wants to solve simple problems without wading through a lot of advanced mathematics, a different approach is needed.

Our approach generally parallels the historical development of the subject. This allows us to describe, with less mathematical sophistication, limited ranges of the observed phenomena. Inevitably, this means that as we proceed we find that some of our ideas and viewpoints have to be revised and expanded, paralleling the experiences of the researchers themselves. We shall, however, at each stage attempt to issue certain caveats to the reader so that he will realize the limitations of the methods being discussed.

This chapter is devoted to the generalized Newtonian model, which can describe the shear rate dependence of the viscosity. In §5.1 we describe the model and several useful empiricisms for the non-Newtonian viscosity. Then in §5.2 we illustrate the use of this model by going through seven illustrative examples in detail; these examples deal with problems which are sufficiently simple that analytical solutions may be obtained. For solving somewhat more complicated problems, a variational principle is available, and it is described and illustrated in §5.3. Up to this point it is presumed that the flow is isothermal. Many important industrial applications, however, involve nonisothermal situations, and §5.4 provides a brief introduction to this very large subject. In that section we give six rather long examples, in which the heat-transfer problems can be solved analytically; these examples are highly idealized, and the reader should be made aware of the fact that most nonisothermal processing problems must be solved numerically. Finally in §5.5 we put the generalized Newtonian model into perspective by showing that it is contained in the Criminale–Ericksen–Filbey equation, which is the most general expression for the stress tensor in steady shear flow for a very wide class of fluids.

To avoid encyclopaedic coverage a number of topics have had to be omitted. We cite some of these along with a few references which can provide an introduction to the literature: boundary-layer theory,[1] entrance effects,[2] instability,[3] turbulent flow correlations,[4] agitation and mixing,[5] fiber spinning,[6] mold filling,[7] withdrawal of plates from fluids,[8] pressure dependence of viscosity,[9] and finite element methods.[10]

§5.1 THE GENERALIZED NEWTONIAN FLUID: ITS ORIGIN, USEFULNESS, AND LIMITATIONS

For the industrial chemical engineer, the most important property of macromolecular fluids discussed in Chapter 4 is the non-Newtonian viscosity—that is, the fact that the viscosity of the fluid changes with the shear rate. Since for some fluids the viscosity can change by a factor of 10, 100, or even 1000, it is evident that such an enormous change cannot be ignored in pipe flow calculations, lubrication problems, design of on-line viscometers, extruder operation, and polymer processing calculations. Therefore it is not surprising that one of the earliest empiricisms to be introduced was Newton's law of viscosity modified to allow the viscosity to vary with shear rate (or with shear stress). That is, for the elementary flow $v_x = v_x(y)$, $v_y = 0$, $v_z = 0$, the early rheologists replaced:

Newtonian fluid:
$$\tau_{yx} = -\mu \frac{dv_x}{dy} \qquad \mu = \text{constant for a given temperature, pressure, and composition} \tag{5.1-1}$$

by the empiricism:

Generalized Newtonian fluid:
$$\tau_{yx} = -\eta \frac{dv_x}{dy} \qquad \eta = \text{a function of } |dv_x/dy| \text{ (or of } |\tau_{yx}|) \tag{5.1-2}$$

Absolute value signs have been used at the right, since one would expect the change in viscosity to depend on the magnitude but not on the sign of the shear rate (or shear stress). Having written Eq. 5.1-2 one can then introduce various empiricisms to describe the experimental non-Newtonian viscosity curves.

[1] A. Acrivos, M. J. Shah, and E. E. Petersen, *A.I.Ch.E. Journal*, *6*, 312–317 (1960); J. L. White and A. B. Metzner, *ibid.*, *11*, 324–330 (1965); S. Y. Lee and W. F. Ames, *ibid.*, *12*, 700–708 (1966); V. G. Fox, L. E. Erickson, and L. T. Fan, *ibid.*, *15*, 327–333 (1969).
[2] S. S. Chen, L. T. Fan, C. L. Hwang, *A.I.Ch.E. Journal*, *16*, 293–299 (1970); N. D. Sylvester and S. L. Rosen, *ibid.*, *16*, 964–972 (1970).
[3] J. P. Tordella, *Rheology* (F. R. Eirich, Ed.), Volume 5, Chapter 2, Academic Press, New York (1969). J. R. A. Pearson, *Ann. Rev. Fluid Mech.*, *8*, 163–181 (1976).
[4] A. B. Metzner and J. C. Reed, *A.I.Ch.E. Journal*, *1*, 434–440 (1955); R. G. Shaver and E. W. Merrill, *ibid.*, *5*, 181–188 (1959); D. W. Dodge and A. B. Metzner, *ibid.*, *5*, 189–204 (1959); D. W. McEachern, *ibid.*, *15*, 885–889 (1969); H. Rubin and C. Elata, *ibid.*, *17*, 990–996 (1971).
[5] A. B. Metzner and R. E. Otto, *A.I.Ch.E. Journal*, *3*, 3–10 (1957); E. S. Godleski and J. C. Smith, *ibid.*, *8*, 617–620 (1961); D. W. Hubbard and F. F. Calvetti, *ibid.*, *18*, 663–665 (1972).
[6] J. R. A. Pearson and Y. T. Shah, *Ind. Eng. Chem. Fundamentals*, *13*, 134–138 (1974); J. R. A. Pearson, Y. T. Shah, and R. D. Mhaskar, *ibid.*, *15*, 31–37 (1976).
[7] J. L. Berger and C. G. Gogos, *Polym. Eng. and Sci.*, *13*, 102–112 (1973); P.-C. Wu, C. F. Huang, and C. G. Gogos, *ibid.*, *14*, 223–230 (1974).
[8] C. Gutfinger and J. A. Tallmadge, *A.I.Ch.E. Journal*, *11*, 403–413 (1965); J. A. Tallmadge, *ibid.*, 16, 925–930 (1970).
[9] N. Galili, R. Takserman-Krozer, and Z. Rigbi, *Rheol. Acta*, *14*, 550–567 (1975).
[10] K. Palit and R. T. Fenner, *A.I.Ch.E. Journal*, *18*, 628–633 (1972).

We now wish to extend the ideas above to any arbitrary flow. First, for an incompressible Newtonian fluid, for any flow field $\boldsymbol{v} = \boldsymbol{v}(x,y,z,t)$ we have:

Incompressible Newtonian fluid:

$$\boldsymbol{\tau} = -\mu\dot{\boldsymbol{\gamma}} \qquad \begin{array}{l}\mu = \text{a constant}\\ \text{for a given}\\ \text{temperature,}\\ \text{pressure, and}\\ \text{composition}\end{array} \tag{5.1–3}$$

in which $\dot{\boldsymbol{\gamma}}$ is the rate-of-deformation tensor $\nabla \boldsymbol{v} + (\nabla \boldsymbol{v})^\dagger$ [cf. Eq. 1.2–2]. To include the idea of a non-Newtonian viscosity, we write:[1]

Incompressible generalized Newtonian fluid:

$$\boldsymbol{\tau} = -\eta\dot{\boldsymbol{\gamma}} \qquad \begin{array}{l}\eta \text{ is a function of}\\ \text{the scalar invariants}\\ \text{of } \dot{\boldsymbol{\gamma}} \text{ (or of } \boldsymbol{\tau})\end{array} \tag{5.1–4}$$

If the non-Newtonian viscosity, a scalar, is to depend on the components of the tensor $\dot{\boldsymbol{\gamma}}$ (or $\boldsymbol{\tau}$), then it must depend only on those particular combinations of components of the tensor that are scalars.[2] As described in §A.3 we may select as three independent combinations:

$$I_{\dot{\gamma}} = \textstyle\sum_i \dot{\gamma}_{ii} \tag{5.1–5}$$

$$II_{\dot{\gamma}} = \textstyle\sum_i \sum_j \dot{\gamma}_{ij}\dot{\gamma}_{ji} \tag{5.1–6}$$

$$III_{\dot{\gamma}} = \textstyle\sum_i \sum_j \sum_k \dot{\gamma}_{ij}\dot{\gamma}_{jk}\dot{\gamma}_{ki} \tag{5.1–7}$$

For an incompressible fluid $I_{\dot{\gamma}} = 2(\nabla \cdot \boldsymbol{v}) = 0$. For shearing flows $III_{\dot{\gamma}}$ turns out to be zero (see Problem 5B.4); since, as we will point our later, Eq. 5.1–4 should be used only for shearing flows, or at least flows that are very nearly shearing, omitting $III_{\dot{\gamma}}$ from any further consideration is not a serious restriction. Hence η will be taken to depend only on $II_{\dot{\gamma}}$. Actually we shall prefer to use $\dot{\gamma}$, the magnitude of $\dot{\boldsymbol{\gamma}}$, instead of $II_{\dot{\gamma}}$:

$$\dot{\gamma} = \sqrt{\tfrac{1}{2}\textstyle\sum_i \sum_j \dot{\gamma}_{ij}\dot{\gamma}_{ji}} = \sqrt{\tfrac{1}{2}II_{\dot{\gamma}}} \tag{5.1–8}$$ [3]

Hence we will write $\eta = \eta(\dot{\gamma})$. Some empiricisms use the stress as the independent variable and, then, we write $\eta = \eta(\tau)$ where τ is the magnitude of $\boldsymbol{\tau}$.

Equation 5.1–4 with $\eta = \eta(\dot{\gamma})$ or $\eta = \eta(\tau)$ will then be the starting point for all of the calculations in this chapter. Its principal usefulness is for calculating flow rates and shearing forces in steady-state shear flows, such as:

- **a.** *Tube flow*
- **b.** *Axial annular flow*
- **c.** *Tangential annular flow*
- **d.** *Helical annular flow*
- **e.** *Flow between parallel planes*
- **f.** *Flow between rotating disks*
- **g.** *Cone-and-plate flow*

The generalized Newtonian fluid model cannot describe normal-force phenomena (Weissenberg rod-climbing effect, jet swell, induced secondary flows), any time-dependent phenomena

[1] M. Reiner, *Deformation, Strain, and Flow*, Interscience, New York (1960).

[2] K. Hohenemser and W. Prager, *ZAMM*, *12*, 216–226 (1932).

[3] In taking the square root one must affix the proper sign so that $\dot{\gamma}$ will be positive!

generally associated with viscoelasticity (recoil, stress relaxation, stress overshoot, orthogonal rheometer flow), or phenomena connected with flows other than steady shear (elongational flow, squeezing flow). It is thus important to recognize that, although the generalized Newtonian fluid is a useful empiricism for an important class of industrially interesting flows, it has severe limitations.

Here we have introduced the generalized Newtonian fluid as a simple extension of the Newtonian fluid, based on experimental observations. In Chapter 8 we shall see that rather general continuum mechanics arguments give exactly Eq. 5.1–4 with $\eta = \eta(\dot{\gamma})$ for the first term of a general expression for steady-state shear flows. The higher terms give information about the normal stresses in such flows. Hence although Eq. 5.1–4 was originally proposed as an empiricism, it has since been legitimized by modern continuum mechanics theories.

Although we state that Eq. 5.1–4 gives correct results for flow rates and shearing forces in steady shear flows, we hasten to add that engineers have not hesitated to apply Eq. 5.1–4 to somewhat more complicated flows and systems slowly varying with time. A thorough assessment of the errors inherent in such calculations is not available, but one feels intuitively that such a practice probably represents good engineering empiricism. We shall give a few examples of this presently.

Let us now turn to the empiricisms for $\eta(\dot{\gamma})$. From Chapter 4 we know what the shapes of the η vs. $\dot{\gamma}$ curves look like. Although for some problems one can feed in the raw data for $\eta(\dot{\gamma})$, it is often useful to make calculations and derivations for specific analytical representations of $\eta(\dot{\gamma})$ that are known to describe the experimental data with sufficient accuracy. Many such empiricisms are available, and we make no attempt at completeness. Some of the most popular ones are:

a. The "Power Law"[4] of Ostwald and de Waele

In almost all industrial problems the descending linear region of the plot of log η vs. log $\dot{\gamma}$ (see §4.3) is the most important region. In fact, for many inexpensive viscometers and many fluids, it is almost impossible to obtain data for the horizontal region of the $\eta(\dot{\gamma})$ curve. The tilted straight line can be described by a "power law":

$$\eta = m\dot{\gamma}^{n-1} \tag{5.1–9}$$

in which m (with units of $\mathrm{N \cdot s^n/m^2}$) and the dimensionless quantity n are constants characteristic of each polymer and each polymer solution. This simple expression is without doubt the most well-known and widely used empiricism in engineering work; many specific flow problems and heat transfer problems have been solved using it and the results have proven to be useful. The power law does not describe the portion of the viscosity curve near $\dot{\gamma} = 0$ where $\eta = \eta_0$, but this region is usually not too important in engineering design. One cannot construct a "time constant" from the constants m and n; this is a serious drawback if one wishes to tie together steady-state flow phenomena with transient viscoelastic phenomena for the description of which a characteristic time constant is necessary. Clearly, when $n = 1$ and $m = \mu$ one obtains the Newtonian fluid. If $n < 1$, the fluid is said to be "pseudoplastic" and if $n > 1$, the fluid is called "dilatant."[5] Most macromolecular fluids are pseudoplastic and values of n in the range of 0.15 to 0.6 are common (see Table 5.1–1).

[4] W. Ostwald, *Kolloid-Z.*, *36*, 99–117 (1925); A. de Waele, *Oil and Color Chem. Assoc. Journal*, *6*, 33–88 (1923).

[5] M. Reiner, *op. cit.*, pp. 306–308, and others use the term "dilatancy" to describe the change in volume of granular masses necessitated by a distortion. The standard example is the apparent drying of wet sand when stepped on.

TABLE 5.1-1

Power-Law Parameters for Aqueous Solutions[a]

Solution	Temperature (K)	$m(\text{N}\cdot\text{s}^n/\text{m}^2)$	n (—)
0.5% Hydroxyethylcellulose	293	0.84	0.5088
	313	0.30	0.5949
	333	0.136	0.645
2.0% Hydroxyethylcellulose	293	93.5	0.189
	313	59.7	0.223
	333	38.5	0.254
1.0% Poly(ethylene oxide)	293	0.994	0.532
	313	0.706	0.544
	333	0.486	0.599

[a] R. M. Turian, Ph.D. Thesis, University of Wisconsin, Madison (1964), pp. 142–148.

b. The "Truncated Power Law" of Spriggs[6]

In order to utilize the simplicity of the power law and still avoid the objection that a characteristic time is not included among the parameters, one can write:

$$\eta = \eta_0 \qquad \dot{\gamma} \leq \dot{\gamma}_0 \tag{5.1-10}$$

$$\eta = \eta_0 \left(\frac{\dot{\gamma}}{\dot{\gamma}_0}\right)^{n-1} \qquad \dot{\gamma} \geq \dot{\gamma}_0 \tag{5.1-11}$$

in which there are three constants: η_0, a zero-shear-rate viscosity, $(1/\dot{\gamma}_0)$, a characteristic time, and n, a dimensionless power law index. This model then contains the horizontal asymptote for small $\dot{\gamma}$ and the "power law" asymptote for large $\dot{\gamma}$.

c. The Eyring Model[7]

From the theory of rate processes, Eyring showed that:

$$\eta = t_0 \tau_0 \left(\frac{\text{arcsinh}\; t_0 \dot{\gamma}}{t_0 \dot{\gamma}}\right) \tag{5.1-12}$$

in which τ_0 is a characteristic stress and t_0 is a characteristic time. This expression was the result of the first attempt to obtain a rough molecular explanation for non-Newtonian viscosity. It was not until many years later that refined statistical mechanical theories could describe non-Newtonian viscosity, and then only for very dilute solutions. Sutterby[8]

[6] T. W. Spriggs, private communication (1965).

[7] J. F. Kincaid, H. Eyring, and A. E. Stearn, *Chem. Revs.*, *28*, 301–365 (1941); F. H. Ree, T. Ree, and H. Eyring. *Ind. Eng. Chem.*, *50*, 1036–1040 (1958).

[8] J. L. Sutterby, *Trans. Soc. Rheol.*, *9*:*2*, 227–241 (1965); *A.I.Ch.E. Journal*, *12*, 63–68 (1966).

modified the Eyring model by raising the quantity in parentheses to a power A; this provides additional flexibility. Another useful empiricism is obtained by adding a constant to the right side of Eq. 5.1–12; this three-constant "Powell–Eyring model" has been found to fit viscosity data for polymeric fluids very well.[9]

d. The Carreau Model[10]

A four parameter model that has the useful properties of the truncated power law model but that does not have a discontinuous first derivative is:

$$\frac{\eta - \eta_\infty}{\eta_0 - \eta_\infty} = [1 + (\lambda\dot{\gamma})^2]^{(n-1)/2} \tag{5.1–13}$$

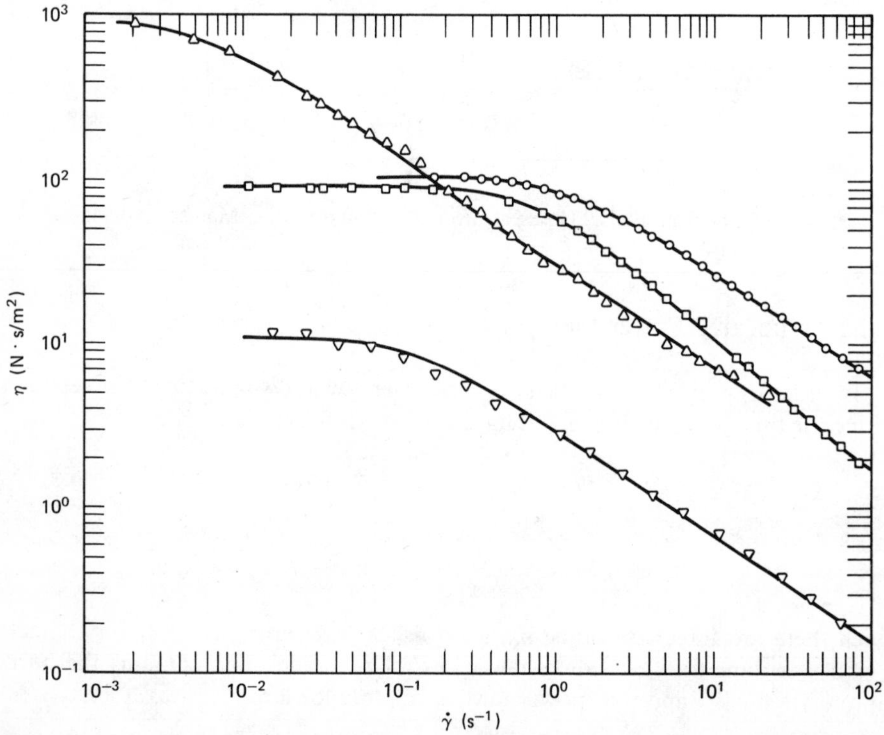

FIGURE 5.1–1. Non-Newtonian viscosity of three polymer solutions and a soap solution as fitted by the Carreau viscosity equation in Eq. 5.1–13 [R. B. Bird, O. Hassager, and S. I. Abdel-Khalik, *A.I.Ch.E. Journal*, *20*, 1041–1066 (1974).]

△ 2% polyisobutylene in Primol 355; data of J. D. Huppler, E. Ashare, and L. A. Holmes, *Trans. Soc. Rheol.*, 11, 159–179 (1968): $\eta_0 = 9.23 \times 10^2$ N·s/m^2, $\eta_\infty = 1.50 \times 10^{-1}$ N·s/m^2, $\lambda = 191$ s, $n = 0.358$.

○ 5% polystyrene in Aroclor 1242; data of E. Ashare, Ph.D. Thesis, University of Wisconsin, Madison (1968): $\eta_0 = 1.01 \times 10^2$ N·s/m^2, $\eta_\infty = 5.9 \times 10^{-2}$ N·s/m^2, $\lambda = 0.84$ s, $n = 0.380$.

▽ 0.75% polyacrylamide (Separan–30) in a 95/5 mixture by weight of water and glycerin; data of B. D. Marsh (1967), as cited by P. J. Carreau, I. F. Macdonald, and R. B. Bird, *Chem. Eng. Sci.*, *23*, 901–911 (1968): $\eta_0 = 10.6$ N·s/m^2, $\eta_\infty = 10^{-2}$ N·s/m^2, $\lambda = 8.04$ s., $n = 0.364$.

□ 7% aluminum soap in decalin and *m*-cresol; data of J. D. Huppler, E. Ashare, and L. A. Holmes, *loc. cit.*: $\eta_0 = 89.6$ N·s/m^2, $\eta_\infty = 10^{-2}$ N·s/m^2, $\lambda = 1.41$ s, $n = 0.200$.

[9] R. E. Powell and H. Eyring, *Nature*, *154*, 427–428 (1944); D. L. Salt, N. W. Ryan, and E. B. Christiansen, *J. Coll. Sci.*, *6*, 146–154 (1951). A temperature-dependent Powell–Eyring model has been used by E. B. Christiansen and S. J. Kelsey, *Chem. Eng. Sci.*, *28*, 1099–1113 (1973).

[10] P. J. Carreau, Ph.D. Thesis, University of Wisconsin, Madison (1968).

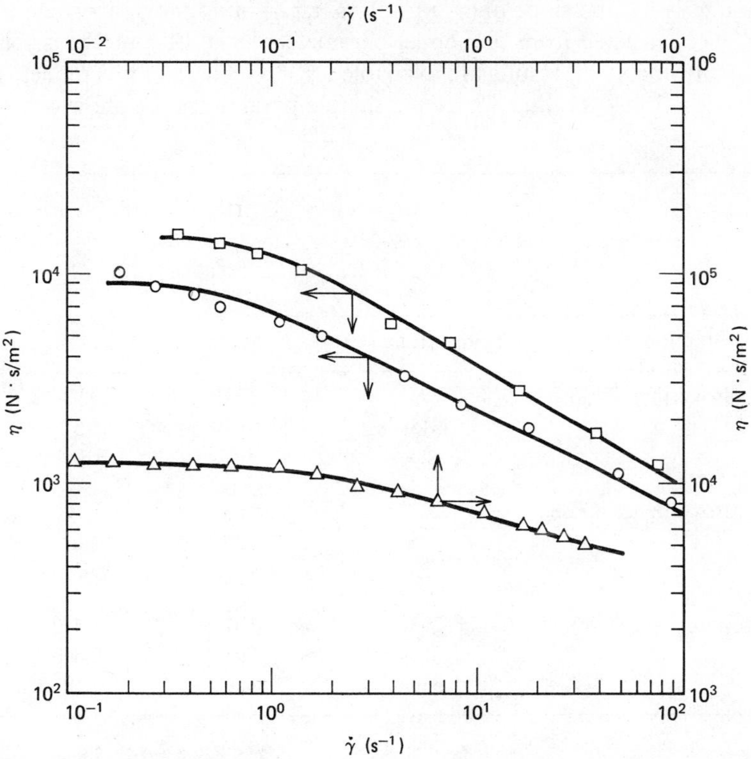

FIGURE 5.1–2. Non-Newtonian viscosity of three polymer melts as described by the Carreau viscosity equation (Eq. 5.1–13). [S. I. Abdel-Khalik, O. Hassager, and R. B. Bird, *Polym. Eng. Sci.*, *14*, 859–867 (1974).]

□ Polystyrene at 453 K; data of T. F. Ballenger, I.-J. Chen, J. W. Crowder, G. E. Hagler, D. C. Bogue, and J. L. White, *Trans. Soc. Rheol.*, *15*, 195–215 (1971): $\eta_0 = 1.48 \times 10^4$ N·s/m^2, $\eta_\infty = 0$, $\lambda = 1.04$ s, $n = 0.398$.

○ High-density polyethylene at 443 K; data of Ballenger, *et al.*, *loc. cit.*: $\eta_0 = 8.92 \times 10^3$ N·s/m^2, $\eta_\infty = 0$, $\lambda = 1.58$ s, $n = 0.496$.

△ Phenoxy-A at 485 K; data of B. D. Marsh as cited by P. J. Carreau, I. F. Macdonald, and R. B. Bird, *Chem. Eng. Sci.*, *23*, 901–911 (1968): $\eta_0 = 1.24 \times 10^4$ N·s/m^2, $\eta_\infty = 0$, $\lambda = 7.44$ s, $n = 0.728$.

in which η_0 is the zero-shear-rate viscosity, η_∞ is the infinite-shear-rate viscosity, λ is a time constant, and n is the dimensionless power-law index. The physical significance of n in Eqs. 5.1–9, 11, and 13 is the same: $(n - 1)$ is the slope of the straight line portion ("power-law region") when $\log \eta$ is plotted against $\log \dot{\gamma}$. For most polymer solutions and polymer melts n lies between 0 and 1. Some examples of data curve-fits are given in Figs. 5.1–1 and 5.1–2.

e. The Ellis Model[11]

The four models above are examples of $\eta = \eta(\dot{\gamma})$. As an illustration of $\eta = \eta(\tau)$ we cite the Ellis model for which:

$$\frac{\eta_0}{\eta} = 1 + \left(\frac{\tau}{\tau_{1/2}}\right)^{\alpha - 1} \qquad (5.1\text{–}14)$$

in which η_0 is the zero-shear-rate viscosity, $\tau_{1/2}$ is the value of the shear stress at which

[11] S. B. Ellis (see M. Reiner, *op. cit.*, p. 246).

$\eta = \eta_0/2$, and $\alpha - 1$ is the slope of $(\eta_0/\eta) - 1$ vs. $\tau/\tau_{1/2}$ on log-log paper. A time constant $\eta_0/\tau_{1/2}$ can be constructed from the model parameters. The Ellis model is relatively easy to use, and many analytical results are available for it.[12] The α in this model is equivalent to the $1/n$ in Eqs. 5.1–9, 11, and 13. Some sample parameters are shown in Table 5.1–2.

TABLE 5.1–2

Ellis Model Parameters for Aqueous Solutions[a]

Solution	Temperature (K)	η_0(N·s/m²)	$\tau_{1/2}$ (N/m²)	α (—)
0.5% Hydroxyethylcellulose	293	0.22	4.93	2.073
	313	0.040	16.5	2.268
	333	0.018	20.0	2.206
2.0% Hydroxyethylcellulose	293	19.85	114	4.454
	313	10.40	84.6	3.920
	333	3.93	83.3	3.699
1.0% Poly(ethylene oxide)	293	2.37	0.412	1.886
	313	0.97	0.660	1.896
	333	0.365	1.18	1.908

[a] R. M. Turian, Ph.D. Thesis, University of Wisconsin, Madison (1964), pp. 164–167.

f. The Bingham Fluid[13]

No listing of empirical models would be complete without giving the Bingham model for a fluid with a yield stress τ_0. If this yield stress is exceeded the viscosity is finite:

$$\eta = \infty \qquad \tau \leq \tau_0 \tag{5.1-15}$$

$$\eta = \mu_0 + \frac{\tau_0}{\dot{\gamma}} \qquad \tau \geq \tau_0 \tag{5.1-16}$$

The model thus has two parameters: τ_0 with dimensions of force per unit area, and μ_0 with dimensions of viscosity. It is primarily used for slurries and pastes. These parameters can be related empirically to the volume fraction of solids ϕ, the particle diameter D_p, and the viscosity of the Newtonian fluid in which the particles are suspended μ_s by:[14]

$$\tau_0 = 312.5 \frac{\phi^3}{D_p{}^2} \tag{5.1-17}$$

$$\mu_0 = \mu_s \exp\left[\phi\left(\frac{5}{2} + \frac{14}{\sqrt{D_p}}\right)\right] \tag{5.1-18}$$

[12] S. Matsuhisa and R. B. Bird, *A.I.Ch.E. Journal*, *11*, 588–595 (1965).

[13] E. C. Bingham, *Fluidity and Plasticity*, McGraw-Hill, New York (1922), pp. 215–218; also *U.S. Bureau of Standards Bulletin*, *13*, 309–353 (1916).

[14] D. G. Thomas, *A.I.Ch.E. Journal*, *7*, 431–437 (1961); *9*, 310–316 (1963).

Here D_p is given in μm and τ_0 in N/m^2. Several dozen other empirical models have been proposed in the rheological literature, but it is not necessary to make a complete listing here.

§5.2 ISOTHERMAL FLOW PROBLEMS

The procedure for the solution of elementary flow problems for generalized Newtonian fluids is exactly the same as for Newtonian fluids, except that the mathematics is more awkward because of the additional complexity introduced by the non-Newtonian viscosity function. In this section we give a series of examples to illustrate the solution procedure.

EXAMPLE 5.2–1 Flow of a Power-Law Fluid in a Straight Circular Tube and in a Slightly Tapered Tube

Rework Example 1.2–3 for the Ostwald-de Waele power-law formula for non-Newtonian viscosity.

SOLUTION

As before we postulate a solution of the form $v_z = v_z(r)$, $v_\theta = 0$, $v_r = 0$, and $p = p(r,\theta,z)$. The z-component of the equation of motion, in terms of $\boldsymbol{\tau}$, is:

$$0 = -\frac{d\mathscr{P}}{dz} - \frac{1}{r}\frac{d}{dr}(r\tau_{rz}) \tag{5.2–1}$$

or using the arguments given in Example 1.2–3:

$$\frac{1}{r}\frac{d}{dr}(r\tau_{rz}) = \frac{\mathscr{P}_0 - \mathscr{P}_L}{L} \tag{5.2–2}$$

This may be integrated to give:

$$\tau_{rz} = \frac{(\mathscr{P}_0 - \mathscr{P}_L)r}{2L} + \frac{C_1}{r} \tag{5.2–3}$$

The constant C_1 has to be zero since one does not expect to have an infinite stress at the tube axis. Equation 5.2–3 then is the result of the application of the equation of motion (i.e., conservation of momentum). Equation 5.2–3 may be written in the alternate form:

$$\tau_{rz} = \tau_R \cdot \frac{r}{R} \tag{5.2–4}$$

where τ_R is the wall shear stress; that is, $\tau_R = \tau_{rz}$ at $r = R$.

Next we have to use the power-law equation for the stress, as given by Eqs. 5.1–4 and 5.1–9. In the latter equation $\dot{\gamma}$ is given by $(-dv_z/dr)$. Then:

$$\begin{aligned}\tau_{rz} &= -\eta\frac{dv_z}{dr} = -m\dot{\gamma}^{n-1}\frac{dv_z}{dr}\\ &= m\left(-\frac{dv_z}{dr}\right)^n\end{aligned} \tag{5.2–5}$$

Combination of Eqs. 5.2–4 and 5.2–5 then gives the differential equation for v_z:

$$m\left(-\frac{dv_z}{dr}\right)^n = \tau_R \cdot \frac{r}{R} \tag{5.2–6}$$

Taking the nth root of both sides and integrating the first order separable equation gives:

$$v_z = -\left(\frac{\tau_R}{mR}\right)^{1/n} \frac{r^{(1/n)+1}}{(1/n)+1} + C_2 \tag{5.2–7}$$

The constant C_2 is evaluated by requiring that v_z be zero at $r = R$. One then gets finally:

$$v_z = \left(\frac{\tau_R}{m}\right)^{1/n} \frac{R}{(1/n)+1}\left[1 - \left(\frac{r}{R}\right)^{(1/n)+1}\right] \tag{5.2–8}$$

For $n < 1$ this gives a velocity profile that is flatter than the parabolic profile in Eq. 1.2–35 for Newtonian fluids (see Fig. 5.2–1).

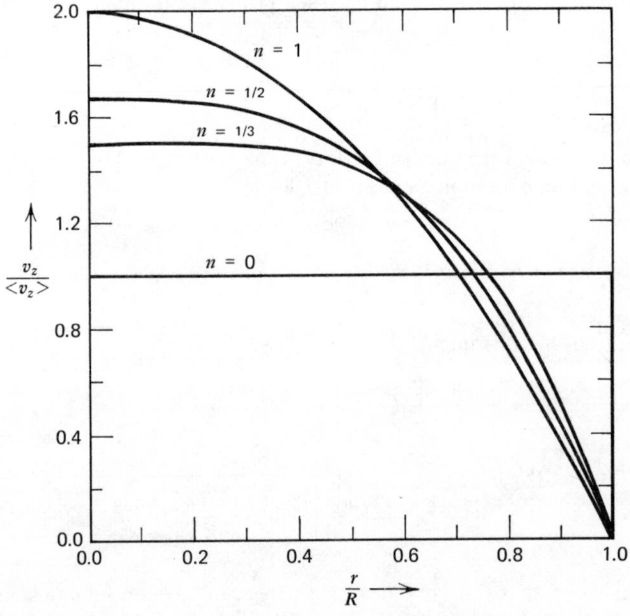

FIGURE 5.2–1. Tube flow velocity profiles for a power-law fluid. Note that the profiles become increasingly flatter as n decreases; $n = 0$ corresponds to plug flow. The Newtonian (parabolic) profile is shown as $n = 1$.

It is then easy to get the volume rate of flow Q:

$$\begin{aligned} Q &= \int_0^{2\pi}\int_0^R v_z r\, dr\, d\theta \\ &= 2\pi R^2 \int_0^1 v_z \cdot \frac{r}{R}\, d\left(\frac{r}{R}\right) \\ &= \frac{\pi R^3}{(1/n)+3}\left(\frac{\tau_R}{m}\right)^{1/n} \\ &= \frac{\pi R^3}{(1/n)+3}\left[\frac{(\mathscr{P}_0 - \mathscr{P}_L)R}{2mL}\right]^{1/n} \end{aligned} \tag{5.2–9}$$

For $n = 1$ and $m = \mu$ this reduces to the Hagen-Poiseuille equation for Newtonian fluids.[1] In Table 5.2–1 we give a summary of the tube flow and slit flow results for several common non-Newtonian viscosity models.

For the tapered tube whose radius decreases linearly from R_0 at $z = 0$ to R_L at $z = L$, we get:

$$\mathscr{P}_0 - \mathscr{P}_L = \frac{2mL}{3n}\left[\frac{Q}{\pi}\left(\frac{1}{n} + 3\right)\right]^n \left(\frac{R_L^{-3n} - R_0^{-3n}}{R_0 - R_L}\right) \tag{5.2–10}$$

for the relation between pressure drop and volume rate of flow.

Sutterby[2] found that the use of a generalized Newtonian fluid model described adequately the Q vs. $\mathscr{P}_0 - \mathscr{P}_L$ relationship for the slow flow of polymer solutions in a converging tube. For very fast flow, on the other hand, the data were well described by the results of an ideal (inviscid) fluid calculation; this is perhaps not too surprising since the inviscid fluid corresponds to $\mathrm{Re} \to \infty$.

The flow of non-Newtonian fluids in tapered tubes has also been studied by Oka and Murata.[3]

EXAMPLE 5.2–2 Thickness of a Film of Polymer Solution Flowing Down an Inclined Plate

Obtain an expression for the thickness of a film of polymer solution as it flows down an inclined plate making an angle α with the vertical. Use the power-law fluid model for the derivation. Take the origin of coordinates to be such that $x = 0$ at the film surface and $x = \delta$ at the plate; the film extends along the plate from $z = 0$ to $z = L$.

SOLUTION

From the equation of motion we get:

$$0 = -\frac{d\tau_{xz}}{dx} + \rho g \cos\alpha \tag{5.2–11}$$

When this is integrated, using the boundary condition that $\tau_{xz} = 0$ at the liquid-air interface at $x = 0$, we get:

$$\tau_{xz} = \rho g x \cos\alpha \tag{5.2–12}$$

We know intuitively that v_z decreases with increasing x, so that $\dot{\gamma}$ will be taken to be $(-dv_z/dx)$ in order to insure that $\dot{\gamma}$ be a positive quantity. Then the power law gives the following expression for τ_{xz}:

$$\begin{aligned}\tau_{xz} &= -\eta\frac{dv_z}{dx} = -m\left(-\frac{dv_z}{dx}\right)^{n-1}\frac{dv_z}{dx} \\ &= +m\left(-\frac{dv_z}{dx}\right)^n\end{aligned} \tag{5.2–13}$$

[1] Laminar-turbulent transition has been studied by D. W. Dodge and A. B. Metzner, *A.I.Ch.E. Journal*, *5*, 189–204 (1959); they found that the laminar-turbulent transition occurred in the modified Reynolds number range $2100 < \mathrm{Re}_n < 3100$, where $\mathrm{Re}_n = (D^n\langle v_z\rangle^{2-n}\rho/m)(3 + (1/n))^{-n}2^{3-n}$ for power-law fluids. For other studies see N. W. Ryan and M. M. Johnson, *A.I.Ch.E. Journal*, *5*, 433–435 (1959); R. W. Hanks, *ibid.*, *9*, 306–309 (1963); D. M. Meter, *ibid.*, *10*, 881–884 (1964).

[2] J. L. Sutterby, *Trans. Soc. Rheol.*, *9:2*, 227–241 (1965).

[3] S. Oka, *Zairyō*, *14*, 241–244 (1965); S. Oka and T. Murata, *Japan. J. of Appl. Phys.*, *8*, 5–8 (1969); S. Oka, *Biorheology*, *10*, 207–212 (1973).

TABLE 5.2–1

Flow Rates through Slits and Tubes

Model (Constants)	Volume Rate of Flow Q in Thin Slit (Width W, Thickness $2B$, Length L) $\tau_{xz}\vert_{x=B} \equiv \tau_B = (\mathscr{P}_0 - \mathscr{P}_L)B/L$		Volume Rate of Flow Q in Circular Tube (Radius R, Length L) $\tau_{rz}\vert_{r=R} \equiv \tau_R = (\mathscr{P}_0 - \mathscr{P}_L)R/2L$	
Power law (m,n)	$\dfrac{2WB^2}{(1/n)+2}\left(\dfrac{\tau_B}{m}\right)^{1/n}$	(A)	$\dfrac{\pi R^3}{(1/n)+3}\left(\dfrac{\tau_R}{m}\right)^{1/n}$	(F)
Ellis $(\eta_0, \tau_{1/2}, \alpha)$	$\dfrac{2WB^2\tau_B}{3\eta_0}\left[1+\dfrac{3}{\alpha+2}\left(\dfrac{\tau_B}{\tau_{1/2}}\right)^{\alpha-1}\right]$	(B)	$\dfrac{\pi R^3\tau_R}{4\eta_0}\left[1+\dfrac{4}{\alpha+3}\left(\dfrac{\tau_R}{\tau_{1/2}}\right)^{\alpha-1}\right]$	(G)
Truncated power law $(\eta_0, \dot{\gamma}_0, n)$	$\dfrac{2WB^2\tau_B}{[(1/n)+2]\eta_0}\left[\left(\dfrac{\eta_0\dot{\gamma}_0}{\tau_B}\right)^{1-(1/n)}+\frac{1}{3}\left(\dfrac{1}{n}-1\right)\left(\dfrac{\eta_0\dot{\gamma}_0}{\tau_B}\right)^3\right]$ (when $\tau_B \gg \eta_0\dot{\gamma}_0$)	(C)	$\dfrac{\pi R^3\tau_R}{[(1/n)+3]\eta_0}\left[\left(\dfrac{\eta_0\dot{\gamma}_0}{\tau_R}\right)^{1-(1/n)}+\frac{1}{4}\left(\dfrac{1}{n}-1\right)\left(\dfrac{\eta_0\dot{\gamma}_0}{\tau_R}\right)^4\right]$ (when $\tau_R \gg \eta_0\dot{\gamma}_0$)	(H)
Eyring (τ_0, t_0)	$\dfrac{2WB^2}{t_0}\left(\dfrac{\tau_0}{\tau_B}\right)^2\left[\dfrac{\tau_B}{\tau_0}\cosh\dfrac{\tau_B}{\tau_0}-\sinh\dfrac{\tau_B}{\tau_0}\right]$	(D)	$\dfrac{2\pi R^3}{t_0}\left(\dfrac{\tau_0}{\tau_R}\right)^3\left[\left\{\frac{1}{2}\left(\dfrac{\tau_R}{\tau_0}\right)^2+1\right\}\cosh\dfrac{\tau_R}{\tau_0}-\dfrac{\tau_R}{\tau_0}\sinh\dfrac{\tau_R}{\tau_0}-1\right]$	(I)
Bingham (μ_0, τ_0)	$\dfrac{2WB^2\tau_B}{3\mu_0}\left[1-\frac{3}{2}\left(\dfrac{\tau_0}{\tau_B}\right)+\frac{1}{2}\left(\dfrac{\tau_0}{\tau_B}\right)^3\right]$ (when $\tau_B \geq \tau_0$)	(E)	$\dfrac{\pi R^3\tau_R}{4\mu_0}\left[1-\frac{4}{3}\left(\dfrac{\tau_0}{\tau_R}\right)+\frac{1}{3}\left(\dfrac{\tau_0}{\tau_R}\right)^4\right]$ (when $\tau_R \geq \tau_0$)	(J)

Combination of Eqs. 5.2–12 and 5.2–13 then gives:

$$-\frac{dv_z}{dx} = \left(\frac{\rho g x}{m}\cos\alpha\right)^{1/n} \qquad (5.2\text{–}14)$$

Integration with boundary condition that $v_z = 0$ at $x = \delta$ gives:

$$v_z = \left(\frac{\rho g \delta}{m}\cos\alpha\right)^{1/n} \frac{\delta}{(1/n)+1}\left[1-\left(\frac{x}{\delta}\right)^{(1/n)+1}\right] \qquad (5.2\text{–}15)$$

for the velocity distribution.

Integration over the cross section of flow (thickness δ and width W) gives for the volume rate of flow:

$$Q = \frac{W\delta^2}{(1/n)+2}\left(\frac{\rho g \delta}{m}\cos\alpha\right)^{1/n} \qquad (5.2\text{–}16)$$

Solution for the film thickness then gives:

$$\delta = \left(\frac{m}{\rho g\cos\alpha}\right)^{1/(2n+1)}\left(\frac{Q[(1/n)+2]}{W}\right)^{n/(2n+1)} \qquad (5.2\text{–}17)$$

This shows how the film thickness depends on the feed rate Q, the fluid properties m and n, and the angle of inclination α.

EXAMPLE 5.2–3 Plane Couette Flow

A macromolecular fluid is confined to the space between two horizontal planes ($x = 0$ and $x = B$) the upper one of which is moving in the positive z-direction with a constant speed v_0. In addition there is a pressure gradient in the z-direction, the pressure at $z = 0$ being p_0 and that at $z = L$ being p_L, with $p_0 > p_L$. Obtain an expression for the volume rate of flow in the z-direction in the slit that results from the combined effects of the motion of the upper plate and the pressure gradient.

Problems of this type arise in diverse processing operations, such as in certain types of extruders, and in various lubrication problems.

SOLUTION

For this flow the equation of motion is:

$$0 = -\frac{dp}{dz} - \frac{d\tau_{xz}}{dx} \qquad (5.2\text{–}18)$$

which may be integrated to give:

$$\tau_{xz} = -\left[\frac{(p_0 - p_L)B}{L}\right](\lambda - \xi) \qquad (5.2\text{–}19)$$

where λ is a constant of integration, and $\xi = x/B$.

Equation 5.1–4 gives

$$\tau_{xz} = -\eta\frac{dv_z}{dx} \qquad (5.2\text{–}20)$$

By way of illustration, we use the power law for η. Two cases have to be distinguished here:

Case I: There is no maximum in the velocity profile $v_z(x)$
Case II: There is a maximum in the velocity profile $v_z(x)$

We consider only Case I, although the final results for Case II will be given.

For Case I, $\eta = m(dv_z/dx)^{n-1}$ and hence Eq. 5.2–20 becomes

$$\tau_{xz} = -m\left(\frac{dv_z}{dx}\right)^n \tag{5.2–21}$$

Combination of Eqs. 5.2–19 and 5.2–21 gives an equation for v_z as a function of x. Integration of this equation and application of the boundary conditions $v_z = 0$ at $x = 0$ and $v_z = v_0$ at $x = B$ gives:

$$\phi(\xi) \equiv \frac{v_z}{v_0} = \frac{\lambda^{s+1} - (\lambda - \xi)^{s+1}}{\lambda^{s+1} - (\lambda - 1)^{s+1}} \tag{5.2–22}$$

where $\xi = x/B$, $s = 1/n$, and λ is a dimensionless parameter given by:

$$\Lambda \equiv \frac{(p_0 - p_L)B}{mL}\left(\frac{B}{v_0}\right)^{1/s}$$

$$= \left[\frac{s+1}{\lambda^{s+1} - (\lambda - 1)^{s+1}}\right]^{1/s} \qquad \textit{Case I} \qquad \Lambda \leq (s+1)^{1/s} \tag{5.2–23}$$

Hence, knowing $\lambda = \lambda(\Lambda,s)$ from Eq. 5.2–23, one can obtain the velocity profile from Eq. 5.2–22.

The volume rate of flow between two planes of width W is then found to be:

$$\frac{Q}{WBv_0} = \int_0^1 \phi\, d\xi = \phi\xi\Big|_0^1 - \int_0^1 \frac{d\phi}{d\xi}\,\xi\, d\xi$$

$$= -(\lambda - 1) + \left(\frac{s+1}{s+2}\right)\frac{\lambda^{s+2} - (\lambda - 1)^{s+2}}{\lambda^{s+1} - (\lambda - 1)^{s+1}} \qquad \textit{Case I} \qquad \Lambda \leq (s+1)^{1/s} \tag{5.2–24}$$

with λ determined from Eq. 5.2–23.

The results analogous to Eqs. 5.2–23 and 5.2–24 for Case II can be shown to be:

$$\Lambda = \left[\frac{s+1}{\lambda^{s+1} - (1 - \lambda)^{s+1}}\right]^{1/s} \qquad \textit{Case II} \qquad \Lambda \geq (s+1)^{1/s} \tag{5.2–25}$$

$$\frac{Q}{WBv_0} = (1 - \lambda) + \left(\frac{s+1}{s+2}\right)\frac{\lambda^{s+2} + (1 - \lambda)^{s+2}}{\lambda^{s+1} - (1 - \lambda)^{s+1}} \qquad \textit{Case II} \qquad \Lambda \geq (s+1)^{1/s} \tag{5.2–26}$$

Hence, whether or not one uses the Case I or Case II formulas depends on whether Λ is larger or smaller than $(s + 1)^{1/s}$. A table of $\lambda = \lambda(\Lambda,s)$ has been prepared by Flumerfelt *et al.*[4] The dimensionless flow rate $\Omega = Q/WBv_0$ is shown in Fig. 5.2–2; this chart is so constructed that it includes the case $p_0 < p_L$ as well as $p_0 > p_L$.

EXAMPLE 5.2–4 Tangential Annular Flow (Bingham Fluid)

Determine the velocity and stress distributions for the tangential flow of a Bingham plastic in an annulus, as functions of the torque $\mathcal{T}$ applied to the outer cylinder. Assume incompressible laminar flow and ignore end effects. The inner and outer radii of the annular gap are κR and R, respectively. Denote the angular velocity of the outer cylinder by W; the inner cylinder is held fixed.

[4] R. W. Flumerfelt, M. W. Pierick, S. L. Cooper, and R. B. Bird, *Ind. Eng. Chem. Fundamentals*, *8*, 354–357 (1969); earlier work was done on this problem by F. W. Kroesser and S. Middleman, *Polym. Eng. Sci.*, *5*, 230–234 (1965) and by Z. Tadmor, *Polym. Eng. Sci.*, *6*, 203–212 (1966).

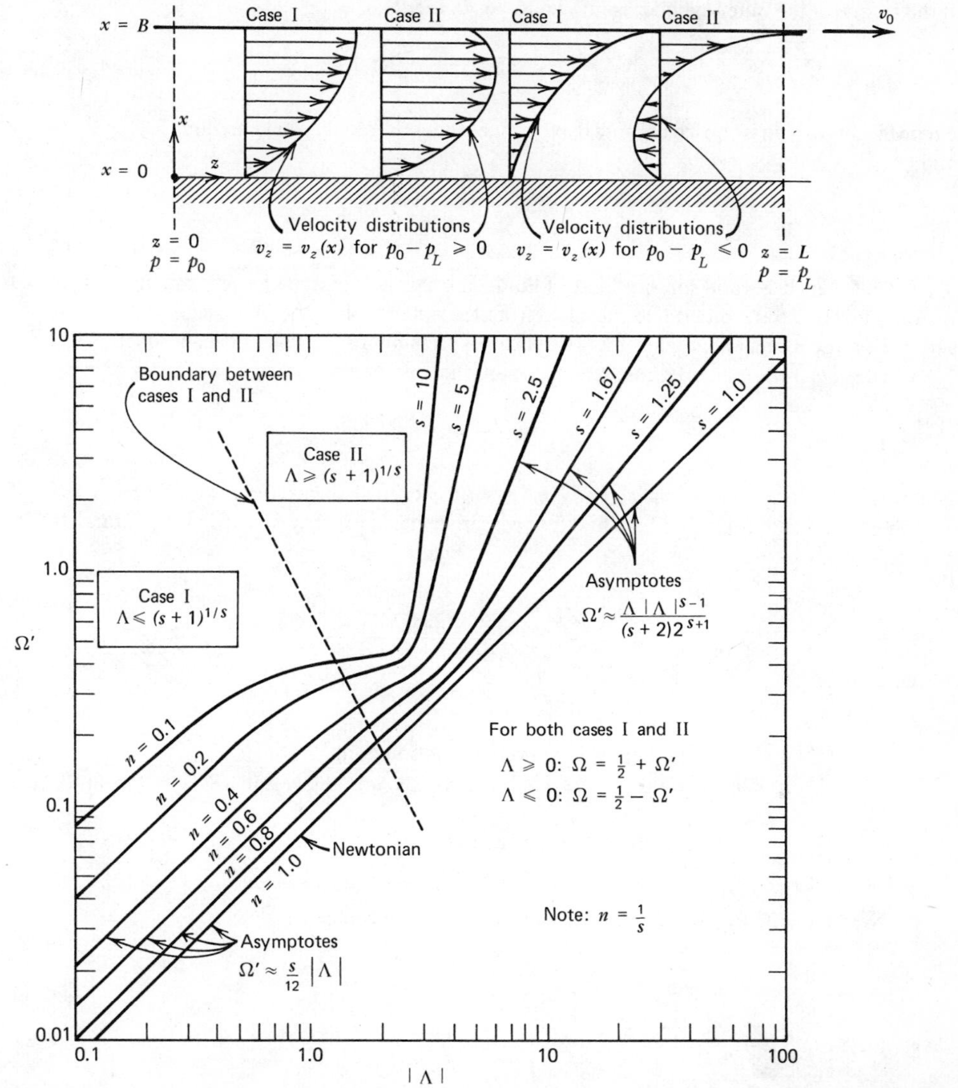

FIGURE 5.2–2. Dimensionless flow rate $\Omega = Q/WBv_0$ as a function of Λ and n (or $1/s$) for generalized Couette flow. [Reprinted with permission from R. W. Flumerfelt, M. W. Pierick, S. L. Cooper, and R. B. Bird, *Ind. Eng. Chem. Fundamentals*, *8*, 354–357 (1969). Copyright by the American Chemical Society.]

SOLUTION

For this system $v_r = v_z = 0$ and $v_\theta = v_\theta(r)$. Hence the only nonvanishing components of $\boldsymbol{\tau}$ are $\tau_{r\theta}$ and $\tau_{\theta r}$, and the θ-equation of motion for steady state is in cylindrical coordinates:

$$0 = \frac{1}{r^2}\frac{d}{dr}(r^2\tau_{r\theta}) \tag{5.2–27}$$

This may be integrated to give:

$$\tau_{r\theta} = \frac{C_1}{r^2} \tag{5.2–28}$$

If the torque at the outer cylinder is known to be $\mathscr{T}$, then:

$$\mathscr{T} = -\tau_{r\theta}\big|_{r=R} \cdot 2\pi RL \cdot R \tag{5.2–29}$$

where the minus sign is chosen because the θ-momentum flux is in the $-r$-direction. Hence $C_1 = -\mathscr{T}/2\pi L$ and:

$$\tau_{r\theta} = -\frac{\mathscr{T}}{2\pi L r^2} \tag{5.2–30}$$

This result, which is valid for any kind of fluid, can also be obtained by recognizing that angular momentum must be transmitted undiminished from the outer to the inner cylinder.

For the Bingham model, the analytical expression to be used depends on the value of $\tau = \sqrt{\frac{1}{2}(\boldsymbol{\tau}:\boldsymbol{\tau})} = \sqrt{\frac{1}{2}\sum_i \sum_j \tau_{ij}{}^2}$. Because $\tau_{r\theta}$ and $\tau_{\theta r}$ are the only nonvanishing components of $\boldsymbol{\tau}$, we have:

$$\tau = \sqrt{\tfrac{1}{2}(\boldsymbol{\tau}:\boldsymbol{\tau})} = \sqrt{\tau_{r\theta}{}^2} = -\tau_{r\theta} \tag{5.2–31}$$

Hence we use Eq. 5.1–16 when $\tau \geq \tau_0$ (i.e., when the critical shear stress is exceeded) and Eq. 5.1–15 when $\tau \leq \tau_0$. From Eq. 5.2–30, we find that we can define a quantity r_0, which is the value of r for which $\tau = \tau_0$:

$$r_0 = \sqrt{\frac{\mathscr{T}}{2\pi\tau_0 L}} \tag{5.2–32}$$

We can then distinguish three situations:

a. If $r_0 \leq \kappa R$, then there will be no fluid motion at all.

b. If $\kappa R < r_0 \leq R$, then there will be viscous flow in the region $\kappa R < r < r_0$ and "plug flow" for $r_0 \leq r \leq R$.

c. If $r_0 \geq R$, then there is flow throughout.

Next we write Eq. 5.1–16 for the system at hand. Inasmuch as

$$\dot{\gamma} = \sqrt{\tfrac{1}{2}(\dot{\boldsymbol{\gamma}}:\dot{\boldsymbol{\gamma}})} = r\frac{d}{dr}\left(\frac{v_\theta}{r}\right) \tag{5.2–33}$$

we get from Eq. 5.1–4:

$$\tau_{r\theta} = -\left\{\mu_0 + \frac{\tau_0}{r\dfrac{d}{dr}\left(\dfrac{v_\theta}{r}\right)}\right\} r\frac{d}{dr}\left(\frac{v_\theta}{r}\right)$$

$$= -\tau_0 - \mu_0 r\frac{d}{dr}\left(\frac{v_\theta}{r}\right) \tag{5.2–34}$$

Since v_θ/r does not decrease with increasing r, $\tau_{r\theta}$ is negative. That is, the θ-component of momentum flows in the $-r$-direction.

By substituting Eq. 5.2–34 into Eq. 5.2–30 and integrating, we get for case ***b***:

$$\frac{v_\theta}{r} = W + \frac{\mathscr{T}}{4\pi L\mu_0 r_0{}^2}\left[1 - \left(\frac{r_0}{r}\right)^2\right] - \frac{\tau_0}{\mu_0}\ln\frac{r}{r_0} \quad \text{for} \quad \kappa R \leq r \leq r_0 \tag{5.2–35}$$

$$\frac{v_\theta}{r} = W \quad \text{for} \quad r_0 \leq r \leq R \tag{5.2–36}$$

in which the boundary condition $v_\theta/r = W$ at $r = r_0$ has been used. For case ***c***, we use the boundary condition that $v_\theta/r = W$ at $r = R$ and get:

$$\frac{v_\theta}{r} = W + \frac{\mathscr{T}}{4\pi L\mu_0 R^2}\left[1 - \left(\frac{R}{r}\right)^2\right] - \frac{\tau_0}{\mu_0}\ln\frac{r}{R} \qquad \text{for} \qquad \kappa R \leq r \leq R \tag{5.2-37}$$

Now for case **c**, if we utilize the condition that at $r = \kappa R$, $v_\theta = 0$, we get:

$$W = \frac{\mathscr{T}}{4\pi L\mu_0 R^2}\left(\frac{1}{\kappa^2} - 1\right) + \frac{\tau_0}{\mu_0}\ln\kappa \tag{5.2-38}$$

This relation between W and $\mathscr{T}$ (known as the Reiner-Riwlin equation[5]) gives a means for determining μ_0 and τ_0 from viscometric data. The tangential annular flow for an Ellis fluid has been given by Matsuhisa and Bird;[6] they also discuss the pressure distribution in this system. The solution for a power-law fluid is given in Problem 5B.3.

EXAMPLE 5.2–5 Axial Annular Flow (Ellis Fluid): An Approximate Treatment

It is desired to obtain the expression for the volume flow rate of a polymer through an annular conduit of length L; the bounding surfaces of the annulus are located at κR and R (with $\kappa < 1$). Use the Ellis fluid model for the derivation.

Since a completely analytical solution for the Ellis model is not possible, use the following approximate procedure: (a) Show how to write the expression for volume rate of flow Q in an annulus for a Newtonian fluid as the expression for Q for plane slit flow multiplied by a "curvature correction factor." (b) Multiply the plane slit flow expression for Q of an Ellis fluid by the same curvature correction factor as was found in (a).

SOLUTION (a) Curvature Correction Factor for Newtonian Fluid

For the Newtonian fluid the volume rate of flow in a coaxial annulus is given by:[7]

$$Q_{\text{ann.}} = \frac{\pi(\mathscr{P}_0 - \mathscr{P}_L)R^4}{8\mu L}\left[(1 - \kappa^4) - \frac{(1 - \kappa^2)^2}{\ln(1/\kappa)}\right] \tag{5.2-39}$$

where $\mathscr{P}_0 - \mathscr{P}_L$ is the modified pressure drop over a length L of conduit. For the flow in a slit of width W, thickness $2B$, and length L ($B \ll W \ll L$) the volume rate of flow is easily found to be[8] (see Eq. 1.2–8):

$$Q_{\text{slit}} = \frac{2}{3}\frac{(\mathscr{P}_0 - \mathscr{P}_L)WB^3}{\mu L} \tag{5.2-40}$$

If in Eq. 5.2–40 we replace W by $2\pi R$ and $2B$ by $R(1 - \kappa)$, we get:

$$Q_{\text{ann.}}\,(\text{approximate}) = \frac{\pi(\mathscr{P}_0 - \mathscr{P}_L)R^4}{6\mu L}(1 - \kappa)^3 \tag{5.2-41}$$

[5] M. Reiner and R. Riwlin, *Kolloid-Z.*, *43*, 1–5 (1927); many more analytical solutions for the Bingham fluid have been given by J. G. Oldroyd, *Proc. Camb. Phil. Soc.*, *43*, 100–105, 383–395, 396–405, 521–532 (1947); *44*, 200–213, 214–228 (1948).
[6] S. Matsuhisa and R. B. Bird, *A.I.Ch.E. Journal*, *11*, 588–595 (1965).
[7] See, for example, R. B. Bird, W. E. Stewart, and E. N. Lightfoot, *Transport Phenomena*, Wiley, New York (1960), pp. 51–54.
[8] R. B. Bird, W. E. Stewart, and E. N. Lightfoot, *op. cit.*, p. 62.

In both Eqs. 5.2–39 and 5.2–41 we set $1 - \kappa$ equal to the small quantity ϵ. Then we divide one by the other to get the Newtonian curvature correction factor C_N:

$$\begin{aligned} C_N &= Q_{\text{ann.}}/Q_{\text{ann.}}\text{ (approximate)} \\ &= \frac{3}{4\epsilon^3}\left\{[1-(1-\epsilon)^4] - \frac{\epsilon^2(2-\epsilon)^2}{\epsilon + \frac{1}{2}\epsilon^2 + \frac{1}{3}\epsilon^3 + \cdots}\right\} \\ &= 1 - \tfrac{1}{2}\epsilon + \frac{1}{60}\epsilon^2 + \frac{1}{120}\epsilon^3 + \cdots \end{aligned} \tag{5.2–42}$$

(b) Approximate Formula for Axial Annular Flow of Ellis Fluid

We now take the slit flow formula for an Ellis fluid from Table 5.2–1 and replace W by $2\pi R$ and $2B$ by $R(1 - \kappa)$ as we did for the Newtonian fluid in (a); we then multiply by C_N:

$$Q_{\text{ann.}} \doteq \frac{\pi(\mathscr{P}_0 - \mathscr{P}_L)R^4}{6\eta_0 L}(1-\kappa)^3\left[1 + \frac{3}{\alpha+2}\left(\frac{(\mathscr{P}_0 - \mathscr{P}_L)R(1-\kappa)}{2\tau_{1/2}L}\right)^{\alpha-1}\right]\cdot C_N \tag{5.2–43}$$

This is an approximate result because the Newtonian curvature correction C_N has been applied. It is believed that this approximate procedure is valid[6] for κ greater than about 0.6; this conclusion was drawn by comparing the exact solution for a power-law fluid[9] with the slit approximation for a power-law fluid multiplied by C_N.

In Figs. 5.2–3 and 5.2–4 we show annular flow experimental data for aqueous solutions of two different polymers.[10] Equation 5.2–43 describes the data quite well. Also, at low flow rates the Ellis model is appreciably better than the power-law model. For annular flow at low flow rates there is an

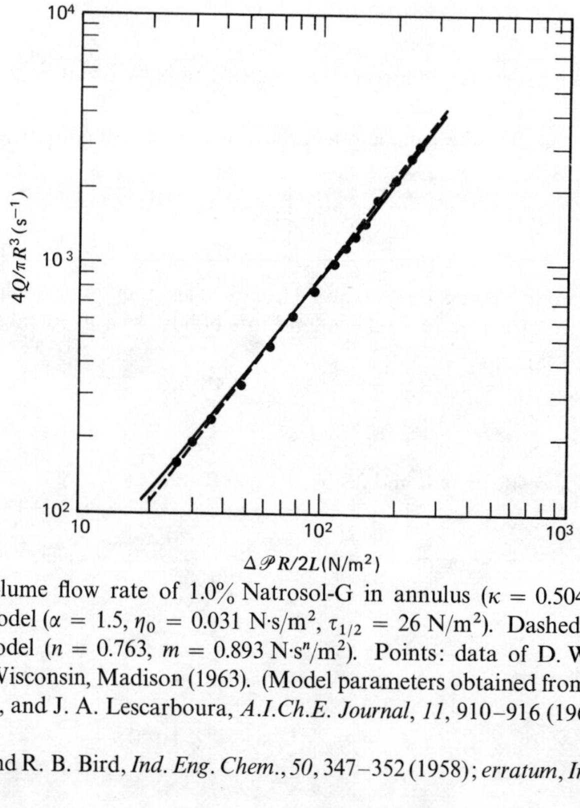

FIGURE 5.2–3. Volume flow rate of 1.0% Natrosol-G in annulus ($\kappa = 0.5043$). Solid curve: Eq. 5.2–43 for the Ellis model ($\alpha = 1.5$, $\eta_0 = 0.031$ N·s/m², $\tau_{1/2} = 26$ N/m²). Dashed curve: exact solution for the power-law model ($n = 0.763$, $m = 0.893$ N·s^n/m²). Points: data of D. W. McEachern, Ph.D. Thesis, University of Wisconsin, Madison (1963). (Model parameters obtained from data of McEachern.) [E. Ashare, R. B. Bird, and J. A. Lescarboura, *A.I.Ch.E. Journal*, *11*, 910–916 (1965).]

[9] A. G. Fredrickson and R. B. Bird, *Ind. Eng. Chem.*, *50*, 347–352 (1958); *erratum*, *Ind. Eng. Chem. Fundamentals 3*, 383 (1964).

[10] E. Ashare, R. B. Bird, and J. A. Lescarboura, *A.I.Ch.E. Journal*, *11*, 910–916 (1965).

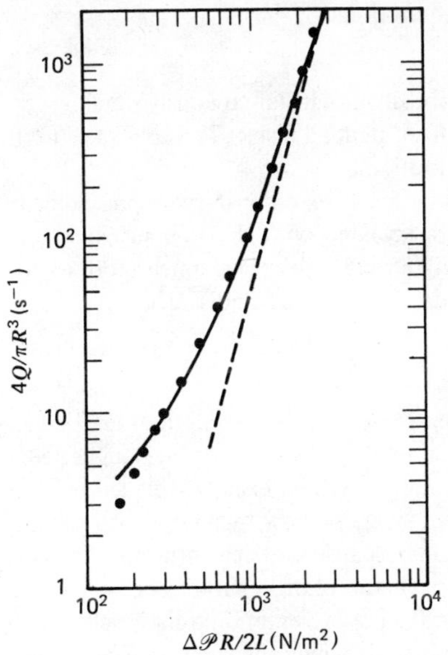

FIGURE 5.2–4. Volume flow rate of 3.5% CMC-70-medium in annulus ($\kappa = 0.624$). Solid curve: Eq. 5.2–43 for the Ellis model ($\alpha = 3.0$, $\eta_0 = 2.27$ N·s/m², $\tau_{1/2} = 152$ N/m²). Dashed curve: exact solution for the power-law model ($n = 0.280$, $m = 60.6$ N·s^n/m²). Points: data of A. G. Fredrickson, Ph.D. Thesis, University of Wisconsin, Madison (1959). (Model parameters obtained from data of Fredrickson.) [E. Ashare, R. B. Bird, and J. A. Lescarboura, *A.I.Ch.E. Journal*, *11*, 910–916 (1965).]

appreciable portion of the cross section in which the velocity gradient is quite small—that is, near the maximum in the velocity distribution. In this region the fact that the power law does not describe the zero-shear-rate viscosity at all does not allow a good description of the system. This inadequacy has also been pointed out by other authors.[11] A similar inadequacy in the power law can be expected to arise in case II of Example 5.2–3.

EXAMPLE 5.2–6 Squeezing Flow between Two Circular Disks

Extend the analysis of Example 1.2–6 to macromolecular fluids by paralleling the previous analysis for a power-law fluid.

SOLUTION

We postulate, as before, that $v_r = v_r(r,z)$, $v_z = v_z(z)$, and $p = p(r)$, and we disregard all inertial terms. Then the starting equations are:

Continuity:
$$\frac{1}{r}\frac{\partial}{\partial r}(rv_r) + \frac{dv_z}{dz} = 0 \qquad (5.2\text{–}44)$$

Motion:
(r-component)
$$0 = -\frac{dp}{dr} - \left(\underline{\frac{1}{r}\frac{\partial}{\partial r}(r\tau_{rr}) - \frac{\tau_{\theta\theta}}{r}} + \frac{\partial \tau_{rz}}{\partial z}\right) \qquad (5.2\text{–}45)$$

Motion:
(z-component)
$$0 = -\left(\underline{\frac{1}{r}\frac{\partial}{\partial r}(r\tau_{rz}) + \frac{\partial \tau_{zz}}{\partial z}}\right) \qquad (5.2\text{–}46)$$

Power law:
$$\tau_{ij} = -m(\tfrac{1}{2}[\underline{\dot\gamma_{rr}^2 + \dot\gamma_{\theta\theta}^2 + \dot\gamma_{zz}^2} + 2\dot\gamma_{rz}^2])^{(n-1)/2}\dot\gamma_{ij} \qquad (5.2\text{–}47)$$

[11] R. D. Vaughn and P. D. Bergman, *Ind. Eng. Chem. Proc. Des. Devel.*, *5*, 44–47 (1966).

We now make several assumptions:

a. It is assumed that locally and instantaneously the flow can be described as the steady-state flow between two fixed parallel disks. This allows us to drop the dashed-underlined terms in the equation of motion.

b. It is assumed that the decrease in non-Newtonian viscosity results predominantly from the velocity gradients associated with the rz-component of the rate-of-deformation tensor. That is, the shearing effects are deemed much more important than the elongation effects. This allows us to drop the dashed underlined terms in the expression for τ_{ij} (and we will need only τ_{rz}).

It is really not known how serious these various assumptions are or over what range of conditions these assumptions are expected to be reasonable. Assumption (a) is suggested by the fact that the corresponding dashed-underlined terms in Eq. 1.2–61 vanish exactly; this assumption corresponds to a "lubrication approximation"—the replacement of a true flow by an approximate flow that is equivalent to a simple shear flow. Assumption (b) is made because one feels intuitively that the omitted terms are smaller than $\dot{\gamma}_{rz}^2$. In any case, the use of these assumptions enables one to parallel exactly the treatment given in Example 1.2–6. Hence we can make the presentation quite brief.

The velocity, pressure, and force equations given in Eqs. 1.2–63, 66, and 68 for the Newtonian fluid, have as their power-law equivalents:

$$v_r = \frac{h^{1+(1/n)}}{1+(1/n)}\left(-\frac{1}{mr^n}\frac{dp}{dr}\right)^{1/n}\left[1-\left(\frac{z}{h}\right)^{1+(1/n)}\right]r \qquad (5.2\text{–}48)$$

$$p - p_a = \frac{(-\dot{h})^n}{h^{2n+1}}\left(\frac{2n+1}{2n}\right)^n \frac{mR^{n+1}}{n+1}\left[1-\left(\frac{r}{R}\right)^{n+1}\right] \qquad (5.2\text{–}49)$$

$$F = \frac{(-\dot{h})^n}{h^{2n+1}}\left(\frac{2n+1}{2n}\right)^n \frac{\pi mR^{n+3}}{n+3} \qquad (5.2\text{–}50)$$

This last result is the *Scott equation*,[12,13] which was first developed in connection with measuring the properties of unvulcanized rubber stocks.

The Scott equation has been tested by Leider[14] who measured the time $t_{1/2}$ for the plates to move from an initial separation h_0 to a separation $h_0/2$ for a constant force F in 181 experimental runs. Regarding Eq. 5.2–50 as a differential equation for $h(t)$, one may integrate from $t = 0$ (where $h = h_0$) to $t = t_{1/2}$ (where $h = h_0/2$) to get:

$$\frac{t_{1/2}}{n} = K_n\left(\frac{\pi R^2 m}{F}\right)^{1/n}\left(\frac{R}{h_0}\right)^{1+(1/n)} \qquad (5.2\text{–}51)$$

$$K_n = \left(\frac{2^{1+(1/n)}-1}{2n}\right)\left(\frac{2n+1}{n+1}\right)\left(\frac{1}{n+3}\right)^{1/n} \qquad (5.2\text{–}52)$$

[12] J. R. Scott, *Trans. Inst. Rubber Ind.*, *7*, 169–186 (1931); *10*, 481–493 (1935).

[13] S. Oka, in *Rheology*, F. R. Eirich (Ed.), Academic Press, New York (1960), Chapter 2, Vol. 3 pp. 73–75; A Cameron, *The Principles of Lubrication*, Longmans, Green, and Co., London (1966), pp. 389–392; D. F. Moore, *The Friction and Lubrication of Elastomers*, Pergamon, New York (1972); A. B. Metzner, *Rheol. Acta*, *10*, 434–444 (1971); M. L. DeMartine and E. L. Cussler, *J. Pharm. Sci.*, *64*, 976–982 (1975); G. Brindley, J. M. Davies, and K. Walters, *J. Non-Newtonian Fluid Mech.*, *1*, 19–37 (1976).

[14] P. J. Leider, *Rheology Research Center Report No. 22*, November 1973, University of Wisconsin, Madison; *Ind. Eng. Chem, Fundamentals*, *13*, 342–346 (1974).

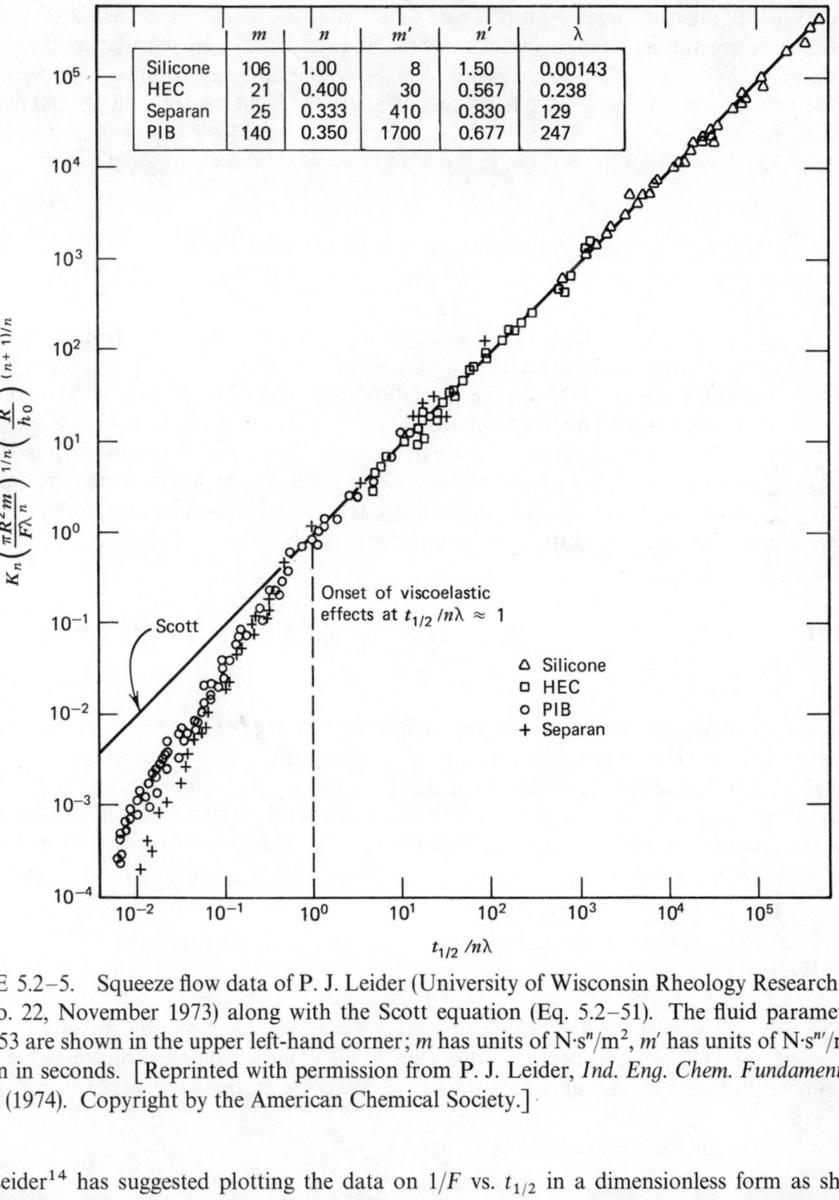

FIGURE 5.2–5. Squeeze flow data of P. J. Leider (University of Wisconsin Rheology Research Center Rept. No. 22, November 1973) along with the Scott equation (Eq. 5.2–51). The fluid parameters for Eq. 5.2–53 are shown in the upper left-hand corner; m has units of $N{\cdot}s^n/m^2$, m' has units of $N{\cdot}s^{n'}/m^2$, and λ is given in seconds. [Reprinted with permission from P. J. Leider, *Ind. Eng. Chem. Fundamentals*, *13*, 342–346 (1974). Copyright by the American Chemical Society.]

Leider[14] has suggested plotting the data on $1/F$ vs. $t_{1/2}$ in a dimensionless form as shown in Fig. 5.2–5. To render the half-time dimensionless, a time-constant λ for the fluid has been introduced. It was remarked in §5.1 that the power-law model contains no time constant. Consequently data from another property have to be introduced. If we make use of the fact that both the viscosity and first normal stress coefficient have a prominent power-law region, that is,

$$\eta = m\dot{\gamma}^{n-1} \quad \text{and} \quad \Psi_1 = m'\dot{\gamma}^{n'-2} \tag{5.2–53}$$

then a time constant can be constructed as follows:

$$\lambda = \left(\frac{m'}{2m}\right)^{1/(n'-n)} \tag{5.2–54}$$

When this choice of time constant is made, it is found that the Scott equation describes the system quite well down to $t_{1/2} = n\lambda$. Below that value there is a systematic deviation associated with the viscoelasticity of the fluid; specifically deviations result because of a prominent stress-overshoot effect that cannot be described by any generalized Newtonian fluid model. However, for $t_{1/2}$ greater than $n\lambda$ the power law seems to work well and the various assumptions seem *a posteriori* to be appropriate. The success of the Scott equation indicates that the squeeze flow experiment can be useful for determining fluid parameters from measurements of F and $t_{1/2}$. From Eq. 5.2–51 we see that a log-log plot of (Fh_0/R^3) vs. $(R/h_0t_{1/2})$ will, for large $t_{1/2}$, give n from the slope of the straight line and m from an intercept:

$$\log\left(\frac{Fh_0}{R^3}\right) = \log\left[\pi m(nK_n)^n\right] + n \log\left(\frac{R}{h_0 t_{1/2}}\right) \tag{5.2–55}$$

From a log-log plot of $(R^2/F)^{1/n}(R/h_0)^{1+(1/n)}$ vs. $t_{1/2}/n$ one can get λ from the break in the curve at $t_{1/2}/n = \lambda$. Explicit illustrations have been given by Leider.[14]

This example has illustrated the use of a generalized Newtonian fluid for a flow that is not a steady state shear flow. In other words, we have violated the restriction placed on the model. Nonetheless, in the limit of slow squeezing flows experiments show that we can apply the model. Experimental data were then used to define the limits of applicability of the model in terms of a characteristic time λ for each fluid. Hence the generalized Newtonian fluid model has suggested a suitable form for the correlation, and the experimental data have established the correlation quantitatively.

EXAMPLE 5.2–7 Enhancement of Axial Annular Flow by Rotating Inner Annulus[15] (Helical Flow of a Power-Law Fluid)

Here we consider the axial flow in a very thin annulus where the inner cylinder radius is R and the gap width is b, which is much smaller than R. The flow is driven by a pressure gradient. We want to investigate the way in which the flow rate changes if the inner cylinder is made to rotate. This is the only illustrative example in which we have more than one nonvanishing component of $\dot{\gamma}$ so that the flows in the two directions are coupled through the dependence of the non-Newtonian viscosity on the magnitude of $\dot{\gamma}$ defined in Eq. 5.1–8.

SOLUTION

The system is sketched in Fig. 5.2–6*a*. Because of the thinness of the slit, curvature can be neglected and the original problem becomes equivalent to the plane-slit problem shown in Fig. 5.2–6*b*; in this figure we also show the coordinate system we use. We postulate that $v_z = v_z(x)$, $v_y = v_y(x)$, $v_x = 0$, and $p = p(z)$.

For these postulates the equations of motion become:

y-component:
$$0 = -\frac{d}{dx}\tau_{xy} \tag{5.2–56}$$

z-component:
$$0 = \frac{\mathscr{P}_0 - \mathscr{P}_L}{L} - \frac{d}{dx}\tau_{xz} \tag{5.2–57}$$

where $\mathscr{P}_0 - \mathscr{P}_L$ is the modified pressure drop between $z = 0$ and $z = L$. The components of the stress

[15] This problem has been studied experimentally and theoretically by A. C. Dierckes, Jr., and W. R. Schowalter, *Ind. Eng. Chem. Fundamentals*, *5*, 263–271 (1966); numerical calculations for the Oldroyd model viscosity function (Eq. 8.1–7) have been made by J. G. Savins and G. C. Wallick, *A.I.Ch.E. Journal*, *12*, 357–363 (1966). See also B. D. Coleman, H. Markovitz, and W. Noll, *Viscometric Flows of Non-Newtonian Fluids*, Springer, New York (1966), pp. 37–41.

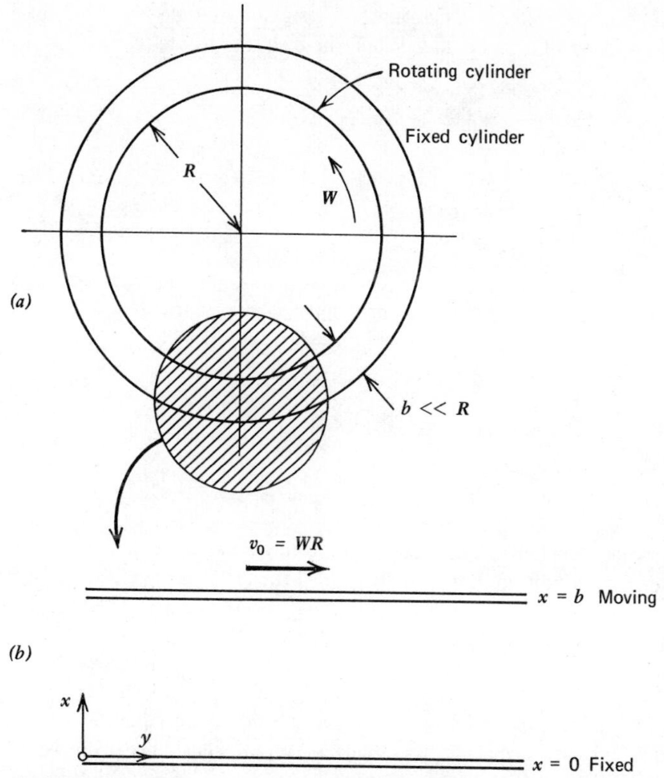

FIGURE 5.2–6. (*a*) Helical flow in a thin annulus; the fluid flows axially because of a pressure gradient and tangentially because of the rotating inner cylinder; (*b*) Coordinate system to be used for the equivalent problem neglecting curvature.

tensor for the power-law fluid are:

$$\tau_{xy} = -\eta(\dot{\gamma})\dot{\gamma}_{xy} = -m\dot{\gamma}^{n-1}\dot{\gamma}_{xy} \tag{5.2–58}$$

$$\tau_{xz} = -\eta(\dot{\gamma})\dot{\gamma}_{xz} = -m\dot{\gamma}^{n-1}\dot{\gamma}_{xz} \tag{5.2–59}$$

in which the magnitude of the rate-of-deformation tensor is:

$$\dot{\gamma} = \sqrt{\left(\frac{dv_y}{dx}\right)^2 + \left(\frac{dv_z}{dx}\right)^2} \tag{5.2–60}$$

It is convenient to use dimensionless quantities:

$$\bar{x} = \frac{x}{b} \qquad \bar{v}_i = \frac{v_i}{v_0} \qquad a = \frac{\mathscr{P}_0 - \mathscr{P}_L}{mL}\frac{b^{n+1}}{v_0{}^n} \tag{5.2–61}$$

and further to abbreviate $d\bar{v}_y/d\bar{x}$ by Y and $d\bar{v}_z/d\bar{x}$ by Z. Then the equations of motion, in terms of the velocity gradients, become:

$$\frac{d}{d\bar{x}}[(Y^2 + Z^2)^{(n-1)/2}Y] = 0 \tag{5.2–62}$$

$$\frac{d}{d\bar{x}}[(Y^2 + Z^2)^{(n-1)/2}Z] = -a \tag{5.2–63}$$

These equations may be integrated at once, and we designate by the symbols C_1 and C_2 the constants of integration that appear on the right side. When these equations are then solved for Y and Z, we get:

$$Y = \frac{d\bar{v}_y}{d\bar{x}} = [C_1^2 + (C_2 - a\bar{x})^2]^{(1-n)/2n} C_1 \tag{5.2–64}$$

$$Z = \frac{d\bar{v}_z}{d\bar{x}} = [C_1^2 + (C_2 - a\bar{x})^2]^{(1-n)/2n}(C_2 - a\bar{x}) \tag{5.2–65}$$

From this point on we specialize to $n = \frac{1}{3}$, since for this choice an analytical solution can be obtained; for other choices of n numerical integrations would have to be performed.

Integration of Eqs. 5.2–64 and 5.2–65 with $n = \frac{1}{3}$ gives:

$$\bar{v}_y = \int_0^{\bar{x}} C_1[C_1^2 + (C_2 - a\bar{x})^2]\, d\bar{x} \tag{5.2–66}$$

$$\bar{v}_z = \int_0^{\bar{x}} (C_2 - a\bar{x})[C_1^2 + (C_2 - a\bar{x})^2]\, d\bar{x} \tag{5.2–67}$$

in which the integration constants have been set equal to zero since both velocity components are zero at $\bar{x} = 0$. The boundary condition that $\bar{v}_z = 0$ at $\bar{x} = 1$ then leads to:

$$C_2 = \frac{a}{2} \tag{5.2–68}$$

The boundary condition that $\bar{v}_y = 1$ at $\bar{x} = 1$ leads to the cubic equation $C_1^3 + \frac{1}{12}a^2 C_1 - 1 = 0$, which has only one real root, according to Decartes' rule of signs. That root is:

$$C_1 = A_+ + A_- \tag{5.2–69}$$

in which:

$$A_\pm = \sqrt[3]{\frac{1}{2} \pm \sqrt{\frac{1}{4} + \frac{1}{27}\left(\frac{a^2}{12}\right)^3}} \tag{5.2–70}$$

The final expressions for the velocity profiles are then:

$$\bar{v}_y = C_1\left[C_1^2\bar{x} + \frac{a^2}{12}(3\bar{x} - 6\bar{x}^2 + 4\bar{x}^3)\right] \tag{5.2–71}$$

$$\bar{v}_z = \frac{a}{2}\left[\left(C_1^2 + \frac{a^2}{4}\right)\bar{x} - \frac{1}{2}a^2\bar{x}^2 + \frac{1}{3}a^2\bar{x}^3\right] - a\left[\left(C_1^2 + \frac{a^2}{4}\right)\frac{\bar{x}^2}{2} - \frac{1}{3}a^2\bar{x}^3 + \frac{1}{4}a^2\bar{x}^4\right] \tag{5.2–72}$$

and the volume rate of flow axially through the annulus is:

$$\begin{aligned} Q &= 2\pi R \int_0^b v_z\, dx = 2\pi R^2 b W \int_0^1 \bar{v}_z\, d\bar{x} \\ &= \frac{\pi R^2 b W a^3}{40}\left[1 + \frac{20}{3}\left(\frac{C_1}{a}\right)^2\right] \end{aligned} \tag{5.2–73}$$

in which $a = (b\,\Delta\mathscr{P}/mL)(b/WR)^{1/3}$. The result is somewhat easier to interpret when the quantity in brackets is written as a series expansion in inverse powers of a by using the asymptotic expansion $C_1 \approx (12/a^2) - (12/a^2)^4 + \cdots$ for large values of a:

$$\begin{aligned} Q &= \frac{\pi R^2 b W a^3}{40}\left[1 + \frac{960}{a^6} + \cdots\right] \\ &= \frac{\pi R b^2}{40}\left(\frac{b\,\Delta\mathscr{P}}{mL}\right)^3\left[1 + 960\left(\frac{WR}{b}\right)^2\left(\frac{mL}{b\,\Delta\mathscr{P}}\right)^6 + \cdots\right] \end{aligned} \tag{5.2–74}$$

in which $\Delta\mathscr{P} = \mathscr{P}_0 - \mathscr{P}_L$. This shows clearly that the flow in the axial direction is enhanced because of the imposed shearing in the tangential direction, since this additional shearing causes the viscosity to be lowered. Note that the correction term is very sensitive to the slit width, which enters as the inverse eighth power, and the pressure gradient, which appears to the minus sixth power.

§5.3 ISOTHERMAL FLOW PROBLEMS BY THE VARIATIONAL METHOD

In the design of viscometers and other equipment for measuring material functions, the experimentalist takes great pains to construct his apparatus so that it is geometrically simple and therefore easy to analyze theoretically. In industrial problems, on the other hand, flow systems are often geometrically complex and hence not amenable to simple theoretical analysis. This means that approximate or numerical procedures are mandatory.

For generalized Newtonian fluids we have available to us a variational principle that can be used as the basis for an approximate method. It is not appropriate here to insert a long discourse on the theory of variational calculus, since excellent discussions of the subject are available elsewhere.[1–3] Instead we present the variational principle and then show by illustrative examples how to use it. In addition to the fact that the variational method utilizes a bounding principle, it has the advantage over finite difference techniques that one obtains analytical expressions rather than tabulations of numerical results. This feature often makes it an attractive alternative.

We consider the flow of an incompressible, generalized Newtonian fluid in a volume V bounded by a surface S. The latter consists of two nonoverlapping regions: S_π, that portion on which forces are specified; and S_v that portion on which the velocity is specified. We consider only steady-state flows for which the $[\boldsymbol{v} \cdot \nabla\boldsymbol{v}]$ term in the equation of motion, the "acceleration term," is either identically zero or else negligible. For such problems there exists the following variational principle:[4–8]

Of all the possible motions within V that are compatible with the equation of continuity $(\nabla \cdot \boldsymbol{v}) = 0$ and the prescribed boundary conditions, the motion that minimizes:

$$J = \int_V \left[\int_0^{\dot\gamma} \eta\dot\gamma\, d\dot\gamma - \rho(\boldsymbol{v}\cdot\boldsymbol{g})\right] dV + \int_{S_\pi} ([\boldsymbol{\pi}\cdot\boldsymbol{v}]\cdot\boldsymbol{n})\, dS \qquad (5.3\text{–}1)$$

will be the steady-state motion. [For flow in conduits of uniform cross section S, the surface integral in J becomes $-\Delta p \int_S v\, dS$ where v is the axial velocity and Δp is the (positive) pressure drop driving the fluid through the conduit.]

To use this principle we pick a "reasonable" expression for $\boldsymbol{v}(x,y,z)$ containing variational parameters $a, b, c, d \ldots$. This expression for $\boldsymbol{v}(x,y,z)$ is then substituted into Eq. 5.3–1 and then $\partial J/\partial a$, $\partial J/\partial b$, $\partial J/\partial c$, $\partial J/\partial d \ldots$ are all set equal to zero to determine $a, b, c, d \ldots$. These

[1] F. B. Hildebrand, *Methods of Applied Mathematics*, Prentice-Hall, Englewood Cliffs, N.J. (1952), Chapter 2.

[2] R. Weinstock, *Calculus of Variations*, McGraw-Hill, New York (1952).

[3] R. S. Schechter, *The Variational Method in Engineering*, McGraw-Hill, New York (1967).

[4] J. Pawlowski, *Kolloid-Z.*, *138*, 6–11 (1954).

[5] Y. Tomita, *Bull. Jap. Soc. Mech. Eng.*, *2*, 469–474 (1959); *Reorojii: Hisenkei Ryūtai no Rikigaku*, Corona, Tokyo (1975), pp. 273–282.

[6] R. B. Bird, *Phys. Fluids*, *3*, 539–541 (1960); *comment*: *5*, 502 (1962).

[7] M. W. Johnson, Jr. *Phys. Fluids*, *3*, 871–878 (1960); *Trans. Soc. Rheol.*, *5*, 9–21 (1961) has given both minimum and maximum principles. See also J. C. Slattery, *Chem. Eng. Sci.*, *19*, 801–806 (1964); R. W. Flumerfelt and J. C. Slattery, *ibid.*, *20*, 157–163 (1965).

[8] R. S. Schechter, *A.I.Ch.E. Journal*, 7, 445–448 (1961).

values of $a, b, c, d \ldots$ are then those which minimize J. In this way we find the "best" solution of the proposed "reasonable" expression for $\mathbf{v}(x,y,z)$. Another analytical form for $\mathbf{v}(x,y,z)$ may then be tried, and if this second form gives a lower value of J, then it is to be preferred. The procedure may be continued indefinitely. This method can require considerable numerical computation, possibly equivalent to that needed for a finite-difference calculation.

EXAMPLE 5.3–1 Axial Annular Flow of a Power-Law Fluid

As in Example 5.2–5 we consider the axial flow of a fluid in a horizontal annulus of length L formed by cylindrical surfaces at $r = \kappa R$ and $r = R$. Here we study the flow of a power-law fluid for which an analytical solution is known,[9] so that we can have an exact result with which we can compare our approximate variational result. It is desired to find the volume rate of flow Q for the fluid.

SOLUTION

A very simple approximate form for the velocity profile that satisfies the boundary conditions is the expression (see Fig. 5.3–1):

$$v_z = a(1 - |\sigma|^{(1/n)+1}) \tag{5.3–2}$$

where

$$\sigma = \frac{2\xi - (1 + \kappa)}{(1 - \kappa)} \tag{5.3–3}$$

This trial velocity distribution contains a as the variational parameter; $\xi = r/R$ is the dimensionless radial coordinate. The quantity $\dot{\gamma}$ is then:

$$\dot{\gamma} = \pm\frac{dv_z}{dr} = \pm\frac{1}{R}\frac{dv_z}{d\xi} = \pm\left(\frac{2/R}{1 - \kappa}\right)\frac{dv_z}{d\sigma} = \mp\frac{2a\left(\frac{1}{n} + 1\right)}{R(1 - \kappa)}|\sigma|^{1/n} \tag{5.3–4}$$

where the upper sign is for $\sigma < 0$ and the lower for $\sigma > 0$.

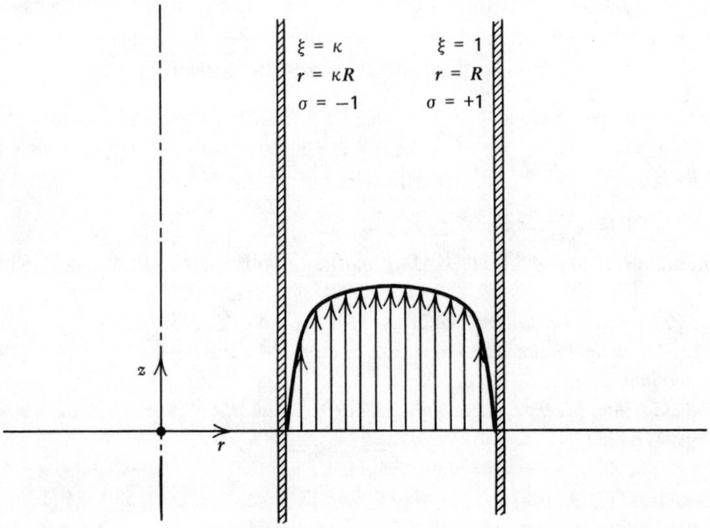

FIGURE 5.3–1. Trial velocity distribution $v_z = a(1 - |\sigma|^{(1/n)+1})$.

The variational function J for this problem becomes:

$$J = 2\pi L \int_{\kappa R}^{R} \frac{m}{n+1} \dot{\gamma}^{n+1} r\, dr - 2\pi\, \Delta p \int_{\kappa R}^{R} v_z r\, dr$$

$$= \frac{2\pi L R^2 m}{n+1} \int_{-1}^{+1} \dot{\gamma}^{n+1} \left[\underline{\left(\frac{1-\kappa}{2}\right)\sigma} + \left(\frac{1+\kappa}{2}\right)\right]\left(\frac{1-\kappa}{2}\right) d\sigma$$

$$- 2\pi\, \Delta p R^2 \int_{-1}^{+1} v_z \left[\underline{\left(\frac{1-\kappa}{2}\right)\sigma} + \left(\frac{1+\kappa}{2}\right)\right]\left(\frac{1-\kappa}{2}\right) d\sigma$$

$$= \frac{\pi L R^2 m}{n+1}(1-\kappa^2)\left(\frac{2a\left(\frac{1}{n}+1\right)}{R(1-\kappa)}\right)^{n+1} \int_0^1 \sigma^{(1/n)+1}\, d\sigma$$

$$-\pi\, \Delta p R^2 (1-\kappa^2) a \int_0^1 (1-\sigma^{(1/n)+1})\, d\sigma$$

$$= \frac{\pi L R^2 m (1-\kappa^2)}{(n+1)\left(\frac{1}{n}+2\right)}\left(\frac{2a\left(\frac{1}{n}+1\right)}{R(1-\kappa)}\right)^{n+1} - \frac{\left(\frac{1}{n}+1\right)}{\left(\frac{1}{n}+2\right)} \pi\, \Delta p R^2 (1-\kappa^2) a \qquad (5.3\text{–}5)$$

The dashed-underlined terms do not contribute since they are odd functions over the interval $(-1, 1)$.

When we set $\partial J/\partial a = 0$ and solve for a we get:

$$a = \left(\frac{\Delta p R}{2mL}\right)^{1/n} \frac{R(1-\kappa)^{(1/n)+1}}{2\left(\frac{1}{n}+1\right)} \qquad (5.3\text{–}6)$$

Then the volumetric flow rate Q is:

$$Q = \pi R^3 \left(\frac{\Delta p R}{2mL}\right)^{1/n} \frac{(1-\kappa)^{(1/n)+2}}{\left(\frac{1}{n}+2\right)} \left(\frac{1+\kappa}{2}\right) \qquad (5.3\text{–}7)$$

Fredrickson and Bird[9] have given a table of dimensionless Q values; when we compare Eq. 5.3–7 with their values we find:

		$Q(\pi R^3)^{-1}\left(\frac{\Delta p R}{2mL}\right)^{-1/n} \frac{((1/n)+2)}{(1-\kappa)^{(1/n)+2}}$	
		Exact	Variational
$n = 0.5$	$\kappa = 0.5$	0.761	0.750
$n = 0.333$	$\kappa = 0.5$	0.765	0.750
$n = 0.333$	$\kappa = 0.2$	0.661	0.600

The variational results are within about 2% of the exact results for $\kappa > 0.5$ and $n > 0.25$, but become worse for smaller κ and n values. Better results could be obtained by using a trial function with two constants a and b instead of the one-constant function in Eq. 5.3–2.

[9] See A. G. Fredrickson and R. B. Bird, *Ind. Eng. Chem.*, *50*, 347–352 (1958); *erratum*: *Ind. Eng. Chem. Fundamentals*, *3*, 383 (1964); see also Problem 5C.1.

A rather detailed variational calculation for the annular flow of a fluid described by the Sutterby model[10] has been carried out by Mitsuishi, Aoyagi, and Soeda.[11] Extensive comparisons between variational calculations and experimental data have also been made for ducts of elliptical, rectangular, and triangular cross-section.[12] A simple empirical technique for estimating flow rates in channels of unusual cross-section has been suggested by Miller.[13] The use of variational methods for flow of various generalized Newtonian fluids around a sphere has been studied by Slattery and coworkers.[14]

EXAMPLE 5.3–2 Estimation of Velocity Distribution for Axial Eccentric Annular Flow (e.g., in a Wire-Coating Device)

A wire of radius R_1 is moving axially with speed v_0 inside a cylindrical cavity of radius R_2 in which a molten polymer is flowing. The wire axis is displaced from the cavity axis by a distance δ. It is desired to estimate the velocity distribution and in particular to determine under what conditions the velocity will be significantly dependent on the variable in the tangential direction (i.e., around the wire). This problem was suggested by a recent study on stability of the wire-coating operation.[15]

SOLUTION

It is appropriate to use bipolar coordinates (see Appendix A). In order to use them it is necessary to relate the geometric quantities R_1, R_2, and δ to the quantities a, ξ_1, and ξ_2 used in bipolar coordinates. This is done in Eqs. A.7–31, 32, and 33. The quantities ξ_1 and ξ_2 specify the locations of the bounding surfaces.

We now want to estimate the velocity profile. We assume a velocity distribution of the form:

$$\frac{v_z}{v_0} = \left(\frac{\xi - \xi_2}{\xi_1 - \xi_2}\right)[1 - A(\xi_1 - \xi)\cos\theta] \tag{5.3–8}$$

in which A will be the variational parameter. This function satisfies the boundary conditions at $\xi = \xi_1$ and $\xi = \xi_2$. For a concentric arrangement A would be zero, but in an eccentric arrangement it is a function of θ. From Fig. A.7–1 it is evident that v_z should depend on a function of θ that is even, and the cosine function satisfies that requirement.

We now wish to determine the optimal value of A by minimizing J; for this problem we are concerned only with the non-Newtonian viscosity term in Eq. 5.3–1. For the flow under consideration (see Table A.7–4):

$$\dot{\gamma} = \sqrt{\tfrac{1}{2}(\dot{\boldsymbol{\gamma}} : \dot{\boldsymbol{\gamma}})} = \sqrt{\left(\frac{X}{a}\frac{\partial v_z}{\partial \xi}\right)^2 + \left(\frac{X}{a}\frac{\partial v_z}{\partial \theta}\right)^2} \tag{5.3–9}$$

in which $X = \cosh\xi + \cos\theta$. For a very large wire speed v_0, the first term under the square-root sign will surely be much larger than the second term. Hence we write:

$$\dot{\gamma} \cong +\frac{X}{a}\frac{\partial v_z}{\partial \xi} \tag{5.3–10}$$

[10] J. L. Sutterby, *Trans. Soc. Rheol.*, *9*:2, 227–241 (1965); *A.I.Ch.E. Journal*, *12*, 63–68 (1966).

[11] N. Mitsuishi, Y. Aoyagi, and H. Soeda, *Kagaku Kōgaku*, *36*, 186–192 (1972).

[12] N. Mitsuishi, Y. Kitayama, and Y. Aoyagi, *Kagaku Kōgaku*, *31*, 570–577 (1967); N. Mitsuishi and Y. Aoyagi, *Memoirs of the Faculty of Engineering*, Kyūshu University, Vol. 28, No. 3, pp. 223–241 (1969); T. Mizushina, N. Mitsuishi, R. Nakamura, *Kagaku Kōgaku*, *28*, 648–652 (1964).

[13] C. Miller, *Ind. Eng. Chem. Fundamentals*, *11*, 524–528 (1972); R. W. Hanks, *Ind. Eng. Chem. Fundamentals*, *13*, 62–66 (1974).

[14] J. C. Slattery, *A.I.Ch.E. Journal*, *8*, 663–667 (1962); M. L. Wasserman and J. C. Slattery, *ibid.*, *10*, 383–388 (1964); S. W. Hopke and J. C. Slattery, *ibid.*, *16*, 224–229, 317–318 (1970).

[15] Z. Tadmor and R. B. Bird, *Polym. Eng. Sci.*, *14*, 124–136 (1974).

The plus sign is chosen since ξ increases from the outer cylinder to the inner cylinder, as does the axial velocity of the fluid. We shall assume the polymer melt viscosity to be described by a power law so that:

$$\begin{aligned}\int_0^{\dot{\gamma}} \eta\dot{\gamma}\, d\dot{\gamma} &= \frac{m}{n+1}\dot{\gamma}^{n+1} \\ &= \frac{m}{n+1}\left(\frac{X}{a}\frac{\partial v_z}{\partial \xi}\right)^{n+1} \\ &= \frac{m}{n+1}\left[\frac{X}{a}\frac{v_0}{\xi_1 - \xi_2}(1 - A(\xi_1 + \xi_2 - 2\xi)\cos\theta)\right]^{n+1}\end{aligned} \tag{5.3–11}$$

Hence the variational integral becomes for a tube of length L:

$$\begin{aligned}J &= L\int_0^{2\pi}\int_{\xi_2}^{\xi_1}\frac{m}{n+1}\left(\frac{X}{a}\frac{\partial v_z}{\partial \xi}\right)^{n+1}\left(\frac{a}{X}\right)^2 d\xi\, d\theta \\ &= \frac{Lmv_0}{(n+1)(\xi_1 - \xi_2)}\int_0^{2\pi}\int_{\xi_2}^{\xi_1}\left(\frac{a}{X}\right)^{n-1}[1 - A(\xi_1 + \xi_2 - 2\xi)\cos\theta]^{n-1}\, d\xi\, d\theta\end{aligned} \tag{5.3–12}$$

When $\partial J/\partial A$ is set equal to zero, it is found that[14] for $R_1 = 0.317$ mm and $R_2 = 0.635$ mm:

δ (mm)	ξ_1	ξ_2	a (mm)	A ($n = 0.5$)	A ($n = 0.75$)	A ($n = 1$)
0.0237	3.6895	2.9982	6.35	0.025	0.008	0
0.1407	1.8184	1.1948	0.953	0.175	0.060	0
0.261	0.8814	0.4812	0.318	0.73	0.25	0

The values of A were obtained by numerical integration. It is thus seen that the θ-dependence becomes more important as the eccentricity increases and as the index n decreases.

§5.4 NONISOTHERMAL FLOW PROBLEMS[1]

In §§5.2 and 5.3 we gave a number of illustrations of how polymer flow problems can be set up and solved; in those sections it was presumed that the fluids were at constant temperature. In this section we give an introduction to the subject of nonisothermal, non-Newtonian flow. We make use of the generalized Newtonian model for describing the stresses in the fluid, and we assume that the thermal conductivity and density of the fluid do not change appreciably with temperature or pressure. The assumption of constant thermal conductivity is not serious since k does not change much with temperature, nor does it

[1] Several review articles are available for this very large subject area: R. L. Pigford, *Chemical Engineering Progress Symposium Series*, No. 17, Vol. 51, 79–92 (1955); A. B. Metzner, *Adv. in Heat Transfer*, *2*, 357–397 (1965); H. H. Winter, *Adv. in Heat Transfer*, *13*, 000–000 (1977). See also *Progress in Heat and Mass Transfer*, Vol. 5 (W. R. Schowalter, A. V. Luikov, W. J. Minkowycz, and N. F. Afgan, Eds.) 1972, which contains many summary articles. The review by G. Astarita and G. C. Sarti, *Theoretical Rheology* (J. F. Hutton, J. R. A. Pearson, and K. Walters, Eds.), Wiley, New York (1974), pp. 123–137, contains a theoretical discussion of the validity of Eq. 5.4–3 for viscoelastic fluids.

appear to change very much with the velocity gradients.[2] The assumption of constant density means that our discussion will be oriented toward forced convection and that free convection will be omitted.[3]

The equations that will form the starting point for the heat-transfer and fluid-flow discussions here are very similar to those in Table 1.2–1:

Continuity:
$$(\nabla \cdot \boldsymbol{v}) = 0 \qquad (5.4\text{–}1)$$
No mass sources within the fluid

Motion:
$$\rho \frac{D\boldsymbol{v}}{Dt} = -\nabla p + [\nabla \cdot \eta\dot{\boldsymbol{\gamma}}] + \rho \boldsymbol{g} \qquad (5.4\text{–}2)$$
Mass per unit volume times acceleration = Pressure force per unit volume + Viscous force per unit volume + Gravitational force per unit volume

Energy:
$$\rho \hat{C}_p \frac{DT}{Dt} = k\,\nabla^2 T + \tfrac{1}{2}\eta(\dot{\boldsymbol{\gamma}} : \dot{\boldsymbol{\gamma}}) \qquad (5.4\text{–}3)$$
Rate of increase of internal energy per unit volume = Rate of addition of energy by heat conduction per unit volume + Rate of conversion of mechanical to thermal energy, per unit volume

In obtaining Eq. 5.4–3 from Eq. 1.1–13 it has been assumed that $\hat{U} = \hat{U}(p,T)$; that is, we assume that $\hat{U}$ does not depend explicitly on any kinematic quantities such as strain and rate-of-strain. This has been a standard assumption in all heat-transfer calculations; although the assumption is generally regarded as reasonable, we know of no thorough experimental study of it.

The non-Newtonian viscosity depends strongly both on velocity gradients and on temperature. Hence the three equations of change are coupled: the temperature distribution cannot be obtained from Eq. 5.4–3 unless the velocity distribution is known, since the latter occurs both in the DT/Dt term and in the viscous heating term; and the velocity distribution cannot be obtained from Eq. 5.4–2 unless the temperature distribution is known, inasmuch as η is greatly influenced by T. It is therefore not surprising that analytical solutions to the above equations can be obtained only by making rather drastic assumptions. Consequently

[2] A. A. Cocci, Jr., and J. J. C. Picot [*Polym. Eng. Sci. 13*, 337–341 (1973)] have found that the thermal conductivity of polydimethylsiloxane samples increased by about 5% as the shear rate went from 0 to 300 s^{-1}.

[3] A theoretical analysis of free-convection heat transfer to non-Newtonian fluids was given by A. Acrivos, *A.I.Ch.E. Journal, 6*, 584–590 (1960). Experimental data were obtained for free convection near a heated vertical plate in an aqueous carboxy-polymethylene solution by I. G. Reilly, C. Tien, and M. Adelman, *Can. J. Chem. Eng., 43*, 157–160 (1965). Numerical calculations for the power-law and Ellis models have been made by H. Ozoe and S. W. Churchill, *A.I.Ch.E. Journal, 18*, 1196–1207 (1972).

most solutions of any real interest will be numerical solutions. One may therefore question the propriety of giving any analytical solutions at all in this section. We feel that a few examples are in order for providing the background for reading some of the publications in the non-Newtonian heat-transfer field. Also analytical solutions can be of value for making order-of-magnitude estimates, for providing the basis for a perturbation-theory solution, for checking numerical programs in limiting cases, and for suggesting dimensionless correlation procedures.

For making nonisothermal flow calculations it is of course necessary to know how the parameters in the rheological equation vary with temperature. For example, if one uses the *power-law* viscosity function, useful empiricisms for the parameters m and n are:[4]

$$m = m^0 e^{-A(T - T_0)/T_0} \tag{5.4-4}$$

$$n = n^0 + \frac{B(T - T_0)}{T_0} \tag{5.4-5}$$

in which T_0 is a reference temperature, and m^0 and n^0 are the values of the parameters at that temperature; the constants A and B are determined for each fluid from experimental data. Often B is found to be quite small and the assumption of constant n is reasonable (see, for example, Table 5.1–1). For the *Ellis model* the following empiricisms seem useful:[4]

$$\eta_0 = \eta_0{}^0 e^{-A'(T - T_0)/T_0} \tag{5.4-6}$$

$$\left(\frac{1}{\alpha}\right) = \left(\frac{1}{\alpha^0}\right) + \frac{B'(T - T_0)}{T_0} \tag{5.4-7}$$

$$\tau_{1/2} = \tau_{1/2}{}^0 e^{C'(T - T_0)/T_0} \tag{5.4-8}$$

in which A', B', and C' are constants for each fluid; here again the constant B' is often sufficiently small that α may be taken to be constant (see Table 5.1–2).

For Newtonian fluids the viscous dissipation term in the energy equation is seldom of importance; notable exceptions are high-speed lubrication problems and boundary-layer heating during space-vehicle reentry. In the polymer field, on the other hand, there are many processes in which the large velocity gradients and very high viscosities conspire to make the viscous heating appreciable, and this in turn has a large effect on the rheological parameters, the flow patterns, and the thermal degradation of the fluids.

We begin by giving several examples of heat-transfer problems with no viscous heating and then several in which viscous heating is included. The power-law is used in all of these examples. The first four examples deal with heat transfer in tubes, and these examples enable one to see how Table 5.4–1 is constructed.[5] Similar results for heat transfer between flat plates are summarized in Table 5.4–2.

[4] R. M. Turian, Ph.D. Thesis, University of Wisconsin, Madison (1964), p. 139 (power law), pp. 161–162 (Ellis model). For some purposes it is convenient to use

$$m^0/m = [1 + C(T - T_0)/T_0]^n \tag{5.4-4a}$$

as suggested by T. Mizushina, R. Itō, Y. Kuriwaki, and K. Yahikozawa, *Kagaku Kōgaku, 31*, 250–255 (1967); here C is a constant, and n is the (constant) exponent in the power-law model.

[5] Tables 5.4–1 and 2 are based on the summary articles by W. J. Beek and R. Eggink, *De Ingenieur, 74*, Ch 81–89 (1962) and J. M. Valstar and W. J. Beek, *ibid., 75*, Ch 1–7 (1963).

TABLE 5.4–1

Asymptotic Results for Nusselt Numbers (Tube Flow)[a]; Nu $= hD/k$

All values are *local* Nu numbers	Constant Wall Temperature. 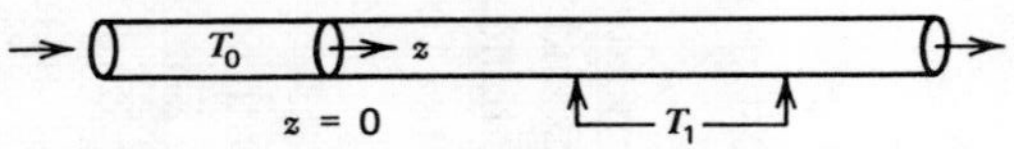			Constant Wall Heat Flux 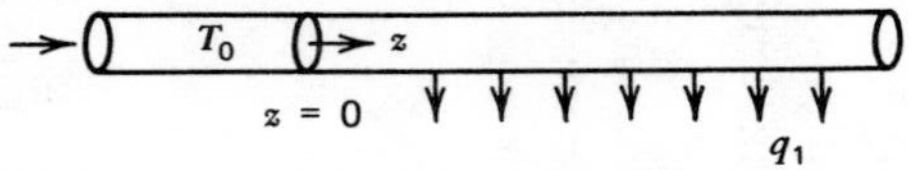		
"Thermal Entrance Region" (small z)	Plug Flow	$\mathrm{Nu} = \frac{1}{\sqrt{\pi}}\left(\frac{\langle v_z\rangle D^2}{\alpha z}\right)^{1/2}$	(A)	Plug Flow	$\mathrm{Nu} = \frac{\sqrt{\pi}}{2}\left(\frac{\langle v_z\rangle D^2}{\alpha z}\right)^{1/2}$	(G)
	Laminar Non-Newtonian Flow	$\mathrm{Nu} = \frac{2}{9^{1/3}\Gamma(\frac{4}{3})}\left[\frac{\langle v_z\rangle D^2}{\alpha z}\left(-\frac{1}{4}\left.\frac{d\phi}{d\xi}\right\|_{\xi=1}\right)\right]^{1/3}$	(B)	Laminar Non-Newtonian Flow	$\mathrm{Nu} = \frac{2\Gamma(\frac{2}{3})}{9^{1/3}}\left[\frac{\langle v_z\rangle D^2}{\alpha z}\left(-\frac{1}{4}\left.\frac{d\phi}{d\xi}\right\|_{\xi=1}\right)\right]^{1/3}$	(H)
	Laminar Newtonian Flow	$\mathrm{Nu} = \frac{2}{9^{1/3}\Gamma(\frac{4}{3})}\left(\frac{\langle v_z\rangle D^2}{\alpha z}\right)^{1/3}$	(C)	Laminar Newtonian Flow	$\mathrm{Nu} = \frac{2\Gamma(\frac{2}{3})}{9^{1/3}}\left(\frac{\langle v_z\rangle D^2}{\alpha z}\right)^{1/3}$	(I)
"Thermally Fully-Developed Flow" (large z)	Plug Flow	$\mathrm{Nu} = 5.772$	(D)	Plug Flow	$\mathrm{Nu} = 8$	(J)
	Laminar Non-Newtonian Flow	$\mathrm{Nu} = \beta_1{}^2$, where β_1 is the *lowest* eigenvalue of $\frac{1}{\xi}\frac{d}{d\xi}\left(\xi\frac{dX_n}{d\xi}\right) + \beta_n{}^2\phi(\xi)X_n = 0;\ X_n(1) = 0$	(E)	Laminar Non-Newtonian Flow	$\mathrm{Nu} = \left[2\int_0^1 \frac{1}{\xi}\left[\int_0^\xi \xi'\phi(\xi')\,d\xi'\right]^2 d\xi\right]^{-1}$	(K)
	Laminar Newtonian Flow	$\mathrm{Nu} = 3.657$	(F)	Laminar Newtonian Flow	$\mathrm{Nu} = \frac{48}{11}$	(L)

NOTE: $\phi(\xi) = v_z/\langle v_z\rangle$ where $\xi = r/R$ and $R = D/2$; for Newtonian fluids $\langle v_z\rangle D^2/\alpha z = \mathrm{RePr}(D/z)$ with $\mathrm{Re} = D\langle v_z\rangle\rho/\mu$

[a] W. J. Beek and R. Eggink, *De Ingenieur, 74,* Ch 85–89 (1962) 31 August.

TABLE 5.4–2

Asymptotic Results for Nusselt Numbers (Thin-Slit Flow)[a]; $\mathrm{Nu} = 4Bh/k$

All values are *local* Nu numbers	Constant Wall Temperature 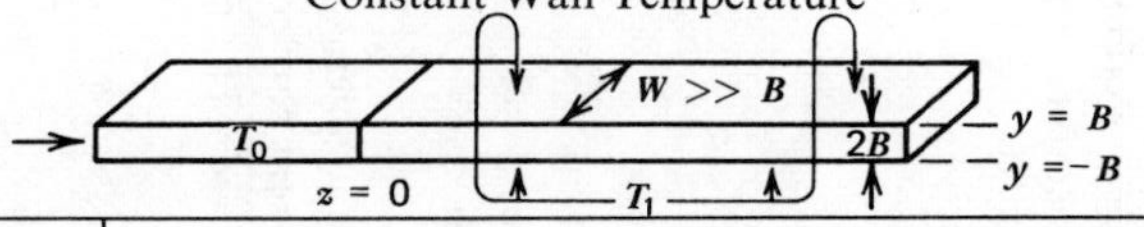		Constant Wall Heat Flux 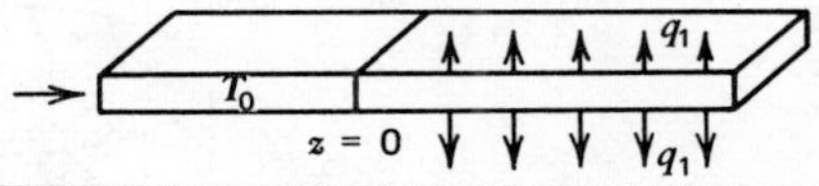	
"Thermal Entrance Region" (small z)	Plug Flow	$\mathrm{Nu} = \frac{4}{\sqrt{\pi}}\left(\frac{\langle v_z\rangle B^2}{\alpha z}\right)^{1/2}$ (A)	Plug Flow	$\mathrm{Nu} = 2\sqrt{\pi}\left(\frac{\langle v_z\rangle B^2}{\alpha z}\right)^{1/2}$ (G)
	Laminar Non-Newtonian Flow	$\mathrm{Nu} = \frac{4}{9^{1/3}\Gamma(\frac{4}{3})}\left[\frac{\langle v_z\rangle B^2}{\alpha z}\left(-\frac{d\phi}{d\sigma}\bigg\vert_{\sigma=1}\right)\right]^{1/3}$ (B)	Laminar Non-Newtonian Flow	$\mathrm{Nu} = \frac{4\Gamma(\frac{2}{3})}{9^{1/3}}\left[\frac{\langle v_z\rangle B^2}{\alpha z}\left(-\frac{d\phi}{d\sigma}\bigg\vert_{\sigma=1}\right)\right]^{1/3}$ (H)
	Laminar Newtonian Flow	$\mathrm{Nu} = \frac{4}{3^{1/3}\Gamma(\frac{4}{3})}\left(\frac{\langle v_z\rangle B^2}{\alpha z}\right)^{1/3}$ (C)	Laminar Newtonian Flow	$\mathrm{Nu} = \frac{4\Gamma(\frac{2}{3})}{3^{1/3}}\left(\frac{\langle v_z\rangle B^2}{\alpha z}\right)^{1/3}$ (I)
"Thermally Fully-Developed Flow" (large z)	Plug Flow	$\mathrm{Nu} = 9.86$ (D)	Plug Flow	$\mathrm{Nu} = 12$ (J)
	Laminar Non-Newtonian Flow	$\mathrm{Nu} = 4\beta_1^{\,2}$, where β_1 is the *lowest* eigenvalue of the equation: $\frac{d^2X_n}{d\sigma^2} + \beta_n^{\,2}\phi(\sigma)X_n = 0;\ X_n(\pm 1) = 0$ (E)	Laminar Non-Newtonian Flow	$\mathrm{Nu} = \left[\frac{1}{4}\int_0^1\left[\int_0^\sigma \phi(\sigma')\,d\sigma'\right]^2 d\sigma\right]^{-1}$ (K)
	Laminar Newtonian Flow	$\mathrm{Nu} = 7.54$ (F)	Laminar Newtonian Flow	$\mathrm{Nu} = \frac{140}{17}$ (L)

NOTE: $\phi(\sigma) = v_z/\langle v_z\rangle$ where $\sigma = y/B$; for Newtonian fluids $\langle v_z\rangle D^2/\alpha z = 4\mathrm{RePr}(B/z)$ with $\mathrm{Re} = 4B\langle v_z\rangle\rho/\mu$

[a] J. M. Valstar and W. J. Beek, *De Ingenieur*, *75*, Ch 1–7 (1963) 4 January.

EXAMPLE 5.4–1 Flow in Tubes with Constant Wall-Heat-Flux[6] (Asymptotic Solution for Large z)

A polymeric fluid is flowing axially in a circular tube of radius R (see Fig. 5.4–1). Before the fluid arrives at the plane $z = 0$, it is at a uniform temperature T_0 and has the fully developed velocity distribution:

$$v_z = v_{\max}\left[1 - \left(\frac{r}{R}\right)^{s+1}\right] \tag{5.4–9}$$

where $s = 1/n$, and $v_{\max}$ is given in Eq. 5.2–8. Then for $z > 0$ there is a constant heat flux q_1 at the wall; if q_1 is positive, heat is leaving the tube and the fluid is cooled, whereas if q_1 is negative, heat is being added as would be the case when the fluid is being warmed by an electrical heating element wrapped

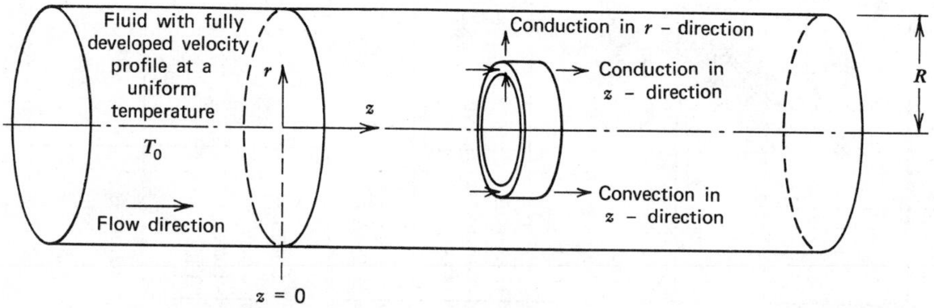

FIGURE 5.4–1. Axial flow in a circular tube with heat transport by conduction and convection.

uniformly about the tube. To obtain an approximate description of this heating or cooling problem, we assume that the power-law constants m and n are not dependent on the temperature; also the temperature dependence of ρ, k, and $\hat{C}_p$ are ignored. The energy equation (Eq. 5.4–3) then becomes, for low flow rates where viscous heating is unimportant:

$$\rho\hat{C}_p v_{\max}\left[1 - \left(\frac{r}{R}\right)^{s+1}\right]\frac{\partial T}{\partial z} = k\left[\frac{1}{r}\frac{\partial}{\partial r}\left(r\frac{\partial T}{\partial r}\right) + \frac{\partial^2 T}{\partial z^2}\right] \tag{5.4–10}$$

Heat convection in the z-direction — Heat conduction in the r-direction — Heat conduction in the z-direction

The three terms here are represented by arrows in Fig. 5.4–1. Except in unusual problems—mainly in liquid metals—the heat conduction in the z-direction is much smaller than the heat convection in the z-direction (the heat transport by fluid flow), so that the last term may be omitted. The problem we want to solve is then:

$$\rho\hat{C}_p v_{\max}\left[1 - \left(\frac{r}{R}\right)^{s+1}\right]\frac{\partial T}{\partial z} = k\frac{1}{r}\frac{\partial}{\partial r}\left(r\frac{\partial T}{\partial r}\right) \tag{5.4–11}$$

[6] This example is an extension to the power-law fluid of §9.8 of R. B. Bird, W. E. Stewart, and E. N. Lightfoot, *Transport Phenomena*, Wiley, New York (1960). The power-law solution was first given by U. Grigull, *Chemie-Ingenieur Technik*, *28*, 553–556 (1956); see also R. B. Bird, *ibid.*, *31*, 569–572 (1959). The problem has also been solved with n = constant and m varying according to Eq. 5.4–4a by Mizushina, Itō, Kuriwaki, and Yahikozawa, *loc. cit.*; see also T. Mizushina and Y. Kuriwaki, *Memoirs of the Faculty of Engineering, Kyōto University, 30*, 511–524 (1968). Further work on the problem has been done by N. Mitsuishi and O. Miyatake, *Kagaku Kōgaku, 32*, 1222–1227 (1968); *Memoirs of the Faculty of Engineering, Kyūshū University*, Vol. *28*, No. 2, 91–107 (1968). The last four papers listed include experimental data.

with the boundary conditions:

$$B.C.\ 1: \quad \text{at } r = 0 \qquad T = \text{finite} \tag{5.4–12}$$

$$B.C.\ 2: \quad \text{at } r = R \qquad -k\left(\frac{\partial T}{\partial r}\right) = q_1 \tag{5.4–13}$$

$$B.C.\ 3: \quad \text{at } z = 0 \qquad T = T_0 \tag{5.4–14}$$

It is convenient to introduce dimensionless quantities:

$$\Theta = \frac{T - T_0}{q_1 R/k}; \qquad \xi = \frac{r}{R}; \qquad \zeta = \frac{\alpha z}{v_{max} R^2} \tag{5.4–15, 16, 17}$$

in which $\alpha = k/\rho\hat{C}_p$ is the thermal diffusivity. The problem may then be restated as:

$$(1 - \xi^{s+1})\frac{\partial \Theta}{\partial \zeta} = \frac{1}{\xi}\frac{\partial}{\partial \xi}\left(\xi \frac{\partial \Theta}{\partial \xi}\right) \tag{5.4–18}$$

with

$$B.C.\ 1: \quad \text{at } \xi = 0 \qquad \Theta = \text{finite} \tag{5.4–19}$$

$$B.C.\ 2: \quad \text{at } \xi = 1 \qquad -\partial\Theta/\partial\xi = 1 \tag{5.4–20}$$

$$B.C.\ 3: \quad \text{at } \zeta = 0 \qquad \Theta = 0 \tag{5.4–21}$$

It is desired to obtain an asymptotic solution to this problem for large ζ.

SOLUTION

Postulate that for large ζ the temperature will have the form:

$$\Theta(\xi,\zeta) = C_0\zeta + \Psi(\xi) \tag{5.4–22}$$

in which C_0 is a constant to be determined from the boundary conditions. This function is, of course, not the complete solution since the third boundary condition cannot be satisfied. Since we are in fact interested only in the behavior at large ζ, we will not try to satisfy the third boundary condition at all. Instead we replace it by:

$$B.C.\ 3': \quad -2\pi R z q_1 = \int_0^{2\pi}\int_0^R \rho\hat{C}_p(T - T_0)v_z r\, dr\, d\theta \tag{5.4–23}$$

or

$$B.C.\ 3': \quad -\zeta = \int_0^1 \Theta(\xi,\zeta)(1 - \xi^{s+1})\xi\, d\xi \tag{5.4–24}$$

This condition is an overall energy balance, stating that the difference between the heat passing through planes $z = 0$ and $z = z$ is equal to the heat passing through the wall at $r = R$.

Substitution of the postulated solution in Eq. 5.4–22 into the partial differential equation then gives the following ordinary differential equation for Ψ:

$$\frac{1}{\xi}\frac{d}{d\xi}\left(\xi \frac{d\Psi}{d\xi}\right) = C_0(1 - \xi^{s+1}) \tag{5.4–25}$$

Integration twice with respect to ξ then gives $\Psi(\xi)$, which can be substituted into Eq. 5.4–22 to yield:

$$\Theta(\xi,\zeta) = C_0\zeta + C_0\left(\frac{\xi^2}{4} - \frac{\xi^{s+3}}{(s+3)^2}\right) + C_1 \ln \xi + C_2 \tag{5.4–26}$$

The constants of integration C_0, C_1, and C_2 are then determined from the boundary conditions; they are found to be:

$$C_1 = 0 \tag{5.4–27}$$

$$C_0 = -\frac{2(s+3)}{(s+1)} \tag{5.4–28}$$

$$C_2 = \frac{4}{s+1}\left[\frac{(s+3)^3 - 8}{16(s+3)(s+5)}\right] \tag{5.4–29}$$

so that finally:

$$\Theta = -\frac{2(s+3)}{(s+1)}\zeta - \frac{2(s+3)}{(s+1)}\left[\frac{\xi^2}{4} - \frac{\xi^{s+3}}{(s+3)^2}\right] + \frac{4}{s+1}\left[\frac{(s+3)^3 - 8}{16(s+3)(s+5)}\right] \tag{5.4–30}$$

This is the temperature distribution for large distances downstream from $z = 0$.

For some engineering calculations it is convenient to work in terms of the local heat-transfer coefficient h, defined by

$$q_1 = h(T_b - T_w) \tag{5.4–31}$$

in which T_w is the local wall temperature and T_b is the local bulk temperature, defined by:

$$T_b(z) = \frac{\int_0^{2\pi}\int_0^R v_z(r)T(r,z)r\,dr\,d\theta}{\int_0^{2\pi}\int_0^R v_z(r)r\,dr\,d\theta} \tag{5.4–32}$$

Usually the dimensionless heat transfer coefficient, or Nusselt number, $\text{Nu} = 2hR/k$ is tabulated. For this problem the Nusselt number is

$$\text{Nu} = \frac{2hR}{k} = \frac{2q_1R}{k(T_b - T_w)} = \frac{2}{\Theta_b - \Theta_w} \tag{5.4–33}$$

in which Θ_b and Θ_w are the dimensionless bulk and wall temperatures defined analogously to Θ in Eq. 5.4–15. When Θ_b and Θ_w are obtained from Eq. 5.4–30, we get finally:

$$\text{Nu} = \frac{8(s+3)(s+5)}{s^2 + 12s + 31} \tag{5.4–34}$$

For the Newtonian fluid s equals 1, and we get $\text{Nu} = \frac{48}{11}$, the entry given in Table 5.4–1. The above derivation can be extended to any arbitrary non-Newtonian viscosity function and the result is given as Eq. K in Table 5.4–1 (see also Problem 5C.5). The result in Eq. 5.4–34 is valid for values of $\langle v_z\rangle D^2/\alpha z$ less than about 10, where $D = 2R$ is the tube diameter.[7]

EXAMPLE 5.4–2 Flow in Tubes with Constant Wall Heat Flux[8] (Asymptotic Solution for Small z)

Repeat the problem of Example 5.4–1, but obtain an asymptotic solution for the region of small z.

[7] The limit of applicability of the asymptotic formula can be found only by comparing this limiting solution with the complete solution given, for example, by N. Mitsuishi and O. Miyatake, *Memoirs of the Faculty of Engineering, Kyūshū University*, Vol. 28, No. 2, pp. 91–107 (1968), Fig. 10; this figure also shows how the Nusselt number increases with n because of the temperature dependence of the parameter m.

[8] R. B. Bird, W. E. Stewart, and E. N. Lightfoot, *op. cit.*, pp. 363–364.

SOLUTION

For small z, the heat removal (if q_1 is positive) or heat penetration (if q_1 is negative) is restricted to a thin shell near the wall, so that the following three approximations lead to results that are accurate in the limit as $z \to 0$:

i. Curvature effects may be ignored and the problem treated as though the wall were flat; the distance from the wall will be designated by $y = R - r$.

ii. The fluid may be regarded as extending from the (flat) heat-transfer surface, $y = 0$, to $y = \infty$.

iii. The velocity profile may be regarded as linear, with a slope given by the velocity gradient at the wall; hence from Eq. 5.2–8, written in terms of y, we find that near the wall $v_z(y) = v_0 y/R$, where $v_0 = (\tau_R/m)^s R = v_{\max}(s+1) = \langle v_z \rangle (s+3)$.

With these assumptions Eq. 5.4–11 becomes:

$$v_0 \frac{y}{R}\frac{\partial T}{\partial z} = \alpha \frac{\partial^2 T}{\partial y^2} \tag{5.4–35}$$

Inasmuch as the boundary condition on $r = R$ is given in terms of the heat flux rather than the temperature, it is easier to work with the differential equation for $q_y = -k(\partial T/\partial y)$, obtained by dividing Eq. 5.4–35 by y and then differentiating with respect to y:

$$v_0 \frac{1}{R}\frac{\partial q_y}{\partial z} = \alpha \frac{\partial}{\partial y}\left(\frac{1}{y}\frac{\partial q_y}{\partial y}\right) \tag{5.4–36}$$

The introduction of the following dimensionless quantities facilitates the solution:

$$\psi = \frac{q_y}{q_1}, \qquad \sigma = \frac{y}{R}, \qquad \zeta = \frac{\alpha z}{v_0 R^2} \tag{5.4–37, 38, 39}$$

Then the problem statement becomes:

$$\frac{\partial \psi}{\partial \zeta} = \frac{\partial}{\partial \sigma}\left(\frac{1}{\sigma}\frac{\partial \psi}{\partial \sigma}\right) \tag{5.4–40}$$

B.C. 1: at $\zeta = 0$ $\quad \psi = 0$ (5.4–41)

B.C. 2: at $\sigma = 0$ $\quad \psi = 1$ (5.4–42)

B.C. 3: at $\sigma = \infty$ $\quad \psi = 0$ (5.4–43)

This problem can be solved by the "method of combination of variables." One postulates that $\psi = \psi(\chi)$, where the new independent variable χ is:

$$\chi = \frac{\sigma}{\sqrt[3]{9\zeta}} \tag{5.4–44}$$

Then the problem becomes:

$$\chi \frac{d^2\psi}{d\chi^2} + (3\chi^3 - 1)\frac{d\psi}{d\chi} = 0 \tag{5.4–45}$$

with $\psi(0) = 1$ and $\psi(\infty) = 0$. The solution is:

$$\psi = \frac{\int_\chi^\infty \chi e^{-\chi^3}\, d\chi}{\int_0^\infty \chi e^{-\chi^3}\, d\chi} = \frac{3}{\Gamma(\frac{2}{3})}\int_\chi^\infty \chi e^{-\chi^3}\, d\chi \tag{5.4–46}$$

from which the temperature profile may be obtained by integration:

$$\int_{T}^{T_0} dT = -\frac{1}{k}\int_{y}^{\infty} q_y \, dy \tag{5.4-47}$$

In dimensionless quantities this becomes:

$$\begin{aligned}\Theta(\sigma,\zeta) &= \frac{T - T_0}{q_1 R/k} = \sqrt[3]{9\zeta}\int_{\chi}^{\infty} \psi \, d\chi \\ &= \frac{3\sqrt[3]{9\zeta}}{\Gamma(\frac{2}{3})}\left[\frac{e^{-\chi^3}}{3} - \chi \int_{\chi}^{\infty} \chi e^{-\chi^3} \, d\chi\right]\end{aligned} \tag{5.4-48}$$

Next we obtain the local Nusselt number, defined in Example 5.4–1, recognizing the fact that the bulk temperature T_b in this thermal inlet region will be virtually indistinguishable from the entrance temperature T_0. Therefore:

$$\text{Nu} = \frac{2hR}{k} = 2\frac{R}{k}\left(\frac{q_1}{T(y = 0) - T_0}\right) = \frac{2}{\Theta(0,\zeta)} \tag{5.4-49}$$

Then when the result in Eq. 5.4–48 is used, we get finally:

$$\text{Nu} = \frac{2\Gamma(\frac{2}{3})}{\sqrt[3]{9\zeta}} = \frac{2\Gamma(\frac{2}{3})}{\sqrt[3]{\dfrac{9\alpha z}{v_0 R^2}}} = \frac{2\Gamma(\frac{2}{3})}{\sqrt[3]{\dfrac{9\alpha z}{(s + 3)\langle v_z\rangle R^2}}} \tag{5.4-50}$$

This result is very nearly the same as the exact solution for $\langle v_z\rangle D^2/\alpha z$ greater than about 30, where $D = 2R$ is the tube diamter.[7]

EXAMPLE 5.4–3 Flow in Tubes with Constant Wall Temperature;[5,9] (Asymptotic Solution for Large z)

Repeat the problem in Example 5.4–1 but with the boundary condition that there is a constant temperature T_1 at the surface $r = R$, for $z > 0$.

SOLUTION

We return to Eq. 5.4–11 and introduce somewhat different dimensionless quantities than we used heretofore:

$$\Theta = \frac{T - T_1}{T_0 - T_1}; \qquad \xi = \frac{r}{R}; \qquad \zeta = \frac{\alpha z}{\langle v_z\rangle R^2} \tag{5.4-51}$$

(Note that the ζ used here is not the same as that in Example 5.4–1 or 5.4–2.) We also use a dimensionless velocity profile defined as follows:

$$\phi(\xi) = \frac{v_z}{\langle v_z\rangle} = \left(\frac{s + 3}{s + 1}\right)(1 - \xi^{s+1}) \tag{5.4-52}$$

[9] B. C. Lyche and R. B. Bird, *Chem. Eng. Sci.*, *6*, 35–41 (1956), gave the solution for constant m and n. The extension to temperature-dependent m and an associated experimental study were given by E. B. Christiansen and S. E. Craig, Jr., *A.I.Ch.E. Journal*, *8*, 154–160 (1962). See also E. B. Christiansen, G. E. Jensen, and F. S. Tao, *ibid.*, *12*, 1196–1202 (1966) and E. B. Christiansen and G. E. Jensen, *ibid.*, *15*, 504–507 (1969).

Then for this constant-wall-temperature situation, the problem statement is:

$$\phi(\xi)\frac{\partial\Theta}{\partial\zeta} = \frac{1}{\xi}\frac{\partial}{\partial\xi}\left(\xi\frac{\partial\Theta}{\partial\xi}\right) \quad (5.4\text{–}53)$$

B.C. 1: at $\zeta = 0$ $\Theta = 1$ (5.4–54)

B.C. 2: at $\xi = 1$ $\Theta = 0$ (5.4–55)

B.C. 3: at $\xi = 0$ $\Theta = \text{finite}$ (5.4–56)

This problem may be attacked by the method of separation of variables. We postulate a solution of the form:

$$\Theta(\xi,\zeta) = X(\xi)Z(\zeta) \quad (5.4\text{–}57)$$

and insert it into the partial differential equation. Then division by XZ gives:

$$\frac{1}{Z}\frac{dZ}{d\zeta} = \frac{1}{\phi X\xi}\frac{d}{d\xi}\left(\xi\frac{dX}{d\xi}\right) \quad (5.4\text{–}58)$$

Both sides are equal to a constant, which we denote by $-\beta^2$. Hence we get two ordinary differential equations:

$$\frac{dZ}{d\zeta} = -\beta^2 Z \quad (5.4\text{–}59)$$

$$\frac{1}{\xi}\frac{d}{d\xi}\left(\xi\frac{dX}{d\xi}\right) + \beta^2\phi X = 0 \quad (5.4\text{–}60)$$

The second of these has boundary conditions: $X'(0) = 0$ and $X(1) = 0$. The X-equation is an eigenvalue problem, and there are many solutions X_i corresponding to the many eigenvalues β_i. The complete solution will then be a linear combination of products of the form given in Eq. 5.4–57:

$$\Theta(\xi,\zeta) = \sum_{i=1}^{\infty} A_i X_i(\xi) e^{-\beta_i^2\zeta} \quad (5.4\text{–}61)$$

The A_i are determined by the requirement that $\Theta = 1$ at $\zeta = 0$. After Eq. 5.4–61 is written for $\zeta = 0$, we multiply both sides by $X_j(\xi)\phi(\xi)\xi\, d\xi$ and integrate from $\xi = 0$ to $\xi = 1$. Then when use is made of the fact that the $X_i(\xi)$ are orthogonal over the range $\xi = 0$ to $\xi = 1$ with respect to the weighting function $\phi(\xi)\xi$, we obtain finally:

$$A_i = \frac{\int_0^1 X_i\phi\xi\, d\xi}{\int_0^1 X_i^2\phi\xi\, d\xi} \quad (5.4\text{–}61a)$$

We now turn to the evaluation of the Nusselt number. The heat transfer coefficient h and Nusselt number Nu are defined as in Eqs. 5.4–31 and 5.4–33. For the constant wall temperature problem, q_1 is not known but can be obtained from the temperature profile using Fourier's law: $q_1 = -k(\partial T/\partial r)_{r=R}$. When this is combined with Eq. 5.4–31, q_1 can be eliminated to obtain a formal expression for h. This can be inserted into the definition of the Nusselt number to give (here $T_w = T_1$):

$$\text{Nu} = \frac{2hR}{k} = -2R\frac{(\partial T/\partial r)_{r=R}}{T_b - T_w}$$

$$= -2\frac{(\partial\Theta/\partial\xi)_{\xi=1}}{\Theta_b} \quad (5.4\text{–}62)$$

Hence, when the temperature profiles are known, the Nusselt number can be calculated. We show in detail how to get Θ_b and $(\partial\Theta/\partial\xi)_{\xi=1}$ from Eq. 5.4–61:

$$\left.\frac{\partial\Theta}{\partial\xi}\right|_{\xi=1} = \sum_{i=1}^{\infty} A_i e^{-\beta_i^2\zeta} X_i'(1) \tag{5.4–63}$$

$$\begin{aligned}
\Theta_b &= \frac{\int_0^1 \phi(\xi)\Theta(\xi,\zeta)\xi\, d\xi}{\int_0^1 \phi(\xi)\xi\, d\xi} \\
&= 2\int_0^1 \phi(\xi)\Theta(\xi,\zeta)\xi\, d\xi \\
&= 2\sum_{i=1}^{\infty} A_i e^{-\beta_i^2\zeta} \int_0^1 \phi(\xi)X_i(\xi)\xi\, d\xi \\
&= -2\sum_{i=1}^{\infty} A_i e^{-\beta_i^2\zeta} \int_0^1 \frac{1}{\beta_i^2}\frac{d}{d\xi}\left(\xi\frac{dX_i}{d\xi}\right) d\xi \\
&= -2\sum_{i=1}^{\infty} A_i e^{-\beta_i^2\zeta} \frac{1}{\beta_i^2}\, \xi\left.\frac{dX_i}{d\xi}\right|_0^1 \\
&= -2\sum_{i=1}^{\infty} \frac{A_i}{\beta_i^2} e^{-\beta_i^2\zeta} X_i'(1)
\end{aligned} \tag{5.4–64}$$

In going from the third to the fourth line, use has been made of Eq. 5.4–60. The Nusselt number is then given by:

$$\text{Nu} = \frac{\sum_{i=1}^{\infty} A_i e^{-\beta_i^2\zeta} X_i'(1)}{\sum_{i=1}^{\infty} (A_i/\beta_i^2) e^{-\beta_i^2\zeta} X_i'(1)} \tag{5.4–65}$$

For large ζ we need only the first term in each sum so that

$$\lim_{\zeta\to\infty} \text{Nu} = \beta_1^2 \tag{5.4–66}$$

Therefore, if we want the local Nusselt number at large ζ, all we need calculate is the first eigenvalue for the boundary-value problem in Eq. 5.4–60. There are several ways of doing this; we illustrate here the use of the method of Stodola and Vianello[10] for the particular case of $n = \frac{1}{2}$ (or $s = 2$).

[10] See, for example, F. B. Hildebrand, *Advanced Calculus for Applications*, Prentice-Hall, Englewood Cliffs, New Jersey (1962), pp. 200–206. To solve the differential equation

$$\frac{d}{dx}\left[p(x)\frac{dy}{dx}\right] + \lambda r(x) y = 0 \tag{5.4–66a}$$

with homogeneous boundary conditions at $x = a$ and $x = b$, do the following: (i) in the λ-term, replace $y(x)$ by a reasonable first guess $y_1(x)$; (ii) solve the resulting differential equation and obtain the solution $y(x) = \lambda f_1(x)$; then repeat (i) with a second guess $y_2(x) = f_1(x)$; solve the resulting differential equation to obtain the solution $y(x) = \lambda f_2(x)$, and continue the process as long as desired. At the nth stage in this process the nth approximation to the lowest eigenvalue λ_1 can be obtained from:

$$\lambda_1^{(n)} = \frac{\int_a^b r(x) f_n(x) y_n(x)\, dx}{\int_a^b r(x)[f_n(x)]^2\, dx} \tag{5.4–66b}$$

For $s = 2$ we have $\phi(\xi) = (\frac{5}{3})(1 - \xi^3)$. As a first guess for the lowest eigenfunction we try $X_1(\xi) = 1 - \xi^3$. Then Eq. 5.4–60 becomes:

$$\frac{d}{d\xi}\left(\xi \frac{dX}{d\xi}\right) = -\tfrac{5}{3}\beta_1{}^2\xi(1 - \xi^3)^2 \tag{5.4–67}$$

This may be integrated to give:

$$\begin{aligned} X(\xi) &= \tfrac{5}{3}\beta_1{}^2\left(\frac{297}{64 \cdot 25} - \frac{1}{4}\xi^2 + \frac{2}{25}\xi^5 - \frac{1}{64}\xi^8\right) \\ &\equiv \beta_1{}^2 f_1(\xi) \end{aligned} \tag{5.4–68}$$

Then, using Eq. 5.4–66b we find:

$$\begin{aligned} [\beta_1{}^2]^{(1)} &= \frac{\int_0^1 \xi(1 - \xi^3)f_1(\xi)(1 - \xi^3)\,d\xi}{\int_0^1 \xi(1 - \xi^3)[f_1(\xi)]^2\,d\xi} \\ &= 3.97 \end{aligned} \tag{5.4–69}$$

This approximate result can be compared with the value Nu = 3.95 obtained by Lyche and Bird[9,5] by a complete numerical solution to the eigenvalue problem. It should be evident that the method of Stodola and Vianello can provide relatively quick estimates of the Nusselt numbers for large z for problems of this type.

EXAMPLE 5.4–4 Flow in Tubes with Constant Wall Temperature;[11–13] (Asymptotic Solution for Small z)

Repeat the problem of Example 5.4–3, but with the requirement to find a solution appropriate for small z.

SOLUTION

The procedure here is very similar to that used in Example 5.4–2. We neglect curvature, approximate the velocity distribution by a linear profile, and imagine that the fluid extends from the heat transfer surface all the way out to infinity. We introduce a dimensionless distance from the wall $\sigma = 1 - \xi$ as before. Then the problem statement in Eqs. 5.4–53 to 56 can be rewritten as:

$$\left.\frac{d\phi}{d\sigma}\right|_{\sigma=0} \sigma \frac{\partial \Theta}{\partial \zeta} = \frac{\partial^2 \Theta}{\partial \sigma^2} \tag{5.4–70}$$

B.C. 1: at $\zeta = 0$ $\quad \Theta = 1$ (5.4–71)

B.C. 2: at $\sigma = 0$ $\quad \Theta = 0$ (5.4–72)

B.C. 3: at $\sigma = \infty$ $\quad \Theta = 1$ (5.4–73)

[11] R. B. Bird, W. E. Stewart, and E. N. Lightfoot, *op. cit.*, pp. 307–308, 349–350.

[12] R. L. Pigford, *C.E.P. Symposium Series*, No. 17, *51*, 79–92 (1955).

[13] The analogous mass transfer problem was discussed by H. Kramers and P. J. Kreyger, *Chem. Eng. Sci.*, *6*, 42–48 (1956).

For the power-law fluid $(d\phi/d\sigma)_{\sigma=0}$ is just $(s + 3)$, as may be seen from Eq. 5.4–52. We now postulate that a solution can be found of the form $\Theta = \Theta(\chi)$, where χ is a new variable:

$$\chi = \frac{\sigma}{\sqrt[3]{9\zeta/(s + 3)}} \tag{5.4–74}$$

Then the problem becomes:

$$\frac{d^2\Theta}{d\chi^2} + 3\chi^2 \frac{d\Theta}{d\chi} = 0 \tag{5.4–75}$$

with $\Theta(0) = 0$ and $\Theta(\infty) = 1$. The solution is:

$$\Theta = \frac{1}{\Gamma(\frac{4}{3})} \int_0^{\chi} e^{-\bar{\chi}^3} d\bar{\chi} \tag{5.4–76}$$

The Nusselt number may now be obtained from Eq. 5.4–62; since for this problem z is quite small, Θ_b will not differ very much from unity. Hence we get:

$$\begin{aligned} \text{Nu} &= +2 \left.\frac{\partial\Theta}{\partial\sigma}\right|_{\sigma=0} = 2 \left.\frac{d\Theta}{d\chi}\right|_{\chi=0} \left(\frac{\partial\chi}{\partial\sigma}\right) \\ &= \frac{2}{\Gamma(\frac{4}{3})} \sqrt[3]{\frac{s + 3}{9\zeta}} = \frac{2}{\Gamma(\frac{4}{3})} \sqrt[3]{\frac{(s + 3)\langle v_z\rangle R^2}{9\alpha z}} \end{aligned} \tag{5.4–77}$$

This is just a special case of the result given in Eq. B of Table 5.4–1.

EXAMPLE 5.4–5 Flow in a Circular Die with Viscous Heating[14]

A molten plastic enters a horizontal circular die of radius R and length L at a temperature T_0. The fluid velocity is sufficiently high that viscous heating is known to be important. Find the temperature distribution in the die assuming (i) the die wall is maintained at temperature T_0; (ii) the fluid is described adequately by a power-law viscosity function with m and n constant, (iii) the fluid has a fully developed velocity profile at the die entrance, (iv) density, heat capacity, and thermal conductivity do not change with temperature or pressure.

What is the maximum temperature predicted for a polyethylene melt that enters the die at 463 K, when the wall shear stress is 2×10^5 N/m^2? The melt has the following physical properties:

$$\rho\hat{C}_p = 1.80 \times 10^6 \text{ J/m}^3\text{·K}$$

$$k = 4.184 \times 10^{-2} \text{ W/m·K}$$

$$n = 0.50$$

$$m = 6.9 \times 10^3 \text{ N·s}^{1/2}/\text{m}^2$$

The radius and length of the die are 0.0394 cm and 1.28 cm, respectively.

[14] Flow of a Newtonian fluid in a circular tube with viscous heating was first solved according to the method given here by H. C. Brinkman, *Appl. Sci. Res.*, *A2*, 120–124 (1951), and the extension to power-law fluids was worked out by R. B. Bird, *J. Soc. Plastics Eng.*, *11*, 35–40 (1955). Compressibility effects were considered by H. L. Toor, *Ind. Eng. Chem.*, *48*, 922–926 (1956), and temperature-dependent viscosity effects by J. B. Lyon and R. E. Gee, *Ind. Eng. Chem.*, *49*, 956–960 (1957), by E. A. Kearsley, *Trans. Soc. Rheol.*, *6*, 253–261 (1962), and by P. C. Sukanek, *Chem. Eng. Sci.*, *26*, 1775–1776 (1971). In the last two publications cited analytical solutions are given for large z, for the Newtonian and power-law fluids, respectively.

SOLUTION

We postulate that $v_z = v_z(r)$, $p = p(z)$, and $T = T(r,z)$. Then the equation of continuity is satisfied identically and the equation of motion (Eq. 5.4–2) and the equation of energy (Eq. 5.4–3) are:

Motion:
$$0 = -\frac{dp}{dz} + \frac{1}{r}\frac{d}{dr}\left(\eta r \frac{dv_z}{dr}\right) \tag{5.4–78}$$

Energy:
$$\rho \hat{C}_p v_z \frac{\partial T}{\partial z} = k \frac{1}{r}\frac{\partial}{\partial r}\left(r \frac{\partial T}{\partial r}\right) + \eta \left(\frac{dv_z}{dr}\right)^2 \tag{5.4–79}$$

Axial heat conduction has been neglected in the energy equation. For the power-law fluid we have to insert $\eta = m(-dv_z/dr)^{n-1}$, and then according to Example 5.2–1

$$v_z = v_{\max}\left[1 - \left(\frac{r}{R}\right)^{s+1}\right] \tag{5.4–80}$$

where $v_{\max} = [(p_0 - p_L)R/2mL]^s[R/(s+1)]$ and $s = 1/n$.

It is convenient to rewrite Eq. 5.4–79 in dimensionless form by introducing the following dimensionless variables:

Radial coordinate:
$$\xi = \frac{r}{R} \tag{5.4–81}$$

Axial coordinate:
$$\zeta = \frac{kz}{\rho \hat{C}_p v_{\max} R^2} \tag{5.4–82}$$

Temperature rise:
$$\Theta = \frac{k(T - T_0)(s+3)^2}{4mR^2\left(\dfrac{v_{\max}}{R}(s+1)\right)^{(1/s)+1}} \tag{5.4–83}$$

Then the mathematical problem to be solved is:

$$(1 - \xi^{s+1})\frac{\partial \Theta}{\partial \zeta} = \frac{1}{\xi}\frac{\partial}{\partial \xi}\left(\xi \frac{\partial \Theta}{\partial \xi}\right) + \frac{(s+3)^2}{4}\xi^{s+1} \tag{5.4–84}$$

with the boundary conditions:

B.C. 1: at $\zeta = 0$ $\quad \Theta = 0$ (5.4–85)

B.C. 2: at $\xi = 0$ $\quad \Theta =$ finite (5.4–86)

B.C. 3: at $\xi = 1$ $\quad \Theta = 0$ (5.4–87)

Because of the heat-production term, Eq. 5.4–84 cannot be directly solved by the method of separation of variables. Intuitively we know that for large distances down the tube a temperature distribution should be attained that depends only on the radial coordinate. Therefore it seems reasonable to postulate a solution of the form:

$$\Theta(\xi,\zeta) = \Theta_1(\xi) - \Theta_2(\xi,\zeta) \tag{5.4–88}$$

in which Θ_1 is the solution for very large ζ, and Θ_2 is a function that just cancels Θ_1 at $\zeta = 0$, and which tends to zero as ζ becomes very large. The Θ_1-part is easily obtained by setting the left side of Eq. 5.4–84 equal to zero and solving the remaining ordinary differential equation using the boundary conditions at

$\xi = 0$ and $\xi = 1$. This gives:

$$\Theta_1(\xi) = \tfrac{1}{4}(1 - \xi^{s+3}) \tag{5.4–89}$$

Now the results in Eqs. 5.4–88 and 89 can be inserted into Eq. 5.4–84, and in this way an equation is obtained for Θ_2:

$$(1 - \xi^{s+1})\frac{\partial \Theta_2}{\partial \zeta} = \frac{1}{\xi}\frac{\partial}{\partial \xi}\left(\xi \frac{\partial \Theta_2}{\partial \xi}\right) \tag{5.4–90}$$

This equation can be solved by the method of separation of variables. We postulate that $\Theta_2(\xi,\zeta) = X(\xi)Z(\zeta)$ and this leads to:

$$\frac{1}{Z}\frac{dZ}{d\zeta} = \frac{1}{X}\frac{1}{1 - \xi^{s+1}}\frac{1}{\xi}\frac{d}{d\xi}\left(\xi \frac{dX}{d\xi}\right) = -a \tag{5.4–91}$$

in which a is the separation constant. This procedure thus leads to two separate ordinary differential equations for $X(\xi)$ and $Z(\zeta)$. The X-equation has to be solved with the boundary conditions that $X(\xi) = 0$ at $\xi = 1$, and $X'(\xi) = 0$ at $\xi = 0$. This boundary-value problem has an infinite set of solutions $X_k(\xi)$ corresponding to the infinite set of eigenvalues a_k. It may be shown directly from the differential equation that the $X_k(\xi)$ satisfy the orthogonality conditions:

$$\int_0^1 (1 - \xi^{s+1})\xi X_k(\xi) X_l(\xi)\, d\xi = 0 \qquad (k \neq l) \tag{5.4–92}$$

Because of the linearity of the original partial differential equation, the expression for $\Theta_2(\xi,\zeta)$ must be given by a superposition of products of functions of ξ and functions of ζ. The final expression for Θ must therefore have the form:

$$\Theta(\xi,\zeta) = \tfrac{1}{4}(1 - \xi^{s+3}) - \sum_{k=1}^{\infty} B_k X_k(\xi) e^{-a_k \zeta} \tag{5.4–93}$$

The B_k have to be determined from the boundary condition at $\zeta = 0$. By multiplying Eq. 5.4–93 (with $\zeta = 0$ and $\Theta = 0$) by $\xi(1 - \xi^{s+1})X_l$ and integrating over ξ from 0 to 1, we get:

$$B_l = \frac{1}{4}\frac{\int_0^1 X_l(1 - \xi^{s+3})(1 - \xi^{s+1})\xi\, d\xi}{\int_0^1 X_l^2(1 - \xi^{s+1})\xi\, d\xi} \tag{5.4–94}$$

The problem is thus solved once the eigenfunctions $X_k(\xi)$ and the eigenvalues a_k are known.

We illustrate what has to be done for the special case that $n = \frac{1}{2}$ or $s = 2$. We seek a solution to the $X(\xi)$ equation in Eq. 5.4–91 in the form of a power series; the ith eigenfunction will thus have the form:

$$X_i(\xi) = \sum_{k=0}^{\infty} b_{ik}\xi^k \tag{5.4–95}$$

in which b_{i0} is arbitrarily chosen to be unity. The coefficient b_{i1} must be zero to satisfy the boundary condition at $\xi = 0$. Substitution of the above postulated solution into the differential equation leads to the following recursion formula: $b_{ik} = -(a_i/k^2)(b_{i,k-2} - b_{i,k-5})$. Hence all of the b_{ik} can be expressed in terms of the eigenvalues a_i, which are at this point still not known:

$$\begin{aligned}
b_{i2} &= -\tfrac{1}{4}a_i & \qquad b_{i5} &= +\frac{1}{25}\,a_i \\
b_{i3} &= 0 & \qquad b_{i6} &= -\frac{1}{2304}\,a_i^3 \\
b_{i4} &= +\frac{1}{64}\,a_i^2 & \qquad b_{i7} &= -\frac{29}{4900}\,a_i^2, \text{ etc.}
\end{aligned} \tag{5.4–96}$$

The boundary condition at $\xi = 1$ then requires that $\sum_{k=0}^{\infty} b_{ik} = 0$; when the b's from Eq. 5.4–96 are

TABLE 5.4–3

Constants for Use in Eq. 5.4–93 When $n = 0.5$[a]

i	a_i	B_i
1	6.582	+0.349
2	39.090	−0.147
3	99.50	+0.057
4	187.9	−0.014[b]

[a] R. B. Bird, *S.P.E. Journal*, *11*, 35–40 (1955).
[b] Estimated value.

substituted into this, we get an algebraic equation for a_i; since this equation is of infinite order, an infinite number of eigenvalues $a_1, a_2, a_3 \ldots$ will be obtained. The first few of these values are shown in Table 5.4–3. Since the X-equation is a second-order differential equation, there must be a second solution in addition to that in Eq. 5.4–95, with the b's provided by Eq. 5.4–96. It can be shown, however, that the second solution becomes infinite at $\xi = 0$ and is hence unacceptable here.

Once we have the eigenvalues a_i and the eigenfunctions X_i, it is a straightforward matter to get the B_i from Eq. 5.4–94. The first few B_i are shown in Table 5.4–3 along with the corresponding eigenvalues.[15] The temperature profiles computed from Eq. 5.4–93 are shown in Fig. 5.4–2. There it can be seen that, at short distances, there is a peak in the temperature profile near the wall where the velocity gradient, and hence the viscous heating, are large. The temperature profiles for a thermally insulated wall are also shown in Fig. 5.4–2 for comparison.

For the particular experimental data given in the problem statement, the maximum temperature will occur at the tube exit, that is at $z = 0.0128$ m. The dimensionless axial distance ζ at the exit is:

$$\begin{aligned}\zeta &= \frac{kz}{\rho \hat{C}_p v_{\max} R^2} = \frac{kz(s+1)m^s}{\rho \hat{C}_p {\tau_R}^s R^3} \\ &= \frac{(4.184 \times 10^{-2})(0.0128)(3)(6.9 \times 10^3)^2}{(1.80 \times 10^6)(2 \times 10^5)^2(0.000394)^3} \\ &= 0.0174 \end{aligned} \tag{5.4–97}$$

From Fig. 5.4–2 we find by interpolation that the maximum dimensionless temperature $\Theta_{\max}$ is about 0.07. Hence from Eq. 5.4–83 we get:

$$\begin{aligned}(T - T_0)_{\max} &= \frac{4mR^2(\tau_R/m)^{s+1}}{k(s+3)^2}\Theta_{\max} \\ &= \frac{4R^2{\tau_R}^3\Theta_{\max}}{25\ m^2\ k} \\ &= \frac{4(3.94 \times 10^{-4})^2(2 \times 10^5)^3(0.07)}{25(6.9 \times 10^3)^2(4.184 \times 10^{-2})} \\ &= 7\ \text{K} \end{aligned} \tag{5.4–98}$$

[15] Note that the X-equations of Eqs. 5.4–60 and 5.4–91 are nearly the same. Since $\phi = v_z/\langle v_z \rangle$ (rather than $v_z/v_{z,\max}$) in Eq. 5.4–60, the eigenvalues in the two examples bear the relation $a_j = (v_{z,\max}/\langle v_z \rangle)\beta_j^2$. Hence the $a_1 = 6.582$ in Table 5.4–3 corresponds to $\beta_1^2 = 3.95$ cited just after Eq. 5.4–69.

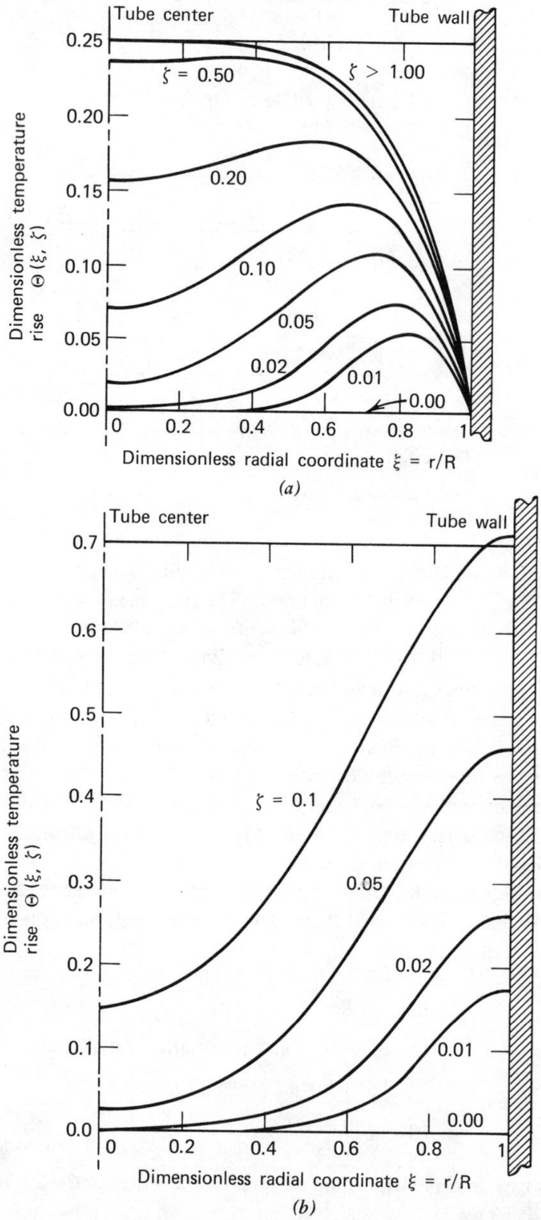

FIGURE 5.4–2. Temperature profiles for tube flow with viscous heating, based on the power-law model with $n = \frac{1}{2}$, and with m and n taken to be constant. (*a*) Constant wall temperature, (*b*) zero wall heat flux. [R. B. Bird, *SPE Journal*, *11*, 35–40 (1955).]

For the insulated wall problem, the temperature rise would have been about 24 K. It is thus evident that viscous heating can produce nonignorable temperature rises in extrusion operations.

We emphasize that the development here has been given for temperature-independent physical properties. The temperature dependence of m, however, cannot really be ignored if the temperature rise is more than a few degrees. Hence the use of the result in Fig. 5.4–2 should really be restricted to estimating flow speeds or pressure drops at which viscous heating just begins to be important.

We conclude with a few comments about the calculation of the eigenvalues a_i. Obtaining the a_i for $i \geq 4$ is tedious by the method outlined above. For $i \geq 4$ the WKBJ (Wentzel-Kramers-Brillouin-Jeffreys) method, widely used in quantum mechanics, is useful for getting both the eigenvalues and eigenfunctions. Applications to non-Newtonian heat-transfer problems have been given by Whiteman and

Drake[16] and by Ziegenhagen.[17] For the problem at hand the application of the WKBJ method gives for any velocity profile v_z:

$$a_i = \left[\frac{(i - \frac{1}{3})\pi}{\int_0^1 \sqrt{v_z(\xi)/v_{z,\,\max}}\, d\xi} \right]^2 \tag{5.4-99}$$

Then for the power law with $s = 2$ (or $n = 0.5$):

$$\begin{aligned} a_i &= \left[\frac{(i - \frac{1}{3})\pi}{\int_0^1 \sqrt{1 - \xi^3}\, d\xi} \right]^2 \\ &= \left[\frac{(i - \frac{1}{3})\pi}{\Gamma(\frac{3}{2})\Gamma(\frac{4}{3})/\Gamma(\frac{11}{6})} \right]^2 \\ &= \left[\frac{(i - \frac{1}{3})\pi}{0.841} \right]^2 \end{aligned} \tag{5.4-100}$$

This gives $a_1 = 6.202$, $a_2 = 38.8$, $a_3 = 99.2$, and $a_4 = 187.6$; by comparing with the exact values in Table 5.4–3 we see that for small i, for which the method is not intended, the results are not too good, but that the error is less than $\frac{1}{2}\%$ for $i = 3$.

EXAMPLE 5.4–6 Viscous Heating in a Cone-and-Plate Viscometer[18]

Repeat the analysis of the problem in Example 1.2–2 for a power-law fluid. Then show how the results can be adapted for use in estimating viscous-heating effects in the cone-and-plate instrument.

SOLUTION (a) Flow between Parallel Plates with the Assumption of Constant Physical Properties

We consider the flow in the z-direction of a fluid between two parallel plates located at $x = 0$ and $x = B$ (cf. Fig. 1.2–1); the lower plate is fixed and the upper one is moving in the positive z-direction with a constant velocity v_0 and $\partial p/\partial z = 0$. For a non-Newtonian fluid the equations of motion and energy that we need for describing the flow are the following simplifications of Eqs. 5.4–2 and 3:

$$0 = \frac{d}{dx}\left(\eta \frac{dv_z}{dx} \right) \tag{5.4-101}$$

$$0 = \frac{d}{dx}\left(k \frac{dT}{dx} \right) + \eta \left(\frac{dv_z}{dx} \right)^2 \tag{5.4-102}$$

into which, for this problem, we substitute $\eta = m(dv_z/dx)^{n-1}$.

We first work through the simplified problem in which m and n, as well as the thermal conductivity, are not varying with temperature. Then from Eq. 5.4–101 we get:

$$v_z = \left(\frac{x}{B} \right) v_0 \tag{5.4-103}$$

[16] I. R. Whiteman and W. B. Drake, *Trans. ASME*, *80*, 728–732 (1958).
[17] A. J. Ziegenhagen, *Int. J. Heat Mass Transfer*, *8*, 499–505 (1965).
[18] The perturbation solution presented here is due to R. M. Turian, *Chem. Eng. Sci.*, *20*, 771–781 (1965); a closed-form solution similar to that discussed in Problem 1C.6 for Newtonian fluids was given by J. Gavis and R. L. Laurence, *Ind. Eng. Chem. Fundamentals*, *7*, 525–527 (1968). In Eq. 5.4–126 where we give (1/20)$\tilde{\text{B}}\text{r}A$, Turian's paper contains erroneously (1/20)$\tilde{\text{B}}\text{r}A/n$; we thank Prof. Turian for correspondence regarding this point.

and then the energy equation becomes:

$$0 = k\left(\frac{d^2T}{dx^2}\right) + m\left(\frac{v_0}{B}\right)^{n+1} \tag{5.4–104}$$

When this is integrated and the boundary conditions that $T = T_0$ at $x = 0, B$ are applied, the final result is:

$$T - T_0 = \frac{mv_0^{n+1}}{2kB^{n-1}}\left[\left(\frac{x}{B}\right) - \left(\frac{x}{B}\right)^2\right] \tag{5.4–105}$$

The maximum temperature in the system occurs at the midplane and is given by:

$$T_{\max} - T_0 = \frac{mv_0^{n+1}}{8kB^{n-1}} \tag{5.4–106}$$

This simple result may be used for making rough estimates of the temperature rise that can be expected in the gap between two moving surfaces in the absence of a pressure gradient.

(b) Flow between Parallel Plates with m Varying with the Temperature, but n and k Constant

We let m vary with temperature according to Eq. 5.4–4 and introduce the following dimensionless variables:

$$\Theta = \frac{T - T_0}{T_0} \qquad \phi = \frac{v_z}{v_0} \qquad \xi = \frac{x}{B} \tag{5.4–107, 108, 109}$$

Then the equations of motion and energy become:

$$e^{-A\Theta}\left(\frac{d\phi}{d\xi}\right)^n = C \tag{5.4–110}$$

$$\frac{d^2\Theta}{d\xi^2} + \text{Br}\, e^{-A\Theta}\left(\frac{d\phi}{d\xi}\right)^{n+1} = 0 \tag{5.4–111}$$

in which $\text{Br} = m^0 v_0^{n+1}/kT_0B^{n-1}$ is the Brinkman number for the power-law fluid, A is the temperature coefficient in Eq. 5.4–4, and C is a constant of integration. Then from Eq. 5.4–110

$$\frac{d\phi}{d\xi} = C^s(1 + a_1\Theta + a_2\Theta^2 + \cdots) \tag{5.4–112}$$

in which $s = 1/n$ and $a_j = (As)^j/j!$. This can be inserted into the energy equation Eq. 5.4–111 to give:

$$\frac{d^2\Theta}{d\xi^2} = -\text{Br}\, C^{s+1}(1 + a_1\Theta + a_2\Theta^2 + \cdots) \tag{5.4–113}$$

Equations 5.4–112 and 113 are solved by a perturbation procedure, using the Brinkman number as the perturbation parameter. We know that the unperturbed problem discussed in (a) has the solution $\phi = \xi$ and $\Theta = \frac{1}{2}\,\text{Br}\,(\xi - \xi^2)$. This suggests that we seek a solution of the form:

$$\phi = \xi + \text{Br}\,\phi_1(\xi) + \text{Br}^2\,\phi_2(\xi) + \cdots \tag{5.4–114}$$

$$\Theta = \tfrac{1}{2}\,\text{Br}\,(\xi - \xi^2) + \text{Br}^2\,\Theta_2(\xi) + \cdots \tag{5.4–115}$$

It is also necessary to expand the integration constant in a similar series:

$$C = C_0 + \text{Br}\, C_1 + \text{Br}^2\, C_2 + \cdots \tag{5.4–116}$$

so that:

$$C^s = C_0{}^s + \text{Br}\,(sC_0^{s-1}C_1) + \text{Br}^2\,(sC_0^{s-1}C_2 + \tfrac{1}{2}s(s-1)C_0^{s-2}C_1{}^2) + \cdots \tag{5.4–117}$$

When these expansions are substituted into Eqs. 5.4–110 and 111, sets of differential equations are obtained by equating equal powers of Br. The resulting differential equations are solved with the boundary conditions that Θ_j and ϕ_j are both zero at $\xi = 0, 1$. In this way, one obtains:

$$\phi = \xi - \tfrac{1}{12}\,\text{Br}\left(\frac{A}{n}\right)(\xi - 3\xi^2 + 2\xi^3) + \cdots \tag{5.4–118}$$

$$\Theta = \tfrac{1}{2}\,\text{Br}\,(\xi - \xi^2) - \tfrac{1}{24}\,\text{Br}^2\left(\frac{A}{n}\right)(n\xi - (n+1)\xi^2 + 2\xi^3 - \xi^4) + \cdots \tag{5.4–119}$$

These results simplify to those given in Eqs. 1.2–27 to 30 when $n = 1$.

(c) Application to the Cone-and-Plate Viscometer

When viscous heating is appreciable, viscometers will give incorrect readings, because of the temperature dependence of the rheological parameters. We now adapt the above results in an approximate way to the cone-and-plate instrument in order to estimate the influence of viscous heating on the measurement of viscosity.

We can make a table of equivalences between the notation used for the parallel plate system and that used in Example 1.2–5 for the cone-and-plate system:

Parallel-Plate System	*Cone-and-Plate System*	
$\xi = \dfrac{x}{B}$	$\bar{\xi} = \dfrac{(\pi/2) - \theta}{\theta_0}$	(5.4–120)
$\phi = \dfrac{v_z}{v_0}$	$\bar{\phi} = \dfrac{v_\phi}{Wr}$	(5.4–121)
$\text{Br} = \dfrac{m^0 v_0^{n+1}}{kT_0 B^{n-1}}$	$\overline{\text{Br}} = \widetilde{\text{Br}}\left(\dfrac{r}{R}\right)^2 = \dfrac{m^0 W^{n+1} R^2}{kT_0\theta_0^{n-1}}\left(\dfrac{r}{R}\right)^2$	(5.4–122)

Recall that the torque required to maintain the rotary motion of the cone is, according to Eq. 1.2–57:

$$\begin{aligned}\mathscr{T} &= 2\pi \int_0^R \tau_{\theta\phi}|_{\theta=\pi/2}\, r^2\, dr \\ &\cong 2\pi \int_0^R \left(-\eta \frac{1}{r}\frac{\partial v_\phi}{\partial \theta}\right)\Bigg|_{\theta=\pi/2} r^2\, dr\end{aligned} \tag{5.4–123}$$

Here an approximate form of $\dot{\gamma}_{\theta\phi}$ is used since $\theta \approx \pi/2$ throughout the gap and $\sin\theta \approx 1$.

Now using the power law:

$$\begin{aligned}\mathscr{T} &= 2\pi R^3 \int_0^1 \left[m\left(-\frac{1}{r}\frac{\partial v_\phi}{\partial \theta}\right)^n\right]_{\theta=\pi/2} \left(\frac{r}{R}\right)^2 d\left(\frac{r}{R}\right) \\ &= 2\pi m^0 R^3 \int_0^1 \left[\left(\frac{d\bar{\phi}}{d\bar{\xi}}\frac{W}{\theta_0}\right)^n\right]_{\bar{\xi}=0} \left(\frac{r}{R}\right)^2 d\left(\frac{r}{R}\right)\end{aligned} \tag{5.4–124}$$

We now assume that $\bar{\phi}$ will be the same function of $\bar{\xi}$ and $\overline{\text{Br}}$ that ϕ is of ξ and Br. Then:

$$\mathscr{T} = \frac{2\pi m^0 W^n R^3}{\theta_0{}^n} \int_0^1 \left[1 - \frac{1}{12}\frac{\widetilde{\text{Br}} A}{n}\left(\frac{r}{R}\right)^2 + \cdots\right]^n \left(\frac{r}{R}\right)^2 d\left(\frac{r}{R}\right) \tag{5.4–125}$$

Use of the binomial expansion and integration then finally gives:

$$\mathscr{T} = \frac{2\pi m^0 W^n R^3}{3\theta_0{}^n}(1 - \tfrac{1}{20}\,\widetilde{\text{Br}}\,A + \cdots) \tag{5.4–126}$$

in which $\widetilde{\text{Br}}$ is defined in Eq. 5.4–122 and A and m^0 are defined in Eq. 5.4–4. Equation 5.4–126 can be used to estimate the effect of viscous heating on the torque reading in the cone-and-plate instrument. Equation 5.4–126 was derived for the case that both the cone and plate surfaces are maintained at temperature T_0. If the plate is maintained at temperature T_0 but the cone is regarded as thermally insulated, then the same result is obtained as above except that $(\frac{1}{20})$ is replaced by $(\frac{1}{5})$.

§5.5 JUSTIFICATION FOR THE USE OF THE GENERALIZED NEWTONIAN FLUID

In this chapter we introduced the generalized Newtonian fluid in an empirical way to account for the enormous changes in viscosity with shear rate observed in many polymer solutions and melts. This simple empirical approach to problem solving has been found to be successful for many problems. To some people the success of the method justifies its use. To other people the use of an empiricism seems unsatisfactory, and they would appreciate knowing if there is any justification for the use of the method. In this instance we can put the method on a firmer basis by showing how the generalized Newtonian fluid comes from rather general continuum mechanical arguments. We do this in §8.5 in some detail, but in this section we offer a brief preview and give the main results.

In Chapter 8 it is shown that for steady-state shearing flows the most general expression for the stress tensor for a very wide class of fluids must be of the form:[1]

$$\boldsymbol{\tau} = -\eta\dot{\boldsymbol{\gamma}} - (\tfrac{1}{2}\Psi_1 + \Psi_2)\{\dot{\boldsymbol{\gamma}}\cdot\dot{\boldsymbol{\gamma}}\} + \tfrac{1}{2}\Psi_1 \frac{\mathscr{D}}{\mathscr{D}t}\dot{\boldsymbol{\gamma}} \tag{5.5–1}$$

Here η, Ψ_1, and Ψ_2 are the material functions defined in §4.3; they are all functions of the magnitude of the rate-of-strain tensor $\dot{\gamma} = \sqrt{\frac{1}{2}(\dot{\boldsymbol{\gamma}} : \dot{\boldsymbol{\gamma}})}$. The derivative $\mathscr{D}/\mathscr{D}t$ is the "corotational time derivative" (also called the "Jaumann derivative");[2] this derivative tells how the components of a tensor change with time as witnessed by an observer translating with the fluid and rotating with it. For a second-order tensor, this derivative is defined as follows:

$$\frac{\mathscr{D}}{\mathscr{D}t}\dot{\boldsymbol{\gamma}} = \frac{\partial}{\partial t}\dot{\boldsymbol{\gamma}} + \{\boldsymbol{v}\cdot\nabla\dot{\boldsymbol{\gamma}}\} + \tfrac{1}{2}(\{\boldsymbol{\omega}\cdot\dot{\boldsymbol{\gamma}}\} - \{\dot{\boldsymbol{\gamma}}\cdot\boldsymbol{\omega}\}) \tag{5.5–2}$$

in which

$$\dot{\boldsymbol{\gamma}} = \nabla\boldsymbol{v} + (\nabla\boldsymbol{v})^\dagger \tag{5.5–3}$$

$$\boldsymbol{\omega} = \nabla\boldsymbol{v} - (\nabla\boldsymbol{v})^\dagger \tag{5.5–4}$$

[1] J. L. Ericksen, in *Viscoelasticity*, J. T. Bergen (Ed.), Academic Press, New York (1960), Eq. 2.12; W. O. Criminale, Jr., J. L. Ericksen, and G. L. Filbey, Jr., *Arch. Rat. Mech. Anal.*, *1*, 410–417 (1958).

[2] G. Jaumann, *Sitzungsberichte Akad. Wiss. Wien*, IIa, *120*, 385–530 (3911); in steady-state flows the $\partial/\partial t$ part of the $\mathscr{D}/\mathscr{D}t$ operator is always omitted.

Equation 5.5–1 is called the *Criminale-Ericksen-Filbey* ("*CEF*") equation;[1] it may be regarded as a special case of the Rivlin-Ericksen equation (see Table 9.4–1, Eq. F).

Equation 5.5–1 is considerably more complicated than Eq. 5.1–4. However, it will be seen that if the terms containing Ψ_1 and Ψ_2 are omitted from the Criminale-Ericksen-Filbey equation, then the generalized Newtonian fluid is recovered. This means that many of the calculations we have made in this chapter for velocity profiles and volume rates of flow will remain valid when we proceed to the more general CEF equation. Therefore what we have learned in this chapter for steady shearing flows does not have to be unlearned as we learn some of the more advanced theories in later chapters.

In this chapter we have applied the generalized Newtonian fluid to some flows that are not steady shearing flows, and this cannot be justified rigorously. Nonetheless generalized Newtonian fluid models will continue to be used outside the region of legitimacy for some time to come, partly because of their simplicity and partly because the computational problems associated with the use of more rigorous equations are so difficult and tedious.

We close the discussion here by showing how Eq. 5.5–1 simplifies for the case of shearing flow between plane parallel plates. All this problem does is to verify that the Ψ_1 and Ψ_2 appearing in the equation are in fact the same as the normal stress coefficients defined in §4.3.

EXAMPLE 5.5–1 Simplification of the CEF Equation for a Steady Simple Shearing Flow

Show how Eq. 5.5–1 simplifies for the shear flow $v_x = \dot{\gamma} y$ between a pair of parallel plates.

SOLUTION

The standard procedure here is to write out all of the pertinent kinematic tensors in matrix form. Then these matrices can be used to write the expression for the stress tensor as a matrix equation. From this equation the desired stresses or combinations of stresses can be obtained easily.

First we write out the velocity-gradient tensor, the rate-of-strain tensor, and the vorticity tensor:

$$\nabla \mathbf{v} = \begin{pmatrix} 0 & 0 & 0 \\ 1 & 0 & 0 \\ 0 & 0 & 0 \end{pmatrix} \dot{\gamma}; \qquad (\nabla \mathbf{v})^\dagger = \begin{pmatrix} 0 & 1 & 0 \\ 0 & 0 & 0 \\ 0 & 0 & 0 \end{pmatrix} \dot{\gamma} \tag{5.5–5, 6}$$

$$\dot{\boldsymbol{\gamma}} = \begin{pmatrix} 0 & 1 & 0 \\ 1 & 0 & 0 \\ 0 & 0 & 0 \end{pmatrix} \dot{\gamma}; \qquad \boldsymbol{\omega} = \begin{pmatrix} 0 & -1 & 0 \\ 1 & 0 & 0 \\ 0 & 0 & 0 \end{pmatrix} \dot{\gamma} \tag{5.5–7, 8}$$

Then the various parts of the corotational derivative can be found (the first two terms are just zero):

$$\{\dot{\boldsymbol{\gamma}} \cdot \dot{\boldsymbol{\gamma}}\} = \begin{pmatrix} 1 & 0 & 0 \\ 0 & 1 & 0 \\ 0 & 0 & 0 \end{pmatrix} \dot{\gamma}^2 \tag{5.5–9}$$

$$\{\boldsymbol{\omega} \cdot \dot{\boldsymbol{\gamma}}\} = \begin{pmatrix} -1 & 0 & 0 \\ 0 & 1 & 0 \\ 0 & 0 & 0 \end{pmatrix} \dot{\gamma}^2 = -\{\dot{\boldsymbol{\gamma}} \cdot \boldsymbol{\omega}\} \tag{5.5–10}$$

When all of these contributions are inserted into Eq. 5.5–1 we find:

$$\begin{pmatrix} \tau_{xx} & \tau_{xy} & \tau_{xz} \\ \tau_{yx} & \tau_{yy} & \tau_{yz} \\ \tau_{zx} & \tau_{zy} & \tau_{zz} \end{pmatrix} = -\eta \begin{pmatrix} 0 & 1 & 0 \\ 1 & 0 & 0 \\ 0 & 0 & 0 \end{pmatrix} \dot{\gamma} - (\tfrac{1}{2}\Psi_1 + \Psi_2) \begin{pmatrix} 1 & 0 & 0 \\ 0 & 1 & 0 \\ 0 & 0 & 0 \end{pmatrix} \dot{\gamma}^2 + \tfrac{1}{2}\Psi_1 \begin{pmatrix} -1 & 0 & 0 \\ 0 & 1 & 0 \\ 0 & 0 & 0 \end{pmatrix} \dot{\gamma}^2 \tag{5.5–11}$$

From this we can obtain directly:

$$\tau_{xy} = \tau_{yx} = -\eta\dot{\gamma} \tag{5.5–12}$$

$$\tau_{xx} - \tau_{yy} = -\Psi_1\dot{\gamma}^2 \tag{5.5–13}$$

$$\tau_{yy} - \tau_{zz} = -\Psi_2\dot{\gamma}^2 \tag{5.5–14}$$

These relations are consistent with the definitions given in Chapter 4. We see thus that the CEF equation provides us with a way of accounting for normal-stress phenomena in steady-state shearing flows, and understanding this equation enlarges to some extent our ability to solve problems and interpret rheological phenomena. We shall return to the CEF equation in Chapter 8.

QUESTIONS FOR DISCUSSION

1. What is the origin of the generalized Newtonian fluid model? Which experiments in Chapters 3 and 4 cannot be described by this model?
2. What empiricisms are available for the function $\eta(\dot{\gamma})$ in the generalized Newtonian fluid? What are the strong and weak points of the "power-law" viscosity function? In what sense do Eqs. 5.1–10 and 11, 13, and 14 "contain" the power-law model?
3. In solving an analytical or numerical problem involving the flow of a polymer, how does one decide which empiricism in §5.1 to use? Does one always need to select a specific functional form for $\eta(\dot{\gamma})$?
4. How does one go from Eqs. 5.1–4 and 9 to Eq. 5.2–5? To Eq. 5.2–13? To Eq. 5.2–21?
5. In connection with Eq. 5.2–31, why are $\tau_{r\theta}$ and $\tau_{\theta r}$ the only nonvanishing components of $\boldsymbol{\tau}$? Draw sketches to describe the "three situations" mentioned just after Eq. 5.2–32. How do you get from Eq. 5.1–4 to Eq. 5.2–34? How do Eqs. 5.2–35, 36, 37, and 38 simplify if $\tau_0 \to 0$?
6. Discuss the origin of Eqs. 5.2–44 to 47 and explain why the dashed-underlined terms may be omitted.
7. Discuss various ways in which the parameters m and n of the power-law model can be deduced from flow experiments.
8. Compare and contrast variational methods and finite-difference techniques for solving non-Newtonian flow problems.
9. How can one distinguish experimentally between "shear thinning" (that is, decrease in viscosity associated with increasing shear rates) and decrease in viscosity due to viscous heating?
10. Would viscous heating be important in a ball-point pen?
11. When a Bingham fluid is sheared at steady state between two parallel plates, which of the velocity distributions below would be possible?

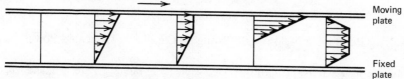

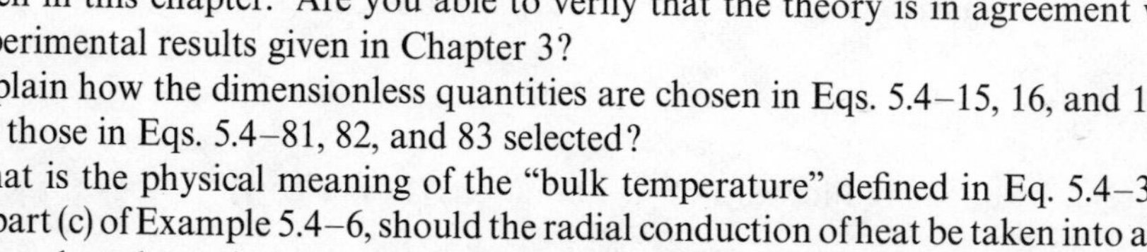

12. Discuss the experiment shown in Fig. 3.1–1 by using some of the analytical results given in this chapter. Are you able to verify that the theory is in agreement with the experimental results given in Chapter 3?
13. Explain how the dimensionless quantities are chosen in Eqs. 5.4–15, 16, and 17. How are those in Eqs. 5.4–81, 82, and 83 selected?
14. What is the physical meaning of the "bulk temperature" defined in Eq. 5.4–32?
15. In part (c) of Example 5.4–6, should the radial conduction of heat be taken into account?
16. When does the variational principle in Eq. 5.3–1 correspond to minimizing the viscous dissipation?
17. Why can we not obtain Eq. A of Table 5.4–1 from Eq. 5.4–77 in the limit $s \to \infty$ (i.e., plug flow)?

PROBLEMS

5A.1 Pipe Flow of a Polyisoprene Solution

A 13.5% (by weight) solution of polyisoprene in isopentane has the following power-law paramete 323 K: $m = 5 \times 10^3$ N·s^n/m^2, and $n = 0.2$. It is being pumped through a pipe that is 10.2 m long anc an internal diameter of 1.3 cm; the flow is known to be laminar. It is desired to build another pipe w length of 20.4 m with the same volume rate of flow and the same pressure drop. What should its ra be?

Answer:

5A.2 Flow of Carboxymethylcellulose Solution in an Annulus

A 3.5% aqueous carboxymethylcellulose solution has Ellis model parameters as follows: $\eta_0 =$ N·s/m^2, $\tau_{1/2} = 152$ N/m^2, and $\alpha = 3.0$. Find the flow rate Q in a horizontal annulus with $R = 2$ cm $\kappa = 0.624$, when $\Delta pR/2L = 500$ N/m^2. Use the approximate relation given in Eq. 5.2–43.

Answer: 154 c

5A.3 Pipe Flow of Polypropylene Melt (Ellis Model)

a. A commercial sample of polypropylene at 403 K has the following Ellis-model paramet $\eta_0 = 1.24 \times 10^4$ N·s/m^2, $\tau_{1/2} = 6.90 \times 10^3$ N/m^2, $\alpha = 2.82$. Calculate the volume rate of flow Q horizontal pipe, 5 cm internal diameter and length 17 m, when the pressure drop is 4.5×10^6 N/m

b. What would be the power-law constants for the fluid? Repeat the calculations in (a) fo power law. Why are the results in (a) and (b) different? Under what conditions would they be expe to be the same?

Answers: (a) 3.87×10^{-6}

(b) 5.92×10^{-7}

5A.4 Pumping a Nylon Melt through a Capillary (Truncated Power Law)[2]

A nylon-6 melt is to be pumped through a capillary of radius 5×10^{-4} m and length $2.5 \times 10^{-}$ with a pressure drop of 2.67×10^7 N/m^2. What volume rate of flow is expected (neglect viscous dissipa as well as entrance and exit effects)? The viscosity data have been fit with a truncated power law, wit following constants: $\eta_0 = 2.50 \times 10^2$ N·s/m^2, $\dot{\gamma}_0 = 800$ s^{-1}, and $n = 0.5$.

Answer: 1.18×10^{-7}

5A.5 Squeezing Flow of Polymer Solutions (Power Law)

One of the polymer solutions studied by Leider[3] in his squeezing-flow experiments was a 0.5% solutio polyacrylamide in glycerine for which the power-law parameters had been measured and found to $m = 25$ N·s^n/m^2, $n = 1/3$; in addition the time constant λ, defined in Eq. 5.2–54, was found to be 1.

This fluid was placed between two circular disks with radius 2.54 cm, with an initial disk separa of $2h_0 = 4.98 \times 10^{-3}$ cm. A 4.07 kg mass was placed on the upper disk and it was found that the t

[1] N. Yamada, N. Kishi, and H. Iizuka, *Kōbunshi Kagaku, 22*, 513–519 (1965).
[2] This problem was taken from W. J. Beek and D. B. Holmes, *Fysisch Technologische Aspecten va Polymeerverwerking*, Laboratorium voor Technische Natuurkunde, Technische Hogeschool Delft (19
[3] P. J. Leider, *Ind. Eng. Chem. Fundamentals, 13*, 342–346 (1974).

for half of the material to be squeezed out was 540 s. Compare this value with the value of $t_{1/2}$ computed from the Scott equation. Verify that the experiment described is indeed within the range of applicability of the Scott equation.

5A.6 Volume Rate of Flow in a Tube-Disk-Slit Assembly

a. The top curve in Fig. 5.1–2 is for a polystyrene melt at 453 K. Using the graph obtain the m and n of the power law appropriate for the power-law region of the data; show all work. Then check the values by obtaining the power-law constants from the Carreau model parameters given in the figure caption.

b. Derive an equation (based on the power-law model) from which the volume rate of flow Q can be obtained for the apparatus shown in Fig. 5A.6, when the pressure drop $p_0 - p_3$ is given. Use the results of Table 5.2–1 and Problem 5B.6 for deriving the expression for Q.

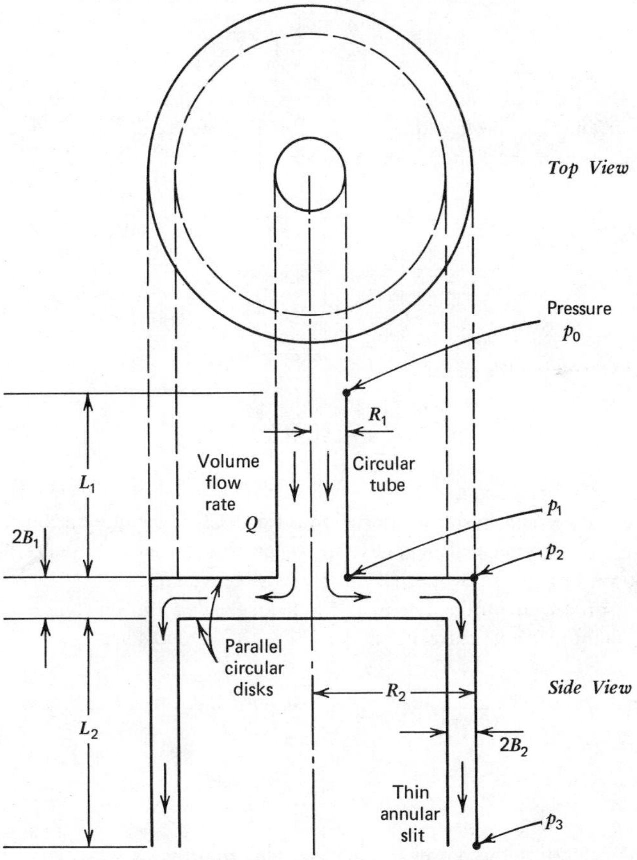

FIGURE 5A.6. Flow downward through a tube, radially between two parallel surfaces and then downward through an annular slit.

c. List the assumptions that are implied in the result obtained in (b).

d. Calculate the volume flow rate in cm³/s for a pressure drop $p_0 - p_3 = 2.1 \times 10^7$ N/m² (which is approximately 210 atm) for the polystyrene melt cited in (a). The dimensions in the device are:

$$R_1 = 0.5 \text{ cm} \quad L_1 = 11 \text{ cm} \quad B_1 = 0.2 \text{ cm}$$
$$R_2 = 7.0 \text{ cm} \quad L_2 = 9 \text{ cm} \quad B_2 = 0.15 \text{ cm}$$

The fluid density is 1 g/cm³.

Answer: 75 cm³/s (approximately)

5B.1 Flow in a Tapered Tube (Power-Law Fluid)

Show how the result in Eq. 5.2–10 is obtained by applying Eq. 5.2–9 locally.

5B.2 Non-Newtonian Flow in Circular Tubes and in Slits

In Example 5.2–1 it is shown how to obtain the volume rate of flow for a power-law fluid through circular tubes. In Table 5.2–1 results are given for the flow of other kinds of "model" fluids in tubes and flat slits. Actually it is not particularly difficult to obtain formal expressions for the flow rates in tubes and slits for any kind of shear-rate dependence of the viscosity.

a. For circular tubes show that the general expression for volume rate of flow can be integrated by parts twice to obtain:

$$Q = 2\pi \int_0^R v_z r\, dr$$

$$= \frac{\pi R^3 \dot{\gamma}_R}{3} - \frac{\pi}{3} \int_0^{\dot{\gamma}_R} r^3\, d\dot{\gamma} \qquad (5B.2\text{–}1)$$

in which $\dot{\gamma} = -dv_z/dr$ and $\dot{\gamma}_R = -dv_z/dr|_{r=R}$. Show further that use of Eq. 5.2–4 and the definition of η in Eq. 5.1–4 enables us to make a change of variable:

$$r = (R/\tau_R)\eta\dot{\gamma} \qquad (5B.2\text{–}2)$$

Show that this leads to:

$$Q = \frac{\pi R^3 \dot{\gamma}_R}{3} - \frac{\pi}{3}\left(\frac{R}{\tau_R}\right)^3 \int_0^{\dot{\gamma}_R} \eta^3 \dot{\gamma}^3\, d\dot{\gamma} \qquad (5B.2\text{–}3)$$

where $\dot{\gamma}_R$ is obtained by solving the equation:

$$\tau_R = \eta(\dot{\gamma}_R)\dot{\gamma}_R \qquad (5B.2\text{–}4)$$

Hence, if one knows $\eta(\dot{\gamma})$ from a cone-and-plate viscometer, for example, then $\dot{\gamma}_R$ can be obtained from Eq. 5B.2–4 in terms of τ_R. When that result is inserted into Eq. 5B.2–3, the integration can be performed to get Q as a function of τ_R. The integrals will have to be performed numerically. Verify Eqs. 5B.2–3 and 4 by inserting the power law and showing that Eq. 5.2–9 is recovered.

b. Repeat the procedure outlined in (a), this time for the flow through a plane slit of width W and thickness $2B$. Show that the results corresponding to Eqs. 5B.2–3 and 4 are:

$$Q = WB^2\dot{\gamma}_B - W\left(\frac{B}{\tau_B}\right)^2 \int_0^{\dot{\gamma}_B} \eta^2\dot{\gamma}^2\, d\dot{\gamma} \qquad (5B.2\text{–}5)$$

$$\tau_B = \eta(\dot{\gamma}_B)\dot{\gamma}_B \qquad (5B.2\text{–}6)$$

Check these results by inserting the power law and showing that the power-law entry in Table 5.2–1 is recovered.

5B.3 Tangential Annular Flow for a Polymer (Power-Law Model)

A polymeric liquid is being sheared in the annular region between two cylindrical surfaces of length L of radii κR and R (with $\kappa < 1$). The inner cylinder is rotating with an angular velocity W, and the outer cylinder is fixed.

a. Show that for this system the θ-component of the equation of motion simplifies to:

$$0 = -\frac{1}{r^2}\frac{d}{dr}(r^2\tau_{r\theta}) \qquad (5B.3\text{–}1)$$

and the $r\theta$-component of $\boldsymbol{\tau}$ for the power-law is:

$$\tau_{r\theta} = m\left(-r\frac{d}{dr}\left(\frac{v_\theta}{r}\right)\right)^n \tag{5B.3-2}$$

b. Show that the equations in (a) along with the appropriate boundary conditions can be solved to give for the velocity distribution:

$$\frac{v_\theta}{Wr} = \frac{(R/r)^{2/n} - 1}{(1/\kappa)^{2/n} - 1} \tag{5B.3-3}$$

c. Obtain an expression for the torque required to turn the inner cylinder.

Answer: (c) $\mathscr{T} = 2\pi(\kappa R)^2 Lm\left(\frac{2W/n}{1-\kappa^{2/n}}\right)^n$

5B.4 Third Invariant of Rate-of-Strain Tensor for Shear Flow

Just after Eq. 5.1–7 it was stated that for shear flows $III_{\dot{\gamma}}$ is identically zero. Verify this comment by considering the flow $v_x = \dot{\gamma}(t)y$, $v_y = 0$, $v_z = 0$.

5B.5 Axial Annular Flow with Inner Cylinder Moving Axially (Power-Law Fluid)

In the system shown in Fig. 5B.5 it is desired to find the velocity distribution in the annular region between A and B for a polymeric fluid. Use the power-law viscosity equation.

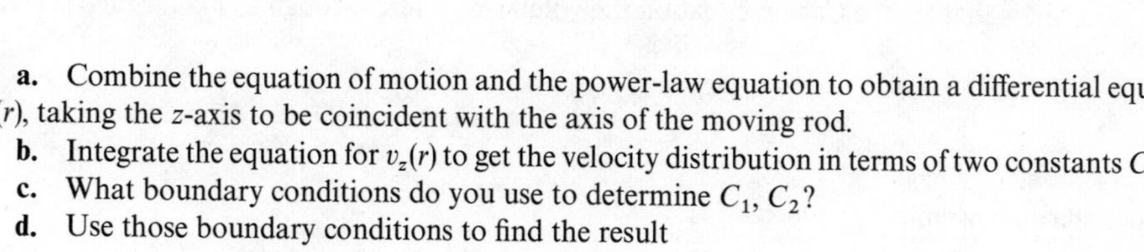

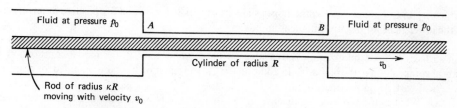

FIGURE 5B.5. Axial annular flow with inner cylinder moving axially.

a. Combine the equation of motion and the power-law equation to obtain a differential equation for $v_z(r)$, taking the z-axis to be coincident with the axis of the moving rod.

b. Integrate the equation for $v_z(r)$ to get the velocity distribution in terms of two constants C_1, C_2.

c. What boundary conditions do you use to determine C_1, C_2?

d. Use those boundary conditions to find the result

$$\frac{v_z(r)}{v_0} = \frac{\xi^{1-s} - 1}{\kappa^{1-s} - 1} \tag{5B.5-1}$$

where $\xi = r/R$ and $s = 1/n$.

e. Show how this simplifies, for the Newtonian fluid, to:

$$\frac{v_z(r)}{v_0} = \frac{\ln \xi}{\ln \kappa} \tag{5B.5-2}$$

f. What is the force acting on the wire in the region between A and B.

g. Is this a "shear flow" according to the definition in §4.1?

5B.6 Radial Flow between Parallel Disks (Power Law)

a. First solve the problem of the power-law fluid flow through a slit of width W and thickness $2B$, and verify the result given in Table 5.2–1.

b. Then solve the problem of slow steady-state radial flow between two fixed parallel disks (see Fig. 1B.10), which are separated by a distance $2B$. Let the inner and outer radii of the disks be R_1 and R_2. Obtain an expression for the volume rate of flow by applying the result of (a) locally in the region between the two disks:

$$Q = \frac{4\pi B^2}{(1/n) + 2}\left(\frac{(p_1 - p_2)B(1 - n)}{m(R_2^{1-n} - R_1^{1-n})}\right)^{1/n} \tag{5B.6–1}$$

This equation has been used by Laurencena and Williams[1] to describe radial flow data for some moderately viscoelastic polymer solutions.

c. Is this a "shear flow" according to the definition in §4.1?

5B.7 Flow in a Falling Film on a Cone (Power Law)

Rework Problem 1B.14(b) for a power-law fluid.

5B.8 Flow of Blood in Tubes (Casson Equation)

To describe the flow of pigment-oil suspensions, Casson[2] proposed the following rheological equation for shearing flows:

$$\sqrt{\pm\tau_{yx}} = \sqrt{\tau_0} + \sqrt{\mu_0}\sqrt{\mp\dot{\gamma}_{yx}} \quad \text{for} \quad \tau_{yx} > \tau_0 \tag{5B.8–1}$$

$$\dot{\gamma}_{yx} = 0 \quad \text{for} \quad \tau_{yx} < \tau_0 \tag{5B.8–2}$$

in which μ_0 and τ_0 are constants; the upper signs in Eq. 5B.8–1 are valid for positive momentum flux, and the lower signs for negative momentum flux.

The Casson equation has proven useful for the description of the flow of blood on both glass and fibrin surfaces.[3] Some information is available on the relation between the model constants and the chemical composition of the blood.[4,5] In addition the Casson equation has been modified for suspensions of spherical particles in polymer solutions.[6]

a. What is the physical significance of the two constants in the equation? To what extent is this equation similar to the Bingham fluid?

b. Show that for the Casson equation the volume flow rate through a circular tube is given by:[3]

$$Q = \frac{\pi R^4(\mathscr{P}_0 - \mathscr{P}_L)}{8\mu_0 L}\left(1 - \tfrac{16}{7}\sqrt{\xi} + \tfrac{4}{3}\xi - \tfrac{1}{21}\xi^4\right) \tag{5B.8–3}$$

in which the dimensionless parameter ξ is

$$\xi = \frac{\tau_0}{(\mathscr{P}_0 - \mathscr{P}_L)R/2L} \tag{5B.8–4}$$

[1] B. R. Laurencena and M. C. Williams, *Trans. Soc. Rheol.*, *18*, 331–355 (1974).
[2] N. Casson, in *Rheology of Disperse Systems* (C. C. Mill, Ed.), Pergamon Press, New York (1959), p. 84.
[3] S. Oka, in *Proc. of the Fourth International Congress on Rheology*, Part 4 (A. L. Copley, Ed.), Wiley, New York (1965), pp. 81–92.
[4] E. W. Merrill, W. G. Margetts, G. R. Cokelet, and E. R. Gilliland, in *Proc. of the Fourth International Congress on Rheology*, Part 4, (A. L. Copley, Ed.), Wiley, New York (1965), pp. 135–143.
[5] E. N. Lightfoot, *Transport Phenomena and Living Systems*, Wiley, New York (1974), pp. 35–38, 430, 438, 440. M. M. Lih, *Transport Phenomena in Medicine and Biology*, Wiley, New York (1975), pp. 378–386.
[6] S. Onogi, T. Matsumoto, and Y. Warashina, *Trans. Soc. Rheol.*, *17*, 175–190 (1973)

5B.9 Distributor Design (Power Law)[7]

In this problem we suggest how the two power-law results in Table 5.2–1 can be combined to get an approximate solution to a rather complicated flow problem. In Fig. 5B.9 we see a distributor system that is supposed to deliver a polymeric fluid with a uniform efflux velocity as the fluid leaves the thin slit between the parallel plates. The circular pipe and parallel plates are horizontal, and the fluid leaves the pipe and enters the parallel-plate system through a thin slit in the tube. The flow between the plates is in the x-direction, because of the presence of vertical dividers that also serve to maintain the spacing between the plates.

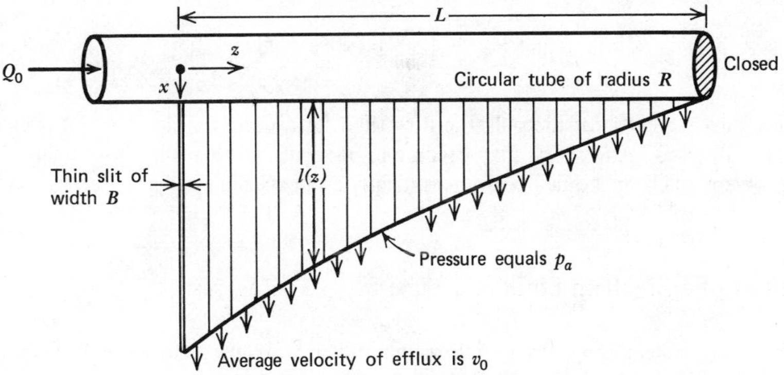

FIGURE 5B.9. Tube of radius R, with slit of width B attached, functioning as a distributor.

a. First consider the flow in the cylindrical tube. Verify that if Q_0 is the volume rate of flow entering the tube, then the volume rate of flow across any plane will be:

$$Q(z) = Q_0\left(1 - \frac{z}{L}\right) \tag{5B.9–1}$$

Then apply the power-law result for a circular tube (Table 5.2–1) locally to obtain the following differential equation for the pressure:

$$Q_0\left(1 - \frac{z}{L}\right) = \frac{\pi R^3}{(1/n) + 3}\left(\frac{R}{2m}\right)^{1/n}\left(-\frac{dp}{dz}\right)^{1/n} \tag{5B.9–2}$$

Integrate the differential equation for $p(z)$ from an arbitrary distance z to the end of the tube $z = L$, and obtain:

$$p - p_a = \frac{2mL}{R}\left(\frac{(1/n) + 3}{\pi R^3}\right)^n \frac{Q_0{}^n}{n + 1}(1 - \zeta)^{n+1} \tag{5B.9–3}$$

where $\zeta = z/L$.

b. Next adapt the slit flow result in Table 5.2–1 to the problem at hand to obtain:

$$v_0 = \frac{B/2}{(1/n) + 2}\left[\frac{(p - p_a)B}{2ml(\zeta)}\right]^{1/n} \tag{5B.9–4}$$

c. From (a) and (b) find the equation of the curve $l(\zeta)$ that will insure uniform efflux:

$$l(\zeta) = \frac{BL}{R(n + 1)}\left(\frac{B}{2\pi R^3}\frac{(1/n) + 3}{(1/n) + 2}\right)^n\left(\frac{Q_0}{v_0}\right)^n(1 - \zeta)^{n+1} \tag{5B.9–5}$$

Note that m does not appear in the result!

[7] This problem was suggested by a conversation with F. H. Ancker, Union Carbide Company, Bound Brook, N.J., who solved a somewhat similar problem by the method used here.

5B.10 Pipe Flow with Sinusoidally Varying Pressure Gradient

A polymer with a power-law exponent $n = 0.5$ is being pumped through a horizontal circular pipe by means of a pressure gradient that is varying sinusoidally about some mean value; that is,

$$\frac{p_0 - p_L}{L} = \alpha + \beta \mathscr{Re}\{e^{i\omega t}\} \tag{5B.10–1}$$

in which α and β are constants.

Show that the time-average volume rate of flow is:[8]

$$\bar{Q} = \frac{\pi R^3}{5}\left(\frac{R}{2m}\right)^2 (\alpha^2 + \tfrac{1}{2}\beta^2) \tag{5B.10–2}$$

Hence the flow rate is greater for the pulsatile flow than for a nonpulsatile flow. Of course, if the frequency is very high viscoelastic effects may become important, and then the above analysis will not be valid. This problem has been studied both theoretically and experimentally by Barnes, Townsend, and Walters.[9]

5B.11 Flow of a Bingham Fluid in a Tube

a. Show that when a Bingham fluid flows in a tube it will have a "plug flow" region in the center with the radius $r_0 = 2\tau_0 L/(\mathscr{P}_0 - \mathscr{P}_L)$.

b. Verify that the velocity distribution is:

$$v_z^{>} = \frac{(\mathscr{P}_0 - \mathscr{P}_L)R^2}{4\mu_0 L}\left[1 - \left(\frac{r}{R}\right)^2\right] - \frac{\tau_0 R}{\mu_0}\left[1 - \left(\frac{r}{R}\right)\right] \qquad r \geq r_0 \tag{5B.11–1}$$

$$v_z^{<} = \frac{(\mathscr{P}_0 - \mathscr{P}_L)R^2}{4\mu_0 L}\left(1 - \frac{r_0}{R}\right)^2 \qquad r \leq r_0 \tag{5B.11–2}$$

c. Obtain the expression for the volume rate of flow given in Table 5.2–1. This result is often referred to as the *Buckingham-Reiner equation.*[10]

5B.12 Viscous Heating between Parallel Plates with One Wall Insulated

Rework the problem given in Example 5.4–6(a), for the power-law fluid with temperature-independent parameters, for the situation that the lower plate is maintained at temperature T_0 whereas the upper plate is thermally insulated.

[8] In order to obtain this result it is convenient to use the following relation:

$$\mathscr{Re}\{w_1\}\mathscr{Re}\{w_2\} = \tfrac{1}{2}[\mathscr{Re}\{w_1 w_2\} + \mathscr{Re}\{w_1 \bar{w}_2\}] \tag{5B.10–1a}$$

in which the overbar indicates a complex conjugate.

[9] H. A. Barnes, P. Townsend, and K. Walters, *Rheol. Acta, 10*, 517–527 (1971); *Nature, 224*, 585–587 (1969). See also Example 8.1–4.

[10] E. Buckingham, *Proc. ASTM, 21*, 1154–1161 (1921); M. Reiner, *Deformation, Strain, and Flow*, Interscience, New York (1960), p. 117. Other Bingham fluid flow problems have been solved by J. G. Oldroyd, *Proc. Camb. Phil. Soc., 43*, 100–105 (1947), *43*, 383–395 (1947), *43*, 396–405 (1947), *43*, 521–532 (1947), *44*, 200–213 (1948), *44*, 214–228 (1948); this series of papers includes problems on steady and unsteady motion as well as a discussion of the proper formulation of the equations for a Bingham material. See also A. Slibar and P. R. Paslay, *Zeits. für angew. Math. u. Mech., 37*, 441–449 (1957); P. R. Paslay and A. Slibar, *Österreichisches Ingenieur-Archiv*, Vol. 10, No. 4, 328–344 (1956); *Petrol. Trans., AIME, 210*, 310–317 (1957); D. B. Clegg and G. Power, *Appl. Sci. Res., A12*, 199–211 (1959).

5B.13 Calculation of Nusselt Number for Flow between Parallel Plates

a. Work through the derivation of the Nusselt number for the flow in a thin slit of thickness $2B$ and width W, analogous to the derivation given in Example 5.4–1 for circular tubes. There is a constant heat flux q_1 at $x = +B$ and $-q_1$ at $x = -B$. Show that the asymptotic expression for the Nusselt number for power-law fluids is:

$$\text{Nu} = \frac{4Bh}{k} = \frac{12(s+4)(2s+5)}{32 + 17s + 2s^2} \tag{5B.13–1}$$

b. Verify that this simplifies to the result in Eq. L of Table 5.4–2 for Newtonian fluids.

c. Show how to get the result above from the general expression in Eq. K of Table 5.4–2.

d. How does the derivation have to be modified if there is a constant heat flux q_1 at one wall and no heat flux at the other?

5B.14 Adjacent Flow of Two Immiscible Fluids[11]

a. Two immiscible polymeric fluids are contained in the space between two parallel planes, located at $x = 0$ and $x = B$, the upper one being in motion in the z-direction with a constant speed v_0, as shown in Fig. 5B.14. Both fluids are describable in terms of power-law parameters, m_I and n_I for fluid "I," and m_{II} and n_{II} for fluid "II." The interface is located at $x = \lambda B$ and there is no pressure gradient in the z-direction. Fluid I is more dense than Fluid II, and it is to be presumed that the flow rate is such that the interface between the fluids remains virtually planar. Find the velocity distribution.

b. (Lengthy!) Repeat the problem in Part (a) when the upper plate is fixed, but there is a pressure gradient in the z-direction.

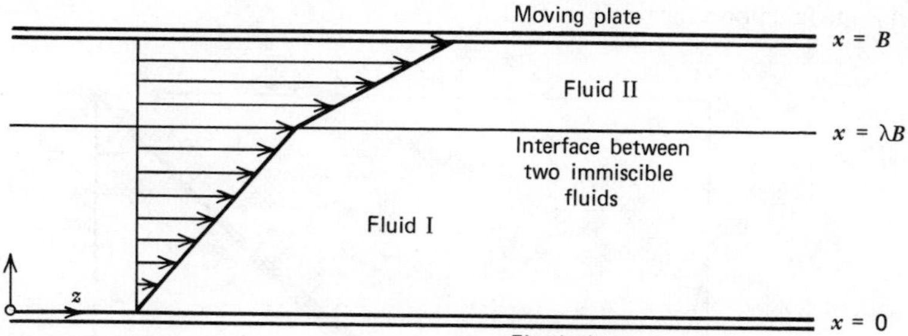

FIGURE 5B.14. Two immiscible power-law fluids being sheared in the region between two parallel plates one of which is in motion.

5C.1 Axial Annular Flow (Power Law Fluid)[1–3]

Obtain an expression for the volume rate of flow of a polymer through an annular conduit of length L; the bounding cylinders of the annulus are located at $r = \kappa R$ and $r = R$ (with $\kappa < 1$). Use the power-law expression for the non-Newtonian viscosity. Let the maximum in the velocity be at the position $r = \lambda R$.

[11] For discussions of the problems of multilayered flows with applications to manufacture of laminated sheets and bicomponent filaments and to mixing processes see W. J. Schrenk, *Plast. Eng.*, *30*, 65–68 (1974); W. J. Schrenk and T. Alfrey, Jr., *S.P.E. Journal*, *29*, 38–42, 43–47 (1973); T. Alfrey, Jr., *Polym. Eng. Sci.*, *9*, 400–404 (1969); W. J. Schrenk and T. Alfrey, Jr., *Polym. Eng. Sci.*, *9*, 393–399 (1969); A. E. Everage, Jr., *Trans. Soc. Rheol.*, *17*, 629–646 (1973). Most of the calculations in these papers were made for Newtonian fluids; in the last reference cited a variational method was employed.

[1] A. G. Fredrickson and R. B. Bird, *Ind. Eng. Chem.*, *50*, 347–352 (1958).

[2] G. C. Wallick and J. G. Savins, *Soc. Petrol. Eng. Journal*, *9*, 311–315 (1969).

[3] A. B. Metzner, "Flow of Non-Newtonian Fluids," in *Handbook of Fluid Dynamics* (V. L. Streeter, Ed.), Section 7, McGraw-Hill, New York (1961).

a. First show that the velocity distributions on the two sides of the maximum velocity are given by:

$$v_z^< = R\left[\frac{(\mathscr{P}_0 - \mathscr{P}_L)R}{2mL}\right]^s \int_\kappa^\xi \left(\frac{\lambda^2}{\xi'} - \xi'\right)^s d\xi' \qquad \kappa \le \xi \le \lambda \tag{5C.1-1}$$

$$v_z^> = R\left[\frac{(\mathscr{P}_0 - \mathscr{P}_L)R}{2mL}\right]^s \int_\xi^1 \left(\xi' - \frac{\lambda^2}{\xi'}\right)^s d\xi' \qquad \lambda \le \xi \le 1 \tag{5C.1-2}$$

in which $s = 1/n$ and $\xi = r/R$.

b. Verify that the requirement that the two expressions above match at the location of the velocity maximum gives the following equation for determining λ as a function of κ and s.

$$\int_\kappa^\lambda \left(\frac{\lambda^2}{\xi} - \xi\right)^s d\xi = \int_\lambda^1 \left(\xi - \frac{\lambda^2}{\xi}\right)^s d\xi \tag{5C.1-3}$$

c. Then show that the volume rate of flow through the annulus is:

$$Q = \pi R^3 \left[\frac{(\mathscr{P}_0 - \mathscr{P}_L)R}{2mL}\right]^s \int_\kappa^1 |\lambda^2 - \xi^2|^{s+1}\xi^{-s}\, d\xi \tag{5C.1-4}$$

This integral can be evaluated once λ is known from Eq. 5C.1–3. The result may be expressed as follows:

$$Q = \pi R^3 \left[\frac{(\mathscr{P}_0 - \mathscr{P}_L)R}{2mL}\right]^s \frac{(1-\kappa)^{s+2}}{s+2}\Upsilon(s,\kappa) \tag{5C.1-5}$$

where $\Upsilon(s,\kappa)$ is the function shown in Fig. 5C.1.

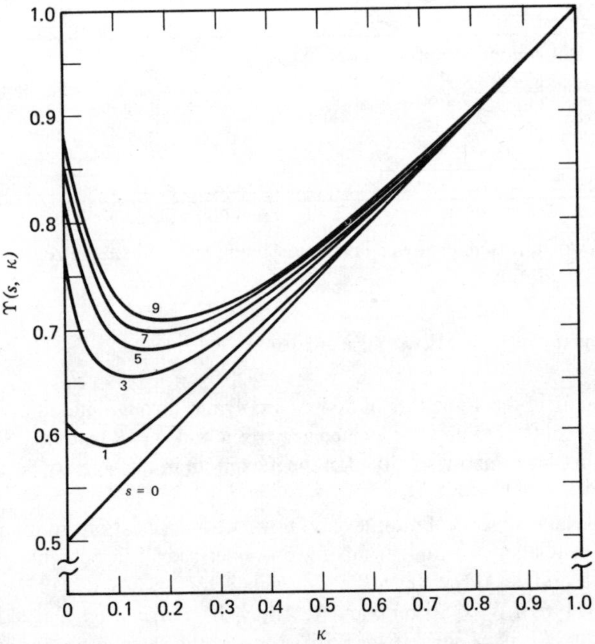

FIGURE 5C.1. The function $\Upsilon(s,\kappa)$ needed for obtaining the volume rate of flow through an annulus for a power-law fluid. A numerical table corresponding to this graph was given by A. G. Fredrickson and R. B. Bird. [Reprinted with permission from A. G. Fredrickson and R. B. Bird, *Ind. Eng. Chem.*, *50*, 347–352 (1958). Copyright by the American Chemical Society.]

5C.2 Flow out of a Tank with an Exit Tube

a. Rework Problem 1C.3(a) for a power-law fluid. Show that the efflux time is:

$$t_{\text{efflux}} = \left(\frac{2mL}{\rho g R_0}\right)^{1/n} \left(\frac{1}{n} + 3\right) \frac{R^2}{R_0{}^3} \left[\frac{(H + L)^{1-(1/n)} - L^{1-(1/n)}}{1 - (1/n)}\right] \tag{5C.2–1}$$

b. Show how this result simplifies to Eq. 1C.3–1 when $n \to 1$ and $m \to \mu$.

5C.3 Axial Annular Flow (Bingham Fluid)

Repeat Problem 5C.1 for the Bingham fluid defined in Eqs. 5.1–15 and 16. Use the following dimensionless quantities introduced by Fredrickson and Bird:[4]

$$T = \frac{2\tau_{rz}L}{(\mathscr{P}_0 - \mathscr{P}_L)R} \tag{5C.3–1}$$

$$T_0 = \frac{2\tau_0 L}{(\mathscr{P}_0 - \mathscr{P}_L)R} \tag{5C.3–2}$$

$$\phi = \left(\frac{2\mu_0 L}{(\mathscr{P}_0 - \mathscr{P}_L)R^2}\right) v_z \tag{5C.3–3}$$

$$\xi = \frac{r}{R} \tag{5C.3–4}$$

a. Show that the momentum-flux distribution and the rheological equation of state may be written:

$$T = \xi - \lambda^2 \xi^{-1} \tag{5C.3–5}$$

$$T = \pm T_0 - \frac{d\phi}{d\xi} \tag{5C.3–6}$$

in which the + sign is used when momentum is being transported in the $+r$-direction and the $-$ sign is used when transport is in the $-r$-direction. Show further that the bounds on the plug-flow region λ_+ and λ_- are given by:

$$\pm T_0 = \lambda_\pm - \left(\frac{\lambda^2}{\lambda_\pm}\right) \tag{5C.3–7}$$

where λ is a constant of integration.

b. Obtain the velocity distribution appropriate for each of the three regions:

$$\phi_- = -T_0(\xi - \kappa) - \tfrac{1}{2}(\xi^2 - \kappa^2) + \lambda^2 \ln \frac{\xi}{\kappa} \qquad \kappa \le \xi \le \lambda_- \tag{5C.3–8}$$

$$\phi_0 = \phi_-(\lambda_-) = \phi_+(\lambda_+) \qquad \lambda_- \le \xi \le \lambda_+ \tag{5C.3–9}$$

$$\phi_+ = -T_0(1 - \xi) + \tfrac{1}{2}(1 - \xi^2) + \lambda^2 \ln \xi \qquad \lambda_+ \le \xi \le 1 \tag{5C.3–10}$$

[4] A. Slibar and P. R. Paslay, *Zeits. angew. Math. u. Mech., 37*, 441–449 (1957); Fredrickson and Bird, *loc. cit.*; see also B. E. Anshus, *Ind. Eng. Chem. Fundamentals, 13*, 162–164 (1974). Helical flow of a Bingham fluid in an annulus has been solved approximately by P. R. Paslay and A. Slibar, *Petroleum Transactions, AIME, 210*, 310–317 (1957).

c. Verify that the determining equation for λ_+ is:

$$2\lambda_+(\lambda_+ - T_0)\ln\frac{\lambda_+ - T_0}{\lambda_+\kappa} - 1 + (T_0 + \kappa)^2 + 2T_0(1 - \lambda_+) = 0 \tag{5C.3–11}$$

d. Show that the volume rate of flow is:

$$Q = \frac{\pi R^4(\mathscr{P}_0 - \mathscr{P}_L)}{8\mu_0 L}\left[(1 - \kappa^4) - 2\lambda_+(\lambda_+ - T_0)(1 - \kappa^2) - \tfrac{4}{3}(1 + \kappa^3)T_0 + \tfrac{1}{3}(2\lambda_+ - T_0)^3 T_0\right]$$

$$\equiv \frac{\pi R^4(\mathscr{P}_0 - \mathscr{P}_L)}{8\mu_0 L}\Omega(\kappa, T_0) \tag{5C.3–12}$$

In Fig. 5C.3 the function Ω is shown as a function of κ and T_0.

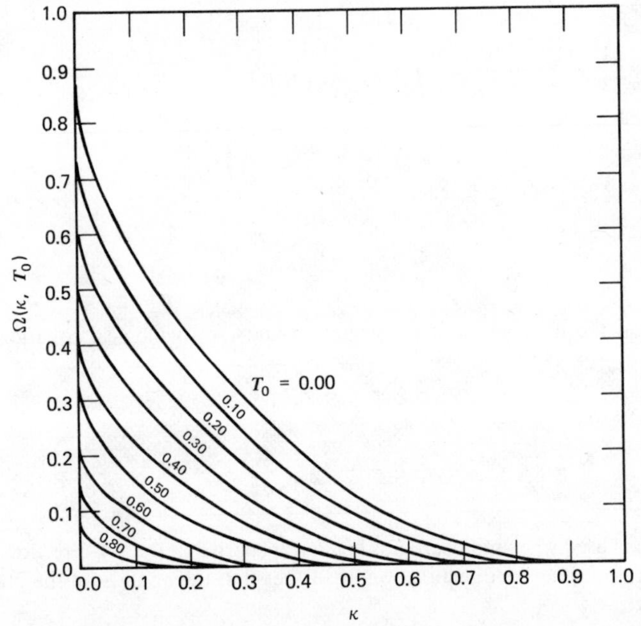

FIGURE 5C.3. The function $\Omega(\kappa, T_0)$ to be used with Eq. 5C.3–12 for computing Bingham flow in an annulus. A numerical table corresponding to this graph was given by Fredrickson and Bird. [Reprinted with permission from A. G. Fredrickson and R. B. Bird, *Ind. Eng. Chem.*, *50*, 347–352 (1958). Copyright by the American Chemical Society.]

5C.4 Falling Cylinder Viscometer (Power Law)

Rework Problem 1C.2 for a power-law fluid. In order to simplify the development, consider only the situation in which the clearance between the cylinder and tube is extremely small so that curvature effects can be completely neglected.

5C.5 Constant-Wall Heat Flux Problem for Arbitrary Velocity Distribution

Rework Example 5.4–1 without making any specific assumption for the velocity profile. That is, instead of introducing Eq. 5.4–9, simply let $v_z/v_{max} = \psi(\xi)$ where $\xi = r/R$.

Show that one obtains the following expressions for the constants C_0, C_1, and C_2 (cf. Eqs. 5.4–27 to 29):

$$C_1 = 0 \tag{5C.5–1}$$

$$C_0 = -\frac{1}{\int_0^1 \psi(\xi)\xi \, d\xi} \tag{5C.5–2}$$

$$C_2 = \frac{\int_0^1 \psi(\xi) \int_0^\xi \frac{1}{\xi'} \int_0^{\xi'} \psi(\xi'')\xi'' \, d\xi'' \, d\xi' \, \xi \, d\xi}{\left[\int_0^1 \psi(\xi)\xi \, d\xi\right]^2} \tag{5C.5–3}$$

Show that this leads finally to the following expression[5] for the local Nusselt number:

$$\frac{1}{\text{Nu}} = \frac{\int_0^1 \frac{1}{\xi}\left[\int_0^\xi \xi'\psi(\xi') \, d\xi'\right]^2 d\xi}{2\left[\int_0^1 \xi\psi(\xi) \, d\xi\right]^2} \tag{5C.5–4}$$

Compare your result with Eq. K in Table 5.4–1.

5C.6 Residence-Time Distribution for Polymeric Fluids

Rework Problem 1C.5 for a polymeric fluid that can be described by a power-law viscosity function. Compare the results with those in Eq. 1C.5–2 and interpret the results.

5C.7 The Rayleigh Problem for a Power-Law Fluid[6]

It is desired to solve the problem given in Problem 1B.16 for a power-law fluid. Use the same dimensionless velocity as before, but define a combined independent variable appropriate to the power-law fluid:

$$\phi_n = \frac{v_x}{v_0} \tag{5C.7–1}$$

$$r = (n+1)^{-1} y(\rho/mtv_0^{n-1})^{1/(n+1)} \tag{5C.7–2}$$

and show that the differential equation for the dimensionless velocity distribution is:

$$\phi_n''(-\phi_n')^{n-1} + (n+1)^n n^{-1} r\phi_n' = 0 \tag{5C.7–3}$$

in which primes denote differentiation with respect to r. Obtain a formal solution to the problem for a pseudoplastic fluid ($n < 1$). Some sample velocity profiles are given in Fig. 5C.7.

It should be borne in mind that in unsteady-state problems of this type, viscoelastic effects will probably be important, and these are not described by the simple power-law model. See §8.1 for the solution to the Rayleigh problem for a viscoelastic fluid.

Answer: $$\phi_n = \left[\frac{(1+n)^n(1-n)}{2n}\right]^{-1/(1-n)} \int_r^\infty (B_n + r^2)^{-1/(1-n)} \, dr$$

where B_n is a constant determined by $\phi_n = 1$ at $r = 0$.

[5] R. N. Lyon, *Chem. Eng. Progr.*, *47*, 75–79 (1951).
[6] R. B. Bird, *A.I.Ch.E. Journal*, *5*, 565–566 (1959).

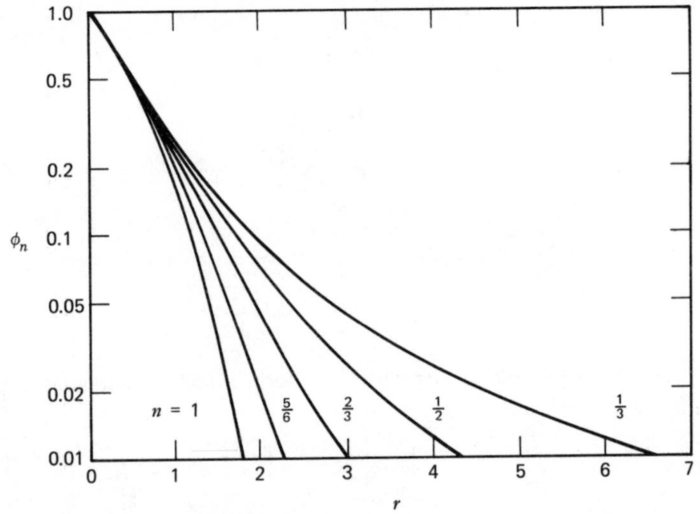

FIGURE 5C.7. Dimensionless velocity profiles for flow of a power-law fluid near a flat surface suddenly set in motion with a constant velocity. [R. B. Bird, *AIChE Journal*, *5*, 565–566 (1959).]

5C.8 Flow of Polymers through Porous Media[7]

The description of the flow of a Newtonian fluid through an isotropic, homogeneous porous medium is based on the classical experiment of Darcy.[8] Darcy's law for one-dimensional flow is

$$v_0 = \frac{k}{\mu}\frac{\Delta p}{L}. \tag{5C.8-1}$$

Here v_0 is the superficial velocity, μ the viscosity, $\Delta p/L$ the pressure gradient, and k the permeability of the medium. The quantity k clearly depends on the structure of the porous medium; despite innumerable investigations, there is no universally accepted expression for k.

A commonly used correlation is the Blake-Kozeny-Carman equation[9]

$$k = \frac{D_p^{\,2}}{180}\frac{\epsilon^3}{(1-\epsilon)^2}. \tag{5C.8-2}$$

Here ϵ is the void fraction of the medium and D_p is the particle diameter. This equation is derived for the "capillary model" of a porous medium. In this model, the medium is represented as a solid permeated by an assemblage of tortuous but continuous channels which have non-uniform cross sections. The model can easily be extended to non-Newtonian fluids; this was first done for the power-law model.[9] Here we develop the analogous result for the Ellis model of Eq. 5.1–14.[7]

a. Equation 5.1–14 can be inserted into the equation of motion for axial flow in a circular tube, and the velocity profile can be obtained. Show that this leads to the following expression for the average velocity:

$$\langle v \rangle = \frac{R\tau_R}{4\eta_0}\left[1 + \frac{4}{\alpha+3}\left(\frac{\tau_R}{\tau_{1/2}}\right)^{\alpha-1}\right] \tag{5C.8-3}$$

[7] T. J. Sadowski and R. B. Bird, *Trans. Soc. Rheol.*, *9:2*, 243–250 (1965); H. C. Park, M. C. Hawley, and R. F. Blanks, *Polymer Engineering and Science*, *15*, 761–773 (1975).

[8] H. P. G. Darcy, *Les fontaines publiques de la ville de Dijon*, Victor Dalmont, Paris (1856); a discussion of this paper is given by M. K. Hubbert, *Trans. A.I.M.E.*, *207*, 222–239 (1956).

[9] R. B. Bird, W. E. Stewart, and E. N. Lightfoot, *Transport Phenomena*, Wiley, New York (1960), §6.4 and Problem 6.Q.

where $\tau_R = R\Delta p/2L$ is the wall shear stress, and R is the tube radius. We now regard the porous medium as a conduit with a complicated cross section with mean hydraulic radius R_h, the latter being the ratio of the cross section to the wetted perimeter; hence, for the flow through the porous medium:

$$\langle v \rangle = \frac{R_h \tau_{Rh}}{2\eta_0}\left[1 + \frac{4}{\alpha + 3}\left(\frac{\tau_{Rh}}{\tau_{1/2}}\right)^{\alpha - 1}\right] \tag{5C.8–4}$$

in which $\tau_{Rh} = R_h\,\Delta p/L$.

b. The mean hydraulic radius can be related to the particle diameter and void fraction by[9] $R_h = D_p\epsilon/6(1 - \epsilon)$; also the average velocity in a pore can be related to the superficial velocity by $v_0 = \langle v \rangle \epsilon$. Show that this leads to

$$v_0 = \frac{R_h \tau_{Rh} \epsilon}{2\eta_0}\left[1 + \frac{4}{\alpha + 3}\left(\frac{\tau_{Rh}}{\tau_{1/2}}\right)^{\alpha - 1}\right] \tag{5C.8–5}$$

with $\tau_{Rh} = (\Delta p/L)\,[D_p\epsilon/6(1 - \epsilon)]$. Because of the error introduced in the mean hydraulic radius concept and the error in L due to tortuosity, the "2" in Eq. 5C.8–5 is replaced by "5" in order to get agreement with experimental Newtonian flow data. This empiricism applies only to the characterization of the porous medium and is independent of the nature of the fluid. There is no reason to expect *a priori* that the same empiricism, if indeed any at all, is applicable to the $4/(\alpha + 3)$ term in Eq. 5C.8–5, which involves a description of the "effective viscosity" of the fluid.

c. Show that when the above replacement is made, Eq. 5C.8–5 can be presented in two useful forms:

i. *Modified Darcy's Law.*

$$v_0 = \frac{k}{\eta_{\text{eff}}}\frac{\Delta p}{L} \tag{5C.8–6}$$

where

$$\frac{1}{\eta_{\text{eff}}} = \frac{1}{\eta_0}\left[1 + \frac{4}{\alpha + 3}\left(\frac{\tau_{Rh}}{\tau_{1/2}}\right)^{\alpha - 1}\right]. \tag{5C.8–7}$$

In Eq. 5C.8–6 the permeability k is the same as that in Eq. 5C.8–1 for Newtonian fluids. This means that one can characterize the medium by using Newtonian fluids and use Darcy's law with an effective viscosity given in Eq. 5C.8–7.

ii. *Modified Friction Factor Expression.*

$$\left(\frac{\Delta p}{\rho v_0^2}\right)\left(\frac{D_p}{L}\right)\left(\frac{\epsilon^3}{1 - \epsilon}\right) = \frac{180}{\left(\dfrac{D_p v_0 \rho}{\eta_0}\right)\left(\dfrac{1}{1 - \epsilon}\right)\left[1 + \dfrac{1}{\alpha + 3}\left(\dfrac{\tau_{Rh}}{\tau_{1/2}}\right)^{\alpha - 1}\right]}. \tag{5C.8–8}$$

This expression is of the form $Y = 180/X$, where Y is a modified friction factor and X is a modified Reynolds number. Sadowski[10] found that most of his data were described by Eq. 5C.8–8; for those runs in which discrepancies were observed, it is believed that elastic effects caused the lack of agreement.

Note: The above analysis is based on the solution of a viscometric flow problem. In such a flow no time-dependent elastic effects are described or expected. In a porous medium, however, elastic effects may occur. As the fluid moves through a tortuous channel in the porous medium, it encounters a capriciously changing cross section. If the fluid relaxation time is small with respect to the time required to go through a contraction or expansion, the fluid will accommodate quickly and no elastic effects would be observed. If, on the other hand, the fluid relaxation time is large with respect to the time to go through a contraction or expansion, then the fluid will not accommodate and elastic effects will be noted.

This suggests that a criterion for the importance of elastic effects will be given by a Deborah number defined by

$$\text{De} = \frac{\eta_0/\tau_{1/2}}{D_p/v_0} = \frac{\text{characteristic time for the fluid}}{\text{characteristic time for the flow system}} \tag{5C.8–9}$$

[10] T. J. Sadowski, *Trans. Soc. Rheol.*, *9*:2, 251–271 (1965); Ph.D. Thesis, University of Wisconsin (1963).

The time D_p/v_0 should give a measure of the time needed to pass a contraction or expansion. On the basis of Sadowski's experiments[10] it appeared that deviations from Eq. 5C.8–8 because of elastic effects seemed to set in at about De = 0.10.

Sadowski[10] found that all of his data could be correlated by modifying Eq. 5C.8–8 by inserting in the square bracket an additional term $-5(\eta_0 v_0/D_p \tau_{1/2})$. This added term accounted empirically for the deviations from Eq. 5C.8–8 due to elastic effects.

For further discussion and conflicting viewpoints see the publications by Middleman and coworkers[11] and by Marshall and Metzner.[12] For a treatment of viscoelastic effects in terms of the motion of a codeformational Maxwell model in a converging-diverging section, see the paper of Wissler.[13]

5D.1 Flow of a Polymeric Fluid through a Square Tube[1]

Obtain an approximate expression for the volume rate of flow of a power-law fluid through a horizonal tube of square cross section. Let the square be described by the lines $x = \pm B$, $y = \pm B$, and let the pressure drop over the tube of length L be $p_0 - p_L$. Use a variational approach and take the trial function to be:

$$v_z = A\left[1 - \left(\frac{x}{B}\right)^2\right]\left[1 - \left(\frac{y}{B}\right)^2\right] \tag{5D.1–1}$$

where A is the variational parameter.

5D.2 Rolling-Ball Viscometer[2,3]

Extend Problem 1D.7 to power-law fluids. Show that

$$\tfrac{1}{3}(2R)^{n+1}(\rho_s - \rho)g \sin\beta = m\left[\pi v_0\left(\frac{2n+1}{n}\right)\right]^n \left(\frac{R}{R-r}\right)^{2n+(1/2)} J_n \tag{5D.2–1}$$

which was derived by Bird and Turian.[2] The quantities J_n are:

$$J_n = 2\int_0^\infty \frac{d\xi}{[I_n(\xi^2)]^n} \tag{5D.2–2}$$

[11] R. H. Christopher and S. Middleman, *Ind. Eng. Chem. Fundamentals*, *4*, 422–426 (1965); N. Y. Gaitonde and S. Middleman, *ibid.*, *6*, 145–147 (1967).

[12] R. J. Marshall and A. B. Metzner, *ibid.*, *6*, 393–400 (1967).

[13] E. H. Wissler, *ibid.*, *10*, 411–417 (1971) see also R. E. Sheffield and A. B. Metzner, *A.I.Ch.E. Journal*, *22*, 736–744 (1976).

[1] For a thorough calculation of this type see R. S. Schechter, *A.I.Ch.E. Journal*, *7*, 445–448 (1961). For other noncircular conduit calculations see N. Mitsuishi and Y. Aoyagi, *J. Chem. Eng. Japan*, *6*, 402–408 (1973)—eccentric annulus; *Chem. Eng. Sci.*, *24*, 309–319 (1969)—rectangular and isosceles triangular ducts; N. Mitsuishi, Y. Kitayama, and Y. Aoyagi, *Kagaku Kōgaku*, *31*, 570–577 (1967)—rectangular and isosceles triangular ducts; N. Mitsuishi, Y. Aoyagi, and H. Soeda, *Kagaku Kōgaku*, *36*, 182–192 (1972)—concentric annulus. In connection with flow through noncircular tubes it must be kept in mind that $\Psi_2 \neq 0$ is a necessary (but not sufficient) condition for secondary flows to exist [J. G. Oldroyd, *Proc. Roy. Soc.*, *A283*, 115–133 (1965)]; these secondary flows are so weak, however, that they do not appreciably influence the Q vs. $\Delta\mathscr{P}$ relation. For application of variational methods to flow around a sphere see A. J. Ziegenhagen, R. B. Bird, and M. W. Johnson, Jr., *Trans. Soc. Rheol.*, *5*, 47–49 (1961); A. J. Ziegenhagen, *Appl. Sci. Res.*, *A14*, 43–56 (1961); S. W. Hopke and J. C. Slattery, *A.I.Ch.E. Journal*, *16*, 224–229 (1970); *ibid.*, *16*, 317–318 (1970).

[2] R. B. Bird and R. M. Turian, *Ind. Eng. Chem. Fundamentals*, *3*, 87 (1964).

[3] J. Šesták and F. Ambros, *Rheol. Acta*, *12*, 70–76 (1973).

where

$$I_n(\alpha) = \int_{-\pi}^{+\pi} (\cos^2 \tfrac{1}{2}\theta + \alpha)^{2+(1/n)}\, d\theta \tag{5D.2–3}$$

An extensive table of J_n has been prepared by Šesták and Ambros.[3] A few sample values are:

n	J_n
0.20	1.8698
0.30	1.5317
0.40	1.2778
0.50	1.0812
0.60	0.9249
0.70	0.7979
0.80	0.6930
0.90	0.6052
1.00	0.5308

Šesták and Ambros made measurements with a rolling-ball viscometer for aqueous solutions of carboxymethylcellulose and polyacrylamide and found Eq. 5D.2–1 to be satisfactory.

CHAPTER 6

THE GENERAL LINEAR VISCOELASTIC FLUID

In Chapter 5 we discussed an expression for the stress tensor that is particularly useful for engineers who must solve problems involving large-deformation flows, both without and with heat transfer. In such problems, as we have seen, the predominant feature of the rheological behavior of the macromolecular fluids is their shear-rate-dependent viscosity.

Although the generalized Newtonian fluid has proven to be of great value in solving problems of engineering interest, its use is strictly speaking limited to steady-state shearing flows. It is generally inappropriate for the description of unsteady flow phenomena, where the elastic response of the polymeric fluid becomes important. In this chapter we want to introduce a rheological equation of state that can describe some of the time-dependent motions of macromolecular fluids, albeit only the very restricted class of flows with very small strains and strain rates—flows in which the fluid never moves very far or very rapidly from its initial configuration.

Why do we spend a whole chapter on such a restricted class of flows? There are several reasons why we need to study this subject, known as "linear viscoelasticity": (1) polymer chemists have evolved several experiments that have enabled them to interrelate structure with the linear mechanical responses; (2) many of the molecular theories of polymeric fluids have been worked out only for the linear viscoelastic effects; and (3) some background in linear viscoelasticity is helpful to proceed to the subject of "nonlinear viscoelasticity", which is treated in Chapters 7, 8, and 9 of this volume. It is this last reason that we shall consider the principal motivation here. For the reader interested in the experiments of linear viscoelasticity, their analysis and molecular interpretations, we recommend the outstanding treatise of Ferry,[1] where a wealth of information and an extensive bibliography are to be found.

In §6.1 we discuss several linear viscoelastic fluid models beginning with the simple Maxwell model containing two constants and ending with the general linear viscoelastic fluid, which contains the "relaxation modulus," $G(t - t')$. Then in §6.2 we work through seven linear viscoelastic problems, which introduce the reader to oscillatory testing, stress relaxation, stress growth, creep, recoil, and wave transmission. These problems are solved using only differential and integral calculus. They can also be solved using Laplace transform; in §6.3 we give two illustrations of this method. Finally in §6.4 we point out that linear viscoelasticity is included in the first term of a much more general "memory integral expansion" for the stress tensor; by examining this expansion we can better understand the limitations which have to be placed on the linear viscoelastic models.

[1] J. D. Ferry, *Viscoelastic Properties of Polymers*, Wiley, New York (1970), Second Edition.

§6.1 THE GENERAL LINEAR VISCOELASTIC FLUID: ITS ORIGIN, USEFULNESS, AND LIMITATIONS

When we stretch a rubber band and then release it, it springs back to its original length. We say that the rubber is "elastic." Inasmuch as rubber is a network polymer with permanent chemical cross-links (see §2.1), it can "remember" perfectly its initial unstretched configuration.[1]

Polymeric fluids also have "elasticity" and "memory" as we saw in the recoil experiment in Chapter 3. We learned there that when the pressure force producing tube flow is removed, the fluid begins retreating in the direction from which it came. The fluid does not return all the way to its original position, since it seems gradually to "forget" where it has come from. For concentrated polymer solutions and for polymer melts this "fading memory" is not surprising, since the structure of such fluids can be thought of as a network formed by temporary physical junctions ("entanglements"), which are continually being destroyed and created (see §2.4). Therefore, if most of the physical junctions have a lifetime of t_0, one would not expect the memory of the fluid to extend much beyond t_0. For very dilute solutions, the memory must be associated with the time needed for the molecules to go from a state of partial alignment ("flow orientation") to a state of random orientation. These ideas will be developed further in Chapters 10 to 15, where the molecular and structural theories are set forth.

We want here to try to capture some of these notions about "elasticity of liquids" and "fading memory" in equations, so that we can obtain a rheological equation of state for macromolecular fluids. No molecular or statistical ideas are used here, only elementary ideas about viscosity and elasticity. The earliest thoughts along these lines are associated with the name of Maxwell, who speculated over a century ago that gases might be viscoelastic. If one writes down the stress in a material sandwiched between two planes of constant y (which are free to move in the x-direction), then the Newtonian liquid with viscosity μ and the Hookean solid[2] with modulus G are described by:

Newton:
$$\tau_{yx} = -\mu \frac{dv_x}{dy} \equiv -\mu\dot{\gamma}_{yx}(t) \tag{6.1-1}$$

Hooke:
$$\tau_{yx} = -G \frac{du_x}{dy} \equiv -G\gamma_{yx}(t_0,t) \tag{6.1-2}$$

Here v_x is the fluid velocity in the x-direction, and u_x is the solid displacement in the x-direction from an equilibrium position at t_0. The quantities γ_{yx} and $\dot{\gamma}_{yx}$ are components of the (infinitesimal) strain tensor and the rate-of-strain (or rate-of-deformation) tensor, respectively; furthermore $\dot{\gamma}_{yx} = \partial\gamma_{yx}/\partial t$. It should be emphasized that Eq. 6.1–2 is valid only for small deformations.

Maxwell[3] recognized that the equation

$$\tau_{yx} + \frac{\mu}{G}\frac{\partial \tau_{yx}}{\partial t} = -\mu\dot{\gamma}_{yx} \tag{6.1-3}$$

[1] For a very readable account of rubber elasticity, the reader should consult L. R. G. Treloar, *The Physics of Rubber Elasticity*, Oxford University Press (1958); the continuum mechanics applicable to large elastic deformations of rubberlike solids is discussed by R. S. Rivlin, in *Rheology*, F. R. Eirich (Ed.), Academic Press, New York (1956), Vol. 1, pp. 351–385.

[2] For an introduction to the classical theory of elasticity, see L. D. Landau and E. M. Lifschitz, *Theory of Elasticity*, Addison-Wesley, Reading (1959).

[3] J. C. Maxwell, *Phil. Trans. Roy. Soc.*, *A157*, 49–88 (1867).

contains the ideas embodied in both Eqs. 6.1–1 and 6.1–2. For steady-state movements Eq. 6.1–3 simplifies to the Newtonian fluid; for rapidly changing stresses, the time-derivative term dominates the left side, and integration with respect to time gives the Hookean solid. This, then, is the simplest expression for the shear stress for a liquid that exhibits both viscosity and elasticity.

We now rewrite Eq. 6.1–3 for an arbitrary tensor component τ_{ij}, and we introduce new notation for the constants, since this notation will fit in better with equations we will develop subsequently: we replace μ by η_0 (a viscosity), and μ/G by λ_0 (a time constant). We may then rewrite the *Maxwell model* in any one of three equivalent forms:

$$\tau_{ij} + \lambda_0 \frac{\partial \tau_{ij}}{\partial t} = -\eta_0 \dot{\gamma}_{ij} \tag{6.1–4}$$

$$\tau_{ij}(t) = -\int_{-\infty}^{t} \left\{ \frac{\eta_0}{\lambda_0} e^{-(t-t')/\lambda_0} \right\} \dot{\gamma}_{ij}(t')\, dt' \tag{6.1–5}$$

$$\tau_{ij}(t) = +\int_{-\infty}^{t} \left\{ \frac{\eta_0}{\lambda_0^{\,2}} e^{-(t-t')/\lambda_0} \right\} \gamma_{ij}(t')\, dt' \tag{6.1–6}$$

where $\gamma_{ij}(t')$ has been introduced as a shorthand for $\gamma_{ij}(t,t')$. To get Eq. 6.1–5 one integrates Eq. 6.1–4 regarding it as a first-order linear differential equation; in so doing one has to introduce the extra condition that τ_{ij} is finite at $t = -\infty$. The function inside the braces in Eq. 6.1–5 is called the *relaxation modulus.* When written in the form of Eq. 6.1–5 the Maxwell model says that the stress at the present time t depends on the rate of strain at time t and also the rate of strain at all past times t', with a weighting factor (the relaxation modulus) that decays exponentially as one goes backwards in time. Thus we see that this form of the Maxwell equation displays explicitly the notion of a fading memory. The fluid remembers very well what it has experienced in the very recent past, but has only a hazy recollection of events in the distant past. It is this form of the stress tensor, in terms of a time integral over the rate-of-strain, which we shall most often use in this book for the representation of viscoelastic phenomena. The form of Maxwell's equation shown in Eq. 6.1–6 is obtained by integrating Eq. 6.1–5 by parts, and using $\gamma_{ij}(t,t') = \int_t^{t'} \dot{\gamma}_{ij}(t'')\, dt''$; note that the reference state for γ_{ij} is taken to be the state at time $t' = t$. In other words, the strain is measured relative to the configuration of the sample at the present time t. Equation 6.1–6 also displays the notion of a fading memory, and the quantity contained within the braces in that equation is referred to as the *memory function.*

In the older literature on viscoelasticity much use was made of mechanical analogies with Eqs. 6.1–1 to 3. The stress-strain relation for the Hookean solid in Eq. 6.1–2 has the same form as the force-displacement relation for a linear spring with force constant G ($F = -GD$). The stress vs. rate-of-strain relation for the Newtonian fluid in Eq. 6.1–1 has the same form as the force vs. rate-of-displacement relation for a dashpot ($F = -\mu\, dD/dt$), the dashpot consisting of a piston that moves in a cylinder containing a Newtonian fluid. The Maxwell model has a stress relation of the same form as that for the force in a spring and dashpot in series ($F + (\mu/G)\, dF/dt = -\mu\, dD/dt$). These mechanical analogs are shown in Fig. 6.1–1. Although many people today sneer at this "spring-and-dashpottery" as a quaint vestige of days gone by, these mechanical models still provide a convenient aid for thinking about the responses of viscoelastic systems. Polymeric molecules have many internal degrees of freedom, and one would not expect that it would be possible to describe the rheological response of a polymeric fluid in terms of anything as simple as a single spring and a single dashpot in series. Therefore it was natural for the early investigators to attempt to simulate polymer behavior by using more and more complicated networks

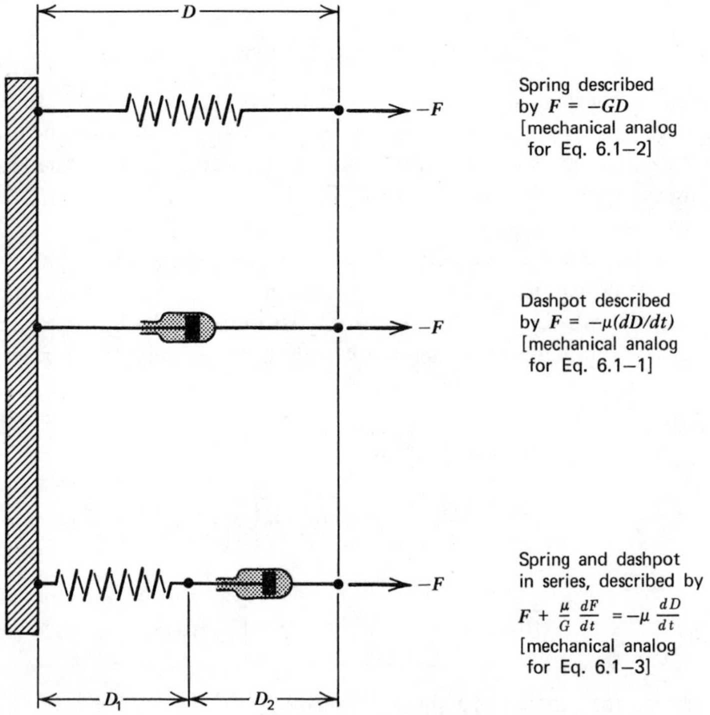

FIGURE 6.1−1. Spring-and-dashpot analogs for rheological equations. Note that for the series arrangement $F = -GD_1$ and $F = -\mu\, dD_2/dt$, since the same force is transmitted through both elements. When the spring relation is differentiated with respect to time, multiplied by μ/G, and added to the dashpot relation we get:

$$F + \frac{\mu}{G}\frac{dF}{dt} = -\mu\left(\frac{dD_1}{dt} + \frac{dD_2}{dt}\right) = -\mu\frac{dD}{dt}$$

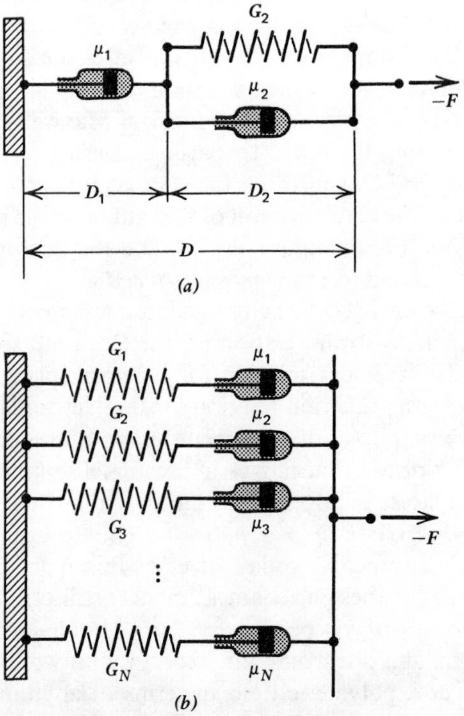

FIGURE 6.1−2. More complicated mechanical models. (*a*) Model that is equivalent to the Jeffreys equation in Eqs. 6.1−7 to 9. (*b*) Model containing N Maxwell elements in parallel, which corresponds to the generalized Maxwell model in Eqs. 6.1−10 to 13.

of springs and dashpots. They found, however, that the stress tensor always comes out in the form of Eq. 6.1–5 or 6, but with a relaxation modulus or memory function that is more and more complicated. For example, the three-element mechanical analog shown in Fig. 6.1–2*a* corresponds to the *Jeffreys model*,[4] with two time constants λ_1 and λ_2, which can be written in any of the three forms:

$$\tau_{ij} + \lambda_1 \frac{\partial \tau_{ij}}{\partial t} = -\eta_0 \left(\dot{\gamma}_{ij} + \lambda_2 \frac{\partial \dot{\gamma}_{ij}}{\partial t} \right) \tag{6.1–7}$$

$$\tau_{ij} = -\int_{-\infty}^{t} \left\{ \frac{\eta_0}{\lambda_1} \left[\left(1 - \frac{\lambda_2}{\lambda_1}\right) e^{-(t-t')/\lambda_1} + 2\lambda_2 \delta(t - t') \right] \right\} \dot{\gamma}_{ij}(t')\, dt' \tag{6.1–8}$$

$$\tau_{ij} = +\int_{-\infty}^{t} \left\{ \frac{\eta_0}{\lambda_1^2} \left[\left(1 - \frac{\lambda_2}{\lambda_1}\right) e^{-(t-t')/\lambda_1} + 2\lambda_2\lambda_1 \frac{\partial}{\partial t'} \delta(t - t') \right] \right\} \gamma_{ij}(t')\, dt' \tag{6.1–9}$$

in which $\delta(t - t')$ is the Dirac delta function.[5] The Jeffreys model has figured prominently in some of the early developments in structural theories, and in addition it has been the starting point for the development of some important nonlinear viscoelastic rheological equations of state. We shall therefore use it in some problems at the end of this chapter by way of illustration.

Another model that we shall have many occasions to refer to is the *generalized Maxwell model*, for which the mechanical analog is shown in Fig. 6.1–2*b*. This model, which has an infinite number of viscosity constants η_k and an infinite number of time constants λ_k, can be thought of as a superposition of an infinite number of Maxwell models. It may be written in any of four alternative ways, two differential forms and two integral forms:

$$\tau_{ij} = \sum_{k=1}^{\infty} \tau_{ij}^{(k)}; \qquad \tau_{ij}^{(k)} + \lambda_k \frac{\partial \tau_{ij}^{(k)}}{\partial t} = -\eta_k \dot{\gamma}_{ij} \tag{6.1–10}$$

$$\left(1 + \sum_{n=1}^{\infty} a_n \frac{\partial^n}{\partial t^n}\right) \tau_{ij} = -\eta_0 \left(1 + \sum_{n=1}^{\infty} b_n \frac{\partial^n}{\partial t^n}\right) \dot{\gamma}_{ij} \tag{6.1–11}$$

$$\tau_{ij} = -\int_{-\infty}^{t} \left\{ \sum_{k=1}^{\infty} \frac{\eta_k}{\lambda_k} e^{-(t-t')/\lambda_k} \right\} \dot{\gamma}_{ij}(t')\, dt' \tag{6.1–12}$$

$$\tau_{ij} = +\int_{-\infty}^{t} \left\{ \sum_{k=1}^{\infty} \frac{\eta_k}{\lambda_k^2} e^{-(t-t')/\lambda_k} \right\} \gamma_{ij}(t')\, dt' \tag{6.1–13}$$

[4] H. Jeffreys, *The Earth*, Cambridge University Press (1929), p. 265.

[5] P. A. M. Dirac, *The Principles of Quantum Mechanics*, Oxford University Press (1947), Third Edition, pp. 58–61. Here we use the definition of M. J. Lighthill, *Fourier Analysis and Generalised Functions*, Cambridge University Press (1964), p. 17, that:

$$\delta(x) = \lim_{n \to \infty} \sqrt{\frac{n}{\pi}}\, e^{-nx^2} \tag{6.1–9a}$$

From this it follows that:

$$\int_{-a}^{a} f(x)\delta(x)\, dx = 2 \int_{0}^{a} f(x)\delta(x)\, dx = f(0) \tag{6.1–9b}$$

$$\int_{-a}^{a} f(x)\delta'(x)\, dx = -f'(0) \tag{6.1–9c}$$

in which $a > 0$ and the prime denotes differentiation with respect to x. Note particularly the occurrence of the factor of 2 in Eq. 6.1–9b when the integral is over the region from 0 to a. This explains the occurrence of the factors of 2 in the δ-function terms in Eqs. 6.1–8 and 9.

It is straightforward to go from Eq. 6.1–10 to Eqs. 6.1–12 and 13 by successive integrations, as was done for the Maxwell model. Recasting the model in the form of Eq. 6.1–11 is somewhat more tedious; the most straightforward procedure is to differentiate Eq. 6.1–12 with respect to t successively to obtain expressions for the higher time derivatives, and then match the coefficients with those in Eq. 6.1–11 (see Problem 6B.4). Note that for steady-state shearing flow $\dot{\gamma}_{yx}$ = constant, and all time derivatives of τ_{yx} are zero; by comparing Eqs. 6.1–10 and 11 for this situation, we find that $\eta_0 = \sum_k \eta_k$.

The generalized Maxwell model has been found to describe quite well the behavior of polymer solutions and polymer melts in linear viscoelastic experiments. Quite a lot is known about the constants η_k and λ_k, both from molecular theories and from comparing empirical rheological equations of state with experimental data. The following empirical equations[6] can be used to reduce the total number of parameters to three (η_0, λ, and α):

$$\eta_k = \frac{\eta_0 \lambda_k}{\sum_k \lambda_k}; \qquad \lambda_k = \frac{\lambda}{k^\alpha} \tag{6.1–14, 15}$$

The Rouse theory for dilute polymer solutions (see Chapter 12) gives a rheological equation of state of the form of Eq. 6.1–10 with η_k proportional to λ_k and with $\lambda_k \sim 1/k^2$ for large k (cf. Eq. 12.3–12). These equations have also been found to describe experimental data for concentrated solutions and melts, with α varying from about 2 to 4.

When we compare Eq. 6.1–5 (for the Maxwell model), Eq. 6.1–8 (for the Jeffreys model), and Eq. 6.1–12 (for the generalized Maxwell model) we find that they are all of the same form: an integral over all past times of a relaxation modulus multiplied by a rate-of-deformation tensor. The only physical ideas that have been included in these models of varying degrees of complexity are those of viscosity and elasticity. An equation that includes all of these models, and many more of course, is the *general linear viscoelastic model*, which may be written in either of two equivalent forms:

$$\boldsymbol{\tau} = -\int_{-\infty}^{t} G(t - t')\dot{\boldsymbol{\gamma}}(t')\, dt' \tag{6.1–16}$$

$$\boldsymbol{\tau} = +\int_{-\infty}^{t} M(t - t')\boldsymbol{\gamma}(t')\, dt' \tag{6.1–17}$$

in which $G(t - t')$ is the *relaxation modulus* and $M(t - t') = \partial G(t - t')/\partial t'$ is the *memory function.* This model will be the starting point for most of the development in this chapter and will also be referred to in connection with nonlinear rheological equations of state and molecular theories. Of course, for getting quantitative answers to problems we shall have to use some specific expression for $G(t - t')$, such as the relaxation modulus for the generalized Maxwell model—the expression between braces in Eq. 6.1–12—along with the empirical relations in Eqs. 6.1–14 and 15.

The general linear viscoelastic model in Eq. 6.1–16 may also be derived by more formal arguments.[7] One assumes that the stress at time t resulting from a step strain at time t' is linear in the strain and multiplied by a decaying function of the elapsed time $t - t'$. An actual flow history may then be regarded as made up of a number of small step strains. By *Boltzmann's superposition principle*[8] it is assumed that the stress contributions from the individual past small step strains at times t' may be added to give the stress at time t. This results in the integral in Eq. 6.1–16. In the linear limit considered in this chapter the Boltz-

[6] T. W. Spriggs, *Chem. Eng. Sci.*, *20*, 931–940 (1965).
[7] A. C. Pipkin, *Lectures on Viscoelasticity Theory*, Springer, New York (1972), pp. 7–12.
[8] L. Boltzmann, *Pogg. Ann. Phys.*, *7*, 624–654 (1876).

mann superposition principle seems reasonable inasmuch as coupling effects between two past deformations must be of second order in the applied deformations and hence be negligible (Problem 6B.7). Experimental evidence has supported the validity of this principle for linear viscoelastic testing, and Eq. 6.1–16 is generally regarded as the correct starting point for the description of the rheology of incompressible viscoelastic liquids in the linear region. It can be used for shear flows, elongational flows, or any other kind of flow. In shear flows, it will not be able to describe the oscillating normal stresses, inasmuch as the normal stresses are second-order effects. They will be dealt with in Chapters 7 to 9.

In Chapter 8 we will obtain Eq. 6.1–16 in another way, by simplifying a rather general result obtained from continuum mechanical arguments. We shall in fact find that Eq. 6.1–16 will appear as the approximate form of the first term in a "memory integral expansion," an expansion in which the first term is a single memory integral, the second term is a double integral, the third term is a triple integral, and so on. Not until we obtain that more general result can we really be precise about the limitations implied in Eq. 6.1–16. We shall find in Chapter 8 that Eq. 6.1–16 is valid as long as the total strain is small (within the memory span of the material) and the velocity gradients are small.

§6.2 SOLUTION OF TIME-DEPENDENT SHEARING PROBLEMS

In the first section it was shown that the elementary concepts of viscosity and elasticity can be combined in a number of ways to develop equations of increasing complexity, culminating with Eq. 6.1–16, which is the most general equation of linear viscoelasticity. In this section we illustrate the use of this equation by solving a series of unsteady-state shearing problems. Most of these problems will appear again in Chapters 7, 8, and 9 where nonlinear effects will be treated; the linear problems discussed here provide a necessary background for the understanding of the nonlinear problems.

Before discussing the transient response phenomena, we mention briefly how Eq. 6.1–16 simplifies for steady-state shear flow. When the fluid has been flowing between parallel plates for a long time with velocity gradient $\dot{\gamma}_{yx}$, then Eq. 6.1–16 becomes:

$$\begin{aligned}\tau_{yx} &= -\int_{-\infty}^{t} G(t - t')\dot{\gamma}_{yx}\, dt' \\ &= -\left[\int_{0}^{\infty} G(s)\, ds\right]\dot{\gamma}_{yx} \\ &\equiv -\eta_0\dot{\gamma}_{yx} \end{aligned} \tag{6.2–1}$$

Here the change of variable $s = t - t'$ has been made. We see that the viscosity is just equal to the integral over the relaxation modulus. The subscript "0" on η_0 indicates that this is the zero-shear-rate viscosity. In the theory of linear viscoelasticity we are able to obtain the viscosity only in this limit of vanishingly small velocity gradients.

EXAMPLE 6.2–1 Small-Amplitude Oscillatory Motion

A polymeric fluid is located in the space between two parallel plates, the upper one of which is made to oscillate with frequency ω in its own plane in the x-direction. As shown in Fig. 4.3–1*b* the velocity profile is assumed to be instantaneously linear,[1] which is a good assumption for highly viscous materials in very narrow slits (cf. Problem 1D.2). Therefore the velocity gradient is changing with time

[1] Inertial effects have been considered by K. Walters and R. A. Kemp, in *Polymer Systems*, R. E. Whetton and R. W. Whorlow (Eds.), Macmillan, London (1968), pp. 237–250, and *Rheol. Acta*, 7, 1–8 (1968).

in the following way:

$$\dot{\gamma}_{yx}(t) = \dot{\gamma}_{yx}{}^0 \cos \omega t \tag{6.2-2}$$

in which we take $\dot{\gamma}_{yx}{}^0$ to be real. Find the time-dependent shear stress τ_{yx} that is needed to maintain this oscillatory motion, and obtain expressions for the real and imaginary parts of the complex viscosity.

SOLUTION

Substitution of the velocity gradient of Eq. 6.2–2 into the rheological equation of state in Eq. 6.1–16 gives:

$$\begin{aligned} \tau_{yx} &= -\int_{-\infty}^{t} G(t - t')\dot{\gamma}_{yx}{}^0 \cos \omega t' \, dt' \\ &= -\dot{\gamma}_{yx}{}^0 \int_0^{\infty} G(s) \cos \omega(t - s) \, ds \\ &= -\left[\int_0^{\infty} G(s) \cos \omega s \, ds\right] \dot{\gamma}_{yx}{}^0 \cos \omega t \\ &\quad -\left[\int_0^{\infty} G(s) \sin \omega s \, ds\right] \dot{\gamma}_{yx}{}^0 \sin \omega t \end{aligned} \tag{6.2-3}$$

Comparison of this result with the definitions of $\eta'(\omega)$ and $\eta''(\omega)$ in Eq. 4.4–3 shows that:

$$\eta'(\omega) = \int_0^{\infty} G(s) \cos \omega s \, ds \tag{6.2-4}$$

$$\eta''(\omega) = \int_0^{\infty} G(s) \sin \omega s \, ds \tag{6.2-5}$$

or, alternatively, we may write the results in terms of the "complex viscosity":

$$\eta^* = \eta' - i\eta'' = \int_0^{\infty} G(s) e^{-i\omega s} \, ds \tag{6.2-6}$$

The relaxation modulus $G(s)$ can be eliminated between Eqs. 6.2–4 and 5 to give the Kramers-Kronig relations (see Problem 6D.2), which interrelate $\eta'(\omega)$ and $\eta''(\omega)$. Finally, we note for future reference that:

$$\lim_{\omega \to 0} \frac{\eta''(\omega)/\omega}{\eta'(\omega)} = \frac{\int_0^{\infty} G(s) s \, ds}{\int_0^{\infty} G(s) \, ds} \tag{6.2-7}$$

We shall find as we go through this section that several other limiting quantities are also equal to the same ratio of integrals.

Let us now see what the expressions for $\eta'(\omega)$ and $\eta''(\omega)$ look like for the particular choice of relaxation modulus given by the generalized Maxwell model (quantity in braces in Eq. 6.1–12):

$$\eta'(\omega) = \sum_{k=1}^{\infty} \frac{\eta_k}{1 + (\lambda_k \omega)^2} \tag{6.2-8}$$

$$\frac{\eta''(\omega)}{\omega} = \sum_{k=1}^{\infty} \frac{\eta_k \lambda_k}{1 + (\lambda_k \omega)^2} \tag{6.2-9}$$

If, in addition, we introduce the expressions for η_k and λ_k given in Eqs. 6.1–14 and 15, these results become:

$$\frac{\eta'}{\eta_0} = \frac{1}{\zeta(\alpha)} \sum_{k=1}^{\infty} \frac{k^{\alpha}}{k^{2\alpha} + (\lambda \omega)^2} \tag{6.2-10}$$

$$\frac{\eta''}{\eta_0} = \frac{\lambda \omega}{\zeta(\alpha)} \sum_{k=1}^{\infty} \frac{1}{k^{2\alpha} + (\lambda \omega)^2} \tag{6.2-11}$$

in which $\zeta(\alpha)$ is the Riemann zeta function.[2] These expressions are not particularly appropriate for manual computation of the functions $\eta'(\omega)$ and $\eta''(\omega)$. Instead we have available[3] an alternate pair of expressions useful for low frequencies ($\omega \ll \lambda^{-1}$):

$$\frac{\eta'}{\eta_0} = 1 - \left[\frac{(\lambda\omega)^2}{\zeta(\alpha)} \sum_{k=1}^{\infty} \frac{1}{k^{\alpha}(k^{2\alpha} + (\lambda\omega)^2)}\right] \tag{6.2–12}$$

$$\frac{\eta''}{\eta_0} = \lambda\omega\left[\frac{\zeta(2\alpha)}{\zeta(\alpha)} - \frac{(\lambda\omega)^2}{\zeta(\alpha)} \sum_{k=1}^{\infty} \frac{1}{k^{2\alpha}(k^{2\alpha} + (\lambda\omega)^2)}\right] \tag{6.2–13}$$

and another pair of asymptotic expressions that are excellent for high frequencies ($\omega \gg \lambda^{-1}$):

$$\frac{\eta'}{\eta_0} \cong \frac{1}{\zeta(\alpha)}\left[\frac{\pi(\lambda\omega)^{(1/\alpha)-1}}{2\alpha \sin((\alpha + 1)\pi/2\alpha)}\right] \tag{6.2–14}$$

$$\frac{\eta''}{\eta_0} \cong \frac{1}{\zeta(\alpha)}\left[\frac{\pi(\lambda\omega)^{(1/\alpha)-1}}{2\alpha \sin(\pi/2\alpha)} - \frac{(\lambda\omega)^{-1}}{2}\right] \tag{6.2–15}$$

These large-frequency expressions are obtained by using the Euler-Maclaurin expansion to convert the sums into integrals (see Problem 6B.5). Equations 6.2–10 and 11 (and their low- and high-frequency equivalents) are useful for curve-fitting data for the small-amplitude oscillatory experiment.

We conclude this illustrative example by reminding the reader that for a Newtonian fluid η' is just a constant (the viscosity), and that η'' is zero; that is, for the Newtonian fluid the shear stress is in phase with the velocity gradient.

EXAMPLE 6.2–2 Stress Relaxation after a Sudden Shearing Displacement

The purpose of this example is to show why the function $G(t - t')$ is called the "relaxation modulus". A polymeric liquid is at rest in the region between two parallel plates for time $t < t_0$. At time $t = t_0$ the upper plate is instantaneously moved slightly in the x-direction, as shown in Fig. 4.3–1*e*. Find the expression for the shear stress at time $t > t_0$.

SOLUTION

To solve this problem we first imagine that the displacement actually occurs during the finite time interval from $t_0 - \epsilon$ to t_0, and then later we let ϵ go to zero. That is, the yx-components of the infinitesimal strain tensor and of the rate-of-strain tensor are as shown in Fig. 4.4–17.

For the displacement occurring in the finite time interval ϵ, Eq. 6.1–16 becomes (for $t > t_0$):

$$\begin{aligned}\tau_{yx}(t) &= -\int_{-\infty}^{t_0-\epsilon} G(t - t')\dot{\gamma}_{yx}(t')\,dt' - \int_{t_0-\epsilon}^{t_0} G(t - t')\dot{\gamma}_{yx}(t')\,dt' - \int_{t_0}^{t} G(t - t')\dot{\gamma}_{yx}(t')\,dt' \\ &= -\frac{\gamma_0}{\epsilon}\int_{t_0-\epsilon}^{t_0} G(t - t')\,dt'\end{aligned} \tag{6.2–16}$$

[2] The Riemann zeta function is defined as

$$\zeta(\alpha) = \sum_{k=1}^{\infty} k^{-\alpha} \tag{6.2–11a}$$

a few sample values being $\zeta(2) = \pi^2/6$, $\zeta(4) = \pi^4/90$, $\zeta(6) = \pi^6/945$ (see *Handbook of Mathematical Functions*, M. Abramowitz and I. A. Stegun (Eds.), Nat. Bur. Stds. Appl. Math. Series No. 55, U.S. Govt. Ptg. Off., Washington, D.C. (1964), p. 807).

[3] T. W. Spriggs, *Chem. Eng. Sci.*, *20*, 931–940 (1965).

since the velocity gradient is zero except during the middle time interval. Now we take the limit as ϵ approaches zero using L'Hospital's rule:

$$\tau_{yx}(t) = \lim_{\epsilon \to 0} (-\gamma_0) \left[\frac{\dfrac{d}{d\epsilon} \displaystyle\int_{t_0 - \epsilon}^{t_0} G(t - t')\, dt'}{\dfrac{d}{d\epsilon}\, \epsilon} \right]$$

$$= -\gamma_0 G(t - t_0) \tag{6.2–17}$$

Thus the function $G(t - t_0)$ describes the way in which the shear stress relaxes after the sudden shearing displacement. If we insert the expression for the relaxation modulus for the generalized Maxwell fluid, we see that the stress dies out as a sum of exponentials. Keep in mind that for a Newtonian fluid there is no delayed stress relaxation—the stress drops instantly to zero as soon as the motion stops.

EXAMPLE 6.2–3 Stress Relaxation after Cessation of Steady Shear Flow

Next we turn our attention to the stress relaxation that occurs in a different type of experiment. Here we envisage a fluid flowing at steady state between two parallel plates, the upper one of which is moving with constant speed in its own plane, for time $t < 0$ (see Fig. 4.3 – 1d). Then at time $t = 0$ the motion of the plate and the fluid is stopped suddenly. It is desired to describe the way in which the stress decays with time after the cessation of the steady shear flow. It is also desired to find the area under the the $\eta^-(t)$ curve.

SOLUTION

For this experiment Eq. 6.1–16 gives for $t < 0$ and $t > 0$:

$$\tau_{yx}(t < 0) = -\eta_0 \dot{\gamma}_0 = -\dot{\gamma}_0 \int_{-\infty}^{t} G(t - t')\, dt' \tag{6.2–18}$$

$$\tau_{yx}(t > 0) = -\eta^- \dot{\gamma}_0 = -\dot{\gamma}_0 \int_{-\infty}^{0} G(t - t')\, dt' \tag{6.2–19}$$

where $\dot{\gamma}_0$ is the shear rate prior to $t = 0$. We now take the ratio of the above two expressions and then integrate it from $t = 0$ to $t = \infty$. This gives for the area under the stress-relaxation curve:

$$\int_0^\infty \frac{\eta^-(t)}{\eta_0}\, dt = \int_0^\infty \frac{\displaystyle\int_{-\infty}^{0} G(t - t')\, dt'}{\displaystyle\int_{-\infty}^{t} G(t - t')\, dt'}\, dt$$

$$= \frac{\displaystyle\int_0^\infty \int_t^\infty G(s)\, ds\, dt}{\displaystyle\int_0^\infty G(s)\, ds}$$

$$= \frac{\displaystyle\int_0^\infty \int_0^s G(s)\, dt\, ds}{\displaystyle\int_0^\infty G(s)\, ds}$$

$$= \frac{\displaystyle\int_0^\infty sG(s)\, ds}{\displaystyle\int_0^\infty G(s)\, ds} \tag{6.2–20}$$

In going from the second to the third line, we interchanged the order of integration so that one integration could be performed. See Fig. 6.2–2a for the graphical interpretation of this integral.

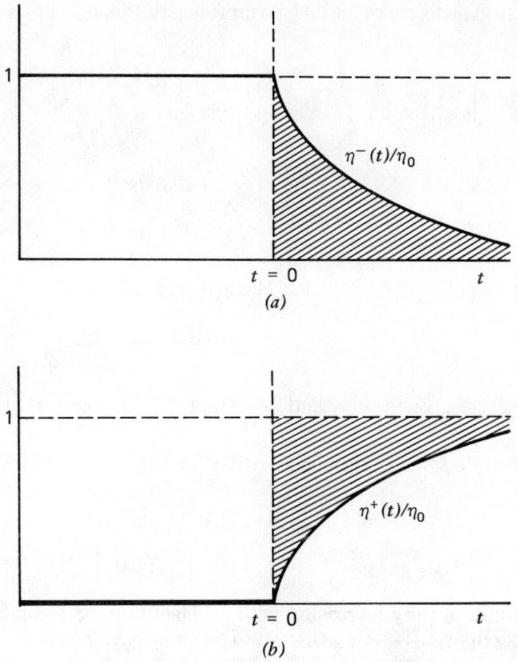

FIGURE 6.2–2. Integrals evaluated in Examples 6.2–3 and 4. (*a*) Integral in Eq. 6.2–20 for stress relaxation is given by the shaded area; (*b*) integral in Eq. 6.2–24 for stress growth is given by the shaded area.

Note further that if we use the relaxation modulus for the generalized Maxwell model, then we find:

$$\frac{\eta^-}{\eta_0} = \frac{\sum_k \eta_k e^{-t/\lambda_k}}{\sum \eta_k} \tag{6.2–21}$$

This result describes the top curve of the set of curves in Fig. 4.4–13, that is, the limiting curve for vanishingly small $\dot{\gamma}_0$. The linear theory cannot describe the dependence of the lower curves on $\dot{\gamma}_0$. To describe those curves we need a nonlinear viscoelastic theory. When we return to the discussion of this experiment in Chapters 7 and 8, we shall find we are able to describe the nonlinear effects.

EXAMPLE 6.2–4 Stress Growth at Inception of Steady Shear Flow

The next flow situation we examine is that shown in Fig. 4.3–1*c*, where a fluid at rest is suddenly made to undergo steady-state shear flow after $t = 0$. Let the velocity gradient for $t > 0$ be $\dot{\gamma}_0$. Find $\eta^+(t)$ and the area under the curve of $1 - (\eta^+/\eta_0)$.

SOLUTION

For this experiment Eq. 6.1–16 gives for $t < 0$ and $t > 0$:

$$\tau_{yx}(t < 0) = 0 \tag{6.2–22}$$

$$\begin{aligned}\tau_{yx}(t > 0) &= -\eta^+\dot{\gamma}_0 \\ &= -\dot{\gamma}_0 \int_0^t G(t - t')\, dt'\end{aligned} \tag{6.2–23}$$

The area between the stress growth curve and its asymptote (see Fig. 6.2–2*b*) is given by:

$$\int_0^\infty \left(1 - \frac{\eta^+(t)}{\eta_0}\right) dt = \int_0^\infty \frac{\int_{-\infty}^0 G(t-t')\,dt'}{\int_{-\infty}^t G(t-t')\,dt'}\,dt$$
$$= \frac{\int_0^\infty \int_t^\infty G(s)\,ds\,dt}{\int_0^\infty G(s)\,ds}$$
$$= \frac{\int_0^\infty sG(s)\,ds}{\int_0^\infty G(s)\,ds} \tag{6.2–24}$$

Thus the area between the stress-growth curve and its asymptote is the same as the area between the stress-relaxation curve and its asymptote.

When the relaxation modulus is specified as that for the generalized Maxwell model, then the stress-growth function is:

$$\frac{\eta^+}{\eta^0} = \frac{\sum_k \eta_k(1 - e^{-t/\lambda_k})}{\sum_k \eta_k} \tag{6.2–25}$$

Note that this curve is monotonically increasing with t. Therefore we cannot, by means of the linear viscoelastic theory, describe the "overshoot effect" shown in Fig. 4.4–9.

EXAMPLE 6.2–5 Constrained Recoil after Cessation of Steady Shear Flow

Next we investigate the system depicted in Fig. 4.3–1*g*. Prior to $t = 0$ the fluid between the two parallel plates is undergoing steady shear flow with velocity gradient $\dot{\gamma}_0$. After $t = 0$, the shear stress is removed so that $\tau_{yx} = 0$. The fluid then recoils and the strain $\gamma_{yx}(0,t)$ can be followed as a function of time. It is presumed that the plate spacing is small enough and the fluid viscous enough that a linear velocity profile is maintained throughout the experiment; that is, inertial effects can be neglected. We wish to find the "ultimate recoil," $\gamma_\infty = \gamma_{yx}(0,t)|_{t=\infty} = \int_0^\infty \dot{\gamma}_{yx}(t)\,dt$ (see Fig. 4.4–21).

SOLUTION

Application of Eq. 6.1–16 to this problem for $t > 0$ gives:

$$0 = -\int_{-\infty}^0 G(t-t')\dot{\gamma}_0\,dt' - \int_0^t G(t-t')\dot{\gamma}_{yx}(t')\,dt' \tag{6.2–26}$$

or, in terms of the variable $s = t - t'$:

$$0 = -\dot{\gamma}_0 \int_t^\infty G(s)\,ds - \int_0^t G(s)\dot{\gamma}_{yx}(t-s)\,ds \tag{6.2–27}$$

Next we integrate over the time from $t = 0$ to $t = \infty$ and this gives:

$$0 = -\dot{\gamma}_0 \int_0^\infty \int_t^\infty G(s)\,ds\,dt - \int_0^\infty \int_0^t G(s)\dot{\gamma}_{yx}(t-s)\,ds\,dt \tag{6.2–28}$$

Next interchange the order of integration to get:

$$0 = -\dot{\gamma}_0 \int_0^\infty \left[\int_0^s dt\right] G(s)\,ds - \int_0^\infty \left[\int_s^\infty \dot{\gamma}_{yx}(t-s)\,dt\right] G(s)\,ds \tag{6.2–29}$$

or

$$0 = -\dot{\gamma}_0 \int_0^\infty sG(s)\,ds - \int_0^\infty \left[\int_0^\infty \dot{\gamma}_{yx}(t')\,dt'\right] G(s)\,ds \tag{6.2–30}$$

The quantity in brackets in Eq. 6.2–30 is just the ultimate recoil γ_∞, for which we then have the final result:

$$\frac{-\gamma_\infty}{\dot{\gamma}_0} = \frac{\int_0^\infty sG(s)\,ds}{\int_0^\infty G(s)\,ds} \tag{6.2–31}$$

Notice that we did not actually solve the integral equation for $\dot{\gamma}_{yx}(t)$ in Eq. 6.2–26, but we were able to extract an integral of the solution, namely, the ultimate recoil. This problem can also be solved by using Laplace transforms, as indicated in Example 6.3–2.

EXAMPLE 6.2–6 Creep after Imposition of Constant Shear Stress

The next experiment we consider is that shown in Fig. 4.3–1f. Prior to $t = 0$ the fluid, contained between two parallel plates, is at rest. After $t = 0$ the fluid sample is subjected to a constant applied shear stress, so that the strain has a response similar to that shown in Fig. 4.4–19. We want to find an expression for the intercept γ_0. Designate the applied shear stress by τ_0 and the ultimate velocity gradient by $\dot{\gamma}_\infty$.

SOLUTION

During the creep process, the rate of strain (or velocity gradient) will be:

$$\dot{\gamma}_{yx} = \frac{d}{dt}\gamma_{yx} \tag{6.2–32}$$

and therefore at any time t the strain will be:

$$\gamma_{yx} = \int_0^t \dot{\gamma}_{yx}(t')\,dt' \tag{6.2–33}$$

For very large time, the strain curve will approach the dashed-line asymptote in Fig. 4.4–19, which is given by the equation:

$$\gamma_{yx}(t \to \infty) = \gamma_0 + \dot{\gamma}_\infty t \tag{6.2–34}$$

If we write Eq. 6.2–33 for $t \to \infty$ and equate the result to Eq. 6.2–34, then we get for the intercept γ_0:

$$\gamma_0 = \int_0^\infty (\dot{\gamma}_{yx}(t') - \dot{\gamma}_\infty)\,dt' \tag{6.2–35}$$

Now we want to get an expression for this quantity by using the rheological equation of state for the linear viscoelastic fluid.

If we write Eq. 6.1–16 for the creep experiment, we have:

$$\tau_0 = -\int_0^t G(t - t')\dot{\gamma}_{yx}(t')\,dt' \tag{6.2–36}$$

But we can also write Eq. 6.1–16 for the steady-state shear flow with $\tau_{yx} = \tau_0$ and $\dot{\gamma}_{yx} = \dot{\gamma}_\infty$:

$$\tau_0 = -\int_{-\infty}^t G(t - t')\dot{\gamma}_\infty\,dt' \tag{6.2–37}$$

We now equate these two expressions for τ_0 to obtain:

$$\dot{\gamma}_\infty \int_0^\infty G(s)\,ds = \int_0^t G(s)\dot{\gamma}_{yx}(t - s)\,ds \tag{6.2–38}$$

This may be rewritten as:

$$\dot{\gamma}_\infty \int_t^\infty G(s)\, ds = \int_0^t G(s)[\dot{\gamma}_{yx}(t-s) - \dot{\gamma}_\infty]\, ds \tag{6.2–39}$$

Next we integrate both sides from $t = 0$ to $t = \infty$ and then interchange the order of integration of s and t; this gives finally:

$$\frac{\gamma_0}{\dot{\gamma}_\infty} = \frac{\int_0^\infty sG(s)\, ds}{\int_0^\infty G(s)\, ds} \tag{6.2–40}$$

Here again we did not solve the integral equation for the velocity gradient, but extracted from the problem only the integral defined in Eq. 6.2–35.

A quantity J_e^0, the *steady-state compliance*, is defined by Eq. 4.4–26:

$$\gamma_0 = -J_e^0 \tau_0 \tag{6.2–41}$$

According to the linear viscoelasticity theory, this quantity is given by:

$$J_e^0 = \frac{\int_0^\infty sG(s)\, ds}{\left[\int_0^\infty G(s)\, ds\right]^2} \tag{6.2–42}$$

This is obtained by combining the results in Eqs. 6.2–37 and 40.

From the above examples we find that certain quantities measured in different shearing experiments are given by the same ratio of integrals: $\int_0^\infty sG(s)\, ds/\int_0^\infty G(s)\, ds$. For future reference we list all of these results in Table 6.2–1. We also include one additional

TABLE 6.2–1

Shear Flow Experiments and "Analogous Quantities"

Experiment	Meaning of Symbols	Measurable Quantity Equal to $\int_0^\infty sG(s)\, ds \Big/ \int_0^\infty G(s)\, ds$
Small-amplitude oscillatory motion	ω = frequency of oscillation	$\lim_{\omega \to 0} \frac{\eta''/\omega}{\eta'}$
Stress relaxation after cessation of steady shear flow	$\dot{\gamma}_0$ = shear rate before stress relaxation	$\lim_{\dot{\gamma}_0 \to 0} \int_0^\infty \frac{\eta^-}{\eta_0}\, dt$
Stress growth after inception of steady shear flow	$\dot{\gamma}_0$ = shear rate during stress growth	$\lim_{\dot{\gamma}_0 \to 0} \int_0^\infty \left(1 - \frac{\eta^+}{\eta_0}\right) dt$
Constrained recoil after cessation of steady shear flow	$\dot{\gamma}_0$ = shear rate before recoil begins	$\lim_{\dot{\gamma}_0 \to 0} \frac{-\gamma_\infty}{\dot{\gamma}_0}$
Creep after application of steady shear stress	$\dot{\gamma}_\infty$ = shear rate when steady state is attained	$\lim_{\tau_0 \to 0} \frac{\gamma_0}{\dot{\gamma}_\infty}$
Steady-state shear flow	$\dot{\gamma}$ = shear rate at steady state	$\lim_{\dot{\gamma} \to 0} \frac{\Psi_1}{2\eta}$

entry—that for the normal stress coefficient in steady-state shear flow—for which nonlinear viscoelasticity theory is required; this result will be obtained in Problem 8B.10, but we include it in the table for completeness.[4]

These "analogies" between various experiments have been the subject of considerable experimentation, and they are particularly useful in providing cross-checks of experimental techniques. The establishment of these interrelations requires no assumptions regarding the relaxation modulus. It must be kept in mind that all of the quantities listed in the right-hand column of Table 6.2–1 are limiting values and thus have to be obtained experimentally by tedious extrapolation processes.

All of the illustrative examples given thus far have assumed that the velocity profiles are linear. Therefore it is not necessary to use the equation of motion in any of the derivations. To conclude this section we give one illustrative example showing how the equation of motion is combined with the linear viscoelastic rheological equation of state to solve a problem.

EXAMPLE 6.2–7 Wave Transmission in a Semi-Infinite Viscoelastic Liquid

A viscoelastic liquid is located in the region $0 \leq y < \infty$. The velocity v_x at the surface $y = 0$ is maintained at $v_x = v_0 \cos \omega t$, where v_0 is the amplitude of the velocity and ω is the frequency. Find the velocity distribution $v_x(y,t)$ throughout the medium, after the initial transients have died out. For the analogous Newtonian-fluid problem see Problem 1D.2.

SOLUTION

We postulate that in the "oscillatory steady state" the velocity distribution will be of the form:

$$v_x = \mathscr{Re}\{v_x{}^0(y)e^{i\omega t}\} \tag{6.2–43}$$

in which $v_x{}^0$ is a complex amplitude with the value $v_x{}^0 = v_0$ at $y = 0$. Equation 6.2–43 describes a situation in which there is a velocity fluctuation at every point in the fluid with frequency ω. However, the phase and amplitude of the oscillation vary with the distance from the oscillating boundary. We further postulate that the shear stress will also be a periodic function:

$$\tau_{yx} = \mathscr{Re}\{\tau_{yx}{}^0(y)e^{i\omega t}\} \tag{6.2–44}$$

in which $\tau_{yx}{}^0(y)$ is a complex function.

The equation of motion for the x-direction is:

$$\rho \frac{\partial v_x}{\partial t} = -\frac{\partial}{\partial y} \tau_{yx} \tag{6.2–45}$$

When we make use of the postulated forms for the velocity and stress we get:

$$\rho i\omega v_x{}^0 = -\frac{d}{dy} \tau_{yx}{}^0 \tag{6.2–46}$$

Next we need an expression for $\tau_{yx}{}^0$, which we get from the rheological equation of state.

[4] The equality of the entries involving $\eta''/\omega\eta'$ and $\Psi_1/2\eta$ has been verified experimentally by T. Kotaka and K. Osaki, *J. Polym. Sci.*, *C15*, 453–479 (1966).

We put the postulated velocity distribution into Eq. 6.1–16 and get:

$$\begin{aligned}\tau_{yx} &= -\int_{-\infty}^{t} G(t-t')\frac{\partial v_x(t')}{\partial y}\,dt' \\ &= -\int_{-\infty}^{t} G(t-t')\,\mathcal{R}e\left\{\frac{dv_x{}^0}{dy}\,e^{i\omega t'}\right\}dt' \\ &= -\mathcal{R}e\left\{\frac{dv_x{}^0}{dy}\int_0^{\infty} G(s)e^{i\omega(t-s)}\,ds\right\} \\ &= -\mathcal{R}e\left\{\frac{dv_x{}^0}{dy}\,e^{i\omega t}\eta^*(\omega)\right\}\end{aligned} \tag{6.2–47}$$

where use has been made of Eq. 6.2–6. Now when Eqs. 6.2–44 and 47 are equated we get the desired expression for $\tau_{yx}{}^0$:

$$\tau_{yx}{}^0 = -\eta^*\frac{dv_x{}^0}{dy} \tag{6.2–48}$$

When this is inserted into Eq. 6.2–46 we get the following differential equation for $v_x{}^0$:

$$\frac{d^2v_x{}^0}{dy^2} - \frac{\rho i\omega v_x{}^0}{\eta^*} = 0 \tag{6.2–49}$$

which has to be solved with the boundary conditions that $v_x{}^0 = v_0$ at $y = 0$, and $v_x{}^0 = 0$ at $y = \infty$. We presume that η^* is known as a function of the frequency, either from experimental data or from some linear viscoelastic model, such as the Jeffreys model or the generalized Maxwell model.

For convenience later in physical interpretation we define $\alpha(\omega)$ and $\beta(\omega)$ by setting

$$\frac{\rho i\omega}{\eta^*} = (\alpha + i\beta)^2 \tag{6.2–50}$$

Then the solution to Eq. 6.2–49 and the concomitant boundary conditions is:

$$v_x{}^0(y) = v_0 e^{-(\alpha+i\beta)y} \tag{6.2–51}$$

Finally, when the result for $v_x{}^0$ in Eq. 6.2–51 is inserted into Eq. 6.2–43, we obtain:

$$v_x(y,t) = v_0 e^{-\alpha y}\cos(\omega t - \beta y) \tag{6.2–52}$$

This shows how the velocity oscillations are attenuated in the y-direction.

§6.3 SOLUTION OF LINEAR VISCOELASTIC PROBLEMS BY LAPLACE TRANSFORMS

In §6.2 we gave the solutions to a variety of simple problems in the linear viscoelasticity of liquids. These problems were solved by elementary methods. Some of them could probably have been solved more quickly or more easily by using Laplace transform techniques. Such methods are very useful in this field, because all of the equations are linear. Problems involving the coupling of the equations of motion of a viscoelastic liquid and the equation of motion of a solid body are obvious candidates for the use of Laplace transforms. Such problems arise in the analysis of certain kinds of viscometers, such as the oscillatory cup-

and-bob viscometer discussed in Problem 1C.8 and Example 4.5–4, and in the study of inertial effects in viscoelastic measurements.

In this section we present the solutions of several problems by means of Laplace transforms. We presume that the reader of this section is already acquainted with the mathematical techniques involved. If not, useful introductions to the subject can be found in various intermediate-level books on applied mathematics.[1]

EXAMPLE 6.3–1 Alternative Formulations of Linear Viscoelastic Models[2,3]

Consider the linear viscoelastic model:

$$\left(1 + a_1 \frac{\partial}{\partial t} + a_2 \frac{\partial^2}{\partial t^2} + \cdots + a_M \frac{\partial^M}{\partial t^M}\right)\tau_{ij} = -\eta_0 \left(1 + b_1 \frac{\partial}{\partial t} + b_2 \frac{\partial^2}{\partial t^2} + \cdots + b_N \frac{\partial^N}{\partial t^N}\right)\dot{\gamma}_{ij} \quad (6.3\text{–}1)$$

a. Obtain the expression for the relaxation modulus corresponding to this model.
b. Show how the expression in (a) simplifies for the special case of the Jeffreys model.

SOLUTION

(a) For the purpose of casting this differential model into the form of the general linear viscoelastic fluid, it is convenient to use a transformation $\mathscr{H}$ that is closely related to the Laplace transform:

$$\mathscr{H}\{f(t)\} \equiv \int_{t_0}^{\infty} e^{-pt} f(t)\, dt = e^{-pt_0}\mathscr{L}\{f(t + t_0)\} \quad (6.3\text{–}2)$$

Note that if $t_0 \geq 0$, then the above is just a statement of the shifting theorem for Laplace transforms. From the definition of $\mathscr{H}$ and the properties of the Laplace transform it is not difficult to see that:

$$\mathscr{H}\left\{\frac{df}{dt}\right\} = p\mathscr{H}\{f\} - f(t_0^+)e^{-pt_0} \quad (6.3\text{–}3)$$

$$\mathscr{H}\left\{\int_{t_0}^{t} g(t - t')f(t')\, dt'\right\} = e^{-pt_0}\mathscr{L}\left\{\int_{t_0}^{t+t_0} g(t + t_0 - t')f(t')\, dt'\right\}$$
$$= e^{-pt_0}\mathscr{L}\{g(t)\}\mathscr{L}\{f(t + t_0)\} = \mathscr{L}\{g(t)\}\mathscr{H}\{f(t)\} \quad (6.3\text{–}4)$$

In Eq. 6.3–3, $f(t_0^+)$ is the limiting value of $f(t)$ as t approaches t_0 from above. Equation 6.3–4 is an $\mathscr{H}$-version of the convolution theorem for Laplace transforms.

Now operate with $\mathscr{H}$ on both sides of Eq. 6.3–1. If we assume that the material is at rest and in a stressfree state for times $t \leq t_0$, so that $e^{-pt_0}\tau_{ij}(t_0^+)$, $e^{-pt_0}(\partial/\partial t)\tau_{ij}(t_0^+), \ldots, e^{-pt_0}(\partial^{M-1}/\partial t^{M-1})\tau_{ij}(t_0^+)$, $e^{-pt_0}\dot{\gamma}_{ij}(t_0^+)$, $e^{-pt_0}(\partial/\partial t)\dot{\gamma}_{ij}(t_0^+), \ldots$, and $e^{-pt_0}(\partial^{N-1}/\partial t^{N-1})\dot{\gamma}_{ij}(t_0^+)$ are all zero even for very large negative values of t_0, then we have:

$$\mathscr{H}\{\tau_{ij}(t)\} = -\eta_0 \left[\frac{1 + b_1 p + b_2 p^2 + \cdots + b_N p^N}{1 + a_1 p + a_2 p^2 + \cdots + a_M p^M}\right]\mathscr{H}\{\dot{\gamma}_{ij}(t)\} \quad (6.3\text{–}5)$$

[1] See, for example, C. R. Wylie, Jr., *Advanced Engineering Mathematics*, McGraw-Hill, New York (1966), Third Edition, Chapter 7; F. B. Hildebrand, *Advanced Calculus for Applications*, Prentice-Hall, Englewood Cliffs, N.J. (1962), Chapter 2.

[2] A. G. Fredrickson, *Principles and Applications of Rheology*, Prentice-Hall, Englewood Cliffs, N.J. (1964), Chapter 6, pp. 127–128.

[3] T. Alfrey, Jr., and E. F. Gurnee, in *Rheology*, F. R. Eirich (Ed.), Academic Press, New York (1956), Vol. 1, Chapter 11.

By comparing Eqs. 6.3–4 and 5, then, we see that:

$$\tau_{ij}(t) = -\int_{t_0}^{t} G(t - t')\dot{\gamma}_{ij}(t')\,dt' \tag{6.3–6}$$

where:

$$G(t) = \eta_0 \mathscr{L}^{-1}\left\{\frac{1 + b_1 p + b_2 p^2 + \cdots + b_N p^N}{1 + a_1 p + a_2 p^2 + \cdots + a_M p^M}\right\} \tag{6.3–7}$$

The general linear viscoelastic fluid model is obtained from Eq. 6.3–6 by letting t_0 go to $-\infty$.

(b) For the Jeffreys model we let $a_1 = \lambda_1$, $b_1 = \lambda_2$ and let all the other a's and b's be zero. Then:

$$\begin{aligned} G(t) &= \eta_0 \mathscr{L}^{-1}\left\{\frac{1 + \lambda_2 p}{1 + \lambda_1 p}\right\} \\ &= \eta_0 \mathscr{L}^{-1}\left\{\frac{\lambda_2}{\lambda_1} + \frac{1 - (\lambda_2/\lambda_1)}{1 + \lambda_1 p}\right\} \\ &= 2\eta_0 \frac{\lambda_2}{\lambda_1}\,\delta(t) + \frac{\eta_0}{\lambda_1}\left(1 - \frac{\lambda_2}{\lambda_1}\right) e^{-t/\lambda_1} \end{aligned} \tag{6.3–8}$$

in agreement with Eq. 6.1–8; note that the definition of the Dirac delta function in footnote 5 of §6.1 has been used.

EXAMPLE 6.3–2 Constrained Recoil after Cessation of Steady Shear Flow

Rework the recoil problem in Example 6.2–5 using Laplace transforms.[4] Use the notation $G_n = \int_0^\infty G(s)s^n\,ds$.

SOLUTION

We begin by rewriting Eq. 6.2–27 as:

$$-\dot{\gamma}_0 G_0 + \dot{\gamma}_0 \int_0^t G(s)\,ds = \int_0^t G(s)\dot{\gamma}_{yx}(t - s)\,ds \tag{6.3–9}$$

We then take the Laplace transform of the entire equation, using the convolution theorem, to get:

$$-\frac{\dot{\gamma}_0 G_0}{p} + \frac{\dot{\gamma}_0 \bar{G}(p)}{p} = \bar{G}(p)\bar{\dot{\gamma}}_{yx}(p) \tag{6.3–10}$$

where $\bar{G} = \mathscr{L}\{G\}$ and $\bar{\dot{\gamma}}_{yx} = \mathscr{L}\{\dot{\gamma}_{yx}\}$. It will be convenient to write $\bar{G}(p)$ as a Taylor expansion in the transform variable p:

$$\begin{aligned} \bar{G}(p) &= \int_0^\infty G(s)e^{-ps}\,ds \\ &= \int_0^\infty G(s)\left[1 - ps + \frac{1}{2!}(ps)^2 - \cdots\right] ds \\ &= G_0 - G_1 p + G_2 p^2 - \cdots \end{aligned} \tag{6.3–11}$$

[4] A. S. Lodge, *Elastic Liquids*, Academic Press, New York (1964), pp. 144–147.

When this is substituted into Eq. 6.3–10, we can solve for $\bar{\dot{\gamma}}_{yx}$:

$$\begin{aligned}\bar{\dot{\gamma}}_{yx} &= \frac{\dot{\gamma}_0}{p} - \frac{\dot{\gamma}_0 G_0}{p\bar{G}(p)} \\ &= \frac{\dot{\gamma}_0}{p} - \frac{\dot{\gamma}_0}{p}\left(1 + \frac{G_1}{G_0}p + \cdots\right) \\ &= -\dot{\gamma}_0 \frac{G_1}{G_0} + O(p) \qquad (6.3\text{–}12)\end{aligned}$$

To find the ultimate recoil we make use of the "final value theorem" in the following way:

$$\begin{aligned}\gamma_\infty &= \lim_{t\to\infty} \int_0^t \dot{\gamma}_{yx}(t')\,dt' \\ &= \lim_{p\to 0} p\mathscr{L}\left\{\int_0^t \dot{\gamma}_{yx}(t')\,dt'\right\} \\ &= \lim_{p\to 0} p\left[\frac{1}{p}\bar{\dot{\gamma}}_{yx}\right] \\ &= \lim_{p\to 0}\left[-\dot{\gamma}_0\frac{G_1}{G_0} + O(p)\right] \\ &= -\dot{\gamma}_0 \frac{\int_0^\infty sG(s)\,ds}{\int_0^\infty G(s)\,ds} \qquad (6.3\text{–}13)\end{aligned}$$

which is the same as the result in Eq. 6.2–31.

§6.4 JUSTIFICATION FOR THE USE OF THE GENERAL LINEAR VISCOELASTIC FLUID

In §6.1 the general linear viscoelastic fluid was introduced in an empirical fashion, following more or less the historical development of the subject. It has served well for the interpretation of experiments and as a framework for presenting and discussing molecular theories. It is natural to ask whether Eq. 6.1–16—the general linear viscoelastic model—can be regarded as a special result obtainable from some more general theory. The answer is "yes," as we shall see in Chapter 8. At this point we give a quick preview of the results that are obtained in Chapter 8 in order to give the reader some perspective.

In Chapter 8 we shall find that there is a rather general "memory integral expansion" capable of describing a wide variety of flows and fluids, the first term of which is very similar indeed to Eq. 6.1–16. This series has the form:[1]

$$\begin{aligned}\boldsymbol{\tau} = &-\int_{-\infty}^t G_I(t-t')\dot{\boldsymbol{\Gamma}}'\,dt' \\ &-\tfrac{1}{2}\int_{-\infty}^t \int_{-\infty}^t G_{II}(t-t', t-t'')\{\dot{\boldsymbol{\Gamma}}'\cdot\dot{\boldsymbol{\Gamma}}'' + \dot{\boldsymbol{\Gamma}}''\cdot\dot{\boldsymbol{\Gamma}}'\}\,dt''\,dt' - \cdots \qquad (6.4\text{–}1)\end{aligned}$$

Here the $\int \cdots dt'$ notation implies integration following a fluid particle. The tensor $\dot{\boldsymbol{\Gamma}}'$ is the velocity gradient that a fluid particle has experienced at some time t' in the past, evaluated

[1] J. D. Goddard, *Trans. Soc. Rheol.*, *11*, 381–399 (1967).

in a coordinate frame rotating with the fluid. It is given by:

$$\dot{\Gamma}'_{ij} \equiv \dot{\Gamma}_{ij}(t,t') = \sum_m \sum_n \Omega_{im}(t,t')\dot{\gamma}_{mn}(t')\Omega_{jn}(t,t') \qquad (6.4\text{–}2)$$

in which the Ω-matrices describe the rotation of the fluid element. At this juncture we make no attempt to explain these rotation matrices completely, since they will be discussed at the beginning of Chapter 7 in some detail. However, for flows in which the vorticity is independent of position and time, the rotation matrix is given by the particularly simple expression involving the vorticity:

$$\Omega_{ij} = \delta_{ij} - \frac{1}{1!}\left(\frac{1}{2}\right)\omega_{ij}(t - t') + \frac{1}{2!}\left(\frac{1}{2}\right)^2 \sum_m \omega_{im}\omega_{mj}(t - t')^2 + \cdots \qquad (6.4\text{–}3)$$

For the present discussion we need not go further into questions of definition. If the fluid never strays very far from its initial configuration, the $\dot{\Gamma}$-tensor simplifies to the $\dot{\gamma}$-tensor; if velocity gradients are always kept quite small, then the second and higher integrals in the series are not important. Hence for small deformations and small deformation gradients, the memory-integral expansion reduces to Eq. 6.1–16 for linear viscoelasticity and $G_I(t - t')$ is seen to be identical to $G(t - t')$.

All of the results obtained in this chapter for the general linear viscoelastic fluid can be obtained also from the first term of the memory-integral expansion in Eq. 6.4–1; in addition the latter is able to describe the shear-rate dependence of the viscosity. If one uses two terms in the expansion, then the behavior of the normal stresses for small-amplitude flows can be found completely. In Chapter 7 we shall give an extensive discussion of the first term in Eq. 6.4–1, and in Chapter 8 the entire series will be examined.

Now that we have given Eq. 6.1–16 and also the series in Eq. 6.4–1, we are in a position to discuss and illustrate the *objectivity principle.* If a rubber band is stretched in the north-south direction, and then gradually rotated toward an east-west orientation, one does not expect that the stress-strain relation for the rubber band will change. That is, the rate of rotation of the rubber band should be immaterial in describing the stress-strain relation. Similarly when we write an equation such as the Maxwell model, we really should do so in such a way that there will be no unwanted dependence of stress on the rotatory motion of the fluid particle as it moves through space. We did not take this precaution in Chapter 6, inasmuch as we agreed at the outset to consider only those flows in which the fluid moves ever so slightly away from its initial position. If we try to apply any of the models of the form of Eq. 6.1–16 to flow systems that do not satisfy this restriction, then we get into trouble. When a rheological equation of state is written in such a form that there is no unwanted dependence of the stresses on the rotatory motion of the fluid, then we say that the equation is "objective." *Equation 6.1–16 is not objective*! The following illustrative example will show how one can run into trouble when the rheological equation of state is not objective.[2]

EXAMPLE 6.4–1 Illustration That the General Linear Viscoelastic Model is not "Objective"

Consider the flow system shown in Fig. 6.4–1, in which a steady-state shear flow experiment is being performed on a rotating turntable. The velocity gradient $\dot{\gamma}$ is presumed to be exceedingly small so that only the single-integral term in Eq. 6.4–1 need be retained. The time variable is chosen in such a way that the $\bar{x}$-axis is lined up with the x-axis at time $t = 0$.

[2] Another example can be found in Problem 7C.2.

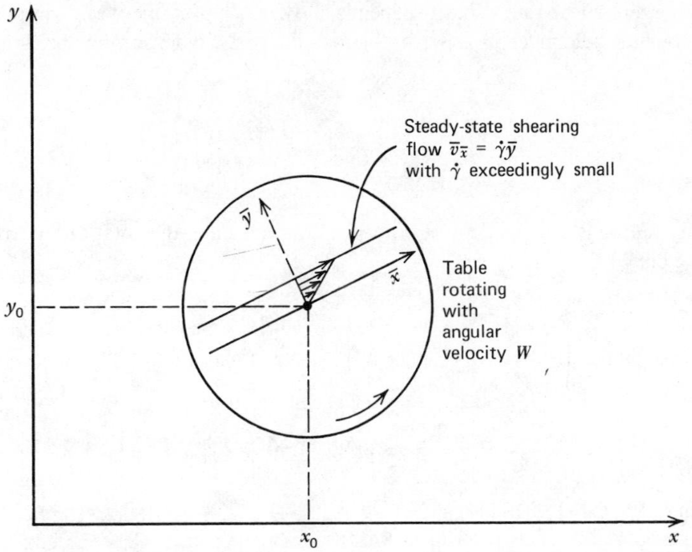

FIGURE 6.4–1. A steady-state shear flow experiment on a rotating turntable. The flow appears to be steady state to an observer standing on the turntable, but the flow appears to be time-dependent to an observer who is not on the turntable.

(a) Use Eq. 6.1–16 to find the components of the stress tensor in the *xyz*-system. Show that one is *not* led to the relation $\eta_0 = \int_0^\infty G(s)\,ds$, but that the viscosity is found to depend on the angular velocity of the turntable! This illustrates that there is a fundamental error in the formulation of the equation for linear viscoelasticity as given in Eq. 6.1–16.

(b) Repeat the derivation using the first term of Eq. 6.4–1, and show that the correct result is obtained.

SOLUTION

(a) The velocity distribution $\mathbf{v}(x,y,t)$ in the flow system as seen by an observer in the *xyz*-system is:

$$v_x = \dot{\gamma}[-(x - x_0)\sin Wt \cos Wt + (y - y_0)\cos^2 Wt] - W(y - y_0) \tag{6.4–4}$$

$$v_y = \dot{\gamma}[-(x - x_0)\sin^2 Wt + (y - y_0)\sin Wt \cos Wt] + W(x - x_0) \tag{6.4–5}$$

The rate-of-strain tensor $\dot{\boldsymbol{\gamma}}$ then has the form

$$\dot{\boldsymbol{\gamma}}(t) = \begin{pmatrix} -\sin 2Wt & \cos 2Wt & 0 \\ \cos 2Wt & \sin 2Wt & 0 \\ 0 & 0 & 0 \end{pmatrix} \dot{\gamma} \tag{6.4–6}$$

Then, according to Eq. 6.1–16 the stresses are given by:

$$\boldsymbol{\tau} = -\dot{\gamma} \int_0^\infty G(s) \begin{pmatrix} -\sin 2W(t - s) & \cos 2W(t - s) & 0 \\ \cos 2W(t - s) & \sin 2W(t - s) & 0 \\ 0 & 0 & 0 \end{pmatrix} ds \tag{6.4–7}$$

At time $t = 0$, when the parallel-plate system is lined up with the x-axis, we get:

$$\tau_{yx} = -\eta_0\dot{\gamma} = -\dot{\gamma} \int_0^\infty G(s) \cos 2Ws\, ds \tag{6.4–8}$$

This tells us that the zero-shear-rate viscosity depends on W, which of course we find impossible to believe.

(b) For the flow field given in Eqs. 6.4–4 and 5, the vorticity tensor may be shown to be:

$$\boldsymbol{\omega}(t) = \begin{pmatrix} 0 & 2W - \dot{\gamma} & 0 \\ \dot{\gamma} - 2W & 0 & 0 \\ 0 & 0 & 0 \end{pmatrix} \tag{6.4–9}$$

Since the vorticity tensor is independent of both position and time, the rotation matrix may be found from Eq. 6.4–3 to be:[3]

$$(\Omega_{ij}) = \begin{pmatrix} \cos\,[(W - \frac{1}{2}\dot{\gamma})(t - t')] & -\sin\,[(W - \frac{1}{2}\dot{\gamma})(t - t')] & 0 \\ \sin\,[(W - \frac{1}{2}\dot{\gamma})(t - t')] & \cos\,[(W - \frac{1}{2}\dot{\gamma})(t - t')] & 0 \\ 0 & 0 & 1 \end{pmatrix} \tag{6.4–10}$$

Then

$$\dot{\Gamma}_{yx} = \dot{\gamma} \cos\,[2Wt - \dot{\gamma}(t - t')] \tag{6.4–11}$$

and the first term of Eq. 6.4–1 gives at $t = 0$:

$$\begin{aligned} \tau_{yx}(t = 0) &= -\dot{\gamma} \int_0^\infty G_I(s) \cos \dot{\gamma}s\, ds \\ &= -\dot{\gamma} \int_0^\infty G_I(s)\, ds + O(\dot{\gamma}^3) \end{aligned} \tag{6.4–12}$$

Therefore, for small $\dot{\gamma}$ we obtain, since $\int_0^\infty G_I(s)\, ds = \eta_0$:

$$\tau_{yx} = -\eta_0 \dot{\gamma} \tag{6.4–13}$$

in agreement with our expectations.

[3] The xx-component in Eq. 6.4–10 is obtained thus:

$$\begin{aligned} \Omega_{xx} &= 1 + \frac{1}{2!}\left(\frac{1}{2}\right)^2 \omega_{xy}\omega_{yx}(t - t')^2 + \frac{1}{4!}\omega_{xy}\omega_{yx}\omega_{xy}\omega_{yx}(t - t')^4 + \cdots \\ &= 1 - \frac{1}{2!}(W - \tfrac{1}{2}\dot{\gamma})^2(t - t')^2 + \frac{1}{4!}(W - \tfrac{1}{2}\dot{\gamma})^4(t - t')^4 + \cdots = \cos\,[(W - \tfrac{1}{2}\dot{\gamma})(t - t')] \end{aligned} \tag{6.4–9a}$$

QUESTIONS FOR DISCUSSION

1. To what extent does the subject material of Chapter 5 overlap that of Chapter 6?
2. What various forms can the equations of linear viscoelasticity assume? Are there still other ways in which they could be written?
3. We have discussed in this chapter certain analogies between the rheological behavior of viscoelastic fluids and the mechanical behavior of springs and dashpots. What analogies are there with electrical circuits?
4. What geological problems was Jeffreys interested in when he discussed viscoelasticity in his book entitled *The Earth*? (cf. footnote 4 in §6.1)
5. Is it possible to turn Eq. 6.1–16 "wrong-side-out"? That is, can the equation be solved for the rate-of-strain tensor?
6. To what extent can the linear viscoelasticity theory of this chapter describe the experiments discussed in Chapters 3 and 4?
7. How does one convert the various expressions in Eqs. 6.1–10 to 13 into one another?
8. What properties does the function $G(t - t')$ in Eq. 6.1–16 have? What is $G(t - t')$ for a Newtonian fluid? A Hookean solid?
9. In the small-amplitude oscillatory motion problem in Example 6.2–1, the velocity profile is assumed to be linear. In Example 6.2–7, on the other hand, we find that the velocity profile near an oscillating wall is far from linear. Explain.
10. How is the complex viscosity related to the relaxation modulus?
11. Sketch the behavior of the functions in Eqs. 6.2–8 and 9, making use of the asymptotic expressions that are known for these functions.
12. Why is $G(t - t')$ called the "relaxation modulus"?
13. What does the G_I term of Eq. 6.4–1 give in place of Eq. 6.4–12 when $t \neq 0$?
14. What is meant when we say that the linear viscoelasticity models are not objective? How serious is this?
15. How are $\alpha(\omega)$ and $\beta(\omega)$ in Eq. 6.2–52 obtained from experimental data on $\eta'(\omega)$ and $\eta''(\omega)$?

PROBLEMS

6A.1 Determination of Relaxation Modulus from Dynamic Viscosity

Compare the following two procedures for obtaining the relaxation modulus $G(s)$ from dynamic viscosity:

a. Equation 6.2–4 can be inverted to give (cf. Eq. 7.5–2):

$$G(s) = \frac{2}{\pi}\int_0^\infty \eta'(\omega)\cos\omega s\,d\omega \tag{6A.1–1}$$

Insert the dynamic viscosity data for 2% polyisobutylene in Primol in Fig. 4.4–3 into the integral and determine $G(s)$.

b. Assume that $G(s)$ has the form given by the expression in braces in Eq. 6.1–12, with $s = t - t'$, and use the relations in Eqs. 6.1–14 and 6.1–15 to reduce the number of parameters to three. Determine η_0, λ, and α by fitting Eqs. 6.2–12 and 14 to the 2% polyisobutylene in Primol data of Fig. 4.4–3. Then plot the curve for $G(s)$.

6A.2 Experimental Test of the Kramers-Kronig Relations

There is a pair of relations (cf. Problem 6D.2 and Table 7.5–2, Eqs. A and B) that enable one to compute η'' from η' data or vice versa; these are known as the Kramers-Kronig relations. For example:

$$\frac{\eta''}{\omega} = \frac{2}{\pi}\int_0^\infty \frac{\eta'(\omega) - \eta'(\varpi)}{\varpi^2 - \omega^2}\,d\varpi \tag{6A.2–1}$$

Use the η' data from Fig. 4.4–3 for 2% polyisobutylene in Primol and see how well you are able to duplicate the η'' data for the same solution given in Fig. 4.4–4.

6B.1 Alternative Forms for the Maxwell and Jeffreys Models

a. Work out the intermediate steps needed to go from Eq. 6.1–4 to Eq. 6.1–5, and then from Eq. 6.1–5 to Eq. 6.1–6.

b. Do the same for the three forms of the Jeffreys model given in Eqs. 6.1–7 through 9.

6B.2 Complex Viscosity for Elementary Linear Models

a. Obtain η' and η'' for the Maxwell model by using each of Eqs. 6.1–4, 5, and 6. Which equation was the easiest to use?

b. Repeat (a) for the Jeffreys model, using Eqs. 6.1–7 to 9.

Hint: In order to use Eq. 6.1–6, you will need to show that if $\dot{\gamma}_{yx} = \dot{\gamma}^0 \mathcal{R}e\{e^{i\omega t}\}$ then

$$\gamma_{yx}(t,t') = \dot{\gamma}^0 \mathcal{R}e\left\{\frac{1}{i\omega}(e^{i\omega t'} - e^{i\omega t})\right\} \tag{6B.2–1}$$

6B.3 Velocity Distribution in a Rotating System

Derive the velocity distribution shown in Eqs. 6.4–4 and 5, and then show how to obtain the rate-of-strain tensor in Eq. 6.4–6 and the vorticity tensor in Eq. 6.4–9.

6B.4 Equivalence of Two Linear Viscoelastic Rheological Equations of State[1]

Consider a rheological equation of state of the form of Eq. 6.1–10 with the sum containing just two terms, so that the polymer is described by four constants η_1, η_2, λ_1, and λ_2. Show that this model can be transformed into the equation:

$$\tau_{ij} + a_1 \frac{\partial \tau_{ij}}{\partial t} + a_2 \frac{\partial^2 \tau_{ij}}{\partial t^2} = -\eta_0 \left(\dot{\gamma}_{ij} + b_1 \frac{\partial \dot{\gamma}_{ij}}{\partial t} \right) \tag{6B.4–1}$$

and obtain the expressions for the new constants a_1, a_2, η_0, and b_1 in terms of the old ones.

Hint: First rewrite the original equation in the form of Eq. 6.1–12. Then find $\partial \tau_{ij}/\partial t$ and $\partial^2 \tau_{ij}/\partial t^2$, and show that:

$$\begin{aligned} \tau_{ij} + a_1 \frac{\partial \tau_{ij}}{\partial t} + a_2 \frac{\partial^2 \tau_{ij}}{\partial t^2} = & -\int_{-\infty}^{t} \left\{ \sum_{k=1}^{2} \left(1 - \frac{a_1}{\lambda_k} + \frac{a_2}{\lambda_k^2}\right) \frac{\eta_k}{\lambda_k} e^{-(t-t')/\lambda_k} \right\} \dot{\gamma}_{ij}(t')\, dt' \\ & - \left(a_1 \sum_{k=1}^{2} \frac{\eta_k}{\lambda_k} - a_2 \sum_{k=1}^{2} \frac{\eta_k}{\lambda_k^2} \right) \dot{\gamma}_{ij}(t) - \left(a_2 \sum_{k=1}^{2} \frac{\eta_k}{\lambda_k} \right) \frac{\partial}{\partial t} \dot{\gamma}_{ij} \end{aligned} \tag{6B.4–2}$$

Show that equating the right sides of Eqs. 6B.4–1 and 2 leads to $a_1 = \lambda_1 + \lambda_2$, $a_2 = \lambda_1 \lambda_2$, $\eta_0 = \eta_1 + \eta_2$, and $b_1 = (\eta_1 \lambda_2 + \eta_2 \lambda_1)/(\eta_1 + \eta_2)$.

6B.5 High-Frequency Expressions for η' and η'' for the Generalized-Maxwell Model

a. First show how to go from Eq. 6.2–10 to Eq. 6.2–14 for $\alpha = 2$ (which corresponds to the Rouse molecular theory of Chapter 12). Make use of the Euler-Maclaurin expansion for converting a sum into an integral:

$$\sum_{k=0}^{m} f(k) = \int_0^m f(t)\, dt + \tfrac{1}{2}[f(0) + f(m)] + \cdots \tag{6B.5–1}$$

Apply this to the sum in Eq. 6.2–10 to get

$$\sum_{k=1}^{\infty} \frac{k^2}{k^4 + (\lambda\omega)^2} = \sum_{k=0}^{\infty} \frac{k^2}{k^4 + (\lambda\omega)^2} \tag{6B.5–2}$$

$$\begin{aligned} &= \int_0^{\infty} \frac{k^2}{k^4 + (\lambda\omega)^2}\, dk + 0 + \cdots \\ &= \frac{1}{2\sqrt{\lambda\omega}} \int_0^{\infty} \frac{\sqrt{t}\, dt}{t^2 + 1} + \cdots \\ &= \frac{1}{2\sqrt{\lambda\omega}} \frac{\pi}{2 \cos(\pi/4)} + \cdots \end{aligned} \tag{6B.5–3}$$

Then the dynamic viscosity becomes:

$$\eta'(\omega) \cong \eta_0 \frac{3\sqrt{2}}{2\pi} \frac{1}{\sqrt{\lambda\omega}} \tag{6B.5–4}$$

b. Derive the results given in Eqs. 6.2–14 and 15 for arbitrary α.

[1] A. J. de Vries, in *Proc. Fourth Int. Congress on Rheology*, E. H. Lee (Ed.), Wiley, New York (1965), Part III, pp. 321–344.

6B.6 The Relaxation Spectrum[2]

Sometimes polymer chemists have chosen to use, instead of the relaxation modulus for the generalized Maxwell model (quantity in braces in Eq. 6.1–12), an expression that implies a "continuum" of exponentials. That is, they write:

$$G(s) = \int_0^\infty \frac{H(\lambda)}{\lambda} e^{-s/\lambda}\, d\lambda = \int_{-\infty}^{+\infty} H(\lambda) e^{-s/\lambda}\, d \ln \lambda \tag{6B.6–1}$$

in which $H(\lambda)$ is called the *relaxation spectrum.* Specifying $H(\lambda)$ is equivalent to specifying $G(t)$. A number of useful empirical formulas for $H(\lambda)$ are available, containing two or three constants.

All of the linear viscoelasticity problems worked in this chapter can be developed in terms of $H(\lambda)$; for example, show that:

$$\eta'(\omega) = \int_0^\infty \frac{H(\lambda)\, d\lambda}{1 + (\lambda\omega)^2} \tag{6B.6–2}$$

$$\frac{\eta''(\omega)}{\omega} = \int_0^\infty \frac{\lambda H(\lambda)\, d\lambda}{1 + (\lambda\omega)^2} \tag{6B.6–3}$$

Interpret the results by comparing them with the expressions for η' and η'' for the Maxwell model.

6B.7 Alternative Derivation of Linear Viscoelasticity[3]

Consider a strain history consisting of a single small step shear strain $\Delta\gamma_1$ at a past time t_1. Mathematically we may say that the strain history has been:

$$\gamma_{yx}(t_1,t) = \Delta\gamma_1 H(t - t_1) \tag{6B.7–1}$$

where $H(x)$, equal to 1 for $x > 0$ and 0 for $x < 0$, is the Heaviside unit step function. By the definition (Eq. 4.4–24) of the relaxation modulus, the shear stress resulting from this shear strain is:

$$\begin{aligned} \tau_{yx}(t) &= -G(t - t_1; \Delta\gamma_1)\, \Delta\gamma_1 \\ &= -G(t - t_1)\, \Delta\gamma_1 + O(\Delta\gamma_1)^3 \end{aligned} \tag{6B.7–2}$$

a. Consider a strain history consisting of two step strains $\Delta\gamma_1$ and $\Delta\gamma_2$ at times t_1 and t_2, respectively. Show that the strain and stress histories may be written:

$$\gamma_{yx}(t_1,t) = \Delta\gamma_1 H(t - t_1) + \Delta\gamma_2 H(t - t_2) \tag{6B.7–3}$$

$$\tau_{yx}(t) = -G(t - t_1)\, \Delta\gamma_1 - G(t - t_2)\, \Delta\gamma_2 + O(\gamma_{yx}{}^3) \tag{6B.7–4}$$

At what order in γ_{yx} may coupling effects between the two step strains occur?

b. Given a strain history $\gamma_{yx}(-\infty,t')$, show that at any time t we may write the strain and the stress as:

$$\begin{aligned} \gamma_{yx}(-\infty,t) &= \int_{-\infty}^{t} H(t - t')\, d\gamma_{yx}(-\infty,t') \\ &= \int_{-\infty}^{t} H(t - t')\dot{\gamma}_{yx}(t')\, dt' \end{aligned} \tag{6B.7–5}$$

[2] J. D. Ferry, *Viscoelastic Properties of Polymers*, Wiley, New York (1970), Second Edition, pp. 63–65.
[3] A. C. Pipkin, *Lectures on Viscoelasticity Theory*, Springer, New York (1972), pp. 10–12.

$$\tau_{yx}(t) = -\int_{-\infty}^{t} G(t - t')\, d\gamma_{yx}(-\infty,t') + O(\gamma_{yx}{}^3)$$

$$= -\int_{-\infty}^{t} G(t - t')\dot{\gamma}_{yx}(t')\, dt' + O(\gamma_{yx}{}^3) \tag{6B.7-6}$$

6B.8 Relaxation Modulus for a Suspension of Rigid Dumbbells

It is shown in Chapter 11 that kinetic theory gives for the relaxation modulus for a dilute suspension of rigid dumbbells:

$$G(t - t') = 2\eta_s\delta(t - t') + nkT[\tfrac{4}{5}\lambda\delta(t - t') + \tfrac{3}{5}e^{-(t-t')/\lambda}] \tag{6B.8-1}$$

Here η_s is the viscosity of the solvent, n is the number of dumbbells per unit volume of solution, and $\lambda = \zeta L^2/12kT$ is a time constant; in the latter, L is the distance between the centers of the two "beads" making up the dumbbell, and ζ is the friction coefficient that gives the Stokes' law resistance of one bead of the dumbbell as it moves through the solvent ($\zeta = 6\pi\eta_s a$, where a is the bead radius).

a. How are the constants in the Jeffreys model of Eq. 6.1–8 related to the kinetic theory parameters which describe a rigid, rodlike macromolecule, idealized as a rigid dumbbell?

b. What are $\eta'(\omega)$ and $\eta''(\omega)$ for a dilute suspension of rigid dumbbells?

6C.1 Complex Viscosity for General Linear Viscoelastic Fluid

a. Use the rheological equation of state in Eq. 6.1–11 and show that $\eta^*/\eta_0 = (P + iQ)/(R + iS)$ or:

$$\frac{\eta'}{\eta_0} = \frac{PR + QS}{R^2 + S^2} \tag{6C.1-1}$$

$$\frac{\eta''}{\eta_0} = \frac{PS - QR}{R^2 + S^2} \tag{6C.1-2}$$

where

$$P(\omega) = 1 - b_2\omega^2 + b_4\omega^4 - \cdots \tag{6C.1-3}$$

$$Q(\omega) = b_1\omega - b_3\omega^3 + b_5\omega^5 - \cdots \tag{6C.1-4}$$

$$R(\omega) = 1 - a_2\omega^2 + a_4\omega^4 - \cdots \tag{6C.1-5}$$

$$S(\omega) = a_1\omega - a_3\omega^3 + a_5\omega^5 - \cdots \tag{6C.1-6}$$

b. It has been shown[1] that if all of the parameters a_n and b_n are expressed in terms of a single time constant λ, thus:

$$a_n = \frac{\pi^{2n}\lambda^n}{(2n + 1)!} = \left(\frac{2n + 3}{3}\right) b_n \tag{6C.1-7}$$

then we get the results in Eqs. 6.2–10 and 11 when $\alpha = 2$. Show that these expressions for a_n and b_n lead to:

$$\eta^* = \frac{3\eta_0}{\pi^2}\left[\frac{\pi\sqrt{i\omega\lambda}\cosh \pi\sqrt{i\omega\lambda} - \sinh \pi\sqrt{i\omega\lambda}}{i\omega\lambda \sinh \pi\sqrt{i\omega\lambda}}\right] \tag{6C.1-8}$$ [2]

[1] T. W. Spriggs and R. B. Bird, *Ind. Eng. Chem. Fundamentals*, *4*, 182–186 (1965).
[2] To get this result use L. B. W. Jolley, *Summation of Series*, Dover, New York (1961), Series No. (124) on p. 22.

or

$$\frac{\eta'}{\eta_0} = \frac{3}{\pi\sqrt{2\lambda\omega}}\left[\frac{\sinh \pi\sqrt{2\lambda\omega} - \sin \pi\sqrt{2\lambda\omega}}{\cosh \pi\sqrt{2\lambda\omega} - \cos \pi\sqrt{2\lambda\omega}}\right] \tag{6C.1-9}$$

$$\frac{\eta''}{\eta_0} = \frac{3}{\pi\sqrt{2\lambda\omega}}\left[\frac{\sinh \pi\sqrt{2\lambda\omega} + \sin \pi\sqrt{2\lambda\omega}}{\cosh \pi\sqrt{2\lambda\omega} - \cos \pi\sqrt{2\lambda\omega}} - \frac{\sqrt{2}}{\pi\sqrt{\lambda\omega}}\right] \tag{6C.1-10}$$

These expressions are then very nearly the same as one obtains from the Rouse theory (see Eqs 12.3–9 and 12).

6D.1 Motion of a Viscoelastic Fluid Pulsating in a Tube[1]

Consider the axial motion of a viscoelastic fluid in a circular tube oscillating about a rest position because of a sinusoidal variation in the pressure gradient:

$$-\frac{\partial p}{\partial z} = \mathcal{Re}\{P^0 e^{i\omega t}\} \tag{6D.1-1}$$

Postulate that all quantities in the system are oscillating sinusoidally but with different phases:

Axial velocity: $$v_z = \mathcal{Re}\{v_z{}^0 e^{i\omega t}\} \tag{6D.1-2}$$

Shear stress: $$\tau_{rz} = \mathcal{Re}\{\tau_{rz}{}^0 e^{i\omega t}\} \tag{6D.1-3}$$

Volume flow rate: $$Q = \mathcal{Re}\{Q^0 e^{i\omega t}\} \tag{6D.1-4}$$

a. Show that the equation of motion becomes:

$$\rho i\omega v_z{}^0 = P^0 + \eta^* \frac{1}{r}\frac{d}{dr}\left(r\frac{dv_z{}^0}{dr}\right) \tag{6D.1-5}$$

b. Verify that the solution of Eq. 6D.1–5 is:

$$v_z{}^0(r;\omega) = \frac{P^0}{\rho i\omega}\left[1 - \frac{J_0(\alpha r)}{J_0(\alpha R)}\right] \tag{6D.1-6}$$

in which R is the tube radius, and $\alpha^2 = -\rho i\omega/\eta^*$.

c. Next show that Q^0 is given by:

$$\begin{aligned} Q^0 &= \frac{\pi R^2 P^0}{\rho i\omega}\left[1 - \frac{2J_1(\alpha R)}{\alpha R J_0(\alpha R)}\right] \\ &= \frac{\pi R^4 P^0}{8\eta^*}\left[1 - \frac{\rho i\omega R^2}{6\eta^*} + \cdots\right] \end{aligned} \tag{6D.1-7}$$

In the second expression, the power-series expansions of the Bessel functions have been used.

[1] This problem was solved for a Maxwell fluid by L. J. F. Broer, *Appl. Sci. Res., A6*, 226–236 (1957), and for the general linear viscoelastic fluid by A. G. Fredrickson, *Principles and Applications of Rheology*, Prentice-Hall, Englewood Cliffs, N.J. (1964), pp. 133 *et seq.*

d. Show that for a Maxwell model the relation between the pressure drop and the volume flow rate is as follows where P^0 is taken to be real:

$$Q = \frac{\pi R^4 P^0}{8\eta_0}\left\{\left[1 + \frac{\rho R^2}{3\eta_0 \lambda}(\lambda\omega)^2 + \cdots\right]\cos\omega t - \left[1 - \frac{\rho R^2}{6\eta_0 \lambda}(1 - (\lambda\omega)^2) + \cdots\right](\lambda\omega)\sin\omega t\right\} \qquad \text{(6D.1–8)}$$

6D.2 Kramers-Kronig Relations[2]

a. By Fourier transformation show that Eqs. 6.2–4 and 5 may be inverted to give:

$$G(s) = \frac{2}{\pi}\int_0^\infty \eta'(\omega)\cos\omega s\, d\omega \qquad (s \geq 0) \qquad \text{(6D.2–1)}$$

$$G(s) = \frac{2}{\pi}\int_0^\infty \eta''(\omega)\sin\omega s\, d\omega \qquad (s > 0) \qquad \text{(6D.2–2)}$$

b. Insert Eqs. 6D.2–1 and 2 for $G(s)$ in Eqs. 6.2–5 and 4, respectively. Show that the results may be written:

$$\eta''(\omega) = \frac{2\omega}{\pi}\int_0^\infty \frac{\eta'(x)}{\omega^2 - x^2}\, dx \qquad \text{(6D.2–3)}$$

$$\eta'(\omega) - \eta'(\infty) = \frac{2}{\pi}\int_0^\infty \frac{x\eta''(x)}{x^2 - \omega^2}\, dx \qquad \text{(6D.2–4)}$$

This may be done by use of the methods of generalized functions.[3] Be careful in treating the behavior of $G(s)$ at $s = 0$.

c. Show that Eqs. 6D.1–3 and 4 may be written in the following forms that are more suited for computations:

$$\eta''(\omega) = \frac{2\omega}{\pi}\int_0^\infty \frac{\eta'(x) - \eta'(\omega)}{\omega^2 - x^2}\, dx \qquad \text{(6D.2–5)}$$

$$\eta'(\omega) - \eta'(\infty) = \frac{2}{\pi}\int_0^\infty \frac{x\eta''(x) - \omega\eta''(\omega)}{x^2 - \omega^2}\, dx \qquad \text{(6D.2–6)}$$

[2] H. A. Kramers, *Atti Congr. Intern. Fisici, Como, 2*, 545–557 (1927); R. de L. Kronig, *J. Opt. Soc. Amer., 12*, 547–557 (1926).

[3] M. J. Lighthill, *Introduction to Fourier Analysis and Generalised Functions*, Cambridge University Press (1964).

CHAPTER 7
QUASILINEAR COROTATIONAL MODELS

The classes of flow problems that can be solved in Chapters 5 and 6 are limited because of the limited ranges of applicability of the generalized Newtonian model (steady-state shear flows) and the general linear viscoelastic model (motions with small deformations and deformation gradients). In order to be able to solve fluid dynamics problems involving flows outside these two classes of motions, we must have a rheological equation of state of greater generality. This chapter is the first of three that deal with the construction of more general rheological equations of state and their applicability to fluid dynamical problem solving.

We take as our starting point the final section of Chapter 6. We found in Example 6.4-1 that the general linear viscoelastic model is not "objective," since it cannot always describe correctly motions in which the fluid undergoes a large deformation from its initial position. In this chapter we revise the linear viscoelastic models of Chapter 6 by rewriting them in such a way that they are "objective." This is done by formulating the rheological equations of state of Chapter 6 in a "corotating reference frame." An observer in this frame will not detect a superposed rigid rotation and the objectivity of the rheological equation of state is thereby insured.[1] The rheological models we obtain by this reformulation of the equations of linear viscoelasticity will be referred to as *quasilinear corotational models*; they are actually nonlinear relations between τ_{ij} and $\dot{\gamma}_{ij}$, but they are linear when written in terms of the analogous quantities defined in the corotational reference frame.

These quasilinear corotational models exhibit a shear-rate-dependent viscosity $\eta(\dot{\gamma})$ and also linear viscoelastic properties such as $\eta'(\omega)$ and $\eta''(\omega)$, thus embracing the main ideas of Chapters 5 and 6. In addition they exhibit normal stresses in steady shear flow, overshoot upon start-up of steady shearing, and other phenomena. These models are not able to describe all rheological phenomena quantitatively, but they can describe many observed effects qualitatively. We thus partly achieve our objective of developing a rheological equation of state that can describe any kind of flow.

However, the improvement in rheological description is inevitably accompanied by increased complexity in the functional form for the stress tensor. Consequently, using the equations of change to solve any nontrivial flow problems for these nonlinear viscoelastic fluid models poses considerable difficulties. Although this may seem discouraging, it must be kept in mind how few hydrodynamics problems can be solved analytically for Newtonian fluids. For polymers with their vastly more complex behavior, numerical solutions will

[1] This idea seems to have been used first for viscoelastic liquids by S. Zaremba, *Bull. Int. Acad. Sci., Cracovie*, 594–614, 614–621 (1903) and then rediscovered by H. Fromm, *Zeits. für angew. Math. u. Mech.*, *25/27*, 146–150 (1947); *28*, 43–54 (1948). See also J. G. Oldroyd, *Proc. Roy. Soc.*, *A245*, 278–297 (1958) and J. D. Goddard and C. Miller, *Rheol. Acta*, *5*, 177–184 (1966) for later discussions of the use of a corotational frame.

generally be needed. Also, in many instances we will have to content ourselves with "solving" problems by shrewd combination of experimental information, analytical solutions for limiting cases, dimensional analysis, and physical insight.

In §7.1 we show how to generate corotational models in a "cookbook" fashion, by using some tables that give the transformations from corotating to fixed reference frames. The reader interested in understanding how the tables were constructed can then read §7.2, whereas the reader who is less interested can omit part or all of that section. In §§7.3 and 7.4 we discuss two specific quasilinear corotational models. Here we show how proposed rheological models are tested by comparing their predictions with experimental data. The rheological equations of state we use in these sections for illustrative purposes are not necessarily the ones we would recommend for serious hydrodynamic calculations.

In §7.5 we give the most general quasilinear corotational model (the Goddard-Miller model), which is equivalent to writing Eq. 6.1–16 in a corotating frame. Here we show how one can obtain approximate interrelations among various material functions and how to attack several fluid mechanics problems. Finally, in §7.6 the use of curvilinear coordinates is discussed.

From the rheological point of view the only new physical idea in this and the following chapters is the formulation of rheological equations of state in a corotating reference frame. However, considerable space is necessarily devoted to testing the rheological models and showing how they are used for solving problems. Chapter 8 extends the discussion to *nonlinear corotational models*, thereby providing more general rheological descriptions and further enlarging our fluid dynamical problem-solving capability.

§7.1 BASIC CONCEPTS OF COROTATIONAL THEORIES

The main purpose of this section is to introduce the corotating reference frame, which is required in this chapter for the description of large-deformation flows of viscoelastic fluids. Up to this point we have used only a fixed frame (or laboratory frame) in which we visualized the flow field to be occurring; it is in this frame that we have formulated and solved the equations of change in Chapter 1. This fixed frame is usually regarded as "nailed down" in the laboratory or in some nonmoving piece of equipment.

With respect to this fixed frame we specify the velocity field by the function $\mathbf{v}(x,y,z,t)$. There are three kinematic tensors[1] that can be defined in terms of the velocity field, and that we will use throughout this and the following chapters:

		*ij*th *Cartesian Component*	
Velocity-gradient tensor:	$\nabla \mathbf{v}$	$\dfrac{\partial}{\partial x_i} v_j$	(7.1–1)
Rate-of-strain tensor:	$\dot{\boldsymbol{\gamma}} = \nabla \mathbf{v} + (\nabla \mathbf{v})^\dagger$	$\dfrac{\partial}{\partial x_i} v_j + \dfrac{\partial}{\partial x_j} v_i$	(7.1–2)
Vorticity tensor:	$\boldsymbol{\omega} = \nabla \mathbf{v} - (\nabla \mathbf{v})^\dagger$	$\dfrac{\partial}{\partial x_i} v_j - \dfrac{\partial}{\partial x_j} v_i$	(7.1–3)

[1] *Caution*: There is considerable diversity in the literature regarding the definitions of these quantities. Some authors define the *ij*th component of $\nabla \mathbf{v}$ to be $\partial v_i/\partial x_j$. Factors of 1/2 are commonly included on the right side of Eqs. 7.1–2 and 3. And some authors use a vorticity tensor that is the negative of ours or 1/2 the negative of ours. One advantage of using Eq. 7.1–2 without the factor of 1/2 is that in the shearing flow $v_x = \dot{\gamma} y$, the $\dot{\gamma}$ appearing here is the same as the $\dot{\gamma}$ that is the magnitude of the tensor $\dot{\boldsymbol{\gamma}}$, namely $\dot{\gamma} = \sqrt{\frac{1}{2}(\dot{\boldsymbol{\gamma}} : \dot{\boldsymbol{\gamma}})}$.

We have already encountered the (symmetric) *rate-of-strain tensor* $\dot{\boldsymbol{\gamma}}$ in earlier chapters, and we know that it describes the rate at which neighboring fluid particles move with respect to one another. The (antisymmetric) *vorticity tensor* $\boldsymbol{\omega}$ is a measure of the local rotation of the fluid.

Instead of the vorticity tensor $\boldsymbol{\omega}$, it is often convenient to use the *angular velocity vector* $\boldsymbol{w}$, which is defined as one-half the fluid "vorticity":

$$\boldsymbol{w} = \tfrac{1}{2}[\nabla \times \boldsymbol{v}] \tag{7.1-4}$$

The relations between $\boldsymbol{\omega}$ and $\boldsymbol{w}$ are as follows:

$$\omega_{ij} = 2 \textstyle\sum_k \epsilon_{ijk} w_k \tag{7.1-5}$$

$$w_i = \tfrac{1}{4} \textstyle\sum_j \sum_k \epsilon_{ijk} \omega_{jk} \tag{7.1-6}$$

We now wish to show that the vector $\boldsymbol{w}$ does represent the local fluid angular velocity.[2]

We can imagine that, in the immediate neighborhood of a point, the fluid will move almost as a rigid body; assume that the angular velocity of the rigid body is $\boldsymbol{W}$. Then the velocity in the neighborhood of the point will be $\boldsymbol{v} = [\boldsymbol{W} \times \boldsymbol{r}]$, where $\boldsymbol{r}$ is a position vector. We now calculate $\boldsymbol{w}$ for this rigid-body motion:

$$\begin{aligned} w_i &= \tfrac{1}{2}[\nabla \times \boldsymbol{v}]_i \\ &= \tfrac{1}{2} \textstyle\sum_j \sum_k \epsilon_{ijk} \dfrac{\partial}{\partial x_j} [\sum_m \sum_n \epsilon_{kmn} W_m x_n] \\ &= \tfrac{1}{2} \textstyle\sum_j \sum_k \sum_m \sum_n \epsilon_{ijk} \epsilon_{kmn} W_m \delta_{jn} \\ &= \tfrac{1}{2} \textstyle\sum_j \sum_k \sum_m \epsilon_{ijk} \epsilon_{mjk} W_m \\ &= W_i \end{aligned} \tag{7.1-7}$$

where, at the very end, use has been made of Eq. A.2-6. We have thus demonstrated that $\boldsymbol{w}$ is the same as the angular velocity vector.

We shall from time to time use the term *homogeneous flow field.* By this we mean that the velocity is linear in the Cartesian coordinates:

$$\boldsymbol{v} = \boldsymbol{v}_0 + [\boldsymbol{\kappa}(t) \cdot \boldsymbol{r}] \tag{7.1-8}$$

where $\boldsymbol{v}_0$ is a constant and $\boldsymbol{\kappa}(t)$ is a tensor independent of the position $\boldsymbol{r}$. For homogeneous flows $\nabla \boldsymbol{v} = \boldsymbol{\kappa}^\dagger$ and $w_i = \tfrac{1}{2} \sum_j \sum_k \epsilon_{ijk} \kappa_{kj}$.

We now introduce the notion of the corotating reference frame that can be used to describe an arbitrary flow field $\boldsymbol{v}(x,y,z,t)$ in the neighborhood of a fluid particle P. As may be seen in Fig. 7.1–1, the corotating frame, with orthogonal unit vectors $\breve{\boldsymbol{\delta}}_1$, $\breve{\boldsymbol{\delta}}_2$, $\breve{\boldsymbol{\delta}}_3$ moves along with the particle P and rotates with the local angular velocity of the fluid $\boldsymbol{w} = \tfrac{1}{2}[\nabla \times \boldsymbol{v}]$. This "corotating" frame is so chosen that at time t it is lined up with the fixed frame. For earlier times t', the corotating frame is "tilted" so that the $\breve{\boldsymbol{\delta}}_i$ are functions of the time t'.

Our interest in this cotranslating and corotating frame arises because of our belief that if we formulate rheological equations of state in this frame, these equations will not

[2] See G. K. Batchelor, *An Introduction to Fluid Dynamics*, Cambridge University Press (1967), p. 81, for a different explanation.

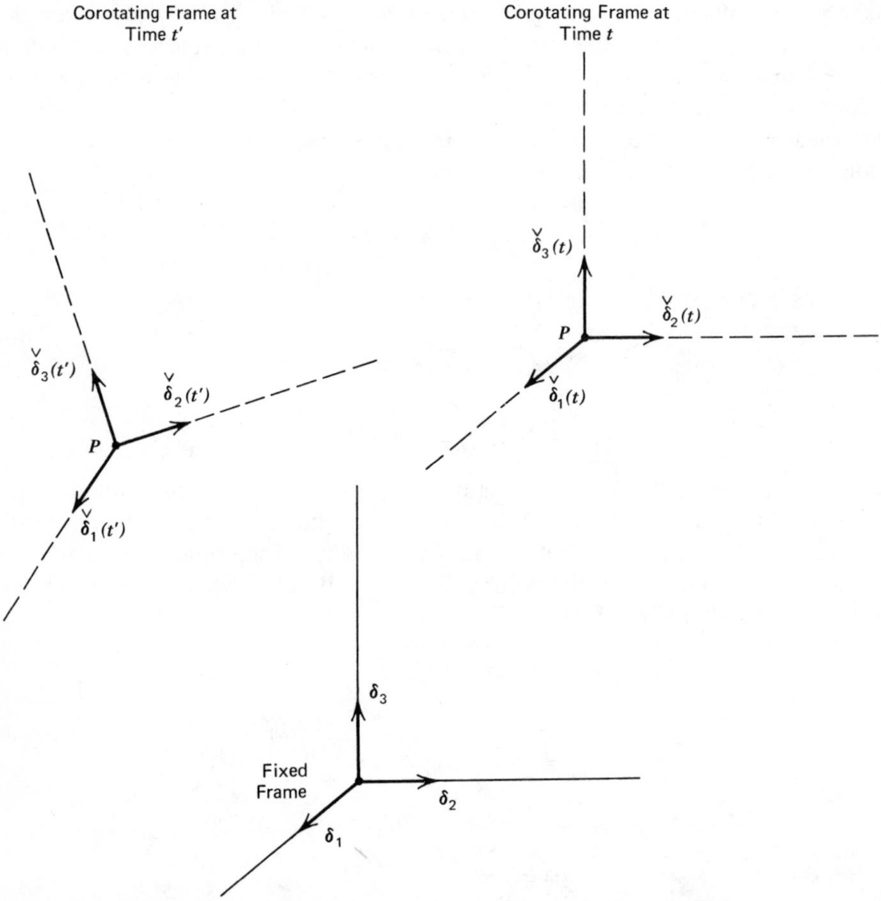

FIGURE 7.1–1. Two frames used to describe the flow near a fluid particle P: the fixed frame and the corotating frame moving along with the particle P. The two frames are lined up at the present time t.

have any unwanted dependence on the local rotation of the fluid. In other words, we will not end up with the unfortunate result we obtained in Example 6.4–1 when we applied the linear viscoelastic model to a large-displacement problem. An observer in the corotating frame will report stresses in terms of the components $\check{\tau}_{ij}$ referred to the cotranslating and corotating frame $\check{\boldsymbol{\delta}}_1\check{\boldsymbol{\delta}}_2\check{\boldsymbol{\delta}}_3$, and he will give the rate of strain in terms of components $\check{\dot{\gamma}}_{ij}$ referred to the cotranslating and corotating frame. Because he is rotating with the fluid, he will report vorticity components $\check{\omega}_{ij} = 0$. The corotating observer will then formulate a rheological equation of state by giving a relation between $\check{\tau}_{ij}$ and $\check{\dot{\gamma}}_{ij}$ that may involve time derivatives and time integrals of these quantities.[3] However, the equations of continuity and motion are generally used in the fixed frame $\boldsymbol{\delta}_1\boldsymbol{\delta}_2\boldsymbol{\delta}_3$ and, therefore, we need the stress tensor components in the fixed frame. Consequently we need transformation laws that

[3] One might wonder whether or not to include spatial derivatives in the rheological equations of state as formulated by the corotating, cotranslating observer. The assumption made here, that the stress in a given fluid particle depends only on the kinematic history of that particle alone (and not on the kinematic history of neighboring particles) seems to have been found applicable to real fluids; no higher level of complexity seems to be warranted at the present time. Fluids obeying this assumption are referred to as (*rheologically*) *simple fluids*.

enable us to transfer a rheological equation of state from the corotating frame into the corresponding equation in the fixed frame. These laws, which are derived in §7.2 are presented in Tables 7.1–1 and 2. Quantities appearing in the rheological equation of state in the corotational formulation will be found in Columns 1 and 1′ in the tables. To get the equation in the fixed frame formulation, these quantities have to be replaced by the corresponding quantities in Columns 2 and 2′. The analogous quantities that would appear in linear viscoelastic rheological equations of state are given in Columns 3 and 3′. (*Warning*: In Tables 7.1–1 and 2 all equations, except those written with boldface symbols, are used only in Cartesian coordinates. See §7.6 for a discussion of curvilinear coordinates.)

In order to use Tables 7.1–1 and 2 one has to know how to apply the $\mathscr{D}/\mathscr{D}t$ operator and how to find the $\Omega_{ij}(t,t')$. Both of these quantities have to do with "subtracting off" the overall rotatory motion of the fluid and hence involve the vorticity. We discuss these quantities briefly here, and more extensively in the next section.

The operator $\mathscr{D}/\mathscr{D}t$ is called the *corotational derivative* or the *Jaumann derivative.*[4] The Jaumann derivative of the components of a second-order tensor $\mathbf{\Lambda}$ is:

$$\frac{\mathscr{D}}{\mathscr{D}t}\Lambda_{ij}(t) = \frac{D}{Dt}\Lambda_{ij}(t) + \tfrac{1}{2}\textstyle\sum_n \left[\omega_{in}(t)\Lambda_{nj}(t) - \Lambda_{in}(t)\omega_{nj}(t)\right] \tag{7.1–9}$$

Hence the Jaumann derivative is made up of two parts: the usual substantial derivative D/Dt, plus an extra term, containing the vorticity tensor, which "corrects for" the rotatory motion of the fluid. We see from the tables that when an observer in the cotranslating and corotating frame reports the time rate of change of stress as $\partial\breve{\tau}_{ij}/\partial t$, the observer in the fixed frame will report the time rate of change as $\mathscr{D}\tau_{ij}/\mathscr{D}t$. Other forms of the $\mathscr{D}/\mathscr{D}t$ operator are given in note (a) of Table 7.1–1; also it is pointed out there that the application of $\mathscr{D}/\mathscr{D}t$ to a tensor product follows the same rule as the differentiation of a product in ordinary calculus.

The Ω_{ij} in Tables 7.1–1 and 2 describe the instantaneous "tilt" of the corotating frame $\breve{\boldsymbol{\delta}}_1\breve{\boldsymbol{\delta}}_2\breve{\boldsymbol{\delta}}_3$ (moving with a given fluid particle) with respect to the fixed frame $\boldsymbol{\delta}_1\boldsymbol{\delta}_2\boldsymbol{\delta}_3$. It is convenient to specify the fluid particle of interest by giving its coordinates at the present time t; thus a fluid particle P that has coordinates x_1, x_2, x_3 at time t is called (x_1,x_2,x_3,t), or simply (x,t) for short. Since the corotating unit vectors $\breve{\boldsymbol{\delta}}_i$ are functions of both the time t' and the particle (x,t) with which they are associated, we write $\breve{\boldsymbol{\delta}}_i = \breve{\boldsymbol{\delta}}_i(x,t,t')$. Usually, however, we will omit the particle dependence and write simply $\breve{\boldsymbol{\delta}}_i(t')$. Similarly we have $\Omega_{ij} = \Omega_{ij}(x,t,t')$, but we will usually write $\Omega_{ij}(t,t')$ for the rotation of the corotating frame between times t and t'. The unit vectors in the two frames are related by the following linear transformation that defines the Ω_{ij}:

$$\breve{\boldsymbol{\delta}}_i(t') = \textstyle\sum_j \Omega_{ij}(t,t')\boldsymbol{\delta}_j \tag{7.1–10}$$

By definition the corotating frame coincides with the nonrotating frame at $t' = t$. Hence at time t we know that $\breve{\boldsymbol{\delta}}_i(t) = \boldsymbol{\delta}_i$, and furthermore that $\Omega_{ij}(t,t) = \delta_{ij}$. The Ω_{ij} are determined by solving a set of differential equations given in note (b) of Table 7.1–1, and clearly the $\Omega_{ij}(t,t')$ depend on the vorticity tensor. For *any flow field* these equations can be solved[5]

[4] G. Jaumann, *Grundlagen der Bewegungslehre*, Leipzig (1905); *Sitzungsberichte Akad. Wiss. Wien, IIa*, *120*, 385–530 (1911). See also W. Prager, *Quart. Appl. Math.*, *18*, 403–407 (1961). The inverse of this "Jaumann differentiation" is called "Jaumann integration," the latter terminology having been introduced by J. D. Goddard and C. Miller, *Rheol. Acta*, *5*, 177–184 (1966).

[5] J. D. Goddard and C. Miller, *loc. cit.*

TABLE 7.1–1

Transformations from Corotating to Fixed Reference Frame: The Stress Tensor and Its Derivatives

(3′) Small deformations	(2′) Fixed components	(1′) Corotational components at t'	(1) Corotational components at t	(2) Fixed components	(3) Small deformations
$\tau_{ij}(t')$	$T'_{ij} \equiv T_{ij}(t,t') = \sum_m \sum_n \Omega_{im}(t,t')\tau_{mn}(t')\Omega_{jn}(t,t')$	$\breve{\tau}_{ij}(t')$	$\breve{\tau}_{ij}(t)$	$\tau_{ij}(t) = \tau_{ij}(t)$	
$\downarrow \frac{\partial}{\partial t'}$	$\downarrow \frac{\partial}{\partial t'}$	$\downarrow \frac{\partial}{\partial t'}$	$\downarrow \frac{\partial}{\partial t}$	$\downarrow \frac{\mathscr{D}}{\mathscr{D}t}$	$\downarrow \frac{\partial}{\partial t}$
$\frac{\partial \tau_{ij}(t')}{\partial t'}$	$\dot{T}'_{ij} \equiv \dot{T}_{ij}(t,t') = \sum_m \sum_n \Omega_{im}(t,t')\frac{\mathscr{D}\tau_{mn}(t')}{\mathscr{D}t'}\Omega_{jn}(t,t')$	$\frac{\partial \breve{\tau}_{ij}(t')}{\partial t'}$	$\frac{\partial \breve{\tau}_{ij}(t)}{\partial t}$	$\frac{\mathscr{D}\tau_{ij}(t)}{\mathscr{D}t}$	$\frac{\partial \tau_{ij}(t)}{\partial t}$
$\downarrow \frac{\partial}{\partial t'}$	$\downarrow \frac{\partial}{\partial t'}$	$\downarrow \frac{\partial}{\partial t'}$	$\downarrow \frac{\partial}{\partial t}$	$\downarrow \frac{\mathscr{D}}{\mathscr{D}t}$	$\downarrow \frac{\partial}{\partial t}$
$\frac{\partial^2 \tau_{ij}(t')}{\partial t'^2}$	$\ddot{T}'_{ij} \equiv \ddot{T}_{ij}(t,t') = \sum_m \sum_n \Omega_{im}(t,t')\frac{\mathscr{D}^2\tau_{mn}(t')}{\mathscr{D}t'^2}\Omega_{jn}(t,t')$	$\frac{\partial^2 \breve{\tau}_{ij}(t')}{\partial t'^2}$	$\frac{\partial^2 \breve{\tau}_{ij}(t)}{\partial t^2}$	$\frac{\mathscr{D}^2\tau_{ij}(t)}{\mathscr{D}t^2}$	$\frac{\partial^2 \tau_{ij}(t)}{\partial t^2}$

$t' = t$

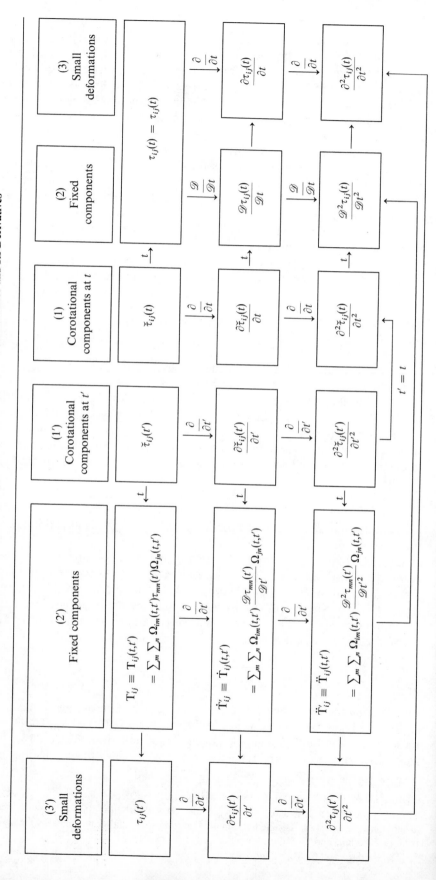

Notes:

a. The Jaumann derivative is defined by

$$\frac{\mathscr{D}}{\mathscr{D}t}\boldsymbol{\tau} = \frac{\partial}{\partial t}\boldsymbol{\tau} + \{\boldsymbol{v}\cdot\nabla\boldsymbol{\tau}\} + \tfrac{1}{2}(\{\boldsymbol{\omega}\cdot\boldsymbol{\tau}\} - \{\boldsymbol{\tau}\cdot\boldsymbol{\omega}\})$$

$$= \frac{D}{Dt}\boldsymbol{\tau} + \tfrac{1}{2}(\{\boldsymbol{\omega}\cdot\boldsymbol{\tau}\} + \{\boldsymbol{\omega}\cdot\boldsymbol{\tau}\}^{\dagger}) \qquad \text{(for } \boldsymbol{\tau} \text{ symmetric)}$$

$$= \frac{D}{Dt}\boldsymbol{\tau} - \{\boldsymbol{w}\times\boldsymbol{\tau}\} - \{\boldsymbol{w}\times\boldsymbol{\tau}\}^{\dagger} \qquad \text{(for } \boldsymbol{\tau} \text{ symmetric)} \tag{A}$$

Note that

$$\frac{\mathscr{D}}{\mathscr{D}t}\{\boldsymbol{\sigma}\cdot\boldsymbol{\tau}\} = \left\{\left(\frac{\mathscr{D}}{\mathscr{D}t}\boldsymbol{\sigma}\right)\cdot\boldsymbol{\tau}\right\} + \left\{\boldsymbol{\sigma}\cdot\left(\frac{\mathscr{D}}{\mathscr{D}t}\boldsymbol{\tau}\right)\right\} \tag{B}$$

b. The $\Omega_{ij}(t,t')$ for a given fluid particle are determined by

$$\frac{\partial}{\partial t'}\Omega_{ij}(t,t') = \tfrac{1}{2}\textstyle\sum_m \Omega_{im}(t,t')\omega_{mj}(t,t') \tag{C}$$

with $\Omega_{ij}(t,t) = \delta_{ij}$. Some solutions of Eq. C are given in Eqs. 7.1–11, 12, and 13. This relation is written for Cartesian coordinates only. It is understood that all tensor components are evaluated at the given fluid particle.

c. The symbol $\overset{t}{\rightarrow}$ is Lodge's "transfer operator" [see A. S. Lodge, *Elastic Liquids*, Academic Press, New York (1964)]; it serves as a reminder that the corotational quantities are referred to (or projected onto) the fixed frame at time t.

d. The time t is regarded as the "present time," the same as the time t occurring in the equations of continuity and motion. The "past time" t' (with $-\infty < t' \leq t$) occurs in rheological equations of state involving integrals. Note that the unprimed columns are special cases of the primed columns in that a quantity in Column 1′ goes over into the corresponding quantity in Column 1 where $t' = t$. Similarly a quantity in Column 2′ goes over into the corresponding quantity in Column 2 when $t' = t$.

TABLE 7.1–2

Transformations from Corotating to Fixed Reference Frame: The Rate-of-Strain Tensor and Its Derivatives

(3′) Small displacements	(2′) Fixed components	(1′) Corotational components at t'	(1) Corotational components at t	(2) Fixed components	(3) Small displacements
$\dot{\gamma}_{ij}(t')$	$\dot{\Gamma}'_{ij} \equiv \dot{\Gamma}'_{ij}(t,t') = \sum_m \sum_n \Omega_{im}(t,t')\dot{\gamma}_{mn}(t')\Omega_{jn}(t,t')$	$\breve{\dot{\gamma}}_{ij}(t')$	$\breve{\dot{\gamma}}_{ij}(t)$	$\dot{\gamma}_{ij}(t)$	$= \dot{\gamma}_{ij}(t)$
$\dfrac{\partial \dot{\gamma}_{ij}(t')}{\partial t'}$	$\ddot{\Gamma}'_{ij} \equiv \ddot{\Gamma}_{ij}(t,t') = \sum_m \sum_n \Omega_{im}(t,t') \dfrac{\mathscr{D}\dot{\gamma}_{mn}(t')}{\mathscr{D}t'} \Omega_{jn}(t,t')$	$\dfrac{\partial \breve{\dot{\gamma}}_{ij}(t')}{\partial t'}$	$\dfrac{\partial \breve{\dot{\gamma}}_{ij}(t)}{\partial t}$	$\dfrac{\mathscr{D}\dot{\gamma}_{ij}(t)}{\mathscr{D}t}$	$\dfrac{\partial \dot{\gamma}_{ij}(t)}{\partial t}$
$\dfrac{\partial^2 \dot{\gamma}_{ij}(t')}{\partial t'^2}$	$\dddot{\Gamma}'_{ij} \equiv \dddot{\Gamma}_{ij}(t,t') = \sum_m \sum_n \Omega_{im}(t,t') \dfrac{\mathscr{D}^2\dot{\gamma}_{mn}(t')}{\mathscr{D}t'^2} \Omega_{jn}(t,t')$	$\dfrac{\partial^2 \breve{\dot{\gamma}}_{ij}(t')}{\partial t'^2}$	$\dfrac{\partial^2 \breve{\dot{\gamma}}_{ij}(t)}{\partial t^2}$	$\dfrac{\mathscr{D}^2\dot{\gamma}_{ij}(t)}{\mathscr{D}t^2}$	$\dfrac{\partial^2 \dot{\gamma}_{ij}(t)}{\partial t^2}$

$t' = t$

Notes:

a. The tensor $\dot{\mathbf{\Gamma}}(t,t')$ may be expanded in a Taylor series about $t' = t$ to give:

$$\dot{\mathbf{\Gamma}}(t,t') = \sum_{n=0}^{\infty} \left.\frac{\partial^n \dot{\mathbf{\Gamma}}}{\partial t'^n}\right|_{t'=t} \frac{(t'-t)^n}{n!}$$

$$= \sum_{n=0}^{\infty} \left(\frac{\mathscr{D}^n}{\mathscr{D}t^n}\dot{\boldsymbol{\gamma}}(t)\right)\frac{(t'-t)^n}{n!}$$

$$= \dot{\boldsymbol{\gamma}}(t) - (t-t')\frac{\mathscr{D}}{\mathscr{D}t}\dot{\boldsymbol{\gamma}} + \frac{1}{2!}(t-t')^2\frac{\mathscr{D}^2}{\mathscr{D}t^2}\dot{\boldsymbol{\gamma}} - \frac{1}{3!}(t-t')^3\frac{\mathscr{D}^3}{\mathscr{D}t^3}\dot{\boldsymbol{\gamma}} + \cdots \quad \text{(A)}$$

b. For the relations between the

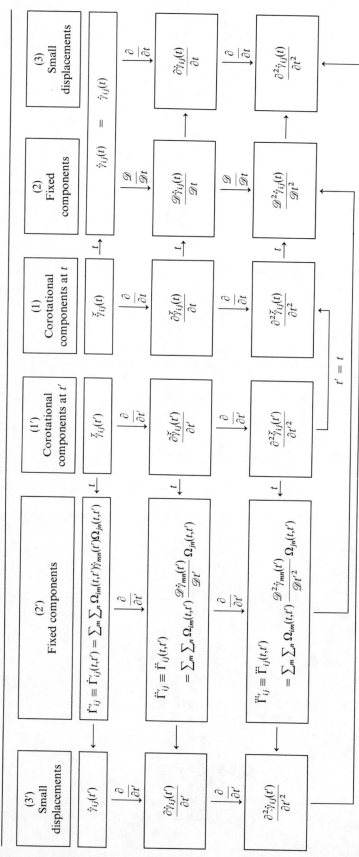

to give:

$$\begin{aligned}\Omega_{ij}(t,t') = \delta_{ij} &+ (-\tfrac{1}{2}) \int_{t'}^{t} \omega_{ij}(t,t'')\, dt'' \\ &+ (-\tfrac{1}{2})^2 \int_{t'}^{t} \int_{t''}^{t} \sum_m \omega_{im}(t,t''')\omega_{mj}(t,t'')\, dt'''\, dt'' \\ &+ (-\tfrac{1}{2})^3 \int_{t'}^{t} \int_{t''}^{t} \int_{t'''}^{t} \sum_m \sum_n \omega_{im}(t,t^{iv})\omega_{mn}(t,t''')\omega_{nj}(t,t'')\, dt^{iv}\, dt'''\, dt'' \\ &+ \cdots \end{aligned} \tag{7.1–11}$$

where $\omega_{ij}(t,t'') = \omega_{ij}(x,t,t'')$ is the vorticity at the particle (x,t) at past time t''. Note that the integration in Eq. 7.1–11 is to be performed following a fluid element, since (x,t) is held constant in the integration. This solution is easily verified by substitution into Eq. C of Table 7.1–1. For *flows with constant vorticity following a fluid particle* Eq. 7.1–11 simplifies to:

$$\begin{aligned}\Omega_{ij}(t,t') &= \delta_{ij} - \tfrac{1}{2}\omega_{ij}(t - t') + \frac{1}{2!}\frac{1}{2^2} \sum_m \omega_{im}\omega_{mj}(t - t')^2 \\ &\quad - \frac{1}{3!}\frac{1}{2^3} \sum_m \sum_n \omega_{im}\omega_{mn}\omega_{nj}(t - t')^3 + \cdots \\ &\equiv e^{-(1/2)\omega_{ij}(t-t')} \end{aligned} \tag{7.1–12}$$

where the exponential notation used here has to be interpreted as a matrix shorthand notation. For *homogeneous rectilinear unsteady-state shearing flows* of the form $v_x = \dot{\gamma}(t)y$, $v_y = 0$, $v_z = 0$ we have:

$$\mathbf{\Omega}(t,t') = \begin{bmatrix} \cos\left[\tfrac{1}{2}\int_{t'}^{t} \dot{\gamma}(t'')\, dt''\right] & \sin\left[\tfrac{1}{2}\int_{t'}^{t} \dot{\gamma}(t'')\, dt''\right] & 0 \\ -\sin\left[\tfrac{1}{2}\int_{t'}^{t} \dot{\gamma}(t'')\, dt''\right] & \cos\left[\tfrac{1}{2}\int_{t'}^{t} \dot{\gamma}(t'')\, dt''\right] & 0 \\ 0 & 0 & 1 \end{bmatrix} \tag{7.1–13}$$

See Problem 7B.4 for an elementary derivation of this equation. For complex flows the determination of Ω_{ij} will be a difficult problem.

In §§7.2 through 7.5 only problems involving Cartesian coordinates are discussed. We purposely restrict these sections in this way so that attention can be focused on the problem of corotating frames and on the physical understanding of the rheological phenomena involved. The extra complications introduced by curvilinear coordinates are postponed to §7.6.

In summary, throughout this chapter and in Chapter 8 we shall follow a two-step procedure for generating rheological equations of state:

I. We formulate a rheological equation of state in a corotating reference frame by giving a relation among $\check{\tau}_{ij}$ and $\check{\gamma}_{ij}$, their products, their time derivatives, or their time integrals, and containing no other variables. The rheological equation of state constructed in that way cannot contain any unwanted dependence on a superposed rotation (cf. Example 6.4–1).

II. After formulating the rheological equation of state in the corotating reference frame, we transform it into a fixed frame so that it can be used in conjunction with the equations of continuity, motion, and energy that are normally written in the fixed frame. To do this the list of transformations in Tables 7.1–1 and 2 is sufficient.

It must be emphasized that following the above procedure does not necessarily lead to a rheological equation of state that describes real fluids. This procedure simply guarantees that the rheological equation of state thus produced will be "objective." We shall see, however, that application of the above method does lead us to some equations that are capable of describing many rheological phenomena.

§7.2 OBJECTIVITY AND THE TRANSFORMATION LAWS BETWEEN REFERENCE FRAMES[1]

The purpose of this section is twofold: to derive the relations presented in Tables 7.1–1 and 2 for transforming quantities between the corotating and fixed reference frames, and to define more precisely the concept of an objective rheological equation of state and verify that rheological equations of state formulated by the corotating observer are objective. We proceed with these objectives in three steps. First, we look at the problem of an arbitrary change in reference frame and derive transformation laws for kinematical and dynamical quantities for such a change. Second, we specialize these results to the corotating frame introduced in §7.1. Third, we define objectivity and illustrate the procedure for determining whether or not a given tensor or rheological equation of state is objective.

Arbitrary Changes of Reference Frame

We begin by considering two frames, the *fixed* O-frame with unit vectors $\boldsymbol{\delta}_1$, $\boldsymbol{\delta}_2$, $\boldsymbol{\delta}_3$ and the *moving* $\bar{O}$-frame with time-dependent unit vectors $\bar{\boldsymbol{\delta}}_1$, $\bar{\boldsymbol{\delta}}_2$, $\bar{\boldsymbol{\delta}}_3$, as shown in Fig. 7.2–1. First we have to describe the displacement of the $\bar{O}$-frame from the O-frame, and the tilt of

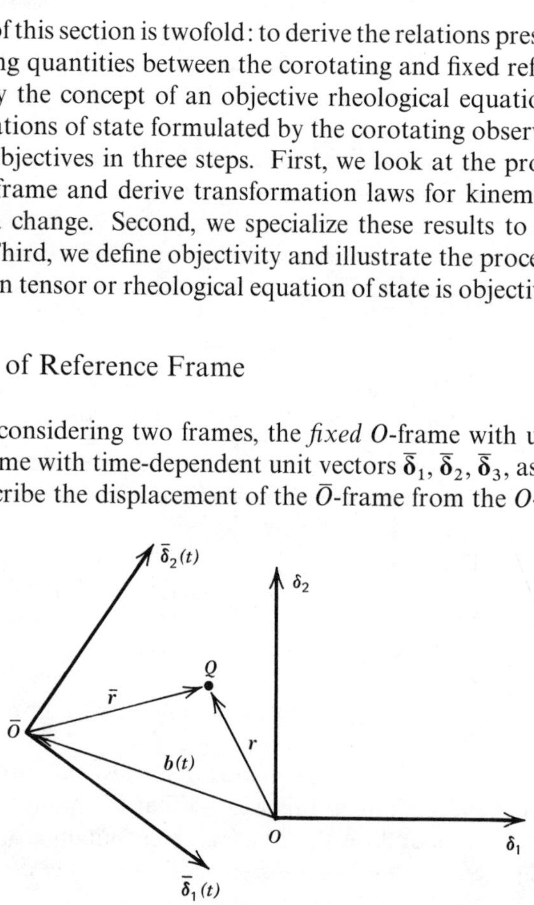

FIGURE 7.2–1. Two reference frames O and $\bar{O}$. A fluid particle Q is also shown. It may be located by position vector $\boldsymbol{r}$ in the O-frame or $\bar{\boldsymbol{r}}$ in the $\bar{O}$-frame. For simplicity $\bar{\boldsymbol{\delta}}_3$ and $\boldsymbol{\delta}_3$ are both taken to be perpendicular to the plane of the paper.

one frame with respect to the other. The instantaneous location of $\bar{O}$ is given by the vector $\boldsymbol{b}(t)$, and the instantaneous orientation of the $\bar{O}$-frame can be specified by giving the unit vectors $\bar{\boldsymbol{\delta}}_i$ in terms of the $\boldsymbol{\delta}_j$:

$$\bar{\boldsymbol{\delta}}_i(t) = \sum_j \mathrm{P}_{ij}(t)\boldsymbol{\delta}_j \tag{7.2–1}$$

[1] In this section only Cartesian coordinates and components are used. See §7.6 for a discussion of curvilinear coordinates.

Note that P_{ij} is the cosine of the angle[2] between $\bar{\boldsymbol{\delta}}_i$ and $\boldsymbol{\delta}_j$. Equation 7.2–1 can be inverted to give:

$$\boldsymbol{\delta}_i = \sum_j P_{ji}(t)\bar{\boldsymbol{\delta}}_j(t) \tag{7.2–2}$$

and the $P_{ij}(t)$ have the property that

$$\sum_m P_{im}P_{jm} = \sum_m P_{mi}P_{mj} = \delta_{ij} \tag{7.2–3}$$

That is, the P_{ij} are the elements of an orthogonal matrix.

We now wish to discuss the coordinates describing the instantaneous location of a fluid particle Q with respect to the two frames (see Fig. 7.2–1). The vectors $\boldsymbol{r}$ and $\bar{\boldsymbol{r}}$ are related by[3]

$$\bar{\boldsymbol{r}} = \boldsymbol{r} - \boldsymbol{b} \tag{7.2–4}$$

We now write $\bar{\boldsymbol{r}}$ in terms of its components in the $\bar{O}$-frame, and the other two vectors in terms of their components in the O-frame:

$$\sum_i \bar{\boldsymbol{\delta}}_i\bar{x}_i = \sum_j \boldsymbol{\delta}_j(x_j - b_j) \tag{7.2–5}$$

Then if Eq. 7.2–2 is used on the right side of Eq. 7.2–5 we get

$$\sum_i \bar{\boldsymbol{\delta}}_i\bar{x}_i = \sum_j \sum_i P_{ij}\bar{\boldsymbol{\delta}}_i(x_j - b_j) \tag{7.2–6}$$

whence

$$\boxed{\bar{x}_i = \sum_j P_{ij}(x_j - b_j)} \tag{7.2–7}$$

and

$$\boxed{x_i = \sum_j P_{ji}\bar{x}_j + b_i} \tag{7.2–8}$$

Equations 7.2–7 and 8 give the relations between the coordinates used by the observers in the two frames to locate a particle Q in their respective frames at some instant of time.

So far we have considered quantities that depend on the relative location of the particle Q and the two frames just at one particular instant. Hence we have not had to concern ourselves with the relative motion of the two frames. Next we wish to obtain the relation between the velocity components used by observers in the two frames to describe the flow. To do this we first consider the relative motion of the two frames.

At any instant the location and orientation of the $\bar{O}$-frame with respect to the O-frame is given by the vector $\boldsymbol{b}$ (with components b_i in the O-frame) and by the matrix P_{ij} of direction cosines. Hence if we know the functions $b_i(t)$ and $P_{ij}(t)$ for all t, we know the relative position and orientation of the two frames at all times. We assume that the functions $b_i(t)$ and $P_{ij}(t)$ are differentiable and denote their first derivative with respect to t by $\dot{b}_i$ and $\dot{P}_{ij}$ respectively.

We now define the velocity of the particle Q *with respect to a given frame* as the first time derivative of the position vector of the particle Q in that frame. Then the velocity

[2] The P_{ij} can also be expressed in terms of three "Euler angles." See, for example, Eqs. 13.1–3 and 4 and Fig. 13.1–2. The P_{ij} are "rho's," standing for "rhotation."

[3] The vector $\boldsymbol{r}$ has components x_1, x_2, x_3; sometimes we designate the set of components simply as x.

components seen by observers in the two frames are given by:

$$v_i \equiv \dot{x}_i \tag{7.2–9}$$

$$\bar{v}_i \equiv \dot{\bar{x}}_i \tag{7.2–10}$$

The v_i are understood to be functions of x_i and t, whereas the $\bar{v}_i$ are functions of $\bar{x}_i$ and t. At any time x_i and $\bar{x}_i$ are related by Eqs. 7.2–7 and 8. In these equations all of the quantities are functions of time, and their first time derivatives have been introduced. Hence at any time we may take the time derivative of Eq. 7.2–7 to find:

$$\dot{\bar{x}}_i = \sum_j \mathrm{P}_{ij}(\dot{x}_j - \dot{b}_j) + \sum_j \dot{\mathrm{P}}_{ij}(x_j - b_j) \tag{7.2–11}$$

We now insert Eqs. 7.2–9 and 10 in Eq. 7.2–11. Furthermore we define the fluid velocity at time t and at position x_i in the O-frame $v_i(x,t)$ as the velocity of the particle Q that happened to have coordinates x_i at time t, with similar remarks for the $\bar{O}$-frame. Then since the particle Q is arbitrary we find for the entire flow field:

$$\boxed{\bar{v}_i(\bar{x},t) = \sum_j \mathrm{P}_{ij}(v_j(x,t) - \dot{b}_j) + \sum_j \dot{\mathrm{P}}_{ij}(x_j - b_j)} \tag{7.2–12}$$

We thus now have the velocity field $\bar{v}_i(\bar{x}_1,\bar{x}_2,\bar{x}_3,t)$ as seen by an observer in the $\bar{O}$-frame in terms of the fixed observer's $v_i(x_1,x_2,x_3,t)$ and the known motion of the $\bar{O}$-frame ($\dot{b}_i(t)$ specifying the instantaneous translation at t and $\dot{\mathrm{P}}_{ij}(t)$ specifying the instantaneous rotation). Once again it is understood that the $\bar{x}_i$ and the x_i are related by Eqs. 7.2–7 and 8.

The relation between the velocity gradients is then:

$$\begin{aligned}
\frac{\partial}{\partial \bar{x}_i}\bar{v}_j(\bar{x},t) &= \sum_k \frac{\partial}{\partial x_k}\sum_l (\mathrm{P}_{jl}v_l(x,t) + \dot{\mathrm{P}}_{jl}x_l)\frac{\partial}{\partial \bar{x}_i}x_k \\
&= \sum_k \sum_l \left(\mathrm{P}_{jl}\frac{\partial}{\partial x_k}v_l + \dot{\mathrm{P}}_{jl}\delta_{kl}\right)\mathrm{P}_{ik} \\
&= \sum_k \sum_l \mathrm{P}_{ik}\left(\frac{\partial}{\partial x_k}v_l\right)\mathrm{P}_{jl} + \sum_k \mathrm{P}_{ik}\dot{\mathrm{P}}_{jk}
\end{aligned} \tag{7.2–13}$$

From this we can now find the corresponding transformations for the rate-of-strain tensor and the vorticity tensor at any point in the flow field:

$$\boxed{\begin{aligned}\bar{\dot{\gamma}}_{ij}(\bar{x},t) &\equiv \frac{\partial \bar{v}_j}{\partial \bar{x}_i} + \frac{\partial \bar{v}_i}{\partial \bar{x}_j} \\ &= \sum_k \sum_l \mathrm{P}_{ik}(t)\dot{\gamma}_{kl}(x,t)\mathrm{P}_{jl}(t)\end{aligned}} \tag{7.2–14}$$

$$\boxed{\begin{aligned}\bar{\omega}_{ij}(\bar{x},t) &\equiv \frac{\partial \bar{v}_j}{\partial \bar{x}_i} - \frac{\partial \bar{v}_i}{\partial \bar{x}_j} \\ &= \sum_k \sum_l \mathrm{P}_{ik}(t)\omega_{kl}(x,t)\mathrm{P}_{jl}(t) + 2\sum_k \mathrm{P}_{ik}(t)\dot{\mathrm{P}}_{jk}(t)\end{aligned}} \tag{7.2–15}$$

Note that the $\dot{\gamma}_{ij}$ obey the usual transformation rule for a coordinate rotation. The components ω_{ij} of the vorticity tensor, however, require additional terms in the transformation rule. This is to be expected because $\boldsymbol{\omega}$ contains information about fluid rotation whereas $\dot{\boldsymbol{\gamma}}$ does not.

Now that we know how components of various kinematical tensors in the two frames are related, let us turn to the transformation rule for the stress tensor. Recall that the total stress tensor $\boldsymbol{\pi}$ is defined in terms of the force per unit area $\boldsymbol{\pi}_n$ transmitted across a surface with unit normal $\boldsymbol{n}$:

$$\boldsymbol{\pi}_n = [\boldsymbol{n} \cdot \boldsymbol{\pi}] = [\boldsymbol{n} \cdot (p\boldsymbol{\delta} + \boldsymbol{\tau})] \tag{7.2–16}$$

Since the force $\boldsymbol{\pi}_n$ is a quantity that is independent of frame, we expect the components of $\boldsymbol{\pi}$ and $\boldsymbol{\tau}$ in the two frames to be interrelated by the usual transformations associated with a coordinate rotation (cf. Eq. 7.2–14):

$$\bar{\tau}_{ij}(\bar{x},t) = \textstyle\sum_m \sum_n \mathrm{P}_{im}(t)\tau_{mn}(x,t)\mathrm{P}_{jn}(t) \tag{7.2–17}$$

Here it is understood that τ_{mn} are components of $\boldsymbol{\tau}$ referred to the $\boldsymbol{\delta}_1\boldsymbol{\delta}_2\boldsymbol{\delta}_3$-frame.

Finally, we inquire about the transformation laws for time derivatives of tensors such as the rate-of-strain and stress tensors, since such quantities are sure to enter some rheological equations of state. We begin by looking at the time rate of change of a scalar function as seen by observers in the two frames and then use the same method for components of tensors of interest. Imagine then a scalar function that has the functional form $s(x,t)$ as seen by the fixed observer and that appears in the arbitrary frame to have the functional form $\bar{s}(\bar{x},t)$. The two functions s and $\bar{s}$ describe the same scalar quantity in space, so that

$$\bar{s}(\bar{x},t) = s(x,t) \tag{7.2–18}$$

provided that each is evaluated at the same place, that is, that $\bar{x}_i$ and x_i are related by Eqs. 7.2–7 and 8. We now compute the partial time derivative of Eq. 7.2–18 in the arbitrary frame; this requires the $\bar{x}_i$ to be held constant in performing the differentiation. Since s does not depend explicitly on the $\bar{x}_i$, it is necessary to apply the chain rule in differentiating the right side of the equation. Thus at any time t

$$\begin{aligned}
\frac{\partial}{\partial t}\bar{s}(\bar{x},t) &= \frac{\partial}{\partial t}s(x,t) + \textstyle\sum_k \dfrac{\partial}{\partial x_k}s(x,t)\dfrac{\partial}{\partial t}x_k(\bar{x},t) \\
&= \frac{\partial s}{\partial t} + \textstyle\sum_k [\sum_j \dot{\mathrm{P}}_{jk}\bar{x}_j + \dot{b}_k]\dfrac{\partial s}{\partial x_k} \qquad \text{(from Eq. 7.2–8)} \\
&= \frac{\partial}{\partial t}s(x,t) + \textstyle\sum_k [\underline{\sum_j \sum_l \dot{\mathrm{P}}_{jk}(t)\mathrm{P}_{jl}(t)(x_l - b_l(t))} + \dot{b}_k(t)]\dfrac{\partial}{\partial x_k}s(x,t)
\end{aligned} \tag{7.2–19}$$

The dashed-underlined term drops out at $\bar{x}_i = 0$. Later we need only those quantities evaluated at the origin of the rotating coordinate system.

In a similar fashion we can take the time derivative of the left and right sides of Eqs. 7.2–14 and 17; from Eq. 7.2–19 these are:

$$\boxed{\begin{aligned}
\frac{\partial}{\partial t}\bar{\dot{\gamma}}_{ij}(\bar{x},t) = \textstyle\sum_k \sum_l \Bigg(&\mathrm{P}_{ik}\left(\dfrac{\partial}{\partial t}\dot{\gamma}_{kl}(x,t)\right)\mathrm{P}_{jl} + \dot{\mathrm{P}}_{ik}\dot{\gamma}_{kl}\mathrm{P}_{jl} + \mathrm{P}_{ik}\dot{\gamma}_{kl}\dot{\mathrm{P}}_{jl}\Bigg) \\
&+ \textstyle\sum_k \sum_l \sum_m [\underline{\sum_n \sum_p \dot{\mathrm{P}}_{nm}\mathrm{P}_{np}(x_p - b_p)} + \dot{b}_m]\mathrm{P}_{ik}\left(\dfrac{\partial \dot{\gamma}_{kl}(x,t)}{\partial x_m}\right)\mathrm{P}_{jl}
\end{aligned}} \tag{7.2–20}$$

$$\frac{\partial}{\partial t}\bar{\tau}_{ij}(\bar{x},t) = \sum_k \sum_l \left(\mathrm{P}_{ik} \left(\frac{\partial}{\partial t} \tau_{kl}(x,t) \right) \mathrm{P}_{jl} + \dot{\mathrm{P}}_{ik}\tau_{kl}\mathrm{P}_{jl} + \mathrm{P}_{ik}\tau_{kl}\dot{\mathrm{P}}_{jl} \right)$$
$$+ \sum_k \sum_l \sum_m \underline{[\sum_n \sum_p \dot{\mathrm{P}}_{nm}\mathrm{P}_{np}(x_p - b_p) + \dot{b}_m]}\mathrm{P}_{ik} \left(\frac{\partial \tau_{kl}(x,t)}{\partial x_m} \right) \mathrm{P}_{jl} \qquad (7.2\text{–}21)$$

As before, the dashed-underlined terms vanish at $\bar{x}_i = 0$. We have retained the dashed-underlined terms in the above results for later use in Eq. 7.2–45, which shows that corotational time derivatives are objective operations. Higher derivatives can be found by repeating this process. This completes our derivation of laws for transforming quantities between a fixed frame and any frame that is moving in an arbitrary way with respect to the fixed one. We can now specialize these results to obtain transformation laws relating corotating and fixed vector and tensor components.

The Corotating Frame

The corotating frame was introduced in §7.1 as a frame that translates with a fluid particle and rotates with the local angular velocity of the fluid at that particle. In general, the fluid velocity and angular velocity vary from particle to particle, so that in the corotating formalism we associate a different corotating frame with each particle. The corotating frame at any particle is used to describe kinematical and dynamical tensors at that particle only. Whereas in the earlier part of this section there was only one arbitrary frame that we were relating to the fixed frame, there is now an infinite number of corotating frames, and this makes it necessary to specify which particle a given corotating frame is moving and rotating with. Before we used the letter "Q" to label a fluid particle; but in the following we find it necessary to use a different method for labeling particles. We motivate our choice of a labeling convention as follows: Suppose we want to calculate the stress tensor $\boldsymbol{\tau}(x,t)$ at some point x_i at the present time t. It is found that in rheologically complex fluids the current stress in a fluid element depends not only on the current rate of deformation but also on the deformation that the fluid element has experienced in the past (see §8.3). Equation 6.4–1 is an example of an equation that contains this physical idea. We thus need to be able to describe kinematical events that happened in the past to the particle at x_i at time t. This suggests that we label particles by their coordinates at the present time t. The particle at which we want to calculate the stress tensor is denoted by (x,t). If we give x and t specific meanings as reference quantities, we must use different symbols to represent the space-fixed coordinates and time at any time other than t. We use x_i' and t' for this purpose; at $t' = t$, the coordinates x_i' and x_i have the same meaning. As an illustration of this notation, $\boldsymbol{\tau}(x',t')$ denotes the stress tensor at the fixed point x_i' at time t'.

We now proceed to specialize the general transformation laws, previously obtained for an arbitrary frame, to the corotational frame associated with fluid particle (x,t). The transformation rules in Eqs. 7.2–14, 15, 17, 20, and 21 involve the b_i and the P_{ij} and their time derivatives. Let us consider each of these quantities in turn. Figure 7.2–2 will be of help in understanding the notation.

The components of the vector $\boldsymbol{b}(t')$ used for locating the origin of the arbitrary frame are now identified with the coordinates of particle (x,t) at time t', which are given by the *displacement functions*[4] $x_i' = x_i'(x,t,t')$. The time rate of change of $\boldsymbol{b}(t')$ with t' is then just the

[4] The displacement functions play a key role in §§9.1 to 3; it is always assumed that the inverse relation $x_i = x_i(x',t,t')$ exists.

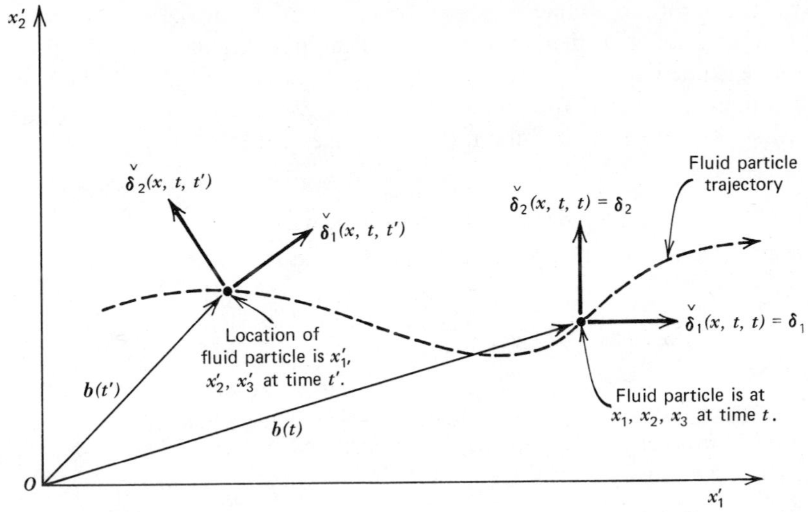

FIGURE 7.2–2. Fixed frame with origin at O, showing the corotating frame, with unit vectors $\check{\boldsymbol{\delta}}_1, \check{\boldsymbol{\delta}}_2, \check{\boldsymbol{\delta}}_3$, that moves with a fluid particle (x,t) and rotates with the instantaneous, local angular velocity of the fluid.

velocity of particle (x,t) at t' relative to the fixed frame:

$$\begin{aligned} \dot{b}_i(t') &= \frac{\partial}{\partial t'} x_i'(x,t,t') \\ &= v_i(x,t,t') \\ &= v_i(x',t')\big|_{x_i'=x_i'(x,t,t')} \end{aligned} \tag{7.2–22}$$

The velocity components are used here in the same sense as they were in Eq. 7.2–9; now, however, we can use the particle labels to show explicitly that the coordinates are to be differentiated at constant particle. Whereas before $\dot{b}_i$ depended only on t', now $v_i(x,t,t')$ carries a "fluid particle label" (x,t). Note that we can write physical quantities in two ways. For example, when we write $\mathbf{v}(x,t,t')$ we mean "the velocity at time t' of the fluid particle (x,t)"; on the other hand $\mathbf{v}(x',t')|_{x_i'=x_i'(x,t,t')}$ means "we look at the fluid velocity field at t' and then select from it just the fluid velocity at the particular point x_i' where particle (x,t) is located at time t'." If the entire field $\mathbf{v}(x',t')$ is known, then the displacement functions can be used to get $\mathbf{v}(x,t,t')$ for all particles (x,t), and vice versa.

To describe the instantaneous tilt of the corotating frame we give the rotation matrix (P_{ij}) a special symbol, namely (Ω_{ij}); note $\Omega_{ij} = \Omega_{ij}(x,t,t')$ gives the tilt at t' of the corotating frame moving with particle (x,t). The unit vectors in this corotating frame are $\check{\boldsymbol{\delta}}_i(x,t,t')$. By making appropriate changes in notation in Eqs. 7.2–1 and 2 we obtain relations between unit vectors in the corotating and fixed frames:

$$\check{\boldsymbol{\delta}}_i(x,t,t') = \textstyle\sum_j \Omega_{ij}(x,t,t')\boldsymbol{\delta}_j \tag{7.2–23}$$

$$\boldsymbol{\delta}_i = \textstyle\sum_j \Omega_{ji}(x,t,t')\check{\boldsymbol{\delta}}_j(x,t,t') \tag{7.2–24}$$

The Ω_{ij} are elements of an orthogonal matrix and must obey a relation similar to Eq. 7.2–3.

Now the corotating frame possesses two features that make the Ω_{ij} special: (1) the time rate of change of Ω_{ij} is determined by the local fluid angular velocity; and (2) the corotating frame is lined up with the fixed frame at time t (see Fig. 7.2–2). These features can be used to obtain a differential equation and initial condition from which Ω_{ij} can be calculated. First, we note that the time rates of change of the corotating unit vectors $\breve{\boldsymbol{\delta}}_i$ are given by:

$$\frac{\partial}{\partial t'}\breve{\boldsymbol{\delta}}_i(x,t,t') = [\boldsymbol{w}(x,t,t') \times \breve{\boldsymbol{\delta}}_i(x,t,t')]$$

$$= -\tfrac{1}{2}[\boldsymbol{\omega}(x,t,t') \cdot \breve{\boldsymbol{\delta}}_i(x,t,t')] \qquad (7.2\text{–}25)$$

$$\breve{\boldsymbol{\delta}}_i(x,t,t) = \boldsymbol{\delta}_i \qquad (7.2\text{–}26)$$

Here $\boldsymbol{w}(x,t,t')$ and $\boldsymbol{\omega}(x,t,t')$ are the fluid angular velocity vector and vorticity tensor at particle (x,t). The two forms of Eq. 7.2–25 can be interrelated by using Eq. 7.1–6. We now use Eq. 7.2–23 to replace $\breve{\boldsymbol{\delta}}_i$ by $\boldsymbol{\delta}_j$ in Eq. 7.2–25:

$$\sum_j \frac{\partial}{\partial t'}\Omega_{ij}(x,t,t')\boldsymbol{\delta}_j = -\tfrac{1}{2}[\boldsymbol{\omega}(x,t,t') \cdot \textstyle\sum_n \Omega_{in}(x,t,t')\boldsymbol{\delta}_n]$$

$$= -\tfrac{1}{2}[\textstyle\sum_j \sum_k \boldsymbol{\delta}_j\boldsymbol{\delta}_k\omega_{jk} \cdot \sum_n \Omega_{in}\boldsymbol{\delta}_n]$$

$$= -\tfrac{1}{2}\textstyle\sum_j \sum_k \sum_n \boldsymbol{\delta}_j\delta_{kn}\omega_{jk}\Omega_{in}$$

$$= -\tfrac{1}{2}\textstyle\sum_j \boldsymbol{\delta}_j(\sum_n \Omega_{in}\omega_{jn}) \qquad (7.2\text{–}27)$$

Taking the jth component of each side we see that the Ω_{ij} must satisfy the following differential equation:

$$\frac{\partial}{\partial t'}\Omega_{ij}(x,t,t') = \tfrac{1}{2}\textstyle\sum_n \Omega_{in}(x,t,t')\omega_{nj}(x,t,t') \qquad (7.2\text{–}28)$$

$$\Omega_{ij}(x,t,t) = \delta_{ij} \qquad (7.2\text{–}29)$$

Equation 7.2–29 corresponds to Eq. 7.2–26, and states that at time t the corotating frame is aligned with the fixed frame.

Now that we have the quantities corresponding to b_i and P_{ij}, we are in a position to adapt the results of Eqs. 7.2–14, 15, 17, 20, and 21 for the corotating frame; we do this by adding the particle label x,t and replacing the $\bar{x},t$ used earlier by $\breve{x},t'$. We are interested in the transformation rules *only at the origin of the corotating frame*, that is, at the particle (x,t). In this way we obtain the expressions for the components of the rate-of-deformation tensor at particle (x,t) as seen by the corotating observer:

$$\breve{\gamma}_{ij}(x,t,\breve{x},t')|_{\breve{x}_i=0} \equiv \breve{\gamma}_{ij}(x,t,t')$$

$$= \textstyle\sum_k \sum_l \Omega_{ik}(x,t,t')\dot{\gamma}_{kl}(x',t')\Omega_{jl}(x,t,t')|_{x_i'=x_i'(x,t,t')}$$

$$= \textstyle\sum_k \sum_l \Omega_{ik}(x,t,t')\dot{\gamma}_{kl}(x,t,t')\Omega_{jl}(x,t,t') \qquad (7.2\text{–}30)^5$$

in which the Ω_{ik} are obtained from the solution of Eqs. 7.2–28 and 29. For the vorticity

[5] Note that Eq. 7.2–30 can be inverted to give $\dot{\gamma}_{ij}$ in terms of the $\breve{\gamma}_{mn}$:

$$\dot{\gamma}_{ij}(x,t,t') = \textstyle\sum_m \sum_n \Omega_{mi}(x,t,t')\breve{\gamma}_{mn}(x,t,t')\Omega_{nj}(x,t,t') \qquad (7.2\text{–}30a)$$

tensor,

$$\begin{aligned}
\breve{\omega}_{ij}(x,t,\breve{x},t')|_{\breve{x}_i=0} &\equiv \breve{\omega}_{ij}(x,t,t') \\
&= \sum_k \sum_l \Omega_{ik}(x,t,t')\omega_{kl}(x',t')\Omega_{jl}(x,t,t')|_{x'_i=x'_i(x,t,t')} \\
&\quad + 2\sum_k \Omega_{ik}(x,t,t')\frac{\partial}{\partial t'}\Omega_{jk}(x,t,t') \\
&= \sum_k \Omega_{ik}[\sum_l \omega_{kl}(x',t')|_{x'_i=x'_i(x,t,t')}\Omega_{jl} \\
&\quad + 2\cdot\tfrac{1}{2}\sum_l \Omega_{jl}\omega_{lk}(x,t,t')] \qquad \text{(from Eq. 7.2–28)} \\
&= \sum_k \sum_l \Omega_{ik}(x,t,t')[\omega_{kl}(x',t')|_{x'_i=x'_i(x,t,t')} \\
&\quad - \omega_{kl}(x,t,t')]\Omega_{jl}(x,t,t') \\
&= 0
\end{aligned} \tag{7.2–31}$$

Hence the vorticity tensor components $\breve{\omega}_{ij}$ seen by the corotating observer right at particle (x,t) are zero; this is, of course, to be expected, since the corotating frame rotates with the local fluid angular velocity. Although the corotating observer sees zero vorticity at (x,t), at neighboring points that would in general not be so; this point, though interesting, does not affect our use of the corotating formalism.

In order to obtain an expression for the time rate of change of the corotating rate-of-strain tensor at the particle (x,t), we specialize the result given in Eq. 7.2–20 to that particle and use Eq. 7.2–28:

$$\begin{aligned}
\frac{\partial}{\partial t'}\breve{\dot{\gamma}}_{ij}(x,t,\breve{x},t')|_{\breve{x}_i=0} &\equiv \frac{\partial}{\partial t'}\breve{\dot{\gamma}}_{ij}(x,t,t') \\
&= \sum_k \sum_l \Bigg[\Omega_{ik}\left(\frac{\partial}{\partial t'}\dot{\gamma}_{kl}(x',t')\right)\Omega_{jl} \\
&\quad + \left(\frac{\partial}{\partial t'}\Omega_{ik}\right)\dot{\gamma}_{kl}\Omega_{jl} + \Omega_{ik}\dot{\gamma}_{kl}\left(\frac{\partial}{\partial t'}\Omega_{jl}\right) \\
&\quad + \Omega_{ik}\sum_m v_m(x',t')\left(\frac{\partial}{\partial x'_m}\dot{\gamma}_{kl}(x',t')\right)\Omega_{jl}\Bigg]\Bigg|_{x'_i=x'_i(x,t,t')} \\
&= \sum_k \sum_l \Omega_{ik}\Bigg[\frac{D}{Dt'}\dot{\gamma}_{kl}(x',t') \\
&\quad + \tfrac{1}{2}\sum_m (\omega_{km}(x',t')\dot{\gamma}_{ml}(x',t') - \dot{\gamma}_{km}(x',t')\omega_{ml}(x',t'))\Bigg]\Bigg|_{x'_i=x'_i(x,t,t')} \Omega_{jl} \\
&= \sum_k \sum_l \Omega_{ik}\left(\frac{\mathscr{D}}{\mathscr{D}t'}\dot{\gamma}_{kl}(x',t')\Big|_{x'_i=x'_i(x,t,t')}\right)\Omega_{jl} \\
&\equiv \sum_k \sum_l \Omega_{ik}\left(\frac{\mathscr{D}}{\mathscr{D}t'}\dot{\gamma}_{kl}(x,t,t')\right)\Omega_{jl}
\end{aligned} \tag{7.2–32}$$

In the last lines of Eq. 7.2–32 we have introduced the *corotational* (or *Jaumann*) *derivative* of the components of a second-order tensor. These are defined for an arbitrary tensor $\mathbf{\Lambda}$ by

$$\frac{\mathscr{D}}{\mathscr{D}t'}\Lambda_{ij}(x',t') \equiv \frac{D}{Dt'}\Lambda_{ij}(x',t') + \tfrac{1}{2}\sum_n [\omega_{in}(x',t')\Lambda_{nj}(x',t') - \Lambda_{in}(x',t')\omega_{nj}(x',t')] \tag{7.2–33}$$

Note that Eq. 7.2–32 could have been obtained directly from Eq. 7.2–30 by taking $\partial/\partial t'$ of the expression for $\check{\gamma}_{ij}(x,t,t')$ given in that result. In doing this, a term $\partial\dot{\gamma}_{kl}(x,t,t')/\partial t'$ would arise in place of the $(D\dot{\gamma}_{kl}(x',t')/Dt')|_{x_i'=x_i'(x,t,t')}$ term that appears in Eq. 7.2–32; that these are equivalent is easily seen by an application of the chain rule.[6]

By taking $\partial/\partial t'$ of Eq. 7.2–32 we can obtain:

$$\frac{\partial^2}{\partial t'^2}\,\check{\gamma}_{ij}(x,t,t') = \textstyle\sum_k \sum_l \Omega_{ik}\left(\dfrac{\mathscr{D}^2}{\mathscr{D}t'^2}\,\dot{\gamma}_{kl}(x,t,t')\right)\Omega_{jl} \tag{7.2–34}$$

Higher time derivatives are obtained in the same way.

When $\check{\gamma}_{ij}$ and its time derivatives are evaluated at the present time t we have:

$$\check{\gamma}_{ij}(x,t,t')|_{t'=t} = \dot{\gamma}_{ij}(x,t) \tag{7.2–35}$$

$$\frac{\partial}{\partial t'}\,\check{\gamma}_{ij}(x,t,t')|_{t'=t} = \frac{\mathscr{D}}{\mathscr{D}t}\,\dot{\gamma}_{ij}(x,t) \tag{7.2–36}$$

$$\frac{\partial^2}{\partial t'^2}\,\check{\gamma}_{ij}(x,t,t')|_{t'=t} = \frac{\mathscr{D}^2}{\mathscr{D}t^2}\,\dot{\gamma}_{ij}(x,t) \tag{7.2–37}$$

These relations are tabulated on the right side of Table 7.1–2. The corresponding transformation rules for the stress tensor are obtained in exactly the same way and are given on the right side of Table 7.1–1.

We now turn our attention to obtaining the more complicated transformation rules on the left sides of Tables 7.1–1 and 2. Let us define a tensor with components $\dot{\Gamma}_{ij}$ by[7]

$$\begin{aligned}\dot{\Gamma}_{ij}(x,t,t';t_0) &\equiv \textstyle\sum_k \sum_l \sum_m \sum_n \Omega_{ki}(x,t,t_0)\Omega_{km}(x,t,t')\dot{\gamma}_{mn}(x,t,t')\Omega_{ln}(x,t,t')\Omega_{lj}(x,t,t_0)\\ &= \textstyle\sum_k \sum_l \Omega_{ki}(x,t,t_0)\check{\gamma}_{kl}(x,t,t')\Omega_{lj}(x,t,t_0)\end{aligned} \tag{7.2–38}$$

[6] As noted after Eq. 7.2–22 we can regard physical quantities such as $\dot{\gamma}_{ij}$ as functions either of particle labels and time, $\dot{\gamma}_{ij}(x,t,t')$, or of spatial coordinates and time, $\dot{\gamma}_{ij}(x',t')$. By definition a partial time derivative of such a quantity expressed as a function of particles and time is always equal to a substantial derivative of the corresponding function of position and time:

$$\begin{aligned}\frac{\partial}{\partial t'}\,\dot{\gamma}_{ij}(x,t,t') &= \left(\frac{\partial}{\partial t'}\,\dot{\gamma}_{ij}(x',t')\right)_{x,t}\\ &= \frac{\partial}{\partial t'}\,\dot{\gamma}_{ij}(x',t') + \textstyle\sum_m \dfrac{\partial x_m'(x,t,t')}{\partial t'}\dfrac{\partial}{\partial x_m'}\,\dot{\gamma}_{ij}(x',t')\\ &= \frac{\partial}{\partial t'}\,\dot{\gamma}_{ij}(x',t') + \textstyle\sum_m v_m(x',t')\dfrac{\partial}{\partial x_m'}\,\dot{\gamma}_{ij}(x',t')\\ &= \frac{D}{Dt'}\,\dot{\gamma}_{ij}(x',t')\end{aligned} \tag{7.2–33a}$$

where it is understood that both sides are evaluated at the same position (i.e., that x_i' and (x,t) are related through the displacement functions). Thus we can write Eq. 7.2–33:

$$\begin{aligned}\frac{\mathscr{D}}{\mathscr{D}t'}\,\Lambda_{ij}(x,t,t') &= \frac{\partial}{\partial t'}\,\Lambda_{ij}(x,t,t') + \tfrac{1}{2}\textstyle\sum_n [\omega_{in}(x,t,t')\Lambda_{nj}(x,t,t')\\ &\quad - \Lambda_{in}(x,t,t')\omega_{nj}(x,t,t')]\end{aligned} \tag{7.2–33b}$$

[7] The motivation for introducing $\dot{\Gamma}_{ij}$ is as follows: In subsequent sections we shall want to reformulate the linear viscoelastic models of Chapter 6 in the corotating frame. Suppose the corotating observer uses

Equation 7.2–38 tells how the components of the rate-of-strain tensor at (x,t) as seen by the corotating observer at time t' are transformed[8] to the fixed frame at a time t_0. Almost always[9] the time t_0 at which the transformation is performed is the same as the present time t. Since at time t the corotating frame is lined up with the fixed frame, $\Omega_{ij}(x,t,t) = \delta_{ij}$ so that

$$\begin{aligned}\dot{\Gamma}_{ij}(x,t,t') &\equiv \dot{\Gamma}_{ij}(x,t,t';t)\\ &= \textstyle\sum_m \sum_n \Omega_{im}(x,t,t')\dot{\gamma}_{mn}(x,t,t')\Omega_{jn}(x,t,t')\\ &= \breve{\dot{\gamma}}_{ij}(x,t,t')\end{aligned} \tag{7.2–39}$$

It is then easy to see that

$$\begin{aligned}\ddot{\Gamma}_{ij}(x,t,t') &\equiv \frac{\partial}{\partial t'}\dot{\Gamma}_{ij}(x,t,t') = \frac{\partial}{\partial t'}\breve{\dot{\gamma}}_{ij}(x,t,t')\\ &= \textstyle\sum_k \sum_l \Omega_{ik}\left(\dfrac{\mathscr{D}}{\mathscr{D}t'}\dot{\gamma}_{kl}(x,t,t')\right)\Omega_{kl}\end{aligned} \tag{7.2–40}$$

Similar results hold for higher derivatives of $\dot{\Gamma}_{ij}$. These transformation rules are tabulated on the left side of Table 7.1–2. The corresponding rules for the stress tensor are found on the left side of Table 7.1–1. This completes our discussion of the transformation laws in Tables 7.1–1 and 2. We now turn to a discussion of objectivity.

Objective Tensors and Rheological Equations of State

In this final part of §7.2 we define the concept of objectivity. Then we show that rheological equations of state formulated in the corotational formalism are objective.

---

an integral model (e.g., Eq. 6.1–16) to compute the stress at his particle (x,t) at time t_0. He would write

$$\breve{\tau}_{mn}(x,t,t_0) = -\int_{-\infty}^{t_0} G(t_0 - t')\breve{\dot{\gamma}}_{mn}(x,t,t')\,dt' \tag{7.2–38a}$$

Now if we want the stress components in the fixed frame we use $\Omega_{ij}(x,t,t_0)$ which gives the tilt of the corotating frame at t_0 (cf. Eq. 7.2–30a):

$$\begin{aligned}\tau_{ij}(x,t,t_0) &= \textstyle\sum_m \sum_n \Omega_{mi}(x,t,t_0)\breve{\tau}_{mn}(x,t,t_0)\Omega_{nj}(x,t,t_0)\\ &= -\int_{-\infty}^{t_0} G \textstyle\sum_m \sum_n \Omega_{mi}(x,t,t_0)\breve{\dot{\gamma}}_{mn}(x,t,t')\Omega_{nj}(x,t,t_0)\,dt'\\ &= -\int_{-\infty}^{t_0} G\,\dot{\Gamma}_{ij}(x,t,t';t_0)\,dt'\end{aligned} \tag{7.2–38b}$$

Thus $\dot{\Gamma}_{ij}$ is a description of past rate of strain in the fixed frame for use in integral equations for computing τ_{ij} at some other time, t_0.

[8] Equation 7.2–38 can also be written as

$$\breve{\dot{\gamma}}_{ij}(x,t,t') \overset{t_0}{\rightarrow} \dot{\Gamma}_{ij}(x,t,t';t_0) \tag{7.2–38c}$$

by using the $\overset{t_0}{\rightarrow}$ operation introduced in note (c) to Table 7.1–1. The symbol $\overset{t_0}{\rightarrow}$ is called a "transfer operator"; it was introduced by A. S. Lodge, *Elastic Liquids*, Academic Press, New York (1964), Chapter 12. The t_0 over the arrow denotes the time at which the corotational quantity is referred to the fixed coordinate system.

[9] An exception might be in the calculation of recoil, for example.

We begin with two definitions:

1. An *objective tensor* is a tensor whose components as seen in any two reference frames are interrelated by the usual transformation rules for a coordinate rotation.

Given any two reference frames, we can always take one to be fixed and the other to be translating and rotating relative to it. Thus we can show that a tensor $\boldsymbol{\sigma}$ is objective by showing that its components in the fixed frame and arbitrarily moving frame (σ_{ij} and $\bar{\sigma}_{ij}$, respectively) discussed at the beginning of this section are related at any instant t by (cf. Eq. 7.2–14):

$$\bar{\sigma}_{ij} = \sum_k \sum_l \mathrm{P}_{ik}(t)\sigma_{kl}\mathrm{P}_{jl}(t) \tag{7.2–41}$$

2. An *objective rheological equation of state* is one that is constructed from objective tensors.

By looking back at Eqs. 7.2–14 and 17 we see immediately that the rate-of-strain and stress tensors at the present time t are objective tensors. However, the velocity gradient tensor, the vorticity tensor, and partial time derivatives of the rate-of-strain and stress tensors, all evaluated at t, are not objective. An example of a rheological equation of state that is not objective is the general linear viscoelastic fluid model; in §6.4 we showed by counterexample that it did not satisfy the definition given above. This we did by choosing two particular reference frames and using the general linear viscoelastic fluid model to predict the viscosity in each frame. The results were different for the two frames, and this could not happen if the rheological equation of state had been made up of objective tensors. In Example 7.2–1 we give an example of an objective rheological equation of state.

EXAMPLE 7.2–1 Objectivity of the Second-Order Fluid

Under conditions of very slow and very slowly changing flow, formal continuum mechanical arguments (§8.4) lead to a rheological equation of state called the second-order fluid:

$$\tau_{ij} = -\alpha_1\dot{\gamma}_{ij} + \alpha_2 \frac{\mathscr{D}}{\mathscr{D}t}\dot{\gamma}_{ij} - \alpha_{11} \sum_m \dot{\gamma}_{im}\dot{\gamma}_{mj} \tag{7.2–42}$$

Prove that the second-order fluid is an objective rheological equation of state.

SOLUTION

In the arbitrarily translating and rotating reference frame, the rheological equation of state for the second-order fluid has the same form as Eq. 7.2–42, except that quantities in the arbitrary frame replace those from the fixed frame:

$$\bar{\tau}_{ij} = -\alpha_1\bar{\dot{\gamma}}_{ij} + \alpha_2 \overline{\frac{\mathscr{D}}{\mathscr{D}t}\dot{\gamma}_{ij}} - \alpha_{11} \sum_m \bar{\dot{\gamma}}_{im}\bar{\dot{\gamma}}_{mj} \tag{7.2–43}$$

We already know that $\boldsymbol{\tau}(t)$ and $\dot{\boldsymbol{\gamma}}(t)$ are objective tensors, so we focus our attention on the last two terms. First, consider the last term:

$$\begin{aligned}
\sum_m \bar{\dot{\gamma}}_{im}\bar{\dot{\gamma}}_{mj} &= \sum_m \sum_k \sum_l \sum_n \sum_p \mathrm{P}_{ik}\dot{\gamma}_{kl}\mathrm{P}_{ml}\mathrm{P}_{mn}\dot{\gamma}_{np}\mathrm{P}_{jp} \\
&= \sum_k \sum_l \sum_n \sum_p \mathrm{P}_{ik}\dot{\gamma}_{kl}\delta_{ln}\dot{\gamma}_{np}\mathrm{P}_{jp} \\
&= \sum_k \sum_l \sum_p \mathrm{P}_{ik}\dot{\gamma}_{kl}\dot{\gamma}_{lp}\mathrm{P}_{jp}
\end{aligned} \tag{7.2–44}$$

Hence, by comparing this result with Eq. 7.2–41 we see that the tensor $\{\dot{\boldsymbol{\gamma}} \cdot \dot{\boldsymbol{\gamma}}\}$ is objective. Finally we examine the Jaumann derivative term:

$$
\begin{aligned}
\overline{\frac{\mathscr{D}}{\mathscr{D}t}\dot{\gamma}_{ij}} &= \frac{\partial}{\partial t}\bar{\dot{\gamma}}_{ij} + \textstyle\sum_k \bar{v}_k \dfrac{\partial}{\partial \bar{x}_k}\bar{\dot{\gamma}}_{ij} + \tfrac{1}{2}\textstyle\sum_k (\bar{\omega}_{ik}\bar{\dot{\gamma}}_{kj} - \bar{\dot{\gamma}}_{ik}\bar{\omega}_{kj}) \\
&= \textstyle\sum_k \sum_l \left[\mathrm{P}_{ik}\left(\dfrac{\partial}{\partial t}\dot{\gamma}_{kl}\right)\mathrm{P}_{jl} + \dot{\mathrm{P}}_{ik}\dot{\gamma}_{kl}\mathrm{P}_{jl} + \mathrm{P}_{ik}\dot{\gamma}_{kl}\dot{\mathrm{P}}_{jl} \right] \\
&\quad + \textstyle\sum_k \sum_l \sum_m [\sum_n \sum_p \dot{\mathrm{P}}_{nm}\mathrm{P}_{np}(x_p - b_p) + \dot{b}_m]\mathrm{P}_{ik}\left(\dfrac{\partial \dot{\gamma}_{kl}}{\partial x_m}\right)\mathrm{P}_{jl} \\
&\quad + \textstyle\sum_k \sum_l [\mathrm{P}_{kl}(v_l - \dot{b}_l) + \dot{\mathrm{P}}_{kl}(x_l - b_l)] \sum_p \dfrac{\partial}{\partial x_p}(\sum_m \sum_n \mathrm{P}_{im}\dot{\gamma}_{mn}\mathrm{P}_{jn})\mathrm{P}_{kp} \\
&\quad + \tfrac{1}{2}\textstyle\sum_k \sum_n \sum_p [(\sum_l \sum_m \mathrm{P}_{il}\omega_{lm}\mathrm{P}_{km} + 2\sum_l \mathrm{P}_{il}\dot{\mathrm{P}}_{kl})(\mathrm{P}_{kn}\dot{\gamma}_{np}\mathrm{P}_{jp}) \\
&\quad - (\mathrm{P}_{in}\dot{\gamma}_{np}\mathrm{P}_{kp})(\textstyle\sum_l \sum_n \mathrm{P}_{kl}\omega_{lm}\mathrm{P}_{jm} + 2\sum_l \mathrm{P}_{kl}\dot{\mathrm{P}}_{jl})] \\
&= \textstyle\sum_k \sum_l \left[\mathrm{P}_{ik}\left(\dfrac{\partial}{\partial t}\dot{\gamma}_{kl}\right)\mathrm{P}_{jl} + \underline{\dot{\mathrm{P}}_{ik}\dot{\gamma}_{kl}\mathrm{P}_{jl} + \mathrm{P}_{ik}\dot{\gamma}_{kl}\dot{\mathrm{P}}_{jl}} \right] \\
&\quad + \textstyle\sum_k \sum_l \sum_m [\underline{-\sum_n \sum_p \mathrm{P}_{nm}\dot{\mathrm{P}}_{np}(x_p - b_p)} + \underline{\dot{b}_m}]\mathrm{P}_{ik}\left(\dfrac{\partial \dot{\gamma}_{kl}}{\partial x_m}\right)\mathrm{P}_{jl} \\
&\quad + \textstyle\sum_k \sum_l \sum_m [\sum_n \sum_p \mathrm{P}_{np}(v_p - \underline{\dot{b}_p})\mathrm{P}_{nm} + \underline{\sum_n \sum_p \dot{\mathrm{P}}_{np}\mathrm{P}_{nm}(x_p - b_p)}]\mathrm{P}_{ik}\left(\dfrac{\partial \dot{\gamma}_{kl}}{\partial x_m}\right)\mathrm{P}_{jl} \\
&\quad + \tfrac{1}{2}\textstyle\sum_l \sum_m [\sum_n \mathrm{P}_{il}(\omega_{ln}\dot{\gamma}_{nm} - \dot{\gamma}_{ln}\omega_{nm})\mathrm{P}_{jm} - 2(\underline{\dot{\mathrm{P}}_{il}\dot{\gamma}_{lm}\mathrm{P}_{jm} + \mathrm{P}_{il}\dot{\gamma}_{lm}\dot{\mathrm{P}}_{jm}})] \\
&= \textstyle\sum_k \sum_l \mathrm{P}_{ik}\left[\dfrac{\partial}{\partial t}\dot{\gamma}_{kl} + \sum_m v_m \dfrac{\partial}{\partial x_m}\dot{\gamma}_{kl} + \tfrac{1}{2}\sum_m (\omega_{km}\dot{\gamma}_{ml} - \dot{\gamma}_{km}\omega_{ml})\right]\mathrm{P}_{jl} \\
&= \textstyle\sum_k \sum_l \mathrm{P}_{ik}\left(\dfrac{\mathscr{D}}{\mathscr{D}t}\dot{\gamma}_{kl}\right)\mathrm{P}_{jl}
\end{aligned}
\tag{7.2–45}
$$

Hence $(\mathscr{D}/\mathscr{D}t)\dot{\boldsymbol{\gamma}}(t)$ is an objective tensor, and therefore the second-order fluid is an objective rheological equation of state. In obtaining the result in Eq. 7.2–45 we have made frequent use of results obtained in the first part of this section; similarly underlined terms in the intermediate steps above are found to cancel.

We conclude our discussion of objectivity by showing that the tensors $\dot{\boldsymbol{\Gamma}}$ and $\mathbf{T}$ that we obtain from the corotational formalism are objective. We illustrate the method for $\dot{\boldsymbol{\Gamma}}$. In order to show that $\dot{\boldsymbol{\Gamma}}$ and its derivatives are objective we must be able to calculate the components of these in both the O- and $\bar{O}$-frames (Fig. 7.2–3). We have already given formulas for determining $\dot{\Gamma}_{ij}, \ddot{\Gamma}_{ij}, \ldots$; in order to determine $\bar{\dot{\Gamma}}_{ij}, \bar{\ddot{\Gamma}}_{ij}, \ldots$ we simply apply the defining formulas, Eqs. 7.2–38 through 40, in the $\bar{O}$-frame. Once these components have been found it is easy to verify that they satisfy Eq. 7.2–41 for any time t.

It is thus necessary to know the tilt of the corotating frame relative to the $\bar{O}$-frame at any past time t'. This can be done formally as follows:

$$
\begin{aligned}
\boldsymbol{\delta}_i(x,t,t') &= \textstyle\sum_j \bar{\Omega}_{ij}(\bar{x},t,t')\bar{\boldsymbol{\delta}}_j(t') \\
&= \textstyle\sum_j \sum_k \bar{\Omega}_{ij}(\bar{x},t,t')\mathrm{P}_{jk}(t')\boldsymbol{\delta}_k
\end{aligned}
\tag{7.2–46}
$$

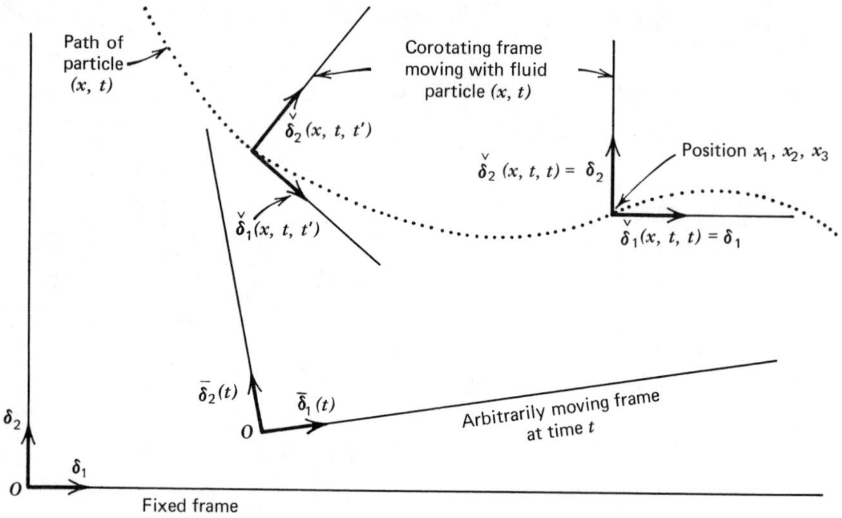

FIGURE 7.2–3. The three reference frames used in the discussion of objectivity.

where $\bar{\Omega}_{ij}(\bar{x},t,t')$ describes the corotating frame moving with the particle $(\bar{x},t)$. Note that a particle can be labeled by either its coordinates $\bar{x}_i$ at time t or its coordinates x_i at time t. Since $(\bar{x},t)$ and (x,t) are used to refer to the same particle, x_i and $\bar{x}_i$ must be related by Eqs. 7.2–7 and 8 (written for time t). Comparing Eq. 7.2–46 with Eq. 7.2–2 gives after a slight rearrangement:

$$\bar{\Omega}_{ij}(\bar{x},t,t')|_{\bar{x}_i=\bar{x}_i(x,t)} = \textstyle\sum_k \Omega_{ik}(x,t,t')\mathrm{P}_{jk}(t') \tag{7.2–47}$$

By taking $\partial/\partial t'$ of Eq. 7.2–47 and using previously developed expressions for $(\partial/\partial t')\Omega_{ik}$ and $\dot{\mathrm{P}}_{ik}$, it is straightforward to show that

$$\frac{\partial}{\partial t'}\bar{\Omega}_{ij}(\bar{x},t,t') = \textstyle\sum_n \bar{\Omega}_{in}(\bar{x},t,t')\bar{\omega}_{nj}(\bar{x},t,t') \tag{7.2–48}$$

In other words, $\bar{\Omega}_{ij}$ is computed from a differential equation in the arbitrary frame in just the same way Ω_{ij} is computed in the fixed frame. The initial condition on $\bar{\Omega}_{ij}$ is different, however:

$$\bar{\Omega}_{ij}(\bar{x},t,t) = \mathrm{P}_{ji}(t) \tag{7.2–49}$$

The components $\bar{\dot{\Gamma}}_{ij}(\bar{x},t,t')$ in the arbitrary frame are (cf. Eq. 7.2–38)

$$\begin{aligned}\bar{\dot{\Gamma}}_{ij}(\bar{x},t,t') &\equiv \textstyle\sum_k\sum_l\sum_m\sum_n \bar{\Omega}_{ki}(\bar{x},t,t)\bar{\Omega}_{km}(\bar{x},t,t')\bar{\dot{\gamma}}_{mn}(\bar{x},t,t')\bar{\Omega}_{ln}(\bar{x},t,t')\bar{\Omega}_{lj}(\bar{x},t,t)\\ &= \textstyle\sum_k\sum_l\sum_m\sum_n\sum_p\sum_q\sum_r\sum_s \mathrm{P}_{ik}(t)\Omega_{kp}(x,t,t')\mathrm{P}_{mp}(t')\mathrm{P}_{mq}(t')\dot{\gamma}_{qr}(x,t,t')\\ &\quad\cdot \mathrm{P}_{nr}(t')\Omega_{ls}(x,t,t')\mathrm{P}_{ns}(t')\mathrm{P}_{jl}(t)\\ &= \textstyle\sum_k\sum_l \mathrm{P}_{ik}(t)\dot{\Gamma}_{kl}(x,t,t')\mathrm{P}_{jl}(t)\end{aligned} \tag{7.2–50}$$

and thus $\dot{\mathbf{\Gamma}}$ is an objective tensor. In particular $\dot{\mathbf{\Gamma}}(x,t,t')$ gives an objective measure at the present time of past rates of strain experienced by particle (x,t). The time derivatives of $\dot{\mathbf{\Gamma}}$ following the fluid are also objective tensors as can be shown by computing these derivatives as seen by the arbitrarily translating and rotating observer. For example

$$\bar{\ddot{\Gamma}}_{ij}(\bar{x},t,t') \equiv \overline{\frac{\partial}{\partial t'}\dot{\Gamma}}_{ij}(\bar{x},t,t')$$

$$= \sum_k \sum_l \sum_m \sum_n \bar{\Omega}_{ki}(\bar{x},t,t)\frac{\partial}{\partial t'}[\bar{\Omega}_{km}(\bar{x},t,t')\bar{\dot{\gamma}}_{mn}(\bar{x},t,t')\bar{\Omega}_{ln}(\bar{x},t,t')]\bar{\Omega}_{lj}(\bar{x},t,t)$$

$$= \sum_k \sum_l \sum_m \sum_n \bar{\Omega}_{ki}(\bar{x},t,t)\bar{\Omega}_{km}(\bar{x},t,t')\left(\overline{\frac{\mathscr{D}}{\mathscr{D}t'}\dot{\gamma}}_{mn}(\bar{x},t,t')\right)\bar{\Omega}_{ln}(\bar{x},t,t')\bar{\Omega}_{lj}(\bar{x},t,t)$$

(cf. Eq. 7.2–32)

$$= \sum_k \sum_l \sum_m \sum_n \mathrm{P}_{ik}(t)\Omega_{km}(x,t,t')\left(\frac{\mathscr{D}}{\mathscr{D}t'}\dot{\gamma}_{mn}(x,t,t')\right)\Omega_{ln}(x,t,t')\mathrm{P}_{jl}(t)$$

(cf. Eqs. 7.2–45, 50)

$$= \sum_k \sum_l \mathrm{P}_{ik}(t)\ddot{\Gamma}_{kl}(x,t,t')\mathrm{P}_{jl}(t) \qquad (7.2\text{–}51)$$

We can conclude from the above that $\ddot{\mathbf{\Gamma}}$ and all higher time derivatives of $\dot{\mathbf{\Gamma}}$ are objective. By paralleling these arguments for **T** (see Table 7.1–1) it is straightforward to show that **T**, $\dot{\mathbf{T}}$, and all higher time derivatives of **T** are objective.

At the particle (x,t), we have shown that tensor quantities seen by the corotating observer at his origin result in objective tensors when transformed into the space-fixed coordinate system. This means that any rheological equation of state formulated by the corotating observer and then transformed into the fixed frame by means of Tables 7.1–1 and 2 is objective; thus the method outlined in §7.1 for constructing objective rheological equations of state is correct. Note that it is not necessary to go through the complete procedure of first formulating a rheological equation of state in the corotating frame and then transforming it to the fixed frame by using Tables 7.1–1 and 2. Instead, the equation can be constructed directly from the objective tensors $\dot{\mathbf{\Gamma}}, \ddot{\mathbf{\Gamma}}, \ldots, \mathbf{T}, \dot{\mathbf{T}}, \ldots$ If these tensors are to be used to describe rate of strain and stress at past times, then they have the forms listed in columns (2′) of the tables. However, if they are used for the present rate of strain and stress, then they appear in the special forms given in columns (2) of the tables. In the remaining part of this chapter we always start in the corotating frame since we feel that this procedure is particularly illustrative of the physical aspects of the methods.

In concluding this section we point out a slight difference in notation between this section and the rest of the book. Here, in the derivation of the transformation laws between frames, we have felt it particularly important to include all functional dependences explicitly in the arguments of various corotating, fixed, and arbitrary reference frame quantities. Since the corotational and fixed components are always evaluated at the particle of interest, we have not felt it necessary to carry this somewhat cumbersome notation into other sections. We use the following abbreviations: $\check{\boldsymbol{\delta}}_i(t') = \check{\boldsymbol{\delta}}_i(x,t,t')$, $\Omega_{ij}(t,t') = \Omega_{ij}(x,t,t')$, $\check{\gamma}_{ij}(t') = \check{\gamma}_{ij}(x,t,t')$, $(\partial/\partial t')\check{\gamma}_{ij}(t') = (\partial/\partial t')\check{\gamma}_{ij}(x,t,t')$, $\dot{\Gamma}_{ij}(t,t') = \dot{\Gamma}_{ij}(x,t,t')$, $(\partial/\partial t')\dot{\Gamma}_{ij}(t,t') = (\partial/\partial t')\dot{\Gamma}_{ij}(x,t,t')$ with similar abbreviations for τ_{ij} and T_{ij}.

§7.3 THE COROTATIONAL JEFFREYS MODEL[1]

In this section we present the first of several applications of the method given in §7.1 for constructing objective rheological equations of state. Because the Jeffreys model of Eq. 6.1–7 is known to describe linear viscoelastic behavior qualitatively, we use it to

[1] J. G. Oldroyd, in *Rheology of Disperse Systems*, Pergamon, London (1959), pp. 1–15; see Eq. 10.

generate a nonlinear viscoelastic model. This is done by rewriting the Jeffreys model in the corotating frame; an alternate form can be obtained by integrating the Jeffreys equation in the corotating frame. This then gives two alternative versions for the Jeffreys model:

$$\breve{\tau}_{ij} + \lambda_1 \frac{\partial \breve{\tau}_{ij}}{\partial t} = -\eta_0 \left(\breve{\dot{\gamma}}_{ij} + \lambda_2 \frac{\partial \breve{\dot{\gamma}}_{ij}}{\partial t} \right) \tag{7.3–1}$$

$$\breve{\tau}_{ij} = -\int_{-\infty}^{t} \left\{ \frac{\eta_0}{\lambda_1} \left[\left(1 - \frac{\lambda_2}{\lambda_1}\right) e^{-(t-t')/\lambda_1} + 2\lambda_2 \delta(t - t') \right] \right\} \breve{\dot{\gamma}}_{ij}(t')\, dt' \qquad (7.3\text{–}2)^2$$

As we shall see presently, in order for the viscosity function to have a physically reasonable shape, λ_2/λ_1 must be between 1/3 and 1. When Tables 7.1–1 and 2 are used, these equations become the *corotational Jeffreys model*:

$$\boldsymbol{\tau} + \lambda_1 \frac{\mathscr{D}}{\mathscr{D}t} \boldsymbol{\tau} = -\eta_0 \left(\dot{\boldsymbol{\gamma}} + \lambda_2 \frac{\mathscr{D}}{\mathscr{D}t} \dot{\boldsymbol{\gamma}} \right) \qquad (7.3\text{–}3)^{3,4}$$

$$\boldsymbol{\tau} = -\int_{-\infty}^{t} \left\{ \frac{\eta_0}{\lambda_1} \left[\left(1 - \frac{\lambda_2}{\lambda_1}\right) e^{-(t-t')/\lambda_1} + 2\lambda_2 \delta(t - t') \right] \right\} \dot{\boldsymbol{\Gamma}}'\, dt' \tag{7.3–4}$$

Here we have used the symbol $\dot{\boldsymbol{\Gamma}}'$, introduced in Table 7.1–2, as a convenient shorthand:

$$\dot{\boldsymbol{\Gamma}}' \equiv \dot{\boldsymbol{\Gamma}}(t,t') = \{\boldsymbol{\Omega}(t,t') \cdot \dot{\boldsymbol{\gamma}}(t') \cdot \boldsymbol{\Omega}^\dagger(t,t')\} \tag{7.3–5}$$

In writing Eqs. 7.3–3 to 5 we have used boldface tensor notation, rather than the Cartesian tensor notation in Eqs. 7.3–1 and 2, so that the indicated operations can be performed in curvilinear coordinates if desired.

In order to evaluate the model, we obtain expressions for the material functions in several different flow situations as described in the following four illustrative examples.

EXAMPLE 7.3–1 Steady Shear Flow of a Corotational Jeffreys Fluid

Consider the steady shear flow of a corotational Jeffreys fluid in the space between two parallel plates, one of which is moving. Let the flow field be $v_x = \dot{\gamma} y$, where $\dot{\gamma}$ is the constant shear rate. Obtain

[2] The factor of 2 in the second term in the relaxation modulus arises because of the definition of the Dirac δ-function given in Eq. 6.1–9b.

[3] An extension of this equation to include higher Jaumann derivatives was suggested by J. G. Oldroyd, in *Second-Order Effects in Elasticity, Plasticity, and Fluid Dynamics*, M. Reiner and D. Abir (Eds.), Macmillan, New York (1964), pp. 520–529. This higher-order equation can be solved explicitly for $\boldsymbol{\tau}$ as shown by R. T. Balmer and J. J. Kauzlarich, in *Developments in Mechanics*, Weiss, Young, Riley, and Rogge (Eds.), Iowa State University Press, Ames (1969), pp. 113–133; see also R. T. Balmer, *Trans. Soc. Rheol.*, *16*, 277–293 (1972).

[4] If we set $\lambda_2 = 0$ and $\lambda_1 = \lambda_0$, then we get the two-constant corotational analog of the Maxwell model of Eq. 6.1–4. This two-constant model was first given by S. Zaremba, *Bull. Int. Acad. Sci. Cracovie*, pp. 594–614 (1903); pp. 614–621 (1903), and rediscovered by H. Fromm, *Zeits. für angew. Math. u. Mech.*, *25/27*, 146–150 (1947); *28*, 43–54 (1948) and T. W. DeWitt, *J. Appl. Phys.*, *26*, 889–894 (1955); it is appropriate to call it the *ZFD-model*. This simple two-constant model can describe qualitatively many observed phenomena in rheology; however it has also serious shortcomings, the most important one of which is that, for simple steady-state shear flow, τ_{yx} increases and then decreases as the shear rate is increased. See Problems 7B.7, 7B.8, 7C.1, 7C.3, 7D.1, and 7D.2 for further information on the ZFD-model.

the expressions for the viscosity and normal stress functions (a) by using Eq. 7.3–3 and (b) by using Eq. 7.3–4.

SOLUTION (a) Use of Differential Form

For the flow under consideration we can obtain the various kinematic tensors needed to find $(\mathscr{D}/\mathscr{D}t)\boldsymbol{\tau}$ and $(\mathscr{D}/\mathscr{D}t)\dot{\boldsymbol{\gamma}}$ (cf. Example 5.5–1):

$$\nabla \boldsymbol{v} = \begin{pmatrix} 0 & 0 & 0 \\ 1 & 0 & 0 \\ 0 & 0 & 0 \end{pmatrix} \dot{\gamma}; \qquad (\nabla \boldsymbol{v})^\dagger = \begin{pmatrix} 0 & 1 & 0 \\ 0 & 0 & 0 \\ 0 & 0 & 0 \end{pmatrix} \dot{\gamma} \tag{7.3–6, 7}$$

$$\dot{\boldsymbol{\gamma}} = \begin{pmatrix} 0 & 1 & 0 \\ 1 & 0 & 0 \\ 0 & 0 & 0 \end{pmatrix} \dot{\gamma}; \qquad \boldsymbol{\omega} = \begin{pmatrix} 0 & -1 & 0 \\ 1 & 0 & 0 \\ 0 & 0 & 0 \end{pmatrix} \dot{\gamma} \tag{7.3–8, 9}$$

Since $\boldsymbol{\tau}$ is independent of x, y, z, and t, then in Eq. A of Table 7.1–1 $\partial\boldsymbol{\tau}/\partial t = \mathbf{0}$ and $\{\boldsymbol{v} \cdot \nabla\boldsymbol{\tau}\} = \mathbf{0}$. Next

$$\{\boldsymbol{\omega} \cdot \boldsymbol{\tau}\} = \begin{pmatrix} 0 & -1 & 0 \\ 1 & 0 & 0 \\ 0 & 0 & 0 \end{pmatrix} \begin{pmatrix} \tau_{xx} & \tau_{xy} & \tau_{xz} \\ \tau_{yx} & \tau_{yy} & \tau_{yz} \\ \tau_{zx} & \tau_{zy} & \tau_{zz} \end{pmatrix} \dot{\gamma} = \begin{pmatrix} -\tau_{yx} & -\tau_{yy} & -\tau_{yz} \\ \tau_{xx} & \tau_{xy} & \tau_{xz} \\ 0 & 0 & 0 \end{pmatrix} \dot{\gamma} \tag{7.3–10}$$

$$\{\boldsymbol{\omega} \cdot \boldsymbol{\tau}\}^\dagger = -\{\boldsymbol{\tau} \cdot \boldsymbol{\omega}\} = \begin{pmatrix} -\tau_{yx} & \tau_{xx} & 0 \\ -\tau_{yy} & \tau_{xy} & 0 \\ -\tau_{yz} & \tau_{xz} & 0 \end{pmatrix} \dot{\gamma} \tag{7.3–11}$$

Hence the corotational time derivatives are:

$$\frac{\mathscr{D}}{\mathscr{D}t}\boldsymbol{\tau} = \begin{pmatrix} -\tau_{yx} & \frac{1}{2}(\tau_{xx} - \tau_{yy}) & -\frac{1}{2}\tau_{yz} \\ \frac{1}{2}(\tau_{xx} - \tau_{yy}) & \tau_{xy} & \frac{1}{2}\tau_{xz} \\ -\frac{1}{2}\tau_{yz} & \frac{1}{2}\tau_{xz} & 0 \end{pmatrix} \dot{\gamma} \tag{7.3–12}$$

$$\frac{\mathscr{D}}{\mathscr{D}t}\dot{\boldsymbol{\gamma}} = \tfrac{1}{2}\{\boldsymbol{\omega} \cdot \dot{\boldsymbol{\gamma}} - \dot{\boldsymbol{\gamma}} \cdot \boldsymbol{\omega}\} = \begin{pmatrix} -1 & 0 & 0 \\ 0 & 1 & 0 \\ 0 & 0 & 0 \end{pmatrix} \dot{\gamma}^2 \tag{7.3–13}$$

We are now in a position to write Eq. 7.3–3 in matrix form for the particular flow field under consideration:

$$\begin{pmatrix} \tau_{xx} & \tau_{xy} & \tau_{xz} \\ \tau_{yx} & \tau_{yy} & \tau_{yz} \\ \tau_{zx} & \tau_{zy} & \tau_{zz} \end{pmatrix} + \lambda_1 \begin{pmatrix} -\tau_{yx} & \frac{1}{2}(\tau_{xx} - \tau_{yy}) & -\frac{1}{2}\tau_{yz} \\ \frac{1}{2}(\tau_{xx} - \tau_{yy}) & \tau_{xy} & \frac{1}{2}\tau_{xz} \\ -\frac{1}{2}\tau_{yz} & \frac{1}{2}\tau_{xz} & 0 \end{pmatrix} \dot{\gamma}$$

$$= -\eta_0 \begin{pmatrix} 0 & 1 & 0 \\ 1 & 0 & 0 \\ 0 & 0 & 0 \end{pmatrix} \dot{\gamma} - \eta_0 \lambda_2 \begin{pmatrix} -1 & 0 & 0 \\ 0 & 1 & 0 \\ 0 & 0 & 0 \end{pmatrix} \dot{\gamma}^2 \tag{7.3–14}$$

Because the stress tensor is symmetrical and because the normal stresses are undetermined to within an isotropic function, the above matrix equation yields five algebraic equations for the measurable stresses:

$$(\tau_{xx} - \tau_{yy}) + \lambda_1(-2\tau_{yx}\dot{\gamma}) = 2\eta_0\lambda_2\dot{\gamma}^2 \quad (7.3\text{–}15)$$

$$(\tau_{yy} - \tau_{zz}) + \lambda_1(\tau_{yx}\dot{\gamma}) = -\eta_0\lambda_2\dot{\gamma}^2 \quad (7.3\text{–}16)$$

$$\tau_{yx} + \lambda_1[\tfrac{1}{2}(\tau_{xx} - \tau_{yy})\dot{\gamma}] = -\eta_0\dot{\gamma} \quad (7.3\text{–}17)$$

$$\tau_{xz} + \lambda_1(-\tfrac{1}{2}\tau_{yz}\dot{\gamma}) = 0 \quad (7.3\text{–}18)$$

$$\tau_{yz} + \lambda_1(+\tfrac{1}{2}\tau_{xz}\dot{\gamma}) = 0 \quad (7.3\text{–}19)$$

From the last two equations it is easy to see that τ_{xz} and τ_{yz} are both zero, as we know they must be for the simple shearing flow considered here (cf. Eq. 4.2–7). Solving the remaining three equations gives:

$$\tau_{yx} = -\eta_0\dot{\gamma}\frac{1 + \lambda_1\lambda_2\dot{\gamma}^2}{1 + \lambda_1^2\dot{\gamma}^2} \quad \text{or} \quad \eta = \eta_0\frac{1 + \lambda_1\lambda_2\dot{\gamma}^2}{1 + \lambda_1^2\dot{\gamma}^2} \quad (7.3\text{–}20)$$

$$\tau_{xx} - \tau_{yy} = -\frac{2\eta_0(\lambda_1 - \lambda_2)\dot{\gamma}^2}{1 + \lambda_1^2\dot{\gamma}^2} \quad \text{or} \quad \Psi_1 = \frac{2\eta_0(\lambda_1 - \lambda_2)}{1 + \lambda_1^2\dot{\gamma}^2} \quad (7.3\text{–}21)$$

$$\tau_{yy} - \tau_{zz} = +\frac{\eta_0(\lambda_1 - \lambda_2)\dot{\gamma}^2}{1 + \lambda_1^2\dot{\gamma}^2} \quad \text{or} \quad \Psi_2 = -\frac{\eta_0(\lambda_1 - \lambda_2)}{1 + \lambda_1^2\dot{\gamma}^2} \quad (7.3\text{–}22)$$

From Eq. 7.3–20 we see that η_0 may be interpreted as the zero-shear-rate viscosity. The viscosity and primary normal stress coefficient given by Eqs. 7.3–20 and 21 are shown in Fig. 7.3–1 for $\lambda_2/\lambda_1 = 2/5$.

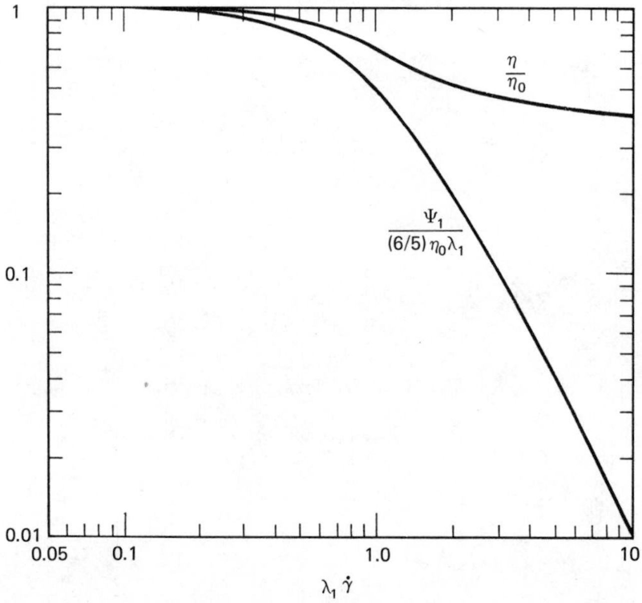

FIGURE 7.3–1. Viscosity and primary normal stress coefficient for the corotational Jeffreys model with $\lambda_2/\lambda_1 = \frac{2}{5}$. (This particular value exactly reproduces the linear viscoelastic properties of a rigid dumbbell solution—see §2.5 and Chapter 11.)

(b) Use of Integral Form

Let us now show how to get the same results by using Eq. 7.3–4. For steady-state shear flows Eq. 7.1–13 simplifies to:

$$\mathbf{\Omega} = \begin{pmatrix} \cos\frac{1}{2}\dot{\gamma}(t-t') & \sin\frac{1}{2}\dot{\gamma}(t-t') & 0 \\ -\sin\frac{1}{2}\dot{\gamma}(t-t') & \cos\frac{1}{2}\dot{\gamma}(t-t') & 0 \\ 0 & 0 & 1 \end{pmatrix} \tag{7.3–23}$$

Then from Eq. 7.3–5 we obtain by doing the tensor multiplication $\dot{\mathbf{\Gamma}}' = \{\mathbf{\Omega}\cdot\dot{\boldsymbol{\gamma}}'\cdot\mathbf{\Omega}^{\dagger}\}$:

$$\dot{\mathbf{\Gamma}}' = \begin{pmatrix} \dot{\gamma}\sin\dot{\gamma}(t-t') & \dot{\gamma}\cos\dot{\gamma}(t-t') & 0 \\ \dot{\gamma}\cos\dot{\gamma}(t-t') & -\dot{\gamma}\sin\dot{\gamma}(t-t') & 0 \\ 0 & 0 & 0 \end{pmatrix} \tag{7.3–24}$$

Therefore Eq. 7.3–4 becomes, for the particular flow field under consideration:

$$\begin{pmatrix} \tau_{xx} & \tau_{xy} & \tau_{xz} \\ \tau_{yx} & \tau_{yy} & \tau_{yz} \\ \tau_{zx} & \tau_{zy} & \tau_{zz} \end{pmatrix} = -\int_0^\infty \left\{\frac{\eta_0}{\lambda_1}\left[\left(1-\frac{\lambda_2}{\lambda_1}\right)e^{-s/\lambda_1} + 2\lambda_2\delta(s)\right]\right\}$$
$$\cdot \begin{pmatrix} \sin\dot{\gamma}s & \cos\dot{\gamma}s & 0 \\ \cos\dot{\gamma}s & -\sin\dot{\gamma}s & 0 \\ 0 & 0 & 0 \end{pmatrix}\dot{\gamma}\,ds \tag{7.3–25}$$

When the integrations in Eq. 7.3–25 are performed, we find that τ_{xz} and τ_{yz} are both zero, and that τ_{yx}, $\tau_{xx} - \tau_{yy}$, and $\tau_{yy} - \tau_{zz}$ are identical with the expressions obtained in part (a).

To what extent are the functions in Eqs. 7.3–20 to 22 in agreement with experimental facts? As can be seen in Fig. 7.3–1 the curves for $\eta(\dot{\gamma})$ and $\Psi_1(\dot{\gamma})$ for $\lambda_2/\lambda_1 = \frac{2}{5}$ have shapes not unlike those found experimentally for many polymeric fluids. Furthermore the second normal stress coefficient is negative, in agreement with laboratory observations; but the model gives a value of -0.5 for the ratio Ψ_2/Ψ_1, whereas measured values of about -0.02 to -0.4 have been reported in low-shear-rate experiments. Note that $\Psi_2/\Psi_1 = -0.5$ implies that there would be no radial variation in the total normal stress exerted on the plate in the cone-and-plate viscometer (see, for example, Eq. 4.5–5), whereas the experimental data shown in Fig. 4A.2 indicate otherwise. Because the calculated material functions cannot reproduce the viscosity function over a large range of the shear rate and because the magnitude of the secondary normal-stress function is unreasonably large, we have to conclude that the corotational Jeffreys model is not entirely satisfactory for the description of steady-state shear flow behavior.

In conclusion, we are now in a position to see why for polymeric fluids some restrictions have to be placed on the value of the ratio λ_2/λ_1. Since most polymer solutions and polymer melts are "shear thinning," we have to require that $\lambda_2/\lambda_1 < 1$. Also, in order to insure that τ_{yx} is a monotone increasing function of shear rate, we have to impose the additional restriction that $\lambda_2/\lambda_1 > \frac{1}{3}$. This restriction means that η_∞/η_0 must always be greater than $\frac{1}{3}$; this is not particularly realistic for concentrated solutions and melts, as can be seen from the experimental data in §4.3.

EXAMPLE 7.3–2 Small-Amplitude Oscillations in a Corotational Jeffreys Fluid

Examine the small-amplitude oscillatory flow, $v_x = \dot{\gamma}(t)y$, with $\dot{\gamma} = \dot{\gamma}^0 \cos\omega t$, where ω is the frequency and $\dot{\gamma}^0$ is a real positive quantity denoting the amplitude of the velocity gradient. Obtain the expressions for $\eta'(\omega)$ and $\eta''(\omega)$. The requirement that the amplitude of the oscillations be small is equivalent to the requirement that $(\dot{\gamma}^0/\omega) \ll 1$.

SOLUTION

Here we use the integral form of the model (see Problem 7C.1 for use of the differential form). For the oscillatory motion, the $\Omega_{ij}(t,t')$ are given by Eq. 7.1–13:

$$\boldsymbol{\Omega} = \begin{bmatrix} \cos\left[\frac{\dot{\gamma}^0}{2\omega}(\sin\omega t - \sin\omega t')\right] & \sin\left[\frac{\dot{\gamma}^0}{2\omega}(\sin\omega t - \sin\omega t')\right] & 0 \\ -\sin\left[\frac{\dot{\gamma}^0}{2\omega}(\sin\omega t - \sin\omega t')\right] & \cos\left[\frac{\dot{\gamma}^0}{2\omega}(\sin\omega t - \sin\omega t')\right] & 0 \\ 0 & 0 & 1 \end{bmatrix} \tag{7.3–26}$$

The rate-of-strain tensor is:

$$\dot{\boldsymbol{\gamma}}(t') = \begin{pmatrix} 0 & 1 & 0 \\ 1 & 0 & 0 \\ 0 & 0 & 0 \end{pmatrix} \dot{\gamma}^0 \cos\omega t' \tag{7.3–27}$$

Then by the tensor multiplication $\dot{\boldsymbol{\Gamma}}' = \{\boldsymbol{\Omega} \cdot \dot{\boldsymbol{\gamma}}' \cdot \boldsymbol{\Omega}^{\dagger}\}$ we obtain:

$$\dot{\boldsymbol{\Gamma}}' = \begin{bmatrix} \sin\left[\frac{\dot{\gamma}^0}{\omega}(\sin\omega t - \sin\omega t')\right] & \cos\left[\frac{\dot{\gamma}^0}{\omega}(\sin\omega t - \sin\omega t')\right] & 0 \\ \cos\left[\frac{\dot{\gamma}^0}{\omega}(\sin\omega t - \sin\omega t')\right] & -\sin\left[\frac{\dot{\gamma}^0}{\omega}(\sin\omega t - \sin\omega t')\right] & 0 \\ 0 & 0 & 0 \end{bmatrix} \dot{\gamma}^0 \cos\omega t'$$

$$\approx \begin{bmatrix} \frac{\dot{\gamma}^0}{\omega}(\sin\omega t - \sin\omega t') & 1 & 0 \\ 1 & -\frac{\dot{\gamma}^0}{\omega}(\sin\omega t - \sin\omega t') & 0 \\ 0 & 0 & 0 \end{bmatrix} \dot{\gamma}^0 \cos\omega t' \tag{7.3–28}$$

The second, approximate, form retains only those terms of first and second order in $\dot{\gamma}^0/\omega$, after expanding sin [ ] and cos [ ] in Taylor series.

Insertion of the second expression of Eq. 7.3–28 into Eq. 7.3–4 enables us to get the stresses. For τ_{yx} we get the following result:

$$\tau_{yx} = -\eta_0\dot{\gamma}^0 \frac{1 + \lambda_1\lambda_2\omega^2}{1 + \lambda_1{}^2\omega^2} \cos\omega t - \eta_0\dot{\gamma}^0 \frac{(\lambda_1 - \lambda_2)\omega}{1 + \lambda_1{}^2\omega^2} \sin\omega t \tag{7.3–29}$$

Comparison of this with the defining equation for η' and η'' (see Eq. 4.4–3) gives:

$$\eta'(\omega) = \frac{\eta_0(1 + \lambda_1\lambda_2\omega^2)}{1 + \lambda_1{}^2\omega^2} \tag{7.3–30}$$

$$\frac{\eta''(\omega)}{\omega} = \frac{\eta_0(\lambda_1 - \lambda_2)}{1 + \lambda_1{}^2\omega^2} \tag{7.3–31}$$

Comparison with Eqs. 7.3–20 and 21 shows that η and η' are identical functions of their arguments, and that $\Psi_1/2$ and η''/ω are identical functions. We know from the discussion in Chapter 4 that for many polymer solutions and melts these pairs of functions do indeed have similar shapes, although the curves do not exactly superpose. Equations 7.3–30 and 31 show that η' decreases with ω and that η'' is proportional to ω for small frequencies. Thus Eqs. 7.3–30 and 31 have qualitatively the correct shapes.

EXAMPLE 7.3–3 Stress Growth at Inception of Steady-State Shear Flow

A fluid between two parallel plates has been at rest for $t < 0$. For $t > 0$ the upper plate moves in the x-direction at constant speed so that the velocity profile $v_x = \dot{\gamma}_0 y$ is maintained. Obtain the expressions for the growth of the shear stress and the normal stresses.

SOLUTION

For this problem $\dot{\Gamma}'_{ij}$ is given by Eq. 7.3–24 for $t' > 0$, and $\dot{\Gamma}'_{ij} = 0$ for $t' < 0$. Therefore the stresses are given by:

$$\begin{pmatrix} \tau_{xx} & \tau_{xy} & 0 \\ \tau_{yx} & \tau_{yy} & 0 \\ 0 & 0 & \tau_{zz} \end{pmatrix} = -\int_0^t \left\{ \frac{\eta_0}{\lambda_1} \left[\left(1 - \frac{\lambda_2}{\lambda_1}\right) e^{-s/\lambda_1} + 2\lambda_2 \delta(s) \right] \right\} \cdot \begin{pmatrix} \sin \dot{\gamma}_0 s & \cos \dot{\gamma}_0 s & 0 \\ \cos \dot{\gamma}_0 s & -\sin \dot{\gamma}_0 s & 0 \\ 0 & 0 & 0 \end{pmatrix} \dot{\gamma}_0 \, ds \tag{7.3–32}$$

When the integrations are performed we obtain:

$$\tau_{yx} = -\eta^+ \dot{\gamma}_0 = -\eta_0 \dot{\gamma}_0 \frac{1 + \lambda_1 \lambda_2 \dot{\gamma}_0^2}{1 + (\lambda_1 \dot{\gamma}_0)^2} - \eta_0 \dot{\gamma}_0 \frac{1 - (\lambda_2/\lambda_1)}{1 + (\lambda_1 \dot{\gamma}_0)^2} (\lambda_1 \dot{\gamma}_0 \sin \dot{\gamma}_0 t - \cos \dot{\gamma}_0 t) e^{-t/\lambda_1} \tag{7.3–33}$$

$$\tau_{xx} - \tau_{yy} = -\Psi_1^+ \dot{\gamma}_0^2 = -2\eta_0 \dot{\gamma}_0^2 \frac{\lambda_1 - \lambda_2}{1 + (\lambda_1 \dot{\gamma}_0)^2} + 2\eta_0 \lambda_1 \dot{\gamma}_0^2 \frac{1 - (\lambda_2/\lambda_1)}{1 + (\lambda_1 \dot{\gamma}_0)^2} \left(\frac{1}{\lambda_1 \dot{\gamma}_0} \sin \dot{\gamma}_0 t + \cos \dot{\gamma}_0 t \right) e^{-t/\lambda_1} \tag{7.3–34}$$

According to Eq. 7.3–33, τ_{yx} attains a maximum at $\dot{\gamma}_0 t = \pi/2$ and then undergoes damped oscillations about the final steady-state value. Similarly, Eq. 7.3–34 indicates that $\tau_{xx} - \tau_{yy}$ attains a maximum at $\dot{\gamma}_0 t = \pi$ and then executes damped oscillations. A few sample curves based on Eqs. 7.3–33 and 34 are shown in Fig. 7.3–2. These curves should be compared with the experimental data shown in §4.4, Experiment (c). It can be seen that the experimental curves do exhibit the "overshoot phenomenon," and furthermore that the maximum in the normal stress difference occurs later than the maximum in the shear stress. However, there seem to be no oscillations in the experimental data, and so we have to conclude that the spurious model-predicted oscillations point to an inherent inadequacy of the corotational Jeffreys model.

EXAMPLE 7.3–4 Stress Relaxation after Cessation of Steady Shear Flow

A fluid between two parallel plates has been flowing for a long time prior to $t = 0$ with a velocity profile $v_x = \dot{\gamma}_0 y$. Suddenly at $t = 0$ the motion is stopped, and the shear and normal stresses relax. Obtain the expressions showing how this relaxation process occurs.

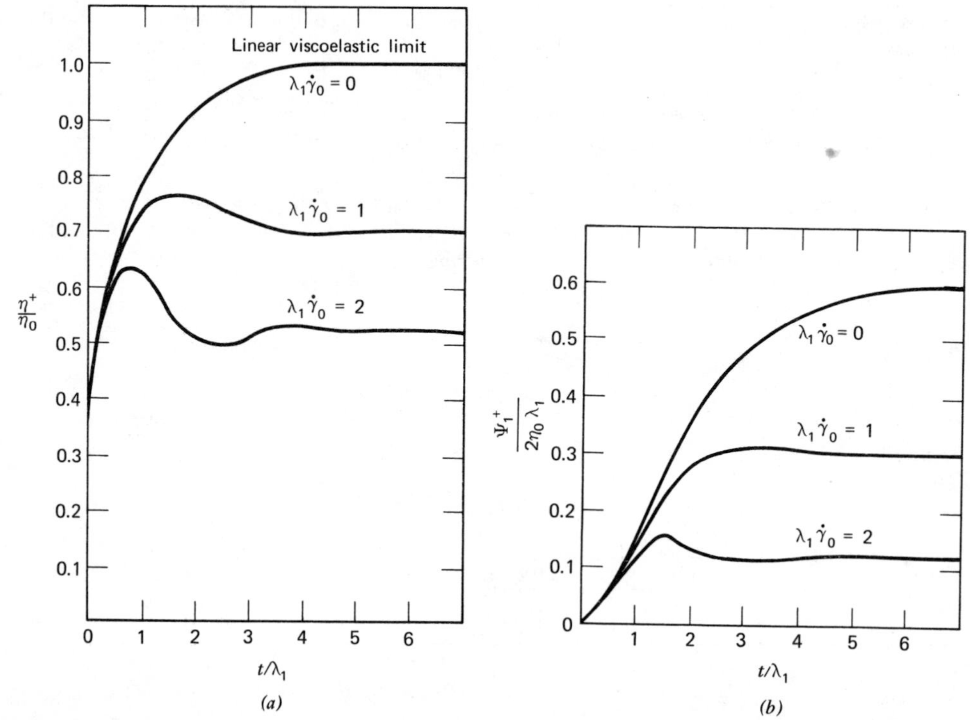

FIGURE 7.3–2. Stress growth curves calculated for the corotational Jeffreys model with $\lambda_2/\lambda_1 = \frac{2}{5}$. (*a*) Growth of the shear stress for several values of $\lambda_1\dot{\gamma}_0$, showing the "overshoot effect." The model gives an instantaneous stress jump to $\eta^+/\eta_0 = \frac{2}{5}$ at time $t = 0$. (*b*) Growth of the primary normal stress difference that also exhibits overshoot. Note that the maxima in the normal stress occur at a later time than the corresponding maxima in the shear stress curves.

SOLUTION

For this problem the $\dot{\Gamma}'_{ij}$ are all zero for $t' > 0$; for $t' < 0$, they are given by Eq. 7.3–24 with t set equal to zero. Therefore the stresses are given by:

$$\begin{pmatrix} \tau_{xx} & \tau_{xy} & 0 \\ \tau_{yx} & \tau_{yy} & 0 \\ 0 & 0 & \tau_{zz} \end{pmatrix} = -\int_t^\infty \left\{ \frac{\eta_0}{\lambda_1} \left[\left(1 - \frac{\lambda_2}{\lambda_1}\right) e^{-s/\lambda_1} + 2\lambda_2 \delta(s) \right] \right\}$$
$$\cdot \begin{pmatrix} -\sin \dot{\gamma}_0(t-s) & \cos \dot{\gamma}_0(t-s) & 0 \\ \cos \dot{\gamma}_0(t-s) & \sin \dot{\gamma}_0(t-s) & 0 \\ 0 & 0 & 0 \end{pmatrix} \dot{\gamma}_0 \, ds \quad (7.3\text{–}35)$$

From this we obtain after performing the indicated integrations:

$$\tau_{yx} = -\eta^- \dot{\gamma}_0 = -\eta_0 \dot{\gamma}_0 \frac{1 - (\lambda_2/\lambda_1)}{1 + (\lambda_1 \dot{\gamma}_0)^2} e^{-t/\lambda_1} \quad (7.3\text{–}36)$$

$$\tau_{xx} - \tau_{yy} = -\Psi_1^- \dot{\gamma}_0^2 = -2\eta_0 \lambda_1 \dot{\gamma}_0^2 \frac{1 - (\lambda_2/\lambda_1)}{1 + (\lambda_1 \dot{\gamma}_0)^2} e^{-t/\lambda_1} \quad (7.3\text{–}37)$$

These results are sketched in Fig. 7.3–3. We find here, as we did for stress growth, that the shear stress has a discontinuity at $t = 0$, whereas the normal stress does not. Thus the η^-/η curves lie below the

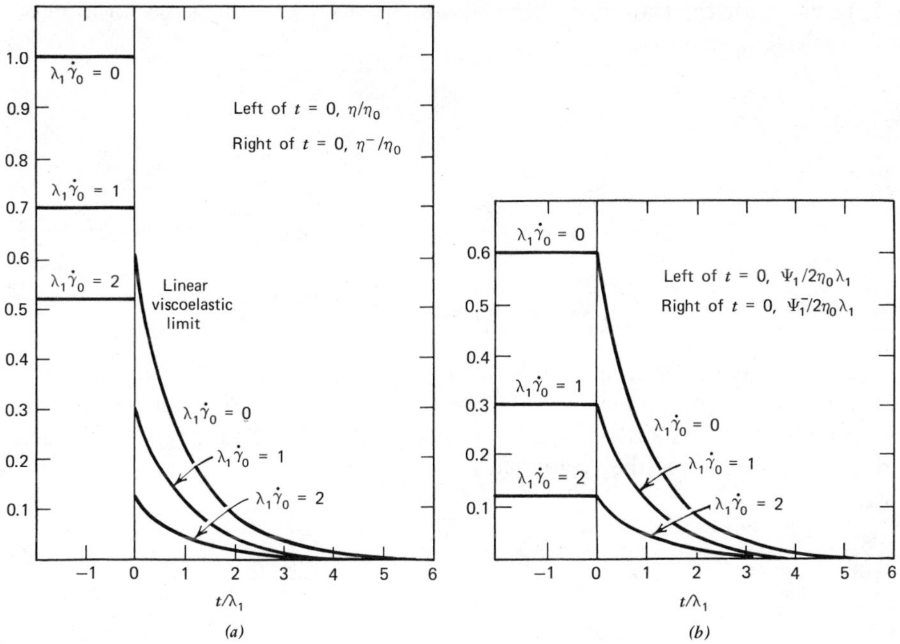

FIGURE 7.3–3. Stress relaxation curves calculated for the corotational Jeffreys model with $\lambda_2/\lambda_1 = \frac{2}{5}$. The lines left of $t = 0$ are the material functions for steady-state shear flow, and the curves to the right of $t = 0$ show the decay of stresses with time. (*a*) Shear stress relaxation, (*b*) Primary normal stress relaxation.

Ψ_1^-/Ψ_1 curves, and this is in agreement with experimental data. It is also known experimentally that both η^-/η and Ψ_1^-/Ψ_1 curves are lowered as $\dot{\gamma}_0$ is increased. Equation 7.3–36 shows this for shear stresses, but Eq. 7.3–37 does not give that kind of behavior for normal stresses. Therefore, once again we find that the model does a pretty good job, but that it is not completely satisfactory.

In this section we have illustrated the generation of a nonlinear viscoelastic model and the testing of the model to see how well it can describe the material functions in various kinds of flows. This "game" has been played by many rheologists in the last 25 years, with many models being proposed and a large number of them being rejected. Unfortunately the testing of these empirical rheological models is often not as thorough as one might desire, since, as we saw in Chapter 4, most material function data are for shear flows, and only limited experimentation has been done for elongational and other nonshearing flows.

§7.4 THE GENERALIZED ZFD MODEL[1]

In the foregoing section we developed an objective nonlinear viscoelastic model. However, the model was not capable of portraying the shear-rate-dependent viscosity function very well. For the solution of some fluid dynamics problems, it is important to describe the viscosity function accurately, with less accuracy needed for the other material functions. In this section we show how to construct a model that gives $\eta(\dot{\gamma})$ fairly realistically.

We begin by writing down a superposition of Maxwell models in the corotating frame, paralleling the corresponding equations in linear viscoelasticity, namely Eqs. 6.1–10

[1] A. J. de Vries, in *Proc. Fourth Int. Congress on Rheology*, E. H. Lee (Ed.), Wiley, New York (1965), Part III, pp. 321–344.

and 12. Then we use the transformation rules of §7.1 to rewrite the superposition of corotational Maxwell models thus:

$$\begin{cases} \boldsymbol{\tau} = \sum_{k=1}^{\infty} \boldsymbol{\tau}_k & (7.4\text{–}1) \\ \boldsymbol{\tau}_k + \lambda_k \frac{\mathscr{D}}{\mathscr{D}t} \boldsymbol{\tau}_k = -\eta_k \dot{\boldsymbol{\gamma}} & (7.4\text{–}2) \end{cases}$$

or

$$\boldsymbol{\tau} = -\int_{-\infty}^{t} \left\{ \sum_{k=1}^{\infty} \frac{\eta_k}{\lambda_k} e^{-(t-t')/\lambda_k} \right\} \dot{\boldsymbol{\Gamma}}' \, dt' \qquad (7.4\text{–}3)$$

These are the differential and integral forms of the *generalized ZFD model* (see footnote 4 of §7.3). For some of the simple flows used for measuring material functions, the solutions for the generalized ZFD model are just obtained by superposing the results for the simple ZFD model. For example the steady shear flow material functions are[1] (cf. Eqs. 7.3–20 to 22 with $\lambda_2 = 0$):

$$\eta = \sum_{k=1}^{\infty} \frac{\eta_k}{1 + (\lambda_k \dot{\gamma})^2} \qquad (7.4\text{–}4)$$

$$\tfrac{1}{2}\Psi_1 = \sum_{k=1}^{\infty} \frac{\eta_k \lambda_k}{1 + (\lambda_k \dot{\gamma})^2} = -\Psi_2 \qquad (7.4\text{–}5)$$

If we introduce the empirical equations in Eqs. 6.1–14 and 15 we can then obtain expressions involving only three parameters: the zero-shear-rate viscosity η_0, a time constant λ, and a dimensionless parameter α that is related to the slope of the log η vs. log $\dot{\gamma}$ curve in the "power law region." We can obtain a rapidly converging expansion for small $\dot{\gamma}$ and an asymptotic expression for large $\dot{\gamma}$ (cf. Eqs. 6.2–12 and 13 and Problem 6B.5):

Low-shear-rate region:

$$\frac{\eta}{\eta_0} = 1 - \frac{(\lambda\dot{\gamma})^2}{\zeta(\alpha)} \sum_{k=1}^{\infty} \frac{1}{k^{\alpha}(k^{2\alpha} + (\lambda\dot{\gamma})^2)} \qquad (7.4\text{–}6)$$

$$\frac{\Psi_1}{2\eta_0\lambda} = -\frac{\Psi_2}{\eta_0\lambda} = \frac{\zeta(2\alpha)}{\zeta(\alpha)} - \frac{(\lambda\dot{\gamma})^2}{\zeta(\alpha)} \sum_{k=1}^{\infty} \frac{1}{k^{2\alpha}(k^{2\alpha} + (\lambda\dot{\gamma})^2)} \qquad (7.4\text{–}7)$$

High-shear-rate region:

$$\frac{\eta}{\eta_0} \cong \frac{1}{\zeta(\alpha)} \left[\frac{\pi(\lambda\dot{\gamma})^{(1/\alpha)-1}}{2\alpha \sin((\alpha+1)\pi/2\alpha)} \right] \qquad (7.4\text{–}8)$$

$$\frac{\Psi_1}{2\eta_0\lambda} = -\frac{\Psi_2}{\eta_0\lambda} \cong \frac{1}{\zeta(\alpha)} \left[\frac{\pi(\lambda\dot{\gamma})^{(1/\alpha)-2}}{2\alpha \sin(\pi/2\alpha)} - \frac{(\lambda\dot{\gamma})^{-2}}{2} \right] \qquad (7.4\text{–}9)$$

Equations 7.4–6 and 8 are able to fit viscosity curves rather well, inasmuch as they describe the "flat regions" near zero shear rate and the "power-law regions" for high shear rates (see Fig. 7.4–1); however, the curves do show some spurious wiggles for large values of α. The normal stress curves in Fig. 7.4–1 and the shear stress growth curves in Fig. 7.4–2 also have qualitatively the right shapes, except for the spurious oscillations in η^+ for large $\dot{\gamma}_0$. According to Eqs. 7.4–8 and 9 the slopes of η and Ψ_1 in the power-law region differ by unity; such behavior is not observed for real fluids as can be seen from the data given in the table at the top of Fig. 5.2–5.

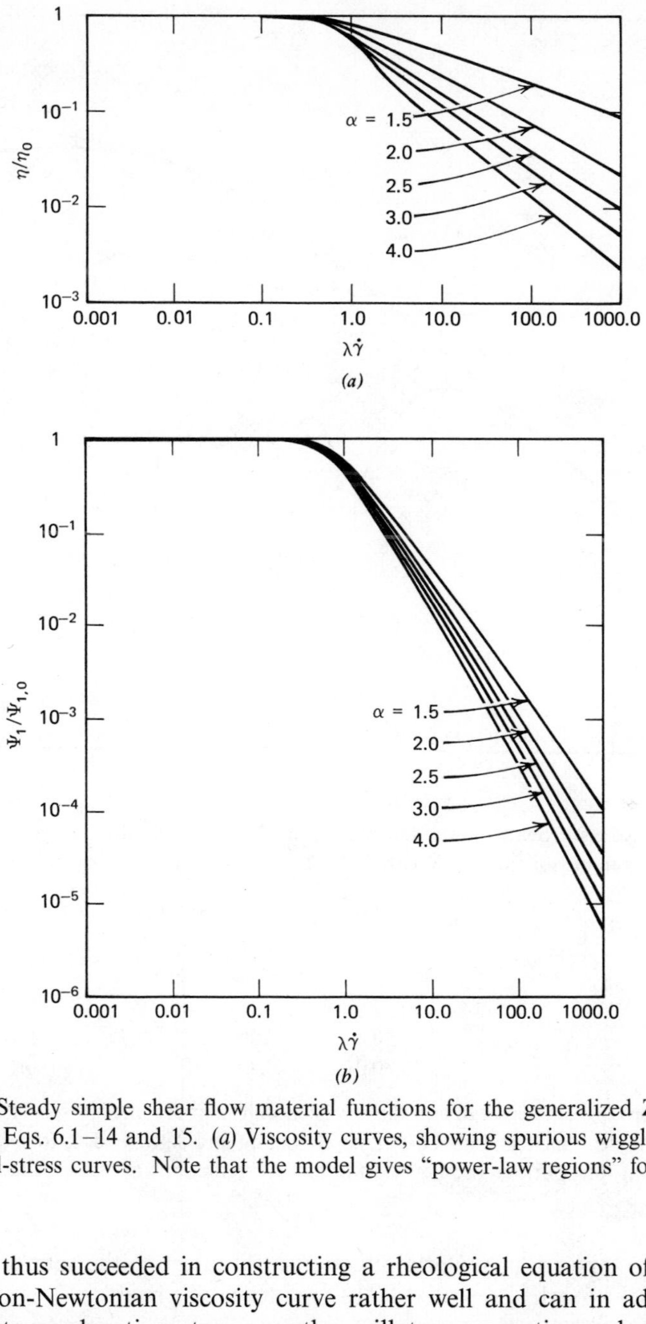

FIGURE 7.4–1. Steady simple shear flow material functions for the generalized ZFD model, with the empiricisms of Eqs. 6.1–14 and 15. (*a*) Viscosity curves, showing spurious wiggle near $\lambda\dot{\gamma} = 1$ for high α; (*b*) Normal-stress curves. Note that the model gives "power-law regions" for both η and Ψ_1.

We have thus succeeded in constructing a rheological equation of state that can reproduce the non-Newtonian viscosity curve rather well and can in addition describe normal stresses, stress relaxation, stress growth, oscillatory properties, and other rheological phenomena *approximately*. Therefore for those problems in which the viscosity plays the dominant role and the various other phenomena are likely to be less important, the generalized ZFD model may prove to be useful. To date, however, no complex flow problems have been solved using this model.

We could go on and give a number of other illustrations of the use of the ZFD model, but nothing new would be learned about the technique of testing rheological equations of state by comparing their predictions for simple flows with measured material functions. In the next section we consider a slightly more general equation obtained by replacing the quantity in braces in Eq. 7.4–3 by an arbitrary function $G(t - t')$. We give a number of

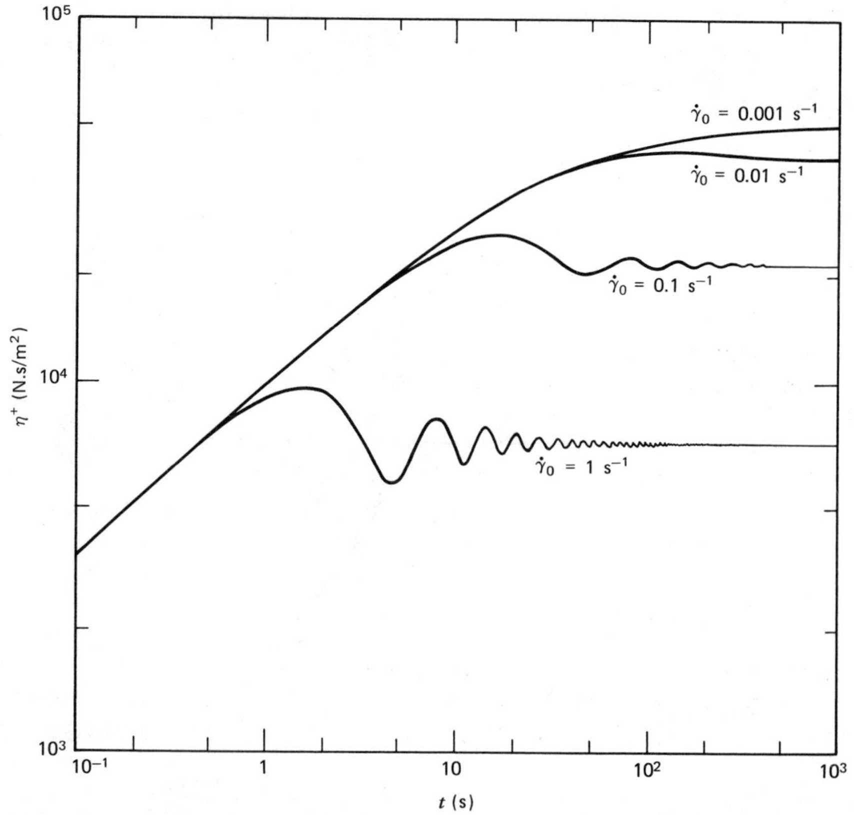

FIGURE 7.4–2. Stress growth on sudden inception of shear flow, for the generalized ZFD model. The calculations are made for Eq. 7.4–3 with

i	λ_i(s)	η_i (N·s/m²)
1	100	1.903×10^4
2	10	2.306×10^4
3	1	6.241×10^3
4	0.1	3.299×10^3
5	0.01	3.72×10^2

and with $\eta_i = 0$ for $i \geq 6$. This set of constants was chosen to fit the linear viscoelastic properties of the low-density polyethylene melt in Fig. 4.6–2 [see H. Chang and A. S. Lodge, *Rheol. Acta, 11*, 127–129 (1972)]. Aside from the spurious oscillations the model predictions are in qualitative agreement with experimental data.

expressions there for various material functions. By inserting the sum of exponentials for $G(t - t')$ the results for the generalized ZFD model can be recovered.

§7.5 THE GODDARD-MILLER MODEL[1]

In §6.1 we discussed a sequence of linear viscoelastic models, arriving finally at the general linear viscoelastic model in Eq. 6.1–16. It was also pointed out that Eq. 6.1–16 could be interpreted as a result of applying the Boltzmann superposition principle. In this section we consider the corotational analog of Eq. 6.1–16; this gives us the most general quasilinear corotational model.

[1] J. D. Goddard and C. Miller, *Rheol. Acta, 5*, 177–184 (1966).

We begin by noting that Eq. 7.3–4 for the corotational Jeffreys model, and Eq. 7.4–3 for the generalized ZFD model both have the same form: they are single integrals over the time history of the rate-of-deformation tensor (suitably rotated) with a relaxation modulus that emphasizes the most recent history. These two models, and many others, can then be regarded as special cases of the following quasilinear rheological equation of state:

$$\boldsymbol{\tau} = -\int_{-\infty}^{t} G(t - t')\dot{\boldsymbol{\Gamma}}'\,dt' \tag{7.5–1}$$

Equation 7.5–1 is the *Goddard-Miller equation* and can be thought of as the result of applying the Boltzmann superposition principle in a corotating frame. Note that for irrotational flows (as well as for small-deformation flows) Eq. 7.5–1 becomes identical with Eq. 6.1–16, since by Eq. 7.1–11 $\dot{\boldsymbol{\Gamma}}'$ becomes $\dot{\boldsymbol{\gamma}}(t')$.

There are many useful empirical expressions for $G(t - t')$. The { }-expression in Eq. 7.4–3 (a sum of exponentials) is useful, since it has been widely used in linear viscoelasticity; this expression can be used along with the empiricisms in Eqs. 6.1–14 and 15 in order to reduce the number of constants. In Problem 7B.11 we cite several other empirical expressions for the relaxation modulus, including one that exactly gives the power-law viscosity $\eta = m\dot{\gamma}^{n-1}$.

We summarize the expressions[2] for a number of material functions for the Goddard-Miller equation in Table 7.5–1. The results in §§7.3 and 7.4 can be obtained by inserting the appropriate particular form of the relaxation modulus. From this table we may draw a number of conclusions about the quasilinear corotational models that fit into the form of Eq. 7.5–1:

1. All models of this form imply that if one has determined $G(t - t')$ from linear viscoelastic measurements, then the nonlinear behavior is known. Extensive experimentation has shown this not to be the case.[3] However, if $G(t - t')$ is determined from the steady shear flow properties, the model may be useful for describing those flows in which one feels intuitively that the details of the elastic responses will not be too important.
2. All models of the form of Eq. 7.5–1 give the secondary normal stress function to be minus one-half the primary normal stress function, whereas a factor of -0.02 to -0.4 would be more appropriate. Furthermore the primary normal stress function will always be intimately related to the viscosity function, so that the model does not allow for two fluids with identical $\eta(\dot{\gamma})$ but different $\Psi_1(\dot{\gamma})$. On the other hand, normal stresses do represent higher-order effects, and for some hydrodynamic problems an accurate description of them may not be necessary.
3. All models of the Goddard-Miller form give an elongational viscosity that is completely independent of the elongation rate. The available data on this property are not sufficiently abundant to make it possible to decide on the adequacy of this result. The model gives no "overshoot" in elongational stress growth, and this does seem to be in agreement with the scanty data available.
4. The models of the form of Eq. 7.5–1 seem to be able to describe reasonably well the stress relaxation after cessation of steady shear flow. For stress growth they show the experimentally observed "overshoot effect" but they give spurious oscillations for high shear rates.

[2] R. B. Bird, O. Hassager, and S. I. Abdel-Khalik, *A.I.Ch.E. Journal*, *20*, 1041–1066 (1974), Section 6.
[3] P. J. Carreau, I. F. Macdonald, and R. B. Bird, *Chem. Eng. Sci.*, *23*, 901–911 (1968).

TABLE 7.5–1

Summary of Material Functions for the Goddard-Miller Equation (Eq. 7.5–1)

Experiments	Material Functions		Comments
Small-amplitude oscillatory motion $v_x = \dot{\gamma}^0 y \mathscr{R}e\{e^{i\omega t}\}$	$\eta^*(\omega) = \eta' - i\eta'' = \int_0^\infty G(s)e^{-i\omega s}\,ds$	(A)	This result is identical to that in Eqs. 6.2–4 and 5. Here $\tau_{yx} = \mathscr{R}e\{\tau^0 e^{i\omega t}\}$ and $\tau^0 = -\eta^*\gamma^0$.
Steady-state shear flow $v_x = \dot{\gamma}y$	$\eta^\#(\dot{\gamma}) = \eta - i\frac{1}{2}\dot{\gamma}\Psi_1 = \int_0^\infty G(s)e^{-i\dot{\gamma}s}\,ds$	(B)	In addition the model gives $\Psi_2 = -\frac{1}{2}\Psi_1$. The zero-shear-rate viscosity is $\eta_0 = \int_0^\infty G(s)\,ds$.
Stress growth at sudden inception of the steady shear flow $v_x = \dot{\gamma}_0 y$ at $t = 0$	$\eta^+(t;\dot{\gamma}_0) = \int_0^t G(s)\cos\dot{\gamma}_0 s\,ds$ $\frac{1}{2}\dot{\gamma}_0\Psi_1^+(t;\dot{\gamma}_0) = \int_0^t G(s)\sin\dot{\gamma}_0 s\,ds$	(C) (D)	η^+ and Ψ_1^+ are defined for $t \geq 0$ by $\tau_{yx} = -\eta^+\dot{\gamma}_0$, $\tau_{xx} - \tau_{yy} = -\Psi_1^+\dot{\gamma}_0{}^2$. *Note:* $\Psi_2^+ = -\frac{1}{2}\Psi_1^+$.
Stress relaxation after sudden cessation of the steady shear flow $v_x = \dot{\gamma}_0 y$ at $t = 0$	$\eta^-(t;\dot{\gamma}_0) = \int_t^\infty G(s)\cos\dot{\gamma}_0(s - t)\,ds$ $\frac{1}{2}\dot{\gamma}_0\Psi_1^-(t;\dot{\gamma}_0) = \int_t^\infty G(s)\sin\dot{\gamma}_0(s - t)\,ds$	(E) (F)	η^- and Ψ_1^- are defined for $t \geq 0$ by $\tau_{yx} = -\eta^-\dot{\gamma}_0$, $\tau_{xx} - \tau_{yy} = -\Psi_1^-\dot{\gamma}_0{}^2$. *Note:* $\Psi_2^- = -\frac{1}{2}\Psi_1^-$.

Steady shear flow with superposed small-amplitude oscillations $v_x = \dot{\gamma}(t)y$ where $\dot{\gamma}(t) = \dot{\gamma}_m + \mathscr{Re}\{\dot{\gamma}^0 e^{i\omega t}\}$ where $\dot{\gamma}_m$ is constant.	$\eta'_{\|\|}(\omega;\dot{\gamma}_m) = \int_0^\infty G(s)[\cos \dot{\gamma}_m s \cos \omega s - (\dot{\gamma}_m/\omega) \sin \dot{\gamma}_m s \sin \omega s]\, ds$ (G) $\eta''_{\|\|}(\omega;\dot{\gamma}_m) = \int_0^\infty G(s)[\cos \dot{\gamma}_m s \sin \omega s - (\dot{\gamma}_m/\omega)(\sin \dot{\gamma}_m s)(1 - \cos \omega s)]\, ds$ (H)	The shear stress is of the form: $\tau_{yx} = \tau_m + \mathscr{Re}\{\tau^0 e^{i\omega t}\}$ where $\tau_m = -\eta(\dot{\gamma}_m)\dot{\gamma}_m$ $\tau^0 = -\eta^*_{\|\|}(\omega;\dot{\gamma}_m)\dot{\gamma}^0$
Steady elongational flow $v_z = \dot{\epsilon}z$ $v_x = -\frac{1}{2}\dot{\epsilon}x$ $v_y = -\frac{1}{2}\dot{\epsilon}y$	$\bar{\eta}(\dot{\epsilon}) = 3 \int_0^\infty G(s)\, ds = 3\eta_0$ (I)	$\bar{\eta}$ is defined by $\tau_{zz} - \tau_{xx} = -\bar{\eta}\dot{\epsilon}$ *Note*: $\bar{\eta}$ is independent of $\dot{\epsilon}$. Result is identical with linear viscoelastic result.
Elongational stress growth at sudden inception of steady elongational flow at $t = 0$ $v_z = +\dot{\epsilon}_0 z$ $v_x = -\frac{1}{2}\dot{\epsilon}_0 x$ $v_y = -\frac{1}{2}\dot{\epsilon}_0 y$	$\bar{\eta}^+(t;\dot{\epsilon}_0) = 3 \int_0^t G(s)\, ds$ (J)	$\bar{\eta}^+$ is defined by $\tau_{zz} - \tau_{xx} = -\bar{\eta}^+\dot{\epsilon}_0$ Result is identical with the linear viscoelastic result.

In the next chapter we shall see that Eq. 7.5–1 is, in fact, just the first term in a "memory integral expansion" and that some of the shortcomings mentioned above can be corrected for by using higher terms in the series. However, inclusion of additional terms gives rheological equations of state that are of such complexity that there is some doubt as to their usefulness in making fluid dynamics calculations.

Before leaving the Goddard-Miller equation we want to discuss some interrelations that can be obtained between various material functions; these relations do not contain the $G(t - t')$ function and hence are model-independent. Recall that in Chapter 6 we were able to obtain a number of relations that connected material functions from various experiments. First we note that the two equations contained in Eq. B of Table 7.5–1 can be inverted by Fourier transform to give:

$$G(s) = \frac{2}{\pi} \int_0^\infty \eta(\dot{\gamma}) \cos \dot{\gamma} s \, d\dot{\gamma} \qquad (s \geq 0) \tag{7.5–2}$$

$$G(s) = \frac{1}{\pi} \int_0^\infty \dot{\gamma} \Psi_1(\dot{\gamma}) \sin \dot{\gamma} s \, d\dot{\gamma} \qquad (s > 0) \tag{7.5–3}$$

Similar expressions can be written for η' and η'' by using Eq. A of Table 7.5–1. If the Goddard-Miller equation gave a perfect representation of the rheological behavior of the fluid, then G could in principle[4] be determined equally well from data on any one of the properties η, Ψ_1, η', or η''; however Eq. 7.5–1 cannot describe simultaneously both steady shear flow and oscillatory properties arbitrarily well. Hence in determining $G(t - t')$ one has to make a choice as to which property one wishes most to describe accurately. If $G(t - t')$ is determined from Eq. 7.5–2, then the viscosity can in principle be reproduced with arbitrarily good precision.

Once we have Eqs. 7.5–2 and 3 we can insert these expressions into any of the relations in Table 7.5–1. In this way many interesting relations can be obtained among the measurable functions,[2] and these relations do not contain $G(t - t')$. We give a sampling of these interrelations in Table 7.5–2. Equation A in this table has been tested extensively.[2,5] By examining the data for viscosity and normal stresses of six polymer melts, three polymer solutions, and one soap solution, with a wide range of chemical and physical properties, it was found that Eq. A of Table 7.5–2 was consistently lower than the experimental values by a factor of about 2 for solutions and about 3 for melts. This is regarded as satisfactory agreement, inasmuch as the Ψ_1 function changed by more than a factor of 10^3 as a function of the shear rate.

Data by Christiansen and Leppard[6] on poly (ethylene oxide) and polyacrylamide solutions have been used to test Eqs. H, I, and J, and very good agreement was reported. Note that these equations suggest that no new information would be obtained by measuring the primary normal stress difference in a small-amplitude oscillatory experiment, if measurements of the shear stresses are already available.

[4] For information on performing the inversion, see J. D. Ferry and K. Ninomiya, in *Viscoelasticity* J. T. Bergen (Ed.), Academic Press, New York (1960), pp. 55–75; R. S. Marvin, *ibid.*, pp. 27–54.

[5] S. I. Abdel-Khalik, O. Hassager, and R. B. Bird, *Polym. Eng. Sci.*, *14*, 859–867 (1974).

[6] E. B. Christiansen and W. R. Leppard, *Trans. Soc. Rheol.*, *18*, 65–86 (1974); note that Eq. H had been verified experimentally earlier by L. C. Akers and M. C. Williams, *J. Chem. Phys.*, *51*, 3834–3841 (1969) and H. Endō and M. Nagasawa, *J. Polymer Sci.*, Part A-2, *8*, 371–381 (1970). Equation H was first obtained by A. S. Lodge [in *Phénomènes de rélaxation et de fluage en rhéologie non-linéaire*, Éditions du C.R.N.S., Paris (1961), p. 51] and later by M. C. Williams and R. B. Bird [*Ind. Eng. Chem. Fundamentals*, *3*, 42–49 (1964)]. T. W. Spriggs [Ph.D. Thesis, University of Wisconsin, Madison (1966)] first obtained Eqs. I and J from the first two terms of Eq. 9.5–1.

TABLE 7.5–2

Relations among Material Functions as Given by the Goddard-Miller Model[2] (Eq. 7.5–1)

Relations between steady shear flow functions:

$$\Psi_1(\dot{\gamma}) = \frac{4}{\pi}\int_0^\infty \frac{\eta(\dot{\gamma}) - \eta(\dot{\gamma}')}{\dot{\gamma}'^2 - \dot{\gamma}^2}\, d\dot{\gamma}' \tag{A}$$

$$\eta(\dot{\gamma}) - \eta_\infty = \frac{1}{\pi}\int_0^\infty \frac{\dot{\gamma}'^2\Psi_1(\dot{\gamma}') - \dot{\gamma}^2\Psi_1(\dot{\gamma})}{\dot{\gamma}'^2 - \dot{\gamma}^2}\, d\dot{\gamma}' \tag{B}$$

Relation between normal stress and the relaxing shear stress:

$$\Psi_1(\dot{\gamma}) = 2\int_0^\infty \eta^-(t;\dot{\gamma})\, dt \tag{C}$$

Normal stress growth in terms of shear stress growth:

$$\Psi_1^+(t;\dot{\gamma}) = \frac{2}{\dot{\gamma}}\int_0^{\dot{\gamma}} E(t;\dot{\gamma}')\, d\dot{\gamma}' \tag{D}$$

where

$$E(t;\dot{\gamma}) = t\eta^+(t;\dot{\gamma}) - \int_0^t \eta^+(t';\dot{\gamma})\, dt' \tag{E}$$

Relations among stress-growth functions and stress-relaxation functions:

$$(\eta - \eta^+) = \eta^- \cos\dot{\gamma}t - \tfrac{1}{2}\dot{\gamma}\Psi_1^- \sin\dot{\gamma}t \tag{F}$$

$$\tfrac{1}{2}\dot{\gamma}(\Psi_1 - \Psi_1^+) = \eta^- \sin\dot{\gamma}t + \tfrac{1}{2}\dot{\gamma}\Psi_1^- \cos\dot{\gamma}t \tag{G}$$

Relations among small-amplitude oscillatory motion functions:

$$\Psi_1{}^d(\omega) = \frac{\eta''(\omega)}{\omega} \tag{H}$$

$$\Psi_1'(\omega) = \frac{-\eta''(\omega) + \eta''(2\omega)}{\omega} \tag{I}$$

$$\Psi_1''(\omega) = \frac{\eta'(\omega) - \eta'(2\omega)}{\omega} \tag{J}$$

EXAMPLE 7.5–1 Elongational Stress Growth for Goddard-Miller Fluid

A cylindrical specimen of a very viscous fluid is at rest before time $t = 0$. The axis of the specimen coincides with the z-axis of a Cartesian coordinate system. Beginning at $t = 0$ the sample is elongated at a constant elongational rate $dv_z/dz = \dot{\epsilon}_0$. Obtain an expression for $\bar{\eta}^+$.

SOLUTION

For the flow $v_z = \dot{\epsilon}_0 z$, $v_x = -\frac{1}{2}\dot{\epsilon}_0 x$, $v_y = -\frac{1}{2}\dot{\epsilon}_0 y$, the rate-of-strain tensor is:

$$\dot{\boldsymbol{\gamma}} = \begin{pmatrix} -1 & 0 & 0 \\ 0 & -1 & 0 \\ 0 & 0 & 2 \end{pmatrix} \dot{\epsilon}_0 \qquad t > 0 \tag{7.5–4}$$

and the vorticity tensor is zero. Therefore $\dot{\boldsymbol{\Gamma}}'$ in Eq. 7.5–1 is equal to $\dot{\boldsymbol{\gamma}}'$. Then the stresses during the elongational growth will be:

$$\begin{pmatrix} \tau_{xx} & \tau_{xy} & \tau_{xz} \\ \tau_{yx} & \tau_{yy} & \tau_{yz} \\ \tau_{zx} & \tau_{zy} & \tau_{zz} \end{pmatrix} = -\int_0^t G(s) \begin{pmatrix} -1 & 0 & 0 \\ 0 & -1 & 0 \\ 0 & 0 & 2 \end{pmatrix} \dot{\epsilon}_0 \, ds \tag{7.5–5}$$

Therefore we get:

$$\tau_{zz} - \tau_{xx} = -3 \int_0^t G(s)\dot{\epsilon}_0 \, ds \tag{7.5–6}$$

and hence (cf. Eq. 4.6–7):

$$\bar{\eta}^+ = 3 \int_0^t G(s) \, ds \tag{7.5–7}$$

This shows that the stresses increase monotonically until a steady-state value $\bar{\eta} = 3\eta_0$ is attained. It has been found, however, difficult to attain steady state in these kinds of experiments as pointed out in §4.6.

EXAMPLE 7.5–2 Integro-Differential Equation for the Dynamics of a Gas Bubble Suspended in a Viscoelastic Fluid[7]

Derive an equation describing the radius as a function of time for a gas bubble suspended in a macromolecular fluid, when the pressure in the gas bubble is a function of time. Use the Goddard-Miller model to describe the mechanical response of the fluid.

SOLUTION

Consider a gas bubble that is spherical in shape; its radius will be called $R(t)$. The viscoelastic fluid outside the bubble is incompressible. If we put the origin of coordinates at the center of the bubble,

[7] This example is based on the publications of H. S. Fogler and J. D. Goddard, *Phys. Fluids*, *13*, 1135–1141 (1970); *J. Appl. Phys.*, *42*, 259–263 (1971). The earliest theoretical treatment on gas-bubble dynamics seems to be that of Lord Rayleigh, *Phil. Mag.*, *34*, 94–98 (1917), who considered the collapse of a spherical cavity in an inviscid fluid, with no surface tension. For this case Eq. 7.5–19 has the solution:

$$t = \sqrt{\frac{3}{2}\frac{\rho R_0{}^2}{p_\infty}} \int_x^1 \frac{x^{3/2} \, dx}{\sqrt{1 - x^3}} \tag{7.5–8a}$$

where $x = R/R_0$ and $R = R_0$ at $t = 0$, at which time p_G suddenly becomes zero. The time for the complete collapse of the bubble is

$$t_c = \frac{\Gamma(\frac{5}{6})\Gamma(\frac{1}{2})}{\sqrt{6}\,\Gamma(\frac{4}{3})} \sqrt{\frac{\rho R_0{}^2}{p_\infty}} = 0.915 \sqrt{\frac{\rho R_0{}^2}{p_\infty}} \tag{7.5–8b}$$

For further discussions of bubble collapse see H. Lamb, *Hydrodynamics*, Cambridge University Press (1932), Sixth Edition, p. 122, and G. K. Batchelor, *An Introduction to Fluid Dynamics*, Cambridge University Press (1967), pp. 486–490. A treatment of the mass-transfer process coupled with the rheological problem has been given in a paper on the dissolution of a gas bubble in a viscoelastic fluid by E. Zana and L. G. Leal, *Ind. Eng. Chem. Fundamentals*, *14*, 175–182 (1975). An earlier paper on vapor-bubble growth in viscoelastic liquids is that by J. R. Street, *Trans. Soc. Rheol.*, *12*, 103–131 (1968).

then the velocity of the fluid will have the form $v_r = v_r(r,t)$, $v_\theta = 0$, $v_\phi = 0$. The equation of continuity is then:

$$\frac{1}{r^2}\frac{\partial}{\partial r}(r^2 v_r) = 0 \tag{7.5–8}$$

from which:

$$r^2 v_r = \text{a function of time} \tag{7.5–9}$$

The function of time can be evaluated at the bubble surface as $R^2\dot{R}$, and hence:

$$v_r(r,t) = \frac{R^2\dot{R}}{r^2} \tag{7.5–10}$$

where the dot indicates differentiation with respect to time.

The equation of motion for the fluid external to the bubble is:

$$\rho\left(\frac{\partial v_r}{\partial t} + v_r\frac{\partial}{\partial r}v_r\right) = -\frac{\partial p}{\partial r} - [\nabla\cdot\boldsymbol{\tau}]_r \tag{7.5–11}$$

When the velocity distribution given in Eq. 7.5–10 is inserted into the equation of motion and the resulting equation integrated from $r = R(t)$ to $r = \infty$, we get:

$$R\ddot{R} + \tfrac{3}{2}\dot{R}^2 = \frac{1}{\rho}(p_L - p_\infty) - \frac{1}{\rho}\int_R^\infty [\nabla\cdot\boldsymbol{\tau}]_r\,dr \tag{7.5–12}$$

where p_L is the pressure in the liquid phase at $r = R$, and p_∞ is the pressure far from the bubble (at the same elevation).

At the gas-liquid interface:[8]

$$(p + \tau_{rr})_G = (p + \tau_{rr})_L + \frac{2\sigma}{R} \tag{7.5–13}$$

where σ is the interfacial tension, and the subscripts G and L refer to the gas and liquid phases, respectively. It is customary to neglect $(\tau_{rr})_G$ since it is usually less important than p_G.

The quantity $[\nabla\cdot\boldsymbol{\tau}]_r$ is given by (see Appendix A):

$$[\nabla\cdot\boldsymbol{\tau}]_r = \frac{\partial\tau_{rr}}{\partial r} + \frac{2\tau_{rr} - \tau_{\theta\theta} - \tau_{\phi\phi}}{r} \tag{7.5–14}$$

For this purely radial flow, $\boldsymbol{\omega} = \mathbf{0}$ and Eq. 7.5–1 simplifies to:

$$\boldsymbol{\tau} = -\int_{-\infty}^{t} G(t - t')\dot{\boldsymbol{\gamma}}(t')\,dt' \tag{7.5–15}$$

and, hence, because of the incompressibility condition, tr $\boldsymbol{\tau} = 0$. Therefore Eq. 7.5–14 simplifies to:

$$[\nabla\cdot\boldsymbol{\tau}]_r = \frac{\partial\tau_{rr}}{\partial r} + 3\frac{\tau_{rr}}{r} \tag{7.5–16}$$

[8] For a discussion of the forces at interfaces see L. D. Landau and E. M. Lifshitz, *Fluid Mechanics*, Addison-Wesley, Reading, Massachusetts, (1959), pp. 230–244.

Insertion of Eqs. 7.5–13 and 16 into Eq. 7.5–12 gives:

$$R\ddot{R} + \tfrac{3}{2}\dot{R}^2 = \frac{p_G - p_\infty}{\rho} - \frac{2\sigma}{\rho R} - \frac{3}{\rho}\int_R^\infty \frac{\tau_{rr}}{r}\,dr \tag{7.5–17}$$

where we have assumed that $(\tau_{rr})_L$ is zero at $r = \infty$. From Eq. 7.5–15 we know that:

$$\begin{aligned}\tau_{rr}(r,t) &= -\int_{-\infty}^{t} G(t - t')\dot{\gamma}_{rr}(r',t')\,Dt' \\ &= -\int_{-\infty}^{t} G(t - t')\left(-\frac{4(R')^2\dot{R}'}{(r')^3}\right)Dt' \\ &= 4\int_{-\infty}^{t} G(t - t')\frac{(R')^2\dot{R}'}{r^3 + (R')^3 - R^3}\,dt'\end{aligned} \tag{7.5–18}$$

where R' is the bubble radius at t'. In the first two lines of Eq. 7.5–18 we have written the integrand as a function of arbitrary spatial location r' and time t'; it is easiest to compute $\dot{\gamma}_{rr}$ in this way. Since particle designation (r,t) does not appear explicitly in this form of the integrand, we indicate integration following a particle by $\int \cdots Dt'$; this is sometimes called "substantial integration" (cf. Eq. 9.2–6). Then in going from the second to third line, the particle label is introduced by means of $(r')^3 = r^3 + (R')^3 - R^3$, which is a consequence of Eq. 7.5–10. This last form is most convenient for performing the integration. When this expression for τ_{rr} is substituted into Eq. 7.5–17 we get finally:

$$R\ddot{R} + \tfrac{3}{2}\dot{R}^2 = \frac{1}{\rho}(p_G - p_\infty) - \frac{2\sigma}{\rho R} - \frac{12}{\rho}\int_{-\infty}^{t} \frac{G(t - t')\dot{R}'(R')^2 \ln (R'/R)}{(R')^3 - R^3}\,dt' \tag{7.5–19}$$

which is the desired integrodifferential equation for $R(t)$.

Fogler and Goddard[7] have used Eq. 7.5–19 with (cf. Eq. 7.3–4):

$$G(t - t') = \frac{\eta_0}{\lambda_1}\left[\left(1 - \frac{\lambda_2}{\lambda_1}\right)e^{-(t-t')/\lambda_1} + 2\lambda_2\delta(t - t')\right] \tag{7.5–20}$$

to study theoretically the collapse of cavities in viscoelastic fluids. They considered several limiting cases analytically and obtained computer solutions for other cases. They concluded that elastic effects may play an important role in cavitation in viscoelastic liquids, particularly when $\lambda_1/(R_0\sqrt{\rho/p_\infty})$ is large, R_0 being the initial bubble radius. This dimensionless group is the ratio of a characteristic time of the fluid to a characteristic collapse time for the bubble (such a ratio of time constants is sometimes referred to as a Deborah number[9]). Specifically, they demonstrated that for large values of $\lambda_1/(R_0\sqrt{\rho/p_\infty})$, a spherical cavity may either collapse or else display oscillations about an equilibrium radius, depending on whether or not the ratio of ambient pressure to the elastic modulus of the fluid exceeds a certain critical value. Even for $\lambda_1/(R_0\sqrt{\rho/p_\infty}) \approx 1$, the ultimate collapse of a bubble is delayed until after several cycles of contraction and expansion have taken place.

[9] M. Reiner, *Physics Today*, Jan. 17, 1964 (p. 62). The importance of the time-constant ratio has been discussed by R. B. Bird, *Can. Journal Chem. Eng., 43*, 161–168 (1965), and by A. B. Metzner, J. L. White, and M. M. Denn, *A.I.Ch.E. Journal, 12*, 863–865 (1966). For a critique of various definitions of Deborah numbers, see R. R. Huilgol, *Trans. Soc. Rheol., 19*, 297–306 (1975). Dimensional analysis and dimensionless groups have been discussed by G. Astarita and G. Marrucci, *Principles of Non-Newtonian Fluid Mechanics*, McGraw-Hill, New York (1974), Chapter 7, and by Y. Tomita, *Hisenkei Ryūtai no Rikigaku*, Corona, Tokyo (1975), pp. 348–349.

EXAMPLE 7.5–3 Estimation of Extrudate Swell for Polymer Melts[5]

Use Tanner's equation for extrudate swell (Eq. 3.5–1) and Eq. A of Table 7.5–2 to develop a generalized chart for estimating the swell ratio D_e/D for polymer melts from a knowledge of the viscosity vs. shear-rate data.

SOLUTION

From the discrepancies between Eq. A of Table 7.5–2 and the experimental data, we know that a useful semiempirical result is:

$$\Psi_1(\dot{\gamma}) = K \cdot \frac{4}{\pi} \int_0^\infty \frac{\eta(\dot{\gamma}) - \eta(\dot{\gamma}')}{\dot{\gamma}'^2 - \dot{\gamma}^2} \, d\dot{\gamma}' \tag{7.5–21}$$

with $K = 2$ for solutions and $K = 3$ for melts. If now we represent the viscosity function by the Carreau equation (Eq. 5.1–13), then Eq. 7.5–21 can be used to prepare the dimensionless chart shown in Fig. 7.5–1; this chart is useful for *estimating* Ψ_1 in the absence of experimental data. Combination of this chart for

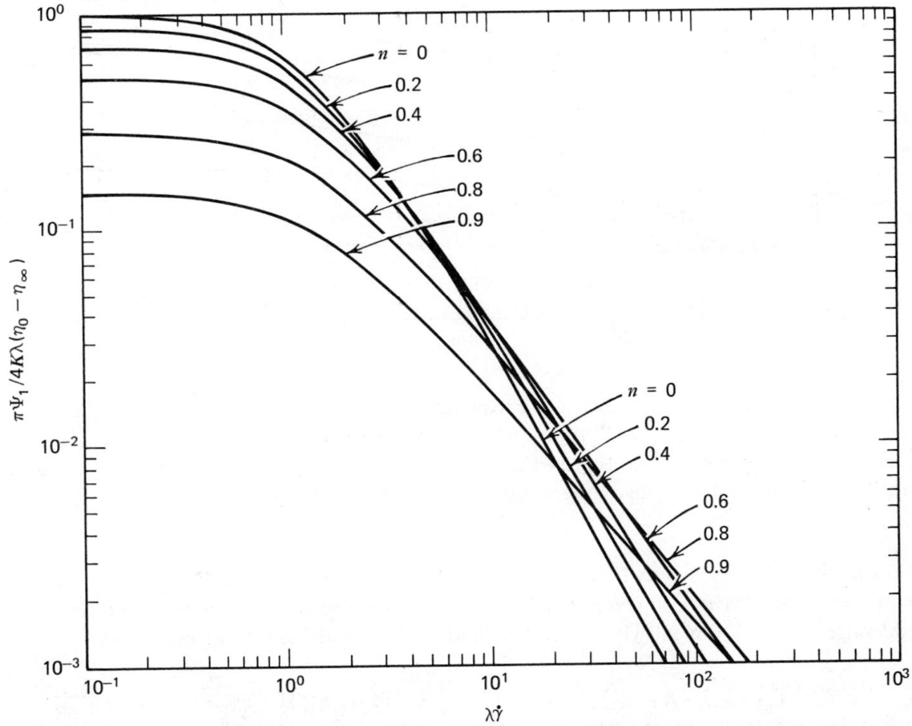

FIGURE 7.5–1. Dimensionless chart for estimating $\Psi_1(\dot{\gamma})$ from viscosity data, when the viscosity data can be described by the Carreau viscosity equation (Eq. 5.1–13):

$$\frac{\eta - \eta_\infty}{\eta_0 - \eta_\infty} = \frac{1}{[1 + (\lambda\dot{\gamma})^2]^{(1-n)/2}}$$

where η_0, η_∞, λ, and n are constants characteristic of the fluid. From limited experimental data it seems that K is about 2 for solutions and about 3 for melts. [S. I. Abdel-Khalik, O. Hassager, and R. B. Bird, *Polym. Eng. Sci.*, *14*, 859–867 (1974).]

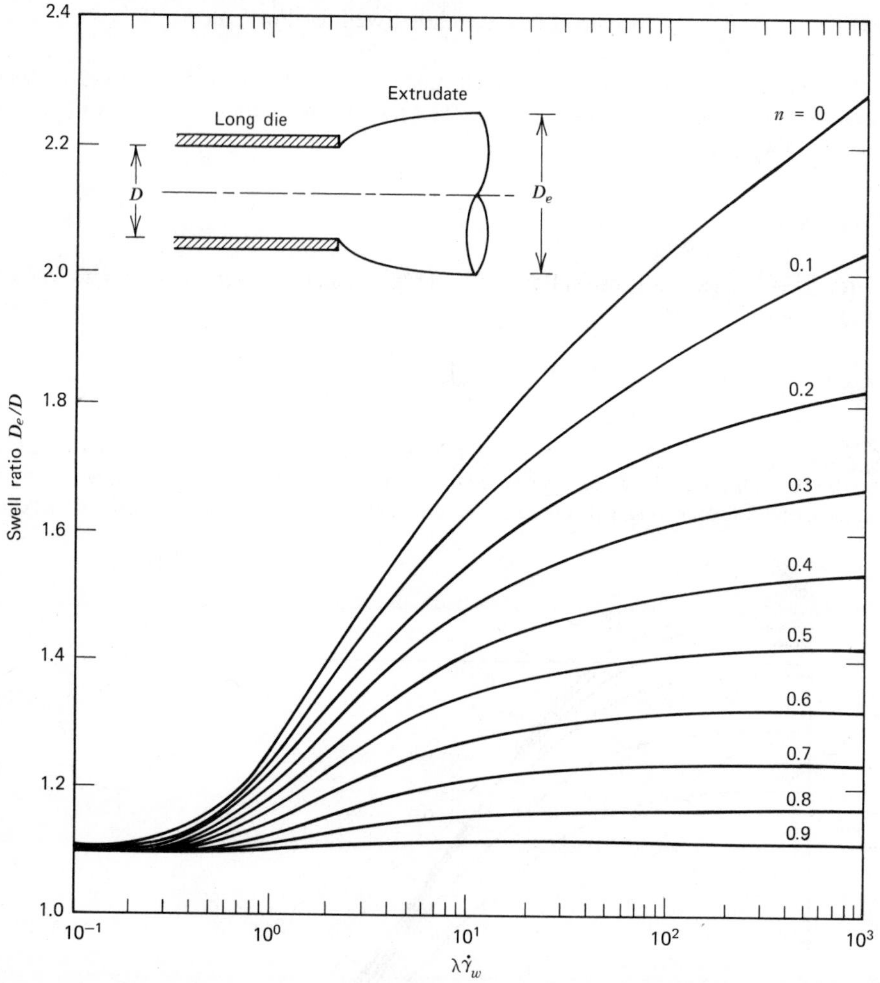

FIGURE 7.5–2. Dimensionless chart for extrudate swell of polymer melts, with the diameter ratio D_e/D plotted vs. a dimensionless wall shear rate. This plot was made by combining Eq. 7.5–21 for $K = 3$ with Tanner's extrudate swell equation in Eq. 3.5–1. For the definition of the parameter n see caption to Fig. 7.5–1. [S. I. Abdel-Khalik, O. Hassager, and R. B. Bird, *Polym. Eng. Sci.*, *14*, 859–867 (1974).]

$K = 3$ with Eq. 3.5–1 then gives the set of curves in Fig. 7.5–2, from which D_e/D may be *estimated* for a melt once λ and n have been determined from the viscosity curve. This chart has been found to be in moderately good agreement with the rather limited data available for polymer melts.

§7.6 COROTATIONAL RHEOLOGICAL EQUATIONS OF STATE IN CURVILINEAR COORDINATES

Thus far in Chapter 7 we have avoided use of curvilinear coordinates (except for Example 7.5–2), in order to put the emphasis on the physical ideas and the method of formulating rheological equations of state. Many problems in fluid dynamics, however, are most conveniently formulated in cylindrical, spherical, or other orthogonal, curvilinear coordinates. Therefore it is appropriate to say a few words about difficulties encountered in curvilinear coordinate systems.

In the *differential rheological equations of state* we encounter the Jaumann derivative at the present time t. The evaluation of this derivative presents no special problems, since the various operations in the Jaumann derivative are explained in Appendix A, and the $\{\boldsymbol{v} \cdot \nabla\boldsymbol{\tau}\}$ operation is given in Tables A.7–1 to 4 for several curvilinear coordinate systems.

In the *integral rheological equations of state*, however, we have to concern ourselves with the fact that as we follow a fluid particle from t' to t, the triad of unit vectors describing the curvilinear coordinate system changes orientation. Hence, instead of the situation described by Fig. 7.1–1, we have to consider that shown in Fig. 7.6–1. Once again we use a triad of unit vectors $\check{\boldsymbol{\delta}}_i$ that rotate with the fluid and that, at time t, coincide with a fixed triad $\boldsymbol{\delta}_i$. We also introduce another triad $\mathring{\boldsymbol{\delta}}_i$ that describes the local orientation of the curvilinear coordinate system fixed in space.

We now let the relations between these various triads be given by:

$$\check{\boldsymbol{\delta}}_i(t') = [\boldsymbol{\delta}_i \cdot \boldsymbol{\Omega}(t,t')] \tag{7.6–1}$$

$$\mathring{\boldsymbol{\delta}}_i(t') = [\boldsymbol{\delta}_i \cdot \boldsymbol{\Lambda}(t,t')] \tag{7.6–2}$$

That is, Eq. 7.6–1 is the same as Eq. 7.1–10 written in tensor notation, and Eq. 7.6–2 tells what unit vectors for the space-fixed curvilinear coordinate system the fluid particle sees at time t'. The two orthogonal tensors $\boldsymbol{\Omega}$ and $\boldsymbol{\Lambda}$ tell how the unit vectors $\boldsymbol{\delta}_i$ have to be turned in order to give $\check{\boldsymbol{\delta}}_i$ and $\mathring{\boldsymbol{\delta}}_i$, respectively, when the fluid particle is at its location at time t'. Keep in mind that $\boldsymbol{\Omega}$ is determined by the rotating fluid motion (a physical process), whereas $\boldsymbol{\Lambda}$ is determined by the choice of the fixed curvilinear coordinate system (a mathematical method for describing locations in the space through which the fluid is flowing). Although it is not necessary, it is generally convenient to choose the $\boldsymbol{\delta}_i$-triad so that it coincides with the $\mathring{\boldsymbol{\delta}}_i$-triad at time t. We shall do that in the illustrative example.

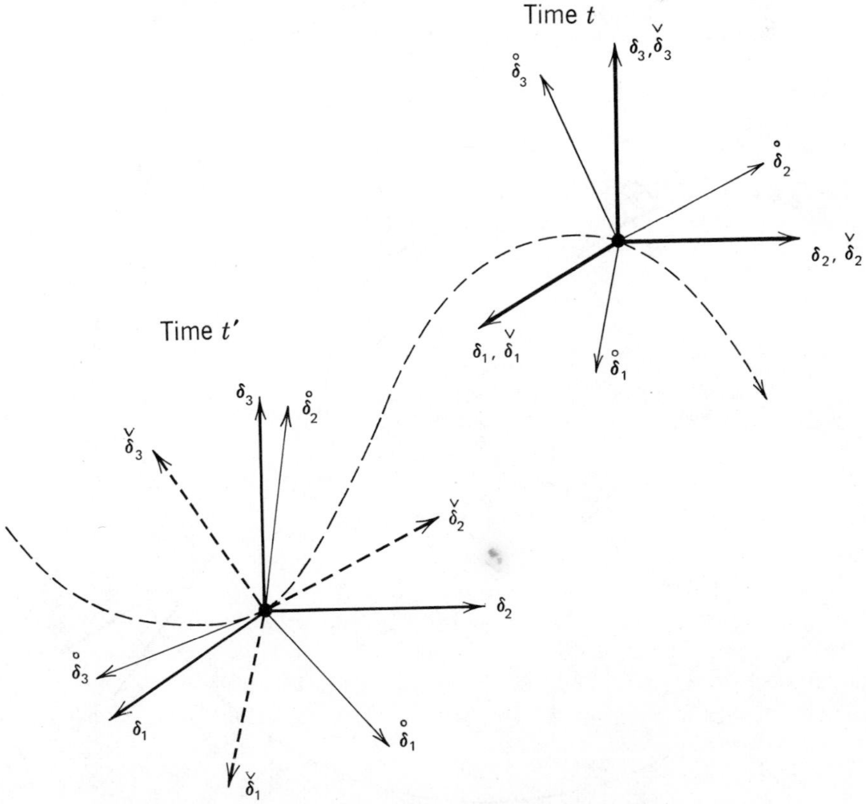

FIGURE 7.6–1. Triads of unit vectors at a fluid particle at times t' and t.

In calculating τ from, for example, the Goddard-Miller model

$$\tau(t) = -\int_{-\infty}^{t} G(t - t')\dot{\Gamma}(x,t,t')\,dt' \tag{7.6–3}$$

one has to get the tensor $\dot{\Gamma}(x,t,t')$ following the fluid element by using Eq. 7.3–5. For a specific curvilinear flow it is usually natural to evaluate the components of the $\dot{\gamma}(t')$ tensor referred to the $\mathring{\delta}_i$-frame at time t'. One then can use Eq. 7.6–2 to obtain the components of $\dot{\gamma}(t')$ in terms of the δ_i-frame. Thereafter it is straightforward to get the components of $\dot{\Gamma}(x,t,t')$ and $\tau(t)$ referred to the fixed curvilinear coordinate system.

Warning: In using matrices to represent the components of tensors, one has to be particularly careful that, in matrix multiplications, all matrices contain tensor components that are referred to the same triad of unit vectors.

EXAMPLE 7.6–1 Calculation of Kinematic Quantities in a Curvilinear Flow

A fluid is located in the annular region between two coaxial cylinders, the outer one of which is rotating with a constant angular velocity (see Fig. 7.6–2). For this steady-state flow with $v_\theta = v_\theta(r)$, $v_r = 0$, and $v_z = 0$, find (a) the velocity gradient tensor, (b) the rate-of-strain tensor, (c) the vorticity tensor, (d) the Jaumann derivative of the rate-of-strain tensor, (e) the $\Omega_{ij}(t,t')$, and (f) the $\dot{\Gamma}'_{ij}$.

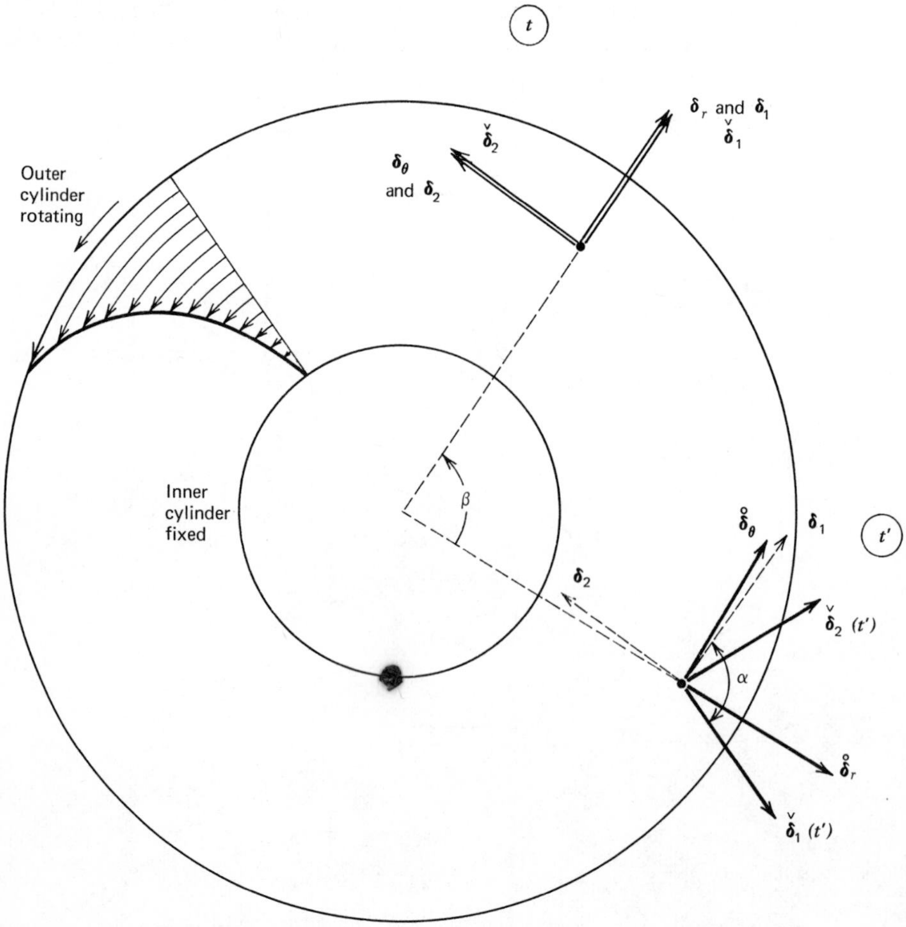

FIGURE 7.6–2. Unit vectors for tangential annular flow.

SOLUTION

(a) From Eqs. S to AA of Table A.7–2 we find for this flow:

$$\nabla \boldsymbol{v} = \begin{pmatrix} 0 & dv_\theta/dr & 0 \\ -v_\theta/r & 0 & 0 \\ 0 & 0 & 0 \end{pmatrix} \tag{7.6–4}$$

Here and in subsequent matrix displays, the matrix elements are the rr-, $r\theta$-, rz-, etc., components of the tensor.

(b) By adding the matrix in Eq. 7.6–4 to its transpose we get:

$$\dot{\boldsymbol{\gamma}} = \begin{pmatrix} 0 & 1 & 0 \\ 1 & 0 & 0 \\ 0 & 0 & 0 \end{pmatrix} r \frac{d}{dr}\left(\frac{v_\theta}{r}\right) \tag{7.6–5}$$

(c) By adding the matrix in Eq. 7.6–4 to the negative of its transpose we find:

$$\boldsymbol{\omega} = \begin{pmatrix} 0 & 1 & 0 \\ -1 & 0 & 0 \\ 0 & 0 & 0 \end{pmatrix} \frac{1}{r}\frac{d}{dr}(rv_\theta) \tag{7.6–6}$$

(d) To get the Jaumann derivative of $\dot{\boldsymbol{\gamma}}$, we have to find the various contributions in Eq. 7.1–9. We note first that $\partial\dot{\boldsymbol{\gamma}}/\partial t = \mathbf{0}$ since the flow is steady state. The $\{\boldsymbol{v} \cdot \nabla\dot{\boldsymbol{\gamma}}\}$ contribution can be obtained by using Table A.7–2; we find that for the flow under consideration:

$$\begin{aligned} \{\boldsymbol{v} \cdot \nabla\dot{\boldsymbol{\gamma}}\} &= \begin{pmatrix} -1 & 0 & 0 \\ 0 & 1 & 0 \\ 0 & 0 & 0 \end{pmatrix} \frac{v_\theta}{r}(\dot{\gamma}_{r\theta} + \dot{\gamma}_{\theta r}) \\ &= \begin{pmatrix} -1 & 0 & 0 \\ 0 & 1 & 0 \\ 0 & 0 & 0 \end{pmatrix} 2\frac{v_\theta}{r} \cdot r \frac{d}{dr}\left(\frac{v_\theta}{r}\right) \end{aligned} \tag{7.6–7}$$

Next, the first of the vorticity terms involves the product:

$$\begin{aligned} \{\boldsymbol{\omega} \cdot \dot{\boldsymbol{\gamma}}\} &= \begin{pmatrix} 0 & 1 & 0 \\ -1 & 0 & 0 \\ 0 & 0 & 0 \end{pmatrix}\begin{pmatrix} 0 & 1 & 0 \\ 1 & 0 & 0 \\ 0 & 0 & 0 \end{pmatrix} \frac{1}{r}\frac{d}{dr}(rv_\theta) \cdot r\frac{d}{dr}\left(\frac{v_\theta}{r}\right) \\ &= \begin{pmatrix} 1 & 0 & 0 \\ 0 & -1 & 0 \\ 0 & 0 & 0 \end{pmatrix} \frac{1}{r}\frac{d}{dr}(rv_\theta) \cdot r\frac{d}{dr}\left(\frac{v_\theta}{r}\right) \end{aligned} \tag{7.6–8}$$

and this is the same as $-\{\dot{\boldsymbol{\gamma}} \cdot \boldsymbol{\omega}\}$. When we add up the various contributions we find:

$$\begin{aligned} \frac{\mathscr{D}}{\mathscr{D}t}\dot{\boldsymbol{\gamma}} &= \begin{pmatrix} 1 & 0 & 0 \\ 0 & -1 & 0 \\ 0 & 0 & 0 \end{pmatrix}\left[\left(-2\frac{v_\theta}{r} + \frac{1}{r}\frac{d}{dr}(rv_\theta)\right) r\frac{d}{dr}\left(\frac{v_\theta}{r}\right)\right] \\ &= \begin{pmatrix} 1 & 0 & 0 \\ 0 & -1 & 0 \\ 0 & 0 & 0 \end{pmatrix}\left[r\frac{d}{dr}\left(\frac{v_\theta}{r}\right)\right]^2 \end{aligned} \tag{7.6–9}$$

(e) For very simple flows such as this one, the easiest way to get the Ω_{ij} is by going back to the definition in Eq. 7.1–10. For the flow under consideration, the local angular velocity of the fluid is:

$$w_z = \tfrac{1}{2}[\nabla \times \boldsymbol{v}]_z$$
$$= \tfrac{1}{2}\frac{1}{r}\frac{d}{dr}(rv_\theta) = \tfrac{1}{2}\omega_{r\theta} \tag{7.6–10}$$

and the angle α through which the $\breve{\boldsymbol{\delta}}_1$ and $\breve{\boldsymbol{\delta}}_2$ vectors move in going from t' to t is just $w_z(t - t')$. Then, using elementary trigonometric arguments and Eq. 7.1–10 one obtains:

$$\boldsymbol{\Omega} = \begin{pmatrix} \cos\alpha & -\sin\alpha & 0 \\ \sin\alpha & \cos\alpha & 0 \\ 0 & 0 & 1 \end{pmatrix} \tag{7.6–11}$$

The same result may be obtained by using Eq. 7.1–12 since this is a flow with constant vorticity, following the fluid particle. To use this equation we have to find the various powers of $\boldsymbol{\omega}$:

$$\boldsymbol{\omega}^2 = \begin{pmatrix} -1 & 0 & 0 \\ 0 & -1 & 0 \\ 0 & 0 & 0 \end{pmatrix}\omega_{r\theta}^{\ 2}$$

$$\boldsymbol{\omega}^3 = \begin{pmatrix} 0 & -1 & 0 \\ 1 & 0 & 0 \\ 0 & 0 & 0 \end{pmatrix}\omega_{r\theta}^{\ 3}$$

$$\boldsymbol{\omega}^4 = \begin{pmatrix} 1 & 0 & 0 \\ 0 & 1 & 0 \\ 0 & 0 & 0 \end{pmatrix}\omega_{r\theta}^{\ 4} \text{ etc.} \tag{7.6–12}$$

in which $\omega_{r\theta} = (1/r)\, d(rv_\theta)/dr$. Then substituting into Eq. 7.1–12 we get:

$$\boldsymbol{\Omega} = \begin{pmatrix} 1 & 0 & 0 \\ 0 & 1 & 0 \\ 0 & 0 & 0 \end{pmatrix}\left[1 - \frac{1}{2!}\left(\frac{\omega_{r\theta}}{2}\right)^2 (t - t')^2 + \frac{1}{4!}\left(\frac{\omega_{r\theta}}{2}\right)^4 (t - t')^2 - \cdots\right]$$
$$- \begin{pmatrix} 0 & 1 & 0 \\ -1 & 0 & 0 \\ 0 & 0 & 0 \end{pmatrix}\left[\frac{1}{1!}\left(\frac{\omega_{r\theta}}{2}\right)(t - t') - \frac{1}{3!}\left(\frac{\omega_{r\theta}}{2}\right)^3 (t - t')^3 + \cdots\right] + \begin{pmatrix} 0 & 0 & 0 \\ 0 & 0 & 0 \\ 0 & 0 & 1 \end{pmatrix}$$
$$= \begin{pmatrix} \cos\tfrac{1}{2}\omega_{r\theta}(t - t') & -\sin\tfrac{1}{2}\omega_{r\theta}(t - t') & 0 \\ +\sin\tfrac{1}{2}\omega_{r\theta}(t - t') & \cos\tfrac{1}{2}\omega_{r\theta}(t - t') & 0 \\ 0 & 0 & 1 \end{pmatrix} \tag{7.6–13}$$

which agrees with Eq. 7.6–11. The elements in the matrix in Eq. 7.6–13 are the rr, $r\theta$, rz, etc. components.

(f) The $\boldsymbol{\Lambda}$-tensor, describing the change of the cylindrical unit vectors as we follow the particle, is also obtainable by elementary trigonometric arguments. It is:

$$\boldsymbol{\Lambda} = \begin{pmatrix} \cos\beta & -\sin\beta & 0 \\ \sin\beta & \cos\beta & 0 \\ 0 & 0 & 1 \end{pmatrix} \tag{7.6–14}$$

in which $\beta = (v_\theta/r)(t - t')$.

The rate-of-strain tensor $\dot{\boldsymbol{\gamma}}$ at the past time t' is given by:

$$\dot{\boldsymbol{\gamma}}(t') = (\mathring{\boldsymbol{\delta}}_r\mathring{\boldsymbol{\delta}}_\theta + \mathring{\boldsymbol{\delta}}_\theta\mathring{\boldsymbol{\delta}}_r)\dot{\gamma}_{r\theta} \tag{7.6–15}$$

where the unit vectors are to be evaluated at t', and $\dot{\gamma}_{r\theta} = r d(v_\theta/r)/dr$. Use of Eq. 7.6–2 then gives:

$$\begin{aligned}\dot{\boldsymbol{\gamma}}(t') &= [(\boldsymbol{\delta}_r \cdot \boldsymbol{\Lambda})(\boldsymbol{\delta}_\theta \cdot \boldsymbol{\Lambda}) + (\boldsymbol{\delta}_\theta \cdot \boldsymbol{\Lambda})(\boldsymbol{\delta}_r \cdot \boldsymbol{\Lambda})]\dot{\gamma}_{r\theta} \\ &= [(\boldsymbol{\delta}_r\boldsymbol{\delta}_r - \boldsymbol{\delta}_\theta\boldsymbol{\delta}_\theta)\sin 2\beta + (\boldsymbol{\delta}_r\boldsymbol{\delta}_\theta + \boldsymbol{\delta}_\theta\boldsymbol{\delta}_r)\cos 2\beta]\dot{\gamma}_{r\theta} \\ &= \begin{pmatrix} \sin 2\beta & \cos 2\beta & 0 \\ \cos 2\beta & -\sin 2\beta & 0 \\ 0 & 0 & 0 \end{pmatrix} \dot{\gamma}_{r\theta}\end{aligned} \tag{7.6–16}$$

Then by performing the matrix multiplication corresponding to Eq. 7.3–5, we find:

$$\begin{aligned}\dot{\boldsymbol{\Gamma}}' &= \begin{pmatrix} \cos\alpha & -\sin\alpha & 0 \\ \sin\alpha & \cos\alpha & 0 \\ 0 & 0 & 1 \end{pmatrix} \begin{pmatrix} \sin 2\beta & \cos 2\beta & 0 \\ \cos 2\beta & -\sin 2\beta & 0 \\ 0 & 0 & 0 \end{pmatrix} \begin{pmatrix} \cos\alpha & \sin\alpha & 0 \\ -\sin\alpha & \cos\alpha & 0 \\ 0 & 0 & 1 \end{pmatrix} \dot{\gamma}_{r\theta} \\ &= \begin{pmatrix} -\sin 2(\alpha - \beta) & \cos 2(\alpha - \beta) & 0 \\ \cos 2(\alpha - \beta) & \sin 2(\alpha - \beta) & 0 \\ 0 & 0 & 0 \end{pmatrix} \dot{\gamma}_{r\theta}\end{aligned} \tag{7.6–17}$$

But since $2(\alpha - \beta) = [\omega_{r\theta} - 2(v_\theta/r)](t - t') = \dot{\gamma}_{r\theta}(t - t')$, we get finally:

$$\dot{\boldsymbol{\Gamma}}(r,t,t') = \begin{pmatrix} -\sin \dot{\gamma}_{r\theta}(t - t') & \cos \dot{\gamma}_{r\theta}(t - t') & 0 \\ \cos \dot{\gamma}_{r\theta}(t - t') & \sin \dot{\gamma}_{r\theta}(t - t') & 0 \\ 0 & 0 & 0 \end{pmatrix} \dot{\gamma}_{r\theta} \tag{7.6–18}$$

This matrix display gives the rr, $r\theta$, rz, etc., components, appropriate for use in Eq. 7.6–3.

QUESTIONS FOR DISCUSSION

1. To what extent do the techniques in Chapter 7 enable us to solve problems of the type discussed in Chapters 5 and 6? Does the Goddard-Miller equation, for example, include the generalized Newtonian fluid and the general linear viscoelastic model as special cases?
2. Explain the steps involved in formulating an objective rheological equation of state. How is Table 7.1–1 useful for doing this? Explain the physical or geometrical meaning of the various symbols used in the table.
3. In what sense are the ZFD, corotating Jeffreys, generalized ZFD, and Goddard-Miller models "quasilinear"? In what way do the nonlinearities enter the equations?
4. What do the illustrative examples in §7.3 demonstrate? Why is the equation of motion not used?
5. When should the differential forms of models be used, and when should the integral forms be used? Can all models be written in both forms?
6. What problems are encountered in finding the Ω_{ij} for a complex flow problem?
7. Equations 7.4–8 and 9 suggest that the slopes of η and Ψ_1 in the power-law region should be related. Is that true experimentally?
8. In the problem statement in Example 7.3–2, it is stated that the requirement of small amplitude of oscillation is equivalent to requiring that $\dot{\gamma}^0/\omega \ll 1$. Justify this statement.
9. In what circumstances is it permissible in Eq. 7.5–1 to replace $\dot{\mathbf{\Gamma}}'$ by $\dot{\boldsymbol{\gamma}}'$? To what extent does Eq. 7.5–1 include other rheological equations of state in this chapter?
10. Is it reasonable to say that the two-constant ZFD model plays a role in viscoelastic fluid dynamics that is analogous to that played by the two-constant van der Waals equation in thermodynamics?
11. Show how to get the results in §7.3 for the corotational Jeffreys model from Table 7.5–1 for the Goddard-Miller model.
12. Obtain the matrix in Eq. 7.3–23 without using Eq. 7.1–13; that is, just use the basic definition of $\mathbf{\Omega}$ and elementary geometrical and trigonometrical arguments.
13. Write $\boldsymbol{\omega}$ in Eq. 7.1–5 in matrix form in order to understand better the relations between the components of $\boldsymbol{\omega}$ and of $\boldsymbol{w}$.
14. Discuss the rheological equations of state:

$$\boldsymbol{\tau} = -\eta(\dot{\gamma})\dot{\boldsymbol{\gamma}}$$

$$\boldsymbol{\tau} = -\int_{-\infty}^{t} G(t - t')\dot{\boldsymbol{\gamma}}(t')\,dt'$$

$$\boldsymbol{\tau} = -\int_{-\infty}^{t} G(t - t')\dot{\mathbf{\Gamma}}(x,t,t')\,dt'$$

with regard to physical basis, range of applicability, experimental determination of the unspecified functions (η and G), useful empiricisms for the unspecified functions, and ability to predict or suggest interrelations.

PROBLEMS

7A.1 Determination of Parameters in Generalized ZFD Model

Use the rheological equation of state in Eq. 7.4–3 along with the empiricisms in Eqs. 6.1–14 and 15, so that there are three parameters: η_0, λ, and α. Determine the values of these parameters for the low-density polyethylene melt for which viscosity data are given in Fig. 4.3–2; do this for $T = 388$ K, 443 K, and 513 K.

a. Obtain η_0 directly from the viscosity at low shear rates.
b. Obtain α from the slope of log η vs. log $\dot{\gamma}$ at high shear rates.
c. Obtain λ from the value of $\dot{\gamma}$ at which the low- and high-shear-rate asymptotes intersect.

7A.2 Estimation of the Extrudate Swell Ratio

For the low-density polyethylene melt for which viscosity data are given in Fig. 4.3–2 estimate the extrudate swell for a wall shear rate of 1.02 s^{-1}, and a temperature of 443 K. From what experimentally measured quantities is the wall shear rate found?

7A.3 Estimation of Hole Pressure Error

Given the viscosity data for 5% polystyrene in Aroclor 1242 in Fig. 5.1–1, estimate the hole pressure error p_H in a rectangular die for a wall shear rate of 101 s^{-1}. Use Eq. 3B.2–3 and neglect the contribution of the secondary normal stress difference. Use Fig. 7.5–1 to estimate the primary normal stress coefficient.

7B.1 Corotational Quantities in Table 7.1–1

a. Verify that the second entry in Column (2′) can be obtained from the first entry in that column by performing the operation $\partial/\partial t'$.
b. Verify Eq. B of Table 7.1–1.
c. Verify by substitution that Eq. 7.1–11 satisfies Eq. C of Table 7.1–1.

7B.2 Some Properties of Steady Simple Shear Flow

Goddard and Miller[1] have derived the following kinematic relations for steady shear flow ($v_x = \dot{\gamma}y$):

$$\{\dot{\boldsymbol{\gamma}} \cdot \dot{\boldsymbol{\gamma}} \cdot \dot{\boldsymbol{\gamma}}\} = \dot{\gamma}^2 \dot{\boldsymbol{\gamma}} \tag{7B.2–1}$$

$$\frac{\mathscr{D}^2 \dot{\boldsymbol{\gamma}}}{\mathscr{D}t^2} = -\dot{\gamma}^2 \dot{\boldsymbol{\gamma}} \tag{7B.2–2}$$

$$\left\{\frac{\mathscr{D}\dot{\boldsymbol{\gamma}}}{\mathscr{D}t} \cdot \frac{\mathscr{D}\dot{\boldsymbol{\gamma}}}{\mathscr{D}t}\right\} = \dot{\gamma}^2 \{\dot{\boldsymbol{\gamma}} \cdot \dot{\boldsymbol{\gamma}}\} \tag{7B.2–3}$$

$$\frac{\mathscr{D}}{\mathscr{D}t} \{\dot{\boldsymbol{\gamma}} \cdot \dot{\boldsymbol{\gamma}}\} = 0 \tag{7B.2–4}$$

[1] J. D. Goddard and C. Miller, *Rheol. Acta*, 5, 177–184 (1966).

Verify these relations by using Eqs. 7.3 – 8, 9, and 13 for steady simple shear flow and the definition $\dot{\gamma} = \sqrt{\frac{1}{2}(\dot{\boldsymbol{\gamma}} : \dot{\boldsymbol{\gamma}})}$.

7B.3 Steady Shear Flow Velocity Field as Seen by Three Different Observers

In Fig. 7B.3 we have a steady simple shearing flow of the form $v_1 = \dot{\gamma} x_2$, as described in terms of the fixed coordinate system with origin at O. Consider two additional coordinate frames, one cotranslating with a fluid particle P and one corotating with the fluid particle. By "corotating" we mean that the frame is rotating with the local fluid angular velocity.

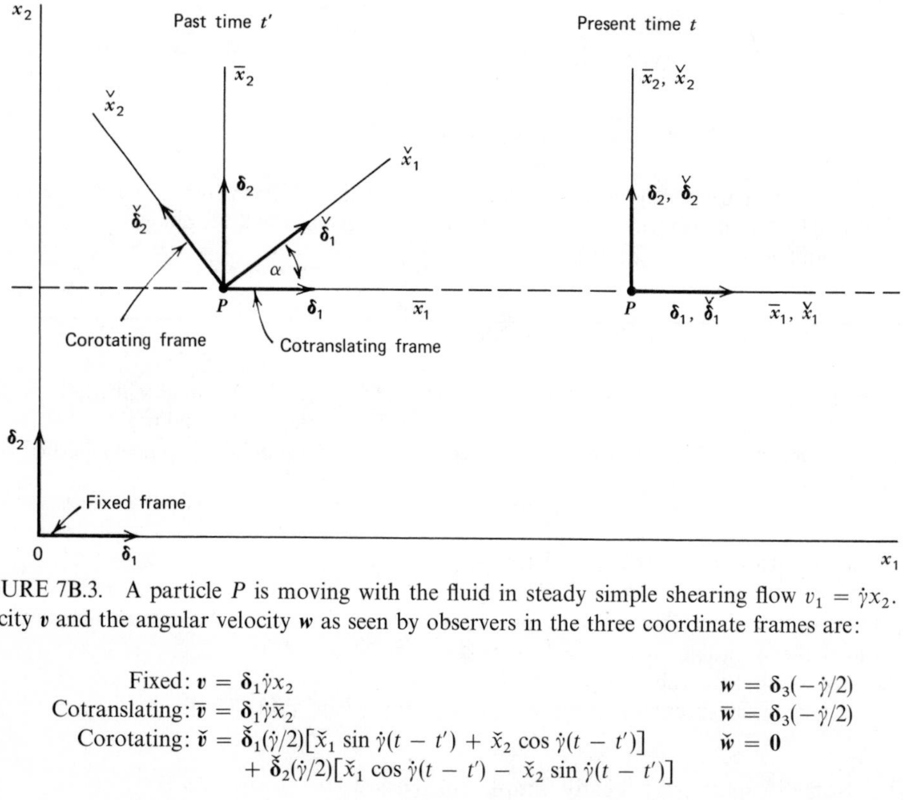

FIGURE 7B.3. A particle P is moving with the fluid in steady simple shearing flow $v_1 = \dot{\gamma} x_2$. The velocity $\boldsymbol{v}$ and the angular velocity $\boldsymbol{w}$ as seen by observers in the three coordinate frames are:

$$\begin{aligned} \text{Fixed: } \boldsymbol{v} &= \boldsymbol{\delta}_1 \dot{\gamma} x_2 & \boldsymbol{w} &= \boldsymbol{\delta}_3(-\dot{\gamma}/2) \\ \text{Cotranslating: } \bar{\boldsymbol{v}} &= \boldsymbol{\delta}_1 \dot{\gamma} \bar{x}_2 & \bar{\boldsymbol{w}} &= \boldsymbol{\delta}_3(-\dot{\gamma}/2) \\ \text{Corotating: } \check{\boldsymbol{v}} &= \check{\boldsymbol{\delta}}_1(\dot{\gamma}/2)[\check{x}_1 \sin \dot{\gamma}(t-t') + \check{x}_2 \cos \dot{\gamma}(t-t')] & \check{\boldsymbol{w}} &= \mathbf{0} \\ &\quad + \check{\boldsymbol{\delta}}_2(\dot{\gamma}/2)[\check{x}_1 \cos \dot{\gamma}(t-t') - \check{x}_2 \sin \dot{\gamma}(t-t')] \end{aligned}$$

a. In the figure caption we list the velocity field $\boldsymbol{v}$ and the angular velocity field $\boldsymbol{w} = \frac{1}{2}[\nabla \times \boldsymbol{v}]$ as seen by observers in all three frames. Verify that these expressions are correct, and then relate them to the more general expressions given in §7.2.

b. Write out the velocity components $\check{v}_1$ and $\check{v}_2$ seen in the corotating frame for the times $t' = t$, $t - (\pi/2\dot{\gamma})$, and $t - (\pi/\dot{\gamma})$. Draw three sketches showing the corotating frame at each of these times. Then on each figure, draw the components $\check{v}_1(t')$ and $\check{v}_2(t')$ at some point in the coordinate system. At the point you chose, also show the resultant vector $\check{\boldsymbol{v}}(t')$ and the space-fixed velocity vector $\boldsymbol{v}$. Interpret these figures.

c. How would the results be modified if the rotating frame were rotating with some angular velocity $\boldsymbol{W}$ different from the local fluid angular velocity $\boldsymbol{w}$?

Hint for Part (a): The velocity field as seen by the corotating observer may be written in the alternative form in terms of $w = w_3$.

$$\check{v}_1 = \dot{\gamma}[-\check{x}_1 \sin w(t-t') + \check{x}_2 \cos w(t-t')] \cos w(t-t') + w\check{x}_2 \tag{7B.3–1}$$

$$\check{v}_2 = \dot{\gamma}[-\check{x}_1 \sin w(t-t') + \check{x}_2 \cos w(t-t')] \sin w(t-t') - w\check{x}_1 \tag{7B.3–2}$$

7B.4 Expression for the Ω_{ij} for Unsteady Shearing Flows

In this problem we derive the matrix expression in Eq. 7.1–13 by elementary methods. We visualize the corotating triad of unit vectors moving along in an *unsteady homogeneous shearing flow*, with the angle α defined as in Fig. 7B.3.

a. Show that for the unsteady shearing flow $v_x = \dot{\gamma}(t)y$ the local angular velocity vector for the fluid is $\boldsymbol{w} = -\frac{1}{2}\dot{\gamma}(t)\boldsymbol{\delta}_z$.

b. Write down a differential equation to describe the change of the angle α with t', by equating $d\alpha/dt'$ to the instantaneous local angular velocity. Integrate that equation with respect to time, using the fact that $\alpha = 0$ at $t' = t$. Obtain finally:

$$\alpha(t,t') = \tfrac{1}{2}\int_{t'}^{t} \dot{\gamma}(t'')\, dt'' \qquad (7B.4\text{–}1)$$

c. Verify that the matrix Ω_{ij} which relates the corotating frame to the fixed frame is given by

$$\boldsymbol{\Omega}(t,t') = \begin{pmatrix} \cos\alpha & \sin\alpha & 0 \\ -\sin\alpha & \cos\alpha & 0 \\ 0 & 0 & 1 \end{pmatrix} \qquad (7B.4\text{–}2)$$

for a shearing flow. Combine the results of (b) and (c) to get Eq. 7.1–13.

7B.5 Radial Flow between Two Parallel Disks

A fluid flows radially in the space between two parallel circular disks (see Fig. 1B.10 and Fig. 7B.5).

a. To what extent is it possible to assume that the velocity distribution is given by $v_r = r^{-1}f(z)$, $v_\theta = 0, v_z = 0$? With this assumption show that the radial location r of a fluid particle at time t is related to its location $r'(r,z,t,t')$ at an earlier time t' by:

$$r = \sqrt{(r')^2 + 2f(t - t')} \qquad (7B.5\text{–}1)$$

b. Show that the angle $\alpha(t,t')$, defined in Fig. 7B.5 is given by:

$$\alpha(t,t') = \left(\frac{f'}{2f}\right)\left(r - \sqrt{r^2 - 2f(t - t')}\right) \qquad (7B.5\text{–}2)$$

provided that $\sqrt{r^2 - 2f(t - t')}$ is greater than r_1.

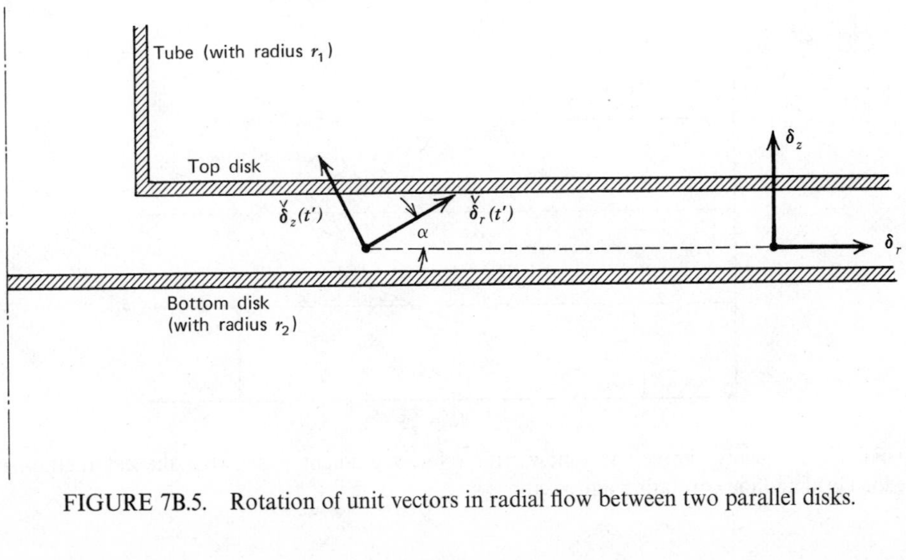

FIGURE 7B.5. Rotation of unit vectors in radial flow between two parallel disks.

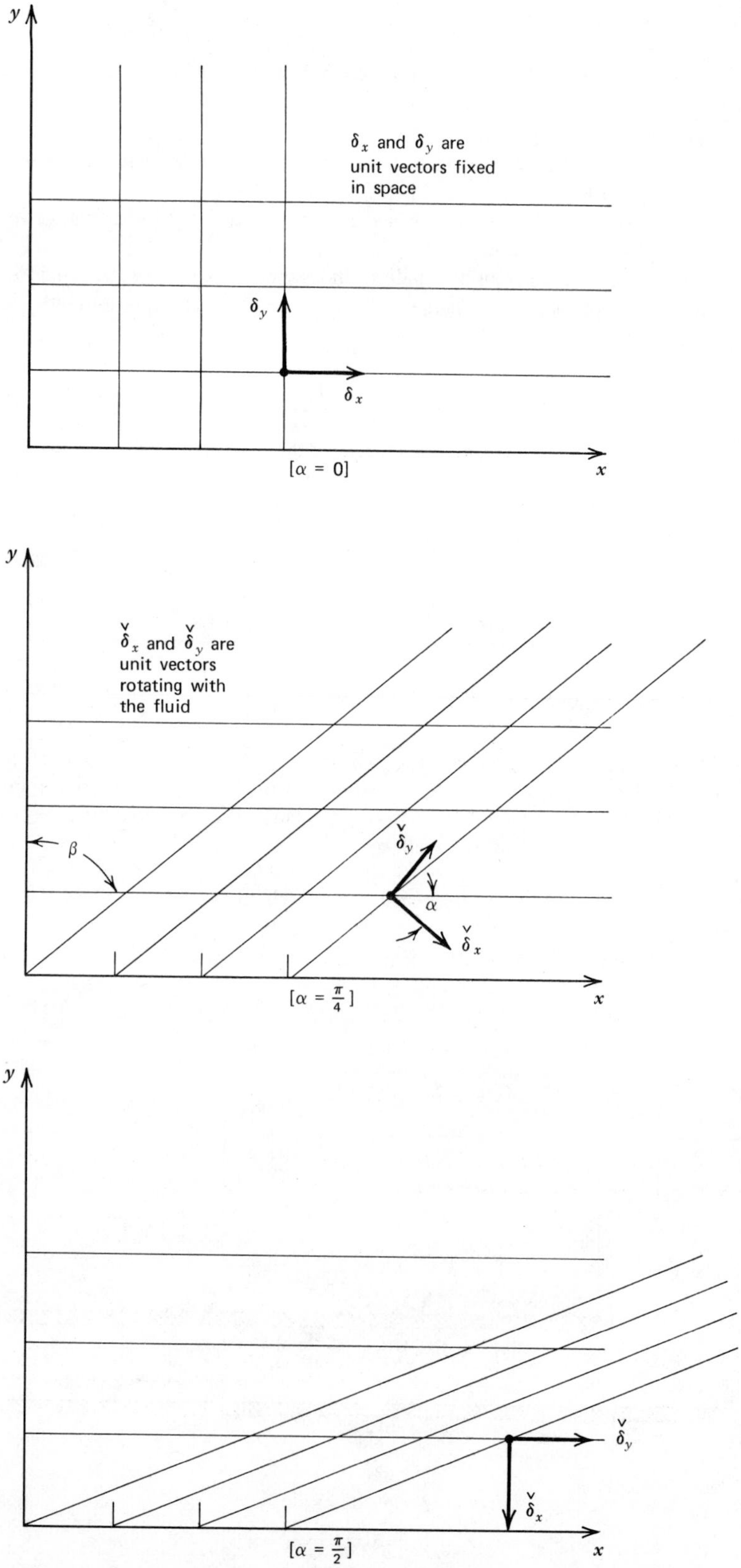

FIGURE 7B.6. Steady simple shear flow with velocity gradient $\dot{\gamma}$, showing the deformation of an embedded grid and the corotating unit vectors.

c. Show also that the Ω_{ij} are given by:

$$\boldsymbol{\Omega} = \begin{pmatrix} \cos\alpha & 0 & \sin\alpha \\ 0 & 1 & 0 \\ -\sin\alpha & 0 & \cos\alpha \end{pmatrix} \tag{7B.5–3}$$

d. Obtain the following expressions for the nonzero components of the tensor $\dot{\boldsymbol{\Gamma}}(r,z,t,t')$:

$$\dot{\Gamma}_{rr} = -\frac{2f(z)\cos^2\alpha}{r^2 - 2f\cdot(t-t')} + \frac{f'(z)\sin 2\alpha}{\sqrt{r^2 - 2f\cdot(t-t')}} \tag{7B.5–4}$$

$$\dot{\Gamma}_{\theta\theta} = \frac{2f(z)}{r^2 - 2f\cdot(t-t')} \tag{7B.5–5}$$

$$\dot{\Gamma}_{zz} = -\frac{2f(z)\sin^2\alpha}{r^2 - 2f\cdot(t-t')} - \frac{f'(z)\sin 2\alpha}{\sqrt{r^2 - 2f\cdot(t-t')}} \tag{7B.5–6}$$

$$\dot{\Gamma}_{rz} = \dot{\Gamma}_{zr} = \frac{f(z)\sin 2\alpha}{r^2 - 2f\cdot(t-t')} + \frac{f'(z)\cos 2\alpha}{\sqrt{r^2 - 2f\cdot(t-t')}} \tag{7B.5–7}$$

Note that in this problem $\dot{\boldsymbol{\Gamma}}'$ is a function of r and z, the present location of a fluid particle, since the flow is inhomogeneous.

e. Describe with words and equations how one would set up the problem of trying to determine the velocity and pressure distributions for one of the models described in this chapter.

f. Verify that the result in Eq. 7B.5–3 does satisfy Eq. C of Table 7.1–1 for the assumed velocity profile.

7B.6 Rotation of a Fluid Element in Simple Shear Flow

Consider a simple shear flow with constant velocity gradient $\dot{\gamma}$ as shown in Fig. 7B.6. At $t = t_0$ a rectangular grid with lines one unit apart is imprinted in the fluid. During the subsequent steady shear flow the imprinted lines deform with the fluid.

At time $t = t_0$ we also introduce a set of unit vectors $\breve{\boldsymbol{\delta}}_x$ and $\breve{\boldsymbol{\delta}}_y$ to coincide with the fixed unit vectors $\boldsymbol{\delta}_x$ and $\boldsymbol{\delta}_y$. During the subsequent motion we require that the unit vectors $\breve{\boldsymbol{\delta}}_x$ and $\breve{\boldsymbol{\delta}}_y$ go along with the fluid; they remain orthogonal and rotate with the local angular velocity of the fluid $\boldsymbol{w} = \frac{1}{2}[\nabla \times \boldsymbol{v}]$ (see Eq. 7.1–4).

At any time t, let α be the angle through which the unit vectors $\breve{\boldsymbol{\delta}}_x$ and $\breve{\boldsymbol{\delta}}_y$ have turned, and let β be the angle through which the grid lines are tilted. Show that $\beta = \arctan 2\alpha$.

7B.7 The ZFD Model for Unsteady Shear Flows[2]

a. Show that for arbitrary unsteady simple shearing flows of the form $v_x = \dot{\gamma}(t)y$, the ZFD model gives the following three simultaneous ordinary differential equations:

$$(\tau_{xx} - \tau_{yy}) + \lambda_0 \frac{d}{dt}(\tau_{xx} - \tau_{yy}) - 2\lambda_0\dot{\gamma}\tau_{yx} = 0 \tag{7B.7–1}$$

$$(\tau_{yy} - \tau_{zz}) + \lambda_0 \frac{d}{dt}(\tau_{yy} - \tau_{zz}) + \lambda_0\dot{\gamma}\tau_{yx} = 0 \tag{7B.7–2}$$

$$\tau_{yx} + \lambda_0 \frac{d}{dt}\tau_{yx} + \tfrac{1}{2}\lambda_0\dot{\gamma}(\tau_{xx} - \tau_{yy}) = -\eta_0\dot{\gamma} \tag{7B.7–3}$$

[2] See footnote 4 in §7.3.

b. From the above set of equations obtain the following second-order ordinary differential equations for the shear stress and the primary normal stress difference:

$$\lambda_0{}^2 \frac{d^2}{dt^2} \tau_{yx} + \left(2\lambda_0 - \frac{\lambda_0{}^2 \ddot{\gamma}}{\dot{\gamma}}\right) \frac{d}{dt} \tau_{yx} + \left[1 + (\lambda_0 \dot{\gamma})^2 - \frac{\lambda_0 \ddot{\gamma}}{\dot{\gamma}}\right] \tau_{yx} = -\eta_0 \dot{\gamma} \tag{7B.7–4}$$

$$\lambda_0{}^2 \frac{d^2}{dt^2} (\tau_{xx} - \tau_{yy}) + \left(2\lambda_0 - \frac{\lambda_0{}^2 \ddot{\gamma}}{\dot{\gamma}}\right) \frac{d}{dt} (\tau_{xx} - \tau_{yy}) + \left[1 + (\lambda_0 \dot{\gamma})^2 - \frac{\lambda_0 \ddot{\gamma}}{\dot{\gamma}}\right] (\tau_{xx} - \tau_{yy}) = -2\eta_0 \lambda_0 \dot{\gamma}^2 \tag{7B.7–5}$$

in which $\ddot{\gamma}$ means $d\dot{\gamma}/dt$.

c. Obtain the results in Eqs. 7.3–20 and 21 (with $\lambda_2 = 0$ and $\lambda_1 = \lambda_0$) for steady shear flow by using Eq. 7B.7–4.

7B.8 Constrained Recoil for the ZFD Fluid

Work the recoil problem of Example 6.2–5 for the ZFD fluid. Start with Eqs. 7B.7–1 to 3.

a. First solve Eq. 7B.7–1 for the normal stress difference:

$$\tau_{xx} - \tau_{yy} = N_0 e^{-t/\lambda_0} \tag{7B.8–1}$$

where $N_0 = -2\eta_0 \lambda_0 \dot{\gamma}_0{}^2/[1 + (\lambda_0 \dot{\gamma}_0)^2]$. Substitute this result into Eq. 7B.7–3 and get an equation that can be solved for $\dot{\gamma}$ to give:

$$\dot{\gamma} = \frac{\partial \gamma(0,t)}{\partial t} = -\frac{\tau_{yx} + \lambda_0 (d\tau_{yx}/dt)}{\eta_0 + \frac{1}{2}\lambda_0 N_0 e^{-t/\lambda_0}} \tag{7B.8–2}$$

b. Obtain the ultimate recoil by performing an integration:

$$\begin{aligned} \gamma_\infty &= -\int_0^\infty \frac{0 + \lambda_0 \tau_0 \cdot 2\delta(t' - 0)}{\eta_0 + \frac{1}{2}\lambda_0 N_0 e^{-t'/\lambda_0}} dt' \\ &= -\lambda \dot{\gamma}_0 \end{aligned} \tag{7B.8–3}$$

7B.9 Corotational Jeffreys Model

In §7.3 various material functions were obtained for the corotational Jeffreys model.

a. Obtain the expression for $\bar{\eta}$ for this model.

b. Attempt to obtain the three constants η_0, λ_1, and λ_2 from some sample steady shear flow data in Chapter 4, and then see how well other material functions can be predicted.

7B.10 Estimation of Maximum Shear Stress in Start-Up of Steady-State Shear Flow

a. Use the Goddard-Miller equation to obtain a formal expression for the maximum value in the function $\eta^+(t;\dot{\gamma}_0)$ which describes the shear-stress growth at the inception of a shear flow ($v_x = 0$ for $t < 0$, and $v_x = \dot{\gamma}_0 y$ for $t > 0$).

b. Next use the relaxation modulus given in part (a) of Problem 7B.11 to obtain an analytical expression for $\eta^+_{\max}(\dot{\gamma}_0)$. Express the result as:

$$\frac{\eta^+_{\max}(\dot{\gamma}_0)}{\eta(\dot{\gamma}_0)} = F(n) \tag{7B.10–1}$$

and find $F(n)$; note that m and $\dot{\gamma}_0$ do not appear in this ratio. (*Note*: In doing this derivation you will obtain an integral that does not seem to be readily available; obtain an approximate expression for the integral by replacing the cosine in the integrand by an appropriate parabola over the range of integration.)

c. Evaluate $F(n)$ for a few sample values of n, including that found for the polystyrene melt in Problem 5A.6(a). Interpret the results in terms of stresses developed in the start-up of polymer processing equipment.

7B.11 Special Forms of the Relaxation Modulus

a. Verify that[3] when the expression

$$G(t - t') = \left(\frac{2m}{\pi}\right)\left[\cos\left(\frac{n\pi}{2}\right)\right]\Gamma(n)(t - t')^{-n} \tag{7B.11–1}$$

is inserted into the Goddard-Miller model, one obtains a viscosity expression given by the power law $\eta = m\dot{\gamma}^{n-1}$. Obtain the analogous expressions for Ψ_1 and $\bar{\eta}$. To what extent are these satisfactory results?

b. Verify that[4]

$$G(t - t') = \left(\frac{2\eta_0}{\sqrt{\pi}\lambda}\right)\left(\frac{2\lambda}{t - t'}\right)^{n/2}\frac{1}{\Gamma(\frac{1}{2}(1 - n))}K_{-n/2}\left(\frac{t - t'}{\lambda}\right) \tag{7B.11–2}$$

leads to the Carreau viscosity expression. $K_{-n/2}$ is a modified Bessel function of the second kind.

c. It has been suggested[5] that

$$G(s) = \frac{\eta_0}{\lambda\Gamma(1 - n)}\left(\frac{\lambda}{s}\right)^n e^{-s/\lambda} \tag{7B.11–3}$$

may be a useful empiricism. Explore some of its properties; what does it give for η, Ψ_1, and $\bar{\eta}$?

7C.1 Small-Amplitude Oscillations in a ZFD Fluid

a. Postulate that in the oscillatory experiment the shear and normal stresses will have the form (for small-amplitude oscillations):

$$\tau_{yx} = \mathcal{Re}\{\tau_{yx}{}^0 e^{i\omega t}\} \tag{7C.1–1}$$

$$\tau_{jj} = \mathcal{Re}\{\tau_{jj}{}^0 e^{2i\omega t}\} + d_j \qquad (j = x,y,z) \tag{7C.1–2}$$

Insert these expressions into the equations in (a) of Problem 7B.7. Then use Eq. 5B.10–1a in order to take the product of the real parts of two complex numbers. Next equate the coefficients of the terms containing $e^{0i\omega t}$, $e^{i\omega t}$, and $e^{2i\omega t}$ to get:

$$(1 + 2i\lambda_0\omega)(\tau_{xx}{}^0 - \tau_{yy}{}^0) - \lambda_0\dot{\gamma}^0\tau_{yx}{}^0 = 0 \tag{7C.1–3}$$

$$(d_x - d_y) - \mathcal{Re}\{\tau_{yx}{}^0\} = 0 \tag{7C.1–4}$$

$$(1 + i\lambda_0\omega)\tau_{yx}{}^0 + \underline{\tfrac{1}{4}\lambda_0\dot{\gamma}^0(\tau_{xx}{}^0 - \tau_{yy}{}^0)} + \underline{\tfrac{1}{2}\lambda_0\dot{\gamma}^0(d_x - d_y)} = -\eta_0\dot{\gamma}^0 \tag{7C.1–5}$$

b. Solve Eq. 7C.1–5 by neglecting the dashed-underlined terms, inasmuch as they would lead to terms of higher order in $\dot{\gamma}^0$. Obtain the expressions for the material functions η', η'', Ψ_1', Ψ_1'', and $\Psi_1{}^d$. Verify that these functions satisfy Eqs. H, I, and J of Table 7.5–2.

[3] W. Heindl and H. Giesekus, *Rheol. Acta, 11*, 152–162 (1972).
[4] A. Co and R. B. Bird, Unpublished Research (1975).
[5] D. J. Segalman, Private Communication (1975).

7C.2 Eccentric Disk Rheometer

Analyze the eccentric disk rheometer of Problem 4B.1 by using the Goddard-Miller model. Both disks rotate with an angular velocity W. The disk centers are displaced by a distance a and the separation between disks is b; we designate the ratio a/b by A.

a. Use the velocity distribution for this system given in Eqs. 4B.1–1 to obtain the matrix representations of $\dot{\boldsymbol{\gamma}}$, $\boldsymbol{\omega}$ and the powers of $\boldsymbol{\omega}$ up through $\boldsymbol{\omega}^5$.

b. Show that for this flow Eq. 7.1–12 gives:

$$\begin{aligned}
\boldsymbol{\Omega}(t,t') &= \boldsymbol{\delta} - \boldsymbol{\Theta}_1\left(\frac{W}{2}(t-t')\right) + \frac{1}{2!}\boldsymbol{\Theta}_2\left(\frac{W}{2}(t-t')\right)^2 \\
&\quad - \frac{1}{3!}\boldsymbol{\Theta}_1\left(\frac{W}{2}(t-t')\right)^3[-(4+A^2)] \\
&\quad + \frac{1}{4!}\boldsymbol{\Theta}_2\left(\frac{W}{2}(t-t')\right)^4[-(4+A^2)] - \cdots \\
&= \boldsymbol{\delta} - \boldsymbol{\Theta}_1(4+A^2)^{-1/2}\sin\left[\left(\frac{W}{2}\right)(4+A^2)^{1/2}(t-t')\right] \\
&\quad + \boldsymbol{\Theta}_2(4+A^2)^{-1}\left\{1-\cos\left[\left(\frac{W}{2}\right)(4+A^2)^{1/2}(t-t')\right]\right\}
\end{aligned} \tag{7C.2–1}$$

in which

$$\boldsymbol{\Theta}_1 = \begin{pmatrix} 0 & 2 & -A \\ -2 & 0 & 0 \\ A & 0 & 0 \end{pmatrix} \tag{7C.2–2}$$

$$\boldsymbol{\Theta}_2 = \begin{pmatrix} -(4+A^2) & 0 & 0 \\ 0 & -4 & 2A \\ 0 & 2A & -A^2 \end{pmatrix} \tag{7C.2–3}$$

c. Show that Eq. 7.5–1 gives:

$$\lim_{A\to 0}\frac{\tau_{xz}}{-AW} = \eta'(W) \tag{7C.2–4}$$

$$\lim_{A\to 0}\frac{\tau_{yz}}{-AW} = \eta''(W) \tag{7C.2–5}$$

where use has to be made of Eq. A of Table 7.5–1. This shows that the measurement of the forces in the x- and y-directions on one disk gives the components of the complex viscosity η^*.

d. Can the same result be obtained by using Eq. 6.1–16? Discuss.

7C.3 The Journal Bearing Problem for a ZFD Fluid

The flow of a Newtonian fluid in a journal bearing is discussed in Problem 1C.9. The same problem has been solved for the ZFD fluid by Fix and Paslay, by use of a perturbation analysis.[1] Work through the details of their analysis.

[1] G. J. Fix and P. R. Paslay, *J. Appl. Mech.*, *34*, 579–582 (1967).

7C.4 Kinematic Tensors for Torsional Flow

For the torsional flow between two parallel disks, the upper one of which is rotating with an angular velocity W, the velocity distribution is $v_\theta = Wr(z/b)$, where b is the distance between the disks. For this flow obtain expressions for the same six kinematic tensors that were worked out in Example 7.6–1.

7D.1 Stress Growth for the ZFD Model

For the stress-growth problem discussed in Example 7.3–3, compute the growth functions for the ZFD model using Laplace transform.

a. Take the Laplace transform of Eqs. 7B.7–1 to 3 for the problem at hand to get:

$$\overline{\tau_{xx} - \tau_{yy}} + \lambda_0 p(\overline{\tau_{xx} - \tau_{yy}}) - 2\lambda_0\dot{\gamma}_0\overline{\tau}_{yx} = 0 \quad (7D.1\text{–}1)$$

$$\overline{\tau_{yy} - \tau_{zz}} + \lambda_0 p(\overline{\tau_{yy} - \tau_{zz}}) + \lambda_0\dot{\gamma}_0\overline{\tau}_{yx} = 0 \quad (7D.1\text{–}2)$$

$$\overline{\tau}_{yx} + \lambda_0 p\overline{\tau}_{yx} + \tfrac{1}{2}\lambda_0\dot{\gamma}_0(\overline{\tau_{xx} - \tau_{yy}}) = -\frac{\eta_0\dot{\gamma}_0}{p} \quad (7D.1\text{–}3)$$

in which p is the transform variable and overlines denote transformed quantities.

b. Solve for the transformed quantities in (a).

c. Invert to obtain:

$$\tau_{yx} = -\frac{\eta_0\dot{\gamma}_0}{1 + (\lambda_0\dot{\gamma}_0)^2}[1 + (\lambda\dot{\gamma}_0 \sin\dot{\gamma}_0 t - \cos\dot{\gamma}_0 t)e^{-t/\lambda_0}] \quad (7D.1\text{–}4)$$

$$\tau_{xx} - \tau_{yy} = -\frac{2\eta_0\lambda_0\dot{\gamma}_0{}^2}{1 + (\lambda_0\dot{\gamma}_0)^2}\left[1 - \left(\frac{1}{\lambda_0\dot{\gamma}_0}\sin\dot{\gamma}_0 t + \cos\dot{\gamma}_0 t\right)e^{-t/\lambda_0}\right] \quad (7D.1\text{–}5)$$

7D.2 Suddenly Accelerated Flow of a ZFD Fluid[1]

A corotational Maxwell model (ZFD fluid) is placed in the region between two parallel plates, located at $y = 0$ and $y = B$. For $t < 0$ the system is at rest, but for $t > 0$ the bottom plate moves in the x-direction with a constant velocity v_0. Denn and Porteous[1] have shown that the velocity-profile development is given by:[2]

$$\frac{v_x}{v_0} = 1 - \frac{y}{B} + \frac{2}{\pi}\sum_{n=1}^{\infty}\frac{\sin n\pi y/B}{n}T_n(t) \quad (7D.2\text{–}1)$$

where

$$T_n(t) = \begin{cases} e^{-t/2\lambda_0}\left[\cosh\dfrac{\sqrt{(a_n\eta_0/\rho B^2)t}}{2\text{ El}} + \dfrac{1}{a_n}\sinh\dfrac{\sqrt{(a_n\eta_0/\rho B^2)t}}{2\text{ El}}\right] & a_n > 0 \\ e^{-t/2\lambda_0}\left[1 + \dfrac{(\eta_0/\rho B^2)t}{2\text{ El}}\right] & a_n = 0 \\ e^{-t/2\lambda_0}\left[\cos\dfrac{\sqrt{-(a_n\eta_0/\rho B^2)t}}{2\text{ El}} - \dfrac{1}{a_n}\sin\dfrac{\sqrt{-(a_n\eta_0/\rho B^2)t}}{2\text{ El}}\right] & a_n < 0 \end{cases} \quad (7D.2\text{–}2)$$

[1] This problem is based on a discussion given by M. M. Denn and K. C. Porteous, *Chem. Eng. Journal*, 2, 280–286 (1971).

[2] H. F. Weinberger, *Partial Differential Equations*, Blaisdell, New York (1965).

Here El is an "elasticity number" defined by:

$$\text{El} = \frac{\eta_0 \lambda_1}{\rho B^2} \tag{7D.2-3}$$

and the a_n are defined by:

$$a_n = 1 - 4\pi^2 n^2 \text{ El} \tag{7D.2-4}$$

Verify this result and then show that the Newtonian solution is recovered when El → 0. Discuss the difference in the character of the solution for the Newtonian and the viscoelastic fluids.

CHAPTER 8
NONLINEAR COROTATIONAL MODELS

In the last chapter we discussed the use of a corotating coordinate frame for constructing objective rheological equations of state. We illustrated the method by writing some of the linear viscoelastic models of Chapter 6 in the corotating frame, thereby obtaining several *quasilinear corotational models*. There is no reason why we should restrict ourselves to rheological equations of state that are linear relations between $\breve{\tau}_{ij}$ and $\breve{\dot{\gamma}}_{ij}$ in the corotating reference frame. In this chapter we remove the restriction of linearity and follow the same procedure used in Chapter 7. This leads to *nonlinear corotational models.*

In the first two sections we present several proposals for creating nonlinear corotational models. In §8.1 we illustrate nonlinear differential models by giving the Oldroyd 8-constant model, which has deservedly enjoyed considerable attention; quite a few fluid dynamics problems have been solved for particular forms of this model. In §8.2 we give several examples of nonlinear integral models. The models of §§8.1 and 2 are, of course, empirical and there is no systematic procedure for improving them.

In §8.3 we give the corotational memory integral expansion, which presents the stress tensor as an infinite series of integrals of increasing complexity. The first term alone gives the Goddard-Miller model, and the double-, triple-, etc., integral terms provide systematic corrections. Although we cannot expect to find the complete expansion immediately useful for solving fluid dynamics problems, we shall see that it is very helpful for gaining a broader perspective of polymer fluid rheology and for systematizing the presentation of useful expressions for the stress tensor in certain simple flows (we call these "reduced rheological equations of state").

Specifically in §8.4 we give the "retarded motion expansion" for slow flows that are steady state or else that vary very slowly in time. And in §8.5 we give the "Criminale-Ericksen-Filbey" or "CEF" equation for steady-state shear flows. Both of these reduced rheological equations of state have been used extensively for interpreting rheological experiments and for making fluid dynamical calculations.

Because of the complexity of the expressions for the stress tensor, the solution of fluid dynamics problems is unavoidably very difficult. Analytical solutions can be obtained for a few simple problems by taking recourse to perturbation methods. For most problems of engineering interest, numerical solutions are necessary.

One very important aspect of fluid dynamics problems is the question of whether or not a flow is stable. In this chapter we do not delve into the subject of instability in viscoelastic fluid dynamics. Much has been published on this topic, but we content ourselves only to cite a few references.[1–7]

[1] H. Giesekus, *Rheol. Acta*, *5*, 239–252 (1966); R. K. Bhatnagar and H. Giesekus, *ibid.*, *9*, 53–60, 412–418 (1970); H. Giesekus and R. K. Bhatnagar, *ibid.*, *10*, 266–274 (1971).

§8.1 NONLINEAR DIFFERENTIAL COROTATIONAL MODELS (THE OLDROYD 8-CONSTANT MODEL)

We take as our starting point the corotational Jeffreys model of Eq. 7.3–3. The $\mathscr{D}/\mathscr{D}t$ terms in the model contain nonlinear terms: terms with products of velocity gradients, and terms with products of stresses and velocity gradients. Oldroyd[1] suggested that a possibly useful generalization of the corotational Jeffreys model could be obtained by adding to the latter all possible terms involving products of the $\check{\tau}_{ij}$ with the $\check{\dot{\gamma}}_{ij}$, and all possible terms quadratic in the $\check{\dot{\gamma}}_{ij}$, with appropriate constants multiplying all these extra terms included. Thus we write down in the corotational frame:

$$\underline{\check{\tau}_{ij} + \lambda_1 \frac{\partial \check{\tau}_{ij}}{\partial t}} + \tfrac{1}{2}\mu_0 \left(\textstyle\sum_k \check{\tau}_{kk}\right)\check{\dot{\gamma}}_{ij}$$
$$- \tfrac{1}{2}\mu_1 \textstyle\sum_k \left(\check{\tau}_{ik}\check{\dot{\gamma}}_{kj} + \check{\dot{\gamma}}_{ik}\check{\tau}_{kj}\right) + \tfrac{1}{2}\nu_1\left(\sum_m \sum_n \check{\tau}_{mn}\check{\dot{\gamma}}_{nm}\right)\delta_{ij}$$
$$= -\eta_0 \left[\underline{\check{\dot{\gamma}}_{ij} + \lambda_2 \frac{\partial \check{\dot{\gamma}}_{ij}}{\partial t}} - \mu_2 \textstyle\sum_k \check{\dot{\gamma}}_{ik}\check{\dot{\gamma}}_{kj} + \tfrac{1}{2}\nu_2\left(\sum_m \sum_n \check{\dot{\gamma}}_{mn}\check{\dot{\gamma}}_{nm}\right)\delta_{ij}\right] \qquad (8.1\text{–}1)$$

The dashed-underlined terms belong to the original corotational Jeffreys model. When transformed into the nonrotating frame, this becomes, according to Tables 7.1–1 and 2:

$$\underline{\boldsymbol{\tau} + \lambda_1 \frac{\mathscr{D}\boldsymbol{\tau}}{\mathscr{D}t}} + \tfrac{1}{2}\mu_0(\operatorname{tr}\boldsymbol{\tau})\dot{\boldsymbol{\gamma}}$$
$$- \tfrac{1}{2}\mu_1\{\boldsymbol{\tau}\cdot\dot{\boldsymbol{\gamma}} + \dot{\boldsymbol{\gamma}}\cdot\boldsymbol{\tau}\} + \tfrac{1}{2}\nu_1(\boldsymbol{\tau}:\dot{\boldsymbol{\gamma}})\boldsymbol{\delta}$$
$$= -\eta_0\left[\underline{\dot{\boldsymbol{\gamma}} + \lambda_2 \frac{\mathscr{D}\dot{\boldsymbol{\gamma}}}{\mathscr{D}t}} - \mu_2\{\dot{\boldsymbol{\gamma}}\cdot\dot{\boldsymbol{\gamma}}\} + \tfrac{1}{2}\nu_2(\dot{\boldsymbol{\gamma}}:\dot{\boldsymbol{\gamma}})\boldsymbol{\delta}\right] \qquad (8.1\text{–}2)$$

Several simplified versions of this model have been used, in which special values are assigned to some of the constants; these are summarized in Table 8.1–1.

There are some restrictions on the choice of the constants that have to be imposed in order that the derived material functions will not be in conflict with experimental data:

1. Since η' is known to decrease with increasing ω, we must impose the requirement that $0 < \lambda_2 < \lambda_1$; see Eq. 8.1–10.
2. Since viscosity is generally a monotone decreasing function of $\dot{\gamma}$, we must require that $0 < \sigma_2 < \sigma_1$ [where $\sigma_i = \lambda_1\lambda_i + \mu_0(\mu_i - \tfrac{3}{2}\nu_i) - \mu_1(\mu_i - \nu_i)$, with $i = 1, 2$]; see Eq. 8.1–7.

[2] F. J. Lockett and R. S. Rivlin, *J. Mécanique*, *7*,475–498 (1968); M. M. Smith and R. S. Rivlin, *ibid.*, *11*, 69–94 (1972).

[3] J. P. Tordella, in *Rheology*, F. R. Eirich (Ed.), Academic Press, New York (1969), Vol. 5, Chapter 2.

[4] M. R. Feinberg and W. R. Schowalter, *Ind. Eng. Chem. Fundamentals*, *8*,332–338 (1969); L. V. McIntire and W. R. Schowalter, *AIChE Journal*, *18*,102–110 (1972).

[5] D. D. Joseph, *Stability of Fluid Motions*, Springer, Berlin (1976).

[6] J. R. A. Pearson, *Ann. Rev. Fluid Mech.*, *8*,163–181 (1976).

[7] C. J. S. Petrie and M. M. Denn, *AIChE Journal*, *22*,209–236 (1976).

[1] J. G. Oldroyd, *Proc. Roy. Soc.*, *A245*,278–297 (1958); *Rheol. Acta*, *1*,337–344 (1961). The parameters $\eta_0, \lambda_1, \lambda_2, \mu_0, \mu_1, \mu_2, \nu_1$, and ν_2 in Eq. 8.1–2 are identical to those used by Oldroyd. See also R. R. Huilgol, *Continuum Mechanics of Viscoelastic Liquids*, Wiley, New York (1975), pp. 191–198.

3. Since τ_{yx} is to be ever-increasing with $\dot{\gamma}$ for steady shear flow, we have to require that $\sigma_2 \geq \frac{1}{9}\sigma_1$.
4. When $\eta(\dot{\gamma})$ and $\eta'(\omega)$ are plotted on the same graph with $\dot{\gamma} = \omega$, the η-curve generally lies above the η'-curve. For this to be true in the region of small $\dot{\gamma}$ and small ω, we have to require that $\sigma_1 - \sigma_2 < \lambda_1(\lambda_1 - \lambda_2)$; see comment at the end of Example 8.1–4.

Consideration of additional material functions and additional flow patterns will in general provide further inequalities.

With eight constants and all the extra terms considerably more variety in rheological response is possible than for the corotational Jeffreys equation. The Oldroyd 8-constant model has been found to be a useful and relatively simple rheological equation of state for making exploratory fluid dynamical calculations.

In the following illustrative examples we work out some of the material functions for this model and show how it has been used to describe several flow systems. In the second example (Example 8.1–2) we show how the model can be recast in an integral form, which leads to a memory-integral expansion. In Problem 8B.3 the connection between the Oldroyd model and the retarded-motion expansion is brought out. And in Problem 11B.9 it is shown how the Oldroyd model can be used as a framework for presenting kinetic theory results (see also Figs. 9.5–1 and 2).

EXAMPLE 8.1–1 Material Functions for the Oldroyd 8-Constant Model

Show how Eq. 8.1–2 simplifies for unsteady shear flow. Then obtain the steady-state shear flow material functions and the small-amplitude oscillatory motion material functions.

SOLUTION

(a) For the unsteady shear flow $v_x(y,t) = \dot{\gamma}(t)y$, Eq. 8.1–2 simplifies to:

$$\left(1 + \lambda_1 \frac{\partial}{\partial t}\right)\tau_{xx} - (\lambda_1 + \mu_1 - \nu_1)\tau_{yx}\dot{\gamma} = \eta_0(\lambda_2 + \mu_2 - \nu_2)\dot{\gamma}^2 \tag{8.1–3}$$

$$\left(1 + \lambda_1 \frac{\partial}{\partial t}\right)\tau_{yy} + (\lambda_1 - \mu_1 + \nu_1)\tau_{yx}\dot{\gamma} = -\eta_0(\lambda_2 - \mu_2 + \nu_2)\dot{\gamma}^2 \tag{8.1–4}$$

$$\left(1 + \lambda_1 \frac{\partial}{\partial t}\right)\tau_{zz} + \nu_1\tau_{yx}\dot{\gamma} = -\eta_0\nu_2\dot{\gamma}^2 \tag{8.1–5}$$

$$\left(1 + \lambda_1 \frac{\partial}{\partial t}\right)\tau_{yx} + \tfrac{1}{2}(\lambda_1 - \mu_1 + \mu_0)\tau_{xx}\dot{\gamma} - \tfrac{1}{2}(\lambda_1 + \mu_1 - \mu_0)\tau_{yy}\dot{\gamma} + \tfrac{1}{2}\mu_0\tau_{zz}\dot{\gamma} = -\eta_0\left(1 + \lambda_2 \frac{\partial}{\partial t}\right)\dot{\gamma} \tag{8.1–6}$$

(b) For *steady-state shear flow* the above equations become a set of four simultaneous algebraic equations. When these are solved we get the following material functions:

$$\frac{\eta}{\eta_0} = \frac{1 + [\lambda_1\lambda_2 + \mu_0(\mu_2 - \frac{3}{2}\nu_2) - \mu_1(\mu_2 - \nu_2)]\dot{\gamma}^2}{1 + [\lambda_1^2 + \mu_0(\mu_1 - \frac{3}{2}\nu_1) - \mu_1(\mu_1 - \nu_1)]\dot{\gamma}^2} \tag{8.1–7}$$

$$\frac{\Psi_1}{2\eta_0\lambda_1} = \frac{\eta(\dot{\gamma})}{\eta_0} - \frac{\lambda_2}{\lambda_1} \tag{8.1–8}$$

TABLE 8.1–1
Models Included in the Oldroyd 8-Constant Model (Eq. 8.1–2)

Name of Model	Number of Constants	Values of Time Constants							Steady-State Shear Flow Material Functions[a]	Elongational Viscosity[b]	References
		λ_1	λ_2	μ_1	μ_2	μ_0	ν_1	ν_2			
Oldroyd 6-constant model	6						0	0	η depends on $\dot{\gamma}$ Ψ_1 depends on $\dot{\gamma}$ Ψ_2 not simply related to Ψ_1	See Eq. 8B.1–3.	Problems 8B.3 and 11B.9
Oldroyd 4-constant model	4			λ_1	λ_2		0	0	η depends on $\dot{\gamma}$ Ψ_1 depends on $\dot{\gamma}$ $\Psi_2 = 0$	See Eq. 8B.1–3.	Example 8.1–4
Oldroyd's "Fluid A"	3			$-\lambda_1$	$-\lambda_2$	0	0	0	$\eta = \eta_0$ $\Psi_1 = 2\eta_0(\lambda_1 - \lambda_2)$ $\Psi_2 = -\Psi_1$	$3\eta_0 \dfrac{1 + \lambda_2\dot{\epsilon} - 2\lambda_1\lambda_2\dot{\epsilon}^2}{1 + \lambda_1\dot{\epsilon} - 2\lambda_1{}^2\dot{\epsilon}^2}$	Example 9.1–1 Problem 9B.8
Oldroyd's "Fluid B"	3			λ_1	λ_2	0	0	0	$\eta = \eta_0$ $\Psi_1 = 2\eta_0(\lambda_1 - \lambda_2)$ $\Psi_2 = 0$	$3\eta_0 \dfrac{1 - \lambda_2\dot{\epsilon} - 2\lambda_1\lambda_2\dot{\epsilon}^2}{1 - \lambda_1\dot{\epsilon} - 2\lambda_1{}^2\dot{\epsilon}^2}$	Example 9.1–1 Problem 9B.8
Williams' 3-constant Oldroyd model	3			λ_1	λ_2	0	$\frac{2}{3}\lambda_1$	$\frac{2}{3}\lambda_2$	η depends on $\dot{\gamma}$ Ψ_1 depends on $\dot{\gamma}$ $\Psi_2 = 0$	$3\eta_0 \dfrac{1 - \lambda_2\dot{\epsilon}}{1 - \lambda_1\dot{\epsilon}}$	Problem 8B.2

Corotational Jeffreys Model	3			0	0	0	0	0	η depends on $\dot{\gamma}$ Ψ_1 depends on $\dot{\gamma}$ $\Psi_2 = -\frac{1}{2}\Psi_1$	$3\eta_0$	§7.3
Denn's modified convected Maxwell model	3		0	λ_1		0	0	0	$\eta = \eta_0(1 - \lambda_1\mu_2\dot{\gamma}^2)$ $\Psi_1 = 2\lambda_1\eta$ $\Psi_2 = 0$	$3\eta_0 \dfrac{1 - \mu_2\dot{\epsilon} - 2\lambda_1\mu_2\dot{\epsilon}^2}{1 - \lambda_1\dot{\epsilon} - 2\lambda_1^2\dot{\epsilon}^2}$	Problem 9B.1
Second-order fluid[c]	3	0		0		0	0	0	$\eta = \eta_0$ $\Psi_1 = -2\eta_0\lambda_2$ $\Psi_2 = \eta_0(\lambda_2 - \mu_2)$	$3\eta_0(1 - \mu_2\dot{\epsilon})$	§8.4
Convected Maxwell model	2		0	λ_1	0	0	0	0	$\eta = \eta_0$ $\Psi_1 = 2\eta_0\lambda_1$ $\Psi_2 = -\Psi_1$	$3\eta_0 \dfrac{1}{1 - \lambda_1\dot{\epsilon} - 2\lambda_1^2\dot{\epsilon}^2}$.	

[a] See Eqs. 8.1–7 to 9 for general expressions.
[b] See Eq. 8B.1–3 for general expression.
[c] For the relation of the Oldroyd model and the third-order fluid see Problem 8B.3.

$$\frac{\Psi_2}{\eta_0\lambda_1} = -\left(1 - \frac{\mu_1}{\lambda_1}\right)\frac{\eta(\dot{\gamma})}{\eta_0} + \frac{\lambda_2}{\lambda_1}\left(1 - \frac{\mu_2}{\lambda_2}\right)$$

$$= -\frac{\Psi_1}{2\eta_0\lambda_1} + \left(\frac{\mu_1}{\lambda_1}\frac{\eta}{\eta_0} - \frac{\mu_2}{\lambda_1}\right) \tag{8.1-9}$$

It is evident that this model does not require the fixed relation $\Psi_2 = -\frac{1}{2}\Psi_1$, which all quasilinear corotational models give. For a particular choice of the eight constants given in Eq. 9.5–3, the predicted material functions are given in Fig. 9.5–1.

(c) For *small-amplitude oscillatory motion*, the nonlinear terms in Eq. 8.1–2 can be omitted, and the material functions are the same as for the corotational Jeffreys model:

$$\frac{\eta'}{\eta_0} = \frac{1 + \lambda_1\lambda_2\omega^2}{1 + (\lambda_1\omega)^2} \tag{8.1-10}$$

$$\frac{\eta''}{\eta_0} = \frac{(\lambda_1 - \lambda_2)\omega}{1 + (\lambda_1\omega)^2} \tag{8.1-11}$$

From this we see that this model does not require that $\eta(\dot{\gamma})$ and $\eta'(\omega)$ be identical functions, nor that $\Psi_1/2$ and η''/ω be identical functions. In addition to these material functions, the elongational viscosity is worked out in Problem 8B.1 and illustrated in Fig. 9.5–2. From all these material-functions results, we can conclude that the Oldroyd model is a considerable improvement over the corotational Jeffreys model; however, the viscosity and other material functions are generally incapable of describing the data for polymeric fluids quantitatively over large ranges of the variables.

EXAMPLE 8.1–2 The Integral Form of the Oldroyd 8-Constant Model

In §§7.3 and 4 we gave the corotational Jeffreys model and the generalized ZFD model in both differential and integral form. Show how to transform Eq. 8.1–2 into an equivalent integral form.

SOLUTION

Because of the nonlinearities in the model when written in the form of Eq. 8.1–1, it is not easy to convert to an integral model. We follow here the iterative procedure suggested by Walters.[2]

If we anticipate that $\check{\tau}_{ij}$ will contain terms linear in the $\check{\dot{\gamma}}_{ij}$, terms quadratic in the $\check{\dot{\gamma}}_{ij}$, etc., then to get the first-order solution one would select the dashed underlined terms of Eq. 8.1–1, all of which contain first-order terms. Using these terms only, Eq. 8.1–1 can be integrated to give:

$$\check{\tau}_{ij} = -\frac{\eta_0\lambda_2}{\lambda_1}\check{\dot{\gamma}}_{ij} - \frac{\eta_0(\lambda_1 - \lambda_2)}{\lambda_1^2}\int_{-\infty}^{t} e^{-(t-t')/\lambda_1}\check{\dot{\gamma}}_{ij}(t')\,dt' \tag{8.1-12}$$

which is just Eq. 7.3–2 in a slightly different form. This expression can be inserted for $\check{\tau}_{ij}$ in the non-underlined terms in Eq. 8.1–1 and the equation can be integrated a second time to give the terms quadratic in the $\check{\dot{\gamma}}_{ij}$:

$$\begin{aligned}\check{\tau}_{ij} = &-\frac{\eta_0\lambda_2}{\lambda_1}\check{\dot{\gamma}}_{ij} - \frac{\eta_0(\lambda_1 - \lambda_2)}{\lambda_1^2}\int_{-\infty}^{t} e^{-(t-t')/\lambda_1}\check{\dot{\gamma}}_{ij}(t')\,dt' \\ &- \frac{\eta_0(\mu_1\lambda_2 - \mu_2\lambda_1)}{\lambda_1^2}\int_{-\infty}^{t} e^{-(t-t')/\lambda_1}\textstyle\sum_k \check{\dot{\gamma}}_{ik}(t')\check{\dot{\gamma}}_{kj}(t')\,dt' \\ &- \frac{\eta_0\mu_1(\lambda_1 - \lambda_2)}{2\lambda_1^3}\int_{-\infty}^{t}\int_{-\infty}^{t'} e^{-(t-t'')/\lambda_1} \\ &\times \textstyle\sum_k \left[\check{\dot{\gamma}}_{ik}(t'')\check{\dot{\gamma}}_{kj}(t') + \check{\dot{\gamma}}_{ik}(t')\check{\dot{\gamma}}_{kj}(t'')\right] dt''\,dt' \\ &+ (\text{terms containing } v_1, v_2)\,\delta_{ij}\end{aligned} \tag{8.1-13}$$

[2] K. Walters, *Zeits. für angew. Math. u. Physik*, *21*, 592–600 (1970).

Note that the μ_0 term does not appear in this approximation because tr $\dot{\boldsymbol{\gamma}} = 0$ for an incompressible fluid. The final term containing ν_1 and ν_2 may be omitted[3] inasmuch as $\breve{\tau}_{ij}$ is undetermined up to an isotropic function. The expression for $\breve{\tau}_{ij}$ in Eq. 8.1–13 can now be put back into Eq. 8.1–1 and the integration process repeated; this iterative scheme clearly leads to multiple integrals of ever-increasing order.

We can now rewrite Eq. 8.1–13 by making two changes: (1) we put all terms linear in $\breve{\dot{\gamma}}_{ij}$ under a single integral, and all terms quadratic in the $\breve{\dot{\gamma}}_{ij}$ in a double integral, etc., and (2) we transform to fixed coordinates. This gives finally:

$$\begin{aligned}\boldsymbol{\tau} = &-\int_{-\infty}^{t} G_I(t-t')\dot{\boldsymbol{\Gamma}}'\,dt' \\ &-\tfrac{1}{2}\int_{-\infty}^{t}\int_{-\infty}^{t} G_{II}(t-t',t-t'')\{\dot{\boldsymbol{\Gamma}}'\cdot\dot{\boldsymbol{\Gamma}}''+\dot{\boldsymbol{\Gamma}}''\cdot\dot{\boldsymbol{\Gamma}}'\}\,dt''\,dt' - \cdots\end{aligned} \tag{8.1–14}$$

in which

$$G_I = \frac{\eta_0}{\lambda_1}\left(1-\frac{\lambda_2}{\lambda_1}\right)e^{-(t-t')/\lambda_1} + 2\eta_0\frac{\lambda_2}{\lambda_1}\delta(t-t') \tag{8.1–15}$$

$$G_{II} = \begin{cases} 0 & \text{if} \quad t' < t'' \le t \\ \left[\dfrac{\eta_0\mu_1(\lambda_1-\lambda_2)}{2\lambda_1{}^3}e^{-(t-t'')/\lambda_1} + \dfrac{2\eta_0(\mu_1\lambda_2-\lambda_1\mu_2)}{\lambda_1{}^2}e^{-(t-t')/\lambda_1}\delta(t'-t'')\right] & \text{if} \quad t' \ge t'' \end{cases} \tag{8.1–16}$$

We are thus led to a *memory-integral expansion* of which we have written out the first two terms. More will be said about such expansions in §8.3. (Note that G_I is identical to the { }- quantity in Eq. 7.3–4; thus the first term in the memory-integral expansion in Eq. 8.1–14 just exactly reproduces the corotational Jeffreys model, as one would expect.)

EXAMPLE 8.1–3 The Rayleigh Problem for an Oldroyd Fluid

A semi-infinite body of viscoelastic liquid with constant ρ is bounded at $y = 0$ by a flat solid surface (the xz-plane). Before $t = 0$ the fluid and the solid surface are at rest; after $t = 0$, the surface moves with constant velocity v_0 in the positive x-direction. Find the velocity distribution. (See Problem 1B.16 for the Newtonian fluid, and Problem 5C.7 for the power-law model.)

SOLUTION

This problem has been solved by Tanner[4] for the Oldroyd model of Eq. 8.1–2 with $\mu_1 = \lambda_1$, $\mu_2 = \lambda_2$, and $\mu_0 = \nu_1 = \nu_2 = 0$. This is then a viscoelastic model with two time constants λ_1 and λ_2, and with a constant viscosity η_0, a constant $\Psi_1 = 2\eta_0(\lambda_1 - \lambda_2)$, and $\Psi_2 = 0$ in steady shear flow (cf. Eqs. 8.1–7 to 9).

We postulate a solution of the form $v_x = v_x(y,t)$. Then the equation of continuity is satisfied identically and the three components of the equation of motion are:

$$\rho\frac{\partial v_x}{\partial t} = -\frac{\partial \tau_{yx}}{\partial y} \tag{8.1–17}$$

$$0 = -\frac{\partial p}{\partial y} - \frac{\partial \tau_{yy}}{\partial y} \tag{8.1–18}$$

$$0 = -\frac{\partial p}{\partial z} \tag{8.1–19}$$

[3] They may be omitted in writing the second-order term. However, they must be retained if one wishes to go through another iteration to get third-order terms.

[4] R. I. Tanner, *Zeits. f. angew. Math. u. Physik*, *13*, 573–580 (1962).

The Oldroyd rheological equation of state becomes for this flow (cf. Eqs. 8.1–3 to 6):

$$\tau_{yx} + \lambda_1 \left(\frac{\partial \tau_{yx}}{\partial t} - \tau_{yy} \frac{\partial v_x}{\partial y} \right) = -\eta_0 \left(\frac{\partial v_x}{\partial y} + \lambda_2 \frac{\partial}{\partial t} \frac{\partial v_x}{\partial y} \right) \tag{8.1–20}$$

$$\tau_{xx} + \lambda_1 \left(\frac{\partial \tau_{xx}}{\partial t} - 2\tau_{yx} \frac{\partial v_x}{\partial y} \right) = +2\eta_0 \lambda_2 \left(\frac{\partial v_x}{\partial y} \right)^2 \tag{8.1–21}$$

$$\tau_{yy} + \lambda_1 \left(\frac{\partial \tau_{yy}}{\partial t} \right) = 0 \tag{8.1–22}$$

Equations 8.1–17 to 22 are the differential equations needed to describe the flow.

From Eq. 8.1–22 we obtain immediately:

$$\tau_{yy} = A(y) e^{-t/\lambda_1} \tag{8.1–23}$$

where $A(y)$ is an arbitrary function of y. But τ_{yy} is known to be zero for $t \leq 0$, so that $A(y)$ has to be taken to be zero. Then the τ_{yy}-term in Eq. 8.1–20 can be omitted and Eqs. 8.1–17 and 20 can be combined to give:

$$\rho \left(\frac{\partial v_x}{\partial t} + \lambda_1 \frac{\partial^2 v_x}{\partial t^2} \right) = \eta_0 \left(\frac{\partial^2 v_x}{\partial y^2} + \lambda_2 \frac{\partial}{\partial t} \frac{\partial^2 v_x}{\partial y^2} \right) \tag{8.1–24}$$

It is convenient to introduce the following dimensionless quantities:

$$V = \frac{v_x}{v_0} \qquad T = \frac{t}{\lambda_1}$$

$$Y = \sqrt{\frac{\rho}{\eta_0 \lambda_1}}\, y \qquad a = \frac{\lambda_2}{\lambda_1} \tag{8.1–25}$$

The mathematical problem that we have to solve then consists of the linear partial differential equation:

$$\frac{\partial V}{\partial T} + \frac{\partial^2 V}{\partial T^2} = \frac{\partial^2 V}{\partial Y^2} + a \frac{\partial^3 V}{\partial T\, \partial Y^2} \tag{8.1–26}$$

along with the following boundary and initial conditions:

$$V = H(T) \qquad \text{at} \qquad Y = 0 \tag{8.1–27}$$

$$V = 0 \qquad \text{as} \qquad Y \to \infty \tag{8.1–28}$$

$$\frac{\partial V}{\partial T} = \frac{\partial^2 V}{\partial T^2} = 0 \qquad \text{for} \qquad T \leq 0 \tag{8.1–29}$$

where $H(T)$ is the Heaviside unit step function, which is zero for $T < 0$ and unity for $T > 0$.

Tanner solved the problem by taking the Laplace transform of Eq. 8.1–26 and using the initial conditions; this gives:

$$(p + p^2)\bar{V} = (1 + ap) \frac{d^2 \bar{V}}{dY^2} \tag{8.1–30}$$

This differential equation can be solved with the boundary conditions that $\bar{V}$ approaches zero as $Y \to \infty$, and $\bar{V} = 1/p$ at $Y = 0$. The result is:

$$\bar{V} = \frac{1}{p} \exp\left(-\sqrt{\frac{p(1 + p)}{1 + ap}}\, Y \right) \tag{8.1–31}$$

Tanner inverted Eq. 8.1–31 by performing a contour integral in the complex plane to obtain the solution:

$$V(Y,T) = \tfrac{1}{2} + \frac{1}{\pi}\int_0^\infty \exp\left[-\sqrt{\frac{u}{2}}\,M(\cos\theta - \sin\theta)Y\right] \cdot \sin\left[uT - \sqrt{\frac{u}{2}}\,M(\cos\theta + \sin\theta)Y\right]\frac{du}{u} \tag{8.1–32}$$

in which $M(u)$ and $\theta(u)$ are defined by:

$$M(u) = \sqrt[4]{\frac{1 + u^2}{1 + a^2u^2}} \tag{8.1–33}$$

$$\theta(u) = \tfrac{1}{2}(\arctan u - \arctan au) \tag{8.1–34}$$

We see that for this very idealized problem and simplified rheological equation of state, an analytical solution can be obtained. For the special case of $a = 1$ (both time constants equal, which, from Eq. 8.1–7 gives $\eta = \eta_0$), Tanner showed that the solution simplifies to the error function expression given in Problem 1B.16 for the Newtonian fluid.

To show how the character of the solution changes as a goes from 0 (the "damped wave equation") to 1 (the "diffusion equation"), Tanner prepared plots such as that shown in Fig. 8.1–1; there the velocity at a fixed position is shown as a function of time. Note how the viscoelastic fluid tends to store elastic energy at short times and then release it at longer times.

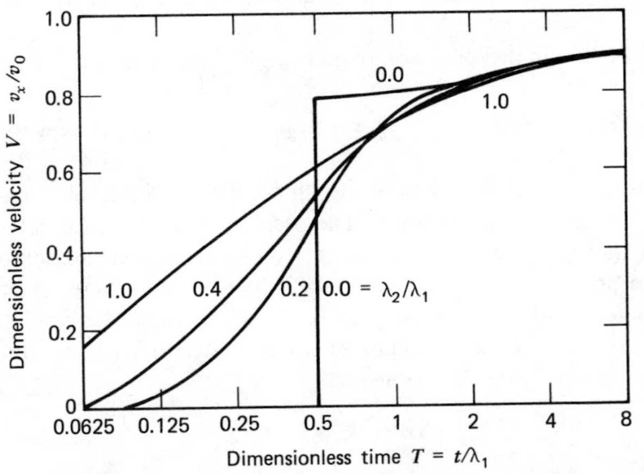

FIGURE 8.1–1. Calculated curves of Tanner for the Rayleigh problem for a viscoelastic fluid. As the parameter a decreases the fluid becomes more and more elastic (i.e., more solidlike). The curves show the velocity variation at $Y = \sqrt{\rho/\eta_0\lambda_1}\,y = 0.5$. [R. I. Tanner, *Z. Angew. Math. Phys.*, *13*, 573–580 (1962).]

EXAMPLE 8.1–4 Tube Flow of an Oldroyd Fluid with a Pulsatile Pressure Gradient[5]

Consider the unsteady flow of a viscoelastic polymeric fluid in a circular tube of radius R. Let the pressure gradient vary sinusoidally about some mean value, so that the z-component of the equation

[5] This problem is patterned closely after the discussion given by K. Walters and P. Townsend, *Proc. Fifth International Congress on Rheology* (S. Onogi, Ed.), University of Tokyo Press, (1970), Vol. 4, pp. 471–483. Other discussions of this problem have been given by H. A. Barnes, P. Townsend, and K. Walters, *Nature, 224*, 585–587 (1969); *Rheol. Acta, 10*, 517–527 (1971). A review article including the pulsatile flow problem as well as flow in a corrugated pipe, flow in a curved pipe, and flow in a noncircular pipe has been prepared by K. Walters, *Progress in Heat and Mass Transfer* (W. R. Schowalter, Ed.), Pergamon Press, New York (1972), Vol. 5, pp. 217–231. Numerical solutions of the pulsatile flow problem, the flow with an impulsively imposed pressure gradient, and the flow with a vibrating pipe were given by P. Townsend, *Rheol. Acta, 12*, 13–18 (1973).

of motion is:

$$\rho \frac{\partial v_z}{\partial t} = \frac{\overline{\Delta p}}{L}(1 + \epsilon\,\mathscr{Re}\{e^{i\omega t}\}) - \frac{1}{r}\frac{\partial}{\partial r}(r\tau_{rz}) \tag{8.1-35}$$

where $\overline{\Delta p}$ is the time-averaged pressure drop over a length L of the tube, ϵ is the (small) amplitude of the pressure-gradient variation, and ω is the frequency of the fluctuations in the pressure drop. If $\epsilon = 0$ (steady-state flow), the velocity distribution is $v_z = v_0(r)$ and the volume rate of flow is Q_0.

We wish to find the mean flow rate $\bar{Q}$ as a function of the mean pressure drop $\overline{\Delta p}$ for a fluid that can be described, at least approximately, by the Oldroyd model of Eq. 8.1–2 with $\mu_1 = \lambda_1$, $\mu_2 = \lambda_2$, and $\nu_1 = \nu_2 = 0$. Such a model has shear-rate-dependent η and Ψ_1, and has $\Psi_2 = 0$; that is, Eq. 8.1–35 has to be solved along with:

$$\boldsymbol{\tau} + \lambda_1\left(\frac{\mathscr{D}\boldsymbol{\tau}}{\mathscr{D}t} - \tfrac{1}{2}\{\boldsymbol{\tau}\cdot\dot{\boldsymbol{\gamma}} + \dot{\boldsymbol{\gamma}}\cdot\boldsymbol{\tau}\}\right) + \tfrac{1}{2}\mu_0(\operatorname{tr}\boldsymbol{\tau})\dot{\boldsymbol{\gamma}} = -\eta_0\left[\dot{\boldsymbol{\gamma}} + \lambda_2\left(\frac{\mathscr{D}\dot{\boldsymbol{\gamma}}}{\mathscr{D}t} - \{\dot{\boldsymbol{\gamma}}\cdot\dot{\boldsymbol{\gamma}}\}\right)\right] \tag{8.1-36}$$

and the boundary conditions

$$v_z = 0 \quad \text{at} \quad r = R$$

$$\frac{\partial v_z}{\partial r} = 0 \quad \text{at} \quad r = 0 \tag{8.1-37}$$

Only a "sinusoidal steady-state" solution is needed, and hence it is not necessary to specify an initial condition.

Follow the perturbation procedure of Walters and Townsend,[5] and seek a solution of the form:

$$v_z(r,t) = v_0 + \epsilon[v_{10} + \mathscr{Re}\{v_{11}e^{i\omega t}\}] + \epsilon^2[v_{20} + \mathscr{Re}\{v_{21}e^{i\omega t}\} + \mathscr{Re}\{v_{22}e^{2i\omega t}\}] + \cdots \tag{8.1-38}$$

in which the v's on the right side are functions of r alone; the v's with a subscript zero are real, whereas the others (contained in $\mathscr{Re}\{\ \}$) are complex. The oscillatory terms in Eq. 8.1–38 will not contribute to $\bar{Q}$; it will turn out that v_{10} is zero and, hence, the lowest-order contribution to $\bar{Q} - Q_0$ will be that involving v_{20}. Hence obtaining $v_{20}(r)$ will be the principal objective here. (Clearly v_{10} must be zero: if it were not, then there would be a term in $\bar{Q}$ proportional to ϵ; and this would mean that $\bar{Q}$ would be different for ϵ positive and for ϵ negative—in other words, $\bar{Q}$ would depend on the phase of the pressure variation—and this is physically unreasonable.)

SOLUTION

First we have to write out Eq. 8.1–36 in matrix form for the flow $v_z = v_z(r,t)$, $v_r = v_\theta = 0$. The rz-component of the equation contains τ_{rr} and tr $\boldsymbol{\tau}$. Hence we also have to select from the matrix form of Eq. 8.1–36 the appropriate equations for τ_{rr} and tr $\boldsymbol{\tau}$. This gives us three equations:

$$\tau_{rz} + \lambda_1\left(\frac{\partial \tau_{rz}}{\partial t} - \underline{\tau_{rr}\dot{\gamma}_{rz}}\right) + \tfrac{1}{2}\mu_0(\operatorname{tr}\boldsymbol{\tau})\dot{\gamma}_{rz} = -\eta_0\left(\dot{\gamma}_{rz} + \lambda_2\frac{\partial \dot{\gamma}_{rz}}{\partial t}\right) \tag{8.1-39}$$

$$\operatorname{tr}\boldsymbol{\tau} + \lambda_1\left(\frac{\partial \operatorname{tr}\boldsymbol{\tau}}{\partial t} - 2\tau_{rz}\dot{\gamma}_{rz}\right) = +2\eta_0\lambda_2\dot{\gamma}_{rz}^{\,2} \tag{8.1-40}$$

$$\tau_{rr} + \lambda_1\frac{\partial \tau_{rr}}{\partial t} = 0 \tag{8.1-41}$$

From the last equation $\tau_{rr} = 0$ and, hence, the dashed-underlined term in Eq. 8.1–39 can be omitted. Hence we have to solve simultaneously Eqs. 8.1–35, 39, and 40. This can be done by expanding τ_{rz}, $\dot{\gamma}_{rz}$, and tr $\boldsymbol{\tau}$ in powers of ϵ as we did in Eq. 8.1–38 for $v_z(r,t)$:

$$\tau_{rz} = \tau_0 + \epsilon[\tau_{10} + \mathscr{Re}\{\tau_{11}e^{i\omega t}\}] + \cdots \tag{8.1–42}$$

$$\dot{\gamma}_{rz} = \dot{\gamma}_0 + \epsilon[\dot{\gamma}_{10} + \mathscr{Re}\{\dot{\gamma}_{11}e^{i\omega t}\}] + \cdots \tag{8.1–43}$$

$$\text{tr}\,\boldsymbol{\tau} = T_0 + \epsilon[T_{10} + \mathscr{Re}\{T_{11}e^{i\omega t}\}] + \cdots \tag{8.1–44}$$

Of course $\dot{\gamma}_0 = dv_0/dr$, $\dot{\gamma}_{10} = dv_{10}/dr$, etc.

These expansions are substituted into Eqs. 8.1–35, 39, and 40, and then terms of equal powers of the parameter ϵ are equated. Equating the terms of zero power of ϵ gives:

$$0 = \frac{\overline{\Delta p}}{L} - \frac{1}{r}\frac{d}{dr}(r\tau_0) \tag{8.1–45}$$

$$\tau_0 + \tfrac{1}{2}\mu_0 T_0\dot{\gamma}_0 = -\eta_0\dot{\gamma}_0 \tag{8.1–46}$$

$$T_0 - 2\lambda_1\tau_0\dot{\gamma}_0 = +2\eta_0\lambda_2\dot{\gamma}_0{}^2 \tag{8.1–47}$$

Combination of these equations and elimination of τ_0 and T_0 give:

$$0 = \frac{\overline{\Delta p}}{L} + \frac{1}{r}\frac{d}{dr}\left(r\dot{\gamma}_0\left\{\eta_0\frac{1 + \mu_0\lambda_2\dot{\gamma}_0{}^2}{1 + \mu_0\lambda_1\dot{\gamma}_0{}^2}\right\}\right) \tag{8.1–48}$$

The quantity in { } is just the non-Newtonian viscosity for the Oldroyd four-constant model being used. Since $\dot{\gamma}_0 = dv_0/dr$, this is a second-order differential equation for $v_0(r)$, which can be solved in parametric form.[5,6] Then τ_0 and T_0 can be obtained as functions of $\dot{\gamma}_0$; these functions are needed later.

Next the terms that involve the first power of ϵ are equated. When this is done it is found that there are some terms independent of time, and other terms sinusoidal in time. Equating the time-independent terms gives three equations containing τ_{10}, T_{10}, and $\dot{\gamma}_{10}$; these equations show that $\dot{\gamma}_{10} = 0$. Since $v_{10} = 0$ at $r = R$, we must accordingly have $v_{10} = 0$ everywhere. Equating the oscillatory terms gives (after removing the $\mathscr{Re}$ operator and the common factors of $e^{i\omega t}$):

$$i\omega\rho v_{11} = \frac{\overline{\Delta p}}{L} - \frac{1}{r}\frac{d}{dr}(r\tau_{11}) \tag{8.1–49}$$

$$(1 + i\lambda_1\omega)\tau_{11} + \tfrac{1}{2}\mu_0(T_0\dot{\gamma}_{11} + \dot{\gamma}_0 T_{11}) = -\eta_0(1 + i\lambda_2\omega)\dot{\gamma}_{11} \tag{8.1–50}$$

$$(1 + i\lambda_1\omega)T_{11} - 2\lambda_1(\tau_0\dot{\gamma}_{11} + \dot{\gamma}_0\tau_{11}) = 4\eta_0\lambda_2\dot{\gamma}_0\dot{\gamma}_{11} \tag{8.1–51}$$

in which τ_0 and T_0 are known as functions of $\dot{\gamma}_0$ from the solution of Eqs. 8.1–45 to 47. By solving simultaneously Eqs. 8.1–50 and 51 for τ_{11} and T_{11}, we get $\tau_{11} = f(\dot{\gamma}_0,\omega)\dot{\gamma}_{11}$ and $T_{11} = F(\dot{\gamma}_0,\omega)\dot{\gamma}_{11}$. Then Eq. 8.1–49 becomes:

$$i\omega\rho v_{11} = \frac{\overline{\Delta p}}{L} - \frac{1}{r}\frac{d}{dr}\left(r\frac{dv_{11}}{dr}f(\dot{\gamma}_0,\omega)\right) \tag{8.1–52}$$

where

$$f(\dot{\gamma}_0,\omega) = -\eta_0\frac{(1 + i\lambda_1\omega)(1 + i\lambda_2\omega) + (3\lambda_2 - \lambda_1 + 2i\lambda_1\lambda_2\omega - \lambda_1{}^2\lambda_2\omega^2)\mu_0\dot{\gamma}_0{}^2 + \mu_0{}^2\lambda_1\lambda_2\dot{\gamma}_0{}^4}{[(1 + i\lambda_1\omega)^2 + \mu_0\lambda_1\dot{\gamma}_0{}^2](1 + \mu_0\lambda_1\dot{\gamma}_0{}^2)} \tag{8.1–53}$$

[6] K. Walters, *Arch. Rat. Mech. Anal.*, *9*, 411–417 (1962); see also M. C. Williams and R. B. Bird, *A.I.Ch.E. Journal*, *8*, 378–382 (1962); see Problem 8C.1.

This second-order ordinary differential equation for the complex function $v_{11}(r)$ has been solved numerically by Walters and Townsend.[5]

In a similar fashion one now proceeds to the terms quadratic in ϵ in Eqs. 8.1–35, 39, and 40. There will be terms independent of t, terms involving $e^{i\omega t}$, and terms involving $e^{2i\omega t}$; each of these groups of terms may be equated separately. The time-independent terms give the following equations:

$$0 = -\frac{1}{r}\frac{d}{dr}(r\tau_{20}) \tag{8.1–54}$$

$$\tau_{20} + \tfrac{1}{2}\mu_0(T_{20}\dot{\gamma}_0 + T_0\dot{\gamma}_{20} + \tfrac{1}{2}\mathcal{Re}\{T_{11}\bar{\dot{\gamma}}_{11}\}) = -\eta_0\dot{\gamma}_{20} \tag{8.1–55}$$

$$T_{20} - 2\lambda_1(\tau_{20}\dot{\gamma}_0 + \tau_0\dot{\gamma}_{20} + \tfrac{1}{2}\mathcal{Re}\{\tau_{11}\bar{\dot{\gamma}}_{11}\}) = 2\eta_0\lambda_2(2\dot{\gamma}_{20}\dot{\gamma}_0 + \tfrac{1}{2}\mathcal{Re}\{\dot{\gamma}_{11}\bar{\dot{\gamma}}_{11}\}) \tag{8.1–56}$$

in which the overbar indicates a complex conjugate. The last two equations can be solved to give τ_{20}

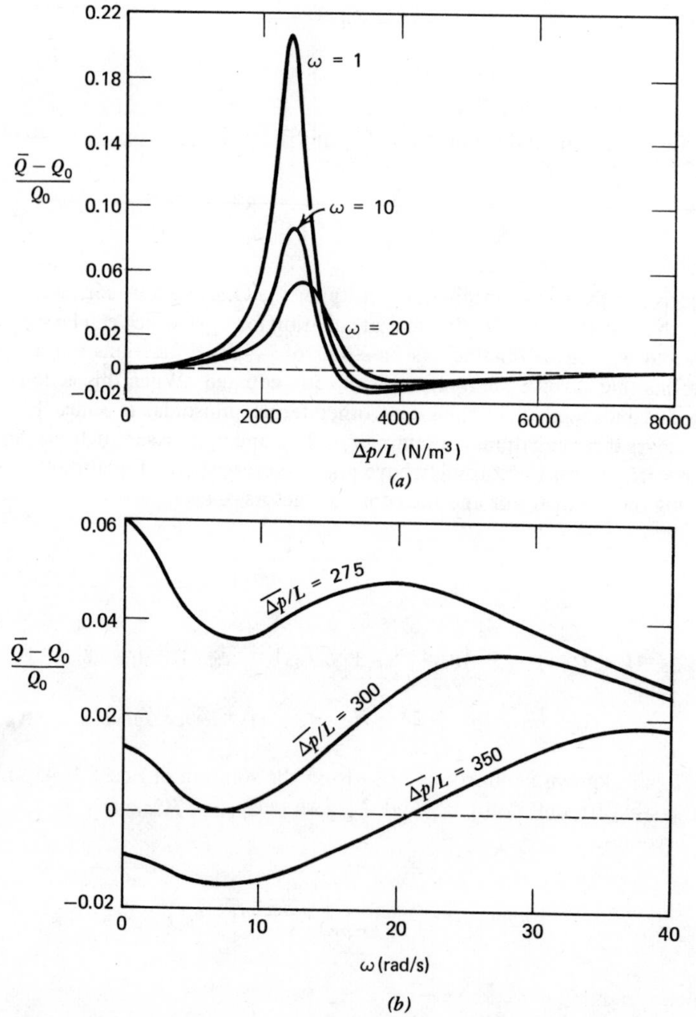

FIGURE 8.1–2. Sample calculated results showing increase (or decrease) in mean flow rate for pulsatile flow. $\eta_0 = 1$ N·s/m², $\lambda_1 = 0.1$ s, $\lambda_2 = 0.02$ s, $\mu_0 = 0.5$ s, $R = 0.25$ cm, $\rho = 1$ g/cm³, $\epsilon = 0.25$. These graphs are based on the Oldroyd model (Eq. 8.1–36) and were calculated by K. Walters and P. Townsend, *Proc. Fifth International Congress on Rheology*, S. Onogi (Ed.), University of Tokyo Press (1970), Vol. 4, pp. 471–483.

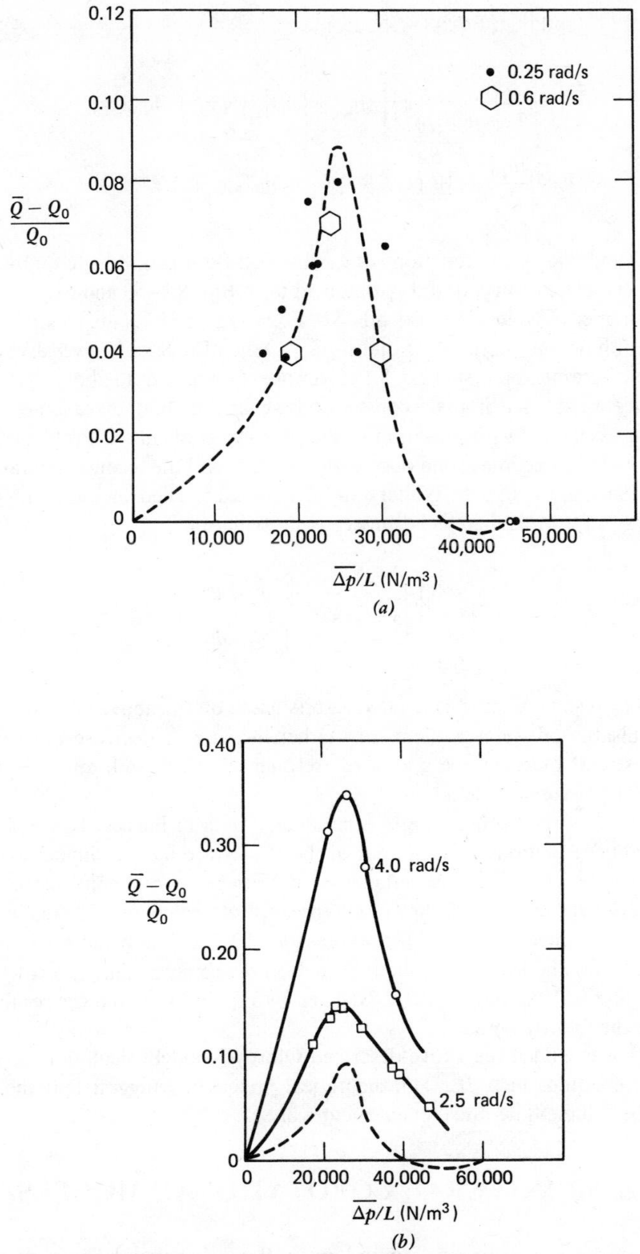

FIGURE 8.1–3. Experimental data on volume-flow-rate enhancement vs. axial pressure gradient for a 1.75% aqueous solution of polyacrylamide. In these experiments $\epsilon = 0.25$ and $R = 0.16$ cm. [H. A. Barnes, P. Townsend, and K. Walters, *Rheol. Acta, 10*, 517–527 (1971).]

(*a*) Low-frequency pulsations of the pressure gradient, showing comparison with dashed curve that results from calculations based on the generalized Newtonian fluid using the experimental $\eta(\dot{\gamma})$ curve;

(*b*) Higher-frequency pulsations, showing the generalized Newtonian model curve, and the increase in enhancement of flow rate as the frequency is raised.

and T_{20} in terms of $\dot{\gamma}_{20}$, and then Eq. 8.1–54 becomes:

$$0 = \frac{1}{r}\frac{d}{dr}\left(r\left[\frac{1}{1+\mu_0\lambda_1\dot{\gamma}_0^2}\right]\left[\{\eta_0 + \tfrac{1}{2}\mu_0(2\lambda_1\dot{\gamma}_0\tau_0 + 4\eta_0\lambda_2\dot{\gamma}_0^2 + T_0)\}\frac{dv_{20}}{dr} + \tfrac{1}{2}\mu_0(\lambda_1\dot{\gamma}_0\mathcal{Re}\{f\} + \eta_0\lambda_2\dot{\gamma}_0 + \tfrac{1}{2}\mathcal{Re}\{F\})\frac{dv_{11}}{dr}\overline{\frac{dv_{11}}{dr}}\right]\right) \tag{8.1–57}$$

In this equation τ_0 and T_0 are the functions of $\dot{\gamma}_0$ obtained from Eqs. 8.1–46 and 47, and f and F are (as mentioned earlier) related to τ_{11} and T_{11} obtained from Eqs. 8.1–50 and 51. The derivative dv_{11}/dr is known from the numerical solution of Eq. 8.1–52. Equation 8.1–57 is thus a second-order differential equation for the real function $v_{20}(r)$. Note that if $\mu_0 = 0$ then Eq. 8.1–36 gives a viscosity independent of the shear rate; furthermore $\mu_0 = 0$ in Eq. 8.1–57 implies (along with the boundary conditions for v_{20}) that $v_{20} = 0$. Hence we see that if η is a constant (no shear thinning), there is no enhancement of the volume rate of flow associated with pulsation of the pressure gradient (cf. Problem 5B.10).

Equation 8.1–57 can be integrated once with no effort, and the second integration was performed numerically, using Simpson's rule, by Walters and Townsend.[5] Then they found $(\bar{Q} - Q_0)/Q_0$, where Q_0 is the volume flow rate when there is no superposed oscillation:

$$\frac{\bar{Q} - Q_0}{Q_0} = \epsilon^2 \frac{\int_0^R v_{20} r\,dr}{\int_0^R v_0 r\,dr} \tag{8.1–58}$$

In Fig. 8.1–2 are shown some sample calculated results based on the above analysis. These calculations suggest that the pulsatile pressure gradient may either increase or decrease the volume rate of flow. They further suggest that there will be a kind of "resonance" effect, with appreciable increase in flow rate near a particular pressure gradient.

In Fig. 8.1–3 are shown some sample experimental data of Barnes, Townsend, and Walters for aqueous polyacrylamide solutions. The shapes of the curves are indeed similar to those obtained by using the Oldroyd model. However, the data show increased flow-rate enhancement as the frequency increases, whereas the Oldroyd model shows just the opposite behavior! This *may* be due to the fact that the calculations were made for $\mu_0 = 5\lambda_1$, whereas $\mu_0 < \lambda_1$ is usually required, since for most fluids the $\eta'(\omega)$ curve lies below the $\eta(\dot{\gamma})$ curve when plotted on the same graph. The calculations have been repeated[7] for a Goddard-Miller model with $G(s)$ given by Eq. 7B.11–3, and the results are qualitatively similar to those for the Oldroyd model.

Townsend[5] has extended the above-described Oldroyd-model calculations to include saw-tooth and square-wave oscillations in $\Delta p/L$. His theoretical predictions suggest that the squarer the wave form, the greater the volume-flow-rate enhancement will be.

§8.2 NONLINEAR INTEGRAL COROTATIONAL MODELS

In the foregoing section we started with the differential form of the corotational Jeffreys model (a quasilinear corotational model) and added some nonlinear terms to obtain the Oldroyd eight-constant model. We could also start out with the integral form of the corotational Jeffreys model and amend it by introducing various types of nonlinearities.

One way to introduce additional nonlinear behavior is to make the relaxation modulus depend on the shear rate through the inclusion of scalar invariants of the rate-of-strain tensor. For example, we could replace Eq. 7.3–4 by:

$$\boldsymbol{\tau} = -\int_{-\infty}^{t}\left\{\frac{\eta_0}{\lambda_0}\left(1 - \frac{\lambda_2}{\lambda_1}\right)f(\dot{\gamma}')e^{-(t-t')/\lambda_1} + 2\lambda_2 g(\dot{\gamma}')\delta(t-t')\right\}\dot{\boldsymbol{\Gamma}}'\,dt' \tag{8.2–1}$$

[7] S. Bhumiratana, Forthcoming Publication.

in which $\dot{\gamma}' \equiv \dot{\gamma}(x,t,t') = \sqrt{\frac{1}{2}(\dot{\boldsymbol{\gamma}}' : \dot{\boldsymbol{\gamma}}')} = \sqrt{\frac{1}{2}(\dot{\boldsymbol{\Gamma}}' : \dot{\boldsymbol{\Gamma}}')}$. If now we choose:

$$f(\dot{\gamma}) = \sqrt{1 + (\lambda_1\dot{\gamma})^2}\sqrt{1 + (\lambda_2\dot{\gamma})^2} \tag{8.2–2}$$

$$g(\dot{\gamma}) = \sqrt{\frac{1 + (\lambda_2\dot{\gamma})^2}{1 + (\lambda_1\dot{\gamma})^2}} \tag{8.2–3}$$

then we get the following two results:

$$\frac{\eta}{\eta_0} = \sqrt{\frac{1 + (\lambda_2\dot{\gamma})^2}{1 + (\lambda_1\dot{\gamma})^2}} \tag{8.2–4}$$

$$\left|\frac{\eta^*}{\eta_0}\right| = \sqrt{\frac{1 + (\lambda_2\omega)^2}{1 + (\lambda_1\omega)^2}} \tag{8.2–5}$$

In this way we obtain exactly the Cox-Merz relation,[1] namely that η vs. $\dot{\gamma}$ should be the same function as $|\eta^*|$ vs. ω (see Eq. 4.7–7). It is known that the Cox-Merz relation describes, at least approximately, the data for many polymeric fluids.

Another way to introduce additional nonlinear behavior is to write:

$$\boldsymbol{\tau} = -\int_{-\infty}^{t} \{\ldots\}[\dot{\boldsymbol{\Gamma}}' + h(\dot{\gamma}')\{\dot{\boldsymbol{\Gamma}}' \cdot \dot{\boldsymbol{\Gamma}}'\}]\, dt' \tag{8.2–6}$$

in which $\{\ldots\}$ is the relaxation modulus in Eq. 7.3–4. If now we choose

$$h(\dot{\gamma}) = \frac{(1 - 2n)(\lambda_1 - \lambda_2)}{1 + (\lambda_1\dot{\gamma})^2} \tag{8.2–7}$$

in which n is an arbitrary constant, then we find that η and Ψ_1 are still as given in Eqs. 7.3–20 and 21, but that $\Psi_2 = -n\Psi_1$. In this way we have provided a means for obtaining any value for the ratio Ψ_2/Ψ_1. We know the ratio is about -0.02 to -0.4 for many fluids.

The empiricisms suggested in Eqs. 8.2–1 and 8.2–6 are special cases of the LeRoy-Pierrard equation:[2]

$$\begin{aligned} \boldsymbol{\tau} = -\int_{-\infty}^{t} &[\alpha_1(t - t', II_{\dot{\gamma}}(t'), III_{\dot{\gamma}}(t'))\dot{\boldsymbol{\Gamma}}(t') \\ &+ \alpha_2(t - t', II_{\dot{\gamma}}(t'), III_{\dot{\gamma}}(t'))\{\dot{\boldsymbol{\Gamma}}(t') \cdot \dot{\boldsymbol{\Gamma}}(t')\}]\, dt' \end{aligned} \tag{8.2–8}$$

in which α_1 and α_2 are functions of the second and third invariants of the $\dot{\boldsymbol{\gamma}}$-tensor, $II_{\dot{\gamma}} = \operatorname{tr} \dot{\boldsymbol{\gamma}}^2$ and $III_{\dot{\gamma}} = \operatorname{tr} \dot{\boldsymbol{\gamma}}^3$. Models of this type have not received much attention in the research literature.

§8.3 THE COROTATIONAL MEMORY INTEGRAL EXPANSION

In Example 8.1–2 we found that when we transformed the Oldroyd 8-constant model into integral form, we obtained Eq. 8.1–14, which gives the stress tensor as a series in which the nth term involves n-fold integrals. Similarly, in Eq. 2.5–3 a kinetic theory

[1] W. P. Cox and E. H. Merz, *J. Polym. Sci.*, *28*, 619–622 (1958); see also §4.7.
[2] P. LeRoy and J. M. Pierrard, *Rheol. Acta*, *12*, 449–454 (1973).

expression was cited for the stress tensor of a dilute solution of rigid dumbbells and in §11.5 we put this result into the form of Eq. 8.1–14. These findings suggest that the most general rheological equation of state we shall need is:

$$\begin{aligned}\boldsymbol{\tau}(x,t) = &-\int_{-\infty}^{t} G_I(t-t')\dot{\boldsymbol{\Gamma}}'\,dt' \\ &-\tfrac{1}{2}\int_{-\infty}^{t}\int_{-\infty}^{t} G_{II}(t-t',t-t'')\{\dot{\boldsymbol{\Gamma}}'\cdot\dot{\boldsymbol{\Gamma}}''+\dot{\boldsymbol{\Gamma}}''\cdot\dot{\boldsymbol{\Gamma}}'\}\,dt''\,dt' \\ &-\tfrac{1}{2}\int_{-\infty}^{t}\int_{-\infty}^{t}\int_{-\infty}^{t}[2G_{III}(t-t',t-t'',t-t''')(\dot{\boldsymbol{\Gamma}}':\dot{\boldsymbol{\Gamma}}'')\dot{\boldsymbol{\Gamma}}''' \\ &+G_{IV}(t-t',t-t'',t-t''')\{\dot{\boldsymbol{\Gamma}}'\cdot\dot{\boldsymbol{\Gamma}}''\cdot\dot{\boldsymbol{\Gamma}}'''+\dot{\boldsymbol{\Gamma}}'''\cdot\dot{\boldsymbol{\Gamma}}''\cdot\dot{\boldsymbol{\Gamma}}'\}]\,dt'''\,dt''\,dt' - \cdots\end{aligned} \tag{8.3–1}$$

which we call the *corotational memory integral expansion.* The kernel functions G_I, G_{II}, G_{III}, etc., are then characteristic functions for each polymeric liquid.

This same result can be obtained by means of continuum mechanical arguments. One begins by postulating that the stresses in a given fluid element depend only on the kinematic history of that fluid element and not on the kinematic history of any adjacent fluid element. Fluids obeying this postulate are called "(rheologically) simple fluids."[1] The word-definition of the simple fluid may be expressed in terms of mathematical notation symbolizing the fact that the stress tensor is a "functional"[2] of the rate-of-strain history of the element. The introduction of the notation of functionals and a discussion of the theory of functionals are outside the scope of an elementary text. However, we note that, just as sufficiently well-behaved functions can be expanded in Taylor series, so also sufficiently well-behaved functionals can be expanded in Fréchet series.[3] Equation 8.3–1 is, in fact, the result of such a Fréchet expansion of a tensor functional. This particular expansion was first presented by Goddard,[4] who patterned his derivation after earlier derivations by Green and Rivlin[5] and Coleman and Noll;[6] these earlier derivations used a codeformational rather than a corotational framework as will be discussed in Chapter 9.

Note that $\boldsymbol{\tau}$ is only part of the total stress tensor $\boldsymbol{\pi} = p\boldsymbol{\delta} + \boldsymbol{\tau}$; we take $\boldsymbol{\tau}$ to be that part of the stress tensor that vanishes at equilibrium. One could write an equation somewhat

[1] The terminology "simple fluid" is due to W. Noll, *Arch. Rat. Mech. Anal., 2*, 197–226 (1958). This and other papers on modern continuum mechanics have been reprinted in *The Rational Mechanics of Materials*, C. Truesdell (Ed.), Gordon and Breach, New York (1965). A proof that Oldroyd's "general elasticoviscous fluid" includes the "simple fluid" was given by A. S. Lodge and J. H. Stark, *Rheol. Acta, 11*, 119–126 (1972); for Oldroyd's criticism of the "simple fluid" see J. G. Oldroyd, *Proc. Roy. Soc., A283*, 115–133 (1965).

[2] V. Volterra, *Theory of Functionals and of Integro-Differential Equations*, Dover, New York (1959), first published in 1930. For concise but readable discussions of the application of functional analysis in continuum mechanics, see R. S. Rivlin, in *Research Frontiers in Fluid Dynamics*, R. J. Seeger and G. Temple (Eds.), Wiley, New York (1965), and A. S. Lodge, *Body Tensor Fields in Continuum Mechanics*, Academic Press, New York (1974), pp. 155–157.

[3] M. Fréchet, *Ann. de l'École Normale Sup.*, 3rd Series, Vol. 27 (1910).

[4] J. D. Goddard, *Trans. Soc. Rheol., 11*, 381–399 (1967). Our G_I, G_{II}, G_{III}, G_{IV} correspond to Goddard's $\psi_1^{(1)}$, $\frac{1}{2}\psi_2^{(2)}$, $\frac{1}{4}\psi_1^{(3)}$, $\frac{1}{4}\psi_3^{(3)}$, respectively, where $\psi_1^{(3)} \equiv \psi_{1,1,1}^{(3)} + \psi_{1,1,3}^{(3)} + \psi_{1,3,3}^{(3)}$, $\psi_2^{(2)} \equiv \psi_{2,2}^{(2)}$, and $\psi_3^{(3)} \equiv \psi_{3,3,3}^{(3)}$.

[5] B. D. Coleman and W. Noll, *Rev. Mod. Phys., 33*, 239–249 (1961); *Ann. N. Y. Acad. Sci., 89*, 672–714 (1961).

[6] A. E. Green and R. S. Rivlin, *Arch. Rat. Mech. Anal., 1*, 1–21 (1957); this paper is also reproduced in the reprint volume cited in footnote 1.

more general than Eq. 8.3–1 by including in the second-order terms an integral over $(\dot{\boldsymbol{\Gamma}}' : \dot{\boldsymbol{\Gamma}}'')\boldsymbol{\delta}$ and similar isotropic contributions in the higher-order terms. However, such terms are of little rheological interest because of the lack of uniqueness in defining p for an incompressible fluid. It is therefore common practice among rheologists to omit isotropic terms in $\boldsymbol{\tau}$, and we adopt that procedure in Volume 1; in Volume 2, however, all isotropic terms arising in the kinetic theory derivations are systematically retained. For compressible fluids, of course, p is defined by the thermodynamic equation of state, and all isotropic terms in the memory-integral expansion must be retained, even a term of first order of the form $-\int_{-\infty}^{t} K_I(t - t')\,\mathrm{tr}\,\dot{\boldsymbol{\Gamma}}'\,dt'\,\boldsymbol{\delta}$.

As far as we know the "simple fluid" model is adequate for describing polymer fluid rheology. To date there have been no indications that more complicated models are needed. Hence, at the present time Eq. 8.3–1 ought to be sufficient for polymer fluid dynamics. Some readers will look aghast at Eq. 8.3–1 and say that substituting this expression for $\boldsymbol{\tau}$ into the equation of motion will result in numerical problems far too complex for current computers, and besides that we know very little about the kernel functions G_I, G_{II}, G_{III}, etc. Nagging questions can also be raised about the rapidity of the convergence of the expansion. These are, of course, legitimate concerns, but there are, in fact, many ways in which Eq. 8.3–1 is quite helpful:

1. *It provides a series in which the first term describes many rheological phenomena.* The first term gives the Goddard-Miller model (Eq. 7.5–1) that in turn contains the general linear viscoelastic fluid of Eq. 6.1–16. We know that the Goddard-Miller model can describe the viscosity, primary normal stress coefficient, stress growth, and stress relaxation semiquantitatively.

2. *It provides a systematic improvement on the Goddard-Miller model.* In Table 8.3–1 we list the material functions for the corotational memory-integral expansion. It can be seen that the second term allows the undesirable relation $\Psi_2 = -\frac{1}{2}\Psi_1$ to be corrected. The second term also allows for deviations from $\bar{\eta} = 3\eta_0$. The third term permits $\eta(\dot{\gamma})$ to have a shape different from $\eta'(\omega)$, and $\frac{1}{2}\Psi_1\dot{\gamma}$ to have a shape different from $\eta''(\omega)$. A preliminary attempt has been made to study the predictions of a truncated expansion, including the triple-integral terms with a simple choice of the G's, and the procedure appears promising.[7]

3. *It provides a starting point for obtaining reduced rheological equations of state.* In §8.4 we use the corotational memory integral expansion to get a simplified equation for the stress tensor for flows that are slow and slowly varying in time. In §8.5 a similar development is made for steady shear flows. These "reduced rheological equations of state" are relatively simple and useful.

4. *It provides a framework for presenting kinetic theory results.* In classical statistical mechanics, thermodynamics provides the framework into which statistical mechanical results can be put. Similarly, the continuum mechanics supplies a convenient framework for presenting molecular theory results. Molecular theories have thus far fitted into the simple fluid hypothesis.

5. *It provides a mechanism for organizing and summarizing rheological models.* Figure 8.3–1 can serve as a road map through the "jungle" of rheological equations of state that have been discussed in Chapters 5 to 8. The utility of the restrictions placed on various models can be better understood in terms of this chart, in which the corotational memory integral expansion clearly

[7] R. B. Bird, O. Hassager, and S. I. Abdel-Khalik, *A.I.Ch.E. Journal, 20*, 1041–1066 (1975), §10.

TABLE 8.3–1
Material Functions from the Corotational Memory-Integral Expansion[a]

Steady Shear Flow

$$\eta = \int_0^\infty G_I(s) \cos \dot{\gamma}s \, ds + \dot{\gamma}^2 \int_0^\infty \int_0^\infty \int_0^\infty [2G_{III}(s,s',s'') \cos \dot{\gamma}s \cos \dot{\gamma}(s' - s'') + G_{IV}(s,s',s'') \cos \dot{\gamma}(s'' - s' + s)] \, ds'' \, ds' \, ds + \cdots \quad \text{(A)}^{b}$$

$$\tfrac{1}{2}\dot{\gamma}\Psi_1 = \int_0^\infty G_I(s) \sin \dot{\gamma}s \, ds + \dot{\gamma}^2 \int_0^\infty \int_0^\infty \int_0^\infty [2G_{III}(s,s',s'') \sin \dot{\gamma}s \cos \dot{\gamma}(s' - s'') + G_{IV}(s,s',s'') \sin \dot{\gamma}(s'' - s' + s)] \, ds'' \, ds' \, ds + \cdots \quad \text{(B)}$$

$$\dot{\gamma}\Psi_2 = -\int_0^\infty G_I(s) \sin \dot{\gamma}s \, ds + \dot{\gamma} \int_0^\infty \int_0^\infty G_{II}(s,s') \cos \dot{\gamma}(s - s') \, ds' \, ds + \cdots \quad \text{(C)}$$

Small-Amplitude Oscillatory Shearing Motion[c]

$$\eta' = \int_0^\infty G_I(s) \cos \omega s \, ds \quad \text{(D)}$$

$$\eta'' = \int_0^\infty G_I(s) \sin \omega s \, ds \quad \text{(E)}$$

$$\Psi_1' = \frac{1}{\omega} \int_0^\infty G_I(s)(\sin 2\omega s - \sin \omega s) \, ds \quad \text{(F)}$$

$$\Psi_1'' = \frac{1}{\omega} \int_0^\infty G_I(s)(\cos \omega s - \cos 2\omega s) \, ds \quad \text{(G)}$$

$$\Psi_1{}^d = \frac{1}{\omega} \int_0^\infty G_I(s) \sin \omega s \, ds \quad \text{(H)}$$

$$\Psi_2' = -\tfrac{1}{2}\Psi_1' + \tfrac{1}{2} \int_0^\infty \int_0^\infty G_{II}(s,s') \cos \omega(s + s') \, ds' \, ds \quad \text{(I)}$$

$$\Psi_2'' = -\tfrac{1}{2}\Psi_1'' + \tfrac{1}{2} \int_0^\infty \int_0^\infty G_{II}(s,s') \sin \omega(s + s') \, ds' \, ds \quad \text{(J)}$$

$$\Psi_2{}^d = -\tfrac{1}{2}\Psi_1{}^d + \tfrac{1}{2} \int_0^\infty \int_0^\infty G_{II}(s,s') \cos \omega(s - s') \, ds' \, ds \quad \text{(K)}$$

Steady Elongational Flow[d]

$$\bar{\eta} = 3 \int_0^\infty G_I(s) \, ds + 3\dot{\epsilon} \int_0^\infty \int_0^\infty G_{II}(s,s') \, ds' \, ds + 9\dot{\epsilon}^2 \int_0^\infty \int_0^\infty \int_0^\infty (2G_{III}(s,s',s'') + G_{IV}(s,s',s'') \, ds'' \, ds' \, ds + \cdots \quad \text{(L)}$$

[a] J. D. Goddard, *Trans. Soc. Rheol.*, *11*, 381–399 (1967); see also R. B. Bird, O. Hassager, and S. I. Abdel-Khalik, *A.I.Ch.E. Journal 20*, 1041–1065 (1974).

[b] Note that $\eta_0 = \int_0^\infty G_I(s) \, ds$.

[c] Note that Eqs. H, I, and J of Table 7.5–2 are valid for interrelating the functions in Eqs. D to H in this table.

[d] Note that the first term of Eq. L gives the "Trouton relation," $\bar{\eta}(0) = 3\eta_0$.

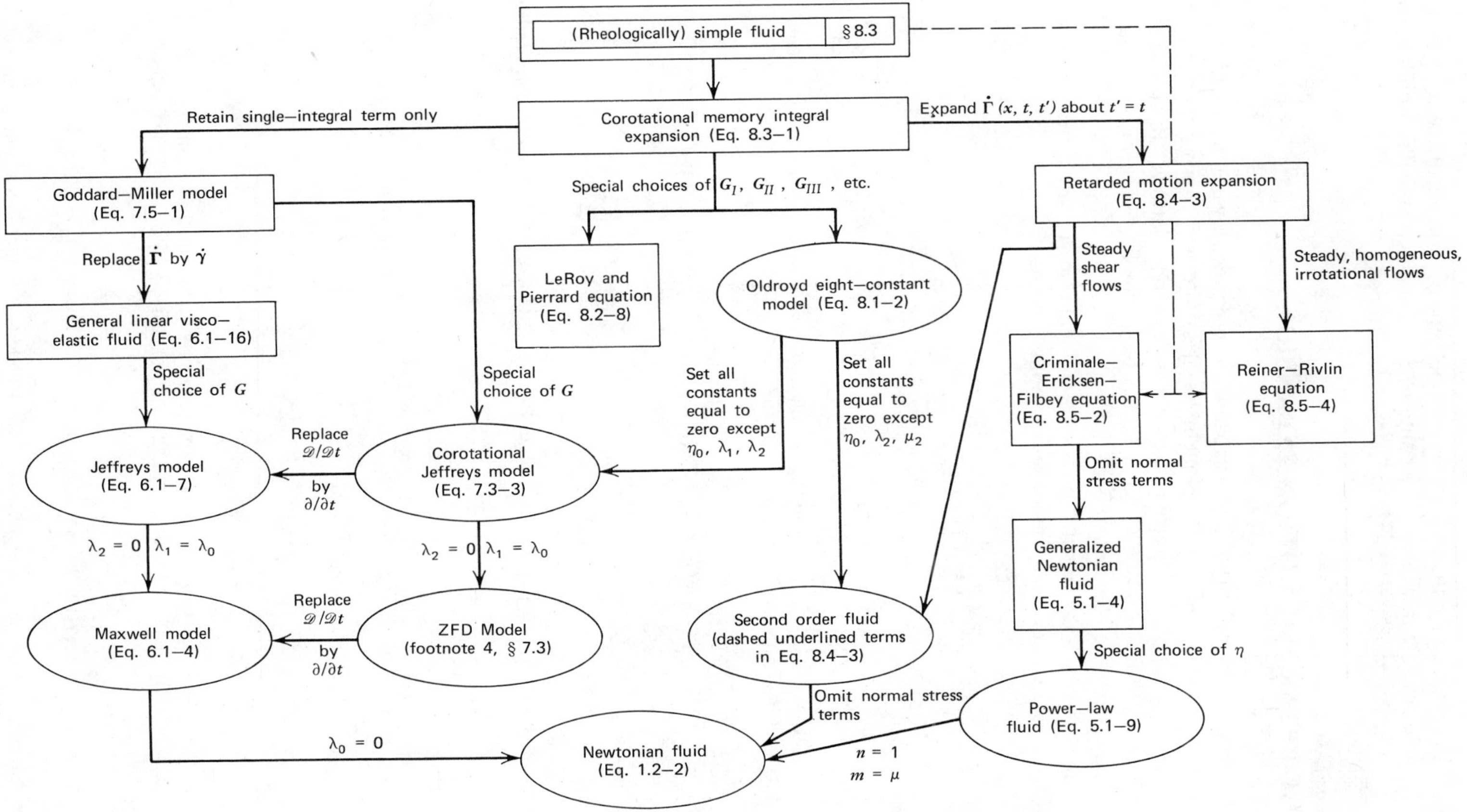

FIGURE 8.3–1. Relations among rheological equations of state.

occupies a prominent position. Models enclosed in an ellipse contain a small number of constants; those contained in a single rectangle contain functions, and the double rectangle indicates a functional.

§8.4 THE RETARDED-MOTION EXPANSION

The corotational memory-integral expansion is not presently useful for hydrodynamic calculations because of its complexity. Nonetheless it is quite useful for providing a common starting point for a variety of approximate and asymptotic relations. In this section we consider the asymptotic form of Eq. 8.3–1 that is useful for flows in which there are only slight departures from the Newtonian behavior.

For flows in which the velocity gradient is changing slowly with time, it is reasonable to expand the tensor $\dot{\mathbf{\Gamma}}(x,t,t')$ in a Taylor series about the time $t' = t$, thus:

$$\dot{\mathbf{\Gamma}}(x,t,t') = \dot{\mathbf{\Gamma}}(x,t,t) + \left.\frac{\partial \dot{\mathbf{\Gamma}}(x,t,t')}{\partial t'}\right|_{t'=t}(t'-t) + \frac{1}{2!}\left.\frac{\partial^2 \dot{\mathbf{\Gamma}}(x,t,t')}{\partial t'^2}\right|_{t'=t}(t'-t)^2 + \cdots \tag{8.4–1}$$

Note that $(\partial/\partial t')\dot{\mathbf{\Gamma}}(x,t,t')$ gives the time rate of change of $\dot{\mathbf{\Gamma}}$ following a fluid particle. Then using Table 7.1–2 we can rewrite this as:

$$\dot{\mathbf{\Gamma}}(x,t,t') = \dot{\boldsymbol{\gamma}}(x,t) - \frac{\mathscr{D}\dot{\boldsymbol{\gamma}}}{\mathscr{D}t}\cdot(t-t') + \frac{1}{2!}\frac{\mathscr{D}^2\dot{\boldsymbol{\gamma}}}{\mathscr{D}t^2}\cdot(t-t')^2 - \frac{1}{3!}\frac{\mathscr{D}^3\dot{\boldsymbol{\gamma}}}{\mathscr{D}t^3}\cdot(t-t')^3 + \cdots \tag{8.4–2}$$

When this is substituted into Eq. 8.3–1, we get (omitting isotropic terms as we agreed to do in §8.3):

$$\boxed{\begin{aligned}\boldsymbol{\tau} = &-\alpha_1\dot{\boldsymbol{\gamma}} + \alpha_2\frac{\mathscr{D}}{\mathscr{D}t}\dot{\boldsymbol{\gamma}} - \alpha_{11}\{\dot{\boldsymbol{\gamma}}\cdot\dot{\boldsymbol{\gamma}}\} \\ &- \alpha_3\frac{\mathscr{D}^2}{\mathscr{D}t^2}\dot{\boldsymbol{\gamma}} + \alpha_{12}\left\{\dot{\boldsymbol{\gamma}}\cdot\left(\frac{\mathscr{D}}{\mathscr{D}t}\dot{\boldsymbol{\gamma}}\right) + \left(\frac{\mathscr{D}}{\mathscr{D}t}\dot{\boldsymbol{\gamma}}\right)\cdot\dot{\boldsymbol{\gamma}}\right\} \\ &- \alpha_{1:11}(\dot{\boldsymbol{\gamma}}:\dot{\boldsymbol{\gamma}})\dot{\boldsymbol{\gamma}} + \cdots\end{aligned}} \tag{8.4–3}$$[1]

[1] Among the third-order terms there is one containing $(\dot{\boldsymbol{\gamma}}:\dot{\boldsymbol{\gamma}})\dot{\boldsymbol{\gamma}}$ and another containing $\{\dot{\boldsymbol{\gamma}}\cdot\dot{\boldsymbol{\gamma}}\cdot\dot{\boldsymbol{\gamma}}\}$. The latter may be eliminated in favor of an additional term of the form $(\dot{\boldsymbol{\gamma}}:\dot{\boldsymbol{\gamma}})\dot{\boldsymbol{\gamma}}$ by using the Cayley-Hamilton theorem (Eq. A.3–25):

$$\dot{\boldsymbol{\gamma}}^3 - \dot{\boldsymbol{\gamma}}^2\,\mathrm{tr}\,\dot{\boldsymbol{\gamma}} + \tfrac{1}{2}\dot{\boldsymbol{\gamma}}[(\mathrm{tr}\,\dot{\boldsymbol{\gamma}})^2 - \mathrm{tr}\,\dot{\boldsymbol{\gamma}}^2] - \boldsymbol{\delta}[\tfrac{1}{6}(\mathrm{tr}\,\dot{\boldsymbol{\gamma}})^3 - \tfrac{1}{2}\,\mathrm{tr}\,\dot{\boldsymbol{\gamma}}\,\mathrm{tr}\,\dot{\boldsymbol{\gamma}}^2 + \tfrac{1}{3}\,\mathrm{tr}\,\dot{\boldsymbol{\gamma}}^3] = \mathbf{0} \tag{8.4–3a}$$

and the fact that $\mathrm{tr}\,\dot{\boldsymbol{\gamma}} = 0$ for an incompressible fluid. In addition the term containing the unit tensor $\boldsymbol{\delta}$ can be discarded inasmuch as $\boldsymbol{\tau}$ can be measured only to within an additive isotropic function. In connection with Eq. 8.4–3 see H. Giesekus, *Zeits. f. angew. Math. u. Mech., 42*, 32–61 (1962); *Rheol. Acta, 3*, 59–71 (1963).

where the α's are constants related to the G's as follows:

$$\alpha_n = \frac{1}{(n-1)!}\int_0^\infty G_I(s)s^{n-1}\,ds$$

$$\alpha_{11} = \int_0^\infty \int_0^\infty G_{II}(s,s')\,ds'\,ds$$

$$\alpha_{12} = \tfrac{1}{2}\int_0^\infty \int_0^\infty G_{II}(s,s')(s+s')\,ds'\,ds$$

$$\alpha_{1:11} = \int_0^\infty \int_0^\infty \int_0^\infty [G_{III}(s,s',s'') + \tfrac{1}{2}G_{IV}(s,s',s'')]\,ds''\,ds'\,ds \qquad (8.4\text{–}4)$$

$$\vdots$$

Equation 8.4–3 is called the *retarded motion expansion*, meaning that the flow is both slow and slowly varying in time. The first term contains the coefficient α_1 (which is, in fact, just the zero-shear-rate viscosity) and is linear in the velocity gradients. The terms containing α_2 and α_{11} contain terms that are quadratic in the velocity gradients, and if we include all the dashed-underlined terms we have what is customarily referred to as the *second-order fluid*.[2] The remaining terms (up to $+\cdots$) contain terms that are cubic in the velocity gradient, and when we include all the terms shown explicitly in Eq. 8.4–3 we get a *third-order fluid*. The second-, third-, and fourth-order fluids have been used for a variety of exploratory calculations for steady-state flows.

Because of its simplicity the second-order fluid has achieved particular prominence and has been widely used. We must emphasize, however, that the second-order fluid has a constant viscosity $\eta = \alpha_1$ and constant normal stress coefficients $\Psi_1 = 2\alpha_2$ and $\Psi_2 = \alpha_{11} - \alpha_2$. For the small-amplitude oscillatory experiment, it gives $\eta' = \alpha_1$ and $\eta'' = \alpha_2\omega$, the validity of which is limited to vanishingly small frequencies. This would be expected, inasmuch as the retarded-motion expansion is applicable only to flows that are slowly changing in time (see Problem 9C.2). The main usefulness of the second-order fluid has been for getting qualitative ideas about the effects of the elasticity of fluids, the directions of secondary flows, and an estimation of the hole-pressure error.

For plane creeping flows $[v_x = v_x(x,y),\ v_y = v_y(x,y),\ v_z = 0]$ of the second-order fluid we have available to us the *Giesekus-Tanner theorem*: Any plane creeping Newtonian velocity field with given velocity boundary conditions is also a solution for the second-order

[2] The second-order fluid is usually given in the literature in terms of the kinematic tensors $\gamma^{(n)}$ (see Chapter 9) thus:

$$\tau = -\beta_1\gamma^{(1)} + \beta_2\gamma^{(2)} - \beta_{11}\{\gamma^{(1)}\cdot\gamma^{(1)}\} \qquad (8.4\text{–}4a)$$

The β's and α's are interrelated by

$$\alpha_1 = \beta_1; \qquad \alpha_2 = \beta_2; \qquad \alpha_{11} = \beta_{11} - \beta_2 \qquad (8.4\text{–}4b)$$

and

$$\eta = \beta_1; \qquad \Psi_1 = 2\beta_2; \qquad \Psi_2 = -2\beta_2 + \beta_{11} \qquad (8.4\text{–}4c)$$

The two types of expansions and their interrelation have been studied extensively by H. Giesekus, *Zeits. f. angew. Math. u. Mech.*, *42*, 32–61 (1962); see also Problem 9B.3. Kinetic theory expressions have been derived for the α's or β's for dilute solutions of rigid dumbbells (Eq. 11.5–29), finitely extendable nonlinear elastic dumbbells (Eqs. 10.5–18 to 23), and other macromolecular models (Table 13.4–1). See also Table 2.5–1.

incompressible fluid with the same boundary conditions.[3] This means that for flows with the same velocity boundary conditions the plane-creeping flow of a Newtonian fluid will be indistinguishable from that for a second-order fluid, as far as the flow pattern is concerned; however, the pressure distribution for the second-order fluid, p, will differ from that for the Newtonian fluid, p_N. If the latter is known, the former may easily be obtained from the *Giesekus-Tanner-Pipkin equation*[4] for plane, incompressible, creeping flows:

$$p = p_N - \frac{\alpha_2}{\alpha_1}\frac{Dp_N}{Dt} + (\alpha_{11} - \tfrac{1}{2}\alpha_2)\dot{\gamma}^2 \tag{8.4-5}$$

in which $\dot{\gamma} = \sqrt{\frac{1}{2}(\dot{\boldsymbol{\gamma}}:\dot{\boldsymbol{\gamma}})} = \sqrt{\frac{1}{2}\sum_i\sum_j \dot{\gamma}_{ij}{}^2}$

When a Newtonian fluid and a second-order fluid flow in the same system, pressure gauges along the walls will register different values for the two fluids. Tanner[3] cites the example of flow between two planes intersecting at an angle $2\theta_0$ with a line sink of strength Q at the junction. The creeping flow solution for Newtonian flow (and hence for second-order fluids) is:

$$v_r = \frac{2Q(\cos 2\theta - \cos 2\theta_0)}{r(\sin 2\theta_0 - 2\theta_0 \cos 2\theta_0)} \qquad v_\theta = 0 \tag{8.4-6}$$

In this flow the measurable normal stresses on the wall due to viscous effects are proportional to $\eta Q/r^2$, whereas the "elastic" stresses are proportional to $\alpha_2 Q^2/r^4$ (that is, they are related to the primary normal stress function Ψ_1). If inertia terms were included as a perturbation, then they would be proportional to $\rho Q^2/r^2$. This indicates that the elastic terms would be dominant near the vicinity of the sink.

EXAMPLE 8.4–1 The Journal-Bearing Problem[5]

In Problem 1C.9 the flow of a Newtonian fluid in a simplified journal-bearing system is discussed. Rework the same problem for the second-order fluid in order to assess the role of the normal stresses.

SOLUTION

The velocity distribution and pressure distribution for the Newtonian fluid in the system in Fig. 1C.9 are given in Eqs. 1C.9–7, 9, and 21. Let us now write down the components of the rate-of-strain tensor and indicate the order of magnitude of the various contributions in terms of the quantity $\delta = c/r_1 \ll 1$; in doing this we make use of the fact that B and Y will be regarded as having a magnitude

[3] The theorem given here is sometimes referred to as Tanner's theorem [R. I. Tanner, *Phys. Fluids*, *9*, 1246–1247 (1968)]. It is a special case of a more general theorem by H. Giesekus, *Rheol. Acta*, *3*, 59–71 (1963); see particularly Hilfsatz 2.4 on p. 63. The uniqueness aspects of the plane-creeping-flow theorem have been discussed by R. R. Huilgol, *SIAM J. Appl. Math.*, *24*, 226–233 (1973). See also Problem 8C.6.
[4] Equation 8.4–5 is a special case of Eq. 31a of H. Giesekus, *Rheol. Acta*, *3*, 59–71 (1963); it was obtained independently by R. I. Tanner and A. C. Pipkin, *Trans. Soc. Rheol.*, *13*, 471–484 (1969); A. C. Pipkin, *Lectures on Viscoelasticity Theory*, Springer, New York (1972), pp 149–153. See also Problem 8D.1. For a further discussion see R. R. Huilgol, *Continuum Mechanics of Viscoelastic Liquids*, Wiley, New York (1975), pp. 228–237.
[5] This example is based on the publication of J. M. Davies and K. Walters, in *Rheology of Lubricants*, Applied Science Publishers, (1973). In this paper results are also given for the third-order fluid and for the Oldroyd model. See also R. I. Tanner, *Australian J. Appl. Sci.*, *14*, 129–136 (1963); A. B. Metzner, *Rheol. Acta*, *10*, 434–444 (1971).

approximately equal to that of c:

$$\dot{\gamma}_{rr} = 2\frac{\partial v_r}{\partial r} = 2\frac{\partial v_r}{\partial Y} \tag{8.4–7}$$

$$O(1)$$

$$\dot{\gamma}_{r\theta} = \frac{1}{r}\frac{\partial v_r}{\partial \theta} + r\frac{\partial}{\partial r}\left(\frac{v_\theta}{r}\right) \approx \frac{1}{r_1}\frac{\partial v_r}{\partial \theta} + \frac{\partial v_\theta}{\partial Y} \tag{8.4–8}$$

$$O(\delta) \qquad O\left(\frac{1}{\delta}\right)$$

$$\dot{\gamma}_{\theta\theta} = 2\left(\frac{1}{r}\frac{\partial v_\theta}{\partial \theta} + \frac{v_r}{r}\right) \approx 2\left(\frac{1}{r_1}\frac{\partial v_\theta}{\partial \theta} + \frac{v_r}{r_1}\right) \tag{8.4–9}$$

$$O(1) \qquad O(\delta)$$

Then the quantity $\dot{\gamma}^2$ in Eq. 8.4–5 is:

$$\dot{\gamma}^2 = \tfrac{1}{2}\textstyle\sum_i \sum_j \dot{\gamma}_{ij}{}^2 \approx \left(\dfrac{\partial v_\theta}{\partial Y}\right)^2 = [Yf(\theta) + g(\theta)]^2 \tag{8.4–10}$$

And the Dp_N/Dt-term in Eq. 8.4–5 is:

$$\begin{aligned} -\frac{\alpha_2}{\alpha_1}\frac{Dp_N}{Dt} &= -\frac{\alpha_2}{\alpha_1}(\boldsymbol{v}\cdot\boldsymbol{\nabla}p_N) \approx -\frac{\alpha_2}{\alpha_1}\frac{v_\theta}{r_1}\frac{\partial}{\partial \theta}p_N \\ &= -\frac{\alpha_2}{\alpha_1}(\tfrac{1}{2}Y^2 f + Yg + W_1 r_1)\cdot\alpha_1 f \end{aligned} \tag{8.4–11}$$

Then, from Eq. 8.4–5 we can get the pressure distribution at the rotating journal surface, $Y = 0$:

$$\begin{aligned} p|_{Y=0} = p_0 &+ \frac{6\alpha_1 W_1 r_1{}^2 a \sin\theta\cdot(c + \frac{1}{2}a\cos\theta)}{(c^2 + \frac{1}{2}a^2)(c + a\cos\theta)^2} \\ &- \alpha_2 W_1 r_1 f(\theta) + (\alpha_{11} - \tfrac{1}{2}\alpha_2)[g(\theta)]^2 \end{aligned} \tag{8.4–12}$$

in which we have used the Newtonian pressure distribution in Eq. 1C.9–21.

Next we need the components of the stress τ_{rr} and $\tau_{r\theta}$, evaluated at $Y = 0$, which arise in the integrals for the torque and force in Eqs. 1C.9–18, 19, and 20. For the second-order fluid (dashed-underlined terms in Eq. 8.4–3) the pertinent stress components are:

$$\begin{aligned} \tau_{rr} = &-\alpha_1\dot{\gamma}_{rr} - \alpha_{11}(\dot{\gamma}_{rr}{}^2 + \dot{\gamma}_{r\theta}{}^2) \\ &+ \alpha_2\left[\left(v_r\frac{\partial}{\partial r} + \frac{v_\theta}{r}\frac{\partial}{\partial \theta}\right)\dot{\gamma}_{rr}\right. \\ &\left.+ \dot{\gamma}_{r\theta}\left(r\frac{\partial}{\partial r}\left(\frac{v_\theta}{r}\right) - \frac{1}{r}\frac{\partial v_r}{\partial \theta}\right)\right] \\ \approx\ &(-\alpha_{11} + \alpha_2)\left(\frac{\partial v_\theta}{\partial Y}\right)^2 + O\left(\frac{1}{\delta}\right) \\ =\ &(-\alpha_{11} + \alpha_2)(Yf + g)^2 + O\left(\frac{1}{\delta}\right) \\ \xrightarrow{Y=0}\ &(-\alpha_{11} + \alpha_2)g^2 \end{aligned} \tag{8.4–13}$$

$$\begin{aligned}
\tau_{r\theta} = \; & -\alpha_1 \dot{\gamma}_{r\theta} + \alpha_2 \left[\left(v_r \frac{\partial}{\partial r} + \frac{v_\theta}{r} \frac{\partial}{\partial \theta} \right) \dot{\gamma}_{r\theta} \right. \\
& \left. + \tfrac{1}{2}(\dot{\gamma}_{\theta\theta} - \dot{\gamma}_{rr}) \left(r \frac{\partial}{\partial r} \left(\frac{v_\theta}{r} \right) - \frac{1}{r} \frac{\partial v_r}{\partial \theta} \right) \right] \\
\approx \; & -\alpha_1 \frac{\partial v_\theta}{\partial Y} + \alpha_2 \left[\left(v_r \frac{\partial}{\partial Y} + \frac{v_\theta}{r_1} \frac{\partial}{\partial \theta} \right) \frac{\partial v_\theta}{\partial Y} \right. \\
& \left. + \tfrac{1}{2} \left(\frac{2}{r_1} \frac{\partial v_\theta}{\partial \theta} - 2 \frac{\partial v_r}{\partial Y} \right) \left(\frac{\partial v_\theta}{\partial Y} - \frac{1}{r_1} \frac{\partial v_r}{\partial \theta} \right) \right] \\
\xrightarrow{Y=0} \; & -\alpha_1 g(\theta) + \alpha_2 W_1 g'(\theta)
\end{aligned} \tag{8.4–14}$$

Now that we know the pressure and the stress components at the rotating journal surface, we can evaluate the integrals for the torque and the components of the force (see Eqs. 1C.9–18 to 20) that the fluid exerts on the journal. When the indicated integrations are performed, we obtain finally:

$$\mathscr{T} = -\frac{2\pi\eta L W_1 r_1^3 (c^2 + 2a^2)}{\sqrt{c^2 - a^2}\,(c^2 + \frac{1}{2}a^2)} \tag{8.4–15}$$

$$F_y = -\frac{6\pi\eta L W_1 r_1^3 a}{\sqrt{c^2 - a^2}\,(c^2 + \frac{1}{2}a^2)} \tag{8.4–16}$$

$$F_x = \frac{2\pi\Psi_1 L W_1^2 r_1^3 a (c^4 - \frac{1}{8}a^2 c^2 + \frac{1}{4}a^4)}{(c^2 - a^2)^{3/2}\,(c^2 + \frac{1}{2}a^2)^2} \tag{8.4–17}$$

in which we have replaced α_1 by η, and α_2 by $\frac{1}{2}\Psi_1$, in order to emphasize the roles of viscosity and primary normal stress coefficient. The derivation shows that the magnitude of the force $F = (F_x^2 + F_y^2)^{1/2}$ for the second-order fluid is greater than the resultant force for the corresponding Newtonian fluid. This has been interpreted by Davies and Walters[5] to mean that lubricants with polymeric additives will support greater loads and therefore result in reduced wear; this seems to be in qualitative agreement with experimental observations. It must be remembered, however, that the analysis here has been drastically over-simplified and, in fact, the main purpose of the example must be regarded as the illustration of hydrodynamical problem solving. There are many other effects that must be considered in doing a complete study of the problem: the role of the shear-rate dependence of the viscosity, the role of viscous dissipation heating, the effects of compressibility of the lubricant, and perhaps others.

EXAMPLE 8.4–2 Hole-Pressure Error[4]

Use the Giesekus-Tanner theorem and the Giesekus-Tanner-Pipkin equation to obtain a theoretical expression for the "hole-pressure error" for a second-order fluid flowing across a transverse slit. The pressure transducer, located at the bottom of a well, gives a pressure reading of p_M (the measured "pressure"), whereas the true reading should be given by $(p + \tau_{yy})_W$, the pressure-plus-normal-stress value at the wall. The system is pictured in Fig. 8.4–1.

SOLUTION (a) Newtonian Fluid

For the creeping flow of a Newtonian fluid the velocity field is symmetric about the plane $x = 0$. Therefore in the y-component of the equation of motion

$$0 = -\frac{\partial}{\partial x}\pi_{xy} - \frac{\partial}{\partial y}\pi_{yy} \tag{8.4–18}$$

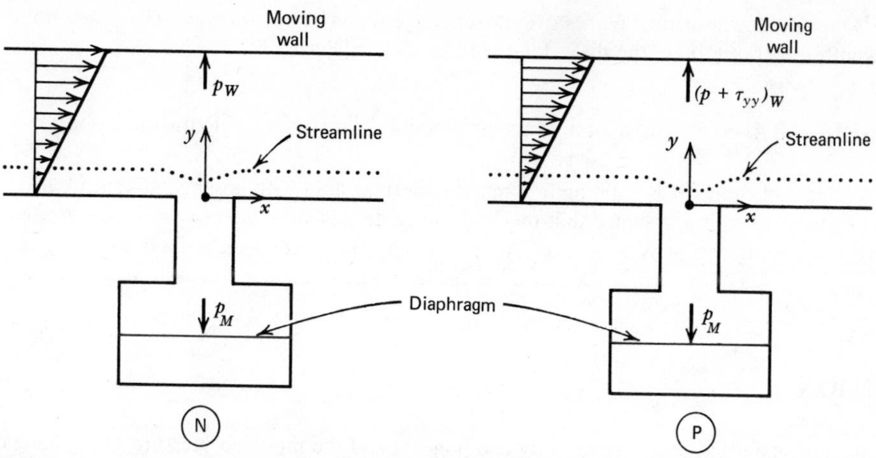

FIGURE 8.4–1. Flow over a transverse slit, with a pressure transducer which gives a pressure reading p_M. For the Newtonian fluid (N) there is no problem for slow flow where $p_M = p_W$. For the polymeric fluid (P) $p_M \neq (p + \tau_{yy})_W$.

the term $\partial\pi_{xy}/\partial x$ vanishes for $x = 0$, and π_{yy} is a constant on the plane $x = 0$. Consequently we conclude that the pressure reading at the wall (p_W) will be identical with the pressure reading at the diaphragm (p_M).

(b) Second-Order Fluid

For the second-order fluid the flow pattern will be identical to that of the Newtonian fluid. At the moving plate the "pressure" reading will be:

$$p_W = (p + \tau_{yy})_W \tag{8.4–19}$$

where p is given by Eq. 8.4–5 and τ_{yy} by the second-order fluid rheological equation of state in Eq. 8.4–3 (dashed-underlined terms only):

$$p = p_N - \frac{\alpha_2}{\alpha_1} v_x \frac{\partial}{\partial x} p_N + (\alpha_{11} - \tfrac{1}{2}\alpha_2)\dot{\gamma}^2 \tag{8.4–20}$$

$$\tau_{yy} = -(\alpha_{11} - \alpha_2)\dot{\gamma}^2 \tag{8.4–21}$$

At the upper wall, the variation of p_N in the flow direction will be very nearly zero when the plates are parallel. The true stress normal to the wall will then be for the second-order fluid:

$$p_W = (p_N + \tfrac{1}{2}\alpha_2\dot{\gamma}^2)_W \tag{8.4–22}$$

At the diaphragm surface, there is no velocity gradient, provided that the slit is deep enough, so that the measured pressure is:

$$p_M = p_N \tag{8.4–23}$$

Hence the hole pressure error $p_H = p_M - p_W$ is:

$$p_H = -\tfrac{1}{2}\alpha_2\dot{\gamma}_W{}^2 = -\tfrac{1}{4}\Psi_1\dot{\gamma}_W{}^2 \tag{8.4–24}$$

This is a special case of the Higashitani-Pritchard equation given in Eq. 3.4–2 (Problem 3B.2). Thus the pressure sensing device in a well will give a reading lower than the true value. Tanner and Pipkin[4] report that the result in Eq. 8.4–24 is in excellent agreement with their experimental measurements for fluids flowing down inclined planes over a transverse slit. The hole-pressure error does not depend on the width-to-depth ratio of the slit as long as it is sufficiently deep.

EXAMPLE 8.4–3 Flow of a Macromolecular Fluid Near a Rotating Sphere[6,7]

A sphere of radius R is rotating in a macromolecular fluid with an angular velocity $\boldsymbol{\delta}_z W$. Find the flow patterns when it is assumed that the fluid can be described by the first- and second-order terms in the retarded motion expansion in Eq. 8.4–3. Specialize the second-order fluid model by requiring that $\Psi_{2,0} = \alpha_{11} - \alpha_2 = 0$; hence the second-order fluid model used will contain two constants $\alpha_1 \equiv \eta_0$ and $\alpha_2 \equiv \frac{1}{2}\Psi_{1,0}$.

SOLUTION

We postulate that the three velocity components and the modified pressure $\mathscr{P}$ are functions of r and θ, but not ϕ. Then the equations of continuity and motion are:

$$\frac{1}{r^2}\frac{\partial}{\partial r}(r^2 v_r) + \frac{1}{r \sin\theta}\frac{\partial}{\partial \theta}(v_\theta \sin\theta) = 0 \tag{8.4–25}$$

$$\rho\left(v_r \frac{\partial v_r}{\partial r} + \frac{v_\theta}{r}\frac{\partial v_r}{\partial \theta} - \frac{{v_\theta}^2 + {v_\phi}^2}{r}\right) = -\frac{\partial \mathscr{P}}{\partial r} - \left(\frac{1}{r^2}\frac{\partial}{\partial r}(r^2 \tau_{rr}) + \frac{1}{r\sin\theta}\frac{\partial}{\partial\theta}(\tau_{r\theta}\sin\theta) - \frac{\tau_{\theta\theta} + \tau_{\phi\phi}}{r}\right) \tag{8.4–26}$$

$$\rho\left(\frac{v_r}{r}\frac{\partial}{\partial r}(r v_\theta) + \frac{v_\theta}{r}\frac{\partial v_\theta}{\partial \theta} - \frac{{v_\phi}^2 \cot\theta}{r}\right) = -\frac{1}{r}\frac{\partial \mathscr{P}}{\partial \theta} - \left(\frac{1}{r^3}\frac{\partial}{\partial r}(r^3 \tau_{r\theta}) + \frac{1}{r\sin\theta}\frac{\partial}{\partial\theta}(\tau_{\theta\theta}\sin\theta) - \frac{\tau_{\phi\phi}\cot\theta}{r}\right) \tag{8.4–27}$$

$$\rho\left(\frac{v_r}{r}\frac{\partial}{\partial r}(r v_\phi) + \frac{v_\theta}{r\sin\theta}\frac{\partial}{\partial\theta}(v_\phi \sin\theta)\right) = -\left(\frac{1}{r^3}\frac{\partial}{\partial r}(r^3 \tau_{r\phi}) + \frac{1}{r\sin^2\theta}\frac{\partial}{\partial\theta}(\tau_{\theta\phi}\sin^2\theta)\right) \tag{8.4–28}$$

The following dimensionless quantities may now be introduced:

$$\begin{aligned}
V_r &= (\rho R/\alpha_1)v_r & T_{ii} &= (\rho R^2/{\alpha_1}^2)\tau_{ii} \qquad i = r, \theta, \phi \\
V_\theta &= (\rho R/\alpha_1)v_\theta & T_{r\theta} &= (\rho R^2/{\alpha_1}^2)\tau_{r\theta} \\
V_\phi &= (1/WR)v_\phi & T_{r\phi} &= (1/\alpha_1 W)\tau_{r\phi} \\
P &= (\rho R^2/{\alpha_1}^2)\mathscr{P} & T_{\theta\phi} &= (1/\alpha_1 W)\tau_{\theta\phi} \\
\xi &= (1/R)r & L &= (WR^2\rho/\alpha_1)^2
\end{aligned} \tag{8.4–29}$$

[6] This example is based on the publication of R. H. Thomas and K. Walters, *Quart. J. Mech. Appl. Math.*, *17*, 39–53 (1964). These authors used the Oldroyd-Walters-Fredrickson rheological equation of state given in Table 9.4–2, but their perturbation analysis gave results equivalent to that of the second-order fluid. For further theoretical development, including third-order terms and effects of external boundaries, see K. Walters and N. D. Waters, *Brit. J. Appl. Phys.*, *14*, 667–671 (1963); *15*, 989–991 (1964); *Rheol. Acta*, *3*, 312–315 (1964).

[7] H. Giesekus, *Rheol. Acta*, *3*, 59–71 (1963) has obtained the solution for the flow of a third-order fluid near a rotating and translating sphere, but neglecting the inertial terms. See also H. Giesekus, *Proc. Fourth International Congress on Rheology*, E. H. Lee and A. L. Copley (Eds.), Wiley, New York (1965), Vol. 1, pp. 249–266. The flow of a third-order fluid past a viscoelastic droplet has been studied by M. G. Wagner and J. C. Slattery, *AIChE Journal*, *17*, 1198–1207 (1971).

Then the equations of continuity and motion become:

$$\frac{1}{\xi^2}\frac{\partial}{\partial \xi}(\xi^2 V_r) + \frac{1}{\xi \sin\theta}\frac{\partial}{\partial \theta}(V_\theta \sin\theta) = 0 \tag{8.4–30}$$

$$V_r\frac{\partial V_r}{\partial \xi} + \frac{V_\theta}{\xi}\frac{\partial V_r}{\partial \theta} - \frac{V_\theta^2 + LV_\phi^2}{\xi} = -\frac{\partial P}{\partial \xi} - \left(\frac{1}{\xi^2}\frac{\partial}{\partial \xi}(\xi^2 T_{rr}) + \frac{1}{\xi \sin\theta}\frac{\partial}{\partial \theta}(T_{r\theta}\sin\theta) - \frac{T_{\theta\theta} + T_{\phi\phi}}{\xi}\right) \tag{8.4–31}$$

$$\frac{V_r}{\xi}\frac{\partial}{\partial \xi}(\xi V_\theta) + \frac{V_\theta}{\xi}\frac{\partial V_\theta}{\partial \theta} - \frac{LV_\phi^2 \cot\theta}{\xi} = -\frac{1}{\xi}\frac{\partial P}{\partial \theta} - \left(\frac{1}{\xi^3}\frac{\partial}{\partial \xi}(\xi^3 T_{r\theta}) + \frac{1}{\xi \sin\theta}\frac{\partial}{\partial \theta}(T_{\theta\theta}\sin\theta) - \frac{T_{\phi\phi}\cot\theta}{\xi}\right) \tag{8.4–32}$$

$$\frac{V_r}{\xi}\frac{\partial}{\partial \xi}(\xi V_\phi) + \frac{V_\theta}{\xi \sin\theta}\frac{\partial}{\partial \theta}(V_\phi \sin\theta) = -\left(\frac{1}{\xi^3}\frac{\partial}{\partial \xi}(\xi^3 T_{r\phi}) + \frac{1}{\xi \sin^2\theta}\frac{\partial}{\partial \theta}(T_{\theta\phi}\sin^2\theta)\right) \tag{8.4–33}$$

The boundary conditions on the above equations are:

$$\begin{aligned} V_r = V_\theta &= 0 && \text{at} \quad \xi = 1 \\ V_\phi &= \sin\theta && \text{at} \quad \xi = 1 \\ V_r = V_\theta = V_\phi &= 0 && \text{at} \quad \xi = \infty \\ P &= 0 && \text{at} \quad \xi = \infty \end{aligned} \tag{8.4–34}$$

When L is small, and terms of order L are neglected, the above equations and boundary conditions have the solution (cf. Problem 1C.7):

$$V_r = V_\theta = P = 0 \tag{8.4–35}$$

$$V_\phi = \frac{\sin\theta}{\xi^2} \tag{8.4–36}$$

so that $T_{r\phi} = T_{\phi r} = (3 \sin\theta)/\xi^3$ are the only nonzero components of the stress tensor. This suggests that we seek a solution to Eqs. 8.4–30 to 33 of the form:

$$V_r = LV_r^{(1)} + L^2 V_r^{(2)} + \cdots \tag{8.4–37}$$

$$V_\theta = LV_\theta^{(1)} + L^2 V_\theta^{(2)} + \cdots \tag{8.4–38}$$

$$V_\phi = \frac{\sin\theta}{\xi^2} + LV_\phi^{(1)} + L^2 V_\phi^{(2)} + \cdots \tag{8.4–39}$$

$$P = LP^{(1)} + L^2 P^{(2)} + \cdots \tag{8.4–40}$$

We now seek a solution to the problem through terms first order in $L = (WR^2\rho/\alpha_1)^2$, which is the square of a Reynolds number.

We now use the dashed underlined terms in Eq. 8.4–3 (with $\alpha_{11} = \alpha_2$), so that:

$$\boldsymbol{\tau} = -\alpha_1 \dot{\boldsymbol{\gamma}} + \alpha_2\{\boldsymbol{v} \cdot \boldsymbol{\nabla}\dot{\boldsymbol{\gamma}} - (\boldsymbol{\nabla}\boldsymbol{v})^\dagger \cdot \dot{\boldsymbol{\gamma}} - \dot{\boldsymbol{\gamma}} \cdot \boldsymbol{\nabla}\boldsymbol{v}\} \tag{8.4–41}$$

This gives, through first order in L, after some rearranging:

$$T_{rr} = -2L\frac{\partial V_r^{(1)}}{\partial \xi} \tag{8.4–42}$$

$$T_{\theta\theta} = -2L\left(\frac{V_r^{(1)}}{\xi} + \frac{1}{\xi}\frac{\partial V_\theta^{(1)}}{\partial \theta}\right) \tag{8.4–43}$$

$$T_{\phi\phi} = -2L\left(\frac{V_r^{(1)}}{\xi} + \frac{V_\theta^{(1)}}{\xi}\cot\theta\right) - 18mL\frac{\sin^2\theta}{\xi^6} \tag{8.4–44}$$

$$T_{r\theta} = -L\left(\frac{1}{\xi}\frac{\partial V_r^{(1)}}{\partial \theta} + \xi\frac{\partial}{\partial \xi}\left(\frac{V_\theta^{(1)}}{\xi}\right)\right) \tag{8.4–45}$$

$$T_{\theta\phi} = -L\frac{\sin\theta}{\xi}\frac{\partial}{\partial\theta}\left(\frac{V_\phi^{(1)}}{\sin\theta}\right) - mL\xi\left(\frac{\partial}{\partial\xi}\frac{\sin\theta}{\xi^3}\right)\left[\frac{1}{\xi}\frac{\partial V_r^{(1)}}{\partial\theta} + 2\xi\frac{\partial}{\partial\xi}\left(\frac{V_\theta^{(1)}}{\xi}\right)\right] \tag{8.4–46}$$

$$T_{r\phi} = \frac{3\sin\theta}{\xi^3} - L\xi\frac{\partial}{\partial\xi}\left(\frac{V_\phi^{(1)}}{\xi}\right) + mL\xi V_r^{(1)}\frac{\partial^2}{\partial\xi^2}\left(\frac{\sin\theta}{\xi^3}\right) - 3mL\xi\left(\frac{\partial}{\partial\xi}\frac{\sin\theta}{\xi^3}\right)\frac{\partial V_r^{(1)}}{\partial\xi} \tag{8.4–47}$$

where the dimensionless parameter:

$$m = \frac{\alpha_2}{\rho R^2} \tag{8.4–48}$$

is a measure of the elastic contribution in the problem; if $m \to 0$, we recover the Newtonian fluid result.

When the above stress expressions and the expansions in Eqs. 8.4–37, 38, 39, and 40 are substituted into Eqs. 8.4–30, 31, and 32, we obtain:

$$\frac{\partial}{\partial\xi}(\xi^2 V_r^{(1)}\sin\theta) + \frac{\partial}{\partial\theta}(\xi V_\theta^{(1)}\sin\theta) = 0 \tag{8.4–49}$$

$$\frac{18m\sin^2\theta}{\xi^7} - \frac{\sin^2\theta}{\xi^5} = -\frac{\partial P}{\partial\xi} + \left(\frac{\partial^2 V_r^{(1)}}{\partial\xi^2} + \frac{2}{\xi}\frac{\partial V_r^{(1)}}{\partial\xi} + \frac{\cot\theta}{\xi^2}\frac{\partial V_r^{(1)}}{\partial\theta} + \frac{1}{\xi^2}\frac{\partial^2 V_r^{(1)}}{\partial\theta^2} - \frac{2V_r^{(1)}}{\xi^2} - \frac{2V_\theta^{(1)}\cot\theta}{\xi^2} - \frac{2}{\xi^2}\frac{\partial V_\theta^{(1)}}{\partial\theta}\right) \tag{8.4–50}$$

$$\frac{18m\cos\theta\sin\theta}{\xi^7} - \frac{\cos\theta\sin\theta}{\xi^5} = -\frac{1}{\xi}\frac{\partial P}{\partial\theta} + \left(\frac{\partial^2 V_\theta^{(1)}}{\partial\xi^2} + \frac{2}{\xi}\frac{\partial V_\theta^{(1)}}{\partial\xi} + \frac{\cot\theta}{\xi^2}\frac{\partial V_\theta^{(1)}}{\partial\theta} + \frac{1}{\xi^2}\frac{\partial^2 V_\theta^{(1)}}{\partial\theta^2} - \frac{V_\theta^{(1)}}{\xi^2\sin^2\theta} + \frac{2}{\xi^2}\frac{\partial V_r^{(1)}}{\partial\theta}\right) \tag{8.4–51}$$

Equation 8.4–49 can be satisfied by introducing a stream function ψ so that:

$$V_r^{(1)} = -\frac{1}{\xi^2\sin\theta}\frac{\partial\psi}{\partial\theta} \tag{8.4–52}$$

$$V_\theta^{(1)} = +\frac{1}{\xi\sin\theta}\frac{\partial\psi}{\partial\xi} \tag{8.4–53}$$

Using the stream function and the operator:

$$D \equiv \frac{\partial^2}{\partial\xi^2} + \frac{1}{\xi^2}\frac{\partial^2}{\partial\theta^2} - \frac{\cot\theta}{\xi^2}\frac{\partial}{\partial\theta} \tag{8.4–54}$$

enables us to write Eqs. 8.4–50 and 51 in a rather compact form. Then P can be eliminated from the two components of the equation of motion by cross-differentiating. This leads to the following partial differential equation for the stream function:

$$D^2\psi = \left(\frac{6}{\xi^5} - \frac{144m}{\xi^7}\right)\sin^2\theta\cos\theta \tag{8.4–55}$$

The solution to this equation which satisfies the boundary conditions at the sphere surface and at infinity is:

$$\psi = \frac{(\xi - 1)^2}{8\xi^3}[(1 - 4m)\xi - 8m]\sin^2\theta\cos\theta \tag{8.4–56}$$

Note that the fundamental character of the flow is altered as m changes, as will be discussed later. Next, from Eqs. 8.4–52 and 53 the velocity components may be obtained to order L.

$$V_r^{(1)} = \frac{(\xi - 1)^2}{8\xi^5}[(1 - 4m)\xi - 8m](3\sin^2\theta - 2) \tag{8.4–57}$$

$$V_\theta^{(1)} = \frac{(\xi - 1)}{4\xi^5}(\xi - 12m)\sin\theta\cos\theta \tag{8.4–58}$$

When the $V_r^{(1)}$ and $V_\theta^{(1)}$ thus obtained are substituted into Eqs. 8.4–46 and 47, the expressions for $T_{\theta\phi}$ and $T_{r\phi}$ are obtained to order L in terms of $V_\phi^{(1)}$, ξ, θ, and the parameter m. These stress components may then be inserted into Eq. 8.4–33 to get a second-order, linear partial differential equation for $V_\phi^{(1)}$. Thomas and Walters[6] succeeded in obtaining a solution to this equation in the form:

$$V_\phi^{(1)} = f_1(\xi)\cdot(\sin\theta) + f_2(\xi)\cdot(\sin^3\theta - \tfrac{4}{5}\sin\theta) \tag{8.4–59}$$

where

$$\begin{aligned} f_1(\xi) &= -\frac{22m^2}{15\xi^8} + \left(\frac{3m}{20} + \frac{9m^2}{5}\right)\frac{1}{\xi^7} - \frac{11}{70}\frac{m}{\xi^6} - \left(\frac{1}{120} + \frac{m}{20} + \frac{m^2}{5}\right)\frac{1}{\xi^5} \\ &\quad + \frac{1}{50\xi^4} + \left(-\frac{1}{80} + \frac{m}{20}\right)\frac{1}{\xi^3} + \left(\frac{1}{1200} + \frac{m}{140} - \frac{2m^2}{15}\right)\frac{1}{\xi^2} \end{aligned} \tag{8.4–60}$$

$$\begin{aligned} f_2(\xi) &= \frac{9m^2}{11\xi^8} - \left(\frac{m}{8} + \frac{3m^2}{2}\right)\frac{1}{\xi^7} + \frac{1}{4}\frac{m}{\xi^6} + \left(\frac{1}{64} - \frac{3m}{8} + \frac{9m^2}{4}\right)\frac{1}{\xi^5} \\ &\quad + \left(\frac{1}{16} - \frac{m}{4}\right)\frac{1}{\xi^3} + \left(-\frac{1}{28}\ln\xi - \frac{5}{64} + \frac{m}{2} - \frac{69m^2}{44}\right)\frac{1}{\xi^4} \end{aligned} \tag{8.4–61}$$

We now have all of the velocity components to order L.

The torque exerted by the fluid on the sphere is then:

$$\begin{aligned} \mathscr{T} &= \int_0^{2\pi}\int_0^{\pi}(-\tau_{\theta\phi})|_{r=R}\cdot R\sin\theta\cdot R^2\sin\theta\,d\theta\,d\phi \\ &= 2\pi\alpha_1 R^3 W\int_0^{\pi}(-T_{r\phi}|_{\xi=1})\sin^2\theta\,d\theta \\ &= 2\pi\alpha_1 R^3 W\int_0^{\pi}\left[3\sin\theta + L\left\{\left(\frac{1}{400} + \frac{3}{140}m - \frac{2}{5}m^2\right)\sin\theta\right.\right. \\ &\quad \left.\left. + \left(\frac{45}{44}m^2 + \frac{85}{224}\right)(\sin^3\theta - \tfrac{4}{5}\sin\theta)\right\}\right]\sin^2\theta\,d\theta \\ &= 8\pi\alpha_1 R^3 W\left[1 + L\left(\frac{1}{1200} + \frac{m}{140} - \frac{2m^2}{15}\right)\right] \end{aligned} \tag{8.4–62}$$ [8]

[8] The coefficient $-2/15$ is in agreement with the creeping flow solution of Giesekus[7] (his Λ_4 coefficient).

where Eq. 8.4–47 and the expressions for $V_r^{(1)}$ and $V_\phi^{(1)}$ have been used. The dimensionless elasticity number, m, is defined in Eq. 8.4–48 and the dimensionless Reynolds-number-squared, L, is given in Eq. 8.4–29.

To illustrate the nature of the secondary flow caused by the inertial and elastic forces, Thomas and Walters[1] plotted projections of the particle paths on a plane containing the axis of rotation. The projection of any streamline is represented approximately by (cf. Eq. 8.4–56):

$$\frac{(\xi - 1)^2}{\xi^3}[(1 - 4m)\xi - 8m] \sin^2\theta \cos\theta = \text{constant} \tag{8.4–63}$$

The projections are quite different depending on the value of the dimensionless parameter $m = \alpha_2/\rho R^2$, which is a measure of the relative strength of the elastic and inertial forces. Three distinct regimes are found:

(i) $0 \leq m \leq 1/12$, in which the inertial flow dominates; here the flow is similar to that of a Newtonian fluid, with the fluid approaching the sphere toward the poles and moving away at the equator (see Figs. 8.4–2*a* and *b*).

(ii) $1/12 < m < 1/4$, in which elastic effects predominate near the sphere, whereas inertial effects are controlling beyond some critical radial coordinate $\xi_{cr} = 8m/(1 - 4m)$; fluid particles inside (outside) this sphere remain inside (outside) during the flow. As can be seen in Fig. 8.4–2*c*, at points labelled N (given by $\xi_N = 12m$ and $\theta_N = \arcsin\sqrt{\frac{2}{3}}$, $v_r = v_\theta = 0$ and the fluid particles move in circles about the axis of rotation.

(iii) $m \geq 1/4$, in which the liquid approaches the sphere at the equator, and then moves away at the poles; here the elastic forces are dominant. As in (ii) at N (given by $\xi_N = 12m$, $\theta_N = \arcsin\sqrt{\frac{2}{3}}$), v_r and v_θ are both zero.

These theoretically-predicted flow patterns have been observed experimentally,[7,9,10] and, in particular the value of $\theta_N = \arcsin\sqrt{\frac{2}{3}} = 54°44'$ has been found to agree quantitatively with the measured value.

Thomas and Walters[6] have further noted that according to Eq. 8.4–62, a plot of $\mathscr{T}/(8\pi\alpha_1 R^3 W)$ vs. L gives at $L = 0$ a slope of $(1/1200) + (1/140)m - (2/15)m^2$. The slope will thus be positive or negative as m is greater or less than 0.11.

In later work by Walters and Waters[6] the above analysis was extended to third-order fluids, for which the generalization of Eq. 8.4–62 is

$$\mathscr{T} = 8\pi\alpha_1 R^3 W\left[1 + L\left(\frac{1}{1200} + \frac{m_1}{140} - \frac{2m_1{}^2}{15} - \frac{12m_2}{5}\right)\right] \tag{8.4–64}$$

in which $L = (WR^2\rho/\alpha_1)^2$, and the dimensionless groups m_1 and m_2 are:

$$m_1 = \frac{2\alpha_2 - \alpha_{11}}{\rho R^2} \tag{8.4–65}$$

$$m_2 = \frac{\alpha_1(\alpha_3 - 2\alpha_{1:11})}{(\rho R^2)^2} \tag{8.4–66}$$

By comparison with Eq. 8B.9–1 we see that m_2 is related to the coefficient of the $\dot{\gamma}^2$-term in the non-Newtonian viscosity function. When Ψ_2 is required to be zero, then m_1 becomes identical to the quantity m defined in Eq. 8.4–48.

[9] K. Walters and J. G. Savins, *Trans. Soc. Rheol.*, *9*:1, 407–416 (1965).
[10] Y. Ide and J. L. White, *J. Appl. Polym. Sci.*, *18*, 2997–3018 (1974).

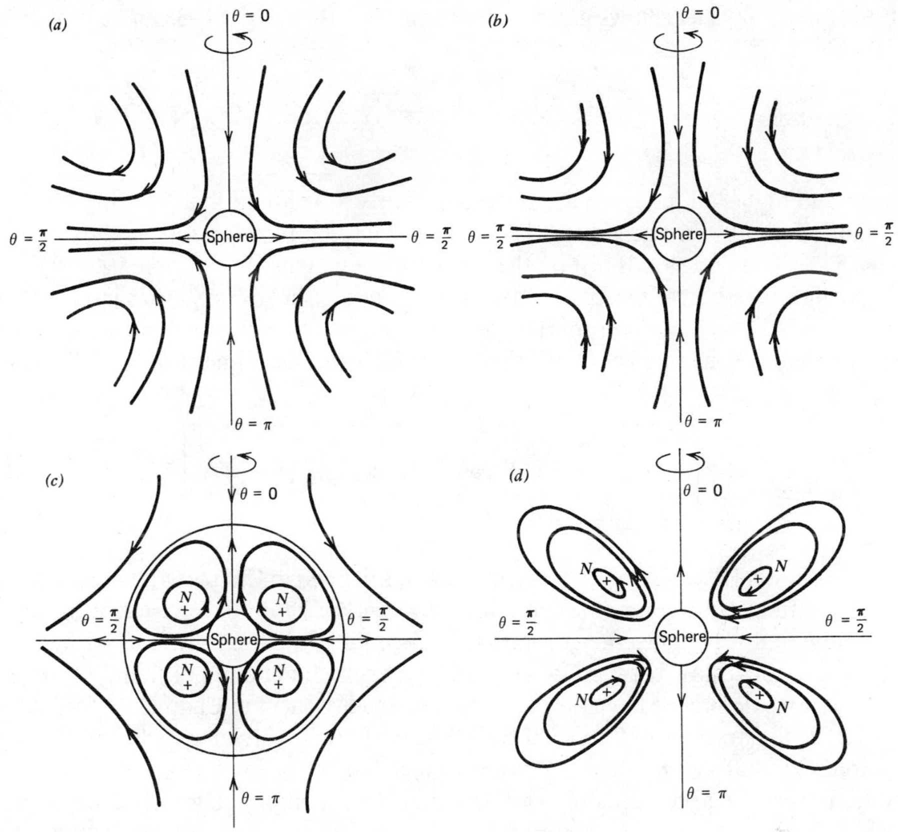

FIGURE 8.4–2. Projections for the particle paths on any plane containing the axis of rotation, near a sphere rotating in a second order fluid. The projections are those calculated for a fluid with $\rho = 1\ \text{g/cm}^3$ and a sphere of radius 3 cm, with

	(a)	(b)	(c)	(d)
α_2	0	0.75	1.5	2.25
m	0	$\frac{1}{12}$	$\frac{1}{6}$	$\frac{1}{4}$

[R. H. Thomas and K. Walters, *Quart. J. of Mech. and Appl. Math.*, *17*, 39–53 (1964), © Oxford University Press, 1964.]

§8.5 SIMPLIFIED EQUATIONS FOR STEADY–STATE FLOWS (THE CRIMINALE-ERICKSEN-FILBEY EQUATION AND THE REINER-RIVLIN EQUATION)

In the last section it was shown that the rheological equation of state for the "simple fluid" can be written as a retarded motion expansion (see Eq. 8.4–3). In this section we show that this expansion can in turn be simplified for two special flows.

a. Steady Shear Flows

Because of the kinematic simplicity of steady shear flows, the expansion in Eq. 8.4–3 can be collapsed[1] into just three terms as we shall describe. If use is made of the kinematical

[1] J. D. Goddard, *Trans. Soc. Rheol.*, *11*, 381–399 (1967), performed the collapsing by a somewhat different procedure; see also H. Giesekus, *Zeits. f. angew. Mech. und Math.*, *1/2*, 32–61 (1962).

relations for steady shear flows given in Problem 7B.2, then Eq. 8.4–3 can be written as:

$$\boldsymbol{\tau} = -(\alpha_1 - \alpha_3\dot{\gamma}^2 + 2\alpha_{1:11}\dot{\gamma}^2)\dot{\boldsymbol{\gamma}} - \alpha_{11}\{\dot{\boldsymbol{\gamma}} \cdot \dot{\boldsymbol{\gamma}}\} + \alpha_2 \frac{\mathscr{D}}{\mathscr{D}t}\dot{\boldsymbol{\gamma}} + \cdots \tag{8.5–1}$$

If fourth and higher order terms are treated similarly, it is found that by using Eqs. 7B.2–1 to 4 additional terms are generated that involve the same three tensors as those appearing in Eq. 8.5–1. Thus $\boldsymbol{\tau}$ depends just on the three kinematic tensors $\dot{\boldsymbol{\gamma}}$, $\dot{\boldsymbol{\gamma}}^2$, and $(\mathscr{D}/\mathscr{D}t)\dot{\boldsymbol{\gamma}}$, and the scalar coefficients of these tensors are all functions of the shear rate $\dot{\gamma}$. (We have omitted all terms multiplied by $\boldsymbol{\delta}$—the isotropic terms—according to the convention we adopted in §8.3.) By this procedure, one finally obtains the Criminale-Ericksen-Filbey (or "CEF") equation:[2]

$$\boldsymbol{\tau} = -\eta\dot{\boldsymbol{\gamma}} - (\tfrac{1}{2}\Psi_1 + \Psi_2)\{\dot{\boldsymbol{\gamma}} \cdot \dot{\boldsymbol{\gamma}}\} + \tfrac{1}{2}\Psi_1 \frac{\mathscr{D}}{\mathscr{D}t}\dot{\boldsymbol{\gamma}} \tag{8.5–2}$$

in which η, Ψ_1, and Ψ_2 are the viscometric functions defined in Chapter 4; these are functions of the magnitude of the rate-of-strain tensor $\dot{\gamma} = \sqrt{\frac{1}{2}(\dot{\boldsymbol{\gamma}} : \dot{\boldsymbol{\gamma}})}$. In §5.5 we cited this equation just to show that it includes the generalized Newtonian fluid, and then we verified that for steady simple shear flow between two parallel plates the functions η, Ψ_1, and Ψ_2 appearing in the equations do indeed have the meanings ascribed to them in Chapter 4. In this section we give a few illustrations of the use of Eq. 8.5–2 for flows involving curvilinear coordinates. We emphasize that Eq. 8.5–2 is restricted to steady-state shear flows, although it will be sensible to apply it approximately to flows that differ somewhat from steady-state shear flows.[3]

b. Steady, Homogeneous, Irrotational Flows

For steady-state flows that are homogeneous and irrotational ($\boldsymbol{\omega} = \mathbf{0}$), all of the corotational derivatives $\mathscr{D}\dot{\boldsymbol{\gamma}}/\mathscr{D}t$, $\mathscr{D}^2\dot{\boldsymbol{\gamma}}/\mathscr{D}t^2, \ldots$ are zero. Then Eq. 8.4–3 gives:

$$\boldsymbol{\tau} = -(\alpha_1 + \alpha_{1:11}II_{\dot{\gamma}})\dot{\boldsymbol{\gamma}} - \alpha_{11}\{\dot{\boldsymbol{\gamma}} \cdot \dot{\boldsymbol{\gamma}}\} + \cdots \tag{8.5–3}$$

Hence we see that the expression for $\boldsymbol{\tau}$ involves products of $\dot{\boldsymbol{\gamma}}$ of increasing order. By repeated application of the Cayley-Hamilton theorem these products may be expressed in terms of $\boldsymbol{\delta}$, $\dot{\boldsymbol{\gamma}}$, and $\{\dot{\boldsymbol{\gamma}} \cdot \dot{\boldsymbol{\gamma}}\}$ with scalar coefficients depending on $I_{\dot{\gamma}}$, $II_{\dot{\gamma}}$, and $III_{\dot{\gamma}}$. For incompressible fluids we omit the isotropic term involving $\boldsymbol{\delta}$ and use the fact that $I_{\dot{\gamma}} = 0$. Then Eq. 8.5–3 becomes:

$$\boldsymbol{\tau} = -f_1(II_{\dot{\gamma}}, III_{\dot{\gamma}})\dot{\boldsymbol{\gamma}} - f_2(II_{\dot{\gamma}}, III_{\dot{\gamma}})\{\dot{\boldsymbol{\gamma}} \cdot \dot{\boldsymbol{\gamma}}\} \tag{8.5–4}$$

[2] W. O. Criminale, Jr., J. L. Ericksen, and G. L. Filbey, Jr., *Arch. Rat. Mech. Anal.*, *1*, 410–417 (1958); see also J. L. Ericksen, in *Viscoelasticity*: *Phenomenological Aspects*, J. T. Bergen (Ed.), Academic Press, New York (1960). The unsteady-state analog of the CEF equation has been derived, but it contains scalar functionals [A. S. Lodge, *Body Tensor Fields in Continuum Mechanics*, Academic Press, New York (1974), Eq. 7.1–11].

[3] All of the viscometric flow problems solved in B. D. Coleman, H. Markovitz, and W. Noll, *Viscometric Flows of Non-Newtonian Fluids*, Springer, New York (1966), can be solved by using Eq. 8.5–2.

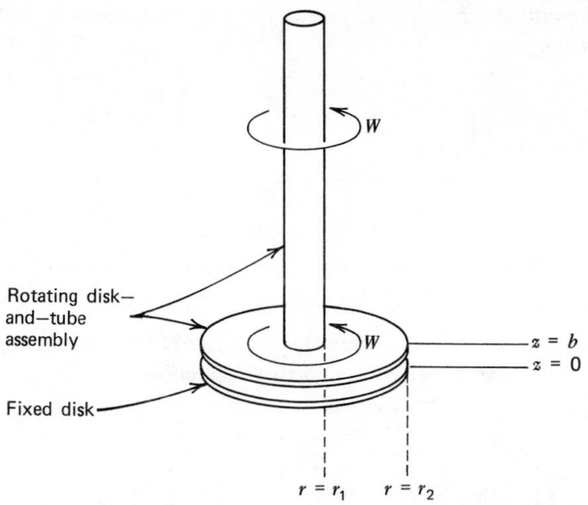

FIGURE 8.5–1. Centripetal pumping in a disk-tube assembly. The viscoelastic fluid is pumped radially inward and up into the tube because of the action of normal stresses.

This is the *Reiner-Rivlin equation.*[4,5,6] When this equation was originally proposed, it was thought to be more generally applicable than it actually is. Hence there are unfortunately many papers in the research literature in which the Reiner-Rivlin equation has been applied incorrectly to flows that are not steady, homogeneous, and irrotational.

EXAMPLE 8.5–1 Centripetal Pumping between Parallel Disks[7]

Two disks are separated by a fixed distance b, and the upper disk is joined to a circular tube (see Fig. 8.5–1). The upper disk is rotated with a constant angular velocity $W\boldsymbol{\delta}_z$. For a Newtonian fluid the rotation of the upper disk would cause a centrifugal force that would tend to suck the fluid downwards out of the tube. For viscoelastic fluids, on the other hand, the normal stresses may cause the fluid to flow inward through the space between the disks and upward through the tube.

The complete calculation of the centripetal pumping is quite difficult, inasmuch as it is not a viscometric flow. However, it is relatively easy to use Eq. 8.5–2 to find out what pressure difference has to be maintained between "1" and "2" in order to prevent flow. Find the height to which the fluid will rise in the tube to balance the normal stresses tending to drive the fluid up the tube.

SOLUTION (a) Newtonian Fluid

Before doing the viscoelastic fluid calculations, it is instructive to solve the problem for the Newtonian fluid, to find the pressure difference required to prevent radially outward flow because of centrifugal effects.

[4] M. Reiner, *Am. J. Math.*, *67*, 350–362 (1945).

[5] R. S. Rivlin, *Proc. Roy. Soc.* (London), *A193*, 260–281 (1948); *Proc. Cambridge Phil. Soc.*, *45*, 88–91 (1949).

[6] K. Weissenberg, *Arch. Sci. Phys. Nat.* (*5*), *140*, 44–106, 130–171 (1935) made an earlier attempt to construct a rheological equation of state starting with the premise that $\boldsymbol{\tau}$ must be some general function of $\dot{\boldsymbol{\gamma}}$. The unsteady-state analog of the Reiner-Rivlin equation has been derived (A. S. Lodge, *op. cit.*, Eq. 7.4–5), but it contains scalar functionals.

[7] Y. Tomita and H. Katō, *Trans. Japan. Soc. Mech. Eng.* (Nippon Kikai Gakkai Ronbunshū), *32*, 241, 1399–1408 (September 1966); B. Maxwell and A. J. Scalora, *Modern Plastics*, *37* (2), 107 (1959); P. A. Good, A. J. Schwartz, and C. W. Macosko, *A.I.Ch.E. Journal*, *20*, 67–73 (1974). Note that where we have 1/6 in Eq. 8.5–11, a more detailed analysis gives 3/20 [K. Stewartson, *Proc. Camb. Phil. Soc.*, *49*, 333–341 (1953)]. For a discussion of tangential flow with slow superimposed radial flow see D. F. James, *Trans. Soc. Rheol.*, *19*, 67–80 (1975).

If we make the postulate that $v_\theta = zf(r)$, $v_r = v_z = 0$, then the r-, θ-, and z-components of the equation of motion become:

$$-\rho \frac{z^2 f^2}{r} = -\frac{\partial p}{\partial r} \tag{8.5–5}$$

$$0 = +\mu \frac{d}{dr}\left(\frac{1}{r}\frac{d}{dr}(rf)\right) \tag{8.5–6}$$

$$0 = -\frac{\partial p}{\partial z} - \rho g \tag{8.5–7}$$

The first of these describes the balance between the centrifugal force and the radial pressure gradient. The second equation can be solved to give the velocity distribution:

$$v_\theta = \frac{Wrz}{b} \tag{8.5–8}$$

Insertion of this into Eq. 8.5–5 then gives:

$$-\rho W^2 r \left(\frac{z}{b}\right)^2 = -\frac{\partial p}{\partial r} \tag{8.5–9}$$

It can now be seen that Eqs. 8.5–7 and 9 are inconsistent, since the two equations give different results for the mixed second derivative $\partial^2 p/\partial r\, \partial z$. This means that the assumption made prior to Eq. 8.5–5 regarding the velocity profile is incorrect, and that strictly speaking secondary flows have to be allowed for. To get an approximate expression for the radial pressure gradient, however, we can obtain the average of Eq. 8.5–9 over z thus:

$$-\frac{\rho W^2 r}{3} = -\frac{d\bar{p}}{dr} \tag{8.5–10}$$

This equation may be integrated to give:

$$\bar{p}_2 - \bar{p}_1 = \tfrac{1}{6}\rho W^2(r_2^2 - r_1^2) \tag{8.5–11}$$

We see that we have to push harder at "2" than at "1" in order to keep the fluid from being thrown out of the gap.

(b) Viscoelastic Fluid

When there is no motion in the r-direction, the r-component of the equation of motion becomes:

$$-\rho \frac{v_\theta^2}{r} = -\frac{\partial p}{\partial r} - \left(\frac{1}{r}\frac{\partial}{\partial r}(r\tau_{rr}) - \frac{\tau_{\theta\theta}}{r} + \frac{\partial \tau_{rz}}{\partial z}\right) \tag{8.5–12}$$

Since the flow is a steady-state shear flow according to the definition in §4.1, the CEF equation may be used to compute the stress components. For this flow the rate-of-strain and vorticity tensors are for the flow field of Eq. 8.5–8 (cf. Eq. 4.5–13):

$$\dot{\boldsymbol{\gamma}} = \begin{pmatrix} 0 & 0 & 0 \\ 0 & 0 & Wr/b \\ 0 & Wr/b & 0 \end{pmatrix} \tag{8.5–13}$$

$$\boldsymbol{\omega} = \begin{pmatrix} 0 & 2Wz/b & 0 \\ -2Wz/b & 0 & -Wr/b \\ 0 & Wr/b & 0 \end{pmatrix} \tag{8.5–14}$$

and hence

$$\{\dot{\gamma} \cdot \dot{\gamma}\} = \begin{pmatrix} 0 & 0 & 0 \\ 0 & (Wr/b)^2 & 0 \\ 0 & 0 & (Wr/b)^2 \end{pmatrix} \tag{8.5–15}$$

$$\{\boldsymbol{\omega} \cdot \dot{\gamma}\} = \begin{pmatrix} 0 & 0 & 2W^2rz/b^2 \\ 0 & -(Wr/b)^2 & 0 \\ 0 & 0 & (Wr/b)^2 \end{pmatrix} \tag{8.5–16}$$

$$\{\boldsymbol{v} \cdot \nabla\dot{\gamma}\} = \begin{pmatrix} 0 & 0 & -W^2rz/b^2 \\ 0 & 0 & 0 \\ -W^2rz/b^2 & 0 & 0 \end{pmatrix} \tag{8.5–17}$$

Substitution of this information into the CEF equation gives for the elements of the stress tensor appearing in Eq. 8.5–12:

$$\tau_{rr} = 0 \tag{8.5–18}$$

$$\tau_{\theta\theta} = -(\Psi_1 + \Psi_2)(Wr/b)^2 \tag{8.5–19}$$

$$\tau_{rz} = 0 \tag{8.5–20}$$

Insertion of these stress-tensor components into Eq. 8.5–12 then gives:

$$\frac{\partial p}{\partial r} = \rho W^2 r \left(\frac{z}{b}\right)^2 - (\Psi_1 + \Psi_2)\frac{W^2}{b^2} r \tag{8.5–21}$$

Then, as before, we average with respect to z to get:

$$\frac{d\overline{p}}{dr} = \frac{\rho W^2 r}{3} - (\Psi_1 + \Psi_2)\frac{W^2}{b^2} r \tag{8.5–22}$$

Finally, integration with respect to r gives:

$$\overline{p}_2 - \overline{p}_1 = \frac{\rho W^2}{6}(r_2{}^2 - r_1{}^2) - \frac{W^2}{b^2}\int_{r_1}^{r_2} (\Psi_1 + \Psi_2) r\, dr \tag{8.5–23}$$

Since $\Psi_1 + \Psi_2$ is positive, the second term will have a sign opposite to the first.

The height of the fluid in the tube that will just balance the pressure difference in Eq. 8.5–23 is then:

$$h = -\frac{W^2}{6g}(r_2{}^2 - r_1{}^2) + \frac{W^2}{\rho g b^2}\int_{r_1}^{r_2} (\Psi_1 + \Psi_2) r\, dr \tag{8.5–24}$$

This is one method for measuring a particular combination of the normal stresses (see Eq. F–1 in Table 4.5–1).

EXAMPLE 8.5–2 Normal Stress Stabilizing Force in Wire-Coating Dies[8]

Here we consider the axial motion of a circular cylinder traveling with speed v_0 inside a circular cylindrical cavity. The cylinder has radius R_1 and the radius of the cavity is R_2. The cylinder axis and

[8] This problem was solved for an Oldroyd model by J. R. Jones, *J. de Méchanique*, *3*, 79–99 (1964); *4*, 121–132 (1965). The treatment given here, based on the CEF equation is due to Z. Tadmor and R. B. Bird, *Polym. Eng. Sci.*, *14*, 124–136 (1974).

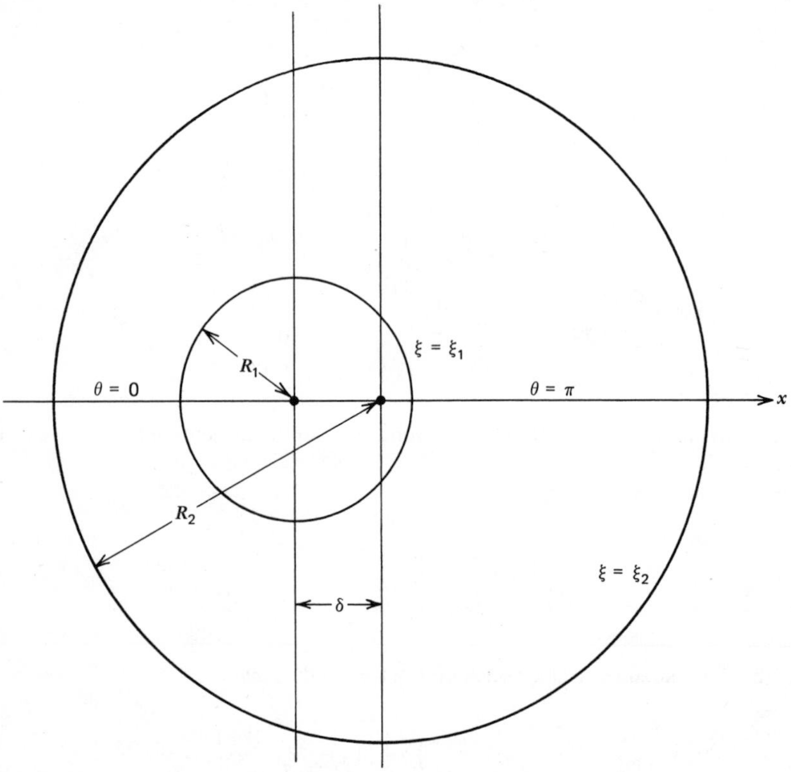

FIGURE 8.5–2. An off-centered wire of radius R_1 moving axially with speed v_0 in a cylindrical cavity of radius R_2 filled with a viscoelastic coating material. The coordinates ξ and θ are coordinates in the bipolar coordinate system (see Fig. A.7–1).

cavity axis are not coincident, but are separated by a small distance δ. Figure 8.5–2 shows the system under consideration, which, because of the eccentric arrangement, is most easily described in terms of bipolar coordinates (see Fig. A.7–1). It is desired to find the lateral force acting on the off-centered cylindrical rod.

This arrangement is of interest in connection with the stabilizing forces acting on an off-center wire in a wire-coating operation, where speeds of up to 25 m/s are encountered. Wire-coating materials in their molten form are viscoelastic liquids, and a salient question is whether the normal stresses in these liquids will help or hinder in the stabilizing of the wire-coating operation. Therefore it is of interest to find out if there is a force on an off-center cylinder that tends to push the cylinder toward the center of the cavity or in the opposite direction.

SOLUTION

We consider here only the situation in which the wire is slightly off center—that is, δ is small compared with R_2. With this assumption we can assume that v_z is a function of ξ alone (the dependence on θ was explored by Tadmor and Bird[8] and found to be negligible), and that the other velocity components v_ξ and v_θ are zero. Furthermore we postulate that the pressure depends on ξ and θ. With these postulates the nonvanishing components of the stress tensor may be found with the help of Table A.7–4 to be:

$$\tau_{\xi\xi} = -\Psi_2 \left(\frac{X}{a} \frac{dv_z}{d\xi} \right)^2 \tag{8.5–25}$$

$$\tau_{\xi z} = -\eta \left(\frac{X}{a} \frac{dv_z}{d\xi} \right) \tag{8.5–26}$$

$$\tau_{zz} = -(\Psi_1 + \Psi_2)\left(\frac{X}{a}\frac{dv_z}{d\xi}\right)^2 \tag{8.5–27}$$

in which $X = \cosh \xi + \cos \theta$. Then the equation of motion has as its ξ-, θ-, and z-components:

$$0 = -\frac{X}{a}\frac{\partial p}{\partial \xi} - \frac{X}{a}\frac{d}{d\xi}\tau_{\xi\xi} + \frac{1}{a}\tau_{\xi\xi}\sinh \xi \tag{8.5–28}$$

$$0 = -\frac{X}{a}\frac{\partial p}{\partial \theta} + \frac{1}{a}\tau_{\xi\xi}\sin \theta \tag{8.5–29}$$

$$0 = -\frac{X}{a}\frac{d}{d\xi}\tau_{\xi z} + \frac{1}{a}\tau_{\xi z}\sinh \xi \tag{8.5–30}$$

Substitution of the expression for $\tau_{\xi z}$ from Eq. 8.5–26 into the third of the equations of motion gives:

$$\frac{d}{d\xi}\left(\eta(\dot{\gamma})\frac{d}{d\xi}v_z\right) = 0 \tag{8.5–31}$$

For the postulated velocity profile $\dot{\gamma} = (X/a)(dv_z/d\xi)$. If it is assumed that $\dot{\gamma}$, and hence η, are nearly constant through the fluid region, then a good approximation for the velocity distribution will be:

$$\frac{v_z}{v_0} = \frac{\xi - \xi_2}{\xi_1 - \xi_2} \tag{8.5–32}$$

where ξ_1 and ξ_2 are the values of ξ corresponding to the radii of the wire and the cylindrical cavity.

Since the pressure is an analytic function:

$$dp = \left(\frac{\partial p}{\partial \xi}\right)d\xi + \left(\frac{\partial p}{\partial \theta}\right)d\theta \tag{8.5–33}$$

The pressure at any point on the surface of the wire can be obtained by substituting Eqs. 8.5–28 and 29 into Eq. 8.5–33 and integrating over θ at constant ξ_1:

$$p(\xi_1,\theta) = p(\xi_1,0) + \int_0^\theta \left(\frac{1}{X}\tau_{\xi\xi}\sin \theta\right)_{\xi=\xi_1} d\theta \tag{8.5–34}$$

where $p(\xi_1,0)$ is an unknown constant.

The force per unit area that the fluid exerts on the cylindrical wire at any point is $[p(\xi_1,\theta) + \tau_{\xi\xi}]\boldsymbol{\delta}_\xi$ where $\boldsymbol{\delta}_\xi$ is the unit vector pointing in the positive ξ-direction. The element of area of the wire on which the force acts is $a\,d\theta\,dz/X$. The component of the force per unit axial distance in the x-direction is then:

$$df_x = [p(\xi_1,\theta) + \tau_{\xi\xi}]|_{\xi=\xi_1}(\boldsymbol{\delta}_\xi \cdot \boldsymbol{\delta}_x)\frac{a\,d\theta}{\cosh \xi_1 + \cos \theta} \tag{8.5–35}$$

Since $\boldsymbol{\delta}_\xi$ can be resolved into components in the x- and y-directions, thus:

$$\boldsymbol{\delta}_\xi = \frac{X}{a}\left[\frac{a\cosh \xi}{\cosh \xi + \cos \theta} - \frac{a\sinh^2 \xi}{(\cosh \xi + \cos \theta)^2}\right]\boldsymbol{\delta}_x + \frac{X}{a}\left[\frac{a\sin \theta \sinh \xi}{(\cosh \xi + \cos \theta)^2}\right]\boldsymbol{\delta}_y \tag{8.5–36}$$

the dot product appearing in Eq. 8.5–35 is:

$$(\boldsymbol{\delta}_\xi \cdot \boldsymbol{\delta}_x) = \frac{(\cosh \xi)(\cosh \xi + \cos \theta) - \sinh^2 \xi}{\cosh \xi + \cos \theta}$$

$$= \frac{1 + \cosh \xi \cos \theta}{\cosh \xi + \cos \theta} \tag{8.5–37}$$

Insertion of this result into Eq. 8.5–35 and integration over θ gives:

$$f_x = 2 \int_0^\pi [p(\xi_1,\theta) + \tau_{\xi\xi}]_{\xi_1} \frac{1 + \cosh \xi_1 \cos \theta}{(\cosh \xi_1 + \cos \theta)^2} a\, d\theta \tag{8.5–38}$$

Next Eq. 8.5–34 is substituted into Eq. 8.5–38. The integral over $p(\xi_1,\theta)$ is zero, and we therefore obtain:

$$f_x = 2a \int_0^\pi \left\{ \left(\int_0^\theta \frac{\tau_{\xi\xi}|_{\xi_1} \sin \theta'}{\cosh \xi_1 + \cos \theta'} d\theta' \right) + \tau_{\xi\xi}|_{\xi_1} \right\} \cdot \frac{1 + \cosh \xi_1 \cos \theta}{(\cosh \xi_1 + \cos \theta)^2} d\theta \tag{8.5–39}$$

Next the normal stress at ξ_1 is obtained by substituting the velocity from Eq. 8.5–32 into Eq. 8.5–25 to give:

$$\tau_{\xi\xi}|_{\xi_1} = -\Psi_2 \left(\frac{\cosh \xi_1 + \cos \theta}{a} \right)^2 \frac{v_0^2}{(\xi_1 - \xi_2)^2} \tag{8.5–40}$$

From the last two equations we then get, if we assume that $\Psi_2(\xi_1)$ depends only weakly on θ:

$$f_x = -\frac{2\Psi_2(\xi_1)}{a} \frac{v_0^2}{(\xi_1 - \xi_2)^2} \int_0^\pi \left\{ \left(\int_0^\theta (\cosh \xi_1 + \cos \theta') \sin \theta' \, d\theta' \right) + (\cosh \xi_1 + \cos \theta)^2 \right\}$$

$$\cdot \frac{1 + \cosh \xi_1 \cos \theta}{(\cosh \xi_1 + \cos \theta)^2} d\theta \tag{8.5–41}$$

After performing the inner integral over θ' we get:

$$f_x = -\frac{2\Psi_2(\xi_1) v_0^2}{a(\xi_1 - \xi_2)^2} \int_0^\pi (1 + \cosh \xi_1 \cos \theta) \left[1 + \frac{(\cosh \xi_1)(1 - \cos \theta) + \frac{1}{2} \sin^2 \theta}{(\cosh \xi_1 + \cos \theta)^2} \right] d\theta \tag{8.5–42}$$

The integral can be shown to be $\pi/2$. Hence the lateral force (per unit length) acting in the x-direction is:

$$f_x = \frac{-\Psi_2(\xi_1) \pi v_0^2}{a(\xi_1 - \xi_2)^2} \tag{8.5–43}$$

where $a(R_1,R_2,\delta)$ is given by Eq. A.7–31. We see that if the secondary normal stress coefficient Ψ_2 is negative (and all the data cited in Chapter 4 indicate that it is), then the force f_x will be positive, and the wire will be self-centering. Note that neither the primary normal stress coefficient nor the viscosity enter in the final expression. Note also that the secondary normal stress coefficient is to be evaluated at ξ_1, where the shear rate is $[(\cosh \xi_1 + \cos \theta)/a][v_0/(\xi_1 - \xi_2)]^2$ (a mean value over θ will have to be taken). The actual importance of this force in wire-coating operations is not known, since the secondary normal stress coefficients have not been measured for the coatings involved at the shear stresses under consideration.

Other forces are at work in wire-coating operations. For example, if the wire axis is not parallel to the axis of the cylindrical cavity there will be appreciable tangential flow of fluid around the wire. Such a flow also gives rise to a stabilizing force. This force has been estimated by Tadmor and Bird.[8]

In this example we have assumed that there is no pressure gradient driving the coating fluid through the cylindrical cavity in the axial direction. The theoretical papers of Jones[8] have included this extra complication.

QUESTIONS FOR DISCUSSION

1. Compare and contrast the corotational Jeffreys model and the Oldroyd eight-constant model as to physical basis, ability to describe material functions, determination of the constants from experimental data, and restrictions that have to be placed on the constants.
2. Why is there no term containing $(\text{tr}\ \dot{\boldsymbol{\gamma}})\dot{\boldsymbol{\gamma}}$ on the right side of Eq. 8.1–2 corresponding to the $(\text{tr}\ \boldsymbol{\tau})\dot{\boldsymbol{\gamma}}$ on the left side?
3. How do the G's have to be selected in the corotational memory integral expansion in order to get the LeRoy and Pierrard equation? What is the physical meaning of this choice?
4. To what extent can one make responsible fluid dynamics calculations knowing only η' and η'' for a fluid?
5. What is the origin of Eq. 8.3–1? How can the G's be determined?
6. What theoretical relations exist among material functions and to what extent can they be trusted? To what extent is Table 8.3–1 helpful in answering this question?
7. Compare and contrast the Oldroyd eight-constant model and the retarded motion expansion as to usefulness for problem solving, determination of the constants from material functions, and interrelation of rheological phenomena.
8. To what extent can you interpret and explain the experiments in Chapter 3 in terms of the various rheological equations of state in Fig. 8.3–1?
9. How do the results in Table 2.5–1 fit into the chart in Fig. 8.3–1?
10. Could cylindrical coordinates be used in solving the problem in Example 8.5–2?
11. What dimensionless groups do you expect to encounter in solving viscoelastic fluid dynamics problems?
12. Is the Goddard-Miller model capable of describing the centripetal pumping in Example 8.5–1?
13. Compare and contrast Example 8.1–4 and Problem 5B.10.

PROBLEMS

8A.1 Rotating Sphere in a Polyisobutylene Solution

Consider a sphere of radius 0.75 cm rotating in the polyisobutylene solution whose viscosity and no[rmal] stress data are shown in Figs. 4.3–3 and 6. At what angular velocity W will the first-order correc[tion] term in Eq. 8.4–62 become 1% of the zero-order term $8\pi\alpha_1 R^3 W$?

8A.2 Height of Rise in Centripetal-Pumping Assembly

For the fluid considered in Problem 8A.1, estimate the height h in Fig. 8.5–1 if $r_1 = 0.2$ cm, $r_2 = 12$ [...] $b = 0.31$ cm, and $W = 5$ rad/s.

8A.3 Estimation of Oldroyd Model Parameters

Consider the four-constant Oldroyd model obtained by setting $\mu_1 = \lambda_1$, $\mu_2 = \lambda_2$, and $\nu_1 = \nu_2 =$ [...] Eq. 8.1–2. Obtain the numerical values of η_0, λ_1, λ_2, and μ_0 by using the data on $\eta(\dot{\gamma})$ and $\eta'(\omega)$ o[f] polyisobutylene in Primol (see Figs. 4.3–3 and 4.4–3). Having obtained the four parameters, plo[t] curve of Ψ_1 given by Eq. 8.1–2 and compare with the experimental data in Fig. 4.3–6. What conclus[ions] do you draw from this comparison?

8A.4 Deviation from Stokes' Law Because of Viscoelasticity of a Macromolecular Fluid

Stokes' law for steady-state, creeping flow around a sphere was given in Example 1.2–4. The analo[gous] problem of solving the equations of continuity and motion (with the inertial terms omitted) fo[r] Oldroyd model of Eq. 8.1–2 was first done by Leslie.[1] In this work μ_1 and μ_2 were set equal to λ_1 [and] λ_2, respectively, and the abbreviations $\sigma_i = \lambda_i\mu_0 + (\lambda_1 - \frac{3}{2}\mu_0)\nu_i$ were used (see the definition of σ after Eq. 8.1–2). Leslie performed a perturbation analysis about the known Newtonian fluid solu[tion] expanding in powers of the dimensionless quantity $v_\infty\lambda_1/R$. His solution is valid for

$$\frac{Rv_\infty\rho}{\eta_0} \ll \frac{v_\infty\lambda_1}{R} < 1 \tag{8A.}$$

Although the details of the derivation are tedious and lengthy, the final result for the drag force quad[ratic] in $v_\infty\lambda_1/R$ is rather simple. This result, as corrected by Giesekus[2] in a later publication, is:

$$F_k = 6\pi\eta_0 R v_\infty\left\{1 - \frac{1}{2275}\left[\left(\frac{401}{11} - 39\frac{\lambda_2}{\lambda_1}\right)\left(1 - \frac{\lambda_2}{\lambda_1}\right) - 471\left(\frac{\sigma_2 - \sigma_1}{\lambda_1{}^2}\right)\right]\left(\frac{v_\infty\lambda_1}{R}\right)^2 + \cdots\right\} \tag{8A}$$

[1] F. M. Leslie, *Quart. J. Mech. Appl. Math.*, *14*, 36–48 (1961), with an appendix by R. I. Tanner.
[2] H. Giesekus, *Rheol. Acta*, *3*, 59–71 (1963), see footnote 16 on p. 69. In this paper the problem [of si]multaneous translation and rotation of a sphere in a viscoelastic liquid is studied. The effect o[f] container boundaries on the movement of a particle through a non-Newtonian fluid has been st[udied] by B. Caswell, *Chem. Eng. Sci.*, *25*, 1167–1176 (1970).

Hence, the deviation from Stokes' law depends on the ratio[3] of a characteristic time of the fluid λ_1 to a characteristic time for the flow process R/v_∞.

a. From Eq. 8A.4–1 show that in applying the above result for F_k to spheres falling in viscoelastic fluids, there is an upper limit to the radius of the sphere that can be used and an upper limit to the speed of descent of the sphere.

b. Design an experiment for testing Eq. 8A.4–2 for a fluid with the following properties:

$$\eta_0 = 1.05\ \text{N·s/m}^2 \qquad \mu_0 = 0.05\ \text{s}$$
$$\lambda_1 = 0.1\ \text{s} \qquad \nu_1 = \nu_2 = 0$$
$$\lambda_2 = 0.02\ \text{s}$$

c. Can Philippoff's bouncing sphere experiment, described in Chapter 3, be understood in the light of Leslie's derivation?

d. Find an expression for the terminal velocity of the falling sphere.

8B.1 Elongational Viscosity for 8-Constant Oldroyd Fluid

a. Write out Eq. 8.1–2 in matrix form for steady-state elongational flow $v_z = \dot{\epsilon}z$, $v_x = -\frac{1}{2}\dot{\epsilon}x$, and $v_y = -\frac{1}{2}\dot{\epsilon}y$, where $\dot{\epsilon}$ is a constant.

b. From the matrix equation obtain a pair of equations for $\tau_{xx} - \tau_{zz}$ and $\text{tr}\,\boldsymbol{\tau} = \tau_{xx} + \tau_{yy} + \tau_{zz}$:

$$(1 - \mu_1\dot{\epsilon})(\tau_{xx} - \tau_{zz}) - (\tfrac{3}{2}\mu_0 - \mu_1)\dot{\epsilon}\ \text{tr}\,\boldsymbol{\tau} = 3\eta_0\dot{\epsilon}(1 - \mu_2\dot{\epsilon}) \tag{8B.1–1}$$

$$(2\mu_1 - 3\nu_1)\dot{\epsilon}(\tau_{xx} - \tau_{zz}) + \text{tr}\,\boldsymbol{\tau} = 3\eta_0\dot{\epsilon}^2(2\mu_2 - 3\nu_2) \tag{8B.1–2}$$

c. Solve the equations in (b) for $\tau_{xx} - \tau_{zz}$ and find that the elongational viscosity $\bar{\eta}$ is:

$$\bar{\eta} = 3\eta_0 \frac{1 - \mu_2\dot{\epsilon} + (\frac{3}{2}\mu_0 - \mu_1)(2\mu_2 - 3\nu_2)\dot{\epsilon}^2}{1 - \mu_1\dot{\epsilon} + (\frac{3}{2}\mu_0 - \mu_1)(2\mu_1 - 3\nu_1)\dot{\epsilon}^2} \tag{8B.1–3}$$

To what extent is this result in accord with known experimental facts? (Note that λ_1 and λ_2 do not appear in this result!)

8B.2 A Three-Constant Oldroyd Model[1]

The Oldroyd eight-constant model in §8.1 has often been used in simplified form by reducing the number of parameters. One simplification is to require that Ψ_2 be zero (this leads to $\lambda_1 = \mu_1$, $\lambda_2 = \mu_2$), and to require arbitrarily that $\boldsymbol{\tau}$ be traceless (this leads to $\nu_1 = \frac{2}{3}\lambda_1$, $\nu_2 = \frac{2}{3}\lambda_2$).

a. Show that this leads to the following expressions for the steady-shear flow functions:

$$\frac{\eta}{\eta_0} = \frac{1 + \frac{2}{3}\lambda_1\lambda_2\dot{\gamma}^2}{1 + \frac{2}{3}(\lambda_1\dot{\gamma})^2} \tag{8B.2–1}$$

[3] A similar ratio, based on the time constant from Ellis model parameters, was successfully used by R. M. Turian, *A.I.Ch.E. Journal*, *13*, 999–1006 (1967), to correlate friction factor data for spheres falling in several aqueous polymer solutions.

[1] M. C. Williams and R. B. Bird, *Phys. Fluids*, *5*, 1126–1127 (1962); the oscillatory normal stresses were given incorrectly in that publication, but were corrected in *Ind. Eng. Chem. Fundamentals*, *3*, 42–49 (1964).

$$\frac{\Psi_1}{2\eta_0\lambda_1} = \frac{1 - (\lambda_2/\lambda_1)}{1 + \frac{2}{3}(\lambda_1\dot{\gamma})^2} \tag{8B.2-2}$$

and that the small-amplitude oscillatory functions are given by Eqs. 8.1–10 and 11. What restrictions have to be placed on the time constants λ_1 and λ_2?

b. Show that $\eta(\dot{\gamma})$ is the same function as $\eta'(c\omega)$, and that $\Psi_1\dot{\gamma}/\sqrt{6}$ vs. $\dot{\gamma}$ is the same function as η'' vs. $c\omega$, with $c = \sqrt{3/2} = 1.24$. The constant c is called the "shift factor," and experimental values[2] lie between about 1.4 and 2.3.

c. An oft-quoted empiricism,[1] known as the *Cox-Merz rule* (cf. §4.7) states that $\eta(\dot{\gamma})$ should be the same function as $|\eta^*|$ vs. ω. To what extent is the three-constant Oldroyd model in agreement with this rule?

8B.3 Third-Order Retarded-Motion Constants Corresponding to the Oldroyd Six-Constant Model

Expand Eq. 8.1–2 (with $v_1 = v_2 = 0$) for slow flow by the following procedure:[3]

a. Recognize that the first-order approximation to Eq. 8.1–2 is $\boldsymbol{\tau} = -\eta_0\dot{\boldsymbol{\gamma}}$, inasmuch as $\mathscr{D}\dot{\boldsymbol{\gamma}}/\mathscr{D}t$ is one order higher in the velocity than $\dot{\boldsymbol{\gamma}}$ is. Insert this first-order expression into the second-order terms in Eq. 8.1–2 and obtain:

$$\boldsymbol{\tau} + \lambda_1 \frac{\mathscr{D}}{\mathscr{D}t}\boldsymbol{\tau} = -\eta_0\left[\dot{\boldsymbol{\gamma}} + \lambda_2\frac{\mathscr{D}}{\mathscr{D}t}\dot{\boldsymbol{\gamma}} + (\mu_1 - \mu_2)\{\dot{\boldsymbol{\gamma}}\cdot\dot{\boldsymbol{\gamma}}\}\right] \tag{8B.3-1}$$

b. Then operate on both sides of Eq. 8B.3–1 with the operator $(1 - \lambda_1\mathscr{D}/\mathscr{D}t)$ and retain only second-order terms. In this way the second-order approximation is obtained:

$$\boldsymbol{\tau} = -\eta_0\left[\dot{\boldsymbol{\gamma}} - (\lambda_1 - \lambda_2)\frac{\mathscr{D}}{\mathscr{D}t}\dot{\boldsymbol{\gamma}} + (\mu_1 - \mu_2)\dot{\boldsymbol{\gamma}}^2\right] \tag{8B.3-2}$$

c. Repeat the process to obtain:

$$\begin{aligned}\boldsymbol{\tau} + \lambda_1\frac{\mathscr{D}}{\mathscr{D}t}\boldsymbol{\tau} = -\eta_0\bigg[\dot{\boldsymbol{\gamma}} &+ \lambda_2\frac{\mathscr{D}}{\mathscr{D}t}\dot{\boldsymbol{\gamma}} + (\mu_1 - \mu_2)\{\dot{\boldsymbol{\gamma}}\cdot\dot{\boldsymbol{\gamma}}\} - \tfrac{1}{2}\mu_1(\lambda_1 - \lambda_2)\frac{\mathscr{D}}{\mathscr{D}t}\{\dot{\boldsymbol{\gamma}}\cdot\dot{\boldsymbol{\gamma}}\} \\ &+ \tfrac{1}{2}(\mu_1 - \mu_0)(\mu_1 - \mu_2)(\dot{\boldsymbol{\gamma}}:\dot{\boldsymbol{\gamma}})\dot{\boldsymbol{\gamma}}\bigg]\end{aligned} \tag{8B.3-3}$$

Then once again apply the operator $(1 - \lambda_1\mathscr{D}/\mathscr{D}t)$ and retain third-order terms. Do not forget the term $\lambda_1(\mathscr{D}^2/\mathscr{D}t^2)\boldsymbol{\tau}$ that now enters, and make use of the Cayley-Hamilton theorem. Then equate the coefficients to those in Eq. 8.4–3 to obtain:

$$\begin{aligned}\alpha_1 &= \eta_0 \\ \alpha_2 &= \eta_0(\lambda_1 - \lambda_2) \\ \alpha_3 &= \eta_0\lambda_1(\lambda_1 - \lambda_2) \\ \alpha_{11} &= \eta_0(\mu_1 - \mu_2) \\ \alpha_{12} &= \eta_0[\lambda_1(\mu_1 - \mu_2) + \tfrac{1}{2}\mu_1(\lambda_1 - \lambda_2)] \\ \alpha_{1:11} &= \tfrac{1}{2}\eta_0(\mu_1 - \mu_0)(\mu_1 - \mu_2)\end{aligned} \tag{8B.3-4}$$

[2] See M. C. Williams, *Chem. Eng. Sci.*, *20*, 693–702 (1965), Figs. 4a and b.
[3] K. Walters, *Zeits. für angew. Math. und Phys.*, *21*, 592–600 (1970), Sec. 5.

d. Invert the result in (c) to obtain:

$$\eta_0 = \alpha_1$$

$$\lambda_1 = \frac{\alpha_3}{\alpha_2}$$

$$\lambda_2 = \frac{\alpha_1\alpha_3 - {\alpha_2}^2}{\alpha_1\alpha_2}$$

$$\mu_1 = 2\left(\frac{\alpha_2\alpha_{12} - \alpha_3\alpha_{11}}{{\alpha_2}^2}\right)$$

$$\mu_2 = \frac{2\alpha_1\alpha_2\alpha_{12} - 2\alpha_1\alpha_3\alpha_{11} - \alpha_{11}{\alpha_2}^2}{\alpha_1{\alpha_2}^2}$$

$$\mu_0 = 2\left(\frac{\alpha_2\alpha_{12}\alpha_{11} - \alpha_3{\alpha_{11}}^2 - \alpha_{1:11}{\alpha_2}^2}{\alpha_{11}{\alpha_2}^2}\right) \tag{8B.3–5}$$

The above comparison shows that slow-flow hydrodynamic problems solved for the Oldroyd model can be taken over at once for the third-order fluid, and vice versa, just by appropriate replacement of constants.[3,4]

8B.4 A Modification of the ZFD Model

a. Recall that the ZFD model requires that $\Psi_2 = -\frac{1}{2}\Psi_1$. This defect of the model can be avoided by using the following modification:

$$\boldsymbol{\tau} = \boldsymbol{\tau}^{(1)} + \boldsymbol{\tau}^{(2)} \tag{8B.4–1}$$

$$\boldsymbol{\tau}^{(1)} + \lambda_0 \frac{\mathscr{D}}{\mathscr{D}t}\boldsymbol{\tau}^{(1)} = -\eta_0\dot{\boldsymbol{\gamma}} \tag{8B.4–2}$$

$$\boldsymbol{\tau}^{(2)} + \lambda_0 \frac{\mathscr{D}}{\mathscr{D}t}\boldsymbol{\tau}^{(2)} = \tfrac{1}{2}b\{\boldsymbol{\tau}^{(1)}\cdot\dot{\boldsymbol{\gamma}} + \dot{\boldsymbol{\gamma}}\cdot\boldsymbol{\tau}^{(1)}\} \tag{8B.4–3}$$

Show that for steady shear flow this model leads to the following results:

$$\frac{\eta}{\eta_0} = \frac{1}{1 + (\lambda_0\dot{\gamma})^2} \tag{8B.4–4}$$

$$\frac{\Psi_1}{2\eta_0\lambda_0} = \frac{1}{1 + (\lambda_0\dot{\gamma})^2} \tag{8B.4–5}$$

$$\frac{\Psi_2}{\eta_0\lambda_0} = -\frac{1 - (b/\lambda_0)}{1 + (\lambda_0\dot{\gamma})^2} \tag{8B.4–6}$$

Note that this same modification could be made in the generalized ZFD model.

b. Consider the integral rheological equation of state given by:

$$\boldsymbol{\tau} = \boldsymbol{\tau}^{(1)} + \boldsymbol{\tau}^{(2)} \tag{8B.4–7}$$

[4] See appendix in the publication by B. Caswell, *Chem. Eng. Sci.*, *25*, 1167–1176 (1970).

where

$$\boldsymbol{\tau}^{(1)} = -\int_{-\infty}^{t} \left\{ \frac{\eta_0}{\lambda_0} e^{-(t-t')/\lambda_0} \right\} \dot{\boldsymbol{\Gamma}}(t')\, dt' \tag{8B.4–8}$$

$$\boldsymbol{\tau}^{(2)} = -\tfrac{1}{2}b \int_{-\infty}^{t} \left\{ \frac{\eta_0}{\lambda_0^2} e^{-(t-t')/\lambda_0} \right\} \left[\int_{t'}^{t} \{\dot{\boldsymbol{\Gamma}}' \cdot \dot{\boldsymbol{\Gamma}}'' + \dot{\boldsymbol{\Gamma}}'' \cdot \dot{\boldsymbol{\Gamma}}'\}\, dt'' \right] dt' \tag{8B.4–9}$$

Show that this model is the same as that given in (a) in differential form. To carry out the proof, first convert the differential and integral forms into the corotating frame and perform the necessary integrations in terms of the corotating components.

8B.5 Tube Flow for CEF Equation

Show that for the axial flow in a circular tube, with $v_z = v_z(r)$, the CEF equation in Eq. 8.5–2 becomes:

$$\begin{pmatrix} \tau_{rr} & \tau_{r\theta} & \tau_{rz} \\ \tau_{\theta r} & \tau_{\theta\theta} & \tau_{\theta z} \\ \tau_{zr} & \tau_{z\theta} & \tau_{zz} \end{pmatrix} = +\eta \begin{pmatrix} 0 & 0 & 1 \\ 0 & 0 & 0 \\ 1 & 0 & 0 \end{pmatrix} \dot{\gamma} - (\tfrac{1}{2}\Psi_1 + \Psi_2) \begin{pmatrix} 1 & 0 & 0 \\ 0 & 0 & 0 \\ 0 & 0 & 1 \end{pmatrix} \dot{\gamma}^2 + \tfrac{1}{2}\Psi_1 \begin{pmatrix} 1 & 0 & 0 \\ 0 & 0 & 0 \\ 0 & 0 & -1 \end{pmatrix} \dot{\gamma}^2 \tag{8B.5–1}$$

in which $\dot{\gamma} = -dv_z/dr$ (why the minus sign?). Write out explicitly the components τ_{rz}, τ_{rr}, τ_{zz}.

8B.6 Helical Annular Flow for CEF Equation

Show how to set up the equations for obtaining the velocity distribution for flow in a horizontal annulus. The inner cylinder of radius κR is moving axially with a velocity v_0 and rotating with an angular velocity W. The outer cylinder of radius R is fixed. Assume that the viscosity function can be described by a power-law function. There is no pressure gradient in the z-direction.

a. Show that the z- and θ-components of the equation of motion are:

$$0 = -\frac{1}{r}\frac{d}{dr}(r\tau_{rz}) \tag{8B.6–1}$$

$$0 = -\frac{1}{r^2}\frac{d}{dr}(r^2\tau_{r\theta}) \tag{8B.6–2}$$

b. Show that the CEF equation gives the following expression for the stress tensor $\boldsymbol{\tau}$:

$$\boldsymbol{\tau} = \begin{pmatrix} \Psi_2(\dot{\gamma}_{r\theta}^2 + \dot{\gamma}_{rz}^2) & \eta\dot{\gamma}_{r\theta} & \eta\dot{\gamma}_{rz} \\ \eta\dot{\gamma}_{r\theta} & (\Psi_1 + \Psi_2)\dot{\gamma}_{r\theta}^2 & (\Psi_1 + \Psi_2)\dot{\gamma}_{rz}\dot{\gamma}_{r\theta} \\ \eta\dot{\gamma}_{rz} & (\Psi_1 + \Psi_2)\dot{\gamma}_{rz}\dot{\gamma}_{r\theta} & (\Psi_1 + \Psi_2)\dot{\gamma}_{rz}^2 \end{pmatrix} \tag{8B.6–3}$$

What are the expressions for the components of the rate-of-strain tensor that are to be inserted into Eq. 8B.6–3?

c. Show that the viscosity function is given by:

$$\eta = m\dot{\gamma}^{n-1} = m(\dot{\gamma}_{r\theta}^2 + \dot{\gamma}_{rz}^2)^{(n-1)/2} \tag{8B.6–4}$$

Note that the viscosity function depends on both velocity gradients. Hence the viscosity used in the axial equation of motion depends on how rapidly the inner cylinder is rotating.

d. Show that the equations of motion can be integrated with respect to r to give:

$$r \frac{dv_z}{dr}\left[\left(\frac{dv_z}{dr}\right)^2 + r^2\left(\frac{d}{dr}\left(\frac{v_\theta}{r}\right)\right)^2\right]^{(n-1)/2} = C_1 \tag{8B.6–5}$$

$$r^3 \frac{d}{dr}\left(\frac{v_\theta}{r}\right)\left[\left(\frac{dv_z}{dr}\right)^2 + r^2\left(\frac{d}{dr}\left(\frac{v_\theta}{r}\right)\right)^2\right]^{(n-1)/2} = C_2 \tag{8B.6–6}$$

Show that for $n = 1$ (Newtonian fluid), the equations for v_z and v_θ can be integrated independently, and that the velocity distribution is just a superposition of the flows found for purely axial flow and purely tangential flow.

e. Show that for $n \neq 1$, Eqs. 8B.6–5 and 6 can be solved for dv_z/dr and $(d/dr)(v_\theta/r)$ and the resulting equations integrated:

$$v_z = \int \left(\frac{C_1}{r}\right)^{1/n}\left[1 + \left(\frac{C_2}{C_1}\frac{1}{r}\right)^2\right]^{(1-n)/2n} dr + C_3 \tag{8B.6–7}$$

$$\frac{v_\theta}{r} = \int \left(\frac{C_2}{C_1}\frac{1}{r^2}\right)\left(\frac{C_1}{r}\right)^{1/n}\left[1 + \left(\frac{C_2}{C_1}\frac{1}{r}\right)^2\right]^{(1-n)/2n} dr + C_4 \tag{8B.6–8}$$

f. What conditions are used to determine the four constants of integration C_1, C_2, C_3, and C_4? To determine them numerical methods have to be used.[5]

8B.7 Centripetal Pumping between Parallel Disks

a. What does Eq. 8.5–24 become for a second-order fluid? To what extent would that result be useful?

b. Assume that $\Psi_2 = -\frac{1}{4}\Psi_1$ and that Ψ_1 may be given by a power law. In that case how can Eq. 8.5–24 be rewritten?

8B.8 Tangential Annular Flow for Second-Order Fluid

Consider the flow in Part (a) of Problem 1B.4, with the inner cylinder stationary. Obtain the pressure distribution in this system for the Newtonian fluid and also for the second-order fluid.

8B.9 Steady Simple Shear Flow of a Third-Order Fluid

Use the retarded-motion expansion in Eq. 8.4–3 to find the material functions for steady shear flow of the form $v_x = \dot{\gamma}y$. Show that the results through terms of third-order are:

$$\eta = \alpha_1 - (\alpha_3 - 2\alpha_{1:11})\dot{\gamma}^2 + \cdots \tag{8B.9–1}$$

$$\Psi_1 = 2\alpha_2 - \cdots \tag{8B.9–2}$$

$$\Psi_2 = (\alpha_{11} - \alpha_2) - \cdots \tag{8B.9–3}$$

Compare these expressions with those for the Oldroyd 8-constant model.

[5] A. C. Dierckes and W. R. Schowalter, *Ind. Eng. Chem. Fundamentals*, 5, 263–271 (1966).

8B.10 Relations among Limiting Values of Material Functions

In Table 6.2–1 a listing is given of "analogous quantities" that may be measured in a variety of rheological experiments. All of the entries there were justified in illustrative examples in §6.2 except for the last one, which involves normal stress. In order to include it we have to prove that:

$$\lim_{\dot{\gamma}\to 0} \frac{\Psi_1}{2\eta} = \frac{\int_0^\infty sG_I(s)\,ds}{\int_0^\infty G_I(s)\,ds} \tag{8B.10–1}$$

Show that this follows directly from Eqs. A and B of Table 8.3–1.

8C.1 Tube Flow of an Oldroyd 8-Constant Fluid[1,2]

Obtain the expression for the volume flow rate vs. pressure drop for the flow of the Oldroyd eight-constant fluid through a circular pipe. Use the following abbreviation for the groups of time constants appearing in Eq. 8.1–7:

$$\sigma_i = \lambda_1\lambda_i + \mu_0(\mu_i - \tfrac{3}{2}\nu_i) - \mu_1(\mu_i - \nu_i) \qquad (i = 1,2) \tag{8C.1–1}$$

Also use the following dimensionless quantities:

$$\Omega = \frac{3\sqrt{\sigma_1}\,Q}{\pi R^3} \tag{8C.1–2}$$

$$X = \sigma_1\dot{\gamma}_R{}^2 \qquad \left(\dot{\gamma}_R = -\left.\frac{dv_z}{dr}\right|_{r=R}\right) \tag{8C.1–3}$$

$$n = \frac{\sigma_2}{\sigma_1} \tag{8C.1–4}$$

Show that

$$\Omega = \left[1 - \frac{1}{2X^2}\left(\frac{1+X}{1+nX}\right)^3 f\right]\sqrt{X} \tag{8C.1–5}$$

in which

$$\begin{aligned} f = {} & \tfrac{1}{2}n^3X^2 - 3n^2(n-1)X + 3n(n-1)(2n-1)\ln(1+X) \\ & - \tfrac{1}{2}X\left(\frac{n-1}{1+X}\right)^2 [6n + (7n-1)X] \end{aligned} \tag{8C.1–6}$$

This gives us, in dimensionless form, the volume rate of flow in terms of the wall shear rate. Show further that the shear rate may be eliminated in favor of the pressure drop through the system by means of the relation:

$$\frac{\sqrt{\sigma_1}}{2\eta_0}\frac{\Delta\mathscr{P}}{L}R = \left(\frac{1+nX}{1+X}\right)\sqrt{X} \tag{8C.1–7}$$

To get Ω in terms of $\Delta\mathscr{P}/L$, Eqs. 8C.1–5, 6, and 7 have to be combined. When this is done numerically the plot in Fig. 8C.1 is obtained. To what extent does the result appear to be realistic?

[1] M. C. Williams and R. B. Bird, *A.I.Ch.E. Journal*, *8*, 378–382 (1962).
[2] K. Walters, *Arch. Rat. Mech. Anal.*, *9*, 411–414 (1962).

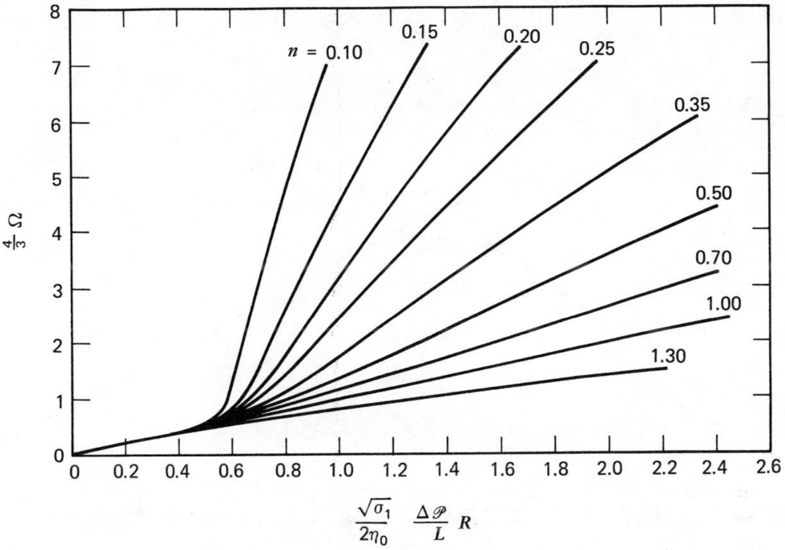

FIGURE 8C.1. Volumetric flow rate vs. pressure difference for an eight-constant Oldroyd liquid in a cylindrical pipe, for various values of $n = \sigma_2/\sigma_1$. The quantities σ_i and Ω are defined in Eqs. 8C.1–1 and 2. [M. C. Williams and R. B. Bird, *A.I.Ch.E. Journal*, *8*, 378–382 (1962).]

8C.2 Spriggs-Bird Two-Constant Model

Spriggs and Bird[3] considered the following nonlinear generalization of Eq. 6.1–11:

$$\left(1 + \sum_{n=1}^{\infty} a_n \mathscr{F}^n\right) \boldsymbol{\tau} = -\eta_0 \left(1 + \sum_{n=1}^{\infty} b_n \mathscr{F}^n\right) \dot{\boldsymbol{\gamma}} \tag{8C.2–1}$$

in which

$$\mathscr{F}\boldsymbol{\tau} = \frac{\mathscr{D}}{\mathscr{D}t}\boldsymbol{\tau} - \tfrac{1}{2}\{\dot{\boldsymbol{\gamma}} \cdot \boldsymbol{\tau} + \boldsymbol{\tau} \cdot \dot{\boldsymbol{\gamma}} - \tfrac{2}{3}(\boldsymbol{\tau} : \dot{\boldsymbol{\gamma}})\boldsymbol{\delta}\} \tag{8C.2–2}$$

By choosing a_n and b_n in a certain way (see Eq. 6C.1–7), the linear viscoelastic material functions of the Rouse theory can be very nearly reproduced.

Show that Eq. 8C.2–1 leads to the following expressions for η and Ψ_1:

$$\frac{\eta}{\eta_0} = \frac{PR + QS}{R^2 + S^2} \tag{8C.2–3}$$

$$\frac{\Psi_1 \dot{\gamma}}{\sqrt{6}\,\eta_0} = \frac{PS - QR}{R^2 + S^2} \tag{8C.2–4}$$

in which P, Q, R, and S are the functions given in Eqs. 6C.1–3 to 6, except that the argument ω is replaced by $\sqrt{\frac{2}{3}}\dot{\gamma}$. This suggests that η and η' should superpose if a "shift factor" of $\sqrt{\frac{2}{3}}$ is applied, and that a similar superposition of η'' and $\Psi_1\dot{\gamma}/\sqrt{6}$ should be possible. One comparison with experimental data is given in Fig. 8C.2; there Eqs. 6C.1–9 and 10 are used for η' and η'', and the same equations with ω replaced by $\sqrt{\frac{2}{3}}\dot{\gamma}$ are used for η and $\Psi_1\dot{\gamma}/\sqrt{6}$, respectively. The only parameters are the zero-shear-rate viscosity η_0 and the time constant λ.

[3] T. W. Spriggs and R. B. Bird, *Ind. Eng. Chem. Fundamentals*, *4*, 182–186 (1965). See also R. Roscoe, *Brit. J. Appl. Phys.*, *15*, 1095–1101 (1964) for a somewhat more general model.

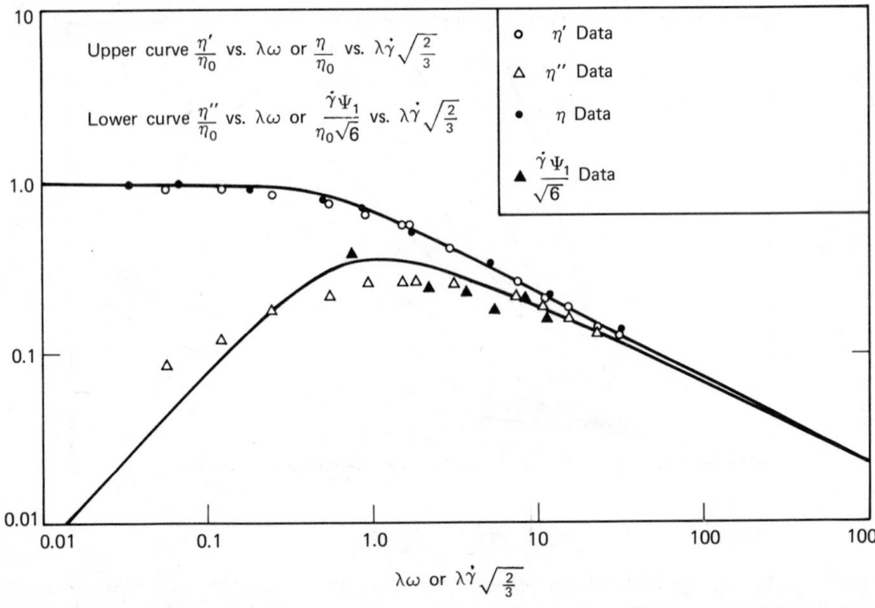

FIGURE 8C.2. Solid curves are functions obtained from the Spriggs-Bird model. The experimental data for η, η', and η'' for 5% polyisobutylene in decalin are taken from T. W. DeWitt, H. Markovitz, F. J. Padden, and L. J. Zapas, *J. Colloid Sci.*, *10*, 174–188 (1955), and the normal stress data are those of H. Markovitz and R. B. Williamson, *Trans. Soc. Rheol.*, *1*, 25–36 (1957). [Reprinted with permission from T. W. Spriggs and R. B. Bird, *Ind. Eng. Chem. Fundamentals*, *4*, 182–186 (1965). Copyright by the American Chemical Society.]

8C.3 A Rheological Equation of State Suggested from Theory of Suspensions

From a kinetic theory study of a dilute suspension of nearly spherical rigid particles, Leal and Hinch[4] have deduced a rheological equation of state of the form:

$$\boldsymbol{\tau} = -c_1\dot{\boldsymbol{\gamma}} - c_2\boldsymbol{\alpha} - c_3\{\dot{\boldsymbol{\gamma}}\cdot\boldsymbol{\alpha} + \boldsymbol{\alpha}\cdot\dot{\boldsymbol{\gamma}} - \tfrac{2}{3}(\boldsymbol{\alpha}:\dot{\boldsymbol{\gamma}})\boldsymbol{\delta}\} \tag{8C.3-1}$$

in which $\boldsymbol{\alpha}$ is a "structure tensor" that obeys the relation:

$$\frac{\mathscr{D}}{\mathscr{D}t}\boldsymbol{\alpha} + c_4\boldsymbol{\alpha} = c_5\dot{\boldsymbol{\gamma}} \tag{8C.3-2}$$

The c_i are positive constants related to the particle shape and concentration. What properties does this rheological equation of state have?

8C.4 Relation between Ψ_1 and η^-

Is Eq. C of Table 7.5–2 valid for the LeRoy and Pierrard equation (see Eq. 8.2–8)? Note that Eq. C of Table 7.5–2 is valid for the dilute suspension of rigid dumbbells as well as for dilute solutions of chain-like bead-spring models (see Example 12.4–1). What conclusions can be drawn from these facts?

[4] L. G. Leal and E. J. Hinch, *J. Fluid Mech.*, *55*, 745–765 (1972); Eqs. 7C.3–1 and 2 are a special case of a model by G. L. Hand, *J. Fluid Mech.*, *13*, 33–46 (1962). For an extended discussion of dilute suspension rheology and the Hand equation, see D. Barthès-Biesel and A. Acrivos, *Int. J. Multiphase Flow*, *1*, 1–24 (1973).

8C.5 Radial Flow between Parallel Disks

Work through the details of the papers by Schwarz and Bruce[5] and by Lee and Williams[5] on radial flow between fixed circular disks. What significant conclusions did the authors draw?

8C.6 The Giesekus-Tanner Theorem for Plane Creeping Flows of Incompressible Second-Order Fluids[6]

In §8.4 a statement of the Giesekus-Tanner theorem was given; here we work through the proof. For the plane flow $v_x(x,y)$, $v_y(x,y)$, $v_z = 0$, we introduce a stream function $\psi(x,y)$ so that $v_x = \partial\psi/\partial y$, $v_y = -\partial\psi/\partial x$.

a. Show that the nonzero stress components for the second-order fluid for plane flow may be written as:

$$\begin{aligned}\tau_{xx} = &-2\alpha_1\psi_{xy} - \alpha_{11}[4\psi_{xy}{}^2 + (\psi_{xx} - \psi_{yy})^2] \\ &+ \alpha_2[2(\psi_y\psi_{xxy} - \psi_x\psi_{xyy}) + (\psi_{xx}{}^2 - \psi_{yy}{}^2)]\end{aligned} \tag{8C.6–1}$$

$$\begin{aligned}\tau_{yy} = &+2\alpha_1\psi_{xy} - \alpha_{11}[4\psi_{xy}{}^2 + (\psi_{yy} - \psi_{xx})^2] \\ &+ \alpha_2[2(\psi_x\psi_{xyy} - \psi_y\psi_{xxy}) + (\psi_{yy}{}^2 - \psi_{xx}{}^2)]\end{aligned} \tag{8C.6–2}$$

$$\begin{aligned}\tau_{yx} = \tau_{xy} = &-\alpha_1(\psi_{yy} - \psi_{xx}) + \alpha_2\left[\left(\psi_y\frac{\partial}{\partial x} - \psi_x\frac{\partial}{\partial y}\right)(\psi_{yy} - \psi_{xx})\right. \\ &\left.+ 2\psi_{xy}(\psi_{xx} + \psi_{yy})\right]\end{aligned} \tag{8C.6–3}$$

where subscripts x and y on the stream function denote partial differentiation.

b. Show that for plane creeping flows the equation of motion may be written in terms of the stream function as:

$$\frac{\partial^2}{\partial x\,\partial y}(\psi_{xx} - \psi_{yy}) + \left(\frac{\partial^2}{\partial y^2} - \frac{\partial^2}{\partial x^2}\right)\psi_{xy} = 0 \tag{8C.6–4}$$

c. By combining the results of (a) and (b) show that

$$\alpha_1\,\nabla^4\psi + (\alpha_{11} + \alpha_2)\left(v_x\frac{\partial}{\partial x} + v_y\frac{\partial}{\partial y}\right)\nabla^4\psi = 0 \tag{8C.6–5}$$

d. Use the result in (c) to verify the Giesekus-Tanner theorem.

8C.7 Extrudate Swell for a Second-Order Fluid[7,8]

The phenomenon of extrudate swell is discussed in §3.5. Let us attempt to describe this effect by using the method of Problem 1D.6.

[5] W. H. Schwarz and C. Bruce, *Chem. Eng. Sci.*, *24*, 399–413 (1969); C. H. Lee and M. C. Williams, *J. Non-Newtonian Fluid Mech.*, *1*, 000–000 (1976).

[6] H. Giesekus, *Rheol. Acta*, *3*, 59–71 (1963); R. I. Tanner, *Phys. Fluids*, *9*, 1246–1247 (1966).

[7] R. B. Bird, R. K. Prud'homme, and M. Gottlieb, Rheology Research Center Report No. 35, University of Wisconsin, Madison, May 1975.

[8] For an excellent summary and critique of errors in the rheology literature pertaining to extrudate swell see J. M. Davies, J. F. Hutton, and K. Walters, *J. Phys. D.: Appl. Phys.*, *6*, 2259–2266 (1973). See also R. R. Huilgol, *Continuum Mechanics of Viscoelastic Liquids*, Wiley, New York (1975), pp. 326–340.

a. Inasmuch as $\langle \pi_{zz} \rangle_1$ and $\langle \pi_{zz} v_z \rangle_1$ are not known and are virtually impossible to estimate, we rewrite them as:

$$\langle \pi_{zz} \rangle_1 = \langle \pi_{zz} - \pi_{rr} \rangle_1 + \langle \pi_{rr} \rangle_1 = -\langle \Psi_1 \dot{\gamma}^2 \rangle_1 + \langle \pi_{rr} \rangle_1 \tag{8C.7–1}$$

$$\langle \pi_{zz} v_z \rangle_1 = \langle (\pi_{zz} - \pi_{rr}) v_z \rangle_1 + \langle \pi_{rr} v_z \rangle_1 = -\langle \Psi_1 \dot{\gamma}^2 v_z \rangle_1 + \alpha \langle \pi_{rr} \rangle_1 \langle v_z \rangle_1 \tag{8C.7–2}$$

in which α is defined as:

$$\alpha = \frac{\langle \pi_{rr} v_z \rangle_1}{\langle \pi_{rr} \rangle_1 \langle v_z \rangle_1} \tag{8C.7–3}$$

Use the r-component of the equation of motion to show that $\alpha = 1$ if $\Psi_2 = 0$.

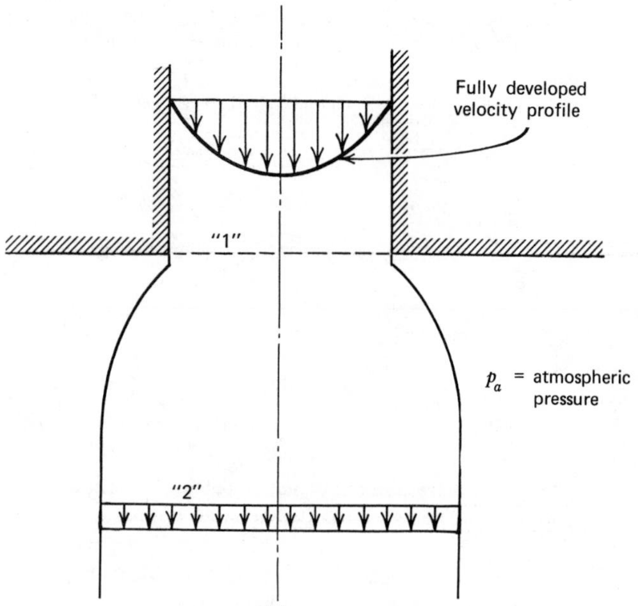

(a)

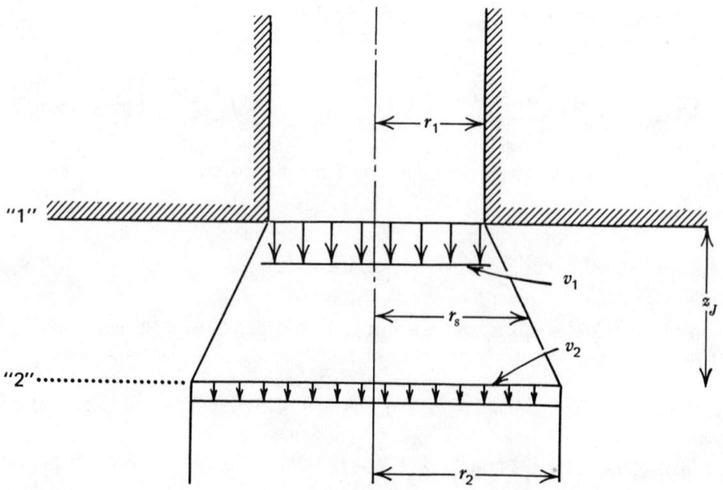

(b)

FIGURE 8C.7. (*a*) Velocity profile rearrangement in extrudate swell (see §3.5 for photograph); (*b*) crude velocity pattern and extrudate shape for trial calculation. The jet surface radius is taken to be $r_s = r_1 + (r_2 - r_1)(z/z_J)$, and z_J is approximated as r_1.

b. Use Eqs. 8C.7–1 and 2 in Eqs. 1D.6–5 and 6 and eliminate $\langle \pi_{rr} \rangle_1$ to obtain:

$$2\alpha \frac{\langle v_z^2 \rangle_1}{\langle v_z \rangle_1^2} - \frac{\langle v_z^3 \rangle_1}{\langle v_z \rangle_1^3} - 2\alpha \frac{S_1}{S_2} + \left(\frac{S_1}{S_2}\right)^2 - \frac{\alpha \langle \Psi_1 \dot{\gamma}^2 \rangle_1}{\frac{1}{2}\rho \langle v_z \rangle_1^2} + \frac{\langle \Psi_1 \dot{\gamma}^2 v_z \rangle_1}{\frac{1}{2}\rho \langle v_z \rangle_1^3} - \frac{p_a(\alpha - 1)}{\frac{1}{2}\rho \langle v_z \rangle_1^2} + e_{ve} = 0 \quad (8C.7–4)$$

in which the dimensionless quantity e_{ve} is defined as:

$$e_{ve} = \frac{-\int_V (\boldsymbol{\tau} : \nabla \boldsymbol{v})\, dV}{\frac{1}{2}\langle v_z \rangle_1^2 \cdot \rho \langle v_z \rangle_1 S_1} \quad (8C.7–5)$$

where V is the volume of the liquid in the jet between planes "1" and "2." Equation 8C.7–4 includes the assumption that there is no velocity profile rearrangement prior to the tube exit.

From Eq. 8C.7–4 we can draw some conclusions:

1. S_2/S_1 depends on Ψ_1 and also slightly on Ψ_2; it also depends on η, which is needed to calculate the various average quantities. In addition it may depend on other rheological parameters that appear in e_{ve}, since a rheological equation of state enters in its evaluation. Hence it appears difficult to try to extract trustworthy Ψ_1 values from extrudate swell.
2. To estimate S_2/S_1 from Eq. 8C.7–4 requires that we be able to estimate e_{ve}. This offers a major obstacle since calculation of e_{ve} necessitates knowing the rheological equation of state and the details of the velocity distribution in the extrudate. Nonetheless Eq. 8C.7–4 may be useful in designing and interpreting extrudate-swell experiments.
3. To illustrate the difficulty we consider a second order fluid with $\Psi_2 = 0$:

$$\boldsymbol{\tau} = -\eta \dot{\boldsymbol{\gamma}} + \tfrac{1}{2}\Psi_1 \left(\dot{\boldsymbol{\gamma}}^2 + \frac{\mathscr{D}}{\mathscr{D}t} \dot{\boldsymbol{\gamma}} \right) \quad (8C.7–6)$$

where η and Ψ_1 are constants. Then if the true flow pattern of Fig. 8C.7*a* is replaced by the fictitious pattern in Fig. 8C.7*b*, with z_J arbitrarily taken to be the same as r_1, show that Eq. 8C.7–4 gives the following equation for S_2/S_1 (lengthy!):

$$\left(\frac{r_1 v_1 \rho}{\eta}\right)\left[\frac{2}{3} - 2\left(\frac{S_1}{S_2}\right) + \left(\frac{S_1}{S_2}\right)^2\right] = \frac{16}{3}\left(\frac{\Psi_1 v_1}{\eta r_1}\right) - 8\left(\sqrt{\frac{S_2}{S_1}} - 1\right)\left(1 + \frac{3}{8}\left(\sqrt{\frac{S_2}{S_1}} - 1\right)\right)^2 \left[1 - \left(\frac{S_1}{S_2}\right)^{3/2}\right]$$
$$- 2\left(\frac{\Psi_1 v_1}{\eta r_1}\right)\left(\sqrt{\frac{S_2}{S_1}} - 1\right)^2 \left[1 - \left(\frac{S_1}{S_2}\right)^3\right] \quad (8C.7–7)$$

Show that in the limit of small Reynolds number $r_1 v_1 \rho/\eta$ and high elasticity number $\Psi_1 v_1/\eta r_1$, the extrudate swell ratio S_2/S_1 is greater than 6.

8D.1 Proof of the Giesekus-Tanner-Pipkin Equation (Eq. 8.4–5)[9]

For creeping flow the equations of motion for incompressible Newtonian and second-order fluids are, in the absence of external forces, respectively:

$$\mathbf{0} = -\nabla p_N + [\nabla \cdot \alpha_1 \dot{\boldsymbol{\gamma}}] \quad (8D.1–1)$$

$$\mathbf{0} = -\nabla p + \left[\nabla \cdot \left(\alpha_1 \dot{\boldsymbol{\gamma}} - \alpha_2 \frac{\mathscr{D}}{\mathscr{D}t}\dot{\boldsymbol{\gamma}} + \alpha_{11}\{\dot{\boldsymbol{\gamma}} \cdot \dot{\boldsymbol{\gamma}}\}\right)\right] \quad (8D.1–2)$$

[9] H. Giesekus, *Rheol. Acta*, *3*, 59–71 (1963); A. C. Pipkin, *Lectures in Viscoelasticity Theory*, Springer, New York (1972), pp. 149–153; R. I. Tanner and A. C. Pipkin, *Trans. Soc. Rheol.*, *13*, 471–484 (1969).

It was shown by the Giesekus-Tanner theorem (see Problem 8C.6) that for *plane* creeping flows, the Newtonian velocity profile satisfies the second-order fluid problem with the same velocity boundary conditions. However, the pressure distributions in the Newtonian fluid (p_N) and in the second-order fluid (p) will be different. We wish here to transform Eq. 8D.1–2 into the form:

$$\mathbf{0} = -\nabla(p + p') + [\nabla \cdot \alpha_1 \dot{\boldsymbol{\gamma}}] \tag{8D.1–3}$$

so that p can be determined from $p = p_N - p'$. Hence it is a question of showing that the divergences of $\mathscr{D}\dot{\boldsymbol{\gamma}}/\mathscr{D}t$ and $\{\dot{\boldsymbol{\gamma}} \cdot \dot{\boldsymbol{\gamma}}\}$ can be written as gradients of scalars.

a. Verify that for any *planar* flow

$$\{\dot{\boldsymbol{\gamma}} \cdot \dot{\boldsymbol{\gamma}}\} = (\boldsymbol{\delta} - \boldsymbol{\delta}_z\boldsymbol{\delta}_z)\dot{\gamma}^2 \tag{8D.1–4}$$

where $\boldsymbol{\delta}$ is the unit tensor, $\boldsymbol{\delta}_z$ is the unit vector in the z-direction which is normal to the plane of flow, and $\dot{\gamma}$ is defined just after Eq. 8.4–5. Then show that

$$[\nabla \cdot \{\dot{\boldsymbol{\gamma}} \cdot \dot{\boldsymbol{\gamma}}\}] = \nabla\dot{\gamma}^2 \tag{8D.1–5}$$

for planar flows.

b. Next show that for *any* flow

$$\left[\nabla \cdot \frac{\mathscr{D}}{\mathscr{D}t}\dot{\boldsymbol{\gamma}}\right] = \frac{D}{Dt}[\nabla \cdot \dot{\boldsymbol{\gamma}}] + [\nabla \boldsymbol{v} : \nabla\dot{\boldsymbol{\gamma}}] + \tfrac{1}{2}[\nabla \cdot \{\boldsymbol{\omega} \cdot \dot{\boldsymbol{\gamma}} - \dot{\boldsymbol{\gamma}} \cdot \boldsymbol{\omega}\}] \tag{8D.1–6}$$

Then use Eq. 8D.1–1 to rewrite this result as:

$$\left[\nabla \cdot \frac{\mathscr{D}}{\mathscr{D}t}\dot{\boldsymbol{\gamma}}\right] = \frac{1}{\alpha_1}\nabla\left(\frac{D}{Dt}p_N\right) - [\nabla \boldsymbol{v} \cdot [\nabla \cdot \dot{\boldsymbol{\gamma}}]] + [\nabla \boldsymbol{v} : \nabla\dot{\boldsymbol{\gamma}}] + \tfrac{1}{2}[\nabla \cdot \{\boldsymbol{\omega} \cdot \dot{\boldsymbol{\gamma}} - \dot{\boldsymbol{\gamma}} \cdot \boldsymbol{\omega}\}] \tag{8D.1–7}$$

Then replace $\nabla \boldsymbol{v}$ by $\frac{1}{2}(\dot{\boldsymbol{\gamma}} + \boldsymbol{\omega})$ and rearrange to get:

$$\left[\nabla \cdot \frac{\mathscr{D}}{\mathscr{D}t}\dot{\boldsymbol{\gamma}}\right] = \frac{1}{\alpha_1}\nabla\left(\frac{D}{Dt}p_N\right) + \tfrac{1}{2}[\dot{\boldsymbol{\gamma}} : \nabla(\dot{\boldsymbol{\gamma}} - \boldsymbol{\omega})] - \tfrac{1}{2}[[\nabla \cdot (\dot{\boldsymbol{\gamma}} - \boldsymbol{\omega})] \cdot \dot{\boldsymbol{\gamma}}] \tag{8D.1–8}$$

Show that the last term is zero [since $(\nabla \cdot \boldsymbol{v}) = 0$] and that the penultimate term can be rearranged as $[\dot{\boldsymbol{\gamma}} : \nabla(\nabla \boldsymbol{v})^\dagger] = \frac{1}{4}\nabla(\dot{\boldsymbol{\gamma}} : \dot{\boldsymbol{\gamma}})$ so that finally:

$$\left[\nabla \cdot \frac{\mathscr{D}}{\mathscr{D}t}\dot{\boldsymbol{\gamma}}\right] = \frac{1}{\alpha_1}\nabla\left(\frac{D}{Dt}p_N\right) + \tfrac{1}{2}\nabla\dot{\gamma}^2 \tag{8D.1–9}$$

c. Use the results of (a) and (b) to obtain Eq. 8.4–5.

CHAPTER 9

CODEFORMATIONAL MODELS

In the preceding two chapters we presented the corotational models. Our motivation for introducing these models was twofold: (i) we wanted to find a rheological equation of state capable of describing the non-Newtonian viscosity of macromolecular fluids as well as their time-dependent linear viscoelastic properties, and (ii) we wanted the rheological equations of state to be objective. The corotational formalism was found to be capable of satisfying these two requirements. In particular we showed that objectivity is automatically insured if we formulate rheological equations of state as relations between a corotating stress tensor $\mathbf{T}(t,t')$ and a corotating rate-of-strain tensor $\dot{\mathbf{\Gamma}}(t,t')$. The reader may wonder whether the corotational framework is the only vantage point from which we may satisfy the two requirements listed above. Indeed the corotational rheological tensors do *not* play such a unique role; rather there exists a wide variety of stress, strain, and rate-of-strain tensors from which objective rheological equations of state may be generated.

In this chapter we introduce the reader to two alternative sets of tensors that we call the "codeformational rheological tensors." With these sets of tensors, both differential and integral codeformational models can be constructed by following the same procedures used in Chapters 7 and 8. We shall see that differential operations in the corotational and codeformational formalisms are simply related. As a result, all of the *differential models* in the two preceding chapters are easily written in terms of codeformational tensors. In fact, most of these models were first proposed in terms of the codeformational formalism, and many of the hydrodynamics problems that we have chosen to discuss previously in conjunction with corotational models were originally solved by using codeformational models. Since the introduction of the codeformational differential models does not supply any new input for problem solving, our primary motivation for including this material is to provide a guide to the research literature, the bulk of which has been written in the codeformational language in recent years.

On the other hand, the tensor quantities involved in corotational and codeformational *integral models* are not related in a simple, closed form. As a result, corotational and codeformational integral models of similar complexity—for example, each quasilinear in its natural basis—give quite different predictions for fluid properties. Thus, the codeformational integral models may be preferred in making certain hydrodynamic calculations, for instance, those closely resembling elongational flow. We give particular emphasis to the codeformational integral models in this chapter and try to bring out the essential differences between them and their corotational counterparts.

In many ways Chapter 9 may be regarded as a parallel to Chapters 7 and 8. Thus in §9.1 we parallel §7.1 in that we describe how the codeformational rheological framework may be used to generate objective rheological equations of state. Without giving any of the mathematical development we complete this section with two tables analogous to Tables 7.1–1 and 2 with interrelations between the codeformational rheological tensors. These tables are arranged in such a way that the reader familiar with Tables 7.1–1 and 2 may also use these tables in a similar handbook fashion to generate objective rheological equations of state and perform manipulations with them. The reader not particularly interested in the mathematical theory of these tables may then proceed to §9.4 and read the remainder of the chapter. On the other hand, the reader desirous of understanding the basis for the codeformational description of rheology should read §§9.2 and 3.

In §9.4 we introduce specific codeformational rheological models and explore their properties in simple flows. Here we give comparisons with experimental data for specific models, and illustrate the process of model testing. In §9.5 we introduce the codeformational analogs of the memory integral expansion presented in Chapter 8. Finally in §9.6 we introduce curvilinear coordinates and solve a simple hydrodynamical problem in curvilinear coordinates.

§9.1 BASIC CONCEPTS OF THE CODEFORMATIONAL FORMALISM

In this section we introduce the main concepts in the codeformational description of rheology. We begin by presenting in Tables 9.1–1 and 2 the basic codeformational quantities:[1] the stress tensors $\boldsymbol{\tau}^{[0]}$ and $\boldsymbol{\tau}_{[0]}$ and their time derivatives, and the strain tensors $\boldsymbol{\gamma}^{[0]}$ and $\boldsymbol{\gamma}_{[0]}$ and their time derivatives. The essence of the codeformational rheological description is that use of the tensors in columns 2 and 2′ of the tables,[2] $\boldsymbol{\tau}^{[n]}$, $\boldsymbol{\tau}_{[n]}$, $\boldsymbol{\tau}^{(n)}$, $\boldsymbol{\tau}_{(n)}$, $\boldsymbol{\gamma}^{[n]}$ . . . , to construct a rheological equation of state guarantees that the resulting equation is objective. A proof of this statement can be found in §9.2. Note that quantities in column 2′ involve a past time t' and should therefore appear inside integrals, whereas quantities from column 2 are evaluated at the present time t and may appear outside integrals. Although these tables are written in general tensor notation in which summations are implied on repeated indices, they may be translated into the familiar Cartesian tensor notation (as used in Tables 7.1–1 and 2) in which summation signs are included explicitly by performing the changes outlined in note (a) to the tables.

In Chapter 7 we used linear viscoelastic models from Chapter 6 as a guide in constructing rheological equations of state. When the linear viscoelastic models were reformulated in terms of the corotational quantities $\mathbf{T}$ and $\dot{\mathbf{\Gamma}}$ by using Tables 7.1–1 and 2, the quasilinear corotational models were obtained. We can follow a similar procedure to construct quasilinear codeformational models. However, in attempting to use Tables 9.1–1 and 2 to replace linear viscoelastic quantities found in columns 3 and 3′ with codeformational tensors in columns 2 and 2′, we discover an interesting fact: there is a choice in the replacement process. For example, $\dot{\boldsymbol{\gamma}}' = \dot{\boldsymbol{\gamma}}(x',t')$ may be replaced by either $\boldsymbol{\gamma}^{[1]\prime} = \boldsymbol{\gamma}^{[1]}(x,t,t')$ or $\boldsymbol{\gamma}'_{[1]} = \boldsymbol{\gamma}_{[1]}(x,t,t')$. That is, there is no unique way in which a given rheological relation, such as the linear viscoelastic model, can be made objective. Rather there are many ways in which this can be done, and all these predict different behavior outside the linear range. Continuum mechanics alone has no way of guiding us in this choice. Only comparisons with experimental data or with molecular theory can yield clues as to which stress and strain tensors will give rheological equations of state that correspond to real fluids.

In order to use Tables 9.1–1 and 2 one must be able to calculate tensors of the kind $\boldsymbol{\Lambda}^{(n)}$ and $\boldsymbol{\Lambda}_{(n)}$, where $\boldsymbol{\Lambda}$ may be $\boldsymbol{\tau}$ or $\dot{\boldsymbol{\gamma}}$, and also the tensors $\boldsymbol{\Delta}$ and $\mathbf{E}$. We shall now define these quantities and tell in what ways the formalism contained in Tables 9.1–1 and 2 is analogous to the corotational formalism.

[1] Tables 9.1–1 and 2 are based on the discussion by J. G. Oldroyd, *Proc. Roy. Soc.*, *A200*, 523–541 (1950).

[2] Note from Tables 9.1–1 and 2 that the difference between parentheses and brackets around a subscript or superscript is very important, and that, for example $\boldsymbol{\tau}_{(1)}$ and $\boldsymbol{\tau}_{[1]}$ have very different meanings.

The tensors $\boldsymbol{\tau}^{(n)}$, $\boldsymbol{\tau}_{(n)}$, $\boldsymbol{\gamma}^{(n)}$, and $\boldsymbol{\gamma}_{(n)}$ are defined by the recurrence relations

$$\boldsymbol{\Lambda}^{(n+1)} = \frac{D}{Dt}\boldsymbol{\Lambda}^{(n)} + \{\boldsymbol{\Lambda}^{(n)} \cdot (\nabla \boldsymbol{v})^{\dagger} + (\nabla \boldsymbol{v}) \cdot \boldsymbol{\Lambda}^{(n)}\} \tag{9.1-1}$$

$$\boldsymbol{\Lambda}_{(n+1)} = \frac{D}{Dt}\boldsymbol{\Lambda}_{(n)} - \{\boldsymbol{\Lambda}_{(n)} \cdot (\nabla \boldsymbol{v}) + (\nabla \boldsymbol{v})^{\dagger} \cdot \boldsymbol{\Lambda}_{(n)}\} \tag{9.1-2}$$

which are valid for any tensor $\boldsymbol{\Lambda}$, and the statements that $\boldsymbol{\tau}^{(0)} = \boldsymbol{\tau}_{(0)} = \boldsymbol{\tau}$, $\boldsymbol{\gamma}^{(1)} = \boldsymbol{\gamma}_{(1)} = \dot{\boldsymbol{\gamma}}$. The operations of increasing the index in the parenthesis by one are the codeformational analogs[3] of the Jaumann differentiation defined in Eq. 7.1–9.

To explain the meaning of the tensors $\boldsymbol{\Delta}$ and $\mathbf{E}$ we first introduce the concept of *displacement functions.* Imagine a fluid flowing in a region with a space-fixed Cartesian coordinate system. We center our attention on a fluid particle P. At the present time t this particle has coordinates x_i where $i = 1, 2, 3$. At a past time t' the particle P had coordinates x_i' where $i = 1, 2, 3$. Now the displacement functions for the particle P are the functions giving the past coordinates of the particle in terms of the present coordinates, the present time, and the past time, that is, the functions (see §7.2):

$$x_i' = x_i'(x,t,t') \tag{9.1-3}$$

Here x is used as an abbreviation for the set of coordinates x_1, x_2, x_3. If we are given the displacement functions x_i' for all values of the present positions then we know the entire flow history for the fluid. The functions in Eq. 9.1–3 may be inverted formally to give

$$x_i = x_i(x',t,t') \tag{9.1-4}$$

It is then convenient to define the tensors $\boldsymbol{\Delta}$ and $\mathbf{E}$ with Cartesian components[4]

$$\Delta_{ij} = \frac{\partial x_i'}{\partial x_j} \tag{9.1-5}$$

$$\mathrm{E}_{ij} = \frac{\partial x_i}{\partial x_j'} \tag{9.1-6}$$

With this notation the entries in columns 2′ may be calculated by matrix multiplications as illustrated in Example 9.4–1. The tensors $\boldsymbol{\Delta}$ and $\mathbf{E}$ may be thought of as deformation gradients; they are the codeformational analogs of the $\boldsymbol{\Omega}$ tensor in the corotational formalism. Note that $\mathbf{E} = \boldsymbol{\Delta}^{-1}$.

Although the most direct way of computing $\boldsymbol{\Delta}$ and $\mathbf{E}$ in Cartesian components is by their definitions in Eqs. 9.1–5 and 6, we note here that formal expansions may be given for

[3] The symbol $\mathfrak{d}/\mathfrak{d}t$ is sometimes used to indicate the codeformational differentiation in both Eqs. 9.1–1 and 2. This symbol is somewhat ambiguous, since the two operations defined in Eqs. 9.1–1 and 2 are very different. For this reason we prefer the notation in Eqs. 9.1–1 and 2.

[4] In curvilinear and nonorthogonal coordinates the mixed components of the tensors $\boldsymbol{\Delta}$ and $\mathbf{E}$ are given by $\Delta^i{}_j = x'^i{}_{,j}$ and $\mathrm{E}^i{}_j = x^i{}_{,j'}$, where the "$,j'$" notation stands for covariant differentiation with respect to x'^j. For the purpose of computing the tensors in column 2′ of Tables 9.1–1 and 2, it is most convenient to substitute the matrix elements $\partial x'^i/\partial x^j$ and $\partial x^i/\partial x'^j$ for the tensor components $\Delta^i{}_j$ and $\mathrm{E}^i{}_j$. This may be done even in curvilinear and nonorthogonal coordinates (see §9.3).

TABLE 9.1–1

Relations among Codeformational Stress Tensors and Their Derivatives[a]

(3′) Small displacements		(2′) Fixed components		(1′) Convected components at t'	(1) Convected components at t		(2) Fixed components		(3) Small displacements
$\tau_{ij}(x,t')$ $\tau^{ij}(x,t')$	$\leftarrow$	$\tau_{ij}^{[0]}(x,t,t') = \Delta^m{}_i \tau_{mn}^{(0)}(x',t')\, \Delta^n{}_j$ $\tau_{[0]}^{ij}(x,t,t') = \mathrm{E}^i{}_m \tau_{(0)}^{mn}(x',t')\mathrm{E}^j{}_n$	$\xleftarrow{t}$	$\hat{\tau}_{ij}(\hat{x},t')$ $\hat{\tau}^{ij}(\hat{x},t')$	$\hat{\tau}_{ij}(\hat{x},t)$ $\hat{\tau}^{ij}(\hat{x},t)$	$\xrightarrow{t}$	$\tau_{ij}^{(0)}(x,t) = \tau_{ij}(x,t)$ $\tau_{(0)}^{ij}(x,t) = \tau^{ij}(x,t)$		
$\downarrow \dfrac{\partial}{\partial t'}$		$\downarrow \dfrac{\partial}{\partial t'}$		$\downarrow \dfrac{\partial}{\partial t'}$	$\downarrow \dfrac{\partial}{\partial t}$		$\downarrow \dfrac{\mathfrak{D}}{\mathfrak{D}t}$		$\downarrow \dfrac{\partial}{\partial t}$
$\dfrac{\partial}{\partial t'}\tau_{ij}(x,t')$ $\dfrac{\partial}{\partial t'}\tau^{ij}(x,t')$	$\leftarrow$	$\tau_{ij}^{[1]}(x,t,t') = \Delta^m{}_i \tau_{mn}^{(1)}(x',t')\, \Delta^n{}_j$ $\tau_{[1]}^{ij}(x,t,t') = \mathrm{E}^i{}_m \tau_{(1)}^{mn}(x',t')\mathrm{E}^j{}_n$	$\xleftarrow{t}$	$\dfrac{\partial \hat{\tau}_{ij}}{\partial t'}$ $\dfrac{\partial \hat{\tau}^{ij}}{\partial t'}$	$\dfrac{\partial \hat{\tau}_{ij}}{\partial t}$ $\dfrac{\partial \hat{\tau}^{ij}}{\partial t}$	$\xrightarrow{t}$	$\tau_{ij}^{(1)}(x,t)$ $\tau_{(1)}^{ij}(x,t)$	$\rightarrow$	$\dfrac{\partial}{\partial t}\tau_{ij}(x,t)$ $\dfrac{\partial}{\partial t}\tau^{ij}(x,t)$
$\downarrow \dfrac{\partial}{\partial t'}$		$\downarrow \dfrac{\partial}{\partial t'}$		$\downarrow \dfrac{\partial}{\partial t'}$	$\downarrow \dfrac{\partial}{\partial t}$		$\downarrow \dfrac{\mathfrak{D}}{\mathfrak{D}t}$		$\downarrow \dfrac{\partial}{\partial t}$
$\dfrac{\partial^2}{\partial t'^2}\tau_{ij}(x,t')$ $\dfrac{\partial^2}{\partial t'^2}\tau^{ij}(x,t')$	$\leftarrow$	$\tau_{ij}^{[2]}(x,t,t') = \Delta^m{}_i \tau_{mn}^{(2)}(x',t')\, \Delta^n{}_j$ $\tau_{[2]}^{ij}(x,t,t') = \mathrm{E}^i{}_m \tau_{(2)}^{mn}(x',t')\mathrm{E}^j{}_n$	$\xleftarrow{t}$	$\dfrac{\partial^2 \hat{\tau}_{ij}}{\partial t'^2}$ $\dfrac{\partial^2 \hat{\tau}^{ij}}{\partial t'^2}$	$\dfrac{\partial^2 \hat{\tau}_{ij}}{\partial t^2}$ $\dfrac{\partial^2 \hat{\tau}^{ij}}{\partial t^2}$	$\xrightarrow{t}$	$\tau_{ij}^{(2)}(x,t)$ $\tau_{(2)}^{ij}(x,t)$	$\rightarrow$	$\dfrac{\partial^2}{\partial t^2}\tau_{ij}(x,t)$ $\dfrac{\partial^2}{\partial t^2}\tau^{ij}(x,t)$

Bottom connectors: (3′) → (3), (2′) → (2), (1′) → (1), each labelled $t' = t$.

[a] See notes at end of Table 9.1–2.

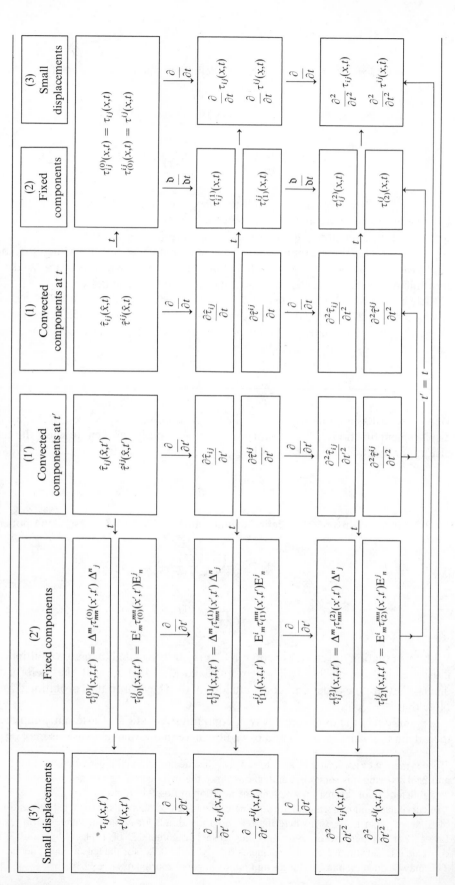

TABLE 9.1–2

Relations among Codeformational Strain Tensors and Their Derivatives

(3′) Small displacements	(2′) Fixed components	(1′) Convected components at t'	(1) Convected components at t	(2) Fixed components	(3) Small displacements
$\gamma_{ij}(x,t,t') = (u_{i,j} + u_{j,i})$ $\gamma^{ij}(x,t,t') = g^{ir}g^{js}(u_{r,s} + u_{s,r})$	$\gamma_{ij}^{[0]}(x,t,t') = +[\Delta^m{}_i g_{mn}(x')\Delta^n{}_j - g_{ij}(x)]$ $\gamma^{ij}_{[0]}(x,t,t') = -[\mathrm{E}^i{}_m g^{mn}(x')\mathrm{E}^j{}_n - g^{ij}(x)]$	$+[\hat{g}_{ij}(\hat{x},t') - \hat{g}_{ij}(\hat{x},t)]$ $-[\hat{g}^{ij}(\hat{x},t') - \hat{g}^{ij}(\hat{x},t)]$			
$\downarrow \frac{\partial}{\partial t'}$	$\downarrow \frac{\partial}{\partial t'}$	$\downarrow \frac{\partial}{\partial t'}$			
$\dot{\gamma}_{ij}(x,t') = (v_{i,j} + v_{j,i})$ $\dot{\gamma}^{ij}(x,t') = g^{ir}g^{js}(v_{r,s} + v_{s,r})$	$\gamma_{ij}^{[1]}(x,t,t') = \Delta^m{}_i \gamma_{mn}^{(1)}(x',t')\Delta^n{}_j$ $\gamma^{ij}_{[1]}(x,t,t') = \mathrm{E}^i{}_m \gamma^{mn}_{(1)}(x',t')\mathrm{E}^j{}_n$	$+\frac{\partial \hat{g}_{ij}}{\partial t'}$ $-\frac{\partial \hat{g}^{ij}}{\partial t'}$	$+\frac{\partial \hat{g}_{ij}}{\partial t}$ $-\frac{\partial \hat{g}^{ij}}{\partial t}$	$\gamma_{ij}^{(1)}(x,t) = \dot{\gamma}_{ij}(x,t)$ $\gamma^{ij}_{(1)}(x,t) = \dot{\gamma}^{ij}(x,t)$	
$\downarrow \frac{\partial}{\partial t'}$	$\downarrow \frac{\partial}{\partial t'}$	$\downarrow \frac{\partial}{\partial t'}$	$\downarrow \frac{\partial}{\partial t}$	$\downarrow \frac{\mathfrak{D}}{\mathfrak{D}t}$	$\downarrow \frac{\partial}{\partial t}$
$\ddot{\gamma}_{ij}(x,t') = \frac{\partial}{\partial t}\dot{\gamma}_{ij}$ $\ddot{\gamma}^{ij}(x,t') = \frac{\partial}{\partial t}\dot{\gamma}^{ij}$	$\gamma_{ij}^{[2]}(x,t,t') = \Delta^m{}_i \gamma_{mn}^{(2)}(x',t')\Delta^n{}_j$ $\gamma^{ij}_{[2]}(x,t,t') = \mathrm{E}^i{}_m \gamma^{mn}_{(2)}(x',t')\mathrm{E}^j{}_n$	$+\frac{\partial^2 \hat{g}_{ij}}{\partial t'^2}$ $-\frac{\partial^2 \hat{g}^{ij}}{\partial t'^2}$	$+\frac{\partial^2 \hat{g}_{ij}}{\partial t^2}$ $-\frac{\partial^2 \hat{g}^{ij}}{\partial t^2}$	$\gamma_{ij}^{(2)}(x,t)$ $\gamma^{ij}_{(2)}(x,t)$	$\ddot{\gamma}_{ij}(x,t)$ $\ddot{\gamma}^{ij}(x,t)$

Horizontal arrows labelled t lead from column (1′) to (2′), from (2′) to (3′), and from (1) to (2); columns (3′), (2′), (1′) are linked to (3), (2), (1) by $t' = t$.

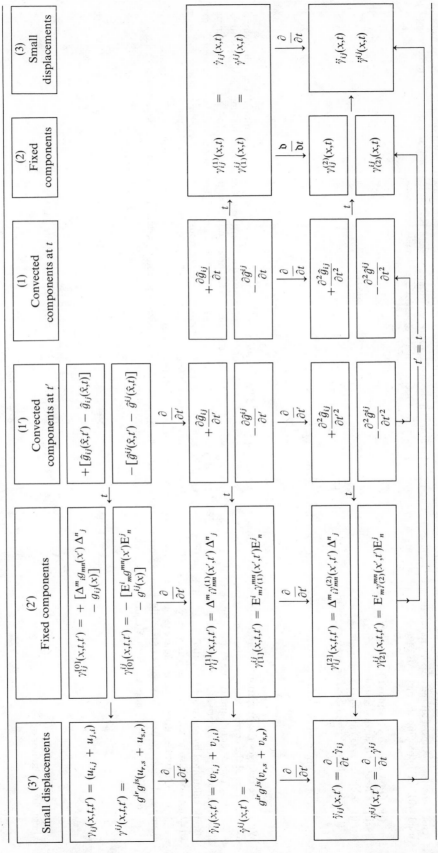

Notes to Tables 9.1–1 and 2.

a. Tables 9.1–1 and 2 are the codeformational analogs of Tables 7.1–1 and 2. They differ from Tables 7.1–1 and 2 in that the notation here is that of general tensor analysis. Almost all the important relations in Tables 9.1–1 and 2 are retained if the space fixed coordinate system is rectangular Cartesian. The tables may be used without employing the methods of general tensor analysis by converting them into rectangular Cartesian coordinates as follows:

(i) Omit columns (1) and (1′) and notes *e* and *g* below.
(ii) Change all superscripts to subscripts.
(iii) Replace $g_{ij}(x)$ and $g_{ij}(x')$ everywhere by δ_{ij}.
(iv) Introduce a summation on every index that occurs twice in a term.

For example, in Cartesian coordinates the first entry in column (2′) of Table 9.1–2 reads

$$\gamma_{ij}^{[0]}(x,t,t') = \textstyle\sum_m \Delta_{mi}\,\Delta_{mj} - \delta_{ij} \tag{A}$$

With these changes the operation for proceeding down in columns (2′) remains partial differentiation with respect to t', and the entries in column (2) are still obtained as the values of the corresponding entries in column (2′) with $t' = t$. As in Tables 7.1–1 and 2 the time t is regarded as the "present time" and the time t' is a "past time."

b. The $\Delta^i{}_j(x,t,t')$ and $\mathrm{E}^i{}_j(x,t,t')$ are defined by $\Delta^i{}_j = x'^i{}_{,j}$ and $\mathrm{E}^i{}_j = x^i{}_{,j'}$ where "$,j'$" denotes covariant differentiation with respect to x'^j. In Cartesian coordinates these reduce to partial derivatives: $\Delta_{ij} = \partial x_i'/\partial x_j$ and $\mathrm{E}_{ij} = \partial x_i/\partial x_j'$. Formal solutions for $\mathbf{\Delta}$ and $\mathbf{E}$ are given in Eqs. 9.1–7 and 8. The tensor entries in columns (2′) may also be calculated by replacing $\Delta^i{}_j$ with $\partial x'^i/\partial x^j$ and $\mathrm{E}^i{}_j$ with $\partial x^i/\partial x'^j$. These latter forms are the most convenient for computations. Here we use the notation that $x'^i = x^i(x,t,t')$; note also $u^i(x,t,t')$ is defined as $u^i = x'^i - x^i$, and $u_i = g_{ij}u^j$; furthermore $v^i(x',t')$ is defined as $(\partial/\partial t')x'^i$, and $v_i = g_{ij}v^j$.

c. Quantities appearing in the same box are components of the same tensor; quantities appearing in different boxes are components of different tensors. For example, $\gamma_{ij}^{[0]}$ and $\gamma_{[0]}^{ij}$ are components of two different tensors $\boldsymbol{\gamma}^{[0]}$ and $\boldsymbol{\gamma}_{[0]}$, whereas $\gamma_{ij}^{(1)}$ and $\gamma_{(1)}^{ij}$ are covariant and contravariant components of the same tensor $\dot{\boldsymbol{\gamma}}$.

d. The tensors $\mathbf{\Lambda}^{(n)}$ and $\mathbf{\Lambda}_{(n)}$, with $\mathbf{\Lambda}$ equal to $\boldsymbol{\tau}$ or $\dot{\boldsymbol{\gamma}}$, are given by the recurrence relations:

$$\mathbf{\Lambda}^{(n+1)} = \frac{D}{Dt}\mathbf{\Lambda}^{(n)} + \{(\nabla \boldsymbol{v})\cdot\mathbf{\Lambda}^{(n)} + \mathbf{\Lambda}^{(n)}\cdot(\nabla \boldsymbol{v})^\dagger\} = \frac{\mathscr{D}}{\mathscr{D}t}\mathbf{\Lambda}^{(n)} + \tfrac{1}{2}\{\dot{\boldsymbol{\gamma}}\cdot\mathbf{\Lambda}^{(n)} + \mathbf{\Lambda}^{(n)}\cdot\dot{\boldsymbol{\gamma}}\} \tag{B}$$

$$\mathbf{\Lambda}_{(n+1)} = \frac{D}{Dt}\mathbf{\Lambda}_{(n)} - \{(\nabla \boldsymbol{v})^\dagger\cdot\mathbf{\Lambda}_{(n)} + \mathbf{\Lambda}_{(n)}\cdot(\nabla \boldsymbol{v})\} = \frac{\mathscr{D}}{\mathscr{D}t}\mathbf{\Lambda}_{(n)} - \tfrac{1}{2}\{\dot{\boldsymbol{\gamma}}\cdot\mathbf{\Lambda}_{(n)} + \mathbf{\Lambda}_{(n)}\cdot\dot{\boldsymbol{\gamma}}\} \tag{C}$$

where, for any tensor $\boldsymbol{\sigma}$:

$$\frac{\mathscr{D}}{\mathscr{D}t}\boldsymbol{\sigma} = \frac{D}{Dt}\boldsymbol{\sigma} + \tfrac{1}{2}\{\boldsymbol{\omega}\cdot\boldsymbol{\sigma} - \boldsymbol{\sigma}\cdot\boldsymbol{\omega}\} \tag{D}$$

The rate-of-strain and vorticity tensors are defined by:

$$\dot{\boldsymbol{\gamma}} = (\nabla \boldsymbol{v}) + (\nabla \boldsymbol{v})^\dagger \tag{E}$$

$$\boldsymbol{\omega} = (\nabla \boldsymbol{v}) - (\nabla \boldsymbol{v})^\dagger \tag{F}$$

and

The operations of increasing the index in a parenthesis by one are the codeformational analogs of the Jaumann differentiation introduced in Chapter 7. Note that the two operations of codeformational differentiation may be expressed in terms of the Jaumann derivative $\mathscr{D}/\mathscr{D}t$ and products involving the rate-of-strain tensor.

e. The quantities Λ_{ij} and Λ^{ij} are the covariant and contravariant components of the second-order tensor $\mathbf{\Lambda}$ with respect to the space fixed coordinate system; Λ_{ij} and Λ^{ij} are related via the components of the metric tensor $g_{ij}(x)$ associated with the space fixed coordinates x^i. Similarly $\hat{\Lambda}_{ij}$ and $\hat{\Lambda}^{ij}$ are related via the components of the metric tensor $\hat{g}_{ij}(\hat{x},t)$ associated with the convected coordinates $\hat{x}^i$. Note that the components of the metric tensor $\hat{g}_{ij}(\hat{x},t)$ are functions of time.

f. Any of the tensors $\Lambda_{ij}^{[n]}$ or $\Lambda_{[n]}^{ij}$ may be expanded in a Taylor series about $t' = t$. For example,

$$\gamma_{ij}^{[0]}(x,t,t') = \sum_{n=0}^{\infty} \frac{(t'-t)^n}{n!} \gamma_{ij}^{[n]}(x,t,t')\big|_{t'=t} = \sum_{n=1}^{\infty} \frac{(t'-t)^n}{n!} \gamma_{ij}^{(n)}(x,t) \tag{H}$$

g. The symbol $\overset{t}{\rightarrow}$ is Lodge's "transfer operator" [see A. S. Lodge, *Elastic Liquids*, Academic Press, New York (1964)]. It serves as a reminder that the convected components are referred to the space fixed coordinate system at time t.

h. The $\gamma_{[1]}$ and $\gamma^{[1]}$ tensors may be expressed in terms of the $\dot{\mathbf{\Gamma}}$ tensor of Chapter 7 by the following expressions derived by Hassager [see R. B. Bird, O. Hassager, and S. I. Abdel-Khalik, *A.I.Ch.E. Journal*, *20*, 1041–1066 (1974)]:

$$\gamma_{[1]} = \{\mathbf{\Theta} \cdot \dot{\mathbf{\Gamma}} \cdot \mathbf{\Theta}^{\dagger}\} \tag{I}$$

$$\gamma^{[1]} = \{\mathbf{\Phi}^{\dagger} \cdot \dot{\mathbf{\Gamma}} \cdot \mathbf{\Phi}\} \tag{J}$$

in which the tensors $\mathbf{\Theta}(x,t,t')$ and $\mathbf{\Phi}(x,t,t') = \mathbf{\Theta}^{-1}$ may be expressed by:

$$\mathbf{\Theta} = \boldsymbol{\delta} + \sum_{n=1}^{\infty} \mathbf{\Theta}_n \tag{K}$$

$$\mathbf{\Theta}_n = \left(\frac{1}{2}\right)^n \int_{t'}^{t} \int_{t'}^{t_1} \cdots \int_{t'}^{t_{n-1}} \{\dot{\mathbf{\Gamma}}_1 \cdot \dot{\mathbf{\Gamma}}_2 \cdot \cdots \dot{\mathbf{\Gamma}}_n\}\, dt_n \cdots dt_2\, dt_1 \tag{L}$$

and

$$\mathbf{\Phi} = \boldsymbol{\delta} + \sum_{n=1}^{\infty} \mathbf{\Phi}_n \tag{M}$$

$$\mathbf{\Phi}_n = \left(\frac{1}{2}\right)^n \int_{t}^{t'} \int_{t}^{t_1} \cdots \int_{t}^{t_{n-1}} \{\dot{\mathbf{\Gamma}}_1 \cdot \dot{\mathbf{\Gamma}}_2 \cdot \cdots \dot{\mathbf{\Gamma}}_n\}\, dt_n \cdots dt_2\, dt_1 \tag{N}$$

in which $\dot{\mathbf{\Gamma}}_i = \dot{\mathbf{\Gamma}}(x,t,t_i)$.

i. These tables are based on the discussion by J. G. Oldroyd, *Proc. Roy. Soc.*, *A200*, 45–63 (1950).

them that are similar to Eq. 7.1–11 for $\boldsymbol{\Omega}$:

$$\boldsymbol{\Delta} = \boldsymbol{\delta} - \int_{t'}^{t} (\nabla \boldsymbol{v})^{\dagger}(x,t,t'')\,dt'' + \int_{t'}^{t}\int_{t''}^{t} \{(\nabla \boldsymbol{v})^{\dagger}(x,t,t'') \cdot (\nabla \boldsymbol{v})^{\dagger}(x,t,t''')\}\,dt'''\,dt'' - \cdots \qquad (9.1\text{–}7)$$

$$\mathbf{E} = \boldsymbol{\delta} + \int_{t'}^{t} (\nabla \boldsymbol{v})^{\dagger}(x,t,t'')\,dt'' + \int_{t'}^{t}\int_{t''}^{t} \{(\nabla \boldsymbol{v})^{\dagger}(x,t,t''') \cdot (\nabla \boldsymbol{v})^{\dagger}(x,t,t'')\}\,dt'''\,dt'' + \cdots \qquad (9.1\text{–}8)$$

In these equations, it can be seen that $(\nabla \boldsymbol{v})^{\dagger} = \frac{1}{2}\{\dot{\boldsymbol{\gamma}} - \boldsymbol{\omega}\}$ occupies somewhat the same role as $\boldsymbol{\omega}$ in the corotational expression. We emphasize that the above are not particularly useful for computing $\boldsymbol{\Delta}$ and $\mathbf{E}$; they are mainly useful in obtaining formal relations among the various kinematic tensors in the codeformational framework.

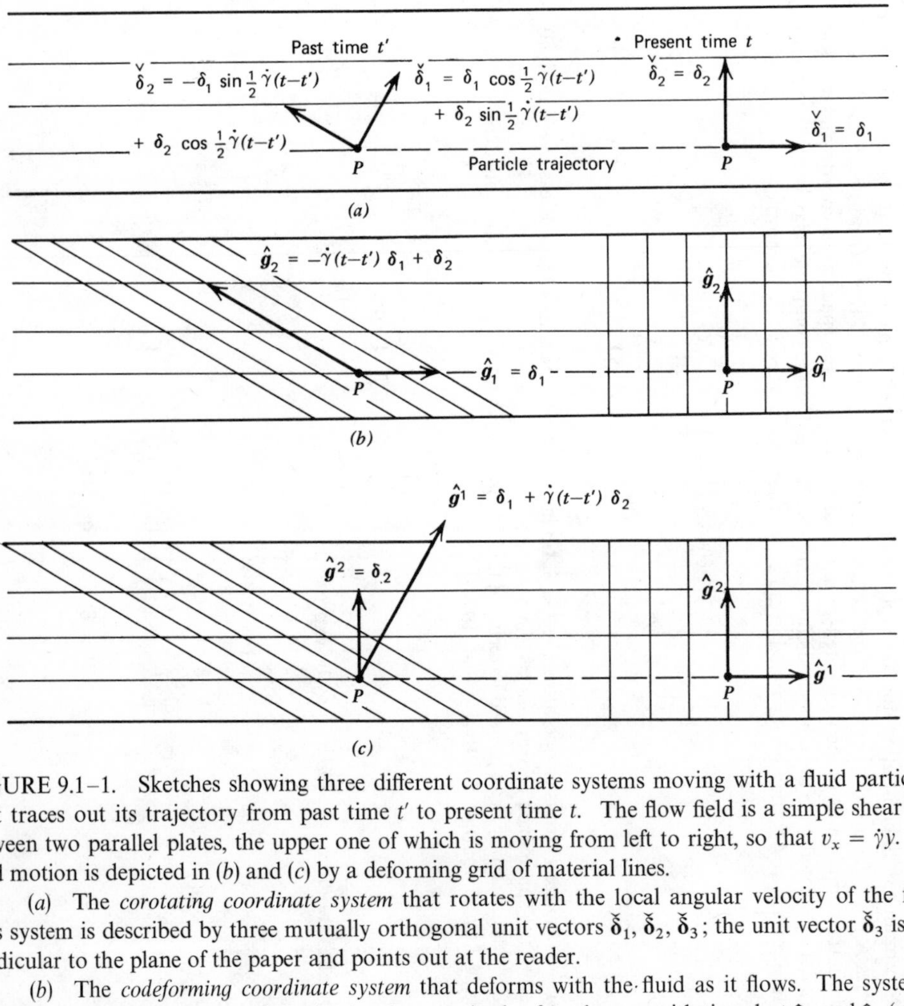

FIGURE 9.1–1. Sketches showing three different coordinate systems moving with a fluid particle P as it traces out its trajectory from past time t' to present time t. The flow field is a simple shear flow between two parallel plates, the upper one of which is moving from left to right, so that $v_x = \dot{\gamma}y$. The fluid motion is depicted in (b) and (c) by a deforming grid of material lines.

(a) The *corotating coordinate system* that rotates with the local angular velocity of the fluid. This system is described by three mutually orthogonal unit vectors $\check{\boldsymbol{\delta}}_1$, $\check{\boldsymbol{\delta}}_2$, $\check{\boldsymbol{\delta}}_3$; the unit vector $\check{\boldsymbol{\delta}}_3$ is perpendicular to the plane of the paper and points out at the reader.

(b) The *codeforming coordinate system* that deforms with the fluid as it flows. The system is described by three "base vectors" $\hat{\boldsymbol{g}}_1, \hat{\boldsymbol{g}}_2, \hat{\boldsymbol{g}}_3$; the magnitude of $\hat{\boldsymbol{g}}_2$ changes with time, but $\hat{\boldsymbol{g}}_1$ and $\hat{\boldsymbol{g}}_3$ (which is perpendicular to the plane of the paper) are of unit length. Note that $\hat{\boldsymbol{g}}_1$, $\hat{\boldsymbol{g}}_2$, and $\hat{\boldsymbol{g}}_3$ are "embedded" in the fluid.

(c) The *coordinate system described by the three "reciprocal base vectors"* $\hat{\boldsymbol{g}}^1$, $\hat{\boldsymbol{g}}^2$, $\hat{\boldsymbol{g}}^3$ as defined by Eq. A.8–2. Here $\hat{\boldsymbol{g}}^3$ (perpendicular to the plane of the paper) and $\hat{\boldsymbol{g}}^2$ are of unit length, but $\hat{\boldsymbol{g}}^1$ (which is perpendicular to the plane of $\hat{\boldsymbol{g}}_2$ and $\hat{\boldsymbol{g}}_3$) has a magnitude that changes with time.

Note that all three coordinate systems are so chosen that they coincide with the triad $\boldsymbol{\delta}_1$, $\boldsymbol{\delta}_2$, $\boldsymbol{\delta}_3$ of a Cartesian coordinate system at time t.

By way of further orientation we compare the corotational and codeformational viewpoints by means of Fig. 9.1–1 and Table 9.1–3. Recall that we began the discussion in Chapter 7 by describing the *corotating unit vectors* $\breve{\boldsymbol{\delta}}_1, \breve{\boldsymbol{\delta}}_2, \breve{\boldsymbol{\delta}}_3$ as a triad of mutually orthogonal unit vectors that rotate with the local fluid angular velocity as described by Eq. A of Table 9.1–3. Here we introduce analogously the *codeforming base vectors* shown in Fig. 9.1–1*b*, whose time rate of change following the particle is described by Eq. B in Table 9.1–3. In the figure we show only an elementary flow; for complex flows the base vectors are defined by Eq. A.8–1 in the codeforming coordinate system, but Eq. B is valid for any flow. It is also customary to introduce the reciprocal base vectors shown in Fig. 9.1–1*c* and described by Eq. C.

In Chapter 7 we described the orientation of the corotating triad of unit vectors by Ω_{ij} defined by Eq. D, the Ω_{ij} being given by the solution to a set of differential equations (Eq. G). Similar definitions can be made in the codeformational description, the quantities Δ_{ij} and E_{ij} fulfilling roles analogous to the Ω_{ij}. It is evident from the table and the figure that to every equation in the corotational formalism there will be two equations in the codeformational formalism. The reason for the dual description in the codeformational viewpoint can be traced to the existence of two sets of base vectors instead of the one set of corotating unit vectors.

In the codeformational description there are two approaches to the subject, the one described in §9.2 and the other in §9.3. In the first approach the quantities $\boldsymbol{\Delta}$ and $\mathbf{E}$ arise

TABLE 9.1–3

Comparison of Corotating Unit Vectors with Codeforming Base Vectors and Associated Kinematic Quantities

$\frac{\partial}{\partial t'}\breve{\boldsymbol{\delta}}_i(x,t,t') = -\frac{1}{2}[\boldsymbol{\omega}(x,t,t')\cdot\breve{\boldsymbol{\delta}}_i(x,t,t')]$ (A) (7.2–25)	$\frac{\partial}{\partial t'}\hat{\boldsymbol{g}}_i(\hat{x},t') = [\hat{\boldsymbol{g}}_i(\hat{x},t')\cdot\nabla\boldsymbol{v}(\hat{x},t')]$ (B) $\frac{\partial}{\partial t'}\hat{\boldsymbol{g}}^i(\hat{x},t') = -[\nabla\boldsymbol{v}(\hat{x},t')\cdot\hat{\boldsymbol{g}}^i(\hat{x},t')]$ (C)
$\breve{\boldsymbol{\delta}}_i(x,t,t') = \sum_j \Omega_{ij}(x,t,t')\boldsymbol{\delta}_j$ $\breve{\boldsymbol{\delta}}_i(x,t,t) = \boldsymbol{\delta}_i$ (D) (7.2–23 and 26)	$\hat{\boldsymbol{g}}_i(\hat{x},t') = \sum_j \boldsymbol{\delta}_j\,\Delta_{ji}(x,t,t')$ $\hat{\boldsymbol{g}}_i(\hat{x},t) = \boldsymbol{\delta}_i$ (E) $\hat{\boldsymbol{g}}^i(\hat{x},t') = \sum_j \mathrm{E}_{ij}(x,t,t')\boldsymbol{\delta}_j$ $\hat{\boldsymbol{g}}^i(\hat{x},t) = \boldsymbol{\delta}_i$ (F)
$\frac{\partial}{\partial t'}\Omega_{ij}(x,t,t') = \frac{1}{2}\sum_n \Omega_{in}(x,t,t')\omega_{nj}(x,t,t')$ $\Omega_{ij}(x,t,t) = \delta_{ij}$ (G) (7.2–28 and 29)	$\frac{\partial}{\partial t'}\Delta_{ij}(x,t,t') = \sum_n (\nabla\boldsymbol{v})_{in}^{\dagger}(x,t,t')\,\Delta_{nj}(x,t,t')$ $\Delta_{ij}(x,t,t) = \delta_{ij}$ (H) (9.2–16 and 17) $\frac{\partial}{\partial t'}\mathrm{E}_{ij}(x,t,t') = -\sum_n \mathrm{E}_{in}(x,t,t')(\nabla\boldsymbol{v})_{nj}^{\dagger}(x,t,t')$ (I) $\mathrm{E}_{ij}(x,t,t) = \delta_{ij}$ (9.2–18 and 19)

naturally in describing the local deformation of the fluid, and two strain tensors $\boldsymbol{\gamma}^{[0]}$ and $\boldsymbol{\gamma}_{[0]}$ are defined in terms of them; to give the time rates of change of $\boldsymbol{\Delta}$ and $\mathbf{E}$, Eqs. H and I are then derived and these relations are needed for obtaining expressions for the rate-of-strain tensor and higher-order kinematic tensors. In the second approach one works directly with the dot products of the base vectors and reciprocal base vectors in Eqs. B and C, $\hat{g}_{ij} = (\hat{\boldsymbol{g}}_i \cdot \hat{\boldsymbol{g}}_j)$ and $\hat{g}^{ij} = (\hat{\boldsymbol{g}}^i \cdot \hat{\boldsymbol{g}}^j)$, which can be seen in Figs. 9.1–1*b* and *c* to give a description of strain that is independent of the local rate of rotation of the fluid. No further use is made of the other entries on the right side of the table; they are presented here merely for reference. Thus the top entries on the right side of the table provide a link with §9.3 whereas the lower entries provide the connection with §9.2. The two approaches of §§9.2 and 9.3 are clearly closely related, and the entries in Table 9.1–3 provide the necessary formulas for interrelating them.

We conclude this section with an example that illustrates the operations involved in manipulating rheological equations of state.

EXAMPLE 9.1–1 Integration of the Differential Forms of Oldroyd's Fluids A and B

The following two possible generalizations of the Jeffreys model of linear viscoelasticity have been proposed by Oldroyd[1] (see Table 8.1–1):

$$\boldsymbol{\tau}^{(0)} + \lambda_1 \boldsymbol{\tau}^{(1)} = -\eta_0(\boldsymbol{\gamma}^{(1)} + \lambda_2 \boldsymbol{\gamma}^{(2)}) \qquad \text{(A)} \quad (9.1\text{–}9)$$

$$\boldsymbol{\tau}_{(0)} + \lambda_1 \boldsymbol{\tau}_{(1)} = -\eta_0(\boldsymbol{\gamma}_{(1)} + \lambda_2 \boldsymbol{\gamma}_{(2)}) \qquad \text{(B)} \quad (9.1\text{–}10)$$

These two models have become known as "Oldroyd's Fluids A and B", respectively. It is desired to rearrange these equations into forms explicit in the stress.

SOLUTION

The rheological tensors in Eqs. 9.1–9 and 10 are defined in columns (2) of Tables 9.1–1 and 2. These tensors are all evaluated at the present time t. For the purpose of manipulating the equations, however, it is easier first to rewrite these tensors as the values at $t' = t$ of the corresponding entries in columns (2′) of Tables 9.1–1 and 2. For instance Oldroyd's fluid A may be expressed as follows:

$$\boldsymbol{\tau}^{[0]}\big|_{t'=t} + \lambda_1 \frac{\partial}{\partial t'} \boldsymbol{\tau}^{[0]}\big|_{t'=t} = -\eta_0 \left(\boldsymbol{\gamma}^{[1]}\big|_{t'=t} + \lambda_2 \frac{\partial}{\partial t'} \boldsymbol{\gamma}^{[1]}\big|_{t'=t} \right) \qquad (9.1\text{–}11)$$

Here we have also used the information given in Tables 9.1–1 and 2 that $\boldsymbol{\tau}^{[1]} = (\partial/\partial t')\boldsymbol{\tau}^{[0]}$ and $\boldsymbol{\gamma}^{[2]} = (\partial/\partial t')\boldsymbol{\gamma}^{[1]}$. We see that Eq. 9.1–11 is a first order, linear, inhomogeneous differential equation in the tensor $\boldsymbol{\tau}^{[0]}$ over the interval $-\infty < t' \leq t$ following a particle. Hence we may solve for $\boldsymbol{\tau}^{[0]}$ using the initial condition that $\boldsymbol{\tau}^{[0]}$ is finite at $t' = -\infty$. If $\boldsymbol{\tau}^{[0]} = \boldsymbol{\tau}$ at $t' = t$, we find that:

$$\begin{aligned} \boldsymbol{\tau}(x,t) &= -\int_{-\infty}^{t} G(t - t')\boldsymbol{\gamma}^{[1]}(x,t,t')\, dt' \\ &= +\int_{-\infty}^{t} M(t - t')\boldsymbol{\gamma}^{[0]}(x,t,t')\, dt' \end{aligned} \qquad \text{(A)} \quad (9.1\text{–}12)$$

Similarly for Oldroyd's fluid B we find:

$$\begin{aligned} \boldsymbol{\tau}(x,t) &= -\int_{-\infty}^{t} G(t - t')\boldsymbol{\gamma}_{[1]}(x,t,t')\, dt' \\ &= +\int_{-\infty}^{t} M(t - t')\boldsymbol{\gamma}_{[0]}(x,t,t')\, dt' \end{aligned} \qquad \text{(B)} \quad (9.1\text{–}13)$$

where[5], in both models,

$$G(t-t') = \frac{\eta_0}{\lambda_1}\left[\left(1-\frac{\lambda_2}{\lambda_1}\right)e^{-(t-t')/\lambda_1} + 2\lambda_2\delta(t-t')\right] \tag{9.1-14}$$

$$M(t-t') = \frac{\eta_0}{\lambda_1^2}\left[\left(1-\frac{\lambda_2}{\lambda_1}\right)e^{-(t-t')/\lambda_1} + 2\lambda_1\lambda_2\frac{\partial}{\partial t'}\delta(t-t')\right] \tag{9.1-15}$$

The integral forms of Oldroyd's fluids A and B have been given both in a rate-of-strain form (involving the $\boldsymbol{\gamma}^{[1]}$ and $\boldsymbol{\gamma}_{[1]}$ tensor, respectively) and in a strain form (involving the $\boldsymbol{\gamma}^{[0]}$ and $\boldsymbol{\gamma}_{[0]}$ tensor, respectively). These forms differ only by an integration by parts. It is worthwhile to point out also that the $\boldsymbol{\gamma}^{[0]}$ and $\boldsymbol{\gamma}_{[0]}$ tensors have no analogs in the corotational formulation. For this reason the corotational Jeffreys model may not be put in a strain integral form similar to the strain integral forms of Oldroyd's fluids A and B.

§9.2 KINEMATICAL AND DYNAMICAL CODE FORMATIONAL TENSORS AND OBJECTIVITY[1]

In the corotational formalism, the basic kinematical quantities are the velocity gradient tensor, the rate-of-deformation tensor, and the vorticity tensor. In Chapter 7 we began by defining these tensors and then proceeded to show how their components as seen by observers in the corotating and fixed frames are related. In the codeformational formalism, on the other hand, the basic kinematical quantities are the displacement functions. From these we define deformation tensors, which are in turn used to define components of finite strain tensors and stress tensors. Time derivatives of the codeformational strain and stress tensors, defined in this way, are then introduced. We next show that all the codeformational tensors in Tables 9.1–1 and 2 are objective. Since all the codeformational rheological tensors can be defined using Cartesian components, the development in this section is restricted to Cartesian coordinates.

Recall from §7.2 that we can label particles by giving their coordinates at the present time t; the particle with coordinates $x_i (i = 1,2,3)$ is denoted by (x,t). The space coordinates of the particle (x,t) at a past time t' we call x_i'. The motion of the material is then given by the *displacement functions*[2] (cf. Eq. 9.1–3):

$$x_i' = x_i'(x,t,t') \tag{9.2-1}$$

which tell the Cartesian coordinates x_i' of each particle (x,t) for all past times t'. Although Eqs. 9.2–1 are very simple in form, they do contain the basis for the codeformational kinematical description. Note that the symbol x_i' appears in the equations with two different meanings: On the left the x_i' denote the space coordinates at time t' of the particle that at time t has space coordinates x_i. On the right, however, the symbol x_i' denotes a function[3], namely the function that expresses the x_i' in terms of the x_i, t, and t'. In the following we

[5] The factor of 2 in the second terms in the relaxation modulus and in the memory function arises because of the definition of the Dirac δ-function given in Eq. 6.1–9b.

[1] The ideas in this section are based to a large extent on the fundamental developments by R. S. Rivlin and J. L. Ericksen, *J. Rat. Mech. Anal., 4*, 323–425 (1955); and by A. E. Green and R. S. Rivlin, *Arch. Rat. Mech. Anal., 1*, 1–21 (1957). We thank Professor M. W. Johnson Jr. for many helpful discussions in connection with this material.

[2] The displacement functions are used in §7.2 to describe the movement of the origin of a corotating frame and are identified with b_i there.

[3] To make this distinction more explicit, a separate symbol is sometimes used to designate the functions x_i'.

assume that the functions in Eqs. 9.2–1 can be inverted to give

$$x_i = x_i(x',t,t') \tag{9.2–2}$$

where x' stands for the set of coordinates x_1', x_2', x_3'. We now proceed to define kinematical quantities used in the codeformational formalism in terms of the displacement functions.

First we consider the Cartesian components of the fluid velocity $\boldsymbol{v}(x',t')$ at the position x' at time t'. These are defined as the velocity components $v_i(x,t,t')$ of the particle (x,t) that happened to pass through the point x' at time t', that is (cf. Eq. 7.2–22):

$$\begin{aligned} v_i(x',t') &= v_i(x,t,t') \\ &= \frac{\partial}{\partial t'} x_i'(x,t,t') \end{aligned} \tag{9.2–3}$$

Note that the velocity components of the particle (x,t) have been defined as the time rate of change of the position coordinates of the particle. Since we have chosen the present time t as the reference time, we must be careful when we define time derivatives at time t. Thus the velocity components of the particle (x,t) at time t are defined by

$$v_i(x,t,t) = \frac{\partial}{\partial t'} x_i'(x,t,t')\big|_{t'=t} \tag{9.2–4}$$

We now define the components of the acceleration of the particle (x,t) by differentiating Eq. 9.2–3 (cf. Eq. 7.2–33a)

$$\begin{aligned} a_i(x',t') &= a_i(x,t,t') \\ &= \frac{\partial}{\partial t'} v_i(x,t,t') \\ &= \left(\frac{\partial}{\partial t'} + \textstyle\sum_k \frac{\partial x_k'}{\partial t'} \frac{\partial}{\partial x_k'}\right) v_i(x',t') \\ &= \left(\frac{\partial}{\partial t'} + \textstyle\sum_k v_k(x',t') \frac{\partial}{\partial x_k'}\right) v_i(x',t') \\ &= \frac{D}{Dt'} v_i(x',t') \end{aligned} \tag{9.2–5}$$

In the last line of Eq. 9.2–5 we have used the "substantial derivative" D/Dt or the time derivative following a fluid particle introduced in §1.1. Also in going from the second to the third line of the equation we have used the chain rule of partial differentiation.

Equation 9.2–3 expresses the velocity components in terms of the displacement functions. Alternatively the displacement functions may be expressed in terms of the velocity field by integration of the equations as follows:

$$\begin{aligned} x_i - x_i' &= \int_{t'}^{t} v_i(x,t,t'')\,dt'' \\ &= \int_{t'}^{t} v_i(x'',t'')\,Dt'' \end{aligned} \tag{9.2–6}$$

In the last line of Eq. 9.2–6 we have introduced the concept of "substantial integration," or time integration following a particle. Note that when the integrand is given as a function of particle and time, dt denotes integration following the particle, whereas Dt is needed when the integrand is given in terms of position and time.

Next we define two "deformation tensors" $\mathbf{\Delta}$ and $\mathbf{E}$. To do this we have to consider a particle that is adjacent to the particle (x,t). At the present time t the components of the vector from the particle (x,t) to the neighboring particle are denoted by dx_i. We therefore label the neighboring particle as $(x + dx,t)$. At a past time t' the vector from the particle (x,t) to the particle $(x + dx,t)$ had Cartesian components dx'_i. We now define a tensor $\mathbf{\Delta}(x,t,t')$ with Cartesian components $\Delta_{ij}(x,t,t')$ by

$$dx'_i = \sum_j \Delta_{ij}(x,t,t')\, dx_j \tag{9.2–7}$$

We see that $\Delta_{ij}(x,t,t')$ may be expressed as

$$\Delta_{ij}(x,t,t') = \frac{\partial x'_i(x,t,t')}{\partial x_j} \tag{9.2–8}$$

We define also the tensor $\mathbf{E}(x,t,t')$ inverse to $\mathbf{\Delta}(x,t,t')$, that is,

$$\mathrm{E}_{ij}(x,t,t') = \frac{\partial x_i(x',t,t')}{\partial x'_j} \tag{9.2–9}$$

In Eq. 9.2–9 $x_i(x',t,t')$ are the functions given in Eq. 9.2–2.

Note that in Cartesian coordinates the matrices (Δ_{ij}) and (E_{ij}) appear in columns 2′ of Tables 9.1–1 and 2. If we know either the tensor $\mathbf{\Delta}$ or $\mathbf{E}$ for all past times t', then we know the entire history of the deformation in the neighborhood of the particle (x,t). It is particularly important to note (see Problem 9D.2) that for an incompressible fluid $\det(\Delta_{ij}) = \det(\mathrm{E}_{ij}) = 1$ for all t' and all particles (x,t).

In addition to containing information about the deformation in the neighborhood of the particle (x,t), the tensors $\mathbf{\Delta}$ and $\mathbf{E}$ also contain information about the local rotation of the material around the particle (x,t). In order to define tensors that do not depend on local rotation, let us now inquire about the distance ds between the particle (x,t) and the neighboring particle $(x + dx,t)$ as a function of time. We see that:

$$\begin{aligned}
[ds(t')]^2 &= \textstyle\sum_i dx'_i\, dx'_i \\
&= \textstyle\sum_i \sum_j \sum_k dx_j\, \Delta_{ij}(x,t,t')\, \Delta_{ik}(x,t,t')\, dx_k \\
&= \textstyle\sum_j \sum_k dx_j (\gamma^{[0]}_{jk}(x,t,t') + \delta_{jk})\, dx_k \\
&= [ds(t)]^2 + \textstyle\sum_j \sum_k dx_j\, \gamma^{[0]}_{jk}(x,t,t')\, dx_k
\end{aligned} \tag{9.2–10}$$

We have here introduced the tensor $\boldsymbol{\gamma}^{[0]}(x,t,t')$ given in column (2′) of Table 9.1–2 in component form; we see that this tensor contains information about how the distances between neighboring particles change with time. It can also be shown that $\boldsymbol{\gamma}^{[0]}$ contains information about the changes in angles between lines from the particle (x,t) to two neighboring particles. In other words $\boldsymbol{\gamma}^{[0]}$ is intimately related to the deformation of the fluid and is independent of superposed rigid rotations. The combination of tensors:

$$\boldsymbol{\delta} + \boldsymbol{\gamma}^{[0]}(x,t,t') = \{\mathbf{\Delta}^\dagger(x,t,t') \cdot \mathbf{\Delta}(x,t,t')\} \tag{9.2–11}$$

is often referred to as the *Cauchy strain tensor*; it is given above with the present configuration as the reference configuration. Similarly the combination:

$$\boldsymbol{\delta} - \boldsymbol{\gamma}_{[0]}(x,t,t') = \{\mathbf{E}(x,t,t') \cdot \mathbf{E}^{\dagger}(x,t,t')\} \tag{9.2–12}$$

is often referred to as the *Finger strain tensor*; here also we have taken the present configuration as the reference configuration. Note that the Finger strain tensor is the inverse of the Cauchy strain tensor, and that for incompressible fluids the determinant of both is unity. The components of the $\boldsymbol{\gamma}^{[0]}$ and $\boldsymbol{\gamma}_{[0]}$ tensors are given in column (2′) of Table 9.1–2; both of these tensors reduce to the infinitesimal strain tensor $\boldsymbol{\gamma}$ of column (3′), used in Chapter 6, for flows with very small displacements (see Problem 9B.5).

We also need to have stress tensors describing the stress at fluid particle (x,t) at time t', referred to the present time t. These tensors, $\boldsymbol{\tau}^{[0]}(x,t,t')$ and $\boldsymbol{\tau}_{[0]}(x,t,t')$, have components given by:

$$\tau_{ij}^{[0]}(x,t,t') = \textstyle\sum_k \sum_l \Delta_{ki}(x,t,t')\tau_{kl}(x',t')\,\Delta_{lj}(x,t,t') \tag{9.2–13}$$

$$\tau_{[0]ij}(x,t,t') = \textstyle\sum_k \sum_l \mathrm{E}_{ik}(x,t,t')\tau_{kl}(x',t')\mathrm{E}_{jl}(x,t,t') \tag{9.2–14}$$

In Eqs. 9.2–13 and 14 $\tau_{kl}(x',t')$ are the Cartesian components of the stress tensor at the position x' at time t'. These two tensors are given in column (2′) of Table 9.1–1. Here, as in other equations in which some terms are expressed as functions of x, t, and t' and others as functions of x' and t', it is understood that all functions are to be evaluated at the same place in space, that is, that x_i' and x_i are related through the displacement functions.

At this point we have defined the four basic codeformational rheological tensors: the strain tensors $\boldsymbol{\gamma}^{[0]}$ and $\boldsymbol{\gamma}_{[0]}$, and the stress tensors $\boldsymbol{\tau}^{[0]}$ and $\boldsymbol{\tau}_{[0]}$. These tensors play the same role in the codeformational point of view as the tensors $\dot{\boldsymbol{\Gamma}}$ and $\mathbf{T}$ do in the corotational framework. As we will show later in the section, the four basic codeformational rheological tensors can be used to describe deformations and stresses in large deformations of macromolecular fluids, without any unwanted dependence on superposed rigid rotations. The basic codeformational tensors as well as the basic corotational tensors, their time derivatives, time integrals, and any combinations of these may therefore be used to construct objective rheological equations of state as we stated in §9.1.

We will now derive expressions for the time derivatives of the four basic codeformational rheological tensors. We begin by considering the time derivatives of the $\boldsymbol{\Delta}$ and $\mathbf{E}$ tensors. The time rate of change of $\Delta_{ij}(x,t,t')$ following the fluid particle (x,t) is:

$$\begin{aligned}
\frac{\partial}{\partial t'}\,\Delta_{ij}(x,t,t') &= \frac{\partial}{\partial t'}\frac{\partial}{\partial x_j}\,x_i'(x,t,t') \\
&= \frac{\partial}{\partial x_j}\frac{\partial}{\partial t'}\,x_i'(x,t,t') \\
&= \frac{\partial}{\partial x_j}\,v_i(x,t,t') \\
&= \textstyle\sum_k \dfrac{\partial x_k'}{\partial x_j}\dfrac{\partial}{\partial x_k'}\,v_i(x',t') \\
&= \textstyle\sum_k (\boldsymbol{\nabla}\boldsymbol{v})_{ki}(x',t')\,\Delta_{kj} \\
&= \textstyle\sum_k (\boldsymbol{\nabla}\boldsymbol{v})_{ki}(x,t,t')\,\Delta_{kj}(x,t,t')
\end{aligned} \tag{9.2–15}$$

We see from Eq. 9.2–15 that $\mathbf{\Delta}$ may alternatively be defined as the solution to the "initial value problem,"

$$\begin{cases} \dfrac{\partial}{\partial t'} \Delta_{ij}(x,t,t') = \sum_k (\nabla \boldsymbol{v})_{ki}(x,t,t')\, \Delta_{kj}(x,t,t') & (9.2\text{–}16) \\ \\ \Delta_{ij}(x,t,t) = \delta_{ij} & (9.2\text{–}17) \end{cases}$$

We note the parallel between Eqs. 9.2–16 and 17 and the initial value problem in Eqs. 7.2–28 and 29. Then, by using the fact that $\mathbf{E} = \mathbf{\Delta}^{-1}$, it may be shown that the tensor $\mathbf{E}$ satisfies a similar initial value problem:

$$\begin{cases} \dfrac{\partial}{\partial t'} \mathrm{E}_{ij}(x,t,t') = -\sum_k \mathrm{E}_{ik}(x,t,t')(\nabla \boldsymbol{v})_{jk}(x,t,t') & (9.2\text{–}18) \\ \\ \mathrm{E}_{ij}(x,t,t) = \delta_{ij} & (9.2\text{–}19) \end{cases}$$

Equations 9.2–16 and 18 may be used to obtain the partial time derivatives with respect to t' of the four basic codeformational rheological tensors. Thus we find for the $\boldsymbol{\gamma}^{[0]}$ tensor,

$$\begin{aligned} \gamma_{ij}^{[1]}(x,t,t') &\equiv \frac{\partial}{\partial t'} \gamma_{ij}^{[0]}(x,t,t') \\ &= \frac{\partial}{\partial t'} [(\textstyle\sum_k \Delta_{ki} \Delta_{kj}) - \delta_{ij}] \\ &= \textstyle\sum_k \left[\left(\dfrac{\partial}{\partial t'} \Delta_{ki}\right) \Delta_{kj} + \Delta_{ki} \left(\dfrac{\partial}{\partial t'} \Delta_{kj}\right)\right] \\ &= \textstyle\sum_k \sum_l \Delta_{ki}[(\nabla \boldsymbol{v})_{lk} + (\nabla \boldsymbol{v})_{kl}]\, \Delta_{lj} \\ &= \textstyle\sum_k \sum_l \Delta_{ki}(x,t,t')\dot{\gamma}_{kl}(x,t,t')\, \Delta_{lj}(x,t,t') \end{aligned} \qquad (9.2\text{–}20)$$

Similarly we find for the derivative of the $\boldsymbol{\gamma}_{[0]}$ tensor:

$$\begin{aligned} \gamma_{[1]ij}(x,t,t') &\equiv \frac{\partial}{\partial t'} \gamma_{[0]ij}(x,t,t') \\ &= \textstyle\sum_k \sum_l \mathrm{E}_{ik}(x,t,t')\dot{\gamma}_{kl}(x,t,t')\mathrm{E}_{jl}(x,t,t') \end{aligned} \qquad (9.2\text{–}21)$$

Then for the time derivatives of the stress tensors:

$$\begin{aligned} \tau_{ij}^{[1]}(x,t,t') &\equiv \frac{\partial}{\partial t'} \tau_{ij}^{[0]}(x,t,t') \\ &= \textstyle\sum_k \sum_l \Delta_{ki}(x,t,t')\tau_{kl}^{(1)}(x,t,t')\, \Delta_{lj}(x,t,t') \end{aligned} \qquad (9.2\text{–}22)$$

and

$$\begin{aligned} \tau_{[1]ij}(x,t,t') &\equiv \frac{\partial}{\partial t'} \tau_{[0]ij}(x,t,t') \\ &= \textstyle\sum_k \sum_l \mathrm{E}_{ik}(x,t,t')\tau_{(1)kl}(x,t,t')\mathrm{E}_{jl}(x,t,t') \end{aligned} \qquad (9.2\text{–}23)$$

In Eqs. 9.2–22 and 23 we have introduced the codeformational derivative operations $^{(1)}$ and $_{(1)}$. These operations are the codeformational analogs of the corotational (or Jaumann) derivative defined in Eq. 7.2–33. In terms of an arbitrary symmetric second order tensor $\mathbf{\Lambda}(x',t')$ we define,[4]

$$\Lambda_{ij}^{(1)}(x',t') = \frac{D}{Dt'}\Lambda_{ij}(x',t') + \sum_l \Lambda_{il}(x',t')(\nabla \mathbf{v})_{jl}(x',t') + \sum_l (\nabla \mathbf{v})_{il}(x',t')\Lambda_{lj}(x',t') \tag{9.2–24}$$

$$\Lambda_{(1)ij}(x',t') = \frac{D}{Dt'}\Lambda_{ij}(x',t') - \sum_l \Lambda_{il}(x',t')(\nabla \mathbf{v})_{lj}(x',t') - \sum_l (\nabla \mathbf{v})_{li}(x',t')\Lambda_{lj}(x',t') \tag{9.2–25}$$

In these definitions, note that: (1) all tensor quantities are evaluated at the same time t' that appears in the substantial derivative D/Dt', and (2) all tensor quantities are evaluated at the same position x'. The process of codeformational differentiation may be repeated any number of times. Notationally we may indicate a repeated codeformational differentiation by adding one to the number in the parentheses. Thus we write in general:

$$\Lambda_{ij}^{(n+1)} = (\Lambda_{ij}^{(n)})^{(1)} \tag{9.2–26}$$

$$\Lambda_{(n+1)ij} = (\Lambda_{(n)ij})_{(1)} \tag{9.2–27}$$

The codeformational or "convected" derivatives may be expressed also in terms of the Jaumann derivative as shown in note (d) to Tables 9.1–1 and 2.

We now return to consider higher derivatives with respect to t' of the four basic codeformational rheological tensors. By repeated differentiation of Eqs. 9.2–20 to 23 and use of Eqs. 9.2–16 and 18, we find in general,

$$\gamma_{ij}^{[n+1]}(x,t,t') = \frac{\partial}{\partial t'}\gamma_{ij}^{[n]} = \sum_k \sum_l \Delta_{ki}(x,t,t')\gamma_{kl}^{(n+1)}(x,t,t')\,\Delta_{lj}(x,t,t') \qquad (n > 0) \tag{9.2–28}$$

$$\gamma_{[n+1]ij}(x,t,t') = \frac{\partial}{\partial t'}\gamma_{[n]ij} = \sum_k \sum_l \mathrm{E}_{ik}(x,t,t')\gamma_{(n+1)kl}(x,t,t')\mathrm{E}_{jl}(x,t,t') \qquad (n > 0) \tag{9.2–29}$$

[4] As a parallel to Eq. 7.2–33b, note that Eqs. 9.2–24 and 25 can be rewritten with Λ_{ij} expressed as a function of particle label and time:

$$\Lambda_{ij}^{(1)}(x,t,t') = \frac{\partial}{\partial t'}\Lambda_{ij}(x,t,t') + \sum_l \Lambda_{il}(x,t,t')(\nabla \mathbf{v})_{jl}(x,t,t') + \sum_l (\nabla \mathbf{v})_{il}(x,t,t')\Lambda_{lj}(x,t,t') = \Lambda_{ij}^{(1)}(x',t')\big|_{x_i'=x_i'(x,t,t')} \tag{9.2–24a}$$

$$\Lambda_{(1)ij}(x,t,t') = \frac{\partial}{\partial t'}\Lambda_{ij}(x,t,t') - \sum_l \Lambda_{il}(x,t,t')(\nabla \mathbf{v})_{lj}(x,t,t') - \sum_l (\nabla \mathbf{v})_{li}(x,t,t')\Lambda_{lj}(x,t,t') = \Lambda_{(1)ij}(x',t')\big|_{x_i'=x_i'(x,t,t')} \tag{9.2–25a}$$

$$\tau_{ij}^{[n+1]}(x,t,t') = \frac{\partial}{\partial t'}\tau_{ij}^{[n]}$$

$$= \sum_k \sum_l \Delta_{ki}(x,t,t')\tau_{kl}^{(n+1)}(x,t,t')\,\Delta_{lj}(x,t,t') \qquad (n \geq 0) \qquad (9.2\text{–}30)$$

$$\tau_{[n+1]ij}(x,t,t') = \frac{\partial}{\partial t'}\tau_{[n]ij}$$

$$= \sum_k \sum_l \mathrm{E}_{ik}(x,t,t')\tau_{(n+1)kl}(x,t,t')\mathrm{E}_{jl}(x,t,t') \qquad (n \geq 0) \qquad (9.2\text{–}31)$$

In Eqs. 9.2–28 to 31 the tensors $\boldsymbol{\gamma}^{(n)}$, $\boldsymbol{\gamma}_{(n)}$, $\boldsymbol{\tau}^{(n)}$, and $\boldsymbol{\tau}_{(n)}$ are defined by the recurrence relations Eqs. 9.2–26 and 27 with the additional information that $\boldsymbol{\gamma}^{(1)} = \boldsymbol{\gamma}_{(1)} = \dot{\boldsymbol{\gamma}}$ and $\boldsymbol{\tau}^{(0)} = \boldsymbol{\tau}_{(0)} = \boldsymbol{\tau}$.

The definitions of the four basic codeformational rheological tensors Eqs. 9.1–10, 12, 13, and 14 and the expressions for their time derivatives with respect to t' given by Eqs. 9.2–24 to 31 contain all the information needed to construct and rearrange rheological equations of state within the codeformational framework. This information is displayed in columns (2′) of Tables 9.1–1 and 2; see notes (a) through (d) to the tables.

It is particularly useful to consider the simplification of the above tensors at $t' = t$ (cf. Eqs. 7.2–35 to 37),

$$\gamma_{ij}^{[n]}(x,t,t')\big|_{t'=t} = \gamma_{ij}^{(n)}(x,t) \qquad (n > 0) \qquad (9.2\text{–}32)$$

$$\gamma_{[n]ij}(x,t,t')\big|_{t'=t} = \gamma_{(n)ij}(x,t) \qquad (n > 0) \qquad (9.2\text{–}33)$$

$$\tau_{ij}^{[n]}(x,t,t')\big|_{t'=t} = \tau_{ij}^{(n)}(x,t) \qquad (n \geq 0) \qquad (9.2\text{–}34)$$

$$\tau_{[n]ij}(x,t,t')\big|_{t'=t} = \tau_{(n)ij}(x,t) \qquad (n \geq 0) \qquad (9.2\text{–}35)$$

These simplifications are arranged in columns (2) of Tables 9.1–1 and 2. We see that the tensors $\boldsymbol{\gamma}^{(n)}$, $\boldsymbol{\gamma}_{(n)}$, $\boldsymbol{\tau}^{(n)}$, and $\boldsymbol{\tau}_{(n)}$ are the nth time derivatives evaluated at the present time t of the four basic codeformational rheological tensors. These tensors may be used in Taylor series expansions of $\boldsymbol{\gamma}^{[0]}$, $\boldsymbol{\gamma}_{[0]}$, $\boldsymbol{\tau}^{[0]}$, and $\boldsymbol{\tau}_{[0]}$ around the present time t. Thus for $\boldsymbol{\gamma}^{[0]}$ we find

$$\gamma_{ij}^{[0]}(x,t,t') = \sum_{n=0}^{\infty} \frac{(t'-t)^n}{n!}\,\gamma_{ij}^{[n]}(x,t,t')\big|_{t'=t}$$

$$= \sum_{n=1}^{\infty} \frac{(t'-t)^n}{n!}\,\gamma_{ij}^{(n)}(x,t) \qquad (9.2\text{–}36)$$

Similar expansions hold for $\boldsymbol{\gamma}_{[0]}$, $\boldsymbol{\tau}^{[0]}$, and $\boldsymbol{\tau}_{[0]}$. The tensors $\boldsymbol{\gamma}^{(n)}$, apparently first given by Dupont,[5] are twice the Oldroyd nth rate-of-strain tensors[6] denoted by $e^{(n)}$, and precisely equal to the nth Rivlin-Ericksen tensors[7] commonly denoted by $\mathbf{A}_n$.

This completes the discussion of the definitions and interrelations between the tensors in columns (2′) and (2) of Tables 9.1–1 and 2. The remainder of this section is devoted to proving that these tensors may be used freely to construct objective rheological equations of state. We shall do this using the general framework presented in §7.2. In §7.2 we showed that the tensors $\dot{\boldsymbol{\Gamma}}$ and $\mathbf{T}$ are objective by showing that their components as viewed by a fixed observer and an arbitrarily translating and rotating observer are related at any time t by the transformation rules for a coordinate rotation (Eq. 7.2–41). In order to determine the components of $\dot{\boldsymbol{\Gamma}}$ (or $\mathbf{T}$) in the arbitrary frame it was necessary to find the rotation matrix

[5] Y. Dupont, *Bull. Sci. Acad. Belgium*, (5), *17*, 441–459 (1931).
[6] J. G. Oldroyd, *Proc. Roy. Soc.*, *A200*, 45–63 (1950).
[7] R. S. Rivlin and J. L. Ericksen, *J. Rat. Mech. Anal.*, *4*, 323–425 (1955).

$\bar{\Omega}_{ij}$ that described the tilt of the corotating frame relative to the arbitrary frame. The corresponding step here is to determine the form of the displacement functions in the arbitrary frame; these can then be used to obtain the components of all codeformational tensors from the definitions given in the preceding text and summarized in Tables 9.1–1 and 2.

The transformation rule for the displacement functions is given directly by Eq. 7.2–7. When this equation is applied at time t we obtain a relation between particle labels in the fixed and arbitrary frames; when it is applied at time t' to the coordinates of a particle (x,t) it gives the relation between displacement functions in the two frames:

$$\bar{x}_i'(\bar{x},t,t') = \sum_j \mathrm{P}_{ij}(t')[x_j'(x,t,t') - b_j(t')] \tag{9.2–37}$$

Similarly the displacement functions for a neighboring particle $(x + dx,t)$ obey the transformation rule

$$\bar{x}_i' + d\bar{x}_i' = \sum_j \mathrm{P}_{ij}(t')[(x_j' + dx_j') - b_j(t')] \tag{9.2–38}$$

where we have suppressed the particle labels for brevity. By subtracting Eq. 9.2–37 from 9.2–38 we obtain a relation for the coordinates of particle $(x + dx,t)$ relative to (x,t), as seen in the two frames:

$$d\bar{x}_i' = \sum_j \mathrm{P}_{ij}(t')\, dx_j' \tag{9.2–39}$$

At time t this becomes

$$d\bar{x}_i = \sum_j \mathrm{P}_{ij}(t)\, dx_j \tag{9.2–40}$$

It is now straightforward to work out the components of the basic codeformational tensors in the arbitrary frame. For example $\mathbf{\Delta}$ is defined by Eq. 9.2–7 and thus we have

$$d\bar{x}_i' = \sum_j \bar{\Delta}_{ij}(\bar{x},t,t')\, d\bar{x}_j \tag{9.2–41}$$

By using Eq. 9.2–39 on the left side of the above and Eq. 9.2–40 on the right side, we find after a slight rearrangement:

$$dx_i' = \sum_j \sum_m \sum_n \mathrm{P}_{mi}(t')\, \bar{\Delta}_{mn}(\bar{x},t,t')\mathrm{P}_{nj}(t)dx_j \tag{9.2–42}$$

A comparison of Eqs. 9.2–42 and 9.2–7 gives:

$$\bar{\Delta}_{ij}(\bar{x},t,t') = \sum_m \sum_n \mathrm{P}_{im}(t')\, \Delta_{mn}(x,t,t')\mathrm{P}_{jn}(t) \tag{9.2–43}$$

Since $\mathrm{E}_{ij} = (\mathbf{\Delta}^{-1})_{ij}$ we find from Eq. 9.2–43 by using the rule for taking the inverse of a product of matrices that the transformation law for $\mathbf{E}$ is

$$\bar{\mathrm{E}}_{ij}(\bar{x},t,t') = \sum_m \sum_n \mathrm{P}_{im}(t)\mathrm{E}_{mn}(x,t,t')\mathrm{P}_{jn}(t') \tag{9.2–44}$$

The transformation laws for the four basic codeformational rheological tensors now follow directly from the definitions in Eqs. 9.2–11 to 9.2–14 and from the transformation rules in Eqs. 7.2–17, 9.2–43 and 9.2–44:

$$\begin{aligned}
\bar{\gamma}_{ij}^{[0]} &\equiv \sum_k \bar{\Delta}_{ki}\, \bar{\Delta}_{kj} - \delta_{ij} \\
&= \sum_k \sum_m \sum_n \sum_r \sum_s \mathrm{P}_{km}(t')\, \Delta_{mn}\mathrm{P}_{in}(t)\mathrm{P}_{kr}(t')\, \Delta_{rs}\mathrm{P}_{js}(t) - \delta_{ij} \\
&= \sum_m \sum_n \sum_s \mathrm{P}_{in}(t)\, \Delta_{mn}\, \Delta_{ms}\mathrm{P}_{js}(t) - \delta_{ij} \\
&= \sum_n \sum_s \mathrm{P}_{in}(t)[\textstyle\sum_m \Delta_{mn}\, \Delta_{ms} - \delta_{ns}]\mathrm{P}_{js}(t) \\
&= \sum_n \sum_s \mathrm{P}_{in}(t)\gamma_{ns}^{[0]}(x,t,t')\mathrm{P}_{js}(t)
\end{aligned} \tag{9.2–45}$$

$$\bar{\gamma}_{[0]ij} = \sum_m \sum_n \mathrm{P}_{im}(t)\gamma_{[0]mn}(x,t,t')\mathrm{P}_{jn}(t) \tag{9.2-46}$$

$$\bar{\tau}_{ij}^{[0]} = \sum_m \sum_n \mathrm{P}_{im}(t)\tau_{mn}^{[0]}(x,t,t')\mathrm{P}_{jn}(t) \tag{9.2-47}$$

$$\bar{\tau}_{[0]ij} = \sum_m \sum_n \mathrm{P}_{im}(t)\tau_{[0]mn}(x,t,t')\mathrm{P}_{jn}(t) \tag{9.2-48}$$

By comparing the above results with Eq. 7.2–41 we see that $\boldsymbol{\gamma}^{[0]}$, $\boldsymbol{\gamma}_{[0]}$, $\boldsymbol{\tau}^{[0]}$, and $\boldsymbol{\tau}_{[0]}$ are all objective tensors and can therefore be used freely to construct objective rheological equations of state. In addition, it is straightforward to show that time derivatives following the fluid of these four tensors are also objective. This can be shown by beginning with the definitions of these derivatives in the arbitrary frame and then transforming each term that appears into the fixed frame as was done for $\ddot{\mathbf{\Gamma}}$ in Eq. 7.2–51. Alternatively, we can take $(\partial/\partial t')_{\bar{x},t}$ of Eqs. 9.2–45 to 9.2–48. From Eq. 7.2–7 we know that this operation is equivalent to taking $(\partial/\partial t')_{x,t}$. In this way, for example, we find

$$\begin{aligned}\bar{\gamma}_{ij}^{[1]} &\equiv \frac{\partial}{\partial t'}\bar{\gamma}_{ij}^{[0]}(\bar{x},t,t') = \sum_m \sum_n \mathrm{P}_{im}(t)\left(\frac{\partial}{\partial t'}\gamma_{mn}^{[0]}(x,t,t')\right)\mathrm{P}_{jn}(t)\\ &= \sum_m \sum_n \mathrm{P}_{im}(t)\gamma_{mn}^{[1]}(x,t,t')\mathrm{P}_{jn}(t)\end{aligned} \tag{9.2-49}$$

Similarly all of the tensors in columns (2′) of Tables 9.1–1 and 2 can be shown to be objective. By evaluating Eqs. 9.2–45 to 9.2–48 and their time derivatives at $t' = t$, we see immediately that $\boldsymbol{\gamma}^{(n)}$, $\boldsymbol{\gamma}_{(n)}$, $\boldsymbol{\tau}^{(n)}$, and $\boldsymbol{\tau}_{(n)}$ are all objective.

In summary we have defined the various codeformational rheological tensors and derived the relations among them. We have further shown that the tensors in columns (2′) of Tables 9.1–1 and 2 (measures of stresses and deformations at a past time) and the tensors in columns (2) (measures of stresses and deformations at the present time) may be used freely to construct objective rheological equations of state. This points out the close analogy between the codeformational formalism presented here and the corotational formalism as presented in §7.2.

We should keep in mind that the entire development in this section for simplicity has been restricted to Cartesian coordinates. All the relations, however, may be generalized to any orthogonal or nonorthogonal coordinate system by coordinate transformations. In this way we obtain precisely the entries in columns (2′) and (2) of Tables 9.1–1 and 2 in the notation of general tensor analysis. In the next section we show how these entries may be obtained alternatively by starting out using nonorthogonal coordinates and general tensor analysis employing the "convected" coordinate formalism.

§9.3 CONVECTED COORDINATE FORMALISM

In this section we present a method for constructing objective rheological equations of state by using a codeforming or "convected" coordinate system.[1] In this respect the methods developed here may be regarded as alternative to those presented in §9.2. Indeed all problems that can be solved using the convected-coordinate formalism may also be solved using the space-fixed coordinate formalism of §9.2. In addition, the convected-coordinate formalism involves the use of nonorthogonal coordinate surfaces, with the

[1] The ideas in this section are based to a large extent on the fundamental development by J. G. Oldroyd, *Proc. Roy. Soc., A200*, 45–63 (1950). Here we restrict our attention to co- and contra-variant convected components; Oldroyd has given a more general presentation that allows for the possibility of mixed components and relative tensors. See also A. S. Lodge, *Elastic Liquids*, Academic Press, London (1964), Chapter 12.

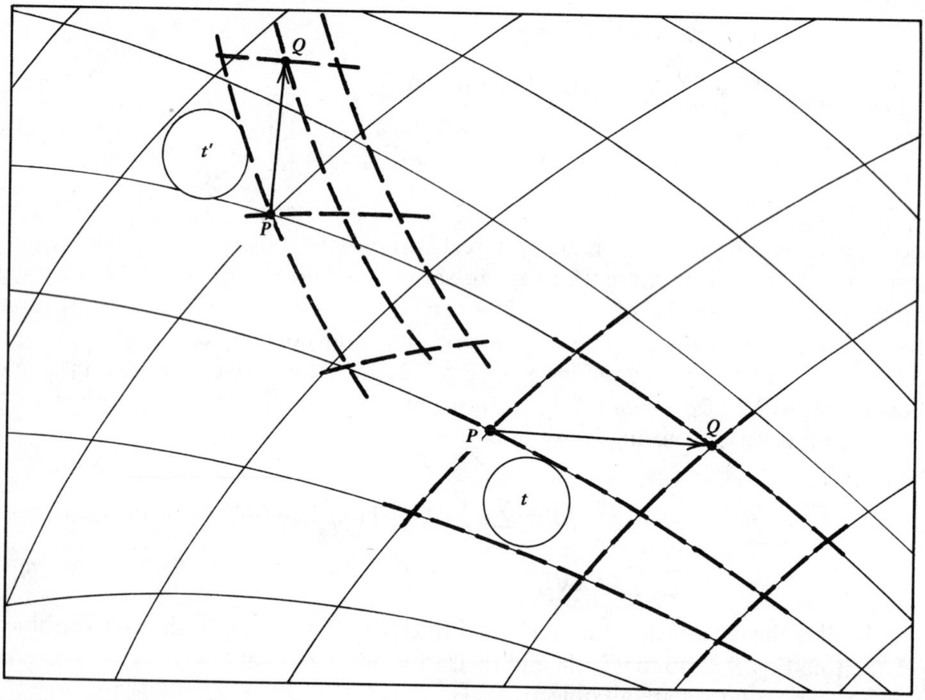

FIGURE 9.3–1. Motion of particles P and Q. Also shown are the space-fixed coordinate curves (———), the convected coordinate curves at time t' (– – – – –), and the convected coordinate curves at time t (– – – – –). For simplicity the figure is restricted to two dimensions. At t' the space coordinates of P are (x'^1, x'^2, x'^3) and the convected coordinates are $(\hat{x}^1, \hat{x}^2, \hat{x}^3)$; at time t the space coordinates of P are (x^1, x^2, x^3), but the convected coordinates are still $(\hat{x}^1, \hat{x}^2, \hat{x}^3)$. The change in shape of the convected coordinate surfaces indicates the deformation the material has undergone. Notice that the convected coordinate surfaces have been chosen coincident with the space fixed coordinate surfaces at time t; we shall often make this choice although in general the convected coordinates at time t may differ from the space coordinates.

consequence that general tensor analysis must be employed. The reader not familiar with general tensor analysis may wish to omit the present section and concentrate on §§7.2 and 9.2, inasmuch as those sections provide sufficient background in the methods for constructing objective rheological equations of state for most purposes. The advanced student, however, will invariably encounter both the space fixed coordinate formalism and the convected formalism in the research literature, and will find §§9.2 and 3 instructive for comparing and contrasting the two methods. In this section then, we will assume that the reader is familiar with nonorthogonal coordinate systems and general tensor analysis to the extent that these are treated in §A.8. All equations will be in general tensor notation; this is indicated by the brackets [( )] around the equation numbers. In particular this involves the use of the implied summation on repeated indices.

Consider now a fluid flowing in a region of space. This region is covered by two coordinate systems (see Fig. 9.3–1): one stationary space fixed coordinate system with coordinates x^i and covariant components of the metric tensor[2] $g_{ij}(x)$, and one moving and deforming "convected" coordinate system with coordinates $\hat{x}^i$ and covariant components

[2] We will here be concerned mainly with components of tensors, and we will construct tensors only from components in the space fixed coordinate system. For a coordinatefree treatment see A. S. Lodge, *Body Tensor Fields in Continuum Mechanics*, Academic Press, New York (1975).

of the metric tensor $\hat{g}_{ij}(\hat{x},t)$. The coordinate surfaces of the convected coordinate system are drawn in the material, and they deform continuously with it. Therefore, the convected coordinate system changes with time, and the components of the metric tensor with respect to the convected coordinate system are functions of time. A particle with convected coordinates $\hat{x}^i$ at any one time has the same convected coordinates at any other time. The convected coordinates may therefore be used as particle labels, and we may speak about the particle $\hat{x}$, meaning the particle with convected coordinates $\hat{x}^i$. By analogy with the development in the previous section the complete deformation history of the fluid may be expressed by the functions:

$$x'^i = x'^i(\hat{x},t') \qquad [(9.3\text{–}1)]$$

for all past times t' up to the present time t and for all particles $\hat{x}$. We assume that Eqs. 9.3–1 have an inverse set:

$$\hat{x}^i = \hat{x}^i(x',t') \qquad [(9.3\text{–}2)]$$

By analogy with Eqs. 9.2–3 and 4 we define the contravariant components of the velocity of the particle $\hat{x}$ at time t' by:

$$v'^i(\hat{x},t') = \frac{\partial}{\partial t'}\, x'^i(\hat{x},t') \qquad [(9.3\text{–}3)]$$

At the present time t, Eqs. 9.3–2 read:

$$\hat{x}^i = \hat{x}^i(x,t) \qquad [(9.3\text{–}4)]$$

Consequently we may eliminate the convected coordinates between Eqs. 9.3–1 and 4 to give:

$$x'^i = x'^i(x,t,t') \qquad [(9.3\text{–}5)]$$

We shall refer to the functions in Eqs. 9.3–5 as the *displacement functions* by analogy with Eqs. 9.2–1. Furthermore we assume that Eqs. 9.3–5 have an inverse set:

$$x^i = x^i(x',t,t') \qquad [(9.3\text{–}6)]$$

Consider now two particles $\hat{x}$ and $(\hat{x} + d\hat{x})$. At a past time t' the distance from $\hat{x}$ to $(\hat{x} + d\hat{x})$ is given by the components of the metric tensor (§A.8):

$$[ds\,(t')]^2 = \hat{g}_{ij}(\hat{x},t')\, d\hat{x}^i\, d\hat{x}^j \qquad [(9.3\text{–}7)]$$

Likewise the angles between lines from the particle $\hat{x}$ to two neighboring particles are also given by $\hat{g}_{ij}(\hat{x},t')$. Thus we may say that the covariant components of the metric tensor with respect to the convected coordinate system describe the shape of a fluid element around $\hat{x}$, and that the functions $\hat{g}_{ij}(\hat{x},t')$ with $-\infty < t' \leq t$ describe the entire deformation history of the fluid in the neighborhood of the particle $\hat{x}$ up to the present time t. Moreover, since the coordinate surfaces that determine $\hat{g}_{ij}(\hat{x},t')$ are embedded in the fluid the $\hat{g}_{ij}(\hat{x},t')$ are unaffected by a superposed rigid body rotation. Notice that in Eq. 9.3–7 the $d\hat{x}^i$ are independent of time, and that the time dependence on the right side of the equation enters only in the $\hat{g}_{ij}(\hat{x},t')$. Therefore we may differentiate Eq. 9.3–7 to get:

$$\frac{d^n}{dt'^n}\,[ds\,(t')]^2 = \left[\frac{\partial^n}{\partial t'^n}\,\hat{g}_{ij}(\hat{x},t')\right] d\hat{x}^i\, d\hat{x}^j \qquad [(9.3\text{–}8)]$$

We see that the time derivatives of the $\hat{g}_{ij}(\hat{x},t')$ govern the time rates of change of the distances from the particle $\hat{x}$ to neighboring particles. The tensor with covariant convected components $(\partial^n/\partial t'^n)\hat{g}_{ij}$ is twice the Oldroyd nth rate-of-strain tensor,[3] and equal to the nth Rivlin-Ericksen tensor[4] $\mathbf{A}_n$ introduced in §9.2. In particular, as we shall see later, the first Rivlin-Ericksen tensor is simply the rate-of-deformation tensor. These statements will become clear in the subsequent development.

The stress tensor at the particle $\hat{x}$ at time t' may be decomposed into components with respect to the convected coordinate system. We shall denote these components by $\hat{\tau}_{ij}(\hat{x},t')$. We assume that the convected components of the stress tensor contain no unwanted dependence on local fluid rotation. The fundamental assumption in the convected coordinate formalism may now be expressed as follows: We may construct objective rheological equations of state for isotropic fluids from the tensor components $\hat{g}_{ij}(\hat{x},t')$ and $\hat{\tau}_{ij}(\hat{x},t')$, their inverses, their time derivatives, and their time integrals in any combination admissible by the laws of general tensor analysis. By combining the results of this section with those of the latter part of §9.2, it is not difficult to verify that this assumption is valid. We must keep in mind, however, that the equation of continuity and the equation of motion will generally be written in terms of physical components in space fixed coordinate systems. Hence we must be able to transform components with respect to convected coordinates into components with respect to fixed coordinates. This is a matter of an ordinary coordinate transformation. Thus by the standard rules for coordinate transformations (Eqs. A.8–46 and 47) we have for the covariant and contravariant convected components of the metric tensor:

$$\hat{g}_{ij}(\hat{x},t') = \frac{\partial x'^k}{\partial \hat{x}^i}\frac{\partial x'^l}{\partial \hat{x}^j}\, g_{kl}(x') \qquad [(9.3\text{–}9)]$$

$$\hat{g}^{ij}(\hat{x},t') = \frac{\partial \hat{x}^i}{\partial x'^k}\frac{\partial \hat{x}^j}{\partial x'^l}\, g^{kl}(x') \qquad [(9.3\text{–}10)]$$

Similarly the transformation laws for the components of the stress tensor are:

$$\hat{\tau}_{ij}(\hat{x},t') = \frac{\partial x'^k}{\partial \hat{x}^i}\frac{\partial x'^l}{\partial \hat{x}^j}\, \tau_{kl}(x',t') \qquad [(9.3\text{–}11)]$$

$$\hat{\tau}^{ij}(\hat{x},t') = \frac{\partial \hat{x}^i}{\partial x'^k}\frac{\partial \hat{x}^j}{\partial x'^l}\, \tau^{kl}(x',t') \qquad [(9.3\text{–}12)]$$

We now consider the time derivatives of the convected components of the metric tensor and of the stress tensor. We operate on both sides of Eq. 9.3–9 with $\partial/\partial t'$ keeping the convected coordinates constant; in so doing we let:

$$\begin{aligned}\frac{\partial}{\partial t'}\frac{\partial x'^k}{\partial \hat{x}^i} &= \frac{\partial}{\partial \hat{x}^i}\frac{\partial}{\partial t'}\, x'^k(\hat{x},t) \\ &= \frac{\partial}{\partial \hat{x}^i}\, v'^k(\hat{x},t) \\ &= \frac{\partial x'^m}{\partial \hat{x}^i}\frac{\partial}{\partial x'^m}\, v'^k(x',t') \qquad [(9.3\text{–}13)]\end{aligned}$$

[3] J. G. Oldroyd, *Proc. Roy. Soc.*, *A200*, 45–63 (1950).
[4] R. S. Rivlin and J. L. Ericksen, *J. Rat. Mech. Anal.*, *4*, 323–425 (1955).

With the aid of Eqs. 9.3–13, we find from Eqs. 9.3–9 that

$$\frac{\partial}{\partial t'}\hat{g}_{ij}(\hat{x},t') = \frac{\partial x'^k}{\partial \hat{x}^i}\frac{\partial x'^l}{\partial \hat{x}^j}\left(\frac{\partial v'^m}{\partial x'^k}g_{ml}(x') + g_{km}(x')\frac{\partial v'^m}{\partial x'^l} + \left(\frac{\partial}{\partial t'}g_{kl}(x'(\hat{x},t'))\right)_{\hat{x}}\right)$$

$$= \frac{\partial x'^k}{\partial \hat{x}^i}\frac{\partial x'^l}{\partial \hat{x}^j}\left(\frac{\partial v'^m}{\partial x'^k}g_{ml} + g_{km}\frac{\partial v'^m}{\partial x'^l} + \frac{\partial g_{kl}}{\partial x'^m}\frac{\partial x'^m}{\partial t'}\right)$$

$$= \frac{\partial x'^k}{\partial \hat{x}^i}\frac{\partial x'^l}{\partial \hat{x}^j}\left(\frac{\partial v'^m}{\partial x'^k}g_{ml} + g_{km}\frac{\partial v'^m}{\partial x'^l} + v'^m\frac{\partial g_{kl}}{\partial x'^m}\right)$$

$$= \frac{\partial x'^k}{\partial \hat{x}^i}\frac{\partial x'^l}{\partial \hat{x}^j}(v'_{l,k} + v'_{k,l})$$

$$= \frac{\partial x'^k}{\partial \hat{x}^i}\frac{\partial x'^l}{\partial \hat{x}^j}\dot{\gamma}_{kl}(x',t') \qquad [(9.3\text{–}14)]$$

Note that in order to differentiate $g_{ij}(x')$ with respect to t' following a particle (constant $\hat{x}$) we have used the chain rule. Also in going from line 3 to line 4 one needs to recognize that the expressions in parentheses obey the transformation law for the covariant components of a second-order tensor. In Cartesian coordinates both of these expressions simplify to $(\partial v'_l/\partial x'_k + \partial v'_k/\partial x'_l)$; and since a tensor equation that is satisfied in one coordinate system is valid in any other coordinate system, the ( ) terms in lines 3 and 4 must be equal. Alternatively, one can show this equality by writing out line 4 and inserting the definitions of the Christoffel symbols. After cancellations are noted, the only terms that remain are those in line 3.

Before we operate on Eq. 9.3–10 let us rewrite it in the form:

$$\frac{\partial x'^i}{\partial \hat{x}^k}\frac{\partial x'^j}{\partial \hat{x}^l}\hat{g}^{kl}(\hat{x},t') = g^{ij}(x') \qquad [(9.3\text{–}15)]$$

We now operate on both sides of Eq. 9.3–15 with $\partial/\partial t'$ while keeping the convected coordinates constant. By using Eqs. 9.3–13 we find that:

$$\frac{\partial x'^i}{\partial \hat{x}^k}\frac{\partial x'^j}{\partial \hat{x}^l}\frac{\partial \hat{g}^{kl}(\hat{x},t')}{\partial t'} = -\frac{\partial x'^m}{\partial \hat{x}^k}\frac{\partial x'^j}{\partial \hat{x}^l}\frac{\partial v'^i}{\partial x'^m}\hat{g}^{kl}(\hat{x},t')$$

$$-\frac{\partial x'^i}{\partial \hat{x}^k}\frac{\partial x'^m}{\partial \hat{x}^l}\frac{\partial v'^j}{\partial x'^m}\hat{g}^{kl}(\hat{x},t') + v'^m\frac{\partial g^{ij}(x')}{\partial x'^m} \qquad [(9.3\text{–}16)]$$

Finally, we introduce Eqs. 9.3–10 into the first two terms on the right-hand side of Eq. 9.3–16. In this way we obtain for the time derivatives of the contravariant components of the metric tensor with respect to the convected coordinate system:

$$\frac{\partial \hat{g}^{ij}}{\partial t'} = -\frac{\partial \hat{x}^i}{\partial x'^k}\frac{\partial \hat{x}^j}{\partial x'^l}\left[g^{ml}\frac{\partial v'^k}{\partial x'^m} + g^{km}\frac{\partial v'^l}{\partial x'^m} - v'^m\frac{\partial g^{kl}}{\partial x'^m}\right]$$

$$= -\frac{\partial \hat{x}^i}{\partial x'^k}\frac{\partial \hat{x}^j}{\partial x'^l}\left[(v'_{n,m} + v'_{m,n})g^{nk}g^{ml}\right]$$

$$= -\frac{\partial \hat{x}^i}{\partial x'^k}\frac{\partial \hat{x}^j}{\partial x'^l}\dot{\gamma}^{kl}(x',t') \qquad [(9.3\text{–}17)]$$

Higher time derivatives may be found by repeated differentiations. In performing these, it must be kept in mind that the space fixed components can depend on t' explicitly as well as through $x'(\hat{x},t')$. As a general result we find:

$$\frac{\partial^n}{\partial t'^n}\,\hat{g}_{ij}(\hat{x},t') = \frac{\partial x'^k}{\partial \hat{x}^i}\frac{\partial x'^l}{\partial \hat{x}^j}\,\gamma_{kl}^{(n)}(x',t') \qquad [(9.3\text{–}18)]$$

$$\frac{\partial^n}{\partial t'^n}\,\hat{g}^{ij}(\hat{x},t') = -\frac{\partial \hat{x}^i}{\partial x'^k}\frac{\partial \hat{x}^j}{\partial x'^l}\,\gamma^{kl}_{(n)}(x',t') \qquad [(9.3\text{–}19)]$$

In these equations $\gamma_{kl}^{(n)}(x',t')$ and $\gamma^{kl}_{(n)}(x',t')$ are given by the conditions that $\gamma^{kl}_{(1)} = \dot{\gamma}^{kl}$ and $\gamma_{kl}^{(1)} = \dot{\gamma}_{kl}$ and the recurrence relations:

$$\gamma_{kl}^{(n+1)} = \frac{\partial}{\partial t}\,\gamma_{kl}^{(n)} + v^m\frac{\partial}{\partial x^m}\,\gamma_{kl}^{(n)} + \frac{\partial v^m}{\partial x^k}\,\gamma_{ml}^{(n)} + \gamma_{km}^{(n)}\,\frac{\partial v^m}{\partial x^l}$$

$$= \frac{\partial}{\partial t}\,\gamma_{kl}^{(n)} + v^m\gamma_{kl,m}^{(n)} + v^m_{,k}\gamma_{ml}^{(n)} + \gamma_{km}^{(n)}v^m_{,l} \qquad [(9.3\text{–}20)]$$

$$\gamma^{kl}_{(n+1)} = \frac{\partial}{\partial t}\,\gamma^{kl}_{(n)} + v^m\frac{\partial}{\partial x^m}\,\gamma^{kl}_{(n)} - \frac{\partial v^k}{\partial x^m}\,\gamma^{ml}_{(n)} - \gamma^{km}_{(n)}\,\frac{\partial v^l}{\partial x^m}$$

$$= \frac{\partial}{\partial t}\,\gamma^{kl}_{(n)} + v^m\gamma^{kl}_{(n),m} - v^k_{,m}\gamma^{ml}_{(n)} - \gamma^{km}_{(n)}v^l_{,m} \qquad [(9.3\text{–}21)]$$

In Eqs. 9.3–20 and 21 we have given two expressions for the recurrence relations. The first forms are the ones that are obtained directly in the differentiation process. These forms are in general easier to use for computational purposes. However each term in the first forms does not in general define the components of a tensor, and it is not obvious that the sum of these terms defines the space fixed components of a tensor. For this purpose the second forms involving covariant derivatives are more useful. To show that the two forms are equivalent we can use either of the methods discussed after Eq. 9.3–14. From the second forms of Eqs. 9.3–20 and 21 we see that the recurrence relations may be written for arbitrary second-order tensors $\mathbf{\Lambda}^{(n)}$ and $\mathbf{\Lambda}_{(n)}$:

$$\mathbf{\Lambda}^{(n+1)} = \frac{D}{Dt}\,\mathbf{\Lambda}^{(n)} + \{(\nabla \boldsymbol{v})\cdot\mathbf{\Lambda}^{(n)} + \mathbf{\Lambda}^{(n)}\cdot(\nabla \boldsymbol{v})^\dagger\} \qquad (9.3\text{–}22)$$

$$\mathbf{\Lambda}_{(n+1)} = \frac{D}{Dt}\,\mathbf{\Lambda}_{(n)} - \{(\nabla \boldsymbol{v})^\dagger\cdot\mathbf{\Lambda}_{(n)} + \mathbf{\Lambda}_{(n)}\cdot(\nabla \boldsymbol{v})\} \qquad (9.3\text{–}23)$$

These operations are defined also in Eqs. 9.2–24 and 25 in Cartesian coordinates.

The relations for the time derivatives of the covariant and contravariant components of the stress tensor with respect to the convected coordinate system are:

$$\frac{\partial^n}{\partial t'^n}\,\hat{\tau}_{ij}(\hat{x},t') = \frac{\partial x'^k}{\partial \hat{x}^i}\frac{\partial x'^l}{\partial \hat{x}^j}\,\tau_{kl}^{(n)}(x',t') \qquad [(9.3\text{–}24)]$$

$$\frac{\partial^n}{\partial t'^n}\hat{\tau}^{ij}(\hat{x},t') = -\frac{\partial \hat{x}^i}{\partial x'^k}\frac{\partial \hat{x}^j}{\partial x'^l}\tau^{kl}_{(n)}(x',t') \qquad [(9.3\text{–}25)]$$

In Eqs. 9.3–24 and 25 the $\tau_{kl}^{(n)}$ and $\tau^{kl}_{(n)}$ are determined by $\tau_{kl}^{(0)} = \tau_{kl}$ and $\tau^{kl}_{(0)} = \tau^{kl}$ with the recurrence relations in Eqs. 9.3–22 and 23.

We have now derived the transformation laws for the time derivatives of the covariant and contravariant components of the metric and stress tensors with respect to the convected coordinate system at an arbitrary past time t'. At the same time this gives us the transformation laws for time integrals of these quantities over past times. Indeed integration is a much simpler operation than differentiation since it involves only additions. In the convected coordinate system we can add the components of any tensor associated with a given particle $\hat{x}$ at different past times t' to obtain the components of a tensor associated with the particle $\hat{x}$. At all past times, however, we may use the transformation rules that we have already derived to rewrite the convected components in terms of fixed components. For example an integral involving the covariant convected components of the stress tensor may be written as follows:

$$\int_{t_1}^{t_2} \hat{\tau}_{ij}(\hat{x},t')\, dt' = \int_{t_1}^{t_2} \frac{\partial x'^k}{\partial \hat{x}^i}\frac{\partial x'^l}{\partial \hat{x}^j}\tau_{kl}(x',t')\, Dt' \qquad [(9.3\text{–}26)]$$

In Eq. 9.3–26 recall that the notation Dt' serves as a reminder that we integrate following the particle $\hat{x}$, that is, the argument x' in τ_{kl} is the location of the particle $\hat{x}$ at time t'.

This completes the discussion of the transformation rules for going from components with respect to the convected coordinates to components with respect to the fixed coordinates. Up to this point we have not specified the convected coordinates in any way. In the solution of actual problems, however, the convected coordinate system (as well as the fixed coordinate system) must be defined explicitly. The convected coordinate system is conveniently defined by specifying that it coincides exactly with the space coordinate system at some time t_0. This could be an initial time prior to which no flow has occurred, or any other well-defined time. Here, however, we shall exclusively use the present time t as the time at which the convected and the space fixed coordinates coincide. This is indicated in Fig. 9.3–1. With this choice the derivatives on the right side of the transformation laws may be rewritten in terms of the displacement functions by use of the chain rule:

$$\begin{aligned}\frac{\partial}{\partial \hat{x}^j} x'^i(\hat{x},t') &= \frac{\partial x^m}{\partial \hat{x}^j}\frac{\partial x'^i}{\partial x^m}\\ &= \delta_j^{\ m}\frac{\partial x'^i}{\partial x^m}\\ &= \frac{\partial}{\partial x^j} x'^i(x,t,t')\end{aligned} \qquad [(9.3\text{–}27)]$$

$$\frac{\partial}{\partial x'^j}\hat{x}^i(x',t') = \frac{\partial}{\partial x'^j} x^i(x',t,t') \qquad [(9.3\text{–}28)]$$

With the aid of Eqs. 9.3–27 and 28 the right sides of all the transformation laws may be written entirely in terms of space fixed coordinates and the displacement functions in Eqs. 9.3–5 and 6. When the right sides of the transformation laws are written in terms of the displacement functions, these define the covariant components of the tensors $\gamma^{[n]}$ and $\tau^{[n]}$, and the contravariant components of the tensors $\boldsymbol{\gamma}_{[n]}$ and $\boldsymbol{\tau}_{[n]}$, with respect to the space fixed coordinate system. For example, we can combine Eq. 9.3–27 with the transformation law for $\hat{g}_{ij}$ in Eqs. 9.3–9 to give[5]

$$\begin{aligned}[\hat{g}_{ij}(\hat{x},t') - \hat{g}_{ij}(\hat{x},t)] &= \left[\frac{\partial x'^m}{\partial x^i}\frac{\partial x'^n}{\partial x^j} g_{mn}(x') - g_{ij}(x)\right] \\ &= [\Delta^m{}_i g_{mn}(x')\,\Delta^n{}_j - g_{ij}(x)] \\ &\equiv \gamma_{ij}^{[0]}(x,t,t') \qquad [(9.3\text{–}29)]\end{aligned}$$

which may be found in the first row of Table 9.1–2. The second and third forms of $\gamma_{ij}^{[0]}$ given above are easily seen to be equal in Cartesian coordinates; and, hence, the equality holds generally. Note, however, that $\Delta^m{}_i \neq \partial x'^m/\partial x^i$ except in Cartesian coordinates. For computations involving curvilinear or nonorthogonal coordinates, the second form, which is suggested by the convected coordinate treatment, provides the most direct method for obtaining $\gamma_{ij}^{[0]}$. To complete the task of constructing the tensor $\boldsymbol{\gamma}^{[0]}$, we associate the above components with the reciprocal base vectors for the space fixed coordinate system:

$$\boldsymbol{\gamma}^{[0]} = \sum_i \sum_j \boldsymbol{g}^i \boldsymbol{g}^j \gamma_{ij}^{[0]} \qquad [(9.3\text{–}30)]$$

Any of the entries in columns (2′) of Tables 9.1–1 and 2 may be found by paralleling the above procedure. The entries in columns (2) are obtained by letting $t' = t$ in corresponding column (2′) entries or by using the recursion formulas in Eqs. 9.3–22 and 23 together with $\boldsymbol{\gamma}^{(1)} = \boldsymbol{\gamma}_{(1)} = \dot{\boldsymbol{\gamma}}$ and $\boldsymbol{\tau}^{(0)} = \boldsymbol{\tau}_{(0)} = \boldsymbol{\tau}$.

§9.4 EMPIRICAL RHEOLOGICAL EQUATIONS OF STATE

As we mentioned in the introduction to this chapter, most rheological model building has historically been done in terms of codeformational, rather than corotational, rheological tensors. Now that we have the codeformational formalism at our disposal, we could parallel the model building of Chapters 7 and 8 with the quantities in Tables 9.1–1 and 2 serving as the basic units for construction. However, little new would be learned by pursuing this line of activity here. Instead, we simply present in Tables 9.4–1 and 2 a brief (and incomplete) listing[1] of some empirical, nonlinear rheological models that have received attention in the literature. These are suggestive of the variety of forms that codeformational model building has led to.

[5] Suppose it is desired to have a measure of strain which can be used to compute stress, for example, at a time t_0 that is different from the time t when the convected and space-fixed coordinate systems are coincident or the time t' at which the strain occurs. If we use the difference in convected metric tensor components suggested in Eq. 9.3–29 and parallel the development leading up to Eq. 7.2–38b, then we are led to define the following space-fixed strain tensor components:

$$\gamma_{ij}^{[0]}(x,t,t';t_0) = \frac{\partial x^k}{\partial x_0{}^i}\frac{\partial x^l}{\partial x_0{}^j}\left[\frac{\partial x'^m}{\partial x^k}\frac{\partial x'^n}{\partial x^l} g_{mn}(x') - g_{kl}(x)\right] \qquad [(9.3\text{–}29a)]$$

where $x_0{}^i = x'^i(x,t,t')|_{t'=t_0}$. Note that $\gamma_{ij}^{[0]}(x,t,t';t)$ is identical to $\gamma_{ij}^{[0]}(x,t,t')$ defined by Eq. 9.3–29. Comparison with Eq. 7.2–38b shows that $\partial x^k/\partial x_0{}^i$ plays the same role here as $\Omega_{mi}(x,t,t_0)$ in that result.

[1] These tables are based on the summaries given by T. W. Spriggs, J. D. Huppler, and R. B. Bird, *Trans. Soc. Rheol.*, *10*, 191–213 (1966), and by M. Yamamoto, *Buttai no Henkeigaku*, Seibundō Shinkōsha, Tokyo (1972).

There are a few general points that need to be made concerning the codeformational models. First, the fact that there are two sets of codeformational tensors instead of the one set in the corotational framework—$\boldsymbol{\gamma}^{(n)}$ and $\boldsymbol{\gamma}_{(n)}$ in place of $\mathscr{D}^n\dot{\boldsymbol{\Gamma}}/\mathscr{D}t^n$, for instance—gives rise to some subtle differences in codeformational models. For example, recall that in generating an objective version of the Jeffreys model in Example 9.1–1, two different models, Oldroyd's fluids A and B, were obtained depending on whether $\partial\dot{\boldsymbol{\gamma}}/\partial t$ was replaced by $\boldsymbol{\gamma}^{(2)}$ or $\boldsymbol{\gamma}_{(2)}$, and so on. Now, even though these models are very similar in appearance, they contain very different fluid behavior. For example, in the rod-climbing experiment, it can be argued that fluid B would climb the rod whereas fluid A would dip near the rod (Problem 9B.8).

Second, some of these models, such as the White-Metzner or the Tanner differential models, have a small number of adjustable constants and functions and may thus be useful for rough engineering calculations. Some of the other models, however, have so much flexibility that it may be difficult to determine the adjustable functions from experimental data.[2]

Third, we must admit that in spite of the abundance of empirical models in the literature, there is no one completely satisfactory model in the sense of being able to describe quantitatively for a given polymeric fluid all of the different material functions presented in Chapter 4. Thus before a rheological model is applied to a hydrodynamics problem, it is important to test the model for simple flow situations related to the flow problem under consideration. We therefore conclude this section with several examples that illustrate the process of model testing and determination of model parameters.

EXAMPLE 9.4–1 Model Testing and Parameter Evaluation

This example is designed primarily to illustrate the process of model testing and evaluation of model parameters. For this purpose we choose the Bird-Carreau model, Eq. D in Table 9.4–2, with the parameter $\epsilon = 0$. The memory function is taken to be:[3]

$$M(t - t', II_{\dot{\gamma}}(t')) = \sum_{k=1}^{\infty} \frac{\eta_k}{\lambda_k^{(2)2}} \frac{e^{-(t-t')/\lambda_k^{(2)}}}{[1 + (\lambda_k^{(1)}\dot{\gamma}(t'))^2]} \tag{9.4–1}$$

where η_k, $\lambda_k^{(1)}$ and $\lambda_k^{(2)}$ are constants and $\dot{\gamma} = (\frac{1}{2}II_{\dot{\gamma}})^{1/2}$ is a measure of the "magnitude" of the tensor $\dot{\boldsymbol{\gamma}}$. Its inclusion in the memory function is designed to account for the destruction of the network structure of the macromolecular fluid due to flow. In order to reduce the number of parameters in the model, we use:[3]

$$\eta_k = \eta_0 \frac{\lambda_k^{(1)}}{\sum_{k=1}^{\infty} \lambda_k^{(1)}} \tag{9.4–2}$$

$$\lambda_k^{(n)} = \lambda_n \left(\frac{2}{k+1}\right)^{\alpha_n} \qquad \text{for } n = 1, 2 \tag{9.4–3}$$

This is a somewhat different empiricism than that used in Eqs. 6.1–14 and 15. The final memory function contains the five parameters: η_0, the zero-shear-rate viscosity; λ_1 and λ_2, time constants; and α_1 and α_2,

[2] For example, B. Bernstein, *Int. J. Non-linear Mech.*, *4*, 183–200 (1969) has suggested that a serious drawback of the K-BKZ model is the inability to determine both of the kernel functions M_1 and M_2 experimentally. In order to overcome this problem, L. J. Zapas, *J. Res. Nat. Bur. Stand.*, *70A*, 525–532 (1966) has proposed a more specific form of the K-BKZ model.

[3] R. B. Bird and P. J. Carreau, *Chem. Eng. Sci.*, *23*, 427–434 (1968); P. J. Carreau, I. F. Macdonald, and R. B. Bird, *Chem. Eng. Sci.*, *23*, 901–911 (1968). This type of rheological equation has been criticized by H. E. van Es and R. M. Christensen, *Trans. Soc. Rheol.*, *17*, 325–330 (1973), using a testing procedure suggested by M. Yamamoto, *Trans. Soc. Rheol.*, *15*, 331–344 (1971).

TABLE 9.4–1
Empirical Differential Models

Name	Reference	Constants or Functions	Equation for Model		Comments
White-Metzner	a	G, $\eta(\dot{\gamma})$	$\boldsymbol{\tau} + \frac{\eta}{G}\boldsymbol{\tau}_{(1)} = -\eta\boldsymbol{\gamma}_{(1)}$	(A)	Modification of Oldroyd B model (cf. Eq. 9.1–10) with $\lambda_2 = 0$.
Tanner	b	λ, $\eta(\dot{\gamma})$	$\boldsymbol{\tau} + \lambda\boldsymbol{\tau}_{(1)} = -\eta\boldsymbol{\gamma}_{(1)}$	(B)	Modification of Oldroyd B model (cf. Eq. 9.1–10) with $\lambda_2 = 0$.
3-Constant Oldroyd	c	$\eta_0, \lambda_1, \lambda_2$	$\boldsymbol{\tau} + \lambda_1(\boldsymbol{\tau}_{(1)} + \frac{1}{3}(\boldsymbol{\tau} : \boldsymbol{\gamma}_{(1)})\boldsymbol{\delta}) = -\eta_0[\boldsymbol{\gamma}_{(1)} + \lambda_2(\boldsymbol{\gamma}_{(2)} + \frac{1}{3}(\boldsymbol{\gamma}_{(1)} : \boldsymbol{\gamma}_{(1)})\boldsymbol{\delta})]$	(C)	Same as Eq. 8.1–2 with $\mu_1 = \lambda_1$, $\mu_2 = \lambda_2$, $\mu_0 = 0$, $\nu_1 = \nu_2 = \frac{2}{3}\lambda_1$.
4-Constant Oldroyd	d	$\eta_0, \lambda_1, \lambda_2, \mu_0$	$\boldsymbol{\tau} + \lambda_1\boldsymbol{\tau}_{(1)} + \frac{1}{2}\mu_0(\text{tr}\,\boldsymbol{\tau})\boldsymbol{\gamma}_{(1)} = -\eta_0(\boldsymbol{\gamma}_{(1)} + \lambda_2\boldsymbol{\gamma}_{(2)})$	(D)	Same as Eq. 8.1–2 with $\mu_1 = \lambda_1$, $\mu_2 = \lambda_2$, $\nu_1 = \nu_2 = 0$; see also Eq. 8.1–36.
Spriggs model	e	$\eta_0, \lambda, \alpha, \epsilon$	$\boldsymbol{\tau} = \sum_{k=1}^{\infty} \boldsymbol{\tau}_k$ $\boldsymbol{\tau}_k + \lambda_k[\boldsymbol{\tau}_{k(1)} - \frac{1}{2}\epsilon\{\boldsymbol{\tau}_k \cdot \boldsymbol{\gamma}_{(1)} + \boldsymbol{\gamma}_{(1)} \cdot \boldsymbol{\tau}_k\} + \frac{1}{3}(1 + \epsilon)(\boldsymbol{\tau}_k : \boldsymbol{\gamma}_{(1)})\boldsymbol{\delta}] = -\eta_k\boldsymbol{\gamma}_{(1)}$ with $\lambda_k = \frac{\lambda}{k^\alpha}$ and $\eta_k = \eta_0 \frac{\lambda_k}{\sum \lambda_k}$	(E)	Superposition of equations for $\boldsymbol{\tau}_k$ based on Eq. 8.1–2 with $\mu_1 = (1 + \epsilon)\lambda_1$, $\nu_2 = \frac{2}{3}(1 + \epsilon)\lambda_1$, $\mu_0 = \lambda_2 = \mu_2 = \nu_2 = 0$.
Rivlin-Ericksen	f	The f_i are functions of the joint invariants of $\boldsymbol{\gamma}^{(1)}$ and $\boldsymbol{\gamma}^{(2)}$. (See discussion after Eq. A.3–24.)	$\boldsymbol{\tau} = f_1\boldsymbol{\gamma}^{(1)} + f_2\boldsymbol{\gamma}^{(2)} + f_3\boldsymbol{\gamma}^{(1)2}$ $+ f_4\boldsymbol{\gamma}^{(2)2} + f_5\{\boldsymbol{\gamma}^{(1)} \cdot \boldsymbol{\gamma}^{(2)} + \boldsymbol{\gamma}^{(2)} \cdot \boldsymbol{\gamma}^{(1)}\}$ $+ f_6\{\boldsymbol{\gamma}^{(1)2} \cdot \boldsymbol{\gamma}^{(2)} + \boldsymbol{\gamma}^{(2)} \cdot \boldsymbol{\gamma}^{(1)2}\}$ $+ f_7\{\boldsymbol{\gamma}^{(1)} \cdot \boldsymbol{\gamma}^{(2)2} + \boldsymbol{\gamma}^{(2)2} \cdot \boldsymbol{\gamma}^{(1)}\}$ $+ f_8\{\boldsymbol{\gamma}^{(1)2} \cdot \boldsymbol{\gamma}^{(2)2} + \boldsymbol{\gamma}^{(2)2} \cdot \boldsymbol{\gamma}^{(1)2}\}$	(F)	The most general expression when it is postulated that $\boldsymbol{\tau}$ depends only on $\boldsymbol{\gamma}^{(1)}$ and $\boldsymbol{\gamma}^{(2)}$; if it is postulated that $\boldsymbol{\tau}$ depends only on $\boldsymbol{\gamma}^{(1)}$, then the Reiner-Rivlin equation is obtained (cf. Eq. 8.5–4).

[a] J. L. White and A. B. Metzner, *J. Appl. Polym. Sci.*, *7*, 1867–1889 (1963).
[b] R. I. Tanner, *ASLE Trans.*, *8*, 179–183 (1965).
[c] M. C. Williams and R. B. Bird, *Phys. Fluids*, *5*, 1126–1127 (1962); *Ind. Eng. Chem. Fundamentals*, *3*, 42–49 (1964).
[d] K. Walters and P. Townsend, *Proceedings of the Fifth International Con*[illegible]

Name	Reference	Equation for Model		Comments
Oldroyd-Walters-Fredrickson	a b c	$\boldsymbol{\tau} = -\int_{-\infty}^{t} G(t-t')\boldsymbol{\gamma}'_{[1]}\,dt'$	(A)	This is a codeformational analog of the Goddard-Miller equation of §7.5.
Lodge's rubberlike liquid	d	$\boldsymbol{\tau} = +\int_{-\infty}^{t} M(t-t')\boldsymbol{\gamma}'_{[0]}\,dt'$	(B)	For a derivation of this kind of model from a molecular network theory, see §15.3. For a derivation from dilute solution theory, see §12.3. This model and the Oldroyd-Walters-Fredrickson model are simply related by an integration by parts.
K–BKZ model	e f	$\boldsymbol{\tau} = +\int_{-\infty}^{t} [M_1(t-t', I_{\gamma'_{[0]}}, II_{\gamma'_{[0]}})\boldsymbol{\gamma}'_{[0]}$ $+ M_2(t-t', I_{\gamma'_{[0]}}, II_{\gamma'_{[0]}})\{\boldsymbol{\gamma}'_{[0]}\cdot\boldsymbol{\gamma}'_{[0]}\}]\,dt'$	(C)	This is the codeformational analog of Eq. 8.2–8, obtainable as an approximation to the codeformational memory-integral expansion.
Bird-Carreau, Bogue-Chen	g h	$\boldsymbol{\tau} = +\int_{-\infty}^{t} M(t-t', II_{\dot{\gamma}}(t'))\left[\left(1+\frac{\epsilon}{2}\right)\boldsymbol{\gamma}'_{[0]} - \left(\frac{\epsilon}{2}\right)\boldsymbol{\gamma}^{[0]'}\right]dt'$	(D)	The Bird-Carreau memory function M is given in Eqs. 9.4–1 to 3.
Tanner-Simmons network rupture model	i	$\boldsymbol{\tau} = +\int_{t-t_R}^{t} M(t-t')\left[\left(1+\frac{\epsilon}{2}\right)\boldsymbol{\gamma}'_{[0]} - \left(\frac{\epsilon}{2}\right)\boldsymbol{\gamma}^{[0]'}\right]dt'$	(E)	t_R is the "network rupture time"; for steady-state shear flow this is defined by $\dot{\gamma}^2 t_R{}^2 = B^2$, where B is a parameter (the "straingth" of the network) characteristic of each fluid. For other flows t_R depends on $I_{\gamma'_{[0]}}$ and $II_{\gamma'_{[0]}}$.
Carreau (Model B)	j	$\boldsymbol{\tau} = \int_{-\infty}^{t} \underset{t''=t'}{\overset{t}{\mathscr{M}}}\,[t-t', II_{\dot{\gamma}}(t'')]\left[\left(1+\frac{\epsilon}{2}\right)\boldsymbol{\gamma}'_{[0]} - \left(\frac{\epsilon}{2}\right)\boldsymbol{\gamma}^{[0]'}\right]dt'$	(F)	$\mathscr{M}$ is a function of $t-t'$ and a functional of $II_{\dot{\gamma}}(t'')$, on $t' \le t'' \le t$. See §15.4 for a discussion of this model, and its relation to the network theory.

[a] J. G. Oldroyd, *Proc. Roy. Soc.* (London), *A200*, 45–63 (1950), Eq. 57.
[b] K. Walters, *Quart. J. Mech. Appl. Math.*, *13*, 444–461 (1960).
[c] A. G. Fredrickson, *Chem. Eng. Sci.*, *17*, 155–166 (1962).
[d] A. S. Lodge, *Trans. Faraday Soc.*, *52*, 120–130 (1956); *Elastic Liquids*, Academic Press, New York (1964), Chapter 6.
[e] A. Kaye, College of Aeronautics, Cranfield, Note No. 134 (1962).
[f] B. Bernstein, E. A. Kearsley, and L. J. Zapas, *Trans. Soc. Rheol.*, *7*, 391–410 (1963).
[g] R. B. Bird, and P. J. Carreau, *Chem. Eng. Sci.*, *23*, 427–434 (1968); P. J. Carreau, I. F. Macdonald, and R. B. Bird, *ibid.*, *23*, 901–911 (1968).
[h] I. Chen and D. C. Bogue, *Trans. Soc. Rheol.*, *16*, 59–78 (1972).
[i] R. I. Tanner and J. M. Simmons, *Chem. Eng. Sci.*, *22*, 1803–1815 (1967); R. I. Tanner, *A.I.Ch.E. Journal*, *15*, 177–183 (1969).
[j] P. J. Carreau, *Trans. Soc. Rheol.*, *16*, 99–129 (1972).

dimensionless parameters. This model is indeed complex, but it is capable of describing quantitatively a large number of phenomena (e.g., non-Newtonian viscosity, primary normal stress coefficient, complex viscosity,[3] stress relaxation); on the other hand it is known to describe other flows only qualitatively (e.g., stress growth at inception of steady shear flow, hysteresis loops). Furthermore, with the simplified memory function given above, the Bird-Carreau model has only a small number of constants, with simple physical meaning, which can be determined from standard material-function measurements.

It is desired to show how to evaluate the five model parameters from experimental data on (a) steady simple shear flow and (b) small-amplitude oscillatory shear flow.

SOLUTION (a) Steady Simple Shear Flow

In order to determine the strain tensor $\boldsymbol{\gamma}_{[0]}$ we first need to know the displacement functions. For the steady shear flow $v_x = \dot{\gamma} y$, $v_y = 0$, $v_z = 0$, these are:

$$x' = x - \dot{\gamma} y(t - t')$$
$$y' = y$$
$$z' = z \tag{9.4-4}$$

These are readily inverted to give $x_i(x', y', z', t, t')$ from which we find:

$$\mathrm{E}_{ij} = \frac{\partial x_i}{\partial x'_j} = \begin{pmatrix} 1 & \dot{\gamma}(t-t') & 0 \\ 0 & 1 & 0 \\ 0 & 0 & 1 \end{pmatrix} \tag{9.4-5}$$

Hence, by matrix multiplication we obtain $\boldsymbol{\gamma}_{[0]}$:

$$\begin{aligned} \gamma_{[0]ij} &= \delta_{ij} - \textstyle\sum_m \mathrm{E}_{im}\mathrm{E}^{\dagger}_{mj} \\ &= \begin{pmatrix} 1 & 0 & 0 \\ 0 & 1 & 0 \\ 0 & 0 & 1 \end{pmatrix} - \begin{pmatrix} 1 & \dot{\gamma}(t-t') & 0 \\ 0 & 1 & 0 \\ 0 & 0 & 1 \end{pmatrix}\begin{pmatrix} 1 & 0 & 0 \\ \dot{\gamma}(t-t') & 1 & 0 \\ 0 & 0 & 1 \end{pmatrix} \\ &= -\begin{pmatrix} \dot{\gamma}^2(t-t')^2 & \dot{\gamma}(t-t') & 0 \\ \dot{\gamma}(t-t') & 0 & 0 \\ 0 & 0 & 0 \end{pmatrix} \end{aligned} \tag{9.4-6}$$

When this strain tensor is inserted into the Bird-Carreau model with the memory function given in the problem statement, the following viscometric functions are obtained:

$$\begin{aligned} \eta &= \sum_{k=1}^{\infty} \frac{\eta_k}{1 + (\lambda_k^{(1)}\dot{\gamma})^2} \\ &\doteq \frac{\pi\eta_0}{\zeta(\alpha_1) - 1} \frac{(2^{\alpha_1}\lambda_1\dot{\gamma})^{(1-\alpha_1)/\alpha_1}}{2\alpha_1 \sin\left(\dfrac{1+\alpha_1}{2\alpha_1}\pi\right)} \qquad \text{(large } \dot{\gamma}\text{)} \end{aligned} \tag{9.4-7}$$

$$\begin{aligned} \Psi_1 &= 2\sum_{k=1}^{\infty} \frac{\eta_k\lambda_k^{(2)}}{1 + (\lambda_k^{(1)}\dot{\gamma})^2} \\ &\doteq \frac{2^{\alpha_2+1}\lambda_2\pi\eta_0}{\zeta(\alpha_1) - 1} \frac{(2^{\alpha_1}\lambda_1\dot{\gamma})^{(1-\alpha_1-\alpha_2)/\alpha_1}}{2\alpha_1 \sin\left(\dfrac{1+\alpha_1-\alpha_2}{2\alpha_1}\pi\right)} \qquad \text{(large } \dot{\gamma}\text{)} \end{aligned} \tag{9.4-8}$$

$$\Psi_2 = 0 \tag{9.4-9}$$

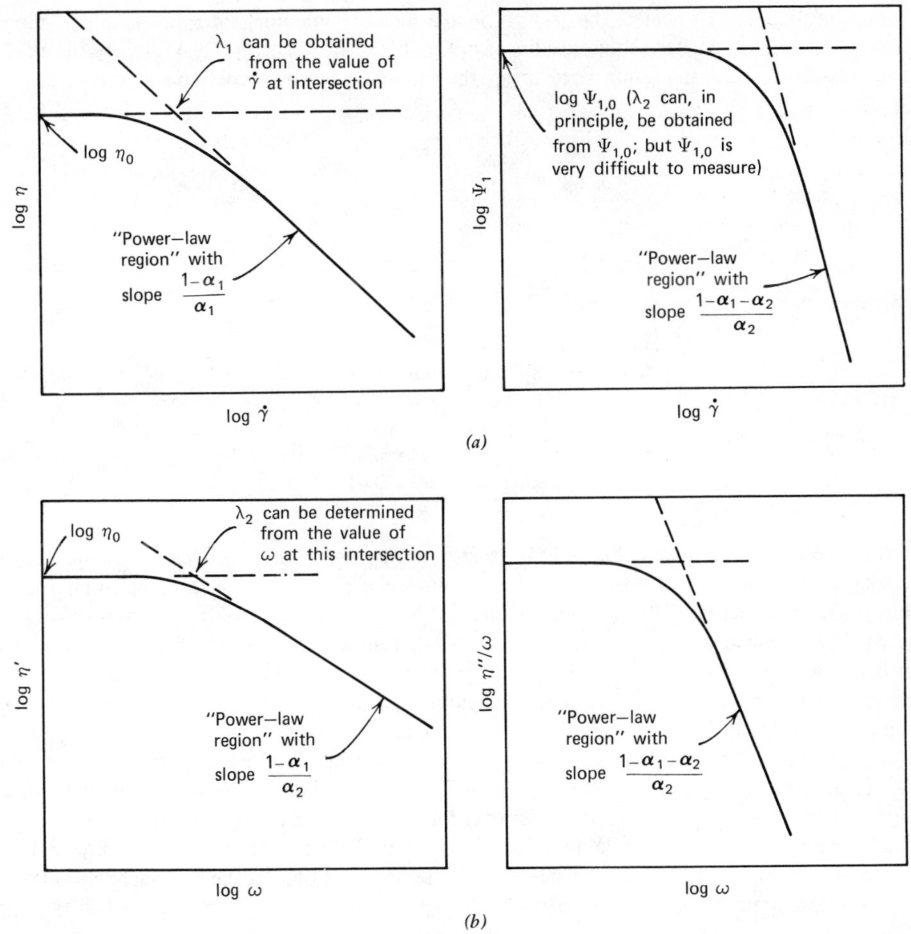

FIGURE 9.4–1. Shear flow material functions for the Bird-Carreau model. (*a*) Sketches of η and Ψ_1 given in Eqs. 9.4–7 and 8 that illustrate how all five model parameters, η_0, λ_1, λ_2, α_1, and α_2, may be obtained by simple geometrical constructions. (*b*) Sketches of η' and η'' from Eqs. 9.4–11 and 12 that show how the parameters η_0, λ_2, α_1, and α_2 may be obtained by simple geometrical constructions.

In order to perform the integrations that lead to Eqs. 9.4–7 and 8 it is convenient to make the substitution $s = t - t'$. The second lines of Eqs. 9.4–7 and 8 are high shear rate asymptotes; these are obtained from the first line expressions by using the Euler-Maclaurin expansion (Problem 6B.5). Also, in these equations $\zeta(\alpha)$ is the Riemann Zeta function for which tables are available. Sketches of η and Ψ_1 are shown in Fig. 9.4–1. It is evident that all five constants can be determined from experimental data on the viscosity function and the primary normal stress coefficient. Note that $\Psi_2 = 0$ for this form of the model; a non-zero Ψ_2 can be obtained by choosing $\epsilon \neq 0$.

(b) Small-Amplitude Oscillatory Shear Flow

For the small-amplitude oscillatory shear flow $v_x = (\dot{\gamma}^0 \cos \omega t)y$ described in Fig. 4.3–1, the inverted displacement functions are $x = x' + (\dot{\gamma}^0 y'/\omega)(\sin \omega t - \sin \omega t')$, $y = y'$, and $z = z'$. From these we can compute $\mathbf{E}$ and then $\boldsymbol{\gamma}_{[0]}$ as in part (a). In this way the yx-component of $\boldsymbol{\gamma}_{[0]}$ is found to be:

$$\gamma_{[0]yx} = -\frac{\dot{\gamma}^0}{\omega}(\sin \omega t - \sin \omega t') \tag{9.4–10}$$

For computing τ_{yx} we can neglect the $\dot{\gamma}^2$ term in the memory function, since in the small-amplitude oscillatory flow we look for terms that are linear in the shear rate. When Eq. 9.4–10 is put in the Bird-Carreau model and the integration performed, the following linear viscoelastic functions are found (cf. Eq. 4.4–3):

$$\eta' = \sum_{k=1}^{\infty} \frac{\eta_k}{1 + (\lambda_k^{(2)}\omega)^2}$$

$$\doteq \frac{\pi\eta_0}{\zeta(\alpha_1) - 1} \frac{(2^{\alpha_2}\lambda_2\omega)^{(1-\alpha_1)/\alpha_2}}{2\alpha_2 \sin\left(\dfrac{1 + 2\alpha_2 - \alpha_1}{2\alpha_2}\pi\right)} \quad \text{(large } \omega\text{)} \tag{9.4–11}$$

$$\frac{\eta''}{\omega} = \sum_{k=1}^{\infty} \frac{\eta_k \lambda_k^{(2)}}{1 + (\lambda_k^{(2)}\omega)^2}$$

$$\doteq \frac{2^{\alpha_2}\lambda_2\pi\eta_0}{\zeta(\alpha_1) - 1} \frac{(2^{\alpha_2}\lambda_2\omega)^{(1-\alpha_1-\alpha_2)/\alpha_2}}{2\alpha_2 \sin\left(\dfrac{1 + \alpha_2 - \alpha_1}{2\alpha_2}\pi\right)} \quad \text{(large } \omega\text{)} \tag{9.4–12}$$

where the second, high-frequency forms have been found by using the Euler-Maclaurin sum formula. These results are shown in Fig. 9.4–1. Here the parameters η_0, λ_2, α_1, and α_2 are seen to have a rather simple significance in terms of the functions η' and η''. Note that λ_1 may not be determined from linear viscoelastic measurements, since it is a parameter associated entirely with nonlinear behavior. For models in which the parameters do not have such a simple geometrical interpretation (as we have obtained in the present example), one will in general have to use a nonlinear least squares analysis for the determination of the model parameters.

The parameters for the Bird-Carreau model discussed here are given in Table 9.4–3 for four polymer solutions, one polymer melt, and one soap solution. These fluids have widely different rheological properties. The zero-shear-rate viscosity varies over four decades, and the time constants vary over three decades. Yet a successful fit of the data was obtained for all fluids for the steady shear flow and small-amplitude oscillatory flow properties considered. This serves to emphasize the thoroughness with which a rheological testing program should be performed.

TABLE 9.4–3

Parameters to Be Used with the Bird-Carreau Model of Example 9.4–1[a]

Fluid	η_0 (N·s/m²)	α_1	α_2	λ_1 (s)	λ_2 (s)	α_2/α_1	λ_1/λ_2
0.9% Hydroxyethyl Cellulose (HEC) in water	1.37	2.60	2.13	0.0981	0.0972	0.82	1.01
0.75% Polyacrylamide (Dow ET 592) in 95/5 H_2O-glycerine	10.4	2.51	1.91	6.93	7.15	0.76	0.97
1.50% Polyacrylamide (Dow ET 592) in 50/50 H_2O-glycerine	280.0	3.01	2.59	45.7	69.8	0.86	0.65
2.0% Polyisobutylene in Primol 355	996.8	2.85	2.33	149.5	170.3	0.82	0.88
7% Aluminum Soap in Decalin and *m*-Cresol	83.1	4.17	3.37	0.838	1.44	0.81	0.582
Phenoxy-A melt at 485 K	12694	1.4	1.1	5.18	6.59	0.79	0.786

[a] P. J. Carreau, I. F. Macdonald, and R. B. Bird, *Chem. Eng. Sci.*, *23*, 901–911 (1968).

The fact that λ_1 and λ_2 are nearly the same suggests that the time constant determined from viscosity will probably be useful for describing elastic effects—at least for some fluids. For extremely difficult flow problems where the prospects of getting an analytical solution seem hopeless, dimensional analysis correlations may be possible. In that case the parameters η_0, α_1, and λ_1 can be used to construct dimensionless groups. Then these groups may be employed to map out flow regimes (as Marsh[4] has done for hysteresis loops) or to get data correlations (such as friction factor charts). Very little effort has been directed along these lines. Because at least three parameters are needed to characterize macromolecular fluids, one can anticipate that the preparation of such correlations will require an enormous effort.

EXAMPLE 9.4–2 Start Up of Elongational Flow for Lodge's Rubberlike Liquid[5]

Obtain an expression for the buildup of stress in a fluid that is at rest for time $t < 0$ and that is suddenly made to undergo an elongational flow with constant elongation rate $\dot{\epsilon}_0$ for $t \geq 0$. In terms of a Cartesian coordinate system x_1, x_2, x_3 (x_1, x_2, and x_3 are convenient symbols for x, y, and z), the velocity field is:

$$v_1 = -\tfrac{1}{2}\dot{\epsilon}(t)x_1$$

$$v_2 = -\tfrac{1}{2}\dot{\epsilon}(t)x_2 \qquad (9.4\text{–}13)$$

$$v_3 = \dot{\epsilon}(t)x_3$$

$$\dot{\epsilon}(t) = \begin{cases} 0 & \text{for} \quad t < 0 \\ \dot{\epsilon}_0 & \text{for} \quad t \geq 0 \end{cases} \qquad (9.4\text{–}14)$$

For $t \geq 0$, the stresses are conveniently described in terms of $\bar{\eta}^+(t;\dot{\epsilon}_0) = (\tau_{11} - \tau_{33})/\dot{\epsilon}_0$. The rheological equation of state to be used for obtaining the stress components is the Lodge (1956) rubberlike liquid, Eq. B in Table 9.4–2. In particular take the memory function to have the form:

$$M(t - t') = \sum_{k=1}^{\infty} \frac{\eta_k}{\lambda_k^2} e^{-(t-t')/\lambda_k} \qquad (9.4\text{–}15)$$

SOLUTION

In order to calculate $\gamma_{[0]}$ appearing in the rheological equation of state, we first need to solve for the displacement functions $x_i' = x_i'(x,t,t')$. From Eqs. 9.4–13 we find for $t' \geq 0$:

$$v_1 = \frac{\partial}{\partial t'} x_1'(x,t,t') = -\tfrac{1}{2}\dot{\epsilon}_0 x_1'(x,t,t') \qquad (t' \geq 0) \qquad (9.4\text{–}16)$$

We integrate this to get:

$$x_1'(x,t,t') = C_1(x,t)e^{-(1/2)\dot{\epsilon}_0 t'} \qquad (t' \geq 0) \qquad (9.4\text{–}17)$$

The function $C_1(x,t)$ may be determined from the initial condition that $x_1' = x_1$ at $t' = t$, that is,

$$x_1 = C_1(x,t)e^{-(1/2)\dot{\epsilon}_0 t} \qquad (9.4\text{–}18)$$

By this method we find the functions:

$$x_1 = x_1' e^{-(1/2)\dot{\epsilon}_0(t-t')}$$

$$x_2 = x_2' e^{-(1/2)\dot{\epsilon}_0(t-t')}$$

$$x_3 = x_3' e^{\dot{\epsilon}_0(t-t')} \qquad (0 \leq t' \leq t) \qquad (9.4\text{–}19)$$

[4] R. B. Bird and B. D. Marsh, *Trans. Soc. Rheol.*, *12*, 479–488 (1968); B. D. Marsh, *Trans. Soc. Rheol.*, *12*, 489–510 (1968).

[5] H. Chang and A. S. Lodge, *Rheol. Acta*, *11*, 127–129 (1972).

For times $t' < 0$ we have:

$$v_1 = \frac{\partial}{\partial t'} x_1'(x,t,t') = 0 \qquad (t' < 0) \tag{9.4–20}$$

Hence

$$x_1' = C_2(x,t) \qquad (t' < 0) \tag{9.4–21}$$

The function $C_2(x,t)$ may be determined by evaluating Eqs. 9.4–19 and 21 at $t' = 0$ and using the continuity of the displacement functions. We find:

$$x_1 = x_1' e^{-(1/2)\dot{\epsilon}_0 t}$$

$$x_2 = x_2' e^{-(1/2)\dot{\epsilon}_0 t}$$

$$x_3 = x_3' e^{\dot{\epsilon}_0 t} \qquad (-\infty < t' \leq 0) \tag{9.4–22}$$

We may now calculate the matrix $E_{ij} = \partial x_i / \partial x_j'$ to get:

$$E_{ij}(t,t') = \begin{pmatrix} e^{-(1/2)\dot{\epsilon}_0(t-t')} & 0 & 0 \\ 0 & e^{-(1/2)\dot{\epsilon}_0(t-t')} & 0 \\ 0 & 0 & e^{\dot{\epsilon}_0(t-t')} \end{pmatrix} \qquad (0 \leq t' \leq t) \tag{9.4–23}$$

$$E_{ij}(t,t') = E_{ij}(t,0) \qquad (-\infty < t' < 0) \tag{9.4–24}$$

where $E_{ij}(t,0)$ is given by Eq. 9.4–23. Hence, using Table 9.1–2 we find:

$$\gamma_{[0]ij}(t,t') = \begin{pmatrix} 1 - e^{-\dot{\epsilon}_0(t-t')} & 0 & 0 \\ 0 & 1 - e^{-\dot{\epsilon}_0(t-t')} & 0 \\ 0 & 0 & 1 - e^{2\dot{\epsilon}_0(t-t')} \end{pmatrix} \qquad (0 \leq t' \leq t) \tag{9.4–25}$$

$$\gamma_{[0]ij}(t,t') = \gamma_{[0]ij}(t,0) \qquad (-\infty < t' < 0) \tag{9.4–26}$$

When the strain tensor in Eqs. 9.4–25 and 26 and the memory function in Eq. 9.4–15 are inserted into Eq. B of Table 9.4–2 and the integration is performed, we find

$$\bar{\eta}^+(t;\dot{\epsilon}_0) = \sum_{k=1}^{\infty} \frac{\eta_k}{\lambda_k^2} \left[\frac{3}{(\lambda_k^{-1} - 2\dot{\epsilon}_0)(\lambda_k^{-1} + \dot{\epsilon}_0)} - \frac{2\lambda_k}{(\lambda_k^{-1} - 2\dot{\epsilon}_0)} e^{-(\lambda_k^{-1} - 2\dot{\epsilon}_0)t} - \frac{\lambda_k}{(\lambda_k^{-1} + \dot{\epsilon}_0)} e^{-(\lambda_k^{-1} + \dot{\epsilon}_0)t} \right] \tag{9.4–27}$$

It is interesting to note that the elongational stress growth function changes character when the imposed elongation rate exceeds the reciprocal of twice the maximum relaxation time. When $\dot{\epsilon}_0 < (2\lambda_{k,\max})^{-1}$ a steady state elongation is approached, but when $\dot{\epsilon}_0 > (2\lambda_{k,\max})^{-1}$ no steady state is predicted. If $\dot{\epsilon}_0 = (2\lambda_k)^{-1}$ for some $k = r$, the integration on t' for that term must be treated separately. We find that the term in Eq. 9.4–27 with $k = r$ must be replaced by:

$$\frac{\eta_r}{\lambda_r^2} (6\dot{\epsilon}_0 t + 1 - e^{-3\dot{\epsilon}_0 t})/(6\dot{\epsilon}_0^2) \tag{9.4–28}$$

The predictions of Eq. 9.4–27 have been compared with Meissner's data for a polyethylene melt by Chang and Lodge[5] (Fig 9.4–2). In their calculations, the memory function was arbitrarily assumed to have five terms of the type shown in Eq. 9.4–15, and a five-term version of Eq. 9.4–27 was consequently used to fit the data. The constants were chosen by matching Eq. 9.4–27 to the data for $\dot{\epsilon}_0 = 0.001\ \mathrm{s}^{-1}$

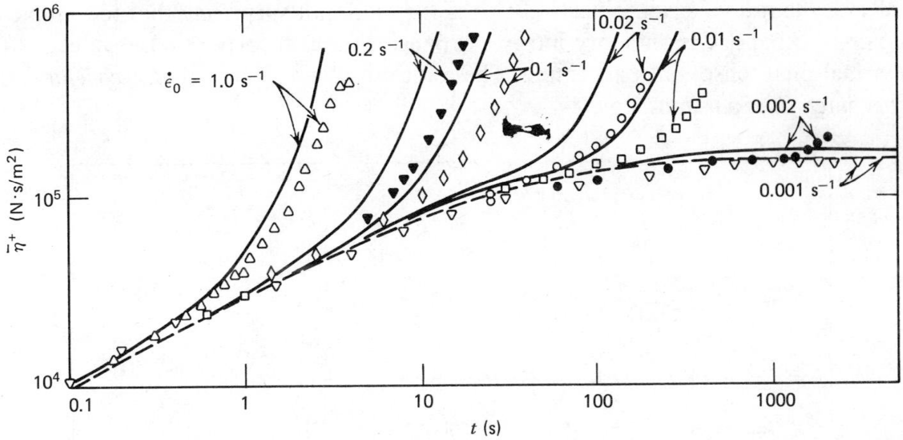

FIGURE 9.4–2. Comparison of calculated and experimental results for $\bar{\eta}^+(t;\dot{\epsilon}_0)$ for a low-density polyethylene melt. △▼◇○□●▽ Experimental data [J. Meissner, *Rheol. Acta.*, *10*, 230–242 (1971).] ———— Predictions of Eqs. 9.4–27 and 28 for Lodge's rubberlike liquid with a five-term memory function, Eq. 9.4–15. The constants for the memory function are tabulated in the caption to Fig. 7.4–2. [H. Chang and A. S. Lodge, *Rheol. Acta*, *11*, 127–129 (1972).]

(which appear to lie in the linear viscoelastic regime) at five times.[6] The rubberlike liquid is seen to describe the upward sweeping nature of the curves rather well. An approximate, truncated version of the corotational memory integral expansion has also been compared to Meissner's data.[7] The latter results do equally well at predicting the rapidly rising part of the curves; in addition, they predict that the data should eventually approach steady-state values. No steady-state values are obtained for the rubberlike liquid at the higher elongation rates. Unfortunately, the data do not extend far enough to distinguish between these predictions.

The comparison given in this example indicates that the Lodge rubberlike liquid is a good model for transient elongational flows, at least for polyethylene melts. On the other hand, the rubberlike liquid predicts a shear-rate-independent viscosity in steady shear flow, and the model should therefore not be used in flow problems where it is judged that non-Newtonian viscosity effects are important.

§9.5 MEMORY INTEGRAL EXPANSIONS[1]

In Chapter 8 we introduced the rheologically simple fluid as a fluid for which the stress in a given fluid element depends only on the deformation history of that fluid element and not on the history of neighboring elements. This concept is expressed mathematically by a tensor functional. If a number of assumptions are made about this functional, it may be expanded into the corotational memory integral expansion, Eq. 8.3–1, in which the deformation history of a fluid element is specified in terms of the rate-of-strain tensor $\dot{\mathbf{\Gamma}}$. At this point it is clear that there are other objective tensors, namely $\boldsymbol{\gamma}_{[1]}$ and $\boldsymbol{\gamma}^{[1]}$, which can be used

[6] A somewhat better fit of the data has been obtained by H. Chang, Ph.D. Thesis, University of Wisconsin, Madison (1973) by fitting the constants in a five-term memory function to data for the two lowest elongation rates. See also A. S. Lodge, *Body Tensor Fields in Continuum Mechanics*, Academic Press, New York (1974), p. 226.

[7] R. B. Bird, O. Hassager, and S. I. Abdel-Khalik, *A.I.Ch.E. Journal*, *20*, 1041–1065 (1974).

[1] The basis for this section is the fundamental development on materials with memory by A. E. Green and R. S. Rivlin, *Arch. Rat. Mech. Anal.*, *1*, 1–21 (1957) with further developments by A. E. Green, R. S. Rivlin, and A. J. M. Spencer, *Arch. Rat. Mech. Anal.*, *3*, 82–92 (1956); A. E. Green and R. S. Rivlin, *Arch. Rat. Mech. Anal.*, *4*, 387–404 (1960); B. D. Coleman and W. Noll, *Rev. Mod. Phys.*, *33*, 239–249 (1961), *errata*: *ibid.*, *36*, 1103 (1964); and A. C. Pipkin, *Rev. Mod. Phys.*, *36*, 1034–1041 (1964).

to describe the rate of strain objectively. It is therefore not surprising that the simple fluid idea can be expanded in memory integral expansions that incorporate the values of these codeformational tensors at past times. These formulas, which we call the *codeformational memory integral expansions*, are:

$$\begin{aligned}\boldsymbol{\tau} = &-\int_{-\infty}^{t} G_1(t - t')\boldsymbol{\gamma}^{[1]'}\,dt' \\ &- \tfrac{1}{2}\int_{-\infty}^{t}\int_{-\infty}^{t} G_2(t - t', t - t'')\{\boldsymbol{\gamma}^{[1]'}\cdot\boldsymbol{\gamma}^{[1]''} \\ &+ \boldsymbol{\gamma}^{[1]''}\cdot\boldsymbol{\gamma}^{[1]'}\}\,dt''\,dt' \\ &- \tfrac{1}{2}\int_{-\infty}^{t}\int_{-\infty}^{t}\int_{-\infty}^{t} [2G_3(t - t', t - t'', t - t''')\boldsymbol{\gamma}^{[1]'}(\boldsymbol{\gamma}^{[1]''} : \boldsymbol{\gamma}^{[1]'''}) \\ &+ G_4(t - t', t - t'', t - t''')\{\boldsymbol{\gamma}^{[1]'}\cdot\boldsymbol{\gamma}^{[1]''}\cdot\boldsymbol{\gamma}^{[1]'''} \\ &+ \boldsymbol{\gamma}^{[1]'''}\cdot\boldsymbol{\gamma}^{[1]''}\cdot\boldsymbol{\gamma}^{[1]'}\}]\,dt'''\,dt''\,dt' - \cdots \end{aligned} \tag{9.5-1}$$

$$\begin{aligned}\boldsymbol{\tau} = &-\int_{-\infty}^{t} G^1(t - t')\boldsymbol{\gamma}'_{[1]}\,dt' - \tfrac{1}{2}\int_{-\infty}^{t}\int_{-\infty}^{t} G^2(t - t', t - t'')\{\boldsymbol{\gamma}'_{[1]}\cdot\boldsymbol{\gamma}''_{[1]} \\ &+ \boldsymbol{\gamma}''_{[1]}\cdot\boldsymbol{\gamma}'_{[1]}\}\,dt''\,dt' - \cdots \end{aligned} \tag{9.5-2}$$

The third-order terms for Eq. 9.5–2 have exactly the same form as those in Eq. 9.5–1. Here we have used the abbreviations $\boldsymbol{\gamma}^{[1]'} = \boldsymbol{\gamma}^{[1]}(x,t,t')$, etc.

As a justification for Eqs. 9.5–1 and 2 we recall that the form of the corotational memory integral expansion was suggested by the series of multiple integrals that resulted from an integration of the Oldroyd eight-constant model. The $\dot{\boldsymbol{\Gamma}}$ tensor appeared inside the integrals because we wrote the Oldroyd model in terms of the corotating rate-of-strain tensor. From §9.1 we know that this model can be rewritten in terms of $\boldsymbol{\tau}^{(n)}$ and $\boldsymbol{\gamma}^{(n)}$ or $\boldsymbol{\tau}_{(n)}$ and $\boldsymbol{\gamma}_{(n)}$. In addition, we have seen in Example 9.1–1 how integration of a model containing $\boldsymbol{\gamma}^{(1)}$ gives rise to an integral over $\boldsymbol{\gamma}^{[1]}$ and similarly for $\boldsymbol{\gamma}_{(1)}$ and $\boldsymbol{\gamma}_{[1]}$. The iterative integration scheme that led to Eq. 8.1–14 can then be used on the codeformational formulations of Oldroyd's eight-constant model. The results are series of the forms in Eqs. 9.5–1 and 2 with the G's given as specific functions involving the model parameters.

It is important to note that Eqs. 9.5–1 and 2 and Eq. 8.3–1 are all equivalent in the sense that they are all expansions representing the same simple fluid postulate. In other words, if we could somehow use the entire series for computations of $\boldsymbol{\tau}$ made with each of the memory integral expansions, we should find that each gives the same answer. From a practical point of view, however, we can only keep a few terms in calculations. This means that the convergence properties[2] of the three expansions are important, and we shall now look at this aspect of Eqs. 8.3–1, 9.5–1 and 2.

In order to compare the convergence properties of the corotational and codeformational memory integral expansions, we need a rheological equation of state from which the G's can be computed and from which exact (meaning not restricted to a particular range of deformation rates) values of fluid properties can be obtained. The kinetic theory for dilute solutions of rigid rodlike particles (§2.5, Chapter 11) provides the needed equations. The first two kernel functions in the corotational memory integral expansion are obtained by comparing Eq. 2.5–3 and Eq. 8.3–1 when both are written for arbitrary, irrotational flows

[2] J. D. Goddard, *Trans. Soc. Rheol.*, *11*, 381–399 (1967).

$(\boldsymbol{\omega} = \mathbf{0})^3$. Similarly, G_1 and G_2 in Eq. 9.5–1 can be determined by a comparison with Eq. 2.5–3 for an irrotational flow.[4] In making this latter comparison, we rewrite $\boldsymbol{\gamma}^{[1]'}$ in terms of $\dot{\boldsymbol{\gamma}}'$ by using Eq. 9.1–7 and the definition of $\boldsymbol{\gamma}^{[1]}$ in Table 9.1–2. Naturally with the G's determined in this way, corotational and codeformational expansions are equivalent through terms of second order in the rate-of-strain tensor; and, furthermore, neither of them is exact outside this range.

For two particular flows, steady simple shear[5] and steady elongation,[6] the behavior of suspensions of rigid rodlike particles has been determined theoretically over the entire range of deformation rates (Chapter 11). We may therefore inquire as to how well the memory integral expansions in terms of the $\dot{\boldsymbol{\Gamma}}$ tensor and the $\boldsymbol{\gamma}^{[1]}$ tensor approximate the exact behavior outside the second-order range. This comparison is shown in Figs. 9.5–1 and 2. We see that the corotational memory integral expansion with two terms approximates the steady shear flow behavior fairly closely over the entire range of shear rates, whereas the codeformational expansion does not. For steady elongational flow the difference between the two expansions is less dramatic, but it still appears that the corotational expansion converges more rapidly than the codeformational expansion. In Figs. 9.5–1 and 2 we have included terms of third order in the rate-of-deformation, which can be obtained in the ways described above for the second order contributions. For additional comparison in these figures, we have also shown the predictions of the Oldroyd eight-constant model Eq. 8.1–2 in which the parameters have been chosen to match exactly, through third-order contributions, all of the steady-state properties from the kinetic theory calculations (see Problem 11B.9):

$$\begin{aligned} \eta_0 &= \eta_s + nkT\lambda & \mu_0 &= -\tfrac{2}{7}\lambda \\ \lambda_1 &= \lambda & \mu_1 &= -\tfrac{1}{7}\lambda \\ \lambda_2 &= \lambda\phi(-3/5) & \mu_2 &= -\tfrac{1}{7}\lambda\phi(21/5) \end{aligned}$$

$$\nu_1 = \nu_2 = 0 \tag{9.5–3}$$

Here $\phi(a) = 1 + a(nkT\lambda)/(\eta_s + nkT\lambda)$, and the kinetic theory parameters are the solvent viscosity η_s, the number density of polymer molecules n, and a molecular time constant λ. Keep in mind that the Oldroyd model contains an infinite number of terms when put into the same form as the memory integral expansions. Also, it must be realized that the comparisons in Figs. 9.5–1 and 2 deal only with two particular steady-state flows, and the convergence properties of the memory-integral expansions may be very different in other flows.

One interesting point that is brought out in the preceding comparisons with molecular theory and that deserves emphasis is that the G's in the memory integral expansions can be completely determined from information about arbitrary, time-dependent, irrotational flows only. This is rather interesting, since once the G's are known the complete rheological character of the fluid is determined. That general, time-dependent irrotational flows contain complete rheological information about any simple fluid can be shown directly from the functional representation of the simple fluid concept[7] (see Problem 9D.1).

[3] S. I. Abdel-Khalik, O. Hassager, and R. B. Bird, *J. Chem. Phys.*, *61*, 4312–4316 (1974). The kernels $G_I - G_{IV}$ are given in §11.5.
[4] R. C. Armstrong and R. B. Bird, *J. Chem. Phys.*, *58*, 2715–2723 (1973).
[5] W. E. Stewart and J. P. Sørensen, *Trans. Soc. Rheol.*, *16*, 1–13 (1972).
[6] R. B. Bird, M. W. Johnson, Jr., and J. F. Stevenson, *Proceedings of the 5th International Congress on Rheology*, S. Onogi (Ed.), Tokyo University Press (1970), Vol. 4, pp. 159–168.
[7] B. D. Coleman and C. Truesdell, *ZAMM*, *45*, 547–551 (1965).

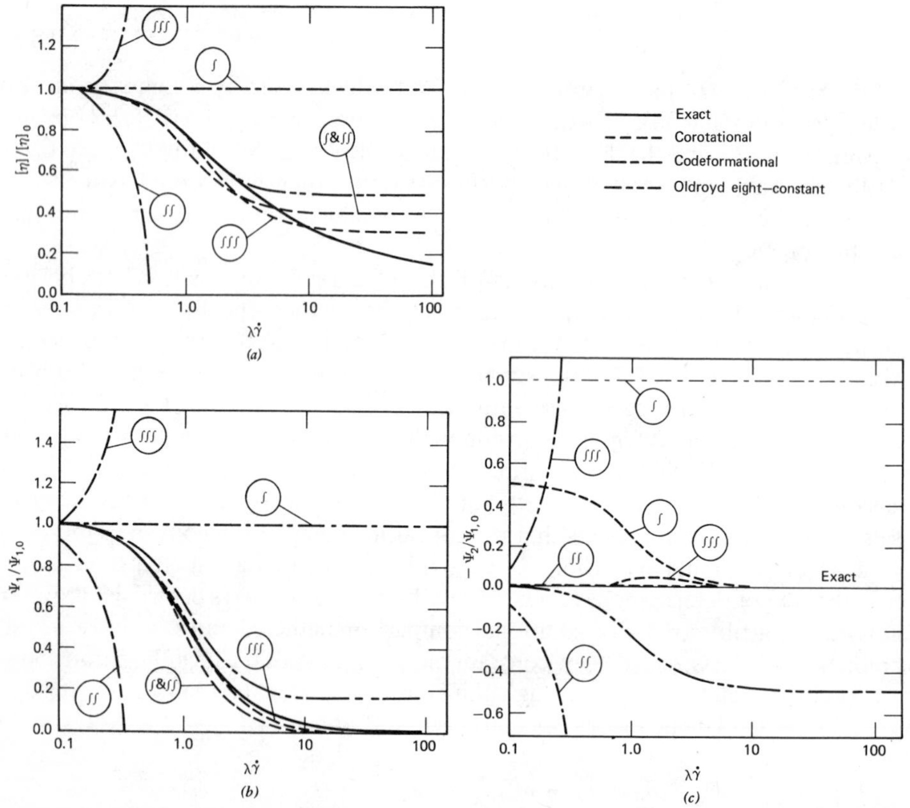

FIGURE 9.5–1. A comparison of exact numerical calculations of the viscometric functions for a dilute solution of rodlike macromolecules with the predictions of a finite number of terms from the corotational memory integral expansion Eq. 8.3–1 and a codeformational memory integral expansion Eq. 9.5–1 and with the predictions of the Oldroyd 8-constant model Eq. 8.1–2. Curves labeled with $\int$ were calculated using one term of the expansion, those labeled with $\int\int$ were obtained using two terms of the expansion, etc. (*a*) Comparison for the intrinsic viscosity $[\eta(\dot{\gamma})]$; (*b*) comparison for the primary normal stress coefficient $\Psi_1(\dot{\gamma})$; (*c*) comparison for the secondary normal stress coefficient $\Psi_2(\dot{\gamma})$. [Adapted from S. I. Abdel-Khalik, O. Hassager, and R. B. Bird, *J. Chem. Phys.*, *61*, 4312–4316 (1974).]

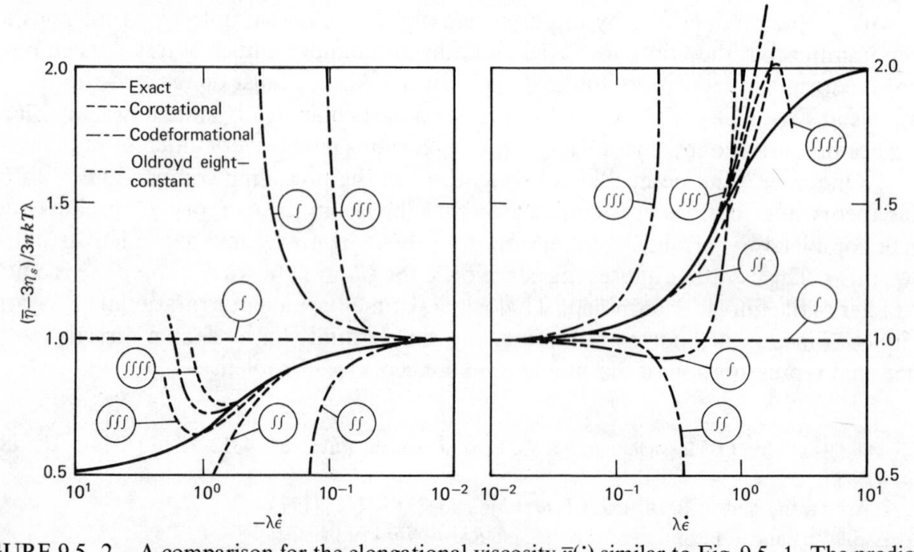

FIGURE 9.5–2. A comparison for the elongational viscosity $\bar{\eta}(\dot{\epsilon})$ similar to Fig. 9.5–1. The prediction of the Oldroyd model approaches 5.20 in each part as $|\lambda\dot{\epsilon}| \to \infty$. [Adapted from S. I. Abdel-Khalik, O. Hassager, and R. B. Bird, *J. Chem. Phys.*, *61*, 4312–4316 (1974).]

The memory integral expansions in terms of any of the $\gamma^{[n]}$ tensors or the $\gamma_{[n]}$ tensors may be used to construct rheological equations of state valid for slow and slowly changing flows, quite similarly to the development in §8.4 for the corotational memory integral expansion. For example we may expand $\gamma^{[1]}$ and $\gamma_{[1]}$ in Taylor series around $t' = t$ as done for $\gamma^{[0]}$ in Eq. 9.2–36. These expansions may then be inserted into Eqs. 9.5–1 and 2 to yield the *retarded motion expansions*:

$$\boxed{\begin{aligned}\tau = &-\beta_1\gamma^{(1)} + \beta_2\gamma^{(2)} - \beta_{11}\{\gamma^{(1)}\cdot\gamma^{(1)}\} \\ &-\beta_3\gamma^{(3)} + \beta_{12}\{\gamma^{(1)}\cdot\gamma^{(2)} + \gamma^{(2)}\cdot\gamma^{(1)}\} \\ &-\beta_{1:11}(\gamma^{(1)}:\gamma^{(1)})\gamma^{(1)} + \cdots\end{aligned}} \tag{9.5–4}$$

$$\boxed{\begin{aligned}\tau = &-\beta^1\gamma_{(1)} + \beta^2\gamma_{(2)} - \beta^{11}\{\gamma_{(1)}\cdot\gamma_{(1)}\} \\ &-\beta^3\gamma_{(3)} - \beta^{12}\{\gamma_{(1)}\cdot\gamma_{(2)} + \gamma_{(2)}\cdot\gamma_{(1)}\} \\ &-\beta^{1:11}(\gamma_{(1)}:\gamma_{(1)})\gamma_{(1)} + \cdots\end{aligned}} \tag{9.5–5}$$ [8]

The β's are related to the G's as follows:

$$\begin{aligned}\beta_n &= \frac{1}{(n-1)!}\int_0^\infty G_1(s)s^{n-1}\,ds \\ \beta_{11} &= \int_0^\infty\int_0^\infty G_2(s,s')\,ds'\,ds \\ \beta_{12} &= \int_0^\infty\int_0^\infty G_2(s,s')(s+s')\,ds'\,ds \\ \beta_{1:11} &= \int_0^\infty\int_0^\infty\int_0^\infty [G_3(s,s',s'') + \tfrac{1}{2}G_4(s,s',s'')]\,ds''\,ds'\,ds \\ &\vdots\end{aligned} \tag{9.5–6}$$

The corresponding relations for β^n, β^{11} ... are obtained by changing all subscripts to superscripts in Eq. 9.5–6. If we include only the dashed-underlined terms in Eq. 9.5–4 or Eq. 9.5–5, we have what is known as the *second-order fluid.* This model was discussed extensively in §8.4. Furthermore, if the expansions are specialized to steady shear flows, the *Criminale-Ericksen-Filbey* (or "*CEF*") *equation* is obtained; for steady-state, homogeneous, potential flows the expansions lead to the *Reiner-Rivlin equation.* These equations were discussed in §8.5.

By way of summary we present Fig. 9.5–3, which shows how the various codeformational rheological equations of state are related. The parallel between this figure and Fig. 8.3–1 is particularly noteworthy; it serves to emphasize that many of the standard

[8] The signs have been chosen so that the coefficients are positive for the FENE dumbbell molecular model [see R. C. Armstrong, Ph.D. Thesis, University of Wisconsin, Madison (1973)]. Note that the sign before β_{12} is + whereas that before β^{12} is −. Relations between the β's are given in Problem 9B.3. Equation 9.5–4 is given to fourth order in Problem 11D.2.

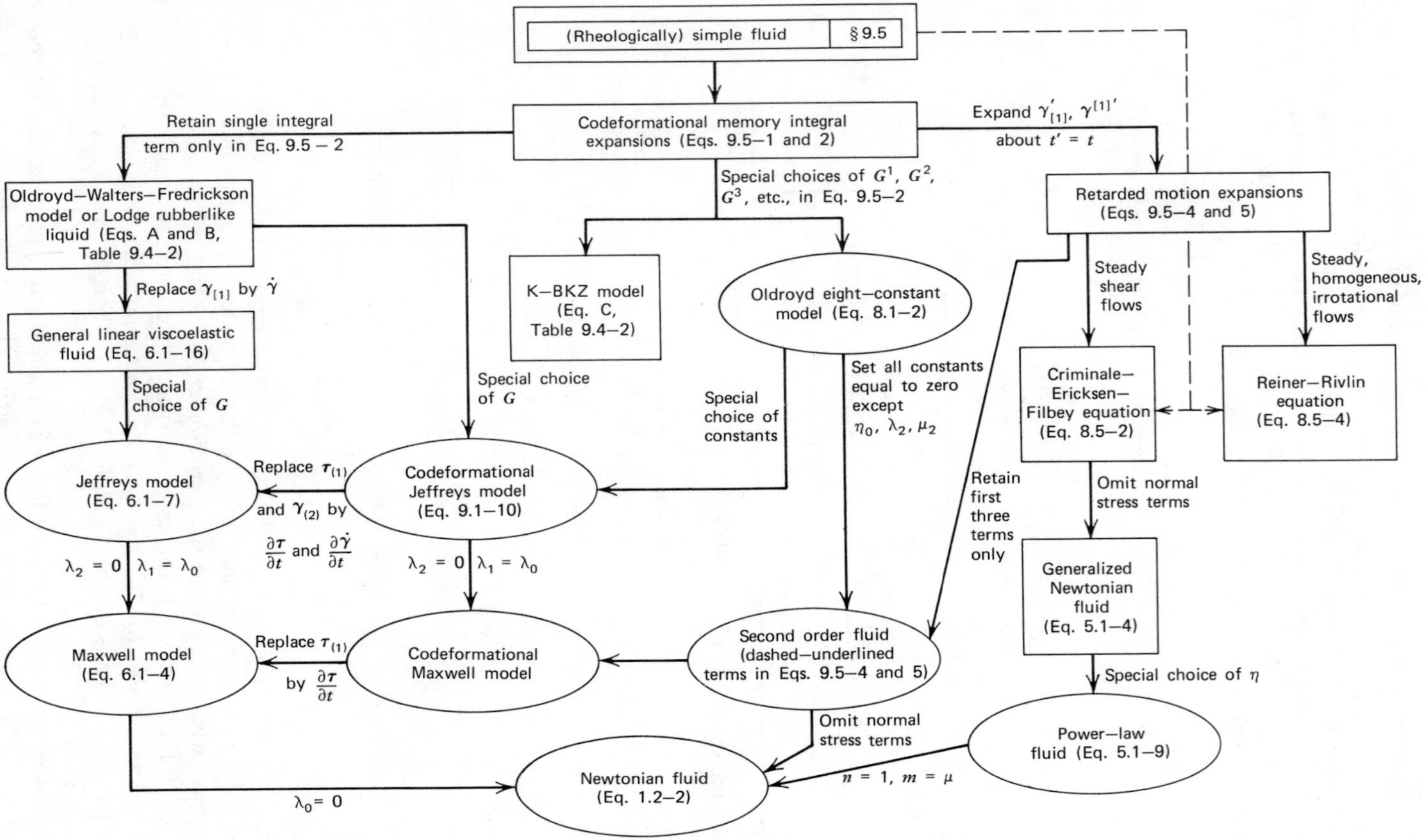

FIGURE 9.5–3. Relations among rheological equations of state.

rheological equations of state can be obtained from either the corotational or the codeformational formalism, and that it is, in fact, very artificial to regard corotational and codeformational models as two separate classes of models.

§9.6 RHEOLOGICAL EQUATIONS OF STATE IN CURVILINEAR COORDINATES

All of the calculations in the two previous sections have for simplicity been made in rectangular Cartesian coordinates. We know, however, that in the solution of flow problems one may often obtain considerable simplification both in the boundary conditions and in the governing equations by the use of curvilinear coordinates. Hence in this section we consider the formulation of codeformational rheological equations of state in cylindrical and spherical coordinates.[1]

The codeformational rheological equations of state in *differential* form involve the tensors $\boldsymbol{\gamma}^{(n)}$, $\boldsymbol{\gamma}_{(n)}$, $\boldsymbol{\tau}^{(n)}$, and $\boldsymbol{\tau}_{(n)}$ defined in note (d) to Tables 9.1–1 and 2. We see that these tensors may all be written in terms of substantial derivatives and products involving the velocity gradient tensor and the stress tensor. Therefore we may use Tables A.7–2 and 3 to express the codeformational derivatives in cylindrical and spherical coordinates. Hence no new problems are encountered with the formulation of differential models in curvilinear coordinates.

The codeformational rheological equations of state in *integral* form involve displacement functions in curvilinear coordinates. By analogy with Eqs. 9.1–3 in Cartesian coordinates we introduce the displacement functions (r',θ',z') and (r',θ',ϕ') as the coordinates at time t' of the particle P that at time t has coordinates (r,θ,z) and (r,θ,ϕ):

$$\begin{cases} r' = r'(r,\theta,z,t,t') \\ \theta' = \theta'(r,\theta,z,t,t') \\ z' = z'(r,\theta,z,t,t') \end{cases} \tag{9.6–1}$$

$$\begin{cases} r' = r'(r,\theta,\phi,t,t') \\ \theta' = \theta'(r,\theta,\phi,t,t') \\ \phi' = \phi'(r,\theta,\phi,t,t') \end{cases} \tag{9.6–2}$$

The velocity components[2] at time t' are given by the displacement functions as follows:

$$\begin{cases} v_r = \dfrac{\partial}{\partial t'} r' \\ v_\theta = r' \dfrac{\partial}{\partial t'} \theta' \\ v_z = \dfrac{\partial}{\partial t'} z' \end{cases} \tag{9.6–3}$$

[1] The reader who wishes to employ more general coordinate systems should read Appendix A and §§9.2 and 3.

[2] In this section by components we mean "physical components" (see Eqs. A.8–10a and b).

$$\begin{cases} v_r = \dfrac{\partial}{\partial t'} r' \\ v_\theta = r' \dfrac{\partial}{\partial t'} \theta' \\ v_\phi = r' \sin\theta' \dfrac{\partial}{\partial t'} \phi' \end{cases} \tag{9.6–4}$$

Here $v_r = v_r(r',\theta',z',t')$, $r' = r'(r,\theta,z,t,t')$ and so on. In some problems we need the relations inverse to Eqs. 9.6–1 expressing the present coordinates and velocities in terms of the coordinates at the past time t'. The strain tensors $\boldsymbol{\gamma}^{[0]}$ and $\boldsymbol{\gamma}_{[0]}$ are given in Tables B.5 and B.6 in rectangular, cylindrical, and spherical coordinates. One may conveniently formulate hydrodynamics problems involving integral models in the strain form by starting with the general expressions for the strain tensor. Considerations of symmetry may then be used to eliminate some terms for the particular problem. This process is illustrated in Example 9.6–1.

EXAMPLE 9.6–1 Inflation of a Spherical Viscoelastic Film[3]

Consider a thin spherical viscoelastic film forming a bubble initially of radius r_0 and thickness Δr_0 such that $\Delta r_0/r_0 \ll 1$ (Fig. 9.6–1). The bubble has been at rest for a long time with its interior filled with a gas of pressure $p_{g,0}$ equal to the surrounding pressure. At time $t = 0$ the external pressure is suddenly removed, and the bubble begins to expand. Calculate the radius of the bubble as a function of time.

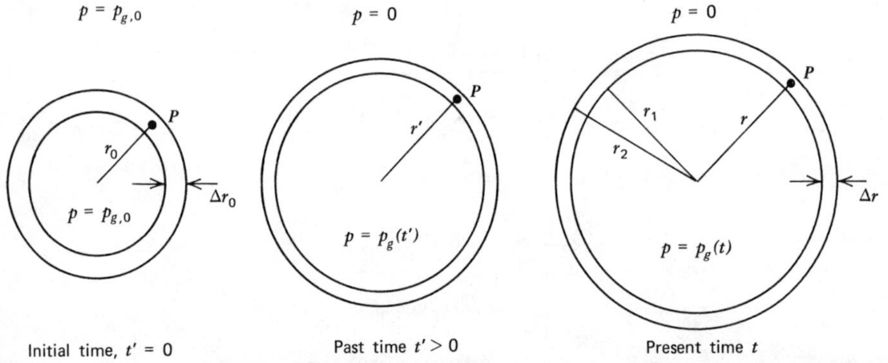

FIGURE 9.6–1. Cross section of a thin viscoelastic film forming a bubble filled with a gas. The bubble is initially at equilibrium with the inside gas pressure equal to the surrounding pressure. At time $t' = 0$ the external pressure is set equal to zero and the bubble expands.

SOLUTION

The liquid in the spherical film undergoes a purely stretching motion. Hence it is important to use a rheological equation of state that gives realistic predictions for flows of this kind. In Example 9.4–2 we found that the Lodge (1956) rubberlike liquid gives good predictions for stress growth in the startup of elongational flow of a low density polyethylene melt. Hence we choose a model of this type. Specifically for illustrative purposes we use the Oldroyd fluid B that is a rubberlike liquid with the special memory function in Eq. 9.1–15. This model does not predict a non-Newtonian viscosity, but since the fluid in

[3] Studies of vapor bubble dynamics in viscoelastic liquids have been done by J. R. Street, *Trans. Soc. Rheol., 12*, 103–131 (1968), and by E. Zana and L. G. Leal, *Ind. Eng. Chem. Fundamentals, 14*, 175–182 (1975).

the thin film does not undergo any shear this may not be a serious defect in this particular flow problem. We write the Oldroyd model B in the form:

$$\tau = -\eta_0 \frac{\lambda_2}{\lambda_1} \dot{\gamma} + \frac{\eta_0}{\lambda_1^2}\left(1 - \frac{\lambda_2}{\lambda_1}\right) \int_{-\infty}^{t} \gamma'_{[0]} e^{-(t-t')/\lambda_1}\, dt' \tag{9.6–5}$$

When $\lambda_2/\lambda_1 = 0$ this simplifies to a codeformational Maxwell model, and when $\lambda_2/\lambda_1 = 1$ we obtain a Newtonian fluid. The shear viscosity of this model is η_0 irrespective of the value of λ_2/λ_1.

We introduce a spherical coordinate system and label the coordinates of a typical particle P by (r,θ,ϕ) at the present time t and by (r',θ',ϕ') at a past time t' (see Fig. 9.6–1). From symmetry considerations we see that the equation of motion in spherical coordinates simplifies to

$$0 = -\frac{\partial p}{\partial r} - \left(\frac{1}{r^2}\frac{\partial}{\partial r}(r^2 \tau_{rr}) - \frac{\tau_{\theta\theta} + \tau_{\phi\phi}}{r}\right) \tag{9.6–6}$$

Hence the pressure drop over the film is

$$\begin{aligned} p_g &= -\int_{r_1}^{r_2} \frac{\partial}{\partial r}(p + \tau_{rr})\, dr \\ &= \int_{r_1}^{r_2} \frac{2\tau_{rr} - \tau_{\theta\theta} - \tau_{\phi\phi}}{r}\, dr \\ &\doteq (2\tau_{rr} - \tau_{\theta\theta} - \tau_{\phi\phi}) \frac{\Delta r}{r} \end{aligned} \tag{9.6–7}$$

In the last line we have introduced the approximation that the stress tensor is constant across the thin film. Let us now calculate the components of the kinematic tensors $\dot{\gamma}$ and $\gamma'_{[0]}$. From Table B.3 we find:

$$\begin{aligned} \dot{\gamma} &= 2\begin{bmatrix} \dfrac{\partial v_r}{\partial r} & 0 & 0 \\ 0 & \dfrac{v_r}{r} & 0 \\ 0 & 0 & \dfrac{v_r}{r} \end{bmatrix} \\ &= 2\begin{pmatrix} -2 & 0 & 0 \\ 0 & 1 & 0 \\ 0 & 0 & 1 \end{pmatrix} \frac{1}{r}\frac{dr}{dt} \end{aligned} \tag{9.6–8}$$

In the last step of Eq. 9.6–8 we have used the incompressibility condition for the viscoelastic liquid in the film in the form tr $\dot{\gamma} = 0$. Note also that we have expressed the velocity component as the time derivative of the displacement function. Similarly from Table B.6 we find:

$$\begin{aligned} \gamma_{[0]} &= \begin{bmatrix} 1 - \left(\dfrac{\partial r}{\partial r'}\right)^2 & 0 & 0 \\ 0 & 1 - \left(\dfrac{r}{r'}\right)^2 & 0 \\ 0 & 0 & 1 - \left(\dfrac{r}{r'}\right)^2 \end{bmatrix} \\ &= \begin{bmatrix} 1 - \left(\dfrac{r'}{r}\right)^4 & 0 & 0 \\ 0 & 1 - \left(\dfrac{r}{r'}\right)^2 & 0 \\ 0 & 0 & 1 - \left(\dfrac{r}{r'}\right)^2 \end{bmatrix} \end{aligned} \tag{9.6–9}$$

In the last step we have again made use of the incompressibility condition for the viscoelastic liquid, this time in the form det $(\boldsymbol{\delta} - \boldsymbol{\gamma}_{[0]}) = 1$.

Before we combine Eqs. 9.6–8 and 9 with Eq. 9.6–7 we need expressions for Δr and $p_g(t)$ as functions of r. First Δr is obtained from the conservation of liquid in the film:

$$r^2\,\Delta r = r_0^{\,2}\,\Delta r_0 \tag{9.6–10}$$

We now approximate the thermodynamic equation of state for the gas inside the bubble by an ideal gas law and assume adiabatic expansion so that

$$p_g(t)V^{\gamma} = p_{g,0}V_0^{\,\gamma} \tag{9.6–11}$$

or

$$p_g(t)r^{3\gamma} = p_{g,0}r_0^{\,3\gamma} \tag{9.6–12}$$

When Eqs. 9.6–5, 8, 9, 10, and 12 are combined with Eq. 9.6–7 the result is an integro-differential equation for $r(t)$. In dimensionless form this equation may be written

$$\frac{dx}{d\tau} = \frac{1}{(\lambda_2/\lambda_1)\beta}\frac{1}{x^{\alpha}} - \frac{(1-(\lambda_2/\lambda_1))x}{6(\lambda_2/\lambda_1)\Lambda^2}\int_{-\infty}^{\tau}\left[\left(\frac{x(\tau)}{x(\tau')}\right)^2 - \left(\frac{x(\tau)}{x(\tau')}\right)^{-4}\right]e^{-(\tau-\tau')/\Lambda}\,d\tau' \tag{9.6–13}$$

where we have introduced the dimensionless quantities,

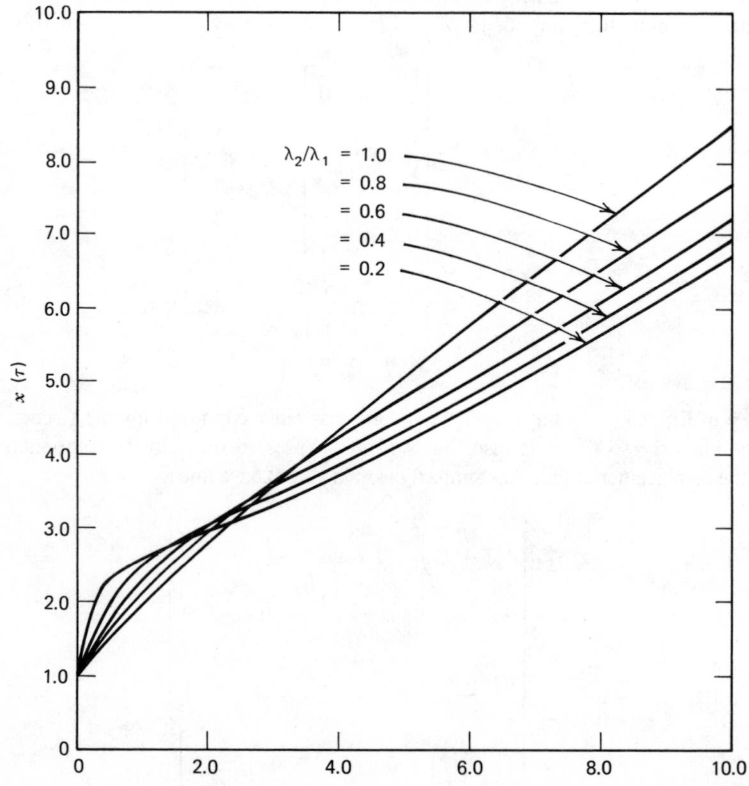

FIGURE 9.6–2. The reduced radius $x(\tau) = r(\tau)/r_0$ versus reduced time $\tau = tp_{g,0}/\eta_0$ as given by the solution of Eq. 9.6–13 for the inflation of a spherical film of an Oldroyd fluid B. The parameters are $\alpha = 0.2$, $\beta = 1.0$, and $\Lambda = 2.0$.

$$x(t') = r(t')/r_0 \tag{9.6–14}$$

$$\Lambda = \lambda_1 p_{g,0} \eta_0^{-1} \tag{9.6–15}$$

$$\tau' = t' p_{g,0} \eta_0^{-1} \tag{9.6–16}$$

$$\alpha = 3\gamma - 4 \tag{9.6–17}$$

$$\beta = 12\, \Delta r_0 / r_0 \tag{9.6–18}$$

Equation 9.6–13 with the condition that $x(\tau') = 1$ for $\tau' \leq 0$ is an initial value problem. For given values of α, β, Λ, and λ_2/λ_1 we may therefore solve numerically for $x(\tau)$ by a forward integration technique. In Fig. 9.6–2 we show the reduced radius $x(\tau)$ as a function of the reduced time τ. We have used the values $\alpha = 0.2$ (corresponding to an ideal diatomic gas), $\beta = 1$, $\Lambda = 2$, and $\lambda_2/\lambda_1 = 0.2$, 0.4, 0.6, 0.8, and 1.0. We see that a viscoelastic film is predicted to expand somewhat faster than a purely viscous film for small times τ, but that for larger times the viscoelastic film is predicted to expand considerably slower than a Newtonian film of the same zero-shear-rate viscosity.

QUESTIONS FOR DISCUSSION

1. What is meant by the term "displacement functions"? Compare and contrast thei in the corotational and codeformational theories.
2. Are the quantities Δ_{ij} and E_{ij} the components of a second order tensor? How are related to each other?
3. What is the physical basis for Oldroyd's "Fluid A" and "Fluid B"? Is there any fu mental reason why the covariant "A" fluid should be preferred over the contrava "B" fluid, or over the analogous corotational model (Eq. 7.3–3)? What is the rel between the "A" and "B" fluids and the Oldroyd eight-constant model discuss Chapter 8?
4. Compare and contrast the viewpoints in §§9.2 and 9.3, as to usability, mathem difficulty, physical basis, use of nonorthogonal coordinates, etc.
5. What is the basis for the models tabulated in Tables 9.4–1 and 2? Is there any *a* reason for preferring these models over the corotational models? Can all of the written in the notation of Chapters 7 and 8? How does one go about determinin constants or functions in these models? How does one decide which model to u a flow calculation?
6. What experiments are presently available for testing empirical rheological mo Which tests are the most crucial?
7. What are the relative merits of the corotational memory-integral expansion (Eq. 8 and the codeformational memory-integral expansions (Eqs. 9.5–1 and 2)? What kinds of memory-integral expansions might be possible?
8. How are the coefficients of Eqs. 8.4–3, 9.5–4, and 9.5–5 related? Will these equations give identical results in all flows, when they are truncated at the second- fluid level?
9. Describe how the data in Fig. 4.4–18 for $G(t;\gamma_0)$ could be used together with the re of Problem 9B.10 to obtain $\eta(\dot{\gamma})$ and $\Psi_1(\dot{\gamma})$ for the 20% polystyrene solution.
10. Is there a codeformational analog to Eq. B of Table 7.1–1?

PROBLEMS

9A.1 Bird-Carreau Model Parameters for Two Polymer Solutions and One Soap Solution

Determine the parameters for the Bird-Carreau model used in Example 9.4–1 for the two polymer solutions and the aluminum soap solution in Figs. 4.3–3 and 6 and Figs. 4.4–3 and 4. To what extent can the parameters for each of these be determined by the methods suggested in Example 9.4–1. Compare graphs of the Bird-Carreau model predictions with the experimental data for η, Ψ_1, η', and η''. How well are the data described?

9B.1 Evaluation of a Proposed Rheological Equation of State

It has been proposed by Ultman and Denn[1] that a useful and relatively simple rheological equation of state, suitable for making hydrodynamic stability calculations, is:

$$\boldsymbol{\tau} + \lambda \boldsymbol{\tau}_{(1)} = -\eta_0 \dot{\boldsymbol{\gamma}} + \nu\{\dot{\boldsymbol{\gamma}} \cdot \dot{\boldsymbol{\gamma}}\} \tag{9B.1–1}$$

in which η_0, λ, and ν are constants.

a. What does this model predict for the steady-state shear flow material functions? Are these results realistic?

b. What does it give for elongational flow properties?

c. What does it give for small-amplitude oscillatory properties?

d. Is ν positive or negative?

e. What limitations (if any) might have to be placed on Sun and Denn's stability calculations[2] because of deficiencies in the model? (*Note*: Questions like this are not always very easy to answer.)

9B.2 Relations among Kinematic Tensors[3]

Show that the $\boldsymbol{\gamma}^{(n)}$ and $\boldsymbol{\gamma}_{(n)}$ are related to the $(\mathscr{D}^n/\mathscr{D}t^n)\dot{\boldsymbol{\gamma}}$ as follows:

$$\boldsymbol{\gamma}^{(1)} = \dot{\boldsymbol{\gamma}} \tag{9B.2–1}$$

$$\boldsymbol{\gamma}^{(2)} = \frac{\mathscr{D}}{\mathscr{D}t}\dot{\boldsymbol{\gamma}} + \{\dot{\boldsymbol{\gamma}} \cdot \dot{\boldsymbol{\gamma}}\} \tag{9B.2–2}$$

$$\boldsymbol{\gamma}^{(3)} = \frac{\mathscr{D}^2}{\mathscr{D}t^2}\dot{\boldsymbol{\gamma}} + \tfrac{3}{2}\frac{\mathscr{D}}{\mathscr{D}t}\{\dot{\boldsymbol{\gamma}} \cdot \dot{\boldsymbol{\gamma}}\} + \{\dot{\boldsymbol{\gamma}} \cdot \dot{\boldsymbol{\gamma}} \cdot \dot{\boldsymbol{\gamma}}\} \tag{9B.2–3}$$

$$\vdots$$

and

$$\boldsymbol{\gamma}_{(1)} = \dot{\boldsymbol{\gamma}} \tag{9B.2–4}$$

$$\boldsymbol{\gamma}_{(2)} = \frac{\mathscr{D}}{\mathscr{D}t}\dot{\boldsymbol{\gamma}} - \{\dot{\boldsymbol{\gamma}} \cdot \dot{\boldsymbol{\gamma}}\} \tag{9B.2–5}$$

$$\boldsymbol{\gamma}_{(3)} = \frac{\mathscr{D}^2}{\mathscr{D}t^2}\dot{\boldsymbol{\gamma}} - \tfrac{3}{2}\frac{\mathscr{D}}{\mathscr{D}t}\{\dot{\boldsymbol{\gamma}} \cdot \dot{\boldsymbol{\gamma}}\} + \{\dot{\boldsymbol{\gamma}} \cdot \dot{\boldsymbol{\gamma}} \cdot \dot{\boldsymbol{\gamma}}\} \tag{9B.2–6}$$

$$\vdots$$

[1] J. S. Ultman and M. M. Denn, *Chem. Eng. Journal*, *2*, 81–89 (1971).
[2] Z. Sun and M. M. Denn, *A.I.Ch.E. Journal*, *18*, 1010–1015 (1972).
[3] H. Giesekus, *Zeits. für angew. Math. u. Mech.*, *42*, 32–61 (1962).

9B.3 Relations among Retarded-Motion Expansion Coefficients

a. Derive the following relations among the coefficients in Eqs. 9.5–4 and 5:

$$\begin{aligned} \beta^1 &= \beta_1 & \beta^{11} &= \beta_{11} - 2\beta_2 \\ \beta^2 &= \beta_2 & \beta^{12} &= -\beta_{12} + 3\beta_3 \\ \beta^3 &= \beta_3 & \beta^{1:11} &= \beta_{1:11} + 3\beta_3 - 2\beta_{12} \\ &\vdots & &\vdots \end{aligned} \tag{9B.3–1}$$

b. Derive the following relations which relate the β's to the α's in the corotational retarded motion expansion:

$$\begin{aligned} \alpha_1 &= \beta_1 & \alpha_{11} &= \beta_{11} - \beta_2 \\ \alpha_2 &= \beta_2 & \alpha_{12} &= \beta_{12} - \tfrac{3}{2}\beta_3 \\ \alpha_3 &= \beta_3 & \alpha_{1:11} &= \beta_{1:11} + \tfrac{1}{2}\beta_3 - \beta_{12} \\ &\vdots & &\vdots \end{aligned} \tag{9B.3–2}$$

The results in Problem 9B.2 will be useful.

9B.4 Relations among Material Functions in Small-Amplitude Oscillatory Shear (for Lodge's Rubberlike Liquid)

Investigate the behavior of the Lodge rubberlike liquid (Eq. B of Table 9.4–2) in small-amplitude oscillatory shearing. Determine whether or not Eqs. H, I, and J of Table 7.5–2 are predicted by this model.

9B.5 Finite and Infinitesimal Strain Tensors

a. Write the expression for $\gamma_{ij}^{[0]}$ in Cartesian coordinates.

b. Into the expression in (a) introduce $x'_m = x_m + u_m$ (what is the meaning of this substitution?) and obtain:

$$\gamma_{ij}^{[0]} = \frac{\partial}{\partial x_i} u_j + \frac{\partial}{\partial x_j} u_i + \text{terms quadratic in } \boldsymbol{u} \tag{9B.5–1}$$

Show that when the particle displacements are extremely small, the quadratic terms can be neglected and the finite strain tensor $\gamma_{ij}^{[0]}$ goes over into the infinitesimal strain tensor γ_{ij}.

c. Repeat the procedure for $\gamma_{[0]ij}$.

9B.6 Base Vectors in Steady Simple Shear Flow

a. Obtain Δ_{ij} and E_{ij} from Eqs. H and I of Table 9.1–3 for the flow $v_x = \dot{\gamma} y$, $v_y = 0$, $v_z = 0$ where $\dot{\gamma}$ is constant. Compare the result for E_{ij} with that in Eq. 9.4–5.

b. Obtain the expressions for $\hat{\boldsymbol{g}}_i$ and $\hat{\boldsymbol{g}}^i$ shown in Fig. 9.1–1 by using Eqs. E and F of Table 9.1–3.

9B.7 Integration of the Differential Forms of Oldroyd's Fluids A and B in Convected Coordinates

Write the rheological models in Eqs. 9.1–9 and 10 in terms of convected components. Integrate these and then convert the results back to space-fixed components. Show how this leads to the results given in Eqs. 9.1–12 to 15.

9B.8 Viscometric Functions and Rod-Climbing for Oldroyd's Fluids A and B

Consider the steady simple shear flow $v_x = \dot{\gamma}y$. For this flow it is desired to obtain the viscometric functions for Oldroyd's fluids A and B and to couple these results with those of Problem 3C.1 to predict the rod-climbing properties of these two models.

a. Use the differential forms of the A and B models (Eqs. 9.1–9 and 10) to show that the steady shear flow properties are

$$\eta = \eta_0 \qquad \text{(A and B)}$$

$$\Psi_1 = 2\eta_0(\lambda_1 - \lambda_2) \qquad \text{(A and B)}$$

$$\Psi_2 = \begin{cases} -2\eta_0(\lambda_1 - \lambda_2) & \text{(A)} \\ 0 & \text{(B)} \end{cases} \tag{9B.8–1}$$

b. Obtain the results in Eq. 9B.8–1 from the rate-of-strain integral forms of these models (Eqs. 9.1–12 and 13). In doing this it is necessary to show that

$$\gamma^{[1]'} = \begin{pmatrix} 0 & \dot{\gamma} & 0 \\ \dot{\gamma} & -2\dot{\gamma}^2(t - t') & 0 \\ 0 & 0 & 0 \end{pmatrix} \tag{9B.8–2}$$

$$\gamma_{[1]'} = \begin{pmatrix} 2\dot{\gamma}^2(t - t') & \dot{\gamma} & 0 \\ \dot{\gamma} & 0 & 0 \\ 0 & 0 & 0 \end{pmatrix} \tag{9B.8–3}$$

c. Now consider the rod-climbing effect for these models. First, rewrite the result[4] from Eq. 3C.1–3 in terms of the material functions

$$\left(r\frac{d}{dr}\pi_{zz}\right) = -\left[\Psi_1 + 2\eta\frac{2\Psi_2 + \dot{\gamma}(d\Psi_2/d\dot{\gamma})}{\eta + \dot{\gamma}(d\eta/d\dot{\gamma})}\right]\dot{\gamma}^2 \tag{9B.8–4}$$

Then show that

$$\left(r\frac{d}{dr}\pi_{zz}\right) = \begin{cases} 6\eta_0(\lambda_1 - \lambda_2)\dot{\gamma}^2 > 0 & \text{(A)} \\ -2\eta_0(\lambda_1 - \lambda_2)\dot{\gamma}^2 < 0 & \text{(B)} \end{cases} \tag{9B.8–5}$$

Interpret.

9B.9 Incompressibility Condition for Infinitesimal Deformations

Use the results of Problems 9B.5 and 9D.2 to prove that the incompressibility condition for infinitesimal deformations may be written in the form,

$$\text{tr}\,\gamma = 0 \tag{9B.9–1}$$

where γ is the infinitesimal strain tensor.

9B.10 Shearing Flows for the K-BKZ Model[5]

It is desired to determine the stress tensor components that are predicted by the K-BKZ model for a simple, time-dependent shearing flow $v_x = \dot{\gamma}(t)y$, $v_y = v_z = 0$. Then from these results, the viscometric functions are to be found.

[4] A.S. Lodge, *Elastic Liquids*, Academic Press, London (1964), pp. 188–194.
[5] B. Bernstein, *Acta Mech.*, *2*, 329–354 (1965); *Int. J. Non-Linear Mech.*, *4*, 183–200 (1969).

a. Use the displacement functions for time-dependent, simple shear flow to show that:

$$\boldsymbol{\gamma}^{[0]\prime} = \begin{pmatrix} 0 & -\gamma' & 0 \\ -\gamma' & \gamma'^2 & 0 \\ 0 & 0 & 0 \end{pmatrix} \tag{9B.10–1}$$

$$\boldsymbol{\gamma}'_{[0]} = -\begin{pmatrix} \gamma'^2 & \gamma' & 0 \\ \gamma' & 0 & 0 \\ 0 & 0 & 0 \end{pmatrix} \tag{9B.10–2}$$

where $\gamma' \equiv \int_{t'}^{t} \dot{\gamma}(t'')\, dt''$ is the magnitude of shear between times t' and t.

b. Verify that both invariants $I_{\gamma_{[0]}}$ and $II_{\gamma_{[0]}}$ can be expressed as simple functions of γ'.

c. By using the form of the K-BKZ model given in Eq. 9C.3–2, substantiate the following expressions for the shear stress and normal stress differences:

$$\tau_{yx} = +\int_{-\infty}^{t} G_*(t - t',\gamma')\gamma'\, dt'$$

$$\tau_{xx} - \tau_{yy} = +\int_{-\infty}^{t} G_*(t - t',\gamma')\gamma'^2\, dt'$$

$$\tau_{yy} - \tau_{zz} = -\int_{-\infty}^{t} L_*(t - t',\gamma')\gamma'^2\, dt' \tag{9B.10–3}$$

where $G_* = -(\bar{M}_1 + \bar{M}_2)$ and $L_* = -\bar{M}_2$. Note that τ_{yx} and $\tau_{xx} - \tau_{yy}$ are not independent functions of shear rate since they are both determined by G_*. The difference $\tau_{yy} - \tau_{zz}$ is given by a different kernel L_*. The transient shear flow data of Christiansen and Leppard[6] suggest that it would be better to have this prediction the other way around, with $\tau_{xx} - \tau_{yy}$ and $\tau_{yy} - \tau_{zz}$ sharing a common kernel and τ_{yx} depending on a second function.

d. Consider the stress relaxation after a sudden shearing displacement experiment (Fig. 4.3–1*e*) in which the shearing displacement γ_0 is applied at $t = 0$. Apply the shear stress expression in Eq. 9B.10–3 to this flow and compare the result with Eq. 4.4–24 to find:

$$G_*(t,\gamma_0) = \frac{\partial G(t;\gamma_0)}{\partial t} \tag{9B.10–4}$$

This part shows how G_* and L_* can be determined experimentally.

e. Specialize the results of part (c) to a steady shear flow with shear rate $\dot{\gamma}$. Show that:

$$\eta = -\int_0^{\infty} G_*(s,\dot{\gamma}s)s\, ds$$

$$\Psi_1 = -\int_0^{\infty} G_*(s,\dot{\gamma}s)s^2\, ds$$

$$\Psi_2 = +\int_0^{\infty} L_*(s,\dot{\gamma}s)s^2\, ds \tag{9B.10–5}$$

Attempts[7,8] to fit experimental data for η and Ψ_1 with the above expressions have yielded different values of G_* depending on whether it was determined from viscosity or primary normal stress coefficient data. The agreement seems to be worst at the highest shear rates.

[6] E. B. Christiansen and W. R. Leppard, *A.I.Ch.E. Journal*, *21*, 999–1006 (1975).

[7] R. I. Tanner and G. Williams, *Trans. Soc. Rheol.*, *14*, 19–38 (1970).

[8] H. -C. Yen and L. V. McIntire, *Trans. Soc. Rheol*, *18*, 495–513 (1974).

9C.1 Convergence of Retarded Motion Expansions

Consider the Lodge (1956) rubberlike liquid in the form:

$$\tau = -\int_{-\infty}^{t} G(t - t')\gamma_{[1]}(t,t')\, dt' \qquad (9C.1\text{–}1)$$

a. Expand $\gamma_{[1]}(t,t')$ in a Taylor series around $t' = t$. Then show that Eq. 9C.1–1 may be written:

$$\tau = \sum_{n=1}^{\infty} (-1)^n \beta^n \gamma_{(n)} \qquad (9C.1\text{–}2)$$

where (cf. Eq. 9.5–6);

$$\beta^n = \frac{1}{(n-1)!}\int_0^{\infty} s^{n-1} G(s)\, ds \qquad (9C.1\text{–}3)$$

b. Now assume that the liquid has been at rest for times $t < 0$ and that a constant shear rate is suddenly imposed at time $t = 0$. Calculate the shear stress growth function $\eta^+(t;\dot{\gamma})$ for times $t > 0$ from Eqs. 9C.1–2 and 3.

c. Specialize Eq. 9C.1–1 to a flow in which the fluid is at rest for times $t < 0$. Then for times $t > 0$ expand $\gamma_{[1]}(t,t')$ in a Taylor series around $t' = t$ to get:

$$\tau = \sum_{n=1}^{\infty} (-1)^n \beta^n(t) \gamma_{(n)} \qquad (9C.1\text{–}4)$$

where

$$\beta^n(t) = \frac{1}{(n-1)!}\int_0^{t} s^{n-1} G(s)\, ds \qquad (9C.1\text{–}5)$$

d. Again consider the stress growth experiment of part (b). Calculate the shear stress growth function $\eta^+(t;\dot{\gamma})$ for times $t > 0$ using Eqs. 9C.1–4 and 5.

e. Do the results of the calculations in parts (b) and (d) agree? Discuss.

9C.2 Stability of Second-Order Fluids

Consider a second-order fluid of density ρ and rheological equation of state given by the dashed-underlined terms in Eq. 9.5–4. The fluid fills the gap between two fixed plates of separation H, as shown in Fig. 9C.2. Initially, at time $t = 0$, the fluid contains a small velocity disturbance of the form $v_1 = U(x_2)$, $v_2 = v_3 = 0$.

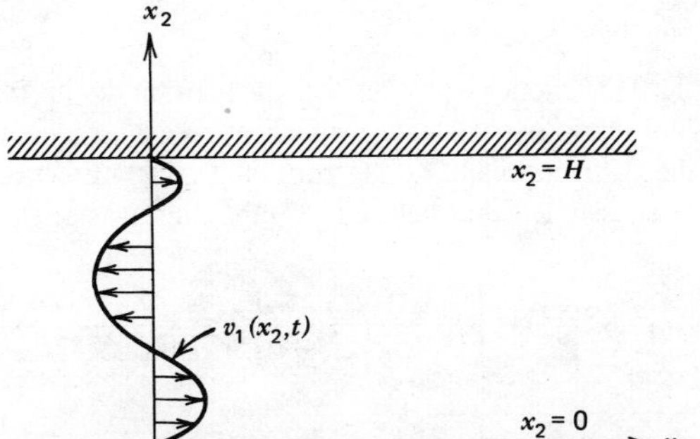

FIGURE 9C.2. A second order fluid is contained in the gap between two parallel plates. At time $t = 0$ the velocity field is of the form $v_1 = U(x_2)$, $v_2 = v_3 = 0$ with the conditions that $U(0) = U(H) = 0$. The plates are fixed at all times $t \geq 0$.

a. Assume that the velocity field for $t \geq 0$ is of the form $v_1 = v_1(x_2,t)$, $v_2 = v_3 = 0$, with the conditions that $v_1(0,t) = v_1(H,t) = 0$. Show that:

$$v_1(x_2,t) = \sum_{n=1}^{\infty} A_n \sin(n\pi x_2/H)e^{-t/\lambda_n} \tag{9C.2-1}$$

where

$$\lambda_n = \frac{\rho H^2 - \beta_2 n^2\pi^2}{\beta_1 n^2 \pi^2} \tag{9C.2-2}$$

and

$$A_n = \frac{2}{H}\int_0^H U(x_2)\sin(n\pi x_2/H)\,dx_2 \tag{9C.2-3}$$

b. Relate the material parameter β_2 to the primary normal stress coefficient Ψ_1 and show that the second order fluid is stable at rest only if $\Psi_1 \leq 0$.

c. Compare the above analysis with experimental observations on polymeric liquids and comment on the suitability of the second order fluid as a rheological equation of state for time-dependent flows.

9C.3 Determination of the Memory Functions in the K-BKZ Model

The purpose of this problem is to show how the memory functions M_1 and M_2 in the K-BKZ model (Eq. C, Table 9.4–2) can be determined from biaxial strain measurements. As a preliminary, it is desired to obtain an alternate form of this model that is often used in the literature for calculations.

a. For incompressible fluids or constant volume flows, the strain tensor $\gamma'_{[0]}$ is constrained by Eq. 9D.2–3. Because of this constraint, there are only two independent scalar invariants of $\gamma'_{[0]}$, and the dependence of the memory functions in the K-BKZ model on $\gamma'_{[0]}$ can thus be represented by including I and II alone as arguments in M_1 and M_2. Verify that these are the only invariants needed by showing that Eq. 9D.2–3 leads to

$$III = 6I - \tfrac{3}{2}I^2 - \tfrac{1}{2}I^3 - \tfrac{3}{2}II + \tfrac{3}{2}(I)(II) \tag{9C.3-1}$$

Here we have used I, II, and III to represent $I_{\gamma'_{[0]}}$, $II_{\gamma'_{[0]}}$ and $III_{\gamma'_{[0]}}$. (In connection with this problem, the discussion of tensor invariants in §A.3 is helpful.)

b. Show that the K-BKZ model in Table 9.4–2 can be recast in the following form:

$$\boldsymbol{\tau} = \int_{-\infty}^{t} [\bar{M}_1(t-t',I,II)\gamma'_{[0]} + \bar{M}_2(t-t',I,II)\gamma^{[0]\prime}]\,dt' + \text{(isotropic terms)} \tag{9C.3-2}$$

where the new memory functions $\bar{M}_1$ and $\bar{M}_2$ are related to M_1 and M_2 by:

$$\bar{M}_1 = M_1 + M_2(I-1)$$
$$\bar{M}_2 = M_2 \tag{9C.3-3}$$

Compare this form of the K-BKZ model with the Bird-Carreau model in Table 9.4–2, Eq. D.

c. Consider a biaxial strain experiment in which at $t = 0$ an incompressible fluid is suddenly stretched by a factor λ_1 in the x-direction and by a factor of λ_2 in the y-direction. For all $t \geq 0$ the sample is held in the new shape. It is assumed that before $t = 0$ the fluid was at equilibrium. Use Eq. 9C.3–2 to show that[1]

$$\bar{W}_1 = (\lambda_1^2 - \lambda_2^2)^{-1}\left[\frac{\lambda_1^2(\tau_{xx}-\tau_{zz})}{(\lambda_1^{-2}\lambda_2^{-2} - \lambda_1^2)} - \frac{\lambda_2^2(\tau_{yy}-\tau_{zz})}{(\lambda_1^{-2}\lambda_2^{-2} - \lambda_2^2)}\right]$$

$$\bar{W}_2 = (\lambda_1^2 - \lambda_2^2)^{-1}\left[\frac{\tau_{yy}-\tau_{zz}}{(\lambda_1^{-2}\lambda_2^{-2} - \lambda_2^2)} - \frac{(\tau_{xx}-\tau_{zz})}{(\lambda_1^{-2}\lambda_2^{-2} - \lambda_1^2)}\right] \tag{9C.3-4}$$

where $\bar{W}_i = \int_{-\infty}^{0} \bar{M}_i(t-t',I,II)\,dt'$.

[1] B. Bernstein, *Acta Mech.*, *2*, 329–354 (1966).

d. Discuss how the results of (c) can be used to determine M_1 and M_2. Experiments like that in (c) have been done on solids, but do not seem possible for liquids. It does not appear possible to do an experiment from which both $\bar{M}_1$ and $\bar{M}_2$ can be determined; this is probably the most serious drawback of the K-BKZ model.

9C.4 Squeezing Flow Between Two Circular Disks for the Tanner Model[2]

Look up the article by Tanner[2] in which he uses the Tanner Model (Eq. B, Table 9.4–1) to solve the squeezing flow problem that was presented in Examples 1.2–6 and 5.2–6 for Newtonian and power-law fluids. Work through the development leading to his Eq. 15. In order to make comparison with Example 5.2–6 easier, η should be regarded as a function of $\dot{\gamma}$, so that Eq. 9 in Tanner's paper is replaced by:

$$\eta = m\dot{\gamma}^{n-1} \tag{9C.4–1}$$

where μ, $k^{1/m}$, and $1/m$ in Tanner's notation correspond to η, m, and n in Eq. 9C.4–1.

Use this result to show that the time $t_{1/2}$ for the initial separation of the plates to be halved under the action of a force F is:

$$\frac{t_{1/2}}{n\lambda} = K_n\left(\frac{\pi R^2 m}{F\lambda^n}\right)^{1/n}\left(\frac{R}{h_0}\right)^{1+(1/n)} - 1 \tag{9C.4–2}$$

How does this result compare with that given in Eq. 5.2–51 and with Leider's experimental data[3] shown in Fig. 5.2–5.

9D.1 Determination of Rheological Behavior from Irrotational Flows[1]

If we describe the deformation history of a fluid element in terms of the values $\gamma^{[1]}(x,t,t')$ takes on in the interval $-\infty < t' < t$, then the simple fluid postulate may be expressed formally in the following way:

$$\boldsymbol{\tau} = \mathop{\mathscr{G}}_{t'=-\infty}^{t'=t} \{\gamma^{[1]}(x,t,t')\} \tag{9D.1–1}$$

in which the symbol $\mathscr{G}$ stands for a tensor "functional."[2] By using the above notation, show that the complete rheological behavior of a simple fluid is determined by its response to an arbitrary, time-dependent, irrotational flow.

To do this, begin by writing $\mathscr{G}$ as the sum of two parts $\mathscr{G}_i$ and $\mathscr{R}$ where $\mathscr{G}_i$ is the response of the fluid in an arbitrary time-dependent, irrotational flow and $\mathscr{R}$ is defined by $\mathscr{G} - \mathscr{G}_i$. Thus for any flow in which $\boldsymbol{\omega} = \mathbf{0}$ we know that $\mathscr{R} = \mathbf{0}$. Now consider the form of the rheological equation of state when written in each of two Cartesian coordinate systems, one of which translates with the fluid element in a space fixed orientation and the other of which translates with the fluid element and also rotates with the local fluid angular velocity. By requiring that the two coordinate systems be coincident at $t' = t$, show how the desired conclusion is reached, namely $\mathscr{R} = \mathbf{0}$.

9D.2 Incompressibility Conditions

Consider an arbitrary deformation that is described by the displacement functions Eqs. 9.2–2. It is desired to establish incompressibility conditions by comparing the mass of material contained in a

[2] R. I. Tanner, *ASLE Trans.*, *8*, 179–183 (1965).
[3] P. J. Leider, *Ind. Eng. Chem. Fundamentals*, *13*, 342–346 (1974).
[1] The argument outlined here was suggested to us by D. J. Segalman.
[2] V. Volterra, *Theory of Functionals and of Integro-Differential Equations*, Dover, New York (1959), first published in 1930. For a concise but readable discussion of the application of functional analysis in continuum mechanics, see R. S. Rivlin, in *Research Frontiers in Fluid Dynamics*, R. J. Seeger and G. Temple (Eds.), Wiley, New York (1965).

volume $V(t)$, the boundaries of which are fluid particles, at the present time t and at an arbitrary past time t'. By a change of variable in the expression for the mass at t', show that:

$$\frac{\rho}{\rho'} = \det \mathbf{\Delta} \tag{9D.2-1}$$

where ρ and ρ' denote the fluid density at t and t'. From this deduce:

$$\det (\boldsymbol{\gamma}^{[0]} + \boldsymbol{\delta}) = 1 \tag{9D.2-2}$$

$$\det (\boldsymbol{\delta} - \boldsymbol{\gamma}_{[0]}) = 1 \tag{9D.2-3}$$

for incompressible fluids.

APPENDIX A

SUMMARY OF VECTOR AND TENSOR NOTATION[1]

The physical quantities encountered in the dynamics of polymeric liquids can be placed in the following categories: *scalars* such as shear rate, temperature, energy, volume, and time; *vectors*, such as velocity, momentum, acceleration, and force; and second order *tensors*, such as the stress, rate-of-strain, and vorticity tensors. We distinguish between these quantities by the following notation:[2]

$$s = \text{scalar (lightface italic)}$$

$$\boldsymbol{v} = \text{vector (boldface italic)}$$

$$\boldsymbol{\tau} = \text{second-order tensor (boldface Greek)}$$

$$\mathsf{B} = \text{tensor of arbitrary order (boldface sans serif)}$$

In addition, boldface Greek symbols with one subscript (such as $\boldsymbol{\delta}_1$ or $\boldsymbol{\tau}_i$) are vectors. For vectors and tensors, several different kinds of multiplication are possible. Some of the operations require the use of special multiplication signs to be defined later: the "single dot" $\cdot$, the "double dot" $:$, and the "cross" $\times$. The parentheses enclosing these special multiplications, or sums of dot or cross multiplications, indicate the type of quantity produced by the multiplication:

$$(\,) = \text{scalar}$$

$$[\,] = \text{vector}$$

$$\{\,\} = \text{tensor}$$

No special significance is attached to the kind of parentheses if the operation enclosed is addition or subtraction, or a multiplication in which $\cdot$, $:$, and $\times$ do not appear. Hence $(\boldsymbol{v} \cdot \boldsymbol{w})$ and $(\boldsymbol{\sigma} : \boldsymbol{\tau})$ are scalars, $[\boldsymbol{v} \times \boldsymbol{w}]$ and $[\boldsymbol{\tau} \cdot \boldsymbol{v}]$ are vectors, and $\{\boldsymbol{\sigma} \cdot \boldsymbol{\tau} + \boldsymbol{\tau} \cdot \boldsymbol{\sigma}\}$ is a tensor. On the other hand, $(\boldsymbol{v} - \boldsymbol{w})$ may be written $[\boldsymbol{v} - \boldsymbol{w}]$ or $\{\boldsymbol{v} - \boldsymbol{w}\}$ when convenient to do so;

[1] This appendix is a modification of Appendix A of *Transport Phenomena*, by R. B. Bird, W. E. Stewart, and E. N. Lightfoot, Wiley, New York (1960).

[2] A convenient notation for blackboard use is: scalar (no underline), vector (single underline), tensor (double underline).

similarly $\boldsymbol{vw}$ may be written $(\boldsymbol{vw})$ or $[\boldsymbol{vw}]$. Actually, scalars can be regarded as zero-order tensors, vectors as first-order tensors. The multiplication signs may be interpreted thus:[3]

Multiplication Sign	*Order of Result*
None	$\sum$
$\times$	$\sum - 1$
$\cdot$	$\sum - 2$
$:$	$\sum - 4$

in which $\sum$ represents the sum of the orders of the quantities being multiplied. For example $s\boldsymbol{\tau}$ is of the order $0 + 2 = 2$, $\boldsymbol{vw}$ is of the order $1 + 1 = 2$, $\boldsymbol{\delta}_1\boldsymbol{\delta}_2$ is of the order $1 + 1 = 2$, $[\boldsymbol{v} \times \boldsymbol{w}]$ is of the order $1 + 1 - 1 = 1$, $(\boldsymbol{\sigma} : \boldsymbol{\tau})$ is of the order $2 + 2 - 4 = 0$, and $\{\boldsymbol{\sigma} \cdot \boldsymbol{\tau}\}$ is of the order $2 + 2 - 2 = 2$.

The basic operations that can be performed on scalar quantities need not be elaborated on here. However, the laws for the algebra of scalars may be used to illustrate three terms that arise in the subsequent discussion of vector operations:

a. For the multiplication of two scalars, r and s, the order of multiplication is immaterial so that the *commutative* law is valid: $rs = sr$.

b. For the successive multiplication of three scalars, q, r, and s, the order in which the multiplications are performed is immaterial, so that the *associative* law is valid: $(qr)s = q(rs)$.

c. For the multiplication of a scalar s by the sum of scalars p, q, and r, it is immaterial whether the addition or multiplication is performed first, so that the *distributive* law is valid: $s(p + q + r) = sp + sq + sr$.

These laws are not generally valid for the analogous vector and tensor operations described in the paragraphs to follow.

§A.1 VECTOR OPERATIONS FROM A GEOMETRICAL VIEWPOINT

In elementary physics courses one is introduced to vectors from a geometrical standpoint. In this section we extend this approach to include the operations of vector multiplication. In §A.2 a parallel analytic treatment is given.

Definition of a Vector and Its Magnitude

A vector $\boldsymbol{v}$ is defined as a quantity of a given magnitude and direction. The magnitude of the vector is designated by $|\boldsymbol{v}|$ or simply by the corresponding lightface symbol v. Two vectors $\boldsymbol{v}$ and $\boldsymbol{w}$ are equal when their magnitudes are equal and when they point in the same direction; they do not have to be collinear or have the same point of origin. If $\boldsymbol{v}$ and $\boldsymbol{w}$ have the same magnitude but point in opposite directions, then $\boldsymbol{v} = -\boldsymbol{w}$.

[3] This interpretation is consistent with the definitions in P. M. Morse and H. Feshbach, *Methods of Theoretical Physics*, McGraw-Hill, New York (1953), Chapter 1.

Addition and Subtraction of Vectors

The addition of two vectors can be accomplished by the familiar parallelogram construction, as indicated by Fig. A. 1–1. Vector addition obeys the following laws:

Commutative: $$(\boldsymbol{v} + \boldsymbol{w}) = (\boldsymbol{w} + \boldsymbol{v}) \tag{A.1–1}$$

Associative: $$(\boldsymbol{v} + \boldsymbol{w}) + \boldsymbol{u} = \boldsymbol{v} + (\boldsymbol{w} + \boldsymbol{u}) \tag{A.1–2}$$

Vector subtraction is performed by reversing the sign of one vector and adding; thus $\boldsymbol{v} - \boldsymbol{w} = \boldsymbol{v} + (-\boldsymbol{w})$. The geometrical construction for this is shown in Fig. A.1–2.

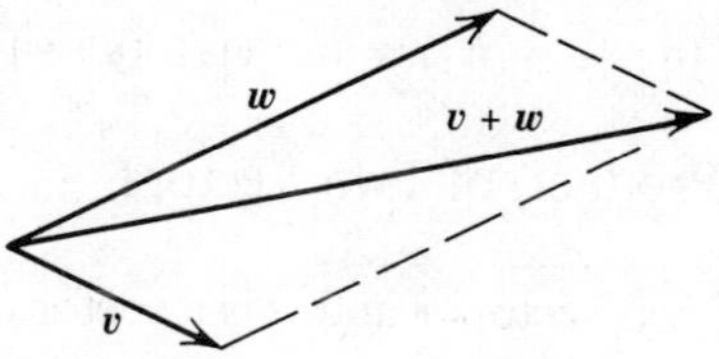

FIGURE A.1–1. Parallelogram construction for adding two vectors.

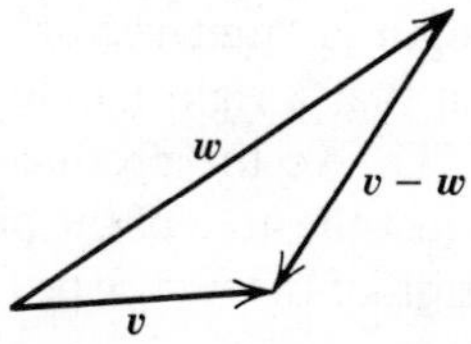

FIGURE A.1–2. Construction for subtracting two vectors.

Multiplication of a Vector by a Scalar

When a vector is multiplied by a scalar, the magnitude of the vector is altered but its direction is not. The following laws are applicable:

Commutative: $$s\boldsymbol{v} = \boldsymbol{v}s \tag{A.1–3}$$

Associative: $$r(s\boldsymbol{v}) = (rs)\boldsymbol{v} \tag{A.1–4}$$

Distributive: $$(q + r + s)\boldsymbol{v} = q\boldsymbol{v} + r\boldsymbol{v} + s\boldsymbol{v} \tag{A.1–5}$$

Scalar Product (or Dot Product) of Two Vectors

The scalar product of two vectors $\boldsymbol{v}$ and $\boldsymbol{w}$ is a scalar quantity defined by

$$(\boldsymbol{v} \cdot \boldsymbol{w}) = vw \cos \phi_{vw} \tag{A.1–6}$$

in which ϕ_{vw} is the angle between the vectors $\boldsymbol{v}$ and $\boldsymbol{w}$. The scalar product is then the magnitude of $\boldsymbol{w}$ multiplied by the projection of $\boldsymbol{v}$ on $\boldsymbol{w}$, or vice versa (Fig. A.1–3). Note that the scalar product of a vector with itself is just the square of the magnitude of the vector:

$$(\boldsymbol{v} \cdot \boldsymbol{v}) = |\boldsymbol{v}|^2 = v^2 \tag{A.1–7}$$

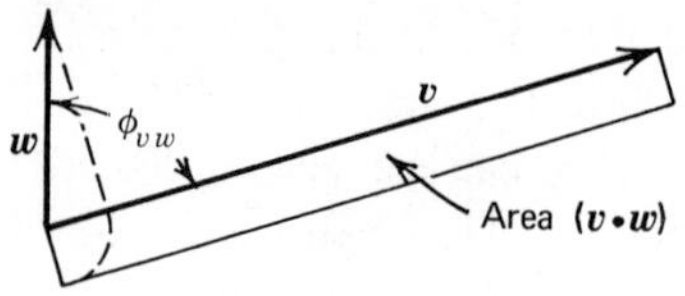

FIGURE A.1–3. The scalar product of two vectors.

The rules governing scalar products are as follows:

Commutative: $$(\boldsymbol{u} \cdot \boldsymbol{v}) = (\boldsymbol{v} \cdot \boldsymbol{u}) \qquad \text{(A.1–8)}$$

Not associative: $$(\boldsymbol{u} \cdot \boldsymbol{v})\boldsymbol{w} \neq \boldsymbol{u}(\boldsymbol{v} \cdot \boldsymbol{w}) \qquad \text{(A.1–9)}$$

Distributive: $$(\boldsymbol{u} \cdot \{\boldsymbol{v} + \boldsymbol{w}\}) = (\boldsymbol{u} \cdot \boldsymbol{v}) + (\boldsymbol{u} \cdot \boldsymbol{w}) \qquad \text{(A.1–10)}$$

Vector Product (or Cross Product) of Two Vectors

The vector product of two vectors $\boldsymbol{v}$ and $\boldsymbol{w}$ is a vector defined by:

$$[\boldsymbol{v} \times \boldsymbol{w}] = \{vw \sin \phi_{vw}\}\boldsymbol{n}_{vw} \qquad \text{(A.1–11)}$$

in which $\boldsymbol{n}_{vw}$ is a vector of unit length (a "unit vector") normal to the plane containing $\boldsymbol{v}$ and $\boldsymbol{w}$ and pointing in the direction that a right-handed screw will move if turned from $\boldsymbol{v}$ toward $\boldsymbol{w}$ through the angle ϕ_{vw}. The vector product is illustrated in Fig. A.1–4. The magnitude of the vector product is just the area of the parallelogram defined by the vectors $\boldsymbol{v}$ and $\boldsymbol{w}$. It follows from the definition of the vector product that:

$$[\boldsymbol{v} \times \boldsymbol{v}] = \boldsymbol{0} \qquad \text{(A.1–12)}$$

Note the following summary of laws governing the vector product operation:

Not commutative: $$[\boldsymbol{v} \times \boldsymbol{w}] = -[\boldsymbol{w} \times \boldsymbol{v}] \qquad \text{(A.1–13)}$$

Not associative: $$[\boldsymbol{u} \times [\boldsymbol{v} \times \boldsymbol{w}]] \neq [[\boldsymbol{u} \times \boldsymbol{v}] \times \boldsymbol{w}] \qquad \text{(A.1–14)}$$

Distributive: $$[\{\boldsymbol{u} + \boldsymbol{v}\} \times \boldsymbol{w}] = [\boldsymbol{u} \times \boldsymbol{w}] + [\boldsymbol{v} \times \boldsymbol{w}] \qquad \text{(A.1–15)}$$

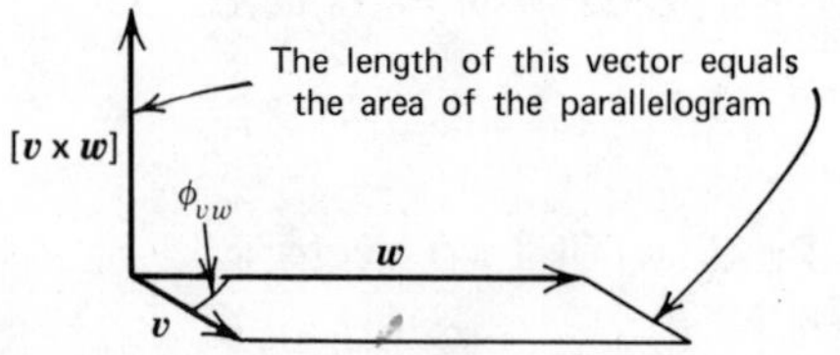

FIGURE A.1–4. The vector product of two vectors.

Multiple Products of Vectors

Somewhat more complicated are multiple products formed by combinations of the multiplication processes just described:

(a) $rs\boldsymbol{v}$ (b) $s(\boldsymbol{v} \cdot \boldsymbol{w})$ (c) $s[\boldsymbol{v} \times \boldsymbol{w}]$
(d) $(\boldsymbol{u} \cdot [\boldsymbol{v} \times \boldsymbol{w}])$ (e) $[\boldsymbol{u} \times [\boldsymbol{v} \times \boldsymbol{w}]]$ (f) $([\boldsymbol{u} \times \boldsymbol{v}] \cdot [\boldsymbol{w} \times \boldsymbol{z}])$
(g) $[[\boldsymbol{u} \times \boldsymbol{v}] \times [\boldsymbol{w} \times \boldsymbol{z}]]$

The geometrical interpretations of the first three of these are straightforward. The magnitude of $(\boldsymbol{u} \cdot [\boldsymbol{v} \times \boldsymbol{w}])$ can easily be shown to represent the volume of a parallelepiped defined by the vectors $\boldsymbol{u}$, $\boldsymbol{v}$, and $\boldsymbol{w}$.

EXERCISES

1. What are the "orders" of the following quantities: $(\boldsymbol{v} \cdot \boldsymbol{w})$, $(\boldsymbol{v} - \boldsymbol{u})\boldsymbol{w}$, $(\boldsymbol{ab} : \boldsymbol{cd})$, $(\boldsymbol{v} \cdot \rho \boldsymbol{wu})$, $[[\boldsymbol{a} \times \boldsymbol{f}] \times [\boldsymbol{b} \times \boldsymbol{g}]]$?
2. Draw a sketch to illustrate the inequality in A.1–9. Are there any special cases for which it becomes an equality?
3. A mathematical surface of area S has an orientation given by a unit normal vector $\boldsymbol{n}$, pointing downstream of the surface. A fluid of density ρ flows through this surface with a velocity $\boldsymbol{v}$. Show that the mass rate of flow through the surface is $w = \rho(\boldsymbol{n} \cdot \boldsymbol{v})S$.
4. The angular velocity $\boldsymbol{W}$ of a rotating solid body is a vector whose magnitude is the rate of angular displacement (radians per second) and whose direction is that in which a right-handed screw would advance if turned in the same direction. The position vector $\boldsymbol{r}$ of a point is that vector from the origin of coordinates to the point. Show that the velocity of any point in a rotating solid body is $\boldsymbol{v} = [\boldsymbol{W} \times \boldsymbol{r}]$. The origin is located on the axis of rotation.
5. A constant force $\boldsymbol{F}$ acts on a body moving with a velocity $\boldsymbol{v}$, which is not necessarily collinear with $\boldsymbol{F}$. Show that the rate at which $\boldsymbol{F}$ does work on the body is $W = (\boldsymbol{F} \cdot \boldsymbol{v})$. [Note that for a variable force $dW = (\boldsymbol{F} \cdot d\boldsymbol{v})$]

§A.2 VECTOR OPERATIONS FROM AN ANALYTICAL VIEWPOINT

In this section a parallel analytical treatment is given to each of the topics presented geometrically in §A.1. In the discussion here we restrict ourselves to rectangular coordinates and label the axes as 1, 2, 3 corresponding to the usual notation of x, y, z; only right-handed coordinates are used.

Many formulas can be expressed compactly in terms of the *Kronecker delta* δ_{ij} and the *permutation symbol* ϵ_{ijk}. These quantities are defined thus:

$$\begin{cases} \delta_{ij} = +1 & \text{if} \quad i = j \qquad \text{(A.2–1)} \\ \delta_{ij} = 0 & \text{if} \quad i \neq j \qquad \text{(A.2–2)} \end{cases}$$

$$\begin{cases} \epsilon_{ijk} = +1 & \text{if} \quad ijk = 123, 231, \text{ or } 312 \qquad \text{(A.2–3)} \\ \epsilon_{ijk} = -1 & \text{if} \quad ijk = 321, 132, \text{ or } 213 \qquad \text{(A.2–4)} \\ \epsilon_{ijk} = 0 & \text{if any two indices are alike} \qquad \text{(A.2–5)} \end{cases}$$

Several relations involving these quantities are useful in proving some vector and tensor identities:

$$\sum_{j=1}^{3} \sum_{k=1}^{3} \epsilon_{ijk}\epsilon_{hjk} = 2\delta_{ih} \tag{A.2–6}$$

$$\sum_{k=1}^{3} \epsilon_{ijk}\epsilon_{mnk} = \delta_{im}\delta_{jn} - \delta_{in}\delta_{jm} \tag{A.2–7}$$

Note that a three-by-three determinant may be written in terms of the ϵ_{ijk}:

$$\begin{vmatrix} a_{11} & a_{12} & a_{13} \\ a_{21} & a_{22} & a_{23} \\ a_{31} & a_{32} & a_{33} \end{vmatrix} = \sum_{i=1}^{3} \sum_{j=1}^{3} \sum_{k=1}^{3} \epsilon_{ijk} a_{1i} a_{2j} a_{3k} \tag{A.2–8}$$

The quantity ϵ_{ijk} thus selects the necessary terms that appear in the determinant and affixes the proper sign to each term.

The Unit Vectors

Let $\boldsymbol{\delta}_1, \boldsymbol{\delta}_2, \boldsymbol{\delta}_3$ be the "unit vectors" (that is, vectors of unit magnitude) in the direction of the 1, 2, 3 axes[1] (Fig. A.2–1). We can use the definitions of the scalar and vector products to tabulate all possible products of each type:

$$\begin{cases} (\boldsymbol{\delta}_1 \cdot \boldsymbol{\delta}_1) = (\boldsymbol{\delta}_2 \cdot \boldsymbol{\delta}_2) = (\boldsymbol{\delta}_3 \cdot \boldsymbol{\delta}_3) = 1 & \text{(A.2–9)} \\ (\boldsymbol{\delta}_1 \cdot \boldsymbol{\delta}_2) = (\boldsymbol{\delta}_2 \cdot \boldsymbol{\delta}_3) = (\boldsymbol{\delta}_3 \cdot \boldsymbol{\delta}_1) = 0 & \text{(A.2–10)} \end{cases}$$

$$\begin{cases} [\boldsymbol{\delta}_1 \times \boldsymbol{\delta}_1] = [\boldsymbol{\delta}_2 \times \boldsymbol{\delta}_2] = [\boldsymbol{\delta}_3 \times \boldsymbol{\delta}_3] = \mathbf{0} & \text{(A.2–11)} \\ [\boldsymbol{\delta}_1 \times \boldsymbol{\delta}_2] = \boldsymbol{\delta}_3; \quad [\boldsymbol{\delta}_2 \times \boldsymbol{\delta}_3] = \boldsymbol{\delta}_1; \quad [\boldsymbol{\delta}_3 \times \boldsymbol{\delta}_1] = \boldsymbol{\delta}_2 & \text{(A.2–12)} \\ [\boldsymbol{\delta}_2 \times \boldsymbol{\delta}_1] = -\boldsymbol{\delta}_3; \quad [\boldsymbol{\delta}_3 \times \boldsymbol{\delta}_2] = -\boldsymbol{\delta}_1; \quad [\boldsymbol{\delta}_1 \times \boldsymbol{\delta}_3] = -\boldsymbol{\delta}_2 & \text{(A.2–13)} \end{cases}$$

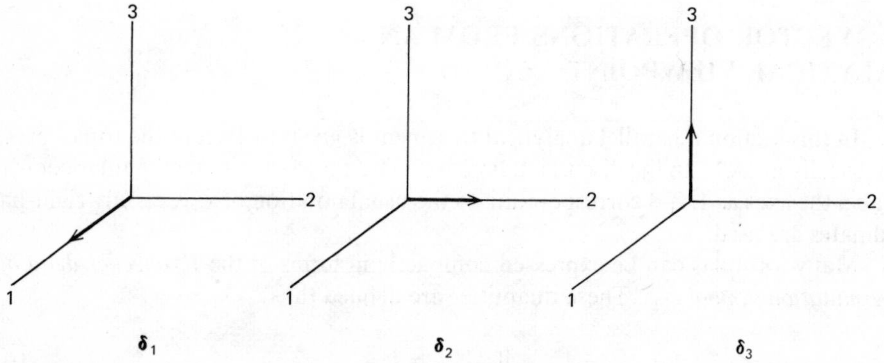

FIGURE A.2–1. The unit vectors $\boldsymbol{\delta}_i$; each vector is of unit magnitude and points in one of the coordinate directions.

All of these relations may be summarized by the following two relations:

$$\boxed{(\boldsymbol{\delta}_i \cdot \boldsymbol{\delta}_j) = \delta_{ij}} \tag{A.2–14}$$

$$\boxed{[\boldsymbol{\delta}_i \times \boldsymbol{\delta}_j] = \sum_{k=1}^{3} \epsilon_{ijk} \boldsymbol{\delta}_k} \tag{A.2–15}$$

[1] In most elementary texts the unit vectors are called $\boldsymbol{i}, \boldsymbol{j}, \boldsymbol{k}$. We prefer to use $\boldsymbol{\delta}_1, \boldsymbol{\delta}_2, \boldsymbol{\delta}_3$ because the components of these vectors are given by the Kronecker delta. That is, the component of $\boldsymbol{\delta}_1$ in the 1-direction is δ_{11} or unity; the component of $\boldsymbol{\delta}_1$ in the two 2-direction is δ_{12} or zero.

in which δ_{ij} is the Kronecker delta and ϵ_{ijk} is the permutation symbol defined in the introduction to this section. These two relations enable us to develop analytic expressions for all the common "dot" and "cross" operations. In the remainder of this section, and in the next section, in developing expressions for vector and tensor operations all we do is to break all vectors up into components and then apply Eqs. A.2–14 and 15.

Definition of a Vector and Its Magnitude

Any vector $\boldsymbol{v}$ can be completely specified by giving the values of its projections v_1, v_2, v_3, on the coordinate axes 1, 2, 3 (Fig. A.2–2). The vector can be constructed by adding vectorially the components multiplied by their corresponding unit vectors:

$$\boldsymbol{v} = \boldsymbol{\delta}_1 v_1 + \boldsymbol{\delta}_2 v_2 + \boldsymbol{\delta}_3 v_3 = \sum_{i=1}^{3} \boldsymbol{\delta}_i v_i \tag{A.2–16}$$

Note that *a vector associates a scalar with each coordinate direction.*[2] It should be emphasized that the v_i are scalars, whereas the $\boldsymbol{\delta}_i v_i$ are vectors that when added together vectorially give $\boldsymbol{v}$.

The magnitude of a vector is given by:

$$|\boldsymbol{v}| = v = \sqrt{v_1^2 + v_2^2 + v_3^2} = \sqrt{\textstyle\sum_i v_i^2} \tag{A.2–17}$$

Two vectors $\boldsymbol{v}$ and $\boldsymbol{w}$ are equal if $v_1 = w_1$, $v_2 = w_2$, and $v_3 = w_3$. Also $\boldsymbol{v} = -\boldsymbol{w}$, if $v_1 = -w_1$, etc.

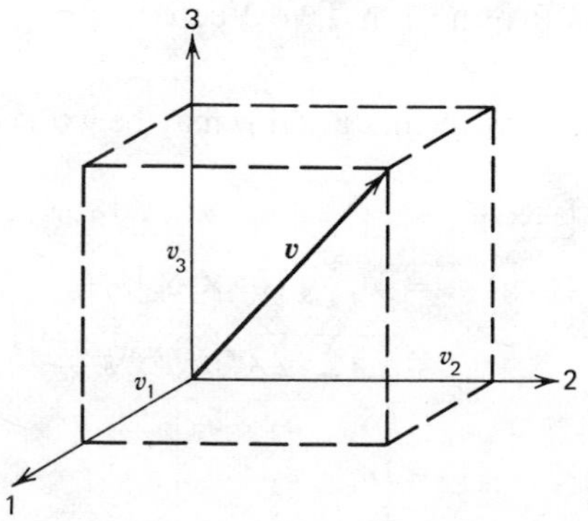

FIGURE A.2–2. The projections of a vector on the coordinate axes 1, 2, and 3.

Addition and Subtraction of Vectors

The addition or subtraction of vectors $\boldsymbol{v}$ and $\boldsymbol{w}$ may be written in terms of components:

$$\boldsymbol{v} + \boldsymbol{w} = \textstyle\sum_i \boldsymbol{\delta}_i v_i + \sum_i \boldsymbol{\delta}_i w_i = \sum_i \boldsymbol{\delta}_i (v_i + w_i) \tag{A.2–18}$$

Geometrically, this corresponds to adding up the projections of $\boldsymbol{v}$ and $\boldsymbol{w}$ on each individual axis and then constructing a vector with these new components. Three or more vectors may be added in exactly the same fashion.

[2] For a discussion of the relation of this definition of a vector to the definition in terms of the rules for transformation of coordinates, see W. Prager, *Mechanics of Continua*, Ginn, Boston (1961).

Multiplication of a Vector by a Scalar

Multiplication of a vector by a scalar corresponds to multiplying each component of the vector by the scalar:

$$s\boldsymbol{v} = s\{\textstyle\sum_i \boldsymbol{\delta}_i v_i\} = \sum_i \boldsymbol{\delta}_i \{s v_i\} \tag{A.2–19}$$

Scalar Product (or Dot Product) of Two Vectors

The scalar product of two vectors $\boldsymbol{v}$ and $\boldsymbol{w}$ is obtained by writing each vector in terms of components according to Eq. A.2–16 and then performing the scalar-product operations on the unit vectors, using Eq. A.2–14:

$$\begin{aligned}(\boldsymbol{v} \cdot \boldsymbol{w}) &= (\{\textstyle\sum_i \boldsymbol{\delta}_i v_i\} \cdot \{\sum_j \boldsymbol{\delta}_j w_j\}) \\ &= \textstyle\sum_i \sum_j (\boldsymbol{\delta}_i \cdot \boldsymbol{\delta}_j) v_i w_j \\ &= \textstyle\sum_i \sum_j \delta_{ij} v_i w_j \\ &= \textstyle\sum_i v_i w_i \end{aligned} \tag{A.2–20}$$

Hence the scalar product of two vectors is obtained by summing the products of the corresponding components of the two vectors. Note that $(\boldsymbol{v} \cdot \boldsymbol{v})$ (sometimes written as $\boldsymbol{v}^2$ or as v^2) is a scalar representing the square of the magnitude of $\boldsymbol{v}$.

Vector Product (or Cross Product) of Two Vectors

The vector product of two vectors $\boldsymbol{v}$ and $\boldsymbol{w}$ may be worked out by using Eqs. A.2–16 and 15:

$$\begin{aligned}[\boldsymbol{v} \times \boldsymbol{w}] &= [\{\textstyle\sum_j \boldsymbol{\delta}_j v_j\} \times \{\sum_k \boldsymbol{\delta}_k w_k\}] \\ &= \textstyle\sum_j \sum_k [\boldsymbol{\delta}_j \times \boldsymbol{\delta}_k] v_j w_k \\ &= \textstyle\sum_i \sum_j \sum_k \epsilon_{ijk} \boldsymbol{\delta}_i v_j w_k \\ &= \begin{vmatrix} \boldsymbol{\delta}_1 & \boldsymbol{\delta}_2 & \boldsymbol{\delta}_3 \\ v_1 & v_2 & v_3 \\ w_1 & w_2 & w_3 \end{vmatrix} \end{aligned} \tag{A.2–21}$$

Here we have made use of Eq. A.2–8. Note that the ith-component of $[\boldsymbol{v} \times \boldsymbol{w}]$ is given by $\sum_j \sum_k \epsilon_{ijk} v_j w_k$; this result finds frequent use in proving vector identities.

Multiple Vector Products

Expressions for the multiple products mentioned in §A.1 can be obtained by using the foregoing analytical expressions for the scalar and vector products. For example, the product $(\boldsymbol{u} \cdot [\boldsymbol{v} \times \boldsymbol{w}])$ may be written:

$$\begin{aligned}(\boldsymbol{u} \cdot [\boldsymbol{v} \times \boldsymbol{w}]) &= \textstyle\sum_i u_i [\boldsymbol{v} \times \boldsymbol{w}]_i \\ &= \textstyle\sum_i \sum_j \sum_k \epsilon_{ijk} u_i v_j w_k \end{aligned} \tag{A.2–22}$$

Then, from Eq. A.2–8, we obtain:

$$(\boldsymbol{u} \cdot [\boldsymbol{v} \times \boldsymbol{w}]) = \begin{vmatrix} u_1 & u_2 & u_3 \\ v_1 & v_2 & v_3 \\ w_1 & w_2 & w_3 \end{vmatrix} \tag{A.2–23}$$

The magnitude of $(\boldsymbol{u} \cdot [\boldsymbol{v} \times \boldsymbol{w}])$ is the volume of a parallelepiped defined by the vectors $\boldsymbol{u}$, $\boldsymbol{v}$, $\boldsymbol{w}$. Furthermore, the vanishing of the determinant is a necessary and sufficient condition that the vectors $\boldsymbol{u}$, $\boldsymbol{v}$, and $\boldsymbol{w}$ be coplanar.

The Position Vector

The usual symbol for the position vector—that is, the vector specifying the location of a point in space—is $\boldsymbol{r}$. The components of $\boldsymbol{r}$ are then x_1, x_2, and x_3, so that

$$\boldsymbol{r} = \textstyle\sum_i \boldsymbol{\delta}_i x_i \tag{A.2–24}$$

This is an irregularity in the notation, since the components have a symbol different from that for the vector. The magnitude of $\boldsymbol{r}$ is usually called $r = \sqrt{x_1^2 + x_2^2 + x_3^2}$.

EXAMPLE A.2–1 Proof of a Vector Identity

The analytical expressions for dot and cross products may be used to prove vector identities; for example, let it be desired to verify the relation

$$[\boldsymbol{u} \times [\boldsymbol{v} \times \boldsymbol{w}]] = \boldsymbol{v}(\boldsymbol{u} \cdot \boldsymbol{w}) - \boldsymbol{w}(\boldsymbol{u} \cdot \boldsymbol{v}) \tag{A.2–25}$$

SOLUTION

Let us rewrite the i-component of the expression on the left side in expanded form:

$$\begin{aligned}
[\boldsymbol{u} \times [\boldsymbol{v} \times \boldsymbol{w}]]_i &= \textstyle\sum_j \sum_k \epsilon_{ijk} u_j [\boldsymbol{v} \times \boldsymbol{w}]_k \\
&= \textstyle\sum_j \sum_k \epsilon_{ijk} u_j \{\sum_l \sum_m \epsilon_{klm} v_l w_m\} \\
&= \textstyle\sum_j \sum_k \sum_l \sum_m \epsilon_{ijk}\epsilon_{klm} u_j v_l w_m \\
&= \textstyle\sum_j \sum_k \sum_l \sum_m \epsilon_{ijk}\epsilon_{lmk} u_j v_l w_m
\end{aligned} \tag{A.2–26}$$

Use may now be made of Eq. A.2–7 to complete the proof:

$$\begin{aligned}
[\boldsymbol{u} \times [\boldsymbol{v} \times \boldsymbol{w}]]_i &= \textstyle\sum_j \sum_l \sum_m (\delta_{il}\delta_{jm} - \delta_{im}\delta_{jl}) u_j v_l w_m \\
&= v_i \textstyle\sum_j \sum_m \delta_{jm} u_j w_m - w_i \sum_j \sum_l \delta_{jl} u_j v_l \\
&= v_i \textstyle\sum_j u_j w_j - w_i \sum_j u_j v_j \\
&= v_i(\boldsymbol{u} \cdot \boldsymbol{w}) - w_i(\boldsymbol{u} \cdot \boldsymbol{v})
\end{aligned} \tag{A.2–27}$$

which is just the i-component of the right side of Eq. A.2–25. In a similar way one may verify such identities as:

$$(\boldsymbol{u} \cdot [\boldsymbol{v} \times \boldsymbol{w}]) = (\boldsymbol{v} \cdot [\boldsymbol{w} \times \boldsymbol{u}]) \tag{A.2–28}$$

$$([\boldsymbol{u} \times \boldsymbol{v}] \cdot [\boldsymbol{w} \times \boldsymbol{z}]) = (\boldsymbol{u} \cdot \boldsymbol{w})(\boldsymbol{v} \cdot \boldsymbol{z}) - (\boldsymbol{u} \cdot \boldsymbol{z})(\boldsymbol{v} \cdot \boldsymbol{w}) \tag{A.2–29}$$

$$[[\boldsymbol{u} \times \boldsymbol{v}] \times [\boldsymbol{w} \times \boldsymbol{z}]] = ([\boldsymbol{u} \times \boldsymbol{v}] \cdot \boldsymbol{z})\boldsymbol{w} - ([\boldsymbol{u} \times \boldsymbol{v}] \cdot \boldsymbol{w})\boldsymbol{z} \tag{A.2–30}$$

EXERCISES

1. Write out the following summations:

   (a) $\sum_{k=1}^{3} k^2$ (b) $\sum_{k=1}^{3} a_k{}^2$ (c) $\sum_{j=1}^{3}\sum_{k=1}^{3} a_{jk}b_{kj}$ (d) $\left(\sum_{j=1}^{3} a_j\right)^2 = \sum_{j=1}^{3}\sum_{k=1}^{3} a_j a_k$

2. A vector $\boldsymbol{v}$ has components $v_x = 1$, $v_y = 3$, $v_z = -5$. A vector $\boldsymbol{w}$ has components $w_x = 2$, $w_y = -1$, $w_z = 1$. Evaluate:
   (a) $(\boldsymbol{v} \cdot \boldsymbol{w})$ (c) The length of $\boldsymbol{v}$ (e) $[\boldsymbol{\delta}_1 \times \boldsymbol{w}]$
   (b) $[\boldsymbol{v} \times \boldsymbol{w}]$ (d) $(\boldsymbol{\delta}_1 \cdot \boldsymbol{v})$ (f) ϕ_{vw}
   (g) $[\boldsymbol{r} \times \boldsymbol{v}]$, where $\boldsymbol{r}$ is the position vector.
3. Evaluate: (a) $([\boldsymbol{\delta}_1 \times \boldsymbol{\delta}_2] \cdot \boldsymbol{\delta}_3)$ (b) $[[\boldsymbol{\delta}_2 \times \boldsymbol{\delta}_3] \times [\boldsymbol{\delta}_1 \times \boldsymbol{\delta}_3]]$.
4. Show that Eq. A. 2–6 is valid for the particular case $i = 1, h = 2$.
   Show that Eq. A. 2–7 is valid for the particular case $i = j = m = 1, n = 2$.
5. Verify Eqs. A.2–28 to 30.
6. Verify that $\sum_{j=1}^{3}\sum_{k=1}^{3} \epsilon_{ijk}\alpha_{jk} = 0$ if $\alpha_{jk} = \alpha_{kj}$.
7. Explain carefully the statement after Eq. A.2–21 that the ith component of $[\boldsymbol{v} \times \boldsymbol{w}]$ is $\sum_j \sum_k \epsilon_{ijk} v_j w_k$.
8. Verify that $([\boldsymbol{v} \times \boldsymbol{w}] \cdot [\boldsymbol{v} \times \boldsymbol{w}]) + (\boldsymbol{v} \cdot \boldsymbol{w})^2 = v^2 w^2$.

§A.3 TENSOR OPERATIONS

In the last section it was shown that expressions could be developed for all common "dot" and "cross" operations for vectors by knowing how to write a vector $\boldsymbol{v}$ as a sum $\sum_i \boldsymbol{\delta}_i v_i$, and by knowing how to manipulate the unit vectors $\boldsymbol{\delta}_i$. In this section we follow a parallel procedure. We write a tensor $\boldsymbol{\tau}$ as a sum $\sum_i \sum_j \boldsymbol{\delta}_i\boldsymbol{\delta}_j\tau_{ij}$, and give formulas for the manipulation of the unit dyads $\boldsymbol{\delta}_i\boldsymbol{\delta}_j$; in this way expressions are developed for the commonly occurring "dot" and "cross" operations for tensors.

The Unit Dyads

The unit vectors $\boldsymbol{\delta}_i$ were defined in the foregoing discussion and then the *scalar products* $(\boldsymbol{\delta}_i \cdot \boldsymbol{\delta}_j)$ and *vector products* $[\boldsymbol{\delta}_i \times \boldsymbol{\delta}_j]$ were given. There is a third kind of product that can be formed with the unit vectors, namely, the *dyadic products* $\boldsymbol{\delta}_i\boldsymbol{\delta}_j$ (written without parenthesis, brackets, or multiplication symbols). According to the rules of notation given in the introduction to Appendix A, the products $\boldsymbol{\delta}_i\boldsymbol{\delta}_j$ are tensors of the second order. Since $\boldsymbol{\delta}_i$ and $\boldsymbol{\delta}_j$ are of unit magnitude, we will refer to the products $\boldsymbol{\delta}_i\boldsymbol{\delta}_j$ as *unit dyads.* Whereas each unit vector in Fig. A.2–1 represents a single coordinate direction, the unit dyads in Fig. A.3–1 represent *ordered* pairs of coordinate directions.

(In physical problems we often work with quantities that require the simultaneous specification of two directions. For example, the flux of x-momentum across a unit area of surface perpendicular to the y-direction is a quantity of this type. Since this quantity is sometimes not the same as the flux of y-momentum perpendicular to the x-direction, it is evident that specifying the two directions is not sufficient; we must also agree upon the order in which the directions are given.)

The dot and cross products of unit vectors were introduced by means of the geometrical definitions of these operations. The analogous operations for the unit dyads

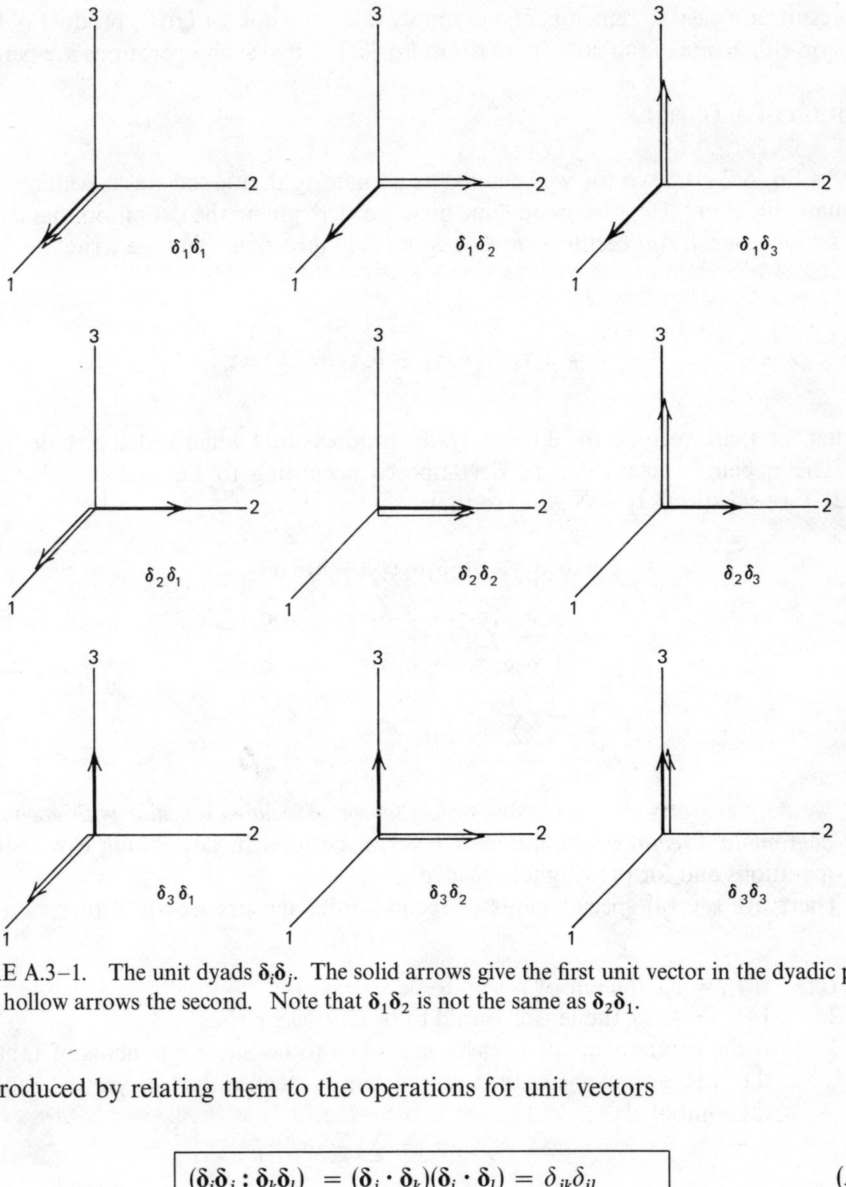

FIGURE A.3–1. The unit dyads $\boldsymbol{\delta}_i\boldsymbol{\delta}_j$. The solid arrows give the first unit vector in the dyadic product and the hollow arrows the second. Note that $\boldsymbol{\delta}_1\boldsymbol{\delta}_2$ is not the same as $\boldsymbol{\delta}_2\boldsymbol{\delta}_1$.

are introduced by relating them to the operations for unit vectors

$$(\boldsymbol{\delta}_i\boldsymbol{\delta}_j : \boldsymbol{\delta}_k\boldsymbol{\delta}_l) = (\boldsymbol{\delta}_j \cdot \boldsymbol{\delta}_k)(\boldsymbol{\delta}_i \cdot \boldsymbol{\delta}_l) = \delta_{jk}\delta_{il} \quad \text{(A.3–1)}$$

$$[\boldsymbol{\delta}_i\boldsymbol{\delta}_j \cdot \boldsymbol{\delta}_k] = \boldsymbol{\delta}_i(\boldsymbol{\delta}_j \cdot \boldsymbol{\delta}_k) = \boldsymbol{\delta}_i\delta_{jk} \quad \text{(A.3–2)}$$

$$[\boldsymbol{\delta}_i \cdot \boldsymbol{\delta}_j\boldsymbol{\delta}_k] = (\boldsymbol{\delta}_i \cdot \boldsymbol{\delta}_j)\boldsymbol{\delta}_k = \delta_{ij}\boldsymbol{\delta}_k \quad \text{(A.3–3)}$$

$$\{\boldsymbol{\delta}_i\boldsymbol{\delta}_j \cdot \boldsymbol{\delta}_k\boldsymbol{\delta}_l\} = \boldsymbol{\delta}_i(\boldsymbol{\delta}_j \cdot \boldsymbol{\delta}_k)\boldsymbol{\delta}_l = \delta_{jk}\boldsymbol{\delta}_i\boldsymbol{\delta}_l \quad \text{(A.3–4)}$$

$$\{\boldsymbol{\delta}_i\boldsymbol{\delta}_j \times \boldsymbol{\delta}_k\} = \boldsymbol{\delta}_i[\boldsymbol{\delta}_j \times \boldsymbol{\delta}_k] = \sum_{l=1}^{3} \epsilon_{jkl}\boldsymbol{\delta}_i\boldsymbol{\delta}_l \quad \text{(A.3–5)}$$

$$\{\boldsymbol{\delta}_i \times \boldsymbol{\delta}_j\boldsymbol{\delta}_k\} = [\boldsymbol{\delta}_i \times \boldsymbol{\delta}_j]\boldsymbol{\delta}_k = \sum_{l=1}^{3} \epsilon_{ijl}\boldsymbol{\delta}_l\boldsymbol{\delta}_k \quad \text{(A.3–6)}$$

These results are easy to remember: one simply takes the dot (or cross) product of the unit vectors on either side of the dot (or cross); in Eq. A.3–1 two such operations are performed.

Definition of a Tensor

In Eq. A.2–16 a vector was defined as a quantity that associates a scalar with each coordinate direction. Here we go to "one higher order" giving the definition that *a (second-order) tensor associates a vector with each coordinate direction,*[1] and we write, analogously to Eq. A.2–16:

$$\boldsymbol{\tau} = \boldsymbol{\delta}_1\boldsymbol{\tau}_1 + \boldsymbol{\delta}_2\boldsymbol{\tau}_2 + \boldsymbol{\delta}_3\boldsymbol{\tau}_3 = \sum_{i=1}^{3} \boldsymbol{\delta}_i\boldsymbol{\tau}_i \tag{A.3–7}$$

Note that the $\boldsymbol{\tau}_i$ are vectors; the $\boldsymbol{\delta}_i\boldsymbol{\tau}_i$ are dyadic products that when added give the tensor $\boldsymbol{\tau}$.

The $\boldsymbol{\tau}_i$ being vectors can be decomposed according to Eq. A.2–16; that is, into Eq. A.3–7 we substitute $\boldsymbol{\tau}_i = \sum_j \boldsymbol{\delta}_j\tau_{ij}$ so that:

$$\begin{aligned}\boldsymbol{\tau} &= \boldsymbol{\delta}_1\boldsymbol{\delta}_1\tau_{11} + \boldsymbol{\delta}_1\boldsymbol{\delta}_2\tau_{12} + \boldsymbol{\delta}_1\boldsymbol{\delta}_3\tau_{13} \\ &\quad + \boldsymbol{\delta}_2\boldsymbol{\delta}_1\tau_{21} + \boldsymbol{\delta}_2\boldsymbol{\delta}_2\tau_{22} + \boldsymbol{\delta}_2\boldsymbol{\delta}_3\tau_{23} \\ &\quad + \boldsymbol{\delta}_3\boldsymbol{\delta}_1\tau_{31} + \boldsymbol{\delta}_3\boldsymbol{\delta}_2\tau_{32} + \boldsymbol{\delta}_3\boldsymbol{\delta}_3\tau_{33} \\ &= \sum_{i=1}^{3}\sum_{j=1}^{3} \boldsymbol{\delta}_i\boldsymbol{\delta}_j\tau_{ij}\end{aligned} \tag{A.3–8}$$

Hence, we may also say that *a (second-order) tensor associates a scalar with each ordered pair of coordinate directions.* Equation A.3–8 can be used in developing expressions for tensor operations and for proving tensor identities.

There are several special kinds of second-order tensors worth noting:

1. If $\tau_{ij} = \tau_{ji}$, the tensor is said to be *symmetric.*
2. If $\tau_{ij} = -\tau_{ji}$, the tensor is said to be *antisymmetric.*
3. If the components of a tensor are taken to be the components of $\boldsymbol{\tau}$, but with the indices transposed, the resulting tensor is called the *transpose* of $\boldsymbol{\tau}$ and given the symbol $\boldsymbol{\tau}^\dagger$:

$$\boldsymbol{\tau}^\dagger = \textstyle\sum_i \sum_j \boldsymbol{\delta}_i\boldsymbol{\delta}_j\tau_{ji} \tag{A.3–9}$$

4. If the components of the tensor are formed by ordered pairs of the components of two vectors $\boldsymbol{v}$ and $\boldsymbol{w}$, the resulting tensor is called the *dyadic product of* $\boldsymbol{v}$ *and* $\boldsymbol{w}$ and given the symbol $\boldsymbol{vw}$:

$$\boldsymbol{vw} = \textstyle\sum_i \sum_j \boldsymbol{\delta}_i\boldsymbol{\delta}_j v_i w_j \tag{A.3–10}$$

   Note that $\boldsymbol{vw} \neq \boldsymbol{wv}$, but that $(\boldsymbol{vw})^\dagger = \boldsymbol{wv}$

5. If the components of the tensor are given by the Kronecker delta δ_{ij}, the resulting tensor is called the *unit tensor* and given the symbol $\boldsymbol{\delta}$:

$$\boldsymbol{\delta} = \textstyle\sum_i \sum_j \boldsymbol{\delta}_i\boldsymbol{\delta}_j\delta_{ij} \tag{A.3–11}$$

[1] Tensors are often defined in terms of the transformation rules; the connections between such a definition and that given above is discussed by W. Prager, *Mechanics of Continua,* Ginn, Boston (1961).

(At this point the reader should carefully review the meanings of δ_{ij}, $\boldsymbol{\delta}_i$, $\boldsymbol{\delta}_i\boldsymbol{\delta}_j$, and $\boldsymbol{\delta}$.)

With the information just given, all of the standard algebraic operations may be summarized.

Addition of Tensors and Dyadic Products

Two tensors are added thus:

$$\begin{aligned}\boldsymbol{\sigma} + \boldsymbol{\tau} &= \textstyle\sum_i \sum_j \boldsymbol{\delta}_i\boldsymbol{\delta}_j\sigma_{ij} + \sum_i \sum_j \boldsymbol{\delta}_i\boldsymbol{\delta}_j\tau_{ij} \\ &= \textstyle\sum_i \sum_j \boldsymbol{\delta}_i\boldsymbol{\delta}_j(\sigma_{ij} + \tau_{ij})\end{aligned} \tag{A.3–12}$$

That is, the sum of two tensors is that tensor whose components are the sums of the corresponding components of the two tensors. The same is true for dyadic products.

Multiplication of a Tensor by a Scalar

Multiplication of a tensor by a scalar corresponds to multiplying each component of the tensor by the scalar:

$$\begin{aligned}s\boldsymbol{\tau} &= s\{\textstyle\sum_i \sum_j \boldsymbol{\delta}_i\boldsymbol{\delta}_j\tau_{ij}\} \\ &= \textstyle\sum_i \sum_j \boldsymbol{\delta}_i\boldsymbol{\delta}_j\{s\tau_{ij}\}\end{aligned} \tag{A.3–13}$$

The same is true for dyadic products.

The Scalar Product (or Double Dot Product) of Two Tensors

Two tensors may be multiplied according to the double dot operation:

$$\begin{aligned}(\boldsymbol{\sigma} : \boldsymbol{\tau}) &= (\{\textstyle\sum_i \sum_j \boldsymbol{\delta}_i\boldsymbol{\delta}_j\sigma_{ij}\} : \{\sum_k \sum_l \boldsymbol{\delta}_k\boldsymbol{\delta}_l\tau_{kl}\}) \\ &= \textstyle\sum_i \sum_j \sum_k \sum_l (\boldsymbol{\delta}_i\boldsymbol{\delta}_j : \boldsymbol{\delta}_k\boldsymbol{\delta}_l)\sigma_{ij}\tau_{kl} \\ &= \textstyle\sum_i \sum_j \sum_k \sum_l \delta_{il}\delta_{jk}\sigma_{ij}\tau_{kl} \\ &= \textstyle\sum_i \sum_j \sigma_{ij}\tau_{ji}\end{aligned} \tag{A.3–14}$$

in which Eq. A.3–1 has been used. Similarly, we may show that:

$$(\boldsymbol{\tau} : \boldsymbol{vw}) = \textstyle\sum_i \sum_j \tau_{ij}v_j w_i \tag{A.3–15}$$

$$(\boldsymbol{uv} : \boldsymbol{wz}) = \textstyle\sum_i \sum_j u_i v_j w_j z_i \tag{A.3–16}$$

The Tensor Product (the Single Dot Product) of Two Tensors

Two tensors may also be multiplied according to the single dot operation:

$$\begin{aligned}\{\boldsymbol{\sigma} \cdot \boldsymbol{\tau}\} &= \{(\textstyle\sum_i \sum_j \boldsymbol{\delta}_i\boldsymbol{\delta}_j\sigma_{ij}) \cdot (\sum_k \sum_l \boldsymbol{\delta}_k\boldsymbol{\delta}_l\tau_{kl})\} \\ &= \textstyle\sum_i \sum_j \sum_k \sum_l \{\boldsymbol{\delta}_i\boldsymbol{\delta}_j \cdot \boldsymbol{\delta}_k\boldsymbol{\delta}_l\}\sigma_{ij}\tau_{kl} \\ &= \textstyle\sum_i \sum_j \sum_k \sum_l \delta_{jk}\boldsymbol{\delta}_i\boldsymbol{\delta}_l\sigma_{ij}\tau_{kl} \\ &= \textstyle\sum_i \sum_l \boldsymbol{\delta}_i\boldsymbol{\delta}_l(\sum_j \sigma_{ij}\tau_{jl})\end{aligned} \tag{A.3–17}$$

That is, the *il*-component of $\{\boldsymbol{\sigma} \cdot \boldsymbol{\tau}\}$ is $\sum_j \sigma_{ij}\tau_{jl}$. Similar operations may be performed with dyadic products. It is common practice to write $\{\boldsymbol{\sigma} \cdot \boldsymbol{\sigma}\}$ as $\boldsymbol{\sigma}^2$, $\{\boldsymbol{\sigma} \cdot \boldsymbol{\sigma}^2\}$ as $\boldsymbol{\sigma}^3$, and so on.

The Vector Product (or Dot Product) of a Tensor with a Vector

When a tensor is dotted into a vector, we get a vector:

$$\begin{aligned} [\boldsymbol{\tau} \cdot \boldsymbol{v}] &= [\{\textstyle\sum_i \sum_j \boldsymbol{\delta}_i\boldsymbol{\delta}_j\tau_{ij}\} \cdot \{\sum_k \boldsymbol{\delta}_k v_k\}] \\ &= \textstyle\sum_i \sum_j \sum_k [\boldsymbol{\delta}_i\boldsymbol{\delta}_j \cdot \boldsymbol{\delta}_k]\tau_{ij}v_k \\ &= \textstyle\sum_i \sum_j \sum_k \boldsymbol{\delta}_i\delta_{jk}\tau_{ij}v_k \\ &= \textstyle\sum_i \boldsymbol{\delta}_i\{\sum_j \tau_{ij}v_j\} \end{aligned} \tag{A.3–18}$$

That is, the ith component of $[\boldsymbol{\tau} \cdot \boldsymbol{v}]$ is $\sum_j \tau_{ij}v_j$. Similarly, the ith component of $[\boldsymbol{v} \cdot \boldsymbol{\tau}]$ is $\sum_j v_j\tau_{ji}$. Clearly, $[\boldsymbol{\tau} \cdot \boldsymbol{v}] \neq [\boldsymbol{v} \cdot \boldsymbol{\tau}]$ unless $\boldsymbol{\tau}$ is symmetric.

Recall that when a vector $\boldsymbol{v}$ is multiplied by a scalar s the resultant vector $s\boldsymbol{v}$ points in the same direction as $\boldsymbol{v}$ but has a different length. But, when $\boldsymbol{\tau}$ is dotted into $\boldsymbol{v}$, the resultant vector $[\boldsymbol{\tau} \cdot \boldsymbol{v}]$ differs from $\boldsymbol{v}$ in *both* length and direction; that is, the tensor $\boldsymbol{\tau}$ "deflects" or "twists" the vector $\boldsymbol{v}$ to form a new vector pointing in a different direction.

The Tensor Product (or Cross Product) of a Tensor with a Vector

When a tensor is crossed with a vector, we get a tensor:

$$\begin{aligned} \{\boldsymbol{\tau} \times \boldsymbol{v}\} &= \{(\textstyle\sum_i \sum_j \boldsymbol{\delta}_i\boldsymbol{\delta}_j\tau_{ij}) \times (\sum_k \boldsymbol{\delta}_k v_k)\} \\ &= \textstyle\sum_i \sum_j \sum_k [\boldsymbol{\delta}_i\boldsymbol{\delta}_j \times \boldsymbol{\delta}_k]\tau_{ij}v_k \\ &= \textstyle\sum_i \sum_j \sum_k \sum_l \epsilon_{jkl}\boldsymbol{\delta}_i\boldsymbol{\delta}_l\tau_{ij}v_k \\ &= \textstyle\sum_i \sum_l \boldsymbol{\delta}_i\boldsymbol{\delta}_l\{\sum_j \sum_k \epsilon_{jkl}\tau_{ij}v_k\} \end{aligned} \tag{A.3–19}$$

Hence, the *il*-component of $\{\boldsymbol{\tau} \times \boldsymbol{v}\}$ is $\sum_j \sum_k \epsilon_{jkl}\tau_{ij}v_k$. Similarly the *lk*-component of $\{\boldsymbol{v} \times \boldsymbol{\tau}\}$ is $\sum_i \sum_j \epsilon_{ijl}v_i\tau_{jk}$.

The Invariants of a Tensor

In §A.2 it was pointed out that a scalar may be formed from a single vector $\boldsymbol{v}$ by forming the product $(\boldsymbol{v} \cdot \boldsymbol{v}) = \sum_i v_iv_i$. From a tensor $\boldsymbol{\tau}$, three independent scalars can be formed by taking the *trace* of (that is, summing the diagonal elements of) $\boldsymbol{\tau}$, $\boldsymbol{\tau}^2$, and $\boldsymbol{\tau}^3$:

$$I_\tau = \text{tr}\,\boldsymbol{\tau} = \textstyle\sum_i \tau_{ii} \tag{A.3–20}$$

$$II_\tau = \text{tr}\,\boldsymbol{\tau}^2 = \textstyle\sum_i \sum_j \tau_{ij}\tau_{ji} \tag{A.3–21}$$

$$III_\tau = \text{tr}\,\boldsymbol{\tau}^3 = \textstyle\sum_i \sum_j \sum_k \tau_{ij}\tau_{jk}\tau_{ki} \tag{A.3–22}$$

These are called *invariants* of the tensor $\boldsymbol{\tau}$, because their values are independent of the choice of coordinate system to which the components of $\boldsymbol{\tau}$ are referred. Other scalars can of course be formed, but they will be combinations of these three. For example, two other widely used invariants are:

$$\overline{II}_\tau = \tfrac{1}{2}(I_\tau^{\,2} - II_\tau) \tag{A.3–23}$$

$$\begin{aligned} \overline{III}_\tau &= \tfrac{1}{6}(I_\tau^{\,3} - 3I_\tau II_\tau + 2III_\tau) \\ &= \det\boldsymbol{\tau} \end{aligned} \tag{A.3–24}$$

It is also possible to form *joint invariants* of two tensors[2] $\boldsymbol{\sigma}$ and $\boldsymbol{\tau}$; these will be tr $\boldsymbol{\sigma}$, tr $\boldsymbol{\tau}$, tr $\boldsymbol{\sigma}^2$, tr $\boldsymbol{\tau}^2$, tr $\boldsymbol{\sigma}^3$, tr $\boldsymbol{\tau}^3$, tr $\{\boldsymbol{\sigma}\cdot\boldsymbol{\tau}\}$, tr $\{\boldsymbol{\sigma}^2\cdot\boldsymbol{\tau}\}$, tr $\{\boldsymbol{\sigma}\cdot\boldsymbol{\tau}^2\}$, and tr $\{\boldsymbol{\sigma}^2\cdot\boldsymbol{\tau}^2\}$.

The Cayley-Hamilton Theorem

A useful theorem using the invariants described above is that of Cayley and Hamilton which states that for a second-order tensor $\boldsymbol{\tau}$:

$$\boldsymbol{\tau}^3 - I_\tau\boldsymbol{\tau}^2 + II_\tau\boldsymbol{\tau} - III_\tau\boldsymbol{\delta} = \mathbf{0} \tag{A.3–25}$$

This allows $\boldsymbol{\tau}^3$ to be expressed in terms of $\boldsymbol{\delta}$, $\boldsymbol{\tau}$, and $\boldsymbol{\tau}^2 = \{\boldsymbol{\tau}\cdot\boldsymbol{\tau}\}$. Extensions of the Cayley-Hamilton theorem have been worked out by Rivlin.[2]

Other Operations

From the foregoing results, it is not difficult to prove the following identities:

$$[\boldsymbol{\delta}\cdot\boldsymbol{v}] = [\boldsymbol{v}\cdot\boldsymbol{\delta}] = \boldsymbol{v} \tag{A.3–26}$$

$$[\boldsymbol{uv}\cdot\boldsymbol{w}] = \boldsymbol{u}(\boldsymbol{v}\cdot\boldsymbol{w}) \tag{A.3–27}$$

$$[\boldsymbol{w}\cdot\boldsymbol{uv}] = (\boldsymbol{w}\cdot\boldsymbol{u})\boldsymbol{v} \tag{A.3–28}$$

$$(\boldsymbol{uv}:\boldsymbol{wz}) = (\boldsymbol{uw}:\boldsymbol{vz}) = (\boldsymbol{u}\cdot\boldsymbol{z})(\boldsymbol{v}\cdot\boldsymbol{w}) \tag{A.3–29}$$

$$(\boldsymbol{\tau}:\boldsymbol{uv}) = ([\boldsymbol{\tau}\cdot\boldsymbol{u}]\cdot\boldsymbol{v}) \tag{A.3–30}$$

$$(\boldsymbol{uv}:\boldsymbol{\tau}) = (\boldsymbol{u}\cdot[\boldsymbol{v}\cdot\boldsymbol{\tau}]) \tag{A.3–31}$$

Magnitude of a Tensor

The magnitude of a tensor is given by

$$\begin{aligned} |\boldsymbol{\tau}| = \tau &= \sqrt{\tfrac{1}{2}(\boldsymbol{\tau}:\boldsymbol{\tau}^\dagger)} \\ &= \sqrt{\tfrac{1}{2}\textstyle\sum_i\sum_j \tau_{ij}^{\,2}} \end{aligned} \tag{A.3–32}$$

Note that for a symmetric tensor $\boldsymbol{\tau}$, $\tau = \sqrt{\tfrac{1}{2}\mathrm{II}_\tau}$.

EXERCISES

1. The components of a symmetrical tensor $\boldsymbol{\tau}$ are:

$$\tau_{xx} = 3 \qquad \tau_{xy} = 2 \qquad \tau_{xz} = -1$$

$$\tau_{yx} = 2 \qquad \tau_{yy} = 2 \qquad \tau_{yz} = 1$$

$$\tau_{zx} = -1 \qquad \tau_{zy} = 1 \qquad \tau_{zz} = 0$$

The components of a vector $\boldsymbol{v}$ are:

$$v_x = 5 \qquad v_y = 3 \qquad v_z = 7$$

[2] R. S. Rivlin, *J. Rat. Mech. Anal.*, *4*, 681–702 (1955); see also R. S. Rivlin, in *Research Frontiers in Fluid Dynamics*, R. J. Seeger and G. Temple (Eds.) Wiley-Interscience, New York (1965), Chapter 5.

Evaluate:

(a) $[\boldsymbol{\tau} \cdot \boldsymbol{v}]$ (b) $[\boldsymbol{v} \cdot \boldsymbol{\tau}]$ (c) $(\boldsymbol{\tau} : \boldsymbol{\tau})$
(d) $(\boldsymbol{v} \cdot [\boldsymbol{\tau} \cdot \boldsymbol{v}])$ (e) $\boldsymbol{v}\boldsymbol{v}$ (f) $[\boldsymbol{\tau} \cdot \boldsymbol{\delta}_1]$
(g) $\{\boldsymbol{\tau} \cdot \boldsymbol{\delta}\}$

2. Evaluate:

(a) $[[\boldsymbol{\delta}_1\boldsymbol{\delta}_2 \cdot \boldsymbol{\delta}_2] \times \boldsymbol{\delta}_1]$ (c) $(\boldsymbol{\delta} : \boldsymbol{\delta})$
(b) $(\boldsymbol{\delta} : \boldsymbol{\delta}_1\boldsymbol{\delta}_2)$ (d) $\{\boldsymbol{\delta} \cdot \boldsymbol{\delta}\}$

3. Verify the identities in Eqs. A.3–26 to 31.
4. If $\boldsymbol{\alpha}$ is symmetrical and $\boldsymbol{\beta}$ is antisymmetrical, show that $(\boldsymbol{\alpha} : \boldsymbol{\beta}) = 0$.
5. Explain carefully the statement after Eq. A.3–17 that the *il*-component of $\{\boldsymbol{\sigma} \cdot \boldsymbol{\tau}\}$ is $\sum_j \sigma_{ij}\tau_{jl}$.

§A.4 THE VECTOR AND TENSOR DIFFERENTIAL OPERATIONS

The vector differential operator $\boldsymbol{\nabla}$, known as "nabla" or "del," is defined in rectangular coordinates as:

$$\boldsymbol{\nabla} = \boldsymbol{\delta}_1 \frac{\partial}{\partial x_1} + \boldsymbol{\delta}_2 \frac{\partial}{\partial x_2} + \boldsymbol{\delta}_3 \frac{\partial}{\partial x_3} = \sum_i \boldsymbol{\delta}_i \frac{\partial}{\partial x_i} \tag{A.4–1}$$

in which the $\boldsymbol{\delta}_i$ are the unit vectors and the x_i are the variables associated with the 1, 2, 3 axes (that is, the x_1, x_2, x_3 are the position coordinates normally referred to as x, y, z). The symbol $\boldsymbol{\nabla}$ is a vector-operator—it has components like a vector but it cannot stand alone; it must operate on a scalar, vector, or tensor function. In this section we summarize the various operations of $\boldsymbol{\nabla}$ on scalars, vectors, and tensors. Just as in §§A.2 and A.3, we decompose vectors and tensors into their components and then use Eqs. A.2–14 and 15, and Eqs. A.3–1 to 6. Keep in mind that this section deals exclusively with rectangular coordinates, for which the unit vectors $\boldsymbol{\delta}_i$ are constants; curvilinear coordinates are discussed in §§A.6, 7. (*Note*: In the kinetic theory discussion in Volume 2 we often use the symbol $\partial/\partial\mathbf{r}$ in lieu of $\boldsymbol{\nabla}$. This notation is particularly convenient in multidimensional spaces.)

The Gradient of a Scalar Field

If s is a scalar function of the variables x_1, x_2, x_3, then the operation of $\boldsymbol{\nabla}$ on s is:

$$\boldsymbol{\nabla} s = \boldsymbol{\delta}_1 \frac{\partial s}{\partial x_1} + \boldsymbol{\delta}_2 \frac{\partial s}{\partial x_2} + \boldsymbol{\delta}_3 \frac{\partial s}{\partial x_3} = \sum_i \boldsymbol{\delta}_i \frac{\partial s}{\partial x_i} \tag{A.4–2}$$

The vector thus constructed from the derivatives of s is designated by $\boldsymbol{\nabla} s$ (or grad s) and is called the *gradient* of the scalar field s. The following properties of the gradient operation should be noted:

Not commutative $$\boldsymbol{\nabla} s \neq s\boldsymbol{\nabla} \tag{A.4–3}$$

Not associative $$(\boldsymbol{\nabla} r)s \neq \boldsymbol{\nabla}(rs) \tag{A.4–4}$$

Distributive $$\boldsymbol{\nabla}(r + s) = \boldsymbol{\nabla} r + \boldsymbol{\nabla} s \tag{A.4–5}$$

The Divergence of a Vector Field

If the vector $\boldsymbol{v}$ is a function of the space variables x_1, x_2, x_3, then a scalar product may be formed with the operator $\boldsymbol{\nabla}$; in obtaining the final form, we use Eq. A.2–14:

$$\begin{aligned}(\boldsymbol{\nabla} \cdot \boldsymbol{v}) &= \left(\left\{\textstyle\sum_i \boldsymbol{\delta}_i \frac{\partial}{\partial x_i}\right\} \cdot \{\textstyle\sum_j \boldsymbol{\delta}_j v_j\}\right) \\ &= \textstyle\sum_i \sum_j (\boldsymbol{\delta}_i \cdot \boldsymbol{\delta}_j) \frac{\partial}{\partial x_i} v_j \\ &= \textstyle\sum_i \sum_j \delta_{ij} \frac{\partial}{\partial x_i} v_j = \sum_i \frac{\partial v_i}{\partial x_i} \end{aligned} \tag{A.4–6}$$

This collection of derivatives of the components of the vector $\boldsymbol{v}$ is called the *divergence* of $\boldsymbol{v}$ (sometimes abbreviated div $\boldsymbol{v}$). Some properties of the divergence operator should be noted:

Not commutative: $$(\boldsymbol{\nabla} \cdot \boldsymbol{v}) \neq (\boldsymbol{v} \cdot \boldsymbol{\nabla}) \tag{A.4–7}$$

Not associative: $$(\boldsymbol{\nabla} \cdot s\boldsymbol{v}) \neq (\boldsymbol{\nabla} s \cdot \boldsymbol{v}) \tag{A.4–8}$$

Distributive: $$(\boldsymbol{\nabla} \cdot \{\boldsymbol{v} + \boldsymbol{w}\}) = (\boldsymbol{\nabla} \cdot \boldsymbol{v}) + (\boldsymbol{\nabla} \cdot \boldsymbol{w}) \tag{A.4–9}$$

The Curl of a Vector Field

A cross product may also be formed between the $\boldsymbol{\nabla}$ operator and the vector $\boldsymbol{v}$, which is a function of the three space variables. This cross product may be simplified by using Eq. A.2–15 and written in a variety of forms:

$$\begin{aligned}[\boldsymbol{\nabla} \times \boldsymbol{v}] &= \left[\left\{\textstyle\sum_j \boldsymbol{\delta}_j \frac{\partial}{\partial x_j}\right\} \times \{\textstyle\sum_k \boldsymbol{\delta}_k v_k\}\right] \\ &= \textstyle\sum_j \sum_k [\boldsymbol{\delta}_j \times \boldsymbol{\delta}_k] \frac{\partial}{\partial x_j} v_k \\ &= \textstyle\sum_i \sum_j \sum_k \epsilon_{ijk} \boldsymbol{\delta}_i \frac{\partial}{\partial x_j} v_k \\ &= \begin{vmatrix} \boldsymbol{\delta}_1 & \boldsymbol{\delta}_2 & \boldsymbol{\delta}_3 \\ \frac{\partial}{\partial x_1} & \frac{\partial}{\partial x_2} & \frac{\partial}{\partial x_3} \\ v_1 & v_2 & v_3 \end{vmatrix} \\ &= \boldsymbol{\delta}_1 \left\{\frac{\partial v_3}{\partial x_2} - \frac{\partial v_2}{\partial x_3}\right\} + \boldsymbol{\delta}_2 \left\{\frac{\partial v_1}{\partial x_3} - \frac{\partial v_3}{\partial x_1}\right\} + \boldsymbol{\delta}_3 \left\{\frac{\partial v_2}{\partial x_1} - \frac{\partial v_1}{\partial x_2}\right\} \end{aligned} \tag{A.4–10}$$

The vector thus constructed is called the *curl* of $\boldsymbol{v}$. Other notations for $[\boldsymbol{\nabla} \times \boldsymbol{v}]$ are curl $\boldsymbol{v}$ and rot $\boldsymbol{v}$, the latter being common in the German literature. The curl operation, like the divergence, is distributive but not commutative or associative. Note that the ith component of $[\boldsymbol{\nabla} \times \boldsymbol{v}]$ is $\sum_j \sum_k \epsilon_{ijk}(\partial v_k/\partial x_j)$.

The Gradient of a Vector Field

In addition to the scalar product $(\boldsymbol{\nabla} \cdot \boldsymbol{v})$ and the vector product $[\boldsymbol{\nabla} \times \boldsymbol{v}]$ one may also form the dyadic product $\boldsymbol{\nabla v}$:

$$\begin{aligned}\boldsymbol{\nabla v} &= \left\{\textstyle\sum_i \boldsymbol{\delta}_i \frac{\partial}{\partial x_i}\right\}\{\textstyle\sum_j \boldsymbol{\delta}_j v_j\} \\ &= \textstyle\sum_i \sum_j \boldsymbol{\delta}_i\boldsymbol{\delta}_j \frac{\partial}{\partial x_i} v_j \qquad \text{(A.4–11)}\end{aligned}$$

This is called the *gradient* of the vector $\boldsymbol{v}$ and is sometimes written grad $\boldsymbol{v}$. It is a second-order tensor whose *ij*-component[1] is $(\partial/\partial x_i)v_j$. Its transpose is:

$$(\boldsymbol{\nabla v})^\dagger = \textstyle\sum_i \sum_j \boldsymbol{\delta}_i\boldsymbol{\delta}_j \frac{\partial}{\partial x_j} v_i \qquad \text{(A.4–12)}$$

whose *ij*-component is $\partial v_i/\partial x_j$. Note that $\boldsymbol{\nabla v} \neq \boldsymbol{v\nabla}$ and $(\boldsymbol{\nabla v})^\dagger \neq \boldsymbol{v\nabla}$.

The Divergence of a Tensor Field

If the tensor $\boldsymbol{\tau}$ is a function of the space variables x_1, x_2, x_3, then a vector product may be formed with operator $\boldsymbol{\nabla}$; in obtaining the final form we use Eq. A.3–3:

$$\begin{aligned}[\boldsymbol{\nabla} \cdot \boldsymbol{\tau}] &= \left[\left\{\textstyle\sum_i \boldsymbol{\delta}_i \frac{\partial}{\partial x_i}\right\} \cdot \left\{\textstyle\sum_j \sum_k \boldsymbol{\delta}_j\boldsymbol{\delta}_k\tau_{jk}\right\}\right] \\ &= \textstyle\sum_i \sum_j \sum_k [\boldsymbol{\delta}_i \cdot \boldsymbol{\delta}_j\boldsymbol{\delta}_k] \frac{\partial}{\partial x_i}\tau_{jk} \\ &= \textstyle\sum_i \sum_j \sum_k \delta_{ij}\boldsymbol{\delta}_k \frac{\partial}{\partial x_i}\tau_{jk} \\ &= \textstyle\sum_k \boldsymbol{\delta}_k \left\{\sum_i \frac{\partial}{\partial x_i}\tau_{ik}\right\} \qquad \text{(A.4–13)}\end{aligned}$$

This is called the *divergence* of the tensor $\boldsymbol{\tau}$, and is sometimes written div $\boldsymbol{\tau}$. The kth component of $[\boldsymbol{\nabla} \cdot \boldsymbol{\tau}]$ is $\sum_i (\partial\tau_{ik}/\partial x_i)$. If $\boldsymbol{\tau}$ is the product $s\boldsymbol{vw}$, then:

$$[\boldsymbol{\nabla} \cdot s\boldsymbol{vw}] = \textstyle\sum_k \boldsymbol{\delta}_k \left\{\sum_i \frac{\partial}{\partial x_i}(s v_i w_k)\right\} \qquad \text{(A.4–14)}$$

The Laplacian of a Scalar Field

If we take the divergence of the gradient of the scalar function s, we obtain

$$\begin{aligned}(\boldsymbol{\nabla} \cdot \boldsymbol{\nabla} s) &= \left(\left\{\textstyle\sum_i \boldsymbol{\delta}_i \frac{\partial}{\partial x_i}\right\} \cdot \left\{\textstyle\sum_j \boldsymbol{\delta}_j \frac{\partial s}{\partial x_j}\right\}\right) \\ &= \textstyle\sum_i \sum_j \delta_{ij} \frac{\partial}{\partial x_i}\frac{\partial s}{\partial x_j} \\ &= \left\{\textstyle\sum_i \frac{\partial^2}{\partial x_i^2} s\right\} \qquad \text{(A.4–15)}\end{aligned}$$

[1] *Caution*: Some authors define the *ij*-component of $\boldsymbol{\nabla v}$ to be $\partial v_i/\partial x_j$.

The collection of differential operators operating on s in the last line is given the symbol $\boldsymbol{\nabla}^2$; hence in rectangular coordinates:

$$(\boldsymbol{\nabla} \cdot \boldsymbol{\nabla}) = \boldsymbol{\nabla}^2 = \frac{\partial^2}{\partial x_1{}^2} + \frac{\partial^2}{\partial x_2{}^2} + \frac{\partial^2}{\partial x_3{}^2} \tag{A.4–16}$$

This is called the *Laplacian* operator. (Some authors use the symbol Δ for the Laplacian operator—this is particularly true in the older German literature; hence $(\boldsymbol{\nabla} \cdot \boldsymbol{\nabla} s)$, $(\boldsymbol{\nabla} \cdot \boldsymbol{\nabla})s$, $\boldsymbol{\nabla}^2 s$, and Δs are all equivalent forms of notation.) The Laplacian operator has only the distributive property, just as the gradient, divergence, and curl.

The Laplacian of a Vector Field

If we take the divergence of the gradient of the vector function $\boldsymbol{v}$, we obtain:

$$\begin{aligned}
[\boldsymbol{\nabla} \cdot \boldsymbol{\nabla} \boldsymbol{v}] &= \left[\left\{\textstyle\sum_i \boldsymbol{\delta}_i \frac{\partial}{\partial x_i}\right\} \cdot \left\{\textstyle\sum_j \sum_k \boldsymbol{\delta}_j \boldsymbol{\delta}_k \frac{\partial}{\partial x_j} v_k\right\}\right] \\
&= \textstyle\sum_i \sum_j \sum_k [\boldsymbol{\delta}_i \cdot \boldsymbol{\delta}_j \boldsymbol{\delta}_k] \frac{\partial}{\partial x_i} \frac{\partial}{\partial x_j} v_k \\
&= \textstyle\sum_i \sum_j \sum_k \delta_{ij} \boldsymbol{\delta}_k \frac{\partial}{\partial x_i} \frac{\partial}{\partial x_j} v_k \\
&= \textstyle\sum_k \boldsymbol{\delta}_k \left(\sum_i \frac{\partial^2}{\partial x_i{}^2} v_k\right)
\end{aligned} \tag{A.4–17}$$

That is, the kth component of $[\boldsymbol{\nabla} \cdot \boldsymbol{\nabla} \boldsymbol{v}]$ is, in rectangular coordinates, just $\boldsymbol{\nabla}^2 v_k$. Alternative notations for $[\boldsymbol{\nabla} \cdot \boldsymbol{\nabla} \boldsymbol{v}]$ are $(\boldsymbol{\nabla} \cdot \boldsymbol{\nabla})\boldsymbol{v}$ and $\boldsymbol{\nabla}^2 \boldsymbol{v}$.

Other Differential Relations

Numerous identities can be proved using the definitions just given:

$$\boldsymbol{\nabla} rs = r\boldsymbol{\nabla} s + s\boldsymbol{\nabla} r \tag{A.4–18}$$

$$(\boldsymbol{\nabla} \cdot s\boldsymbol{v}) = (\boldsymbol{\nabla} s \cdot \boldsymbol{v}) + s(\boldsymbol{\nabla} \cdot \boldsymbol{v}) \tag{A.4–19}$$

$$(\boldsymbol{\nabla} \cdot [\boldsymbol{v} \times \boldsymbol{w}]) = (\boldsymbol{w} \cdot [\boldsymbol{\nabla} \times \boldsymbol{v}]) - (\boldsymbol{v} \cdot [\boldsymbol{\nabla} \times \boldsymbol{w}]) \tag{A.4–20}$$

$$[\boldsymbol{\nabla} \times s\boldsymbol{v}] = [\boldsymbol{\nabla} s \times \boldsymbol{v}] + s[\boldsymbol{\nabla} \times \boldsymbol{v}] \tag{A.4–21}$$

$$[\boldsymbol{\nabla} \cdot \boldsymbol{\nabla} \boldsymbol{v}] = \boldsymbol{\nabla}(\boldsymbol{\nabla} \cdot \boldsymbol{v}) - [\boldsymbol{\nabla} \times [\boldsymbol{\nabla} \times \boldsymbol{v}]] \tag{A.4–22}$$

$$[\boldsymbol{v} \cdot \boldsymbol{\nabla} \boldsymbol{v}] = \tfrac{1}{2}\boldsymbol{\nabla}(\boldsymbol{v} \cdot \boldsymbol{v}) - [\boldsymbol{v} \times [\boldsymbol{\nabla} \times \boldsymbol{v}]] \tag{A.4–23}$$

$$[\boldsymbol{\nabla} \cdot \boldsymbol{v}\boldsymbol{w}] = [\boldsymbol{v} \cdot \boldsymbol{\nabla} \boldsymbol{w}] + \boldsymbol{w}(\boldsymbol{\nabla} \cdot \boldsymbol{v}) \tag{A.4–24}$$

$$(s\boldsymbol{\delta} : \boldsymbol{\nabla} \boldsymbol{v}) = s(\boldsymbol{\nabla} \cdot \boldsymbol{v}) \tag{A.4–25}$$

$$[\boldsymbol{\nabla} \cdot s\boldsymbol{\delta}] = \boldsymbol{\nabla} s \tag{A.4–26}$$

$$[\boldsymbol{\nabla} \cdot s\boldsymbol{\tau}] = [\boldsymbol{\nabla} s \cdot \boldsymbol{\tau}] + s[\boldsymbol{\nabla} \cdot \boldsymbol{\tau}] \tag{A.4–27}$$

EXAMPLE A.4–1 Proof of a Tensor Identity

Prove that for symmetrical $\boldsymbol{\tau}$

$$(\boldsymbol{\tau} : \nabla \boldsymbol{v}) = (\nabla \cdot [\boldsymbol{\tau} \cdot \boldsymbol{v}]) - (\boldsymbol{v} \cdot [\nabla \cdot \boldsymbol{\tau}]) \tag{A.4–28}$$

SOLUTION

First we write out the right side in terms of components:

$$\begin{aligned}(\nabla \cdot [\boldsymbol{\tau} \cdot \boldsymbol{v}]) &= \textstyle\sum_i \frac{\partial}{\partial x_i} [\boldsymbol{\tau} \cdot \boldsymbol{v}]_i \\ &= \textstyle\sum_i \sum_j \frac{\partial}{\partial x_i} \tau_{ij} v_j \end{aligned} \tag{A.4–29}$$

$$\begin{aligned}(\boldsymbol{v} \cdot [\nabla \cdot \boldsymbol{\tau}]) &= \textstyle\sum_j v_j [\nabla \cdot \boldsymbol{\tau}]_j \\ &= \textstyle\sum_j \sum_i v_j \frac{\partial}{\partial x_i} \tau_{ij} \end{aligned} \tag{A.4–30}$$

The left side may be written as:

$$\begin{aligned}(\boldsymbol{\tau} : \nabla \boldsymbol{v}) &= \textstyle\sum_i \sum_j \tau_{ji} \frac{\partial}{\partial x_i} v_j \\ &= \textstyle\sum_i \sum_j \tau_{ij} \frac{\partial}{\partial x_i} v_j \end{aligned} \tag{A.4–31}$$

the second form resulting from using the symmetry of $\boldsymbol{\tau}$. Subtraction of Eq. A.4–30 from Eq. A.4–29 will give Eq. A.4–31.

EXERCISES

1. Perform all the operations in Eq. A.4–6 by writing out all the summations instead of using the $\sum$ notation.
2. A field $\boldsymbol{v}(x,y,z)$ is said to be *irrotational* if $[\nabla \times \boldsymbol{v}] = \mathbf{0}$. Which of the following fields are irrotational:
   (a) $v_x = by \qquad v_y = 0 \qquad v_z = 0$
   (b) $v_x = bx \qquad v_y = 0 \qquad v_z = 0$
   (c) $v_x = by \qquad v_y = bx \qquad v_z = 0$
   (d) $v_x = -by \qquad v_y = bx \qquad v_z = 0$
   Here b is a constant.
3. Evaluate $(\nabla \cdot \boldsymbol{v})$, $\nabla \boldsymbol{v}$ and $[\nabla \cdot \boldsymbol{v}\boldsymbol{v}]$ for the four fields in Exercise 2.
4. A vector $\boldsymbol{v}$ has components
   $$v_i = \sum_{j=1}^{3} \alpha_{ij} x_j$$
   with $\alpha_{ij} = \alpha_{ji}$ and $\sum_{i=1}^{3} \alpha_{ii} = 0$; the α_{ij} are constants. Evaluate $(\nabla \cdot \boldsymbol{v})$, $[\nabla \times \boldsymbol{v}]$, $\nabla \boldsymbol{v}$, $(\nabla \boldsymbol{v})^\dagger$ and $[\nabla \cdot \boldsymbol{v}\boldsymbol{v}]$. (*Hint*: In connection with evaluating $[\nabla \times \boldsymbol{v}]$, see Exercise 6 in §A.2.)
5. Verify that $\nabla^2(\nabla \cdot \boldsymbol{v}) = (\nabla \cdot (\nabla^2 \boldsymbol{v}))$, and that $[\nabla \cdot (\nabla \boldsymbol{v})^\dagger] = \nabla(\nabla \cdot \boldsymbol{v})$.
6. Verify that $(\nabla \cdot [\nabla \times \boldsymbol{v}]) = 0$ and $[\nabla \times \nabla s] = \mathbf{0}$.

7. If $\boldsymbol{r}$ is the position vector (with components x_1, x_2, x_3) and $\boldsymbol{v}$ is any vector, show that:
   (a) $(\nabla \cdot \boldsymbol{r}) = 3$
   (b) $[\nabla \times \boldsymbol{r}] = \boldsymbol{0}$
   (c) $\nabla(\boldsymbol{v} \cdot \boldsymbol{r}) = \boldsymbol{v}$ (where $\boldsymbol{v}$ is a constant)
   (d) $[\boldsymbol{r} \times [\nabla \cdot \boldsymbol{vv}]] = [\nabla \cdot \boldsymbol{v}[\boldsymbol{r} \times \boldsymbol{v}]]$ (where $\boldsymbol{v}$ is a function of position)
8. Develop an alternate expression for $[\nabla \times [\nabla \cdot s\boldsymbol{vv}]]$
9. Is the following relation valid?

$$\tfrac{1}{2}\rho(\nabla \cdot \boldsymbol{v}) = [\nabla \times \rho\boldsymbol{v}] + \tfrac{1}{2}(\nabla \cdot \rho\boldsymbol{v}) - (\boldsymbol{v} \cdot \nabla\rho)$$

10. Write out in full in rectangular coordinates:
   (a) $\dfrac{\partial}{\partial t}\rho\boldsymbol{v} = -[\nabla \cdot \rho\boldsymbol{vv}] - \nabla p - [\nabla \cdot \boldsymbol{\tau}] + \rho\boldsymbol{g}$
   (b) $\boldsymbol{\tau} = -\mu\{\nabla\boldsymbol{v} + (\nabla\boldsymbol{v})^\dagger - \tfrac{2}{3}(\nabla \cdot \boldsymbol{v})\boldsymbol{\delta}\}$
11. If $(\nabla \cdot \boldsymbol{v}) = 0$, then $\boldsymbol{v} = [\nabla \times \boldsymbol{A}]$ where $\boldsymbol{A}$ is some vector. If $[\nabla \cdot \boldsymbol{\tau}] = \boldsymbol{0}$ and $\boldsymbol{\tau}$ is symmetrical, then $\boldsymbol{\tau} = \{\nabla \times \{\nabla \times \boldsymbol{\Phi}\}^\dagger\}$ where $\boldsymbol{\Phi}$ is some symmetrical tensor. Prove these statements.
12. If $\boldsymbol{r}$ is the position vector and r is its magnitude, verify that
   (a) $\nabla \dfrac{1}{r} = -\dfrac{\boldsymbol{r}}{r^3}$
   (b) $\nabla f(r) = \dfrac{1}{r}\dfrac{df}{dr}\boldsymbol{r}$
   (c) $\nabla(\boldsymbol{a} \cdot \boldsymbol{r}) = \boldsymbol{a}$ if $\boldsymbol{a}$ is a constant vector
   (d) $\nabla\nabla r^n = nr^{n-2}[(n-2)r^{-2}\boldsymbol{rr} + \boldsymbol{\delta}]$

§A.5 VECTOR AND TENSOR INTEGRAL THEOREMS

For performing general proofs in continuum physics, several integral theorems are extremely useful.

The Gauss-Ostrogradskii Divergence Theorem

If V is a closed region in space surrounded by a surface S, then[1]

$$\int_V (\nabla \cdot \boldsymbol{v})\, dV = \int_S (\boldsymbol{n} \cdot \boldsymbol{v})\, dS \qquad \text{(A.5–1)}$$

in which $\boldsymbol{n}$ is the outwardly directed unit normal vector. This is known as the *divergence theorem* of Gauss and Ostrogradskii. Two closely allied theorems for scalars and tensors are:

$$\int_V \nabla s\, dV = \int_S \boldsymbol{n}s\, dS \qquad \text{(A.5–2)}$$

$$\int_V [\nabla \cdot \boldsymbol{\tau}]\, dV = \int_S [\boldsymbol{n} \cdot \boldsymbol{\tau}]\, dS \qquad \text{(A.5–3)[2]}$$

[1] The right side may be written in a variety of ways:

$$\int_S (\boldsymbol{v} \cdot \boldsymbol{n})\, dS \quad \text{or} \quad \int_S v_n\, dS \quad \text{or} \quad \int_S (\boldsymbol{v} \cdot \boldsymbol{dS})$$

[2] See P. M. Morse and H. Feshbach, *Methods of Theoretical Physics*, McGraw-Hill, New York (1953), p. 66.

The last relation is also valid for dyadic products $\boldsymbol{vw}$. Note that in all three equations $\boldsymbol{\nabla}$ in the volume integral is just replaced by $\boldsymbol{n}$ in the surface integral; this is a useful mnemonic device.

The Stokes Curl Theorem

If S is a surface bounded by the closed curve C, then

$$\int_S (\boldsymbol{n} \cdot [\boldsymbol{\nabla} \times \boldsymbol{v}])\, dS = \oint_C (\boldsymbol{t} \cdot \boldsymbol{v})\, dC \qquad \text{(A.5–4)}$$

in which $\boldsymbol{t}$ is a unit tangential vector in the direction of integration along C; $\boldsymbol{n}$ is the unit normal vector to S in the direction that a right-hand screw would move if its head were twisted in the direction of integration along C. A similar relation exists for tensors.[2]

$$\int_S [\boldsymbol{n} \cdot \{\boldsymbol{\nabla} \times \boldsymbol{\tau}\}]\, dS = \oint_C [\boldsymbol{t} \cdot \boldsymbol{\tau}]\, dC \qquad \text{(A.5–5)}$$

The Leibnitz Formula for Differentiating a Triple Integral

Let V be a closed moving region in space surrounded by a surface S; let the velocity of any surface element be $\boldsymbol{v}_S$. Then, if $s(x,y,z,t)$ is a scalar function of position and time:

$$\frac{d}{dt}\int_V s\, dV = \int_V \frac{\partial s}{\partial t}\, dV + \int_S s(\boldsymbol{v}_S \cdot \boldsymbol{n})\, dS \qquad \text{(A.5–6)}$$

This is an extension of the *Leibnitz formula* for differentiating an integral; keep in mind that $V = V(t)$ and $S = S(t)$. Equation A.5–6 also applies to vectors and tensors.

EXERCISES

1. Consider the vector field:

$$\boldsymbol{v} = \boldsymbol{\delta}_1 x_1 + \boldsymbol{\delta}_2 x_3 + \boldsymbol{\delta}_3 x_2$$

   Evaluate both sides of Eq. A.5–1 over the region bounded by the planes $x_1 = 0, x_1 = 1$; $x_2 = 0, x_2 = 2$; $x_3 = 0, x_3 = 4$.
2. Use the same vector field to evaluate both sides of Eq. A.5–4 for the face $x_1 = 1$ in Example 1.
3. Consider the time-dependent scalar function:

$$s = x + y + zt$$

   Evaluate both sides of Eq. A.5–6 over the volume bounded by the planes: $x_1 = 0$, $x_1 = t$; $x_2 = 0, x_2 = 2t$; $x_3 = 0, x_3 = 4t$.
4. Use Eq. A.5–5 to show that:

$$2\int_S \boldsymbol{n}\, dS = \oint_C [\boldsymbol{r} \times \boldsymbol{t}]\, dC$$

   where $\boldsymbol{r}$ is the position vector locating a point on C with respect to the origin.

§A.6 VECTOR AND TENSOR ALGEBRA IN CURVILINEAR COORDINATES

Thus far we have considered only rectangular coordinates, x, y, and z. Although formal derivations are usually made in rectangular coordinates, for working problems it is often more natural to use curvilinear coordinates. The two most commonly occurring curvilinear coordinate systems are the *cylindrical* and the *spherical*. In the following we discuss only these two systems, but the method can also be applied to all *orthogonal* coordinate systems, that is, those in which the three families of coordinate surfaces are mutually perpendicular.

We are primarily interested in knowing how to write various differential operations, such as ∇s, $[\nabla \times \boldsymbol{v}]$, and $(\boldsymbol{\tau} : \nabla \boldsymbol{v})$ in curvilinear coordinates. It turns out that we can do this quite simply if we know, for the coordinate system being used, just two things:

(a) The expression for ∇ in curvilinear coordinates.
(b) The spatial derivatives of the unit vectors in curvilinear coordinates.

Hence, we want to focus our attention on these two points. However, in this section we discuss transformation of coordinates and transformation of unit vectors.

Cylindrical Coordinates

In cylindrical coordinates, instead of designating the coordinates of a point by x, y, z, we locate the point by giving the values of r, θ, z. These coordinates[1] are shown in Fig. A.6–1*a*. They are related to the rectangular coordinates by:

$$x = r\cos\theta \qquad \text{(A.6–1)} \qquad\qquad r = +\sqrt{x^2 + y^2} \qquad \text{(A.6–4)}$$

$$y = r\sin\theta \qquad \text{(A.6–2)} \qquad\qquad \theta = \arctan(y/x) \qquad \text{(A.6–5)}$$

$$z = z \qquad \text{(A.6–3)} \qquad\qquad z = z \qquad \text{(A.6–6)}$$

In order to convert derivatives of scalars with respect to x, y, z into derivatives with respect to r, θ, z, the "chain rule" of partial differentiation[2] is used. The derivative operators are

[1] *Caution*: We have chosen to use the familiar r, θ, z-notation for cylindrical coordinates rather than to switch to some less familiar symbols, even though there are two situations in which confusion can arise: (a) Occasionally one has to use cylindrical and spherical coordinates in the same problem, and the symbols r and θ have different meanings in the two systems; (b) occasionally one deals with the position vector $\boldsymbol{r}$ in problems involving cylindrical coordinates, but then the magnitude of $\boldsymbol{r}$ is not the same as the coordinate r, but rather $\sqrt{r^2 + z^2}$. In such situations, as in Fig. A.6–1, we can use overbars for the cylindrical coordinates and write $\bar{r}$, $\bar{\theta}$, $\bar{z}$. For most discussions bars will not be needed.

[2] For example, for a scalar function $\phi(x,y,z) = \psi(r,\theta,z)$

$$\left(\frac{\partial\phi}{\partial x}\right)_{y,z} = \left(\frac{\partial r}{\partial x}\right)_{y,z}\left(\frac{\partial\psi}{\partial r}\right)_{\theta,z} + \left(\frac{\partial\theta}{\partial x}\right)_{y,z}\left(\frac{\partial\psi}{\partial\theta}\right)_{r,z} + \left(\frac{\partial z}{\partial x}\right)_{y,z}\left(\frac{\partial\psi}{\partial z}\right)_{r,\theta}$$

Note that we are careful to use different symbols ϕ and ψ, since ϕ is a different function of x, y, z than ψ is of r, θ, and z!

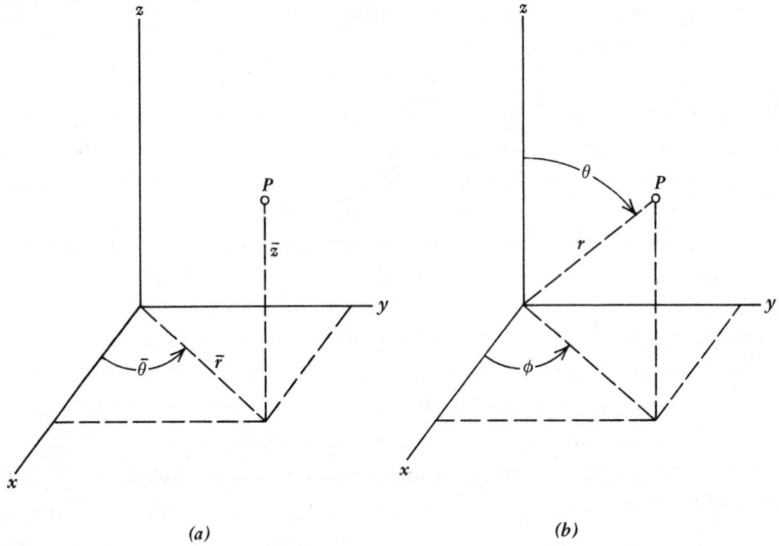

FIGURE A.6–1. (*a*) Cylindrical coordinates with $0 \leq \bar{r} < \infty, 0 \leq \bar{\theta} < 2\pi, -\infty < \bar{z} < \infty$. (*b*) Spherical coordinates with $0 \leq r < \infty, 0 \leq \theta \leq \pi, 0 \leq \phi < 2\pi$. Note that $\bar{r}$ and $\bar{\theta}$ in cylindrical coordinates are not the same as r and θ in spherical coordinates. Note also how the position vector is written in the three coordinate systems:

Rectangular:	$\boldsymbol{r} = \boldsymbol{\delta}_x x + \boldsymbol{\delta}_y y + \boldsymbol{\delta}_z z;$	$r = \sqrt{x^2 + y^2 + z^2}$
Cylindrical:	$\boldsymbol{r} = \boldsymbol{\delta}_r \bar{r} + \boldsymbol{\delta}_z \bar{z};$	$r = \sqrt{\bar{r}^2 + \bar{z}^2}$
Spherical:	$\boldsymbol{r} = \boldsymbol{\delta}_r r;$	$r = r$

See footnote 1 of §A.6 regarding the omission of the bars in cylindrical coordinates.

readily found to be related thus:

$$\frac{\partial}{\partial x} = (\cos\theta)\frac{\partial}{\partial r} + \left(-\frac{\sin\theta}{r}\right)\frac{\partial}{\partial \theta} + (0)\frac{\partial}{\partial z} \tag{A.6–7}$$

$$\frac{\partial}{\partial y} = (\sin\theta)\frac{\partial}{\partial r} + \left(\frac{\cos\theta}{r}\right)\frac{\partial}{\partial \theta} + (0)\frac{\partial}{\partial z} \tag{A.6–8}$$

$$\frac{\partial}{\partial z} = (0)\frac{\partial}{\partial r} + (0)\frac{\partial}{\partial \theta} + (1)\frac{\partial}{\partial z} \tag{A.6–9}$$

With these relations, derivatives of any scalar functions (including, of course, components of vectors and tensors) with respect to x, y, and z can be expressed in terms of derivatives with respect to r, θ, and z.

Having discussed the interrelationship of the coordinates and derivatives in the two coordinate systems, we now turn to the relation between the unit vectors. We begin by noting that the unit vectors $\boldsymbol{\delta}_x$, $\boldsymbol{\delta}_y$, $\boldsymbol{\delta}_z$ (or $\boldsymbol{\delta}_1$, $\boldsymbol{\delta}_2$, $\boldsymbol{\delta}_3$ as we have been calling them) are independent of position—that is, independent of x, y, z. In cylindrical coordinates the unit vectors $\boldsymbol{\delta}_r$ and $\boldsymbol{\delta}_\theta$ will depend on position, as can be seen in Fig. A.6–2. The unit vector $\boldsymbol{\delta}_r$ is a vector of unit length in the direction of increasing r; the unit vector $\boldsymbol{\delta}_\theta$ is a vector of unit length in the direction of increasing θ. Clearly as the point P is moved around on the xy-plane, the directions of $\boldsymbol{\delta}_r$ and $\boldsymbol{\delta}_\theta$ change. Elementary trigonometrical arguments lead

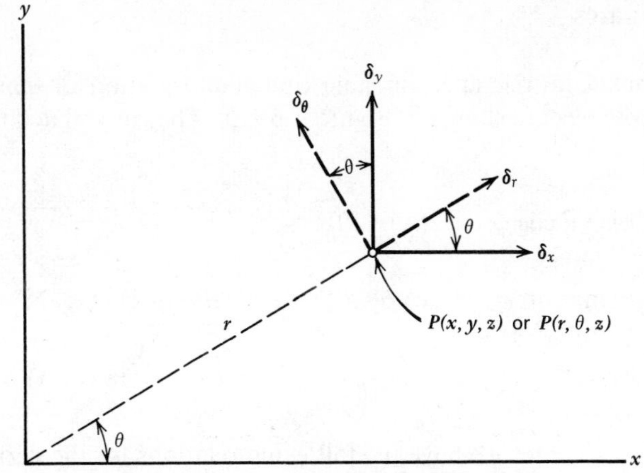

FIGURE A.6–2. Unit vectors in rectangular and cylindrical coordinates; the z-axis and unit vector $\boldsymbol{\delta}_z$ have been omitted for simplicity.

to the following relations:

$$\boldsymbol{\delta}_r = (\cos\theta)\boldsymbol{\delta}_x + (\sin\theta)\boldsymbol{\delta}_y + (0)\boldsymbol{\delta}_z \tag{A.6–10}$$

$$\boldsymbol{\delta}_\theta = (-\sin\theta)\boldsymbol{\delta}_x + (\cos\theta)\boldsymbol{\delta}_y + (0)\boldsymbol{\delta}_z \tag{A.6–11}$$

$$\boldsymbol{\delta}_z = (0)\boldsymbol{\delta}_x + (0)\boldsymbol{\delta}_y + (1)\boldsymbol{\delta}_z \tag{A.6–12}$$

These may be solved for $\boldsymbol{\delta}_x$, $\boldsymbol{\delta}_y$, and $\boldsymbol{\delta}_z$ to give:

$$\boldsymbol{\delta}_x = (\cos\theta)\boldsymbol{\delta}_r + (-\sin\theta)\boldsymbol{\delta}_\theta + (0)\boldsymbol{\delta}_z \tag{A.6–13}$$

$$\boldsymbol{\delta}_y = (\sin\theta)\boldsymbol{\delta}_r + (\cos\theta)\boldsymbol{\delta}_\theta + (0)\boldsymbol{\delta}_z \tag{A.6–14}$$

$$\boldsymbol{\delta}_z = (0)\boldsymbol{\delta}_r + (0)\boldsymbol{\delta}_\theta + (1)\boldsymbol{\delta}_z \tag{A.6–15}$$

The utility of these two sets of relations will be made clear in the next section.

Vectors and tensors can be decomposed into components with respect to cylindrical coordinates just as was done for rectangular coordinates in Eqs. A.2–16 and A.3–8 (i.e., $\boldsymbol{v} = \boldsymbol{\delta}_r v_r + \boldsymbol{\delta}_\theta v_\theta + \boldsymbol{\delta}_z v_z$). Also the multiplication rules for the unit vectors and unit dyads are the same as in Eqs. A.2–14 and 15 and A.3–1 to 6. Consequently the various dot and cross product operations (but *not* the differential operations!) are performed as described in §§A.2 and A.3. For example:

$$(\boldsymbol{v} \cdot \boldsymbol{w}) = v_r w_r + v_\theta w_\theta + v_z w_z \tag{A.6–16}$$

$$\begin{aligned}[\boldsymbol{v} \times \boldsymbol{w}] &= \boldsymbol{\delta}_r(v_\theta w_z - v_z w_\theta) + \boldsymbol{\delta}_\theta(v_z w_r - v_r w_z) \\ &\quad + \boldsymbol{\delta}_z(v_r w_\theta - v_\theta w_r)\end{aligned} \tag{A.6–17}$$

$$\begin{aligned}\{\boldsymbol{\sigma} \cdot \boldsymbol{\tau}\} &= \boldsymbol{\delta}_r\boldsymbol{\delta}_r(\sigma_{rr}\tau_{rr} + \sigma_{r\theta}\tau_{\theta r} + \sigma_{rz}\tau_{zr}) \\ &\quad + \boldsymbol{\delta}_r\boldsymbol{\delta}_\theta(\sigma_{rr}\tau_{r\theta} + \sigma_{r\theta}\tau_{\theta\theta} + \sigma_{rz}\tau_{z\theta}) \\ &\quad + \boldsymbol{\delta}_r\boldsymbol{\delta}_z(\sigma_{rr}\tau_{rz} + \sigma_{r\theta}\tau_{\theta z} + \sigma_{rz}\tau_{zz}) \\ &\quad + \text{etc.}\end{aligned} \tag{A.6–18}$$

Spherical Coordinates

We now tabulate for reference the same kind of information for spherical coordinates r, θ, ϕ. These coordinates are shown in Figure A.6–1*b*. They are related to the rectangular coordinates by:

$$x = r \sin\theta \cos\phi \quad \text{(A.6–19)} \qquad r = +\sqrt{x^2 + y^2 + z^2} \quad \text{(A.6–22)}$$

$$y = r \sin\theta \sin\phi \quad \text{(A.6–20)} \qquad \theta = \arctan\left(\sqrt{x^2 + y^2}/z\right) \quad \text{(A.6–23)}$$

$$z = r \cos\theta \quad \text{(A.6–21)} \qquad \phi = \arctan(y/x) \quad \text{(A.6–24)}$$

For the spherical coordinates we have the following relations for the derivative operators:

$$\frac{\partial}{\partial x} = (\sin\theta\cos\phi)\frac{\partial}{\partial r} + \left(\frac{\cos\theta\cos\phi}{r}\right)\frac{\partial}{\partial\theta} + \left(-\frac{\sin\phi}{r\sin\theta}\right)\frac{\partial}{\partial\phi} \quad \text{(A.6–25)}$$

$$\frac{\partial}{\partial y} = (\sin\theta\sin\phi)\frac{\partial}{\partial r} + \left(\frac{\cos\theta\sin\phi}{r}\right)\frac{\partial}{\partial\theta} + \left(\frac{\cos\phi}{r\sin\theta}\right)\frac{\partial}{\partial\phi} \quad \text{(A.6–26)}$$

$$\frac{\partial}{\partial z} = (\cos\theta)\frac{\partial}{\partial r} + \left(-\frac{\sin\theta}{r}\right)\frac{\partial}{\partial\theta} + (0)\frac{\partial}{\partial\phi} \quad \text{(A.6–27)}$$

The relations between the unit vectors are:

$$\boldsymbol{\delta}_r = (\sin\theta\cos\phi)\boldsymbol{\delta}_x + (\sin\theta\sin\phi)\boldsymbol{\delta}_y + (\cos\theta)\boldsymbol{\delta}_z \quad \text{(A.6–28)}$$

$$\boldsymbol{\delta}_\theta = (\cos\theta\cos\phi)\boldsymbol{\delta}_x + (\cos\theta\sin\phi)\boldsymbol{\delta}_y + (-\sin\theta)\boldsymbol{\delta}_z \quad \text{(A.6–29)}$$

$$\boldsymbol{\delta}_\phi = (-\sin\phi)\boldsymbol{\delta}_x + (\cos\phi)\boldsymbol{\delta}_y + (0)\boldsymbol{\delta}_z \quad \text{(A.6–30)}$$

and

$$\boldsymbol{\delta}_x = (\sin\theta\cos\phi)\boldsymbol{\delta}_r + (\cos\theta\cos\phi)\boldsymbol{\delta}_\theta + (-\sin\phi)\boldsymbol{\delta}_\phi \quad \text{(A.6–31)}$$

$$\boldsymbol{\delta}_y = (\sin\theta\sin\phi)\boldsymbol{\delta}_r + (\cos\theta\sin\phi)\boldsymbol{\delta}_\theta + (\cos\phi)\boldsymbol{\delta}_\phi \quad \text{(A.6–32)}$$

$$\boldsymbol{\delta}_z = (\cos\theta)\boldsymbol{\delta}_r + (-\sin\theta)\boldsymbol{\delta}_\theta + (0)\boldsymbol{\delta}_\phi \quad \text{(A.6–33)}$$

And, finally, some sample operations in spherical coordinates are:

$$\begin{aligned}(\boldsymbol{\sigma} : \boldsymbol{\tau}) = {} & \sigma_{rr}\tau_{rr} + \sigma_{r\theta}\tau_{\theta r} + \sigma_{r\phi}\tau_{\phi r} \\ & + \sigma_{\theta r}\tau_{r\theta} + \sigma_{\theta\theta}\tau_{\theta\theta} + \sigma_{\theta\phi}\tau_{\phi\theta} \\ & + \sigma_{\phi r}\tau_{r\phi} + \sigma_{\phi\theta}\tau_{\theta\phi} + \sigma_{\phi\phi}\tau_{\phi\phi}\end{aligned} \quad \text{(A.6–34)}$$

$$(\boldsymbol{u} \cdot [\boldsymbol{v} \times \boldsymbol{w}]) = \begin{vmatrix} u_r & u_\theta & u_\phi \\ v_r & v_\theta & v_\phi \\ w_r & w_\theta & w_\phi \end{vmatrix} \quad \text{(A.6–35)}$$

That is, the relations (not involving $\boldsymbol{\nabla}$!) given in §§A.2 and 3 can be written directly in terms of spherical components.

EXERCISES

1. Verify the results in Eqs. A.6–7 to 9 and 25 to 27.
2. Verify Eqs. A.6–10 to 15.
3. Verify the results in Eqs. A.6–16, 17, and 18 by using the procedure described above these equations.
4. If $\boldsymbol{r}$ is the instantaneous position vector for a particle, show that the velocity and acceleration of the particle are given by:

$$\boldsymbol{v} = \frac{d}{dt}\boldsymbol{r} = \boldsymbol{\delta}_r \dot{r} + \boldsymbol{\delta}_\theta r\dot{\theta} + \boldsymbol{\delta}_z \dot{z}$$

$$\boldsymbol{a} = \boldsymbol{\delta}_r(\ddot{r} - r\dot{\theta}^2) + \boldsymbol{\delta}_\theta(r\ddot{\theta} + 2\dot{r}\dot{\theta}) + \boldsymbol{\delta}_z \ddot{z}$$

in cylindrical coordinates. The dots indicate time derivatives of the coordinates.

§A.7 DIFFERENTIAL OPERATIONS IN CURVILINEAR COORDINATES

We now turn to the use of the $\boldsymbol{\nabla}$ operator in curvilinear coordinates. As in the previous section, we restrict ourselves primarily to cylindrical and spherical coordinates.

Cylindrical Coordinates

From Eqs. A.6–10, 11, and 12 we can obtain expressions for the spatial derivatives of the unit vectors $\boldsymbol{\delta}_r$, $\boldsymbol{\delta}_\theta$, and $\boldsymbol{\delta}_z$:

$$\frac{\partial}{\partial r}\boldsymbol{\delta}_r = 0 \qquad \frac{\partial}{\partial r}\boldsymbol{\delta}_\theta = 0 \qquad \frac{\partial}{\partial r}\boldsymbol{\delta}_z = 0 \tag{A.7–1}$$

$$\frac{\partial}{\partial \theta}\boldsymbol{\delta}_r = \boldsymbol{\delta}_\theta \qquad \frac{\partial}{\partial \theta}\boldsymbol{\delta}_\theta = -\boldsymbol{\delta}_r \qquad \frac{\partial}{\partial \theta}\boldsymbol{\delta}_z = 0 \tag{A.7–2}$$

$$\frac{\partial}{\partial z}\boldsymbol{\delta}_r = 0 \qquad \frac{\partial}{\partial z}\boldsymbol{\delta}_\theta = 0 \qquad \frac{\partial}{\partial z}\boldsymbol{\delta}_z = 0 \tag{A.7–3}$$

The reader would do well to interpret these derivatives geometrically by considering the way $\boldsymbol{\delta}_r$, $\boldsymbol{\delta}_\theta$, $\boldsymbol{\delta}_z$ change as the location of P is changed in Fig. A.6–2.

We now use the definition of the $\boldsymbol{\nabla}$ operator in Eq. A.4–1, the expressions in Eqs. A.6–13, 14, and 15, and the derivative operators in Eqs. A.6–7, 8, and 9 to obtain the formula for

$\boldsymbol{\nabla}$ in cylindrical coordinates:

$$\begin{aligned}\boldsymbol{\nabla} &= \boldsymbol{\delta}_x \frac{\partial}{\partial x} + \boldsymbol{\delta}_y \frac{\partial}{\partial y} + \boldsymbol{\delta}_z \frac{\partial}{\partial z} \\ &= (\boldsymbol{\delta}_r \cos\theta - \boldsymbol{\delta}_\theta \sin\theta)\left(\cos\theta \frac{\partial}{\partial r} - \frac{\sin\theta}{r}\frac{\partial}{\partial \theta}\right) \\ &\quad + (\boldsymbol{\delta}_r \sin\theta + \boldsymbol{\delta}_\theta \cos\theta)\left(\sin\theta \frac{\partial}{\partial r} + \frac{\cos\theta}{r}\frac{\partial}{\partial \theta}\right) + \boldsymbol{\delta}_z \frac{\partial}{\partial z}\end{aligned} \tag{A.7–4}$$

When this is multiplied out, there is considerable simplification, and we get:

$$\boxed{\boldsymbol{\nabla} = \boldsymbol{\delta}_r \frac{\partial}{\partial r} + \boldsymbol{\delta}_\theta \frac{1}{r}\frac{\partial}{\partial \theta} + \boldsymbol{\delta}_z \frac{\partial}{\partial z}} \tag{A.7–5}$$

for *cylindrical* coordinates. This may be used for obtaining all differential operations in cylindrical coordinates, provided that Eqs. A.7–1, 2, and 3 are used to differentiate any unit vectors on which $\boldsymbol{\nabla}$ operates. This point will be made clear in the subsequent illustrative example.

Spherical Coordinates

The spatial derivatives of $\boldsymbol{\delta}_r$, $\boldsymbol{\delta}_\theta$, and $\boldsymbol{\delta}_\phi$ are given by differentiating Eqs. A.6–28, 29, and 30:

$$\boxed{\frac{\partial}{\partial r}\boldsymbol{\delta}_r = \mathbf{0} \qquad \frac{\partial}{\partial r}\boldsymbol{\delta}_\theta = \mathbf{0} \qquad \frac{\partial}{\partial r}\boldsymbol{\delta}_\phi = \mathbf{0}} \tag{A.7–6}$$

$$\boxed{\frac{\partial}{\partial \theta}\boldsymbol{\delta}_r = \boldsymbol{\delta}_\theta \qquad \frac{\partial}{\partial \theta}\boldsymbol{\delta}_\theta = -\boldsymbol{\delta}_r \qquad \frac{\partial}{\partial \theta}\boldsymbol{\delta}_\phi = \mathbf{0}} \tag{A.7–7}$$

$$\boxed{\frac{\partial}{\partial \phi}\boldsymbol{\delta}_r = \boldsymbol{\delta}_\phi \sin\theta \qquad \frac{\partial}{\partial \phi}\boldsymbol{\delta}_\theta = \boldsymbol{\delta}_\phi \cos\theta \qquad \frac{\partial}{\partial \phi}\boldsymbol{\delta}_\phi = -\boldsymbol{\delta}_r \sin\theta - \boldsymbol{\delta}_\theta \cos\theta} \tag{A.7–8}$$

The use of Eqs. A.6–31, 32, and 33 and Eqs. A.6–25, 26, and 27 gives the following expression for the $\boldsymbol{\nabla}$ operator:

$$\boxed{\boldsymbol{\nabla} = \boldsymbol{\delta}_r \frac{\partial}{\partial r} + \boldsymbol{\delta}_\theta \frac{1}{r}\frac{\partial}{\partial \theta} + \boldsymbol{\delta}_\phi \frac{1}{r\sin\theta}\frac{\partial}{\partial \phi}} \tag{A.7–9}$$

in *spherical* coordinates. This expression may be used for obtaining differential operations in spherical coordinates, provided that Eqs. A.7–6, 7, and 8 are used concomitantly.

Once again we emphasize that one needs only two items of information in order to write the $\boldsymbol{\nabla}$-operations in orthogonal curvilinear coordinates: (1) the expression for $\boldsymbol{\nabla}$, and (2) the expressions for the spatial derivatives of the unit vectors. Thus far we have discussed the two most-used curvilinear coordinates. We now consider briefly the relations for any orthogonal coordinates q_i. Let the relation between Cartesian coordinates and the orthogonal coordinates be given by:

$$x_1 = x_1(q_1,q_2,q_3)$$
$$x_2 = x_2(q_1,q_2,q_3) \qquad \text{or} \qquad x_i = x_i(q_\alpha)$$
$$x_3 = x_3(q_1,q_2,q_3) \tag{A.7–10}$$

These can be solved for the q_α to get the inverse relations $q_\alpha = q_\alpha(x_i)$. Then[1] the unit vectors in rectangular coordinates $\boldsymbol{\delta}_i$ and those in curvilinear coordinates $\boldsymbol{\delta}_\alpha$ are related thus:

$$\boldsymbol{\delta}_\alpha = \sum_i \frac{1}{h_\alpha}\left(\frac{\partial x_i}{\partial q_\alpha}\right)\boldsymbol{\delta}_i = \sum_i h_\alpha \left(\frac{\partial q_\alpha}{\partial x_i}\right)\boldsymbol{\delta}_i \tag{A.7–11}$$

$$\boldsymbol{\delta}_i = \sum_\alpha h_\alpha \left(\frac{\partial q_\alpha}{\partial x_i}\right)\boldsymbol{\delta}_\alpha = \sum_\alpha \frac{1}{h_\alpha}\left(\frac{\partial x_i}{\partial q_\alpha}\right)\boldsymbol{\delta}_\alpha \tag{A.7–12}$$

Consequently, we finally obtain for the derivatives of the unit vectors:

$$\boxed{\frac{\partial \boldsymbol{\delta}_\alpha}{\partial q_\beta} = \frac{\boldsymbol{\delta}_\beta}{h_\alpha}\frac{\partial h_\beta}{\partial q_\alpha} - \delta_{\alpha\beta}\sum_{\gamma=1}^{3}\frac{\boldsymbol{\delta}_\gamma}{h_\gamma}\frac{\partial h_\alpha}{\partial q_\gamma}} \tag{A.7–13}$$

and for the $\boldsymbol{\nabla}$-operator:

$$\boxed{\boldsymbol{\nabla} = \sum_\alpha \frac{\boldsymbol{\delta}_\alpha}{h_\alpha}\frac{\partial}{\partial q_\alpha}} \tag{A.7–14}$$

where

$$h_\alpha^2 = \sum_i \left(\frac{\partial x_i}{\partial q_\alpha}\right)^2 = \left[\sum_i \left(\frac{\partial q_\alpha}{\partial x_i}\right)^2\right]^{-1} \tag{A.7–15}$$

The h_α are called "scale factors."

In Tables A.7–1, 2, 3, and 4 we summarize the differential operations most commonly encountered in rectangular, cylindrical, spherical, and bipolar coordinates.[2] The curvilinear coordinate expressions given there are easily obtained by the method illustrated in the following two examples.

EXAMPLE A.7–1 Differential Operations in Cylindrical Coordinates

Derive expressions for $(\boldsymbol{\nabla} \cdot \boldsymbol{v})$ and $\boldsymbol{\nabla}\boldsymbol{v}$ in cylindrical coordinates.

SOLUTION

(a) We begin by writing $\boldsymbol{\nabla}$ in cylindrical coordinates and decomposing $\boldsymbol{v}$ into its components thus:

$$(\boldsymbol{\nabla} \cdot \boldsymbol{v}) = \left(\left\{\boldsymbol{\delta}_r \frac{\partial}{\partial r} + \boldsymbol{\delta}_\theta \frac{1}{r}\frac{\partial}{\partial \theta} + \boldsymbol{\delta}_z \frac{\partial}{\partial z}\right\} \cdot \left\{\boldsymbol{\delta}_r v_r + \boldsymbol{\delta}_\theta v_\theta + \boldsymbol{\delta}_z v_z\right\}\right) \tag{A.7–16}$$

[1] P. Morse and H. Feshbach, *Methods of Theoretical Physics*, McGraw-Hill, New York (1953), p. 26.
[2] For other coordinate systems see the extensive compilation of P. Moon and D. E. Spencer, *Field Theory Handbook*, Springer, Berlin (1961).

Expanding, we get:

$$(\nabla \cdot v) = \left(\boldsymbol{\delta}_r \cdot \frac{\partial}{\partial r} \boldsymbol{\delta}_r v_r\right) + \left(\boldsymbol{\delta}_r \cdot \frac{\partial}{\partial r} \boldsymbol{\delta}_\theta v_\theta\right) + \left(\boldsymbol{\delta}_r \cdot \frac{\partial}{\partial r} \boldsymbol{\delta}_z v_z\right)$$
$$+ \left(\boldsymbol{\delta}_\theta \cdot \frac{1}{r}\frac{\partial}{\partial \theta} \boldsymbol{\delta}_r v_r\right) + \left(\boldsymbol{\delta}_\theta \cdot \frac{1}{r}\frac{\partial}{\partial \theta} \boldsymbol{\delta}_\theta v_\theta\right) + \left(\boldsymbol{\delta}_\theta \cdot \frac{1}{r}\frac{\partial}{\partial \theta} \boldsymbol{\delta}_z v_z\right)$$
$$+ \left(\boldsymbol{\delta}_z \cdot \frac{\partial}{\partial z} \boldsymbol{\delta}_r v_r\right) + \left(\boldsymbol{\delta}_z \cdot \frac{\partial}{\partial z} \boldsymbol{\delta}_\theta v_\theta\right) + \left(\boldsymbol{\delta}_z \cdot \frac{\partial}{\partial z} \boldsymbol{\delta}_z v_z\right) \quad \text{(A.7–17)}$$

We now use the relations given in Eqs. A.7–1, 2, and 3 to evaluate the derivatives of the unit vectors. This gives

$$(\nabla \cdot v) = (\boldsymbol{\delta}_r \cdot \boldsymbol{\delta}_r)\frac{\partial v_r}{\partial r} + (\boldsymbol{\delta}_r \cdot \boldsymbol{\delta}_\theta)\frac{\partial v_\theta}{\partial r} + (\boldsymbol{\delta}_r \cdot \boldsymbol{\delta}_z)\frac{\partial v_z}{\partial r}$$
$$+ (\boldsymbol{\delta}_\theta \cdot \boldsymbol{\delta}_r)\frac{1}{r}\frac{\partial v_r}{\partial \theta} + (\boldsymbol{\delta}_\theta \cdot \boldsymbol{\delta}_\theta)\frac{1}{r}\frac{\partial v_\theta}{\partial \theta} + (\boldsymbol{\delta}_\theta \cdot \boldsymbol{\delta}_z)\frac{1}{r}\frac{\partial v_z}{\partial \theta}$$
$$+ \frac{v_r}{r}(\boldsymbol{\delta}_\theta \cdot \boldsymbol{\delta}_\theta) + \frac{v_\theta}{r}(\boldsymbol{\delta}_\theta \cdot \{-\boldsymbol{\delta}_r\})$$
$$+ (\boldsymbol{\delta}_z \cdot \boldsymbol{\delta}_r)\frac{\partial v_r}{\partial z} + (\boldsymbol{\delta}_z \cdot \boldsymbol{\delta}_\theta)\frac{\partial v_\theta}{\partial z} + (\boldsymbol{\delta}_z \cdot \boldsymbol{\delta}_z)\frac{\partial v_z}{\partial z} \quad \text{(A.7–18)}$$

Since $(\boldsymbol{\delta}_r \cdot \boldsymbol{\delta}_r) = 1$, $(\boldsymbol{\delta}_r \cdot \boldsymbol{\delta}_\theta) = 0$, etc., the latter simplifies to:

$$(\nabla \cdot v) = \frac{\partial v_r}{\partial r} + \frac{1}{r}\frac{\partial v_\theta}{\partial \theta} + \frac{v_r}{r} + \frac{\partial v_z}{\partial z} \quad \text{(A.7–19)}$$

which is the same as Eq. A of Table A.7–2. The procedure is a bit tedious, but it is straightforward.

(b) Next we examine the dyadic product ∇v:

$$\nabla v = \left\{\boldsymbol{\delta}_r \frac{\partial}{\partial r} + \boldsymbol{\delta}_\theta \frac{1}{r}\frac{\partial}{\partial \theta} + \boldsymbol{\delta}_z \frac{\partial}{\partial z}\right\}\{\boldsymbol{\delta}_r v_r + \boldsymbol{\delta}_\theta v_\theta + \boldsymbol{\delta}_z v_z\}$$
$$= \boldsymbol{\delta}_r\boldsymbol{\delta}_r \frac{\partial v_r}{\partial r} + \boldsymbol{\delta}_r\boldsymbol{\delta}_\theta \frac{\partial v_\theta}{\partial r} + \boldsymbol{\delta}_r\boldsymbol{\delta}_z \frac{\partial v_z}{\partial r} + \boldsymbol{\delta}_\theta\boldsymbol{\delta}_r \frac{1}{r}\frac{\partial v_r}{\partial \theta}$$
$$+ \boldsymbol{\delta}_\theta\boldsymbol{\delta}_\theta \frac{1}{r}\frac{\partial v_\theta}{\partial \theta} + \boldsymbol{\delta}_\theta\boldsymbol{\delta}_z \frac{1}{r}\frac{\partial v_z}{\partial \theta} + \boldsymbol{\delta}_\theta\boldsymbol{\delta}_\theta \frac{v_r}{r} - \boldsymbol{\delta}_\theta\boldsymbol{\delta}_r \frac{v_\theta}{r}$$
$$+ \boldsymbol{\delta}_z\boldsymbol{\delta}_r \frac{\partial v_r}{\partial z} + \boldsymbol{\delta}_z\boldsymbol{\delta}_\theta \frac{\partial v_\theta}{\partial z} + \boldsymbol{\delta}_z\boldsymbol{\delta}_z \frac{\partial v_z}{\partial z}$$
$$= \boldsymbol{\delta}_r\boldsymbol{\delta}_r \frac{\partial v_r}{\partial r} + \boldsymbol{\delta}_r\boldsymbol{\delta}_\theta \frac{\partial v_\theta}{\partial r} + \boldsymbol{\delta}_r\boldsymbol{\delta}_z \frac{\partial v_z}{\partial r} + \boldsymbol{\delta}_\theta\boldsymbol{\delta}_r\left(\frac{1}{r}\frac{\partial v_r}{\partial \theta} - \frac{v_\theta}{r}\right) + \boldsymbol{\delta}_\theta\boldsymbol{\delta}_\theta\left(\frac{1}{r}\frac{\partial v_\theta}{\partial \theta} + \frac{v_r}{r}\right)$$
$$+ \boldsymbol{\delta}_\theta\boldsymbol{\delta}_z \frac{1}{r}\frac{\partial v_z}{\partial \theta} + \boldsymbol{\delta}_z\boldsymbol{\delta}_r \frac{\partial v_r}{\partial z} + \boldsymbol{\delta}_z\boldsymbol{\delta}_\theta \frac{\partial v_\theta}{\partial z} + \boldsymbol{\delta}_z\boldsymbol{\delta}_z \frac{\partial v_z}{\partial z} \quad \text{(A.7–20)}$$

Hence, the rr-component is $\partial v_r/\partial r$, the $r\theta$-component is $\partial v_\theta/\partial r$, etc.

EXAMPLE A.7–2 Differential Operations in Spherical Coordinates

Find the r-component of $[\nabla \cdot \boldsymbol{\tau}]$ in spherical coordinates:

SOLUTION

Using Eq. A. 7–9 we have:

$$[\nabla \cdot \boldsymbol{\tau}]_r = \left[\left\{\boldsymbol{\delta}_r \frac{\partial}{\partial r} + \boldsymbol{\delta}_\theta \frac{1}{r}\frac{\partial}{\partial \theta} + \boldsymbol{\delta}_\phi \frac{1}{r \sin\theta}\frac{\partial}{\partial \phi}\right\} \cdot \{\boldsymbol{\delta}_r\boldsymbol{\delta}_r\tau_{rr} + \boldsymbol{\delta}_r\boldsymbol{\delta}_\theta\tau_{r\theta} + \boldsymbol{\delta}_r\boldsymbol{\delta}_\phi\tau_{r\phi}\right.$$

$$\left. + \boldsymbol{\delta}_\theta\boldsymbol{\delta}_r\tau_{\theta r} + \boldsymbol{\delta}_\theta\boldsymbol{\delta}_\theta\tau_{\theta\theta} + \boldsymbol{\delta}_\theta\boldsymbol{\delta}_\phi\tau_{\theta\phi} + \boldsymbol{\delta}_\phi\boldsymbol{\delta}_r\tau_{\phi r} + \boldsymbol{\delta}_\phi\boldsymbol{\delta}_\theta\tau_{\phi\theta} + \boldsymbol{\delta}_\phi\boldsymbol{\delta}_\phi\tau_{\phi\phi}\}\right]_r \qquad \text{(A.7–21)}$$

We now use Eqs. A. 7–6, 7, 8 and Eq. A. 3–3. Since we want only the r-component, we sort out only those terms which contribute to the coefficient of $\boldsymbol{\delta}_r$:

$$\left[\boldsymbol{\delta}_r \frac{\partial}{\partial r} \cdot \boldsymbol{\delta}_r\boldsymbol{\delta}_r\tau_{rr}\right] = [\boldsymbol{\delta}_r \cdot \boldsymbol{\delta}_r\boldsymbol{\delta}_r]\frac{\partial \tau_{rr}}{\partial r} = \boldsymbol{\delta}_r \frac{\partial \tau_{rr}}{\partial r} \qquad \text{(A.7–22)}$$

$$\left[\boldsymbol{\delta}_\theta \frac{1}{r}\frac{\partial}{\partial \theta} \cdot \boldsymbol{\delta}_\theta\boldsymbol{\delta}_r\tau_{\theta r}\right] = [\boldsymbol{\delta}_\theta \cdot \boldsymbol{\delta}_\theta\boldsymbol{\delta}_r]\frac{1}{r}\frac{\partial}{\partial \theta}\tau_{\theta r} + \text{other term} \qquad \text{(A.7–23)}$$

$$\left[\boldsymbol{\delta}_\phi \frac{1}{r \sin\theta}\frac{\partial}{\partial \phi} \cdot \boldsymbol{\delta}_\phi\boldsymbol{\delta}_r\tau_{\phi r}\right] = [\boldsymbol{\delta}_\phi \cdot \boldsymbol{\delta}_\phi\boldsymbol{\delta}_r]\frac{1}{r \sin\phi}\frac{\partial}{\partial \theta}\tau_{\phi r} + \text{other term} \qquad \text{(A.7–24)}$$

$$\left[\boldsymbol{\delta}_\theta \frac{1}{r}\frac{\partial}{\partial \theta} \cdot \boldsymbol{\delta}_r\boldsymbol{\delta}_r\tau_{rr}\right] = \frac{\tau_{rr}}{r}\left[\boldsymbol{\delta}_\theta \cdot \left\{\frac{\partial}{\partial \theta}\boldsymbol{\delta}_r\right\}\boldsymbol{\delta}_r\right] + \frac{\tau_{rr}}{r}\left[\boldsymbol{\delta}_\theta \cdot \boldsymbol{\delta}_r\left\{\frac{\partial}{\partial \theta}\boldsymbol{\delta}_r\right\}\right]$$

$$= \frac{\tau_{rr}}{r}[\boldsymbol{\delta}_\theta \cdot \boldsymbol{\delta}_\theta\boldsymbol{\delta}_r] = \boldsymbol{\delta}_r \frac{\tau_{rr}}{r} \qquad \text{(A.7–25)}$$

$$\left[\boldsymbol{\delta}_\phi \frac{1}{r \sin\theta}\frac{\partial}{\partial \phi} \cdot \boldsymbol{\delta}_r\boldsymbol{\delta}_r\tau_{rr}\right] = \frac{\tau_{rr}}{r \sin\theta}\left[\boldsymbol{\delta}_\phi \cdot \left\{\frac{\partial}{\partial \phi}\boldsymbol{\delta}_r\right\}\boldsymbol{\delta}_r\right]$$

$$= \frac{\tau_{rr}}{r \sin\theta}[\boldsymbol{\delta}_\phi \cdot \boldsymbol{\delta}_\phi \sin\theta\boldsymbol{\delta}_r]$$

$$= \boldsymbol{\delta}_r \frac{\tau_{rr}}{r} \qquad \text{(A.7–26)}$$

$$\left[\boldsymbol{\delta}_\theta \frac{1}{r}\frac{\partial}{\partial \theta} \cdot \boldsymbol{\delta}_\theta\boldsymbol{\delta}_\theta\tau_{\theta\theta}\right] = \boldsymbol{\delta}_r\left(-\frac{\tau_{\theta\theta}}{r}\right) + \text{other term} \qquad \text{(A.7–27)}$$

$$\left[\boldsymbol{\delta}_\phi \frac{1}{r \sin\theta}\frac{\partial}{\partial \phi} \cdot \boldsymbol{\delta}_\theta\boldsymbol{\delta}_r\tau_{\theta r}\right] = \boldsymbol{\delta}_r \frac{\tau_{\theta r}\cos\theta}{r \sin\theta} \qquad \text{(A.7–28)}$$

$$\left[\boldsymbol{\delta}_\phi \frac{1}{r \sin\theta}\frac{\partial}{\partial \phi} \cdot \boldsymbol{\delta}_\phi\boldsymbol{\delta}_\phi\tau_{\phi\phi}\right] = \boldsymbol{\delta}_r\left(\frac{-\tau_{\phi\phi}}{r}\right) + \text{other terms} \qquad \text{(A.7–29)}$$

Combining the above results we get:

$$[\nabla \cdot \boldsymbol{\tau}]_r = \frac{1}{r^2}\frac{\partial}{\partial r}(r^2\tau_{rr}) + \frac{\tau_{\theta r}}{r}\cot\theta + \frac{1}{r}\frac{\partial}{\partial \theta}\tau_{\theta r}$$

$$+ \frac{1}{r \sin\theta}\frac{\partial \tau_{\phi r}}{\partial \phi} - \frac{\tau_{\theta\theta} + \tau_{\phi\phi}}{r} \qquad \text{(A.7–30)}$$

Note that this expression is correct whether or not $\boldsymbol{\tau}$ is symmetrical.

TABLE A.7–1

Summary of Differential Operations Involving the ∇-Operator in Rectangular Coordinates (x,y,z)

$$(\nabla \cdot \boldsymbol{v}) = \frac{\partial v_x}{\partial x} + \frac{\partial v_y}{\partial y} + \frac{\partial v_z}{\partial z} \quad \text{(A)}$$

$$(\nabla^2 s) = \frac{\partial^2 s}{\partial x^2} + \frac{\partial^2 s}{\partial y^2} + \frac{\partial^2 s}{\partial z^2} \quad \text{(B)}$$

$$\begin{aligned}(\boldsymbol{\tau} : \nabla \boldsymbol{v}) = {} & \tau_{xx}\left(\frac{\partial v_x}{\partial x}\right) + \tau_{xy}\left(\frac{\partial v_x}{\partial y}\right) + \tau_{xz}\left(\frac{\partial v_x}{\partial z}\right) \\ & + \tau_{yx}\left(\frac{\partial v_y}{\partial x}\right) + \tau_{yy}\left(\frac{\partial v_y}{\partial y}\right) + \tau_{yz}\left(\frac{\partial v_y}{\partial z}\right) \\ & + \tau_{zx}\left(\frac{\partial v_z}{\partial x}\right) + \tau_{zy}\left(\frac{\partial v_z}{\partial y}\right) + \tau_{zz}\left(\frac{\partial v_z}{\partial z}\right)\end{aligned} \quad \text{(C)}$$

$$[\nabla s]_x = \frac{\partial s}{\partial x} \quad \text{(D)}$$

$$[\nabla s]_y = \frac{\partial s}{\partial y} \quad \text{(E)}$$

$$[\nabla s]_z = \frac{\partial s}{\partial z} \quad \text{(F)}$$

$$[\nabla \times \boldsymbol{v}]_x = \frac{\partial v_z}{\partial y} - \frac{\partial v_y}{\partial z} \quad \text{(G)}$$

$$[\nabla \times \boldsymbol{v}]_y = \frac{\partial v_x}{\partial z} - \frac{\partial v_z}{\partial x} \quad \text{(H)}$$

$$[\nabla \times \boldsymbol{v}]_z = \frac{\partial v_y}{\partial x} - \frac{\partial v_x}{\partial y} \quad \text{(I)}$$

$$[\nabla \cdot \boldsymbol{\tau}]_x = \frac{\partial \tau_{xx}}{\partial x} + \frac{\partial \tau_{yx}}{\partial y} + \frac{\partial \tau_{zx}}{\partial z} \quad \text{(J)}$$

$$[\nabla \cdot \boldsymbol{\tau}]_y = \frac{\partial \tau_{xy}}{\partial x} + \frac{\partial \tau_{yy}}{\partial y} + \frac{\partial \tau_{zy}}{\partial z} \quad \text{(K)}$$

$$[\nabla \cdot \boldsymbol{\tau}]_z = \frac{\partial \tau_{xz}}{\partial x} + \frac{\partial \tau_{yz}}{\partial y} + \frac{\partial \tau_{zz}}{\partial z} \quad \text{(L)}$$

$$[\nabla^2 \boldsymbol{v}]_x = \frac{\partial^2 v_x}{\partial x^2} + \frac{\partial^2 v_x}{\partial y^2} + \frac{\partial^2 v_x}{\partial z^2} \quad \text{(M)}$$

$$[\nabla^2 \boldsymbol{v}]_y = \frac{\partial^2 v_y}{\partial x^2} + \frac{\partial^2 v_y}{\partial y^2} + \frac{\partial^2 v_y}{\partial z^2} \quad \text{(N)}$$

$$[\nabla^2 \boldsymbol{v}]_z = \frac{\partial^2 v_z}{\partial x^2} + \frac{\partial^2 v_z}{\partial y^2} + \frac{\partial^2 v_z}{\partial z^2} \quad \text{(O)}$$

$$[\boldsymbol{v} \cdot \nabla \boldsymbol{w}]_x = v_x\left(\frac{\partial w_x}{\partial x}\right) + v_y\left(\frac{\partial w_x}{\partial y}\right) + v_z\left(\frac{\partial w_x}{\partial z}\right) \quad \text{(P)}$$

$$[\boldsymbol{v} \cdot \nabla \boldsymbol{w}]_y = v_x\left(\frac{\partial w_y}{\partial x}\right) + v_y\left(\frac{\partial w_y}{\partial y}\right) + v_z\left(\frac{\partial w_y}{\partial z}\right) \quad \text{(Q)}$$

$$[\boldsymbol{v} \cdot \nabla \boldsymbol{w}]_z = v_x\left(\frac{\partial w_z}{\partial x}\right) + v_y\left(\frac{\partial w_z}{\partial y}\right) + v_z\left(\frac{\partial w_z}{\partial z}\right) \quad \text{(R)}$$

TABLE A.7-1 (*Continued*)

$$\{\nabla \boldsymbol{v}\}_{xx} = \frac{\partial v_x}{\partial x} \quad \text{(S)}$$

$$\{\nabla \boldsymbol{v}\}_{xy} = \frac{\partial v_y}{\partial x} \quad \text{(T)}$$

$$\{\nabla \boldsymbol{v}\}_{xz} = \frac{\partial v_z}{\partial x} \quad \text{(U)}$$

$$\{\nabla \boldsymbol{v}\}_{yx} = \frac{\partial v_x}{\partial y} \quad \text{(V)}$$

$$\{\nabla \boldsymbol{v}\}_{yy} = \frac{\partial v_y}{\partial y} \quad \text{(W)}$$

$$\{\nabla \boldsymbol{v}\}_{yz} = \frac{\partial v_z}{\partial y} \quad \text{(X)}$$

$$\{\nabla \boldsymbol{v}\}_{zx} = \frac{\partial v_x}{\partial z} \quad \text{(Y)}$$

$$\{\nabla \boldsymbol{v}\}_{zy} = \frac{\partial v_y}{\partial z} \quad \text{(Z)}$$

$$\{\nabla \boldsymbol{v}\}_{zz} = \frac{\partial v_z}{\partial z} \quad \text{(AA)}$$

$$\{\boldsymbol{v} \cdot \nabla \boldsymbol{\tau}\}_{xx} = (\boldsymbol{v} \cdot \nabla)\tau_{xx} \quad \text{(BB)}$$

$$\{\boldsymbol{v} \cdot \nabla \boldsymbol{\tau}\}_{xy} = (\boldsymbol{v} \cdot \nabla)\tau_{xy} \quad \text{(CC)}$$

$$\{\boldsymbol{v} \cdot \nabla \boldsymbol{\tau}\}_{xz} = (\boldsymbol{v} \cdot \nabla)\tau_{xz} \quad \text{(DD)}$$

$$\{\boldsymbol{v} \cdot \nabla \boldsymbol{\tau}\}_{yx} = (\boldsymbol{v} \cdot \nabla)\tau_{yx} \quad \text{(EE)}$$

$$\{\boldsymbol{v} \cdot \nabla \boldsymbol{\tau}\}_{yy} = (\boldsymbol{v} \cdot \nabla)\tau_{yy} \quad \text{(FF)}$$

$$\{\boldsymbol{v} \cdot \nabla \boldsymbol{\tau}\}_{yz} = (\boldsymbol{v} \cdot \nabla)\tau_{yz} \quad \text{(GG)}$$

$$\{\boldsymbol{v} \cdot \nabla \boldsymbol{\tau}\}_{zx} = (\boldsymbol{v} \cdot \nabla)\tau_{zx} \quad \text{(HH)}$$

$$\{\boldsymbol{v} \cdot \nabla \boldsymbol{\tau}\}_{zy} = (\boldsymbol{v} \cdot \nabla)\tau_{zy} \quad \text{(II)}$$

$$\{\boldsymbol{v} \cdot \nabla \boldsymbol{\tau}\}_{zz} = (\boldsymbol{v} \cdot \nabla)\tau_{zz} \quad \text{(JJ)}$$

where the operator $(\boldsymbol{v} \cdot \nabla) = v_x \frac{\partial}{\partial x} + v_y \frac{\partial}{\partial y} + v_z \frac{\partial}{\partial z}$.

TABLE A.7–2

Summary of Differential Operations Involving the ∇-Operator in Cylindrical Coordinates (r,θ,z)

$$(\nabla \cdot v) = \frac{1}{r}\frac{\partial}{\partial r}(rv_r) + \frac{1}{r}\frac{\partial v_\theta}{\partial \theta} + \frac{\partial v_z}{\partial z} \tag{A}$$

$$(\nabla^2 s) = \frac{1}{r}\frac{\partial}{\partial r}\left(r\frac{\partial s}{\partial r}\right) + \frac{1}{r^2}\frac{\partial^2 s}{\partial \theta^2} + \frac{\partial^2 s}{\partial z^2} \tag{B}$$

$$\begin{aligned}(\tau : \nabla v) = {} & \tau_{rr}\left(\frac{\partial v_r}{\partial r}\right) + \tau_{r\theta}\left(\frac{1}{r}\frac{\partial v_r}{\partial \theta} - \frac{v_\theta}{r}\right) + \tau_{rz}\left(\frac{\partial v_r}{\partial z}\right) \\ & + \tau_{\theta r}\left(\frac{\partial v_\theta}{\partial r}\right) + \tau_{\theta\theta}\left(\frac{1}{r}\frac{\partial v_\theta}{\partial \theta} + \frac{v_r}{r}\right) + \tau_{\theta z}\left(\frac{\partial v_\theta}{\partial z}\right) \\ & + \tau_{zr}\left(\frac{\partial v_z}{\partial r}\right) + \tau_{z\theta}\left(\frac{1}{r}\frac{\partial v_z}{\partial \theta}\right) + \tau_{zz}\left(\frac{\partial v_z}{\partial z}\right)\end{aligned} \tag{C}$$

$$[\nabla s]_r = \frac{\partial s}{\partial r} \tag{D}$$

$$[\nabla s]_\theta = \frac{1}{r}\frac{\partial s}{\partial \theta} \tag{E}$$

$$[\nabla s]_z = \frac{\partial s}{\partial z} \tag{F}$$

$$[\nabla \times v]_r = \frac{1}{r}\frac{\partial v_z}{\partial \theta} - \frac{\partial v_\theta}{\partial z} \tag{G}$$

$$[\nabla \times v]_\theta = \frac{\partial v_r}{\partial z} - \frac{\partial v_z}{\partial r} \tag{H}$$

$$[\nabla \times v]_z = \frac{1}{r}\frac{\partial}{\partial r}(rv_\theta) - \frac{1}{r}\frac{\partial v_r}{\partial \theta} \tag{I}$$

$$[\nabla \cdot \tau]_r = \frac{1}{r}\frac{\partial}{\partial r}(r\tau_{rr}) + \frac{1}{r}\frac{\partial}{\partial \theta}\tau_{\theta r} + \frac{\partial}{\partial z}\tau_{zr} - \frac{\tau_{\theta\theta}}{r} \tag{J}$$

$$[\nabla \cdot \tau]_\theta = \frac{1}{r^2}\frac{\partial}{\partial r}(r^2\tau_{r\theta}) + \frac{1}{r}\frac{\partial}{\partial \theta}\tau_{\theta\theta} + \frac{\partial}{\partial z}\tau_{z\theta} + \frac{\tau_{\theta r} - \tau_{r\theta}}{r} \tag{K}$$

$$[\nabla \cdot \tau]_z = \frac{1}{r}\frac{\partial}{\partial r}(r\tau_{rz}) + \frac{1}{r}\frac{\partial}{\partial \theta}\tau_{\theta z} + \frac{\partial}{\partial z}\tau_{zz} \tag{L}$$

$$[\nabla^2 v]_r = \frac{\partial}{\partial r}\left(\frac{1}{r}\frac{\partial}{\partial r}(rv_r)\right) + \frac{1}{r^2}\frac{\partial^2 v_r}{\partial \theta^2} + \frac{\partial^2 v_r}{\partial z^2} - \frac{2}{r^2}\frac{\partial v_\theta}{\partial \theta} \tag{M}$$

$$[\nabla^2 v]_\theta = \frac{\partial}{\partial r}\left(\frac{1}{r}\frac{\partial}{\partial r}(rv_\theta)\right) + \frac{1}{r^2}\frac{\partial^2 v_\theta}{\partial \theta^2} + \frac{\partial^2 v_\theta}{\partial z^2} + \frac{2}{r^2}\frac{\partial v_r}{\partial \theta} \tag{N}$$

$$[\nabla^2 v]_z = \frac{1}{r}\frac{\partial}{\partial r}\left(r\frac{\partial v_z}{\partial r}\right) + \frac{1}{r^2}\frac{\partial^2 v_z}{\partial \theta^2} + \frac{\partial^2 v_z}{\partial z^2} \tag{O}$$

$$[v \cdot \nabla w]_r = v_r\left(\frac{\partial w_r}{\partial r}\right) + v_\theta\left(\frac{1}{r}\frac{\partial w_r}{\partial \theta} - \frac{w_\theta}{r}\right) + v_z\left(\frac{\partial w_r}{\partial z}\right) \tag{P}$$

$$[v \cdot \nabla w]_\theta = v_r\left(\frac{\partial w_\theta}{\partial r}\right) + v_\theta\left(\frac{1}{r}\frac{\partial w_\theta}{\partial \theta} + \frac{w_r}{r}\right) + v_z\left(\frac{\partial w_\theta}{\partial z}\right) \tag{Q}$$

$$[v \cdot \nabla w]_z = v_r\left(\frac{\partial w_z}{\partial r}\right) + v_\theta\left(\frac{1}{r}\frac{\partial w_z}{\partial \theta}\right) + v_z\left(\frac{\partial w_z}{\partial z}\right) \tag{R}$$

TABLE A.7–2 (*Continued*)

$$\{\nabla \boldsymbol{v}\}_{rr} = \frac{\partial v_r}{\partial r} \tag{S}$$

$$\{\nabla \boldsymbol{v}\}_{r\theta} = \frac{\partial v_\theta}{\partial r} \tag{T}$$

$$\{\nabla \boldsymbol{v}\}_{rz} = \frac{\partial v_z}{\partial r} \tag{U}$$

$$\{\nabla \boldsymbol{v}\}_{\theta r} = \frac{1}{r}\frac{\partial v_r}{\partial \theta} - \frac{v_\theta}{r} \tag{V}$$

$$\{\nabla \boldsymbol{v}\}_{\theta\theta} = \frac{1}{r}\frac{\partial v_\theta}{\partial \theta} + \frac{v_r}{r} \tag{W}$$

$$\{\nabla \boldsymbol{v}\}_{\theta z} = \frac{1}{r}\frac{\partial v_z}{\partial \theta} \tag{X}$$

$$\{\nabla \boldsymbol{v}\}_{zr} = \frac{\partial v_r}{\partial z} \tag{Y}$$

$$\{\nabla \boldsymbol{v}\}_{z\theta} = \frac{\partial v_\theta}{\partial z} \tag{Z}$$

$$\{\nabla \boldsymbol{v}\}_{zz} = \frac{\partial v_z}{\partial z} \tag{AA}$$

$$\{\boldsymbol{v} \cdot \nabla \boldsymbol{\tau}\}_{rr} = (\boldsymbol{v} \cdot \nabla)\tau_{rr} - \frac{v_\theta}{r}(\tau_{r\theta} + \tau_{\theta r}) \tag{BB}$$

$$\{\boldsymbol{v} \cdot \nabla \boldsymbol{\tau}\}_{r\theta} = (\boldsymbol{v} \cdot \nabla)\tau_{r\theta} + \frac{v_\theta}{r}(\tau_{rr} - \tau_{\theta\theta}) \tag{CC}$$

$$\{\boldsymbol{v} \cdot \nabla \boldsymbol{\tau}\}_{rz} = (\boldsymbol{v} \cdot \nabla)\tau_{rz} - \frac{v_\theta}{r}\tau_{\theta z} \tag{DD}$$

$$\{\boldsymbol{v} \cdot \nabla \boldsymbol{\tau}\}_{\theta r} = (\boldsymbol{v} \cdot \nabla)\tau_{\theta r} + \frac{v_\theta}{r}(\tau_{rr} - \tau_{\theta\theta}) \tag{EE}$$

$$\{\boldsymbol{v} \cdot \nabla \boldsymbol{\tau}\}_{\theta\theta} = (\boldsymbol{v} \cdot \nabla)\tau_{\theta\theta} + \frac{v_\theta}{r}(\tau_{r\theta} + \tau_{\theta r}) \tag{FF}$$

$$\{\boldsymbol{v} \cdot \nabla \boldsymbol{\tau}\}_{\theta z} = (\boldsymbol{v} \cdot \nabla)\tau_{\theta z} + \frac{v_\theta}{r}\tau_{rz} \tag{GG}$$

$$\{\boldsymbol{v} \cdot \nabla \boldsymbol{\tau}\}_{zr} = (\boldsymbol{v} \cdot \nabla)\tau_{zr} - \frac{v_\theta}{r}\tau_{z\theta} \tag{HH}$$

$$\{\boldsymbol{v} \cdot \nabla \boldsymbol{\tau}\}_{z\theta} = (\boldsymbol{v} \cdot \nabla)\tau_{z\theta} + \frac{v_\theta}{r}\tau_{zr} \tag{II}$$

$$\{\boldsymbol{v} \cdot \nabla \boldsymbol{\tau}\}_{zz} = (\boldsymbol{v} \cdot \nabla)\tau_{zz} \tag{JJ}$$

where the operator $(\boldsymbol{v} \cdot \nabla) = v_r \frac{\partial}{\partial r} + \frac{v_\theta}{r}\frac{\partial}{\partial \theta} + v_z \frac{\partial}{\partial z}$

TABLE A.7–3

Summary of Differential Operations Involving the ∇-Operator in Spherical Coordinates (r,θ,ϕ)

$$(\nabla \cdot \boldsymbol{v}) = \frac{1}{r^2}\frac{\partial}{\partial r}(r^2 v_r) + \frac{1}{r \sin\theta}\frac{\partial}{\partial \theta}(v_\theta \sin\theta) + \frac{1}{r\sin\theta}\frac{\partial v_\phi}{\partial \phi} \tag{A}$$

$$(\nabla^2 s) = \frac{1}{r^2}\frac{\partial}{\partial r}\left(r^2\frac{\partial s}{\partial r}\right) + \frac{1}{r^2 \sin\theta}\frac{\partial}{\partial \theta}\left(\sin\theta\frac{\partial s}{\partial \theta}\right) + \frac{1}{r^2\sin^2\theta}\frac{\partial^2 s}{\partial \phi^2} \tag{B}$$

$$\begin{aligned}(\boldsymbol{\tau} : \nabla \boldsymbol{v}) = {} & \tau_{rr}\left(\frac{\partial v_r}{\partial r}\right) + \tau_{r\theta}\left(\frac{1}{r}\frac{\partial v_r}{\partial \theta} - \frac{v_\theta}{r}\right) + \tau_{r\phi}\left(\frac{1}{r\sin\theta}\frac{\partial v_r}{\partial \phi} - \frac{v_\phi}{r}\right) \\ & + \tau_{\theta r}\left(\frac{\partial v_\theta}{\partial r}\right) + \tau_{\theta\theta}\left(\frac{1}{r}\frac{\partial v_\theta}{\partial \theta} + \frac{v_r}{r}\right) + \tau_{\theta\phi}\left(\frac{1}{r\sin\theta}\frac{\partial v_\theta}{\partial \phi} - \frac{v_\phi}{r}\cot\theta\right) \\ & + \tau_{\phi r}\left(\frac{\partial v_\phi}{\partial r}\right) + \tau_{\phi\theta}\left(\frac{1}{r}\frac{\partial v_\phi}{\partial \theta}\right) + \tau_{\phi\phi}\left(\frac{1}{r\sin\theta}\frac{\partial v_\phi}{\partial \phi} + \frac{v_r}{r} + \frac{v_\theta}{r}\cot\theta\right)\end{aligned} \tag{C}$$

$$[\nabla s]_r = \frac{\partial s}{\partial r} \tag{D}$$

$$[\nabla s]_\theta = \frac{1}{r}\frac{\partial s}{\partial \theta} \tag{E}$$

$$[\nabla s]_\phi = \frac{1}{r\sin\theta}\frac{\partial s}{\partial \phi} \tag{F}$$

$$[\nabla \times \boldsymbol{v}]_r = \frac{1}{r\sin\theta}\frac{\partial}{\partial \theta}(v_\phi \sin\theta) - \frac{1}{r\sin\theta}\frac{\partial v_\theta}{\partial \phi} \tag{G}$$

$$[\nabla \times \boldsymbol{v}]_\theta = \frac{1}{r\sin\theta}\frac{\partial v_r}{\partial \phi} - \frac{1}{r}\frac{\partial}{\partial r}(r v_\phi) \tag{H}$$

$$[\nabla \times \boldsymbol{v}]_\phi = \frac{1}{r}\frac{\partial}{\partial r}(r v_\theta) - \frac{1}{r}\frac{\partial v_r}{\partial \theta} \tag{I}$$

$$[\nabla \cdot \boldsymbol{\tau}]_r = \frac{1}{r^2}\frac{\partial}{\partial r}(r^2 \tau_{rr}) + \frac{1}{r\sin\theta}\frac{\partial}{\partial \theta}(\tau_{\theta r}\sin\theta) + \frac{1}{r\sin\theta}\frac{\partial}{\partial \phi}\tau_{\phi r} - \frac{\tau_{\theta\theta} + \tau_{\phi\phi}}{r} \tag{J}$$

$$[\nabla \cdot \boldsymbol{\tau}]_\theta = \frac{1}{r^3}\frac{\partial}{\partial r}(r^3 \tau_{r\theta}) + \frac{1}{r\sin\theta}\frac{\partial}{\partial \theta}(\tau_{\theta\theta}\sin\theta) + \frac{1}{r\sin\theta}\frac{\partial}{\partial \phi}\tau_{\phi\theta} + \frac{(\tau_{\theta r} - \tau_{r\theta}) - \tau_{\phi\phi}\cot\phi}{r} \tag{K}$$

$$[\nabla \cdot \boldsymbol{\tau}]_\phi = \frac{1}{r^3}\frac{\partial}{\partial r}(r^3 \tau_{r\phi}) + \frac{1}{r\sin\theta}\frac{\partial}{\partial \theta}(\tau_{\theta\phi}\sin\theta) + \frac{1}{r\sin\theta}\frac{\partial}{\partial \phi}\tau_{\phi\phi} + \frac{(\tau_{\phi r} - \tau_{r\phi}) + \tau_{\phi\theta}\cot\theta}{r} \tag{L}$$

$$\begin{aligned}[\nabla^2 \boldsymbol{v}]_r = {} & \frac{\partial}{\partial r}\left(\frac{1}{r^2}\frac{\partial}{\partial r}(r^2 v_r)\right) + \frac{1}{r^2\sin\theta}\frac{\partial}{\partial \theta}\left(\sin\theta\frac{\partial v_r}{\partial \theta}\right) + \frac{1}{r^2\sin^2\theta}\frac{\partial^2 v_r}{\partial \phi^2} \\ & - \frac{2}{r^2\sin\theta}\frac{\partial}{\partial \theta}(v_\theta\sin\theta) - \frac{2}{r^2\sin\theta}\frac{\partial v_\phi}{\partial \phi}\end{aligned} \tag{M}$$

$$[\nabla^2 \boldsymbol{v}]_\theta = \frac{1}{r^2}\frac{\partial}{\partial r}\left(r^2\frac{\partial v_\theta}{\partial r}\right) + \frac{1}{r^2}\frac{\partial}{\partial \theta}\left(\frac{1}{\sin\theta}\frac{\partial}{\partial \theta}(v_\theta\sin\theta)\right) + \frac{1}{r^2\sin^2\theta}\frac{\partial^2 v_\theta}{\partial \phi^2} + \frac{2}{r^2}\frac{\partial v_r}{\partial \theta} - \frac{2\cot\theta}{r^2\sin\theta}\frac{\partial v_\phi}{\partial \phi} \tag{N}$$

$$[\nabla^2 \boldsymbol{v}]_\phi = \frac{1}{r^2}\frac{\partial}{\partial r}\left(r^2\frac{\partial v_\phi}{\partial r}\right) + \frac{1}{r^2}\frac{\partial}{\partial \theta}\left(\frac{1}{\sin\theta}\frac{\partial}{\partial \theta}(v_\phi\sin\theta)\right) + \frac{1}{r^2\sin^2\theta}\frac{\partial^2 v_\phi}{\partial \phi^2} + \frac{2}{r^2\sin\theta}\frac{\partial v_r}{\partial \phi} + \frac{2\cot\theta}{r^2\sin\theta}\frac{\partial v_\theta}{\partial \phi} \tag{O}$$

$$[\boldsymbol{v} \cdot \nabla \boldsymbol{w}]_r = v_r\left(\frac{\partial w_r}{\partial r}\right) + v_\theta\left(\frac{1}{r}\frac{\partial w_r}{\partial \theta} - \frac{w_\theta}{r}\right) + v_\phi\left(\frac{1}{r\sin\theta}\frac{\partial w_r}{\partial \phi} - \frac{w_\phi}{r}\right) \tag{P}$$

$$[\boldsymbol{v} \cdot \nabla \boldsymbol{w}]_\theta = v_r\left(\frac{\partial w_\theta}{\partial r}\right) + v_\theta\left(\frac{1}{r}\frac{\partial w_\theta}{\partial \theta} + \frac{w_r}{r}\right) + v_\phi\left(\frac{1}{r\sin\theta}\frac{\partial w_\theta}{\partial \phi} - \frac{w_\phi}{r}\cot\theta\right) \tag{Q}$$

$$[\boldsymbol{v} \cdot \nabla \boldsymbol{w}]_\phi = v_r\left(\frac{\partial w_\phi}{\partial r}\right) + v_\theta\left(\frac{1}{r}\frac{\partial w_\phi}{\partial \theta}\right) + v_\phi\left(\frac{1}{r\sin\theta}\frac{\partial w_\phi}{\partial \phi} + \frac{w_r}{r} + \frac{w_\theta}{r}\cot\theta\right) \tag{R}$$

TABLE A.7–3 (*Continued*)

$$\{\nabla \boldsymbol{v}\}_{rr} = \frac{\partial v_r}{\partial r} \tag{S}$$

$$\{\nabla \boldsymbol{v}\}_{r\theta} = \frac{\partial v_\theta}{\partial r} \tag{T}$$

$$\{\nabla \boldsymbol{v}\}_{r\phi} = \frac{\partial v_\phi}{\partial r} \tag{U}$$

$$\{\nabla \boldsymbol{v}\}_{\theta r} = \frac{1}{r}\frac{\partial v_r}{\partial \theta} - \frac{v_\theta}{r} \tag{V}$$

$$\{\nabla \boldsymbol{v}\}_{\theta\theta} = \frac{1}{r}\frac{\partial v_\theta}{\partial \theta} + \frac{v_r}{r} \tag{W}$$

$$\{\nabla \boldsymbol{v}\}_{\theta\phi} = \frac{1}{r}\frac{\partial v_\phi}{\partial \theta} \tag{X}$$

$$\{\nabla \boldsymbol{v}\}_{\phi r} = \frac{1}{r\sin\theta}\frac{\partial v_r}{\partial \phi} - \frac{v_\phi}{r} \tag{Y}$$

$$\{\nabla \boldsymbol{v}\}_{\phi\theta} = \frac{1}{r\sin\theta}\frac{\partial v_\theta}{\partial \phi} - \frac{v_\phi}{r}\cot\theta \tag{Z}$$

$$\{\nabla \boldsymbol{v}\}_{\phi\phi} = \frac{1}{r\sin\theta}\frac{\partial v_\phi}{\partial \phi} + \frac{v_r}{r} + \frac{v_\theta}{r}\cot\theta \tag{AA}$$

$$\{\boldsymbol{v}\cdot\nabla\boldsymbol{\tau}\}_{rr} = (\boldsymbol{v}\cdot\nabla)\tau_{rr} - \left(\frac{v_\theta}{r}\right)(\tau_{r\theta} + \tau_{\theta r}) - \left(\frac{v_\phi}{r}\right)(\tau_{r\phi} + \tau_{\phi r}) \tag{BB}$$

$$\{\boldsymbol{v}\cdot\nabla\boldsymbol{\tau}\}_{r\theta} = (\boldsymbol{v}\cdot\nabla)\tau_{r\theta} + \left(\frac{v_\theta}{r}\right)(\tau_{rr} - \tau_{\theta\theta}) - \left(\frac{v_\phi}{r}\right)(\tau_{\phi\theta} + \tau_{r\phi}\cot\theta) \tag{CC}$$

$$\{\boldsymbol{v}\cdot\nabla\boldsymbol{\tau}\}_{r\phi} = (\boldsymbol{v}\cdot\nabla)\tau_{r\phi} - \left(\frac{v_\theta}{r}\right)\tau_{\theta\phi} + \left(\frac{v_\phi}{r}\right)[(\tau_{rr} - \tau_{\phi\phi}) + \tau_{r\theta}\cot\theta] \tag{DD}$$

$$\{\boldsymbol{v}\cdot\nabla\boldsymbol{\tau}\}_{\theta r} = (\boldsymbol{v}\cdot\nabla)\tau_{\theta r} + \left(\frac{v_\theta}{r}\right)(\tau_{rr} - \tau_{\theta\theta}) - \left(\frac{v_\phi}{r}\right)(\tau_{\theta\phi} + \tau_{\phi r}\cot\theta) \tag{EE}$$

$$\{\boldsymbol{v}\cdot\nabla\boldsymbol{\tau}\}_{\theta\theta} = (\boldsymbol{v}\cdot\nabla)\tau_{\theta\theta} + \left(\frac{v_\theta}{r}\right)(\tau_{r\theta} + \tau_{\theta r}) - \left(\frac{v_\phi}{r}\right)(\tau_{\theta\phi} + \tau_{\phi\theta})\cot\theta \tag{FF}$$

$$\{\boldsymbol{v}\cdot\nabla\boldsymbol{\tau}\}_{\theta\phi} = (\boldsymbol{v}\cdot\nabla)\tau_{\theta\phi} + \left(\frac{v_\theta}{r}\right)\tau_{r\phi} + \left(\frac{v_\phi}{r}\right)[\tau_{\theta r} + (\tau_{\theta\theta} - \tau_{\phi\phi})\cot\theta] \tag{GG}$$

$$\{\boldsymbol{v}\cdot\nabla\boldsymbol{\tau}\}_{\phi r} = (\boldsymbol{v}\cdot\nabla)\tau_{\phi r} - \left(\frac{v_\theta}{r}\right)\tau_{\phi\theta} + \left(\frac{v_\phi}{r}\right)[(\tau_{rr} - \tau_{\phi\phi}) + \tau_{\theta r}\cot\theta] \tag{HH}$$

$$\{\boldsymbol{v}\cdot\nabla\boldsymbol{\tau}\}_{\phi\theta} = (\boldsymbol{v}\cdot\nabla)\tau_{\phi\theta} + \left(\frac{v_\theta}{r}\right)\tau_{\phi r} + \left(\frac{v_\phi}{r}\right)[\tau_{r\theta} + (\tau_{\theta\theta} - \tau_{\phi\phi})\cot\theta] \tag{II}$$

$$\{\boldsymbol{v}\cdot\nabla\boldsymbol{\tau}\}_{\phi\phi} = (\boldsymbol{v}\cdot\nabla)\tau_{\phi\phi} + \left(\frac{v_\phi}{r}\right)(\tau_{r\phi} + \tau_{\phi r}) + \left(\frac{v_\phi}{r}\right)(\tau_{\theta\phi} + \tau_{\phi\theta})\cot\theta \tag{JJ}$$

where the operator $(\boldsymbol{v}\cdot\nabla) = v_r\frac{\partial}{\partial r} + \frac{v_\theta}{r}\frac{\partial}{\partial \theta} + \frac{v_\phi}{r\sin\theta}\frac{\partial}{\partial \phi}$

TABLE A.7–4

Summary of Differential Operations Involving the ∇-Operator in Bipolar Coordinates (ξ,θ,z)

In this table the abbreviation $X = \cosh\xi + \cos\theta$ is used.

$$(\nabla\cdot\boldsymbol{v}) = \frac{X^2}{a}\frac{\partial}{\partial\xi}\left(\frac{v_\xi}{X}\right) + \frac{X^2}{a}\frac{\partial}{\partial\theta}\left(\frac{v_\theta}{X}\right) + \frac{\partial v_z}{\partial z} \tag{A}$$

$$\nabla^2 s = \left(\frac{X}{a}\right)^2\frac{\partial^2 s}{\partial\xi^2} + \left(\frac{X}{a}\right)^2\frac{\partial^2 s}{\partial\theta^2} + \frac{\partial^2 s}{\partial z^2} \tag{B}$$

$$\begin{aligned}(\boldsymbol{\tau}:\nabla\boldsymbol{v}) = {}& \tau_{\xi\xi}\left(\frac{X}{a}\frac{\partial v_\xi}{\partial\xi} + \frac{v_\theta}{a}\sin\theta\right) + \tau_{\xi\theta}\left(\frac{X}{a}\frac{\partial v_\xi}{\partial\theta} + \frac{v_\theta}{a}\sinh\xi\right) + \tau_{\xi z}\left(\frac{\partial v_\xi}{\partial z}\right) + \tau_{\theta\xi}\left(\frac{X}{a}\frac{\partial v_\theta}{\partial\xi} - \frac{v_\xi}{a}\sin\theta\right)\\ &+ \tau_{\theta\theta}\left(\frac{X}{a}\frac{\partial v_\theta}{\partial\theta} - \frac{v_\xi}{a}\sinh\xi\right) + \tau_{\theta z}\left(\frac{\partial v_\theta}{\partial z}\right) + \tau_{z\xi}\left(\frac{X}{a}\frac{\partial v_z}{\partial\xi}\right) + \tau_{z\theta}\left(\frac{X}{a}\frac{\partial v_z}{\partial\theta}\right) + \tau_{zz}\left(\frac{\partial v_z}{\partial z}\right)\end{aligned} \tag{C}$$

$$[\nabla s]_\xi = \frac{X}{a}\frac{\partial s}{\partial\xi} \tag{D}$$

$$[\nabla s]_\theta = \frac{X}{a}\frac{\partial s}{\partial\theta} \tag{E}$$

$$[\nabla s]_z = \frac{\partial s}{\partial z} \tag{F}$$

$$[\nabla\times\boldsymbol{v}]_\xi = \frac{X}{a}\frac{\partial v_z}{\partial\theta} - \frac{\partial v_\theta}{\partial z} \tag{G}$$

$$[\nabla\times\boldsymbol{v}]_\theta = \frac{\partial v_\xi}{\partial z} - \frac{X}{a}\frac{\partial v_z}{\partial\xi} \tag{H}$$

$$[\nabla\times\boldsymbol{v}]_z = \frac{X^2}{a}\frac{\partial}{\partial\xi}\left(\frac{v_\theta}{X}\right) - \frac{X^2}{a}\frac{\partial}{\partial\theta}\left(\frac{v_\xi}{X}\right) \tag{I}$$

$$[\nabla\cdot\boldsymbol{\tau}]_\xi = \frac{X}{a}\frac{\partial\tau_{\xi\xi}}{\partial\xi} + \frac{X}{a}\frac{\partial\tau_{\theta\xi}}{\partial\theta} + \frac{\partial\tau_{z\xi}}{\partial z} + \frac{\tau_{\theta\theta}-\tau_{\xi\xi}}{a}\sinh\xi + \frac{\tau_{\theta\xi}+\tau_{\xi\theta}}{a}\sin\theta \tag{J}$$

$$[\nabla\cdot\boldsymbol{\tau}]_\theta = \frac{X}{a}\frac{\partial\tau_{\xi\theta}}{\partial\xi} + \frac{X}{a}\frac{\partial\tau_{\theta\theta}}{\partial\theta} + \frac{\partial\tau_{z\theta}}{\partial z} + \frac{\tau_{\theta\theta}-\tau_{\xi\xi}}{a}\sin\theta - \frac{\tau_{\theta\xi}+\tau_{\xi\theta}}{a}\sinh\xi \tag{K}$$

$$[\nabla\cdot\boldsymbol{\tau}]_z = \frac{X}{a}\frac{\partial\tau_{\xi z}}{\partial\xi} + \frac{X}{a}\frac{\partial\tau_{\theta z}}{\partial\theta} + \frac{\partial\tau_{zz}}{\partial z} - \frac{\tau_{\xi z}}{a}\sinh\xi + \frac{\tau_{\theta z}}{a}\sin\theta \tag{L}$$

$$[\nabla^2\boldsymbol{v}]_\xi = \left(\frac{X}{a}\right)^2\frac{\partial^2 v_\xi}{\partial\xi^2} + \left(\frac{X}{a}\right)^2\frac{\partial^2 v_\xi}{\partial\theta^2} + \frac{\partial^2 v_\xi}{\partial z^2} - \frac{v_\xi}{a^2}(\sinh^2\xi + \sin^2\theta) + 2\frac{X}{a^2}\left(\frac{\partial v_\theta}{\partial\theta}\sinh\xi + \frac{\partial v_\theta}{\partial\xi}\sin\theta\right) \tag{M}$$

$$[\nabla^2\boldsymbol{v}]_\theta = \left(\frac{X}{a}\right)^2\frac{\partial^2 v_\theta}{\partial\xi^2} + \left(\frac{X}{a}\right)^2\frac{\partial^2 v_\theta}{\partial\theta^2} + \frac{\partial^2 v_\theta}{\partial z^2} - \frac{v_\theta}{a}(\sin^2\theta + \sinh^2\xi) - 2\frac{X}{a^2}\left(\frac{\partial v_\xi}{\partial\xi}\sin\theta + \frac{\partial v_\xi}{\partial\theta}\sinh\xi\right) \tag{N}$$

$$[\nabla^2\boldsymbol{v}]_z = \left(\frac{X}{a}\right)^2\frac{\partial^2 v_z}{\partial\xi^2} + \left(\frac{X}{a}\right)^2\frac{\partial^2 v_z}{\partial\theta^2} + \frac{\partial^2 v_z}{\partial z^2} \tag{O}$$

$$[\boldsymbol{v}\cdot\nabla\boldsymbol{w}]_\xi = v_\xi\left(\frac{X}{a}\frac{\partial w_\xi}{\partial\xi} + \frac{w_\theta}{a}\sin\theta\right) + v_\theta\left(\frac{X}{a}\frac{\partial w_\xi}{\partial\theta} + \frac{w_\theta}{a}\sinh\xi\right) + v_z\frac{\partial w_\xi}{\partial z} \tag{P}$$

$$[\boldsymbol{v}\cdot\nabla\boldsymbol{w}]_\theta = v_\xi\left(\frac{X}{a}\frac{\partial w_\theta}{\partial\xi} - \frac{w_\xi}{a}\sin\theta\right) + v_\theta\left(\frac{X}{a}\frac{\partial w_\theta}{\partial\theta} - \frac{w_\xi}{a}\sinh\xi\right) + v_z\frac{\partial w_\theta}{\partial z} \tag{Q}$$

$$[\boldsymbol{v}\cdot\nabla\boldsymbol{w}]_z = v_\xi\left(\frac{X}{a}\frac{\partial w_z}{\partial\xi}\right) + v_\theta\left(\frac{X}{a}\frac{\partial w_z}{\partial\theta}\right) + v_z\frac{\partial w_z}{\partial z} \tag{R}$$

TABLE A.7–4 (*Continued*)

$$\{\nabla v\}_{\xi\xi} = \frac{X}{a}\frac{\partial v_\xi}{\partial \xi} + \frac{v_\theta}{a}\sin\theta \tag{S}$$

$$\{\nabla v\}_{\xi\theta} = \frac{X}{a}\frac{\partial v_\theta}{\partial \xi} - \frac{v_\xi}{a}\sin\theta \tag{T}$$

$$\{\nabla v\}_{\xi z} = \frac{X}{a}\frac{\partial v_z}{\partial \xi} \tag{U}$$

$$\{\nabla v\}_{\theta\xi} = \frac{X}{a}\frac{\partial v_\xi}{\partial \theta} + \frac{v_\theta}{a}\sinh\xi \tag{V}$$

$$\{\nabla v\}_{\theta\theta} = \frac{X}{a}\frac{\partial v_\theta}{\partial \theta} - \frac{v_\xi}{a}\sinh\xi \tag{W}$$

$$\{\nabla v\}_{\theta z} = \frac{X}{a}\frac{\partial v_z}{\partial \theta} \tag{X}$$

$$\{\nabla v\}_{z\xi} = \frac{\partial v_\xi}{\partial z} \tag{Y}$$

$$\{\nabla v\}_{z\theta} = \frac{\partial v_\theta}{\partial z} \tag{Z}$$

$$\{\nabla v\}_{zz} = \frac{\partial v_z}{\partial z} \tag{AA}$$

$$\{v \cdot \nabla\tau\}_{\xi\xi} = (v \cdot \nabla)\tau_{\xi\xi} + \left(\frac{\tau_{\xi\theta} + \tau_{\theta\xi}}{a}\right)(v_\xi \sin\theta + v_\theta \sinh\xi) \tag{BB}$$

$$\{v \cdot \nabla\tau\}_{\xi\theta} = (v \cdot \nabla)\tau_{\xi\theta} + \left(\frac{\tau_{\theta\theta} - \tau_{\xi\xi}}{a}\right)(v_\xi \sin\theta + v_\theta \sinh\xi) \tag{CC}$$

$$\{v \cdot \nabla\tau\}_{\xi z} = (v \cdot \nabla)\tau_{\xi z} + \frac{\tau_{\theta z}}{a}(v_\xi \sin\theta + v_\theta \sinh\xi) \tag{DD}$$

$$\{v \cdot \nabla\tau\}_{\theta\xi} = (v \cdot \nabla)\tau_{\theta\xi} + \left(\frac{\tau_{\theta\theta} - \tau_{\xi\xi}}{a}\right)(v_\xi \sin\theta + v_\theta \sinh\xi) \tag{EE}$$

$$\{v \cdot \nabla\tau\}_{\theta\theta} = (v \cdot \nabla)\tau_{\theta\theta} - \left(\frac{\tau_{\xi\theta} + \tau_{\theta\xi}}{a}\right)(v_\xi \sin\theta + v_\theta \sinh\xi) \tag{FF}$$

$$\{v \cdot \nabla\tau\}_{\theta z} = (v \cdot \nabla)\tau_{\theta z} - \left(\frac{\tau_{\xi z}}{a}\right)(v_\xi \sin\theta + v_\theta \sinh\xi) \tag{GG}$$

$$\{v \cdot \nabla\tau\}_{z\xi} = (v \cdot \nabla)\tau_{z\xi} + \left(\frac{\tau_{z\theta}}{a}\right)(v_\xi \sin\theta + v_\theta \sinh\xi) \tag{HH}$$

$$\{v \cdot \nabla\tau\}_{z\theta} = (v \cdot \nabla)\tau_{z\theta} - \left(\frac{\tau_{z\xi}}{a}\right)(v_\xi \sin\theta + v_\theta \sinh\xi) \tag{II}$$

$$\{v \cdot \nabla\tau\}_{zz} = (v \cdot \nabla)\tau_{zz} \tag{JJ}$$

where the operator $(v \cdot \nabla) = v_\xi \frac{X}{a}\frac{\partial}{\partial \xi} + v_\theta \frac{X}{a}\frac{\partial}{\partial \theta} + v_z \frac{\partial}{\partial z}$ (KK)

EXERCISES

1. Obtain $(\nabla \cdot \boldsymbol{v})$, $[\nabla \times \boldsymbol{v}]$, and $\nabla \boldsymbol{v}$ in spherical coordinates.
2. Obtain $[\nabla \cdot \boldsymbol{\tau}]$ in cylindrical coordinates.
3. Use Table A.7–2 to write down directly the following quantities in cylindrical coordinates:
   a. $(\nabla \cdot \rho \boldsymbol{v})$ where ρ is a scalar.
   b. $[\nabla \cdot \rho \boldsymbol{v}\boldsymbol{v}]_r$ where ρ is a scalar.
   c. $[\nabla \cdot p\boldsymbol{\delta}]_\theta$ where p is a scalar.
   d. $(\nabla \cdot [\boldsymbol{\tau} \cdot \boldsymbol{v}])$
   e. $[\boldsymbol{v} \cdot \nabla \boldsymbol{v}]_\theta$
   f. $\nabla \boldsymbol{v} + (\nabla \boldsymbol{v})^\dagger$
4. Show that in any orthogonal coordinate system q_1, q_2, q_3:

$$(\nabla \cdot \boldsymbol{v}) = \frac{1}{h_1 h_2 h_3}\left[\frac{\partial}{\partial q_1}(h_2 h_3 v_1) + \frac{\partial}{\partial q_2}(h_3 h_1 v_2) + \frac{\partial}{\partial q_3}(h_1 h_2 v_3)\right]$$

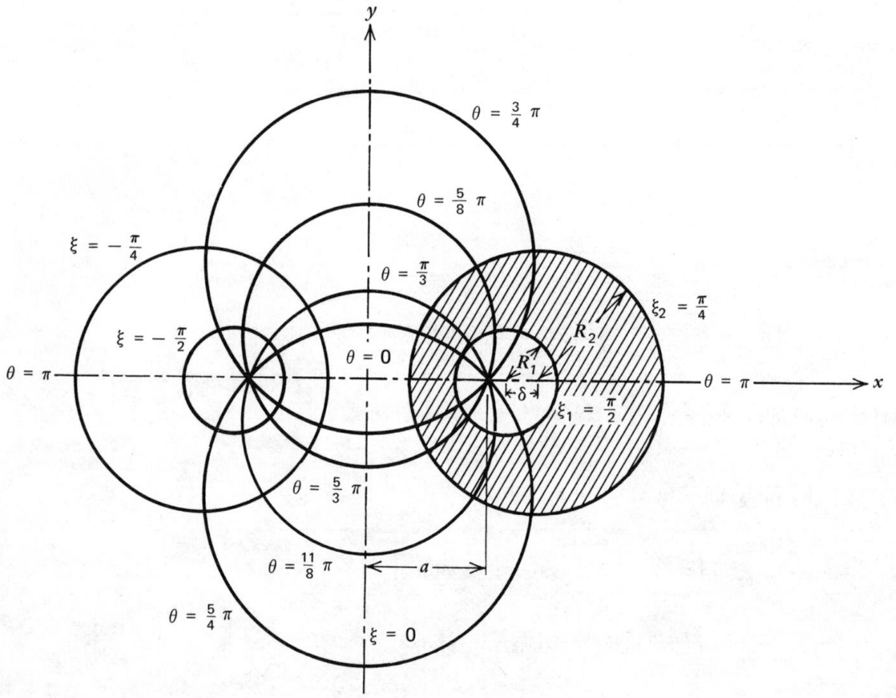

FIGURE A.7–1. Bipolar coordinate system, showing how these coordinates are useful for describing the eccentric annular region between two cylinders of radii R_1 and R_2 with the centers displaced by a distance δ. When R_1, R_2, and δ are given, the parameter a is obtained from

$$\frac{\delta}{R_2} = \sqrt{1 + \left(\frac{a}{R_2}\right)^2} - \sqrt{\left(\frac{R_1}{R_2}\right)^2 + \left(\frac{a}{R_2}\right)^2} \qquad \text{(A.7–31)}$$

and the values ξ_1 and ξ_2 defining the boundaries of the eccentric annular region are given by:

$$\xi_1 = \operatorname{arcsinh} a/R_1; \qquad \xi_2 = \operatorname{arcsinh} a/R_2 \qquad \text{(A.7–32, 33)}$$

A graphical presentation of Eq. A.7–31 is given in Fig. A.7–2.

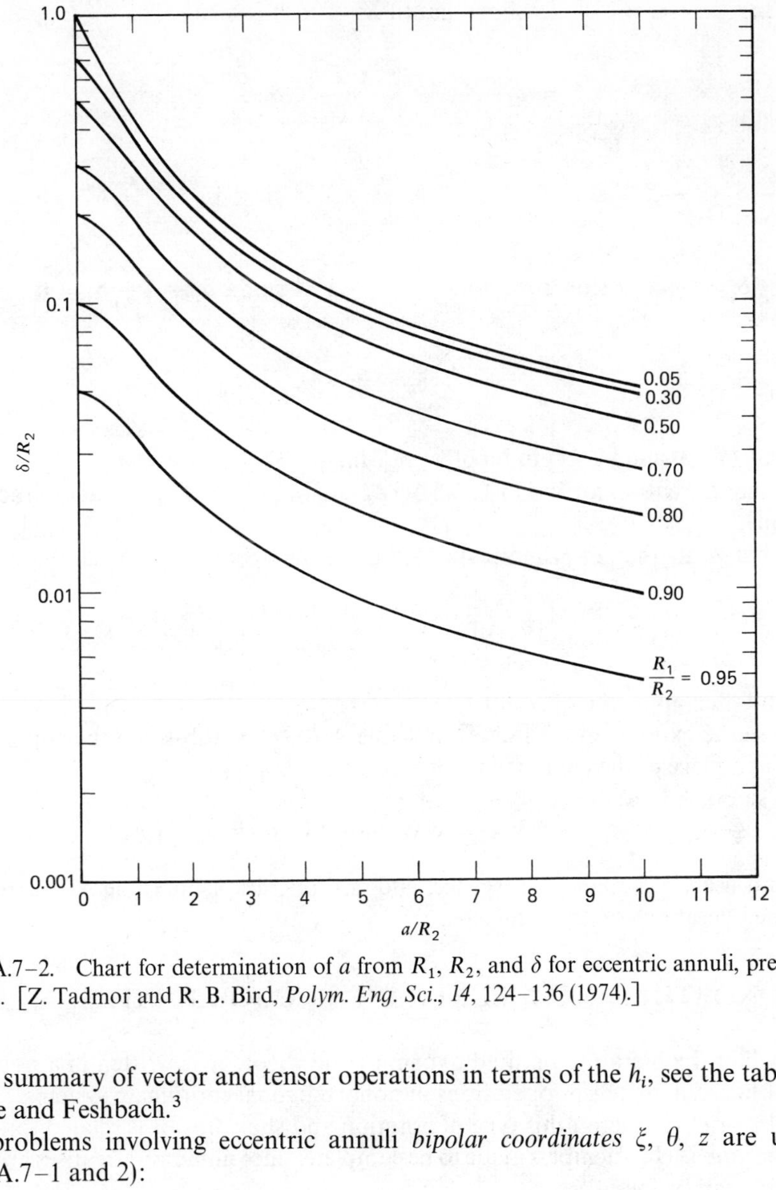

FIGURE A.7–2. Chart for determination of a from R_1, R_2, and δ for eccentric annuli, prepared from Eq. A.7–31. [Z. Tadmor and R. B. Bird, *Polym. Eng. Sci.*, *14*, 124–136 (1974).]

For a summary of vector and tensor operations in terms of the h_i, see the tabulation in Morse and Feshbach.[3]

5. For problems involving eccentric annuli *bipolar coordinates* ξ, θ, z are used[4] (see Figs. A.7–1 and 2):

$$x = \frac{a \sinh \xi}{\cosh \xi + \cos \theta} \qquad \xi = \operatorname{arctanh} \frac{2ax}{a^2 + (x^2 + y^2)}$$

$$y = \frac{a \sin \theta}{\cosh \xi + \cos \theta} \qquad \theta = \arctan \frac{2ay}{a^2 - (x^2 + y^2)}$$

$$z = z \qquad z = z$$

where a is a constant. The coordinates have the following ranges: $-\infty < \xi < +\infty$;

[3] P. M. Morse and H. Feshbach, *op. cit.*, p. 115.
[4] P. M. Morse and H. Feshbach, *op. cit.*, p. 1210.

$0 \leq \theta \leq 2\pi$; $-\infty < z < +\infty$. Show that (if $X = \cosh \xi + \cos \theta$):

$$\nabla = \boldsymbol{\delta}_\xi \frac{X}{a}\frac{\partial}{\partial \xi} + \boldsymbol{\delta}_\theta \frac{X}{a}\frac{\partial}{\partial \theta} + \boldsymbol{\delta}_z \frac{\partial}{\partial z}$$

and

$$\frac{\partial}{\partial \xi}\boldsymbol{\delta}_\xi = -X^{-1}\sin\theta\,\boldsymbol{\delta}_\theta; \qquad \frac{\partial}{\partial \xi}\boldsymbol{\delta}_\theta = +X^{-1}\sin\theta\,\boldsymbol{\delta}_\xi; \qquad \frac{\partial}{\partial \xi}\boldsymbol{\delta}_z = \mathbf{0}$$

$$\frac{\partial}{\partial \theta}\boldsymbol{\delta}_\xi = -X^{-1}\sinh\xi\,\boldsymbol{\delta}_\theta; \qquad \frac{\partial}{\partial \theta}\boldsymbol{\delta}_\theta = +X^{-1}\sinh\xi\,\boldsymbol{\delta}_\xi; \qquad \frac{\partial}{\partial \theta}\boldsymbol{\delta}_z = \mathbf{0}$$

$$\frac{\partial}{\partial z}\boldsymbol{\delta}_\xi = \mathbf{0}; \qquad \frac{\partial}{\partial z}\boldsymbol{\delta}_\theta = \mathbf{0}; \qquad \frac{\partial}{\partial z}\boldsymbol{\delta}_z = \mathbf{0}$$

Evaluate $(\nabla \cdot \boldsymbol{v})$ and $[\nabla \cdot \boldsymbol{\tau}]$ in bipolar coordinates.

6. Verify that the entries for $\nabla^2\boldsymbol{v}$ in Table A.7–2 can be obtained by any one of the following methods:
   (a) First verify that, in cylindrical coordinates the operator $(\nabla \cdot \nabla)$ is

   $$(\nabla \cdot \nabla) = \frac{\partial^2}{\partial r^2} + \frac{1}{r}\frac{\partial}{\partial r} + \frac{1}{r^2}\frac{\partial^2}{\partial \theta^2} + \frac{\partial^2}{\partial z^2}$$

   and then apply the operator to $\boldsymbol{v}$.
   (b) Use the expression for $[\nabla \cdot \boldsymbol{\tau}]$ in Table A.7–2, but substitute the components for $\nabla\boldsymbol{v}$ in place of the components of $\boldsymbol{\tau}$, so as to obtain $[\nabla \cdot \nabla\boldsymbol{v}]$.
   (c) Use Eq. A.4–22

   $$\nabla^2\boldsymbol{v} = \nabla(\nabla \cdot \boldsymbol{v}) - [\nabla \times [\nabla \times \boldsymbol{v}]]$$

   and use the gradient, divergence, and curl operations in Table A.7–2, to evaluate the operations on the right side.

§A.8 NONORTHOGONAL CURVILINEAR COORDINATES

Much of the literature on fluid dynamics and rheology is written in a notation that is capable of describing tensor operations in nonorthogonal coordinate systems. Therefore, we append a brief section on this type of notation and show how it is related to that of the foregoing sections. No attempt is made to be complete, since numerous treatises are available where details can be found.[1]

We consider a curvilinear coordinate system (which may be nonorthogonal) in which the coordinates of a point in space are described by coordinates x^i $(i = 1,2,3)$.[2] The directed

[1] The treatment given here is similar to the development given by A. P. Wills, *Vector Analysis*, Dover Reprint, New York (1958), Chapter 10; see also L. I. Sedov, *Introduction to the Mechanics of a Continuous Medium*, Addison-Wesley, Reading, Massachusetts (1965) and A. J. McConnell, *Applications of Tensor Analysis*, Dover Reprint, New York (1957).

[2] Note that the x^i are nonorthogonal coordinates; they are *not* the same as the x_i (rectangular coordinates) in §§A.1 to A.4. A superscript i is introduced because one customarily adopts the rule that summations will be performed on pairs of repeated indices—one a *super*script and the other a *sub*script. A second rule is that a nonsummed (and hence, nonrepeated) index must appear in every term in the equation in the same position, that is a superscript or as a subscript. In applying these rules we find that a superscript in the denominator is equivalent to a subscript in the numerator—that is, in $\partial s/\partial x^i$ the i is counted as a subscript. One might wonder why it is not usual to write $d\boldsymbol{r} = \sum_i \boldsymbol{g}^i dx_i$ using the reciprocal base vectors defined in Eq. A.8–2. The reason for this is that the dx_i are *not* exact differentials (see Sedov,[1] p. 6).

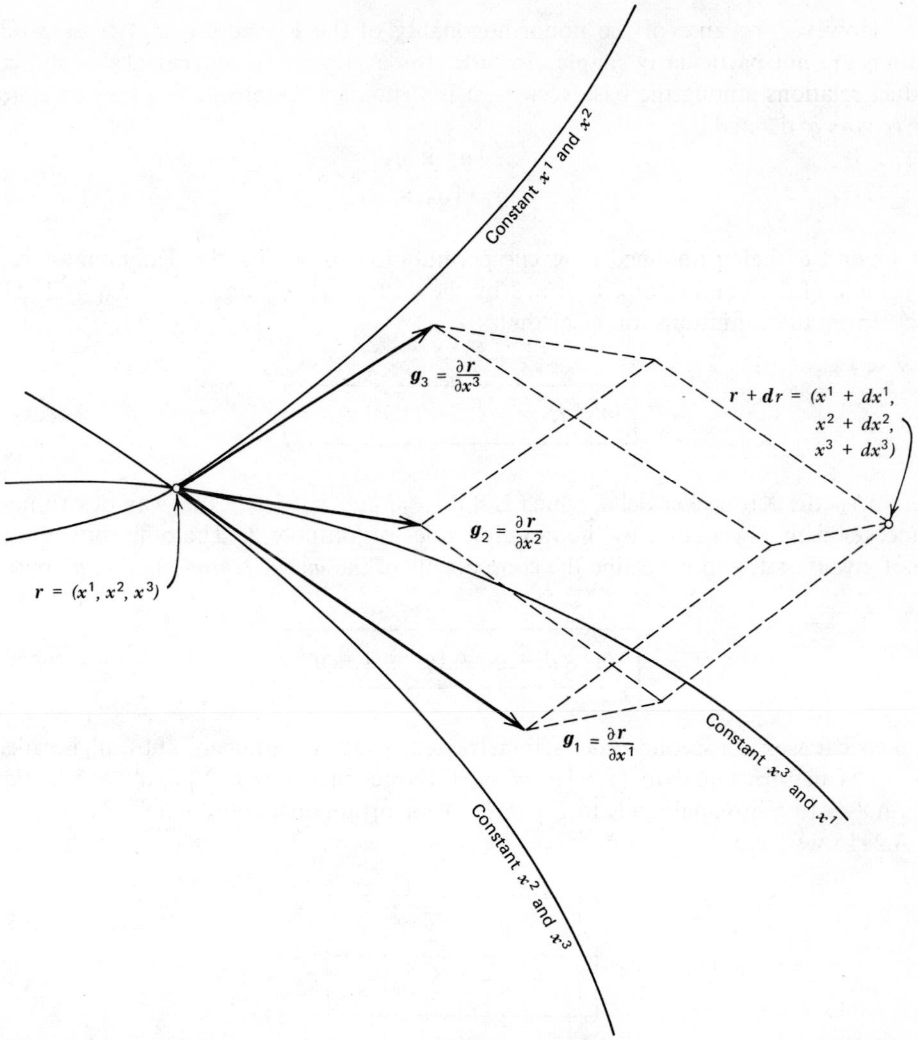

FIGURE A.8–1. The base vectors g_i in a nonorthogonal curvilinear coordinate system. The parallelepiped formed by the g_i has a volume $(g_1 \cdot [g_2 \times g_3]) = \sqrt{g}$.

line segment dr, joining two points a differential distance apart, may be written as[2] (see Fig. A.8–1):

$$dr = \sum_{i=1}^{3} g_i \, dx^i \tag{A.8–1}$$

which serves to define the *base vectors*[3] g_i; that is $g_i = \partial r/\partial x^i$. These base vectors, directed tangentially along the coordinate curves are not necessarily orthogonal nor are they of unit length. However, they play much the same role as the unit vectors δ_i used heretofore, in that all vector and tensor operations can be developed in terms of the dot and cross product operations for the g_i.

[3] The use of a boldface Roman symbol with one subscript as a vector is in violation of the rules given at the beginning of Appendix A. We do this because the closely related metric tensor, defined presently, is always designated in component form by g_{ij}. It would be unwise to tamper with a universally adopted notation such as g_{ij}. Note that for orthogonal coordinates the g_{ii} are just the same as the h_i^2 ("scale factors") in §A.7, and $g_{ij} = 0$ for $i \neq j$.

However, because of the nonorthogonality of the $\boldsymbol{g}_i$, the dot and cross product relations are not particularly simple. In order to develop more convenient dot and cross product relations among the base vectors, it is customary to introduce a set of *reciprocal base vectors* $\boldsymbol{g}^i$ defined by:

$$\boldsymbol{g}^1 = \frac{[\boldsymbol{g}_2 \times \boldsymbol{g}_3]}{(\boldsymbol{g}_1 \cdot [\boldsymbol{g}_2 \times \boldsymbol{g}_3])} \tag{A.8–2}$$

with $\boldsymbol{g}^2$ and $\boldsymbol{g}^3$ being obtained by cyclic permutation; note that the denominator is just the volume of the parallelepiped formed by the vectors $\boldsymbol{g}_1$, $\boldsymbol{g}_2$, and $\boldsymbol{g}_3$ (cf. Eq. A.2–23).

From the definitions it is clear that

$$\boxed{(\boldsymbol{g}^i \cdot \boldsymbol{g}_j) = \delta_j^{\ i}; \quad (\boldsymbol{g}_i \cdot \boldsymbol{g}^j) = \delta_i^{\ j}} \tag{A.8–3,4}$$

where $\delta_j^{\ i}$ is the Kronecker delta, which is 0 if $i \neq j$ and 1 if $i = j$; this way of writing the Kronecker delta is required by the notation rules of footnote 2. The other dot products do not give 0 or 1, and we define the components of the *metric tensor*[4] g_{ij} and g^{ij} by:

$$\boxed{(\boldsymbol{g}_i \cdot \boldsymbol{g}_j) = g_{ij}; \quad (\boldsymbol{g}^i \cdot \boldsymbol{g}^j) = g^{ij}} \tag{A.8–5, 6}$$

The metric tensor is a second-order symmetric tensor; the determinant of the g_{ij} is called g, and $\sqrt{g}$ is in turn equal to $(\boldsymbol{g}_1 \cdot [\boldsymbol{g}_2 \times \boldsymbol{g}_3])$. Note further that $\sum_l g_{kl} g^{lj} = \delta_k^{\ j}$. Equations A.8–3 to 6 are analogous to Eq. A.2–14 for orthogonal coordinates. Analogous to Eq. A.2–15 we have

$$\boxed{[\boldsymbol{g}_i \times \boldsymbol{g}_j] = \textstyle\sum_k \mathscr{E}_{ijk} \boldsymbol{g}^k} \tag{A.8–7}$$

$$\boxed{[\boldsymbol{g}^i \times \boldsymbol{g}^j] = \textstyle\sum_k \mathscr{E}^{ijk} \boldsymbol{g}_k} \tag{A.8–8}$$

Here $\mathscr{E}_{ijk} = \epsilon_{ijk}\sqrt{g}$ and $\mathscr{E}^{ijk} = \epsilon^{ijk}/\sqrt{g}$, where $\epsilon_{ijk} = \epsilon^{ijk}$ are defined in Eqs. A.2–3, 4, and 5. In addition to the operations in Eqs. A.8–3 to 8 there are dyadic operations corresponding to those in Eqs. A.3–1 to 6; for example, $[\boldsymbol{g}_i \cdot \boldsymbol{g}_j \boldsymbol{g}_k] = (\boldsymbol{g}_i \cdot \boldsymbol{g}_j)\boldsymbol{g}_k = g_{ij}\boldsymbol{g}_k$, $[\boldsymbol{g}^i \cdot \boldsymbol{g}_j \boldsymbol{g}_k] = (\boldsymbol{g}^i \cdot \boldsymbol{g}_j)\boldsymbol{g}_k = \delta_j^{\ i}\boldsymbol{g}_k$, etc.

EXAMPLE

In spherical coordinates $x^1 = r$, $x^2 = \theta$, $x^3 = \phi$; the base vectors are $\boldsymbol{g}_1 = \boldsymbol{\delta}_r$, $\boldsymbol{g}_2 = r\boldsymbol{\delta}_\theta$, $\boldsymbol{g}_3 = r \sin\theta\, \boldsymbol{\delta}_\phi$ and the reciprocal base vectors are $\boldsymbol{g}^1 = \boldsymbol{\delta}_r$, $\boldsymbol{g}^2 = r^{-1}\boldsymbol{\delta}_\theta$, $\boldsymbol{g}^3 = (r \sin\theta)^{-1}\boldsymbol{\delta}_\phi$. Since these coordinates are orthogonal the g_{ij} and g^{ij} are zero for $i \neq j$. Furthermore, $g_{rr} = g^{rr} = 1$, $g_{\theta\theta} = r^2$, $g^{\theta\theta} = r^{-2}$, etc.

It is now possible to express vectors and tensors in terms of the base vectors, analogously to Eqs. A.2–16 and A.3–8. However, the fact that we now have two kinds of

[4] Note that from Eqs. A.8–1 and 5 the square of the separation of two adjacent points is:

$$(d\boldsymbol{r} \cdot d\boldsymbol{r}) = \textstyle\sum_i \sum_j g_{ij}\, dx^i\, dx^j \tag{A.8–4a}$$

base vectors ($\boldsymbol{g}_i$ and $\boldsymbol{g}^i$), whereas we previously had only one kind of unit vector, $\boldsymbol{\delta}_i$, means that expansion into components can be made in several ways:

$$\boldsymbol{v} = \sum_i \boldsymbol{g}_i v^i = \sum_i \boldsymbol{g}^i v_i \tag{A.8–9}$$

$$\begin{aligned}\boldsymbol{\tau} &= \sum_i \sum_j \boldsymbol{g}_i \boldsymbol{g}_j \tau^{ij} = \sum_i \sum_j \boldsymbol{g}^i \boldsymbol{g}^j \tau_{ij} \\ &= \sum_i \sum_j \boldsymbol{g}_i \boldsymbol{g}^j \tau^i_{\cdot j} = \sum_i \sum_j \boldsymbol{g}^i \boldsymbol{g}_j \tau_i^{\cdot j}\end{aligned} \tag{A.8–10}$$

[A dot is placed over (under) an index that has been lowered (raised).] The components v_i and v^i are called *covariant* and *contravariant* components of the vector $\boldsymbol{v}$. Similarly τ_{ij}, τ^{ij}, and $\tau_i^{\cdot j}$ (or $\tau^i_{\cdot j}$) are called the *covariant*, *contravariant*, and *mixed* components of the tensor $\boldsymbol{\tau}$. It must be emphasized that these components need not have the same dimensions as the vector ($\boldsymbol{v}$) or tensor ($\boldsymbol{\tau}$) from which they came!

EXAMPLE

Once again we illustrate with spherical coordinates: $v_1 = v_r$, $v_2 = rv_\theta$, $v_3 = r \sin\theta\, v_\phi$ and $v^1 = v_r$, $v^2 = r^{-1}v_\theta$, $v^3 = (r \sin\theta)^{-1}v_\phi$. Note that $\boldsymbol{v} = \boldsymbol{\delta}_r v_r + \boldsymbol{\delta}_\theta v_\theta + \boldsymbol{\delta}_\phi v_\phi$ differs from Eq. A.8–9 only in that the "scale factors" (cf. Eq. A.7–15) are "tucked into" the base vectors and components in such a way that they cancel out properly.[5] The notational advantage in this will appear presently.

It should be added that the covariant and contravariant components are simply related:

$$v_j = \sum_i g_{ij} v^i \tag{A.8–11}$$

as may easily be shown by taking the dot product of $\boldsymbol{g}_j$ with Eq. A.8–9. Similarly

$$\tau^{ij} = \sum_k \sum_l g^{ik} g^{jl} \tau_{kl} \tag{A.8–12}$$

Equations A.8–11 and 12 illustrate the use of the g_{ij} and g^{ij} as "raising and lowering operators."

The various kinds of vector and tensor multiplications can be performed as before; however, because of the several ways of expanding vectors and tensors in their components, these operations may be written in a number of ways. For example, consider the scalar

[5] We note here that whereas v_1, v_2, v_3 are covariant components, and v^1, v^2, v^3 are contravariant components, v_r, v_θ, v_ϕ are *physical components*. Sometimes the notation $v_{\langle 1 \rangle}, v_{\langle 2 \rangle}, v_{\langle 3 \rangle}$ is used for designating physical components. In orthogonal coordinate systems:

$$v_{\langle i \rangle} = \frac{v_i}{h_i} = h_i v^i \qquad \text{(no sum on } i\text{)} \tag{A.8–10a}$$

$$\tau_{\langle ij \rangle} = \frac{\tau_{ij}}{h_i h_j} = h_i h_j \tau^{ij} \qquad \text{(no sums on } i, j\text{)} \tag{A.8–10b}$$

where $h_i = \sqrt{g_{ii}}$ (no sum on i). If $\boldsymbol{v}$ and $\boldsymbol{\tau}$ are the velocity vector and stress tensor, respectively, then the physical components $v_{\langle i \rangle}$ and $\tau_{\langle ij \rangle}$ have dimensions of velocity and stress; the covariant and contravariant components of $\boldsymbol{v}$ and $\boldsymbol{\tau}$ do not necessarily have these dimensions.

product of two vectors:

$$\begin{aligned}(\boldsymbol{v} \cdot \boldsymbol{w}) &= (\{\textstyle\sum_i \boldsymbol{g}_i v^i\} \cdot \{\textstyle\sum_j \boldsymbol{g}^j w_j\}) \\ &= \textstyle\sum_i \sum_j (\boldsymbol{g}_i \cdot \boldsymbol{g}^j) v^i w_j \\ &= \textstyle\sum_i \sum_i \delta_i^{\ j} v^i w_j \\ &= \textstyle\sum_i v^i w_i \end{aligned} \tag{A.8–13}$$

or

$$\begin{aligned}(\boldsymbol{v} \cdot \boldsymbol{w}) &= (\{\textstyle\sum_i \boldsymbol{g}_i v^i\} \cdot \{\textstyle\sum_j \boldsymbol{g}_j w^j\}) \\ &= \textstyle\sum_i \sum_j (\boldsymbol{g}_i \cdot \boldsymbol{g}_j) v^i w^j \\ &= \textstyle\sum_i \sum_j g_{ij} v^i w^j \end{aligned} \tag{A.8–14}$$

Clearly two other combinations may be used. Note that the results are all really the same. The result in Eq. A.8–14 may be written $\sum_i v^i\{\sum_j g_{ij} w^j\} = \sum_i v^i w_i$ using Eq. A.8–11, thereby obtaining the result in Eq. A.8–13.

Other operations proceed similarly:

$$\begin{aligned}[\boldsymbol{\tau} \cdot \boldsymbol{v}] &= [\{\textstyle\sum_i \sum_j \boldsymbol{g}_i \boldsymbol{g}_j \tau^{ij}\} \cdot \{\textstyle\sum_k \boldsymbol{g}^k v_k\}] \\ &= \textstyle\sum_i \sum_j \sum_k [\boldsymbol{g}_i \boldsymbol{g}_j \cdot \boldsymbol{g}^k] \tau^{ij} v_k \\ &= \textstyle\sum_i \sum_j \sum_k \boldsymbol{g}_i \delta_j^{\ k} \tau^{ij} v_k \\ &= \textstyle\sum_i \sum_j \boldsymbol{g}_i \tau^{ij} v_j \\ &= \textstyle\sum_i \boldsymbol{g}_i (\sum_j \tau^{ij} v_j) \end{aligned} \tag{A.8–15}$$

Hence, ith contravariant component of $[\boldsymbol{\tau} \cdot \boldsymbol{v}]$ is $\sum_j \tau^{ij} v_j$.

Cross products may also be performed using the rules in Eqs. A.8–7 and 8:

$$\begin{aligned}[\boldsymbol{v} \times \boldsymbol{w}] &= [\{\textstyle\sum_i \boldsymbol{g}_i v^i\} \times \{\textstyle\sum_j \boldsymbol{g}_j w^j\}] \\ &= \textstyle\sum_i \sum_j [\boldsymbol{g}_i \times \boldsymbol{g}_j] v^i w^j \\ &= \textstyle\sum_i \sum_j \sum_k \mathscr{E}_{ijk} \boldsymbol{g}^k v^i w^j \end{aligned} \tag{A.8–16}$$

so that the kth covariant component of $[\boldsymbol{v} \times \boldsymbol{w}]$ is $\sum_i \sum_j \mathscr{E}_{ijk} v^i w^j$.

Differentiation operations are performed similarly to those described in §A.7. Recall that, in differentiating the unit vectors in curvilinear coordinates, one obtains results of the type shown in Eqs. A.7–1 to 3, 6 to 8, and 13. Note that the spatial derivative of one unit vector gives in general a linear combination of all the unit vectors, the coefficients of which are functions of position. Hence the derivatives of the $\boldsymbol{g}_i$ and $\boldsymbol{g}^i$ will be expected to have the form:

$$\boxed{\frac{\partial}{\partial x^j} \boldsymbol{g}_i = \textstyle\sum_k \left\{ {k \atop ij} \right\} \boldsymbol{g}_k; \quad \frac{\partial}{\partial x^j} \boldsymbol{g}^i = -\textstyle\sum_k \left\{ {i \atop kj} \right\} \boldsymbol{g}^k} \tag{A.8–17, 18}$$

where the coefficients $\left\{ {k \atop ij} \right\}$ are called *Christoffel symbols* (some authors use the notation $\Gamma_{ij}^{\ \ k}$); they are *not* the components of a third-order tensor. By forming appropriate combinations of dot products of the base vectors with Eqs. A.8–17 and 18 it is possible[1] to get explicit expressions for the $\left\{ {k \atop ij} \right\}$ in terms of the g_{ij}:

$$\left\{ {k \atop ij} \right\} = \tfrac{1}{2} \textstyle\sum_l g^{kl} \left(\frac{\partial g_{il}}{\partial x^j} + \frac{\partial g_{jl}}{\partial x^i} - \frac{\partial g_{ij}}{\partial x^l} \right) \tag{A.8–19}$$

Christoffel symbols for three orthogonal coordinate systems are given in Table A.8–1. The $\boldsymbol{\nabla}$ operator is:

$$\boxed{\boldsymbol{\nabla} = \textstyle\sum_i \boldsymbol{g}^i \frac{\partial}{\partial x^i}} \tag{A.8–20}$$

This is analogous to Eq. A.7–14 (note once again that the "scale factors" have been absorbed into the $\boldsymbol{g}^i$). As in §A.7, once we know the expressions for $\boldsymbol{\nabla}$ and for the differentiation of

TABLE A.8–1

Christoffel Symbols and Metric Tensor Components for Several Orthogonal Coordinate Systems

	Christoffel Symbols	Metric Tensor Components[a]
Rectangular	All $\{\ \} = 0$	$g_{xx} = g_{yy} = g_{zz} = 1$ All other g_{ij} are zero.
Cylindrical	$\left\{ {r \atop \theta\theta} \right\} = -r$ $\left\{ {\theta \atop \theta r} \right\} = \left\{ {\theta \atop r\theta} \right\} = \frac{1}{r}$ All other $\{\ \}$ are zero.	$g_{rr} = 1$ $g_{\theta\theta} = r^2$ $g_{zz} = 1$ All other g_{ij} are zero.
Spherical	$\left\{ {r \atop \theta\theta} \right\} = -r$ $\left\{ {r \atop \phi\phi} \right\} = -r \sin^2 \theta$ $\left\{ {\theta \atop \phi\phi} \right\} = -\sin\theta \cos\theta$ $\left\{ {\phi \atop r\phi} \right\} = \left\{ {\phi \atop \phi r} \right\} = \frac{1}{r}$ $\left\{ {\theta \atop r\theta} \right\} = \left\{ {\theta \atop \theta r} \right\} = \frac{1}{r}$ $\left\{ {\phi \atop \theta\phi} \right\} = \left\{ {\phi \atop \phi\theta} \right\} = \cot\theta$ All other $\{\ \}$ are zero.	$g_{rr} = 1$ $g_{\theta\theta} = r^2$ $g_{\phi\phi} = r^2 \sin^2 \theta$ All other g_{ij} are zero.

[a] In orthogonal coordinates g_{ij} and g^{ij} are zero if $i \neq j$; furthermore, when $i = j$, $g^{ii} = (g_{ii})^{-1}$. The g_{ii} (no sum on i) are the same as h_i^2 in §A.7.

the base vectors, any $\boldsymbol{\nabla}$-operations can be worked out. We begin by considering the gradient operations $\boldsymbol{\nabla}s$, $\boldsymbol{\nabla v}$, and $\boldsymbol{\nabla\tau}$. For a *scalar*:

$$\boldsymbol{\nabla}s = \sum_i \boldsymbol{g}^i \frac{\partial s}{\partial x^i} \equiv \sum_i \boldsymbol{g}^i s_{,i} \tag{A.8–21}$$

For a *vector*:

$$\begin{aligned}\boldsymbol{\nabla v} &= \left\{\sum_i \boldsymbol{g}^i \frac{\partial}{\partial x^i}\right\}\left\{\sum_j \boldsymbol{g}^j v_j\right\} \\ &= \sum_i \sum_j \left(\boldsymbol{g}^i\boldsymbol{g}^j \frac{\partial}{\partial x^i} v_j + \boldsymbol{g}^i v_j \frac{\partial}{\partial x^i} \boldsymbol{g}^j\right) \\ &= \sum_i \sum_j \left(\boldsymbol{g}^i\boldsymbol{g}^j \frac{\partial}{\partial x^i} v_j + \boldsymbol{g}^i v_j \left(-\sum_k \begin{Bmatrix} j \\ ki \end{Bmatrix} \boldsymbol{g}^k\right)\right) \\ &= \sum_i \sum_j \boldsymbol{g}^i\boldsymbol{g}^j \left(\frac{\partial}{\partial x^i} v_j - \sum_k \begin{Bmatrix} k \\ ji \end{Bmatrix} v_k\right) \\ &\equiv \sum_i \sum_j \boldsymbol{g}^i\boldsymbol{g}^j v_{j,i}\end{aligned} \tag{A.8–22}$$

or alternatively:

$$\begin{aligned}\boldsymbol{\nabla v} &= \left\{\sum_i \boldsymbol{g}^i \frac{\partial}{\partial x^i}\right\}\left\{\sum_j \boldsymbol{g}_j v^j\right\} \\ &= \sum_i \sum_j \boldsymbol{g}^i\boldsymbol{g}_j \left(\frac{\partial}{\partial x^i} v^j + \sum_k \begin{Bmatrix} j \\ ki \end{Bmatrix} v^k\right) \\ &\equiv \sum_i \sum_j \boldsymbol{g}^i\boldsymbol{g}_j v^j{}_{,i}\end{aligned} \tag{A.8–23}$$

For a second-order *tensor* we have several choices

$$\begin{aligned}\boldsymbol{\nabla\tau} &= \sum_i \sum_j \sum_k \boldsymbol{g}^i\boldsymbol{g}^j\boldsymbol{g}^k \left(\frac{\partial}{\partial x^i} \tau_{jk} - \sum_l \begin{Bmatrix} l \\ ji \end{Bmatrix} \tau_{lk} - \sum_l \begin{Bmatrix} l \\ ki \end{Bmatrix} \tau_{jl}\right) \\ &\equiv \sum_i \sum_j \sum_k \boldsymbol{g}^i\boldsymbol{g}^j\boldsymbol{g}^k \tau_{jk,i}\end{aligned} \tag{A.8–24}$$

$$\begin{aligned}\boldsymbol{\nabla\tau} &= \sum_i \sum_j \sum_k \boldsymbol{g}^i\boldsymbol{g}_j\boldsymbol{g}_k \left(\frac{\partial}{\partial x^i} \tau^{jk} + \sum_l \begin{Bmatrix} j \\ li \end{Bmatrix} \tau^{lk} + \sum_l \begin{Bmatrix} k \\ li \end{Bmatrix} \tau^{jl}\right) \\ &\equiv \sum_i \sum_j \sum_k \boldsymbol{g}^i\boldsymbol{g}_j\boldsymbol{g}_k \tau^{jk}{}_{,i}\end{aligned} \tag{A.8–25}$$

There are two additional choices involving the mixed components.

In the above expressions we have introduced the comma notation for *covariant differentiation*;[6] it is simply a convenient shorthand notation for the components of the gradient operations. The terms in the covariant derivatives containing the Christoffel symbols arise from the differentiation of the unit vectors. [Note that these correspond to the "extra terms" in $\boldsymbol{\nabla v}$ obtained in Eq. A.7–20, namely, $\boldsymbol{\delta}_\theta\boldsymbol{\delta}_\theta(v_r/r) - \boldsymbol{\delta}_\theta\boldsymbol{\delta}_r(v_\theta/r)$].

There are several important properties of covariant differentiation that deserve mention. One, known as *Ricci's lemma*, is that $g_{ij,k} = 0$ and $g^{ij}{}_{,k} = 0$. A second is that the usual differentiation rules of calculus apply: $(v_i w_j)_{,k} = v_i w_{j,k} + v_{i,k} w_j$.

[6] This operation is called covariant differentiation because a subscript (i.e., covariant) index is used. It is possible to define another operation called "contravariant differentiation" that may be obtained by analogous procedures using $\boldsymbol{\nabla} = \sum_i \boldsymbol{g}_i(\partial/\partial x_i)$. This operation, however is seldom used.

Various other ∇ operations can be worked out by the foregoing procedures; for example:[7]

$$(\nabla \cdot \boldsymbol{v}) = \sum_i v^i{}_{,i} \qquad \text{(A.8–26)}$$

$$[\nabla \times \boldsymbol{v}] = \sum_i \sum_j \sum_k \mathscr{E}^{ijk} \boldsymbol{g}_i v_{k,j} \qquad \text{(A.8–27)}$$

$$\nabla^2 s = \sum_i \sum_j g^{ij} s_{,ij} \qquad \text{(A.8–28)}$$

$$[\nabla \cdot \boldsymbol{\tau}] = \sum_i \sum_j \boldsymbol{g}_i \tau^{ji}{}_{,j} \qquad \text{(A.8–29)}$$

Many alternate forms are possible by "juggling" the indices using the raising and lowering operations with the g_{ij} and g^{ij}. For example, $[\nabla \cdot \boldsymbol{\tau}]$ can be written $\sum_i \sum_j \boldsymbol{g}^i \tau^j{}_{.i,j}$ or $\sum_i \sum_j \sum_k \boldsymbol{g}^i g^{jk} \tau_{ji,k}$.

Thus far we have restricted ourselves to a single set of coordinates x^i. Let us now consider a second set of coordinates $\bar{x}^i$. These two sets of coordinates are related to one another by the coordinate transformations

$$\bar{x}^i = \bar{x}^i(x^1, x^2, x^3) \qquad x^i = x^i(\bar{x}^1, \bar{x}^2, \bar{x}^3) \qquad \text{(A.8–30, 31)}$$

which are assumed to be single valued and differentiable. We may then write:

$$d\bar{x}^i = \sum_j \left(\frac{\partial \bar{x}^i}{\partial x^j}\right) dx^j \qquad dx^i = \sum_j \left(\frac{\partial x^i}{\partial \bar{x}^j}\right) d\bar{x}^j \qquad \text{(A.8–32, 33)}$$

Now, if we call $\bar{\boldsymbol{g}}_i$ the base vectors in the barred coordinate system, we may write Eq. A.8–1 in terms of both sets of coordinates:

$$d\boldsymbol{r} = \sum_i \boldsymbol{g}_i \, dx^i \qquad d\boldsymbol{r} = \sum_i \bar{\boldsymbol{g}}_i \, d\bar{x}^i \qquad \text{(A.8–34, 35)}$$

Combining the last two sets of equations we get the transformation rules for the base vectors:

$$\boxed{\boldsymbol{g}_i = \sum_j \frac{\partial \bar{x}^j}{\partial x^i} \bar{\boldsymbol{g}}_j \quad \bar{\boldsymbol{g}}_i = \sum_j \frac{\partial x^j}{\partial \bar{x}^i} \boldsymbol{g}_j} \qquad \text{(A.8–36, 37)}$$

It may further be shown that there are corresponding transformation rules for the reciprocal base vectors:

$$\boxed{\boldsymbol{g}^i = \sum_j \frac{\partial x^i}{\partial \bar{x}^j} \bar{\boldsymbol{g}}^j \quad \bar{\boldsymbol{g}}^i = \sum_j \frac{\partial \bar{x}^i}{\partial x^j} \boldsymbol{g}^j} \qquad \text{(A.8–38, 39)}$$

Once the transformation rules for the base vectors are known the transformation rules for the components of any vector or tensor can easily be written down. Consider a vector $\boldsymbol{v}$, which may be written in either of two forms in both coordinate systems:

$$\boldsymbol{v} = \sum_i \boldsymbol{g}_i v^i = \sum_i \bar{\boldsymbol{g}}_i \bar{v}^i \qquad \boldsymbol{v} = \sum_i \boldsymbol{g}^i v_i = \sum_i \bar{\boldsymbol{g}}^i \bar{v}_i \qquad \text{(A.8–40, 41)}$$

[7] Simpler formulas for orthogonal coordinates can be found in Wills, *op. cit.*, pp. 210–211.

By combining Eq. A.8–40 with Eqs. A.8–36, 37, and by combining Eq. A.8–41 with Eqs. A.8–38, 39, we get:

$$v^i = \sum_j \frac{\partial x^i}{\partial \bar{x}^j} \bar{v}^j \qquad \bar{v}^i = \sum_j \frac{\partial \bar{x}^i}{\partial x^j} v^j \tag{A.8–42, 43}$$

$$v_i = \sum_j \frac{\partial \bar{x}^j}{\partial x^i} \bar{v}_j \qquad \bar{v}_i = \sum_j \frac{\partial x^j}{\partial \bar{x}^i} v_j \tag{A.8–44, 45}$$

These equations are often taken to be the definition of the contravariant and covariant components of a vector. Similar transformation rules may be found for second and higher-order tensors; for example:

$$\bar{\tau}^{ij} = \sum_k \sum_l \frac{\partial \bar{x}^i}{\partial x^k} \frac{\partial \bar{x}^j}{\partial x^l} \tau^{kl} \tag{A.8–46}$$

$$\bar{\tau}_{ij} = \sum_k \sum_l \frac{\partial x^k}{\partial \bar{x}^i} \frac{\partial x^l}{\partial \bar{x}^j} \tau_{kl} \tag{A.8–47}$$

$$\bar{\tau}^i_{.j} = \sum_k \sum_l \frac{\partial \bar{x}^i}{\partial x^k} \frac{\partial x^l}{\partial \bar{x}^j} \tau^k_{.l} \tag{A.8–48}$$

Here, too, these relations are often taken to be the definitions of the contravariant, covariant, and mixed components of a tensor. We have chosen, however, to define vectors and tensors in terms of the base vectors as in Eqs. A.8–9 and 10 and to develop the rules for operating with the base vectors (Eqs. A.8–3 to 8 for the algebraic rules, Eqs. A.8–17, 18 for the differentiation rules, and Eqs. A.8–36 to 39 for the transformation rules). With these rules for the base vectors, the expressions for the components of the vectors and tensors can then be derived. This mode of presentation is that adopted by Wills[1] and Sedov,[1] whereas McConnell[1] uses relations such as Eqs. A.8–42 to 48 as the basic starting point.

EXERCISES

1. Verify that (p. 261 in Wills[1])

$$(\boldsymbol{\nabla} \cdot \boldsymbol{v}) = \frac{1}{\sqrt{g}} \sum_i \frac{\partial}{\partial x^i} (\sqrt{g}\, v^i)$$

$$\boldsymbol{\nabla}^2 s = \frac{1}{\sqrt{g}} \sum_i \sum_j \frac{\partial}{\partial x^i} \left(\sqrt{g}\, g^{ij} \frac{\partial s}{\partial x^j} \right)$$

2. Use the results of (1) to derive the expressions for $(\boldsymbol{\nabla} \cdot \boldsymbol{v})$ and $\boldsymbol{\nabla}^2 s$ given in Tables A.7–2, 3 for cylindrical and spherical coordinates. Note that the vector components v^i in (1) will have to be converted into components referred to the unit vectors as described in the example following Eq. A.8–10.
3. Show that if a second order tensor is symmetric (i.e., $\tau^{ij} = \tau^{ji}$) in one coordinate system, it is symmetric in all other coordinate systems.

§A.9 FURTHER COMMENTS ON VECTOR-TENSOR NOTATION

The bold-face notation used here is called *Gibbs notation.*[1] Also widely used is another notation referred to as Cartesian tensor notation.[2] A few examples suffice to compare the two systems:

Gibbs Notation	Expanded Notation in Terms of Unit Vectors and Unit Dyads	Cartesian Tensor Notation
$(\boldsymbol{v} \cdot \boldsymbol{w})$	$\sum_i v_i w_i$	$v_i w_i$
$[\boldsymbol{v} \times \boldsymbol{w}]$	$\sum_i \sum_j \sum_k \epsilon_{ijk} \boldsymbol{\delta}_i v_j w_k$	$\epsilon_{ijk} v_j w_k$
$[\nabla \cdot \boldsymbol{\tau}]$	$\sum_i \sum_j \boldsymbol{\delta}_i \dfrac{\partial}{\partial x_j} \tau_{ji}$	$\partial_j \tau_{ji}$ (or $\tau_{ji,j}$)
$\nabla^2 s$	$\sum_i \dfrac{\partial^2}{\partial x_i^2} s$	$\partial_i \partial_i s$ (or $s_{,ii}$)
$[\nabla \times [\nabla \times \boldsymbol{v}]]$	$\sum_i \sum_j \sum_k \sum_m \sum_n \boldsymbol{\delta}_i \epsilon_{ijk} \epsilon_{kmn} \dfrac{\partial}{\partial x_j} \dfrac{\partial}{\partial x_m} v_n$	$\epsilon_{ijk} \epsilon_{kmn} \partial_j \partial_m v_n$ (or $\epsilon_{ijk} \epsilon_{kmn} v_{n,mj}$)
$\{\boldsymbol{\tau} \times \boldsymbol{v}\}$	$\sum_i \sum_j \sum_k \sum_l \epsilon_{jkl} \boldsymbol{\delta}_i \boldsymbol{\delta}_l \tau_{ij} v_k$	$\epsilon_{jkl} \tau_{ij} v_k$

The two outer columns are just two different ways of abbreviating the operations described explicitly in the middle column in rectangular coordinates. The rules for converting from one system to another are easy:

To convert from expanded notation to Cartesian tensor notation:

1. Omit all summation signs.
2. Omit all unit vectors and unit dyads.
3. Replace $\partial/\partial x_i$ by ∂_i (or $_{,i}$ following the symbol).

To convert from Cartesian tensor notation to expanded notation:

1. Supply summation signs for all repeated indices.
2. Supply unit vectors and unit dyads for all nonrepeated indices; in each term of a tensor equation the unit vectors must appear in the same order in the unit dyads.
3. Replace ∂_i (or $_{,i}$) by $\partial/\partial x_i$.

The Gibbs notation is compact, easy to read, and devoid of any reference to a particular coordinate system; however, one has to know the meaning of the dot and cross operations and the use of boldface symbols. The Cartesian tensor notation indicates the nature of the operations explicitly in rectangular coordinates, but errors in reading or writing subscripts

[1] J. W. Gibbs, *Vector Analysis*, Dover Reprint, New York (1960).
[2] W. Prager, *Mechanics of Continua*, Ginn, Boston (1961).

can be most aggravating. People who know both systems equally well prefer the Gibbs notation for general discussions and for presenting results, but revert to Cartesian tensor notation for doing proofs of identities.

Similar index notation can also be used in lieu of the expanded notation in §A.8. One simply remembers that base vectors (or reciprocal base vectors) have to be supplied for every nonrepeated contravariant (or covariant) index, and that sums are then implied on all pairs of repeated indices (one subscript and one superscript). For example, the relation

$$[\boldsymbol{v} \times \boldsymbol{w}] = [\nabla \cdot \boldsymbol{\tau}] \tag{A.9–1}$$

in expanded form becomes:

$$\sum_i \sum_j \sum_k \mathscr{E}_{ijk} \boldsymbol{g}^k v^i w^j = \sum_i \sum_k \boldsymbol{g}_k \tau^{ik}{}_{,i} \tag{A.9–2}$$

In abbreviated notation we may simply write down the statement of the equality of the kth components:

$$\mathscr{E}_{ijl} g^{lk} v^i w^j = \tau^{ik}{}_{,i} \tag{A.9–3}$$

or

$$\mathscr{E}_{ijk} v^i w^j = \tau^i{}_{.k,i} \tag{A.9–4}$$

depending on whether we choose to write the equation in contravariant or covariant components. It is in this notation that most tensor books and continuum mechanics research papers are written. Usually the scaffolding provided by the base vectors is carefully removed so that the reader sees nothing but the relations among the components.

Occasionally *matrix notation* is used to display the components of vectors and tensors with respect to designated coordinate systems. For example, when $v_x = \dot{\gamma} y$, $v_y = 0$, $v_z = 0$, $\nabla \boldsymbol{v}$ can be written in two ways:

$$\nabla \boldsymbol{v} = \boldsymbol{\delta}_y \boldsymbol{\delta}_x \dot{\gamma} = \begin{pmatrix} 0 & 0 & 0 \\ \dot{\gamma} & 0 & 0 \\ 0 & 0 & 0 \end{pmatrix} \tag{A.9–5}$$

The second "=" is not really an "equals" sign, but has to be interpreted as "may be displayed as." Note that this notation is somewhat dangerous since one has to infer the unit dyads that are to be multiplied by the matrix elements—in this case, $\boldsymbol{\delta}_x \boldsymbol{\delta}_x$, $\boldsymbol{\delta}_x \boldsymbol{\delta}_y$, etc. If we had used cylindrical coordinates, $\nabla \boldsymbol{v}$ would be represented by the matrix:

$$\nabla \boldsymbol{v} = \begin{pmatrix} \dot{\gamma} \sin\theta \cos\theta & -\dot{\gamma} \sin^2\theta & 0 \\ \dot{\gamma} \cos^2\theta & -\dot{\gamma} \sin\theta \cos\theta & 0 \\ 0 & 0 & 0 \end{pmatrix} \tag{A.9–6}$$

where the matrix elements are understood to be multiplied by $\boldsymbol{\delta}_r \boldsymbol{\delta}_r$, $\boldsymbol{\delta}_r \boldsymbol{\delta}_\theta$, etc., and then added together.

Despite the hazard of misinterpretation and the loose use of "=," the matrix notation enjoys widespread use, the main reason being that the "dot" operations correspond to standard matrix multiplication rules. For example:

$$(\boldsymbol{v} \cdot \boldsymbol{w}) = (v_1 \quad v_2 \quad v_3) \begin{pmatrix} w_1 \\ w_2 \\ w_3 \end{pmatrix} = v_1 w_1 + v_2 w_2 + v_3 w_3 \tag{A.9–7}$$

$$[\boldsymbol{\tau} \cdot \boldsymbol{v}] = \begin{pmatrix} \tau_{11} & \tau_{12} & \tau_{13} \\ \tau_{21} & \tau_{22} & \tau_{23} \\ \tau_{31} & \tau_{32} & \tau_{33} \end{pmatrix} \begin{pmatrix} v_1 \\ v_2 \\ v_3 \end{pmatrix} = \begin{pmatrix} \tau_{11}v_1 + \tau_{12}v_2 + \tau_{13}v_3 \\ \tau_{21}v_1 + \tau_{22}v_2 + \tau_{23}v_3 \\ \tau_{31}v_1 + \tau_{32}v_2 + \tau_{33}v_3 \end{pmatrix} \tag{A.9–8}$$

Of course such matrix multiplications can be performed only when the components are referred to the same unit vectors or base vectors. (We normally enclose the elements of the matrix array in parentheses; for very large arrays, however, brackets are used for ease in typesetting.)

APPENDIX B

COMPONENTS OF THE EQUATION OF MOTION, THE RATE-OF-STRAIN TENSOR, THE VORTICITY TENSOR, AND THE FINITE STRAIN TENSORS IN THREE COORDINATE SYSTEMS

TABLE B.1

Components of the Equations of Motion[a] in Terms of τ

Rectangular Coordinates (x,y,z):

$$\rho\left(\frac{\partial v_x}{\partial t} + v_x \frac{\partial}{\partial x} v_x + v_y \frac{\partial}{\partial y} v_x + v_z \frac{\partial}{\partial z} v_x\right) = -\left[\frac{\partial}{\partial x}\tau_{xx} + \frac{\partial}{\partial y}\tau_{yx} + \frac{\partial}{\partial z}\tau_{zx}\right] - \frac{\partial p}{\partial x} + \rho g_x \quad \text{(B.1–1)}$$

$$\rho\left(\frac{\partial v_y}{\partial t} + v_x \frac{\partial}{\partial x} v_y + v_y \frac{\partial}{\partial y} v_y + v_z \frac{\partial}{\partial z} v_y\right) = -\left[\frac{\partial}{\partial x}\tau_{xy} + \frac{\partial}{\partial y}\tau_{yy} + \frac{\partial}{\partial z}\tau_{zy}\right] - \frac{\partial p}{\partial y} + \rho g_y \quad \text{(B.1–2)}$$

$$\rho\left(\frac{\partial v_z}{\partial t} + v_x \frac{\partial}{\partial x} v_z + v_y \frac{\partial}{\partial y} v_z + v_z \frac{\partial}{\partial z} v_z\right) = -\left[\frac{\partial}{\partial x}\tau_{xz} + \frac{\partial}{\partial y}\tau_{yz} + \frac{\partial}{\partial z}\tau_{zz}\right] - \frac{\partial p}{\partial z} + \rho g_z \quad \text{(B.1–3)}$$

Cylindrical Coordinates (r,θ,z):

$$\rho\left(\frac{\partial v_r}{\partial t} + v_r \frac{\partial v_r}{\partial r} + \frac{v_\theta}{r}\frac{\partial v_r}{\partial \theta} - \frac{v_\theta^2}{r} + v_z \frac{\partial v_r}{\partial z}\right) = -\left[\frac{1}{r}\frac{\partial}{\partial r}(r\tau_{rr}) + \frac{1}{r}\frac{\partial}{\partial \theta}\tau_{\theta r} + \frac{\partial}{\partial z}\tau_{zr} - \frac{\tau_{\theta\theta}}{r}\right] - \frac{\partial p}{\partial r} + \rho g_r \quad \text{(B.1–4)}$$

$$\rho\left(\frac{\partial v_\theta}{\partial t} + v_r \frac{\partial v_\theta}{\partial r} + \frac{v_\theta}{r}\frac{\partial v_\theta}{\partial \theta} + \frac{v_r v_\theta}{r} + v_z \frac{\partial v_\theta}{\partial z}\right) = -\left[\frac{1}{r^2}\frac{\partial}{\partial r}(r^2\tau_{r\theta}) + \frac{1}{r}\frac{\partial}{\partial \theta}\tau_{\theta\theta} + \frac{\partial}{\partial z}\tau_{z\theta} + \frac{\tau_{\theta r} - \tau_{r\theta}}{r}\right]$$

$$- \frac{1}{r}\frac{\partial p}{\partial \theta} + \rho g_\theta \quad \text{(B.1–5)}$$

$$\rho\left(\frac{\partial v_z}{\partial t} + v_r \frac{\partial v_z}{\partial r} + \frac{v_\theta}{r}\frac{\partial v_z}{\partial \theta} + v_z \frac{\partial v_z}{\partial z}\right) = -\left[\frac{1}{r}\frac{\partial}{\partial r}(r\tau_{rz}) + \frac{1}{r}\frac{\partial}{\partial \theta}\tau_{\theta z} + \frac{\partial}{\partial z}\tau_{zz}\right] - \frac{\partial p}{\partial z} + \rho g_z \quad \text{(B.1–6)}$$

Spherical Coordinates (r,θ,ϕ):

$$\rho\left(\frac{\partial v_r}{\partial t} + v_r \frac{\partial v_r}{\partial r} + \frac{v_\theta}{r}\frac{\partial v_r}{\partial \theta} + \frac{v_\phi}{r\sin\theta}\frac{\partial v_r}{\partial \phi} - \frac{v_\theta^2 + v_\phi^2}{r}\right) = -\left[\frac{1}{r^2}\frac{\partial}{\partial r}(r^2\tau_{rr}) + \frac{1}{r\sin\theta}\frac{\partial}{\partial \theta}(\tau_{\theta r}\sin\theta)\right.$$

$$\left. + \frac{1}{r\sin\theta}\frac{\partial}{\partial \phi}\tau_{\phi r} - \frac{\tau_{\theta\theta} + \tau_{\phi\phi}}{r}\right] - \frac{\partial p}{\partial r} + \rho g_r \quad \text{(B.1–7)}$$

$$\rho\left(\frac{\partial v_\theta}{\partial t} + v_r \frac{\partial v_\theta}{\partial r} + \frac{v_\theta}{r}\frac{\partial v_\theta}{\partial \theta} + \frac{v_\phi}{r\sin\theta}\frac{\partial v_\theta}{\partial \phi} + \frac{v_r v_\theta}{r} - \frac{v_\phi^2\cot\theta}{r}\right) = -\left[\frac{1}{r^3}\frac{\partial}{\partial r}(r^3\tau_{r\theta}) + \frac{1}{r\sin\theta}\frac{\partial}{\partial \theta}(\tau_{\theta\theta}\sin\theta)\right.$$

$$\left. + \frac{1}{r\sin\theta}\frac{\partial}{\partial \phi}\tau_{\phi\theta} + \frac{(\tau_{\theta r} - \tau_{r\theta}) - \tau_{\phi\phi}\cot\theta}{r}\right]$$

$$- \frac{1}{r}\frac{\partial p}{\partial \theta} + \rho g_\theta \quad \text{(B.1–8)}$$

$$\rho\left(\frac{\partial v_\phi}{\partial t} + v_r \frac{\partial v_\phi}{\partial r} + \frac{v_\theta}{r}\frac{\partial v_\phi}{\partial \theta} + \frac{v_\phi}{r\sin\theta}\frac{\partial v_\phi}{\partial \phi} + \frac{v_\phi v_r}{r} + \frac{v_\theta v_\phi}{r}\cot\theta\right) = -\left[\frac{1}{r^3}\frac{\partial}{\partial r}(r^3\tau_{r\phi}) + \frac{1}{r\sin\theta}\frac{\partial}{\partial \theta}(\tau_{\theta\phi}\sin\theta)\right.$$

$$\left. + \frac{1}{r\sin\theta}\frac{\partial}{\partial \phi}\tau_{\phi\phi} + \frac{(\tau_{\phi r} - \tau_{r\phi}) + \tau_{\phi\theta}\cot\theta}{r}\right]$$

$$- \frac{1}{r\sin\theta}\frac{\partial p}{\partial \phi} + \rho g_\phi \quad \text{(B.1–9)}$$

[a] In these equations no assumption is made regarding the symmetry of $\boldsymbol{\tau}$.

TABLE B.2

Components of the Equations of Motion for a Newtonian Fluid with Constant Density (ρ) and Constant Viscosity (μ)

Rectangular Coordinates (x,y,z):

$$\rho\left(\frac{\partial v_x}{\partial t} + v_x\frac{\partial v_x}{\partial x} + v_y\frac{\partial v_x}{\partial y} + v_z\frac{\partial v_x}{\partial z}\right) = \mu\left[\frac{\partial^2}{\partial x^2}v_x + \frac{\partial^2}{\partial y^2}v_x + \frac{\partial^2}{\partial z^2}v_x\right] - \frac{\partial p}{\partial x} + \rho g_x \qquad \text{(B.2–1)}$$

$$\rho\left(\frac{\partial v_y}{\partial t} + v_x\frac{\partial v_y}{\partial x} + v_y\frac{\partial v_y}{\partial y} + v_z\frac{\partial v_y}{\partial z}\right) = \mu\left[\frac{\partial^2}{\partial x^2}v_y + \frac{\partial^2}{\partial y^2}v_y + \frac{\partial^2}{\partial z^2}v_y\right] - \frac{\partial p}{\partial y} + \rho g_y \qquad \text{(B.2–2)}$$

$$\rho\left(\frac{\partial v_z}{\partial t} + v_x\frac{\partial v_z}{\partial x} + v_y\frac{\partial v_z}{\partial y} + v_z\frac{\partial v_z}{\partial z}\right) = \mu\left[\frac{\partial^2}{\partial x^2}v_z + \frac{\partial^2}{\partial y^2}v_z + \frac{\partial^2}{\partial z^2}v_z\right] - \frac{\partial p}{\partial z} + \rho g_z \qquad \text{(B.2–3)}$$

Cylindrical Coordinates (r,θ,z):

$$\rho\left(\frac{\partial v_r}{\partial t} + v_r\frac{\partial v_r}{\partial r} + \frac{v_\theta}{r}\frac{\partial v_r}{\partial \theta} - \frac{v_\theta^{\,2}}{r} + v_z\frac{\partial v_r}{\partial z}\right) = \mu\left[\frac{\partial}{\partial r}\left(\frac{1}{r}\frac{\partial}{\partial r}(rv_r)\right) + \frac{1}{r^2}\frac{\partial^2 v_r}{\partial \theta^2} + \frac{\partial^2 v_r}{\partial z^2} - \frac{2}{r^2}\frac{\partial v_\theta}{\partial \theta}\right] - \frac{\partial p}{\partial r} + \rho g_r \qquad \text{(B.2–4)}$$

$$\rho\left(\frac{\partial v_\theta}{\partial t} + v_r\frac{\partial v_\theta}{\partial r} + \frac{v_\theta}{r}\frac{\partial v_\theta}{\partial \theta} + \frac{v_r v_\theta}{r} + v_z\frac{\partial v_\theta}{\partial z}\right) = \mu\left[\frac{\partial}{\partial r}\left(\frac{1}{r}\frac{\partial}{\partial r}(rv_\theta)\right) + \frac{1}{r^2}\frac{\partial^2 v_\theta}{\partial \theta^2} + \frac{\partial^2 v_\theta}{\partial z^2} + \frac{2}{r^2}\frac{\partial v_r}{\partial \theta}\right] - \frac{1}{r}\frac{\partial p}{\partial \theta} + \rho g_\theta \qquad \text{(B.2–5)}$$

$$\rho\left(\frac{\partial v_z}{\partial t} + v_r\frac{\partial v_z}{\partial r} + \frac{v_\theta}{r}\frac{\partial v_z}{\partial \theta} + v_z\frac{\partial v_z}{\partial z}\right) = \mu\left[\frac{1}{r}\frac{\partial}{\partial r}\left(r\frac{\partial v_z}{\partial r}\right) + \frac{1}{r^2}\frac{\partial^2 v_z}{\partial \theta^2} + \frac{\partial^2 v_z}{\partial z^2}\right] - \frac{\partial p}{\partial z} + \rho g_z \qquad \text{(B.2–6)}$$

Spherical Coordinates (r,θ,ϕ):

$$\rho\left(\frac{\partial v_r}{\partial t} + v_r\frac{\partial v_r}{\partial r} + \frac{v_\theta}{r}\frac{\partial v_r}{\partial \theta} + \frac{v_\phi}{r\sin\theta}\frac{\partial v_r}{\partial \phi} - \frac{v_\theta^{\,2} + v_\phi^{\,2}}{r}\right) = \mu\left[\frac{\partial}{\partial r}\left(\frac{1}{r^2}\frac{\partial}{\partial r}(r^2 v_r)\right) + \frac{1}{r^2\sin\theta}\frac{\partial}{\partial \theta}\left(\sin\theta\frac{\partial v_r}{\partial \theta}\right) + \frac{1}{r^2\sin^2\theta}\frac{\partial^2 v_r}{\partial \phi^2} - \frac{2}{r^2\sin\theta}\frac{\partial}{\partial \theta}(v_\theta\sin\theta) - \frac{2}{r^2\sin\theta}\frac{\partial v_\phi}{\partial \phi}\right] - \frac{\partial p}{\partial r} + \rho g_r \qquad \text{(B.2–7)}$$

$$\rho\left(\frac{\partial v_\theta}{\partial t} + v_r\frac{\partial v_\theta}{\partial r} + \frac{v_\theta}{r}\frac{\partial v_\theta}{\partial \theta} + \frac{v_\phi}{r\sin\theta}\frac{\partial v_\theta}{\partial \phi} + \frac{v_r v_\theta}{r} - \frac{v_\phi^{\,2}\cot\theta}{r}\right) = \mu\left[\frac{1}{r^2}\frac{\partial}{\partial r}\left(r^2\frac{\partial v_\theta}{\partial r}\right) + \frac{1}{r^2}\frac{\partial}{\partial \theta}\left(\frac{1}{\sin\theta}\frac{\partial}{\partial \theta}(v_\theta\sin\theta)\right) + \frac{1}{r^2\sin^2\theta}\frac{\partial^2 v_\theta}{\partial \phi^2} + \frac{2}{r^2}\frac{\partial v_r}{\partial \theta} - \frac{2\cot\theta}{r^2\sin\theta}\frac{\partial v_\phi}{\partial \phi}\right] - \frac{1}{r}\frac{\partial p}{\partial \theta} + \rho g_\theta \qquad \text{(B.2–8)}$$

$$\rho\left(\frac{\partial v_\phi}{\partial t} + v_r\frac{\partial v_\phi}{\partial r} + \frac{v_\theta}{r}\frac{\partial v_\phi}{\partial \theta} + \frac{v_\phi}{r\sin\theta}\frac{\partial v_\phi}{\partial \phi} + \frac{v_\phi v_r}{r} + \frac{v_\theta v_\phi}{r}\cot\theta\right) = \mu\left[\frac{1}{r^2}\frac{\partial}{\partial r}\left(r^2\frac{\partial v_\phi}{\partial r}\right) + \frac{1}{r^2}\frac{\partial}{\partial \theta}\left(\frac{1}{\sin\theta}\frac{\partial}{\partial \theta}(v_\phi\sin\theta)\right) + \frac{1}{r^2\sin^2\theta}\frac{\partial^2 v_\phi}{\partial \phi^2} + \frac{2}{r^2\sin\theta}\frac{\partial v_r}{\partial \phi} + \frac{2\cot\theta}{r^2\sin\theta}\frac{\partial v_\theta}{\partial \phi}\right]$$
$$- \frac{1}{r\sin\theta}\frac{\partial p}{\partial \phi} + \rho g_\phi \qquad \text{(B.2–9)}$$

TABLE B.3

Components of the Rate-of-Strain Tensor $\dot{\gamma} = \nabla \boldsymbol{v} + (\nabla \boldsymbol{v})^{\dagger}$

Rectangular Coordinates (x,y,z):

$$\dot{\gamma}_{xx} = 2\frac{\partial v_x}{\partial x} \tag{B.3–1}$$

$$\dot{\gamma}_{yy} = 2\frac{\partial v_y}{\partial y} \tag{B.3–2}$$

$$\dot{\gamma}_{zz} = 2\frac{\partial v_z}{\partial z} \tag{B.3–3}$$

$$\dot{\gamma}_{xy} = \dot{\gamma}_{yx} = \frac{\partial v_y}{\partial x} + \frac{\partial v_x}{\partial y} \tag{B.3–4}$$

$$\dot{\gamma}_{yz} = \dot{\gamma}_{zy} = \frac{\partial v_z}{\partial y} + \frac{\partial v_y}{\partial z} \tag{B.3–5}$$

$$\dot{\gamma}_{zx} = \dot{\gamma}_{xz} = \frac{\partial v_x}{\partial z} + \frac{\partial v_z}{\partial x} \tag{B.3–6}$$

Cylindrical Coordinates (r,θ,z):

$$\dot{\gamma}_{rr} = 2\frac{\partial v_r}{\partial r} \tag{B.3–7}$$

$$\dot{\gamma}_{\theta\theta} = 2\left(\frac{1}{r}\frac{\partial v_\theta}{\partial \theta} + \frac{v_r}{r}\right) \tag{B.3–8}$$

$$\dot{\gamma}_{zz} = 2\frac{\partial v_z}{\partial z} \tag{B.3–9}$$

$$\dot{\gamma}_{r\theta} = \dot{\gamma}_{\theta r} = r\frac{\partial}{\partial r}\left(\frac{v_\theta}{r}\right) + \frac{1}{r}\frac{\partial v_r}{\partial \theta} \tag{B.3–10}$$

$$\dot{\gamma}_{\theta z} = \dot{\gamma}_{z\theta} = \frac{1}{r}\frac{\partial v_z}{\partial \theta} + \frac{\partial v_\theta}{\partial z} \tag{B.3–11}$$

$$\dot{\gamma}_{zr} = \dot{\gamma}_{rz} = \frac{\partial v_r}{\partial z} + \frac{\partial v_z}{\partial r} \tag{B.3–12}$$

Spherical Coordinates (r,θ,ϕ):

$$\dot{\gamma}_{rr} = 2\frac{\partial v_r}{\partial r} \tag{B.3–13}$$

$$\dot{\gamma}_{\theta\theta} = 2\left(\frac{1}{r}\frac{\partial v_\theta}{\partial \theta} + \frac{v_r}{r}\right) \tag{B.3–14}$$

$$\dot{\gamma}_{\phi\phi} = 2\left(\frac{1}{r\sin\theta}\frac{\partial v_\phi}{\partial \phi} + \frac{v_r}{r} + \frac{v_\theta \cot\theta}{r}\right) \tag{B.3–15}$$

$$\dot{\gamma}_{r\theta} = \dot{\gamma}_{\theta r} = r\frac{\partial}{\partial r}\left(\frac{v_\theta}{r}\right) + \frac{1}{r}\frac{\partial v_r}{\partial \theta} \tag{B.3–16}$$

$$\dot{\gamma}_{\theta\phi} = \dot{\gamma}_{\phi\theta} = \frac{\sin\theta}{r}\frac{\partial}{\partial \theta}\left(\frac{v_\phi}{\sin\theta}\right) + \frac{1}{r\sin\theta}\frac{\partial v_\theta}{\partial \phi} \tag{B.3–17}$$

$$\dot{\gamma}_{\phi r} = \dot{\gamma}_{r\phi} = \frac{1}{r\sin\theta}\frac{\partial v_r}{\partial \phi} + r\frac{\partial}{\partial r}\left(\frac{v_\phi}{r}\right) \tag{B.3–18}$$

TABLE B.4

Components of the Vorticity Tensor[a] $\boldsymbol{\omega} = \nabla \boldsymbol{v} - (\nabla \boldsymbol{v})^{\dagger}$

Rectangular Coordinates (x,y,z):

$$\omega_{xy} = -\omega_{yx} = \frac{\partial v_y}{\partial x} - \frac{\partial v_x}{\partial y} \tag{B.4–1}$$

$$\omega_{yz} = -\omega_{zy} = \frac{\partial v_z}{\partial y} - \frac{\partial v_y}{\partial z} \tag{B.4–2}$$

$$\omega_{zx} = -\omega_{xz} = \frac{\partial v_x}{\partial z} - \frac{\partial v_z}{\partial x} \tag{B.4–3}$$

Cylindrical Coordinates (r,θ,z):

$$\omega_{r\theta} = -\omega_{\theta r} = \frac{1}{r}\frac{\partial}{\partial r}(rv_\theta) - \frac{1}{r}\frac{\partial v_r}{\partial \theta} \tag{B.4–4}$$

$$\omega_{\theta z} = -\omega_{z\theta} = \frac{1}{r}\frac{\partial v_z}{\partial \theta} - \frac{\partial v_\theta}{\partial z} \tag{B.4–5}$$

$$\omega_{zr} = -\omega_{rz} = \frac{\partial v_r}{\partial z} - \frac{\partial v_z}{\partial r} \tag{B.4–6}$$

Spherical Coordinates (r,θ,ϕ):

$$\omega_{r\theta} = -\omega_{\theta r} = \frac{1}{r}\frac{\partial}{\partial r}(rv_\theta) - \frac{1}{r}\frac{\partial v_r}{\partial \theta} \tag{B.4–7}$$

$$\omega_{\theta\phi} = -\omega_{\phi\theta} = \frac{1}{r\sin\theta}\frac{\partial}{\partial \theta}(v_\phi \sin\theta) - \frac{1}{r\sin\theta}\frac{\partial v_\theta}{\partial \phi} \tag{B.4–8}$$

$$\omega_{\phi r} = -\omega_{r\phi} = \frac{1}{r\sin\theta}\frac{\partial v_r}{\partial \phi} - \frac{1}{r}\frac{\partial}{\partial r}(rv_\phi) \tag{B.4–9}$$

[a] The diagonal components are all zero.

TABLE B.5

The Strain Tensor $\gamma^{[0]}$ for a General Deformation[a]

Rectangular Coordinates (x,y,z):

$$x' = x'(x,y,z,t,t')$$
$$y' = y'(x,y,z,t,t')$$
$$z' = z'(x,y,z,t,t')$$

$$\gamma_{xx}^{[0]} = \left(\frac{\partial x'}{\partial x}\right)^2 + \left(\frac{\partial y'}{\partial x}\right)^2 + \left(\frac{\partial z'}{\partial x}\right)^2 - 1 \tag{B.5–1}$$

$$\gamma_{yy}^{[0]} = \left(\frac{\partial x'}{\partial y}\right)^2 + \left(\frac{\partial y'}{\partial y}\right)^2 + \left(\frac{\partial z'}{\partial y}\right)^2 - 1 \tag{B.5–2}$$

$$\gamma_{zz}^{[0]} = \left(\frac{\partial x'}{\partial z}\right)^2 + \left(\frac{\partial y'}{\partial z}\right)^2 + \left(\frac{\partial z'}{\partial z}\right)^2 - 1 \tag{B.5–3}$$

Rectangular Coordinates (x,y,z):

$$\gamma_{xy}^{[0]} = \gamma_{yx}^{[0]} = \left(\frac{\partial x'}{\partial x}\right)\left(\frac{\partial x'}{\partial y}\right) + \left(\frac{\partial y'}{\partial x}\right)\left(\frac{\partial y'}{\partial y}\right) + \left(\frac{\partial z'}{\partial x}\right)\left(\frac{\partial z'}{\partial y}\right) \tag{B.5–4}$$

$$\gamma_{yz}^{[0]} = \gamma_{zy}^{[0]} = \left(\frac{\partial x'}{\partial y}\right)\left(\frac{\partial x'}{\partial z}\right) + \left(\frac{\partial y'}{\partial y}\right)\left(\frac{\partial y'}{\partial z}\right) + \left(\frac{\partial z'}{\partial y}\right)\left(\frac{\partial z'}{\partial z}\right) \tag{B.5–5}$$

$$\gamma_{zx}^{[0]} = \gamma_{xz}^{[0]} = \left(\frac{\partial x'}{\partial x}\right)\left(\frac{\partial x'}{\partial z}\right) + \left(\frac{\partial y'}{\partial x}\right)\left(\frac{\partial y'}{\partial z}\right) + \left(\frac{\partial z'}{\partial x}\right)\left(\frac{\partial z'}{\partial z}\right) \tag{B.5–6}$$

Cylindrical Coordinates (r,θ,z):

$$r' = r'(r,\theta,z,t,t')$$
$$\theta' = \theta'(r,\theta,z,t,t')$$
$$z' = z'(r,\theta,z,t,t')$$

$$\gamma_{rr}^{[0]} = \left(\frac{\partial r'}{\partial r}\right)^2 + \left(r'\frac{\partial \theta'}{\partial r}\right)^2 + \left(\frac{\partial z'}{\partial r}\right)^2 - 1 \tag{B.5–7}$$

$$\gamma_{\theta\theta}^{[0]} = \frac{1}{r^2}\left[\left(\frac{\partial r'}{\partial \theta}\right)^2 + \left(r'\frac{\partial \theta'}{\partial \theta}\right)^2 + \left(\frac{\partial z'}{\partial \theta}\right)^2\right] - 1 \tag{B.5–8}$$

$$\gamma_{zz}^{[0]} = \left(\frac{\partial r'}{\partial z}\right)^2 + \left(r'\frac{\partial \theta'}{\partial z}\right)^2 + \left(\frac{\partial z'}{\partial z}\right)^2 - 1 \tag{B.5–9}$$

$$\gamma_{r\theta}^{[0]} = \gamma_{\theta r}^{[0]} = \frac{1}{r}\left[\left(\frac{\partial r'}{\partial r}\right)\left(\frac{\partial r'}{\partial \theta}\right) + r'^2\left(\frac{\partial \theta'}{\partial r}\right)\left(\frac{\partial \theta'}{\partial \theta}\right) + \left(\frac{\partial z'}{\partial r}\right)\left(\frac{\partial z'}{\partial \theta}\right)\right] \tag{B.5–10}$$

$$\gamma_{\theta z}^{[0]} = \gamma_{z\theta}^{[0]} = \frac{1}{r}\left[\left(\frac{\partial r'}{\partial \theta}\right)\left(\frac{\partial r'}{\partial z}\right) + r'^2\left(\frac{\partial \theta'}{\partial \theta}\right)\left(\frac{\partial \theta'}{\partial z}\right) + \left(\frac{\partial z'}{\partial \theta}\right)\left(\frac{\partial z'}{\partial z}\right)\right] \tag{B.5–11}$$

$$\gamma_{zr}^{[0]} = \gamma_{rz}^{[0]} = \left(\frac{\partial r'}{\partial r}\right)\left(\frac{\partial r'}{\partial z}\right) + r'^2\left(\frac{\partial \theta'}{\partial r}\right)\left(\frac{\partial \theta'}{\partial z}\right) + \left(\frac{\partial z'}{\partial r}\right)\left(\frac{\partial z'}{\partial z}\right) \tag{B.5–12}$$

Spherical Coordinates (r,θ,ϕ):

$$r' = r'(r,\theta,\phi,t,t')$$
$$\theta' = \theta'(r,\theta,\phi,t,t')$$
$$\phi' = \phi'(r,\theta,\phi,t,t')$$

$$\gamma_{rr}^{[0]} = \left(\frac{\partial r'}{\partial r}\right)^2 + \left(r'\frac{\partial \theta'}{\partial r}\right)^2 + \left(r'\sin\theta'\frac{\partial \phi'}{\partial r}\right)^2 - 1 \tag{B.5–13}$$

$$\gamma_{\theta\theta}^{[0]} = \frac{1}{r^2}\left[\left(\frac{\partial r'}{\partial \theta}\right)^2 + \left(r'\frac{\partial \theta'}{\partial \theta}\right)^2 + \left(r'\sin\theta'\frac{\partial \phi'}{\partial \theta}\right)^2\right] - 1 \tag{B.5–14}$$

$$\gamma_{\phi\phi}^{[0]} = \frac{1}{(r\sin\theta)^2}\left[\left(\frac{\partial r'}{\partial \phi}\right)^2 + \left(r'\frac{\partial \theta'}{\partial \phi}\right)^2 + \left(r'\sin\theta'\frac{\partial \phi'}{\partial \phi}\right)^2\right] - 1 \tag{B.5–15}$$

$$\gamma_{r\theta}^{[0]} = \gamma_{\theta r}^{[0]} = \frac{1}{r}\left[\left(\frac{\partial r'}{\partial r}\right)\left(\frac{\partial r'}{\partial \theta}\right) + r'^2\left(\frac{\partial \theta'}{\partial r}\right)\left(\frac{\partial \theta'}{\partial \theta}\right) + (r'\sin\theta')^2\left(\frac{\partial \phi'}{\partial r}\right)\left(\frac{\partial \phi'}{\partial \theta}\right)\right] \tag{B.5–16}$$

$$\gamma_{\theta\phi}^{[0]} = \gamma_{\phi\theta}^{[0]} = \frac{1}{r^2\sin\theta}\left[\left(\frac{\partial r'}{\partial \theta}\right)\left(\frac{\partial r'}{\partial \phi}\right) + r'^2\left(\frac{\partial \theta'}{\partial \theta}\right)\left(\frac{\partial \theta'}{\partial \phi}\right) + (r'\sin\theta')^2\left(\frac{\partial \phi'}{\partial \theta}\right)\left(\frac{\partial \phi'}{\partial \phi}\right)\right] \tag{B.5–17}$$

$$\gamma_{\phi r}^{[0]} = \gamma_{r\phi}^{[0]} = \frac{1}{r\sin\theta}\left[\left(\frac{\partial r'}{\partial r}\right)\left(\frac{\partial r'}{\partial \phi}\right) + r'^2\left(\frac{\partial \theta'}{\partial r}\right)\left(\frac{\partial \theta'}{\partial \phi}\right) + (r'\sin\theta')^2\left(\frac{\partial \phi'}{\partial r}\right)\left(\frac{\partial \phi'}{\partial \phi}\right)\right] \tag{B.5–18}$$

[a] For incompressible fluids det $(\boldsymbol{\gamma}^{[0]} + \boldsymbol{\delta}) = 1$.

TABLE B.6

The Strain Tensor $\gamma_{[0]}$ for a General Deformation[a]

Rectangular Coordinates (x,y,z):

$$x = x(x',y',z',t,t')$$
$$y = y(x',y',z',t,t')$$
$$z = z(x',y',z',t,t')$$

$$\gamma_{[0]xx} = 1 - \left[\left(\frac{\partial x}{\partial x'}\right)^2 + \left(\frac{\partial x}{\partial y'}\right)^2 + \left(\frac{\partial x}{\partial z'}\right)^2\right] \tag{B.6–1}$$

$$\gamma_{[0]yy} = 1 - \left[\left(\frac{\partial y}{\partial x'}\right)^2 + \left(\frac{\partial y}{\partial y'}\right)^2 + \left(\frac{\partial y}{\partial z'}\right)^2\right] \tag{B.6–2}$$

$$\gamma_{[0]zz} = 1 - \left[\left(\frac{\partial z}{\partial x'}\right)^2 + \left(\frac{\partial z}{\partial y'}\right)^2 + \left(\frac{\partial z}{\partial z'}\right)^2\right] \tag{B.6–3}$$

$$\gamma_{[0]xy} = \gamma_{[0]yx} = -\left[\left(\frac{\partial x}{\partial x'}\right)\left(\frac{\partial y}{\partial x'}\right) + \left(\frac{\partial x}{\partial y'}\right)\left(\frac{\partial y}{\partial y'}\right) + \left(\frac{\partial x}{\partial z'}\right)\left(\frac{\partial y}{\partial z'}\right)\right] \tag{B.6–4}$$

$$\gamma_{[0]yz} = \gamma_{[0]zy} = -\left[\left(\frac{\partial y}{\partial x'}\right)\left(\frac{\partial z}{\partial x'}\right) + \left(\frac{\partial y}{\partial y'}\right)\left(\frac{\partial z}{\partial y'}\right) + \left(\frac{\partial y}{\partial z'}\right)\left(\frac{\partial z}{\partial z'}\right)\right] \tag{B.6–5}$$

$$\gamma_{[0]zx} = \gamma_{[0]xz} = -\left[\left(\frac{\partial x}{\partial x'}\right)\left(\frac{\partial z}{\partial x'}\right) + \left(\frac{\partial x}{\partial y'}\right)\left(\frac{\partial z}{\partial y'}\right) + \left(\frac{\partial x}{\partial z'}\right)\left(\frac{\partial z}{\partial z'}\right)\right] \tag{B.6–6}$$

Cylindrical Coordinates (r,θ,z):

$$r = r(r',\theta',z',t,t')$$
$$\theta = \theta(r',\theta',z',t,t')$$
$$z = z(r',\theta',z',t,t')$$

$$\gamma_{[0]rr} = 1 - \left[\left(\frac{\partial r}{\partial r'}\right)^2 + \left(\frac{1}{r'}\frac{\partial r}{\partial \theta'}\right)^2 + \left(\frac{\partial r}{\partial z'}\right)^2\right] \tag{B.6–7}$$

$$\gamma_{[0]\theta\theta} = 1 - r^2\left[\left(\frac{\partial \theta}{\partial r'}\right)^2 + \left(\frac{1}{r'}\frac{\partial \theta}{\partial \theta'}\right)^2 + \left(\frac{\partial \theta}{\partial z'}\right)^2\right] \tag{B.6–8}$$

$$\gamma_{[0]zz} = 1 - \left[\left(\frac{\partial z}{\partial r'}\right)^2 + \left(\frac{1}{r'}\frac{\partial z}{\partial \theta'}\right)^2 + \left(\frac{\partial z}{\partial z'}\right)^2\right] \tag{B.6–9}$$

$$\gamma_{[0]r\theta} = \gamma_{[0]\theta r} = -r\left[\left(\frac{\partial r}{\partial r'}\right)\left(\frac{\partial \theta}{\partial r'}\right) + \frac{1}{r'^2}\left(\frac{\partial r}{\partial \theta'}\right)\left(\frac{\partial \theta}{\partial \theta'}\right) + \left(\frac{\partial r}{\partial z'}\right)\left(\frac{\partial \theta}{\partial z'}\right)\right] \tag{B.6–10}$$

$$\gamma_{[0]\theta z} = \gamma_{[0]z\theta} = -r\left[\left(\frac{\partial \theta}{\partial r'}\right)\left(\frac{\partial z}{\partial r'}\right) + \frac{1}{r'^2}\left(\frac{\partial \theta}{\partial \theta'}\right)\left(\frac{\partial z}{\partial \theta'}\right) + \left(\frac{\partial \theta}{\partial z'}\right)\left(\frac{\partial z}{\partial z'}\right)\right] \tag{B.6–11}$$

$$\gamma_{[0]zr} = \gamma_{[0]rz} = -\left[\left(\frac{\partial r}{\partial r'}\right)\left(\frac{\partial z}{\partial r'}\right) + \frac{1}{r'^2}\left(\frac{\partial r}{\partial \theta'}\right)\left(\frac{\partial z}{\partial \theta'}\right) + \left(\frac{\partial r}{\partial z'}\right)\left(\frac{\partial z}{\partial z'}\right)\right] \tag{B.6–12}$$

TABLE B.6 (*Continued*)

Spherical Coordinates (r,θ,ϕ):

$$r = r(r',\theta',\phi',t,t')$$
$$\theta = \theta(r',\theta',\phi',t,t')$$
$$\phi = \phi(r',\theta',\phi',t,t')$$

$$\gamma_{[0]rr} = 1 - \left[\left(\frac{\partial r}{\partial r'}\right)^2 + \left(\frac{1}{r'}\frac{\partial r}{\partial \theta'}\right)^2 + \left(\frac{1}{r' \sin \theta'}\frac{\partial r}{\partial \phi'}\right)^2\right] \quad \text{(B.6–13)}$$

$$\gamma_{[0]\theta\theta} = 1 - r^2\left[\left(\frac{\partial \theta}{\partial r'}\right)^2 + \left(\frac{1}{r'}\frac{\partial \theta}{\partial \theta'}\right)^2 + \left(\frac{1}{r' \sin \theta'}\frac{\partial \theta}{\partial \phi'}\right)^2\right] \quad \text{(B.6–14)}$$

$$\gamma_{[0]\phi\phi} = 1 - r^2 \sin^2 \theta\left[\left(\frac{\partial \phi}{\partial r'}\right)^2 + \left(\frac{1}{r'}\frac{\partial \phi}{\partial \theta'}\right)^2 + \left(\frac{1}{r' \sin \theta'}\frac{\partial \phi}{\partial \phi'}\right)^2\right] \quad \text{(B.6–15)}$$

$$\gamma_{[0]r\theta} = \gamma_{[0]\theta r} = -r\left[\left(\frac{\partial r}{\partial r'}\right)\left(\frac{\partial \theta}{\partial r'}\right) + \frac{1}{r'^2}\left(\frac{\partial r}{\partial \theta'}\right)\left(\frac{\partial \theta}{\partial \theta'}\right) + \frac{1}{(r' \sin \theta')^2}\left(\frac{\partial r}{\partial \phi'}\right)\left(\frac{\partial \theta}{\partial \phi'}\right)\right] \quad \text{(B.6–16)}$$

$$\gamma_{[0]\theta\phi} = \gamma_{[0]\phi\theta} = -r^2 \sin \theta\left[\left(\frac{\partial \theta}{\partial r'}\right)\left(\frac{\partial \phi}{\partial r'}\right) + \frac{1}{r'^2}\left(\frac{\partial \theta}{\partial \theta'}\right)\left(\frac{\partial \phi}{\partial \theta'}\right) + \frac{1}{(r' \sin \theta')^2}\left(\frac{\partial \theta}{\partial \phi'}\right)\left(\frac{\partial \phi}{\partial \phi'}\right)\right] \quad \text{(B.6–17)}$$

$$\gamma_{[0]\phi r} = \gamma_{[0]r\phi} = -r \sin \theta\left[\left(\frac{\partial r}{\partial r'}\right)\left(\frac{\partial \phi}{\partial r'}\right) + \frac{1}{r'^2}\left(\frac{\partial r}{\partial \theta'}\right)\left(\frac{\partial \phi}{\partial \theta'}\right) + \frac{1}{(r' \sin \theta')^2}\left(\frac{\partial r}{\partial \phi'}\right)\left(\frac{\partial \phi}{\partial \phi'}\right)\right] \quad \text{(B.6–18)}$$

[a] For incompressible fluids det $(\boldsymbol{\delta} - \boldsymbol{\gamma}_{[0]}) = 1$.

NOTATION FOR VOLUME 1

Symbols appearing infrequently or only in one section are generally not included in this list. The equation, figure, table, or problem references give the locations where the symbols are first used or where they are defined.

Italic and Roman Symbols

A	Helmholtz free energy	Eq. 2.3–9
a	Parameter in bipolar coordinates	Fig. A.7–1
a_n	Constants in general linear viscoelastic equation	Eq. 6.1–11
a_i	Components of the acceleration vector	Eq. 9.2–5
B	Thickness or half-thickness (of film, slit, etc.)	Eq. 1.2–6 Fig. 1B.9
Br	Brinkman number	Eq. 1.2–17 and Eq. 5.4–111
b_n	Constants in general linear viscoelastic equation	Eq. 6.1–11
C_{ij}	Kramers matrix	Eq. 2.3–20
c	Mass concentration	Eq. 2.4–2
D/Dt	Substantial or material time derivative	Table 1.1–1
$\mathscr{D}/\mathscr{D}t$	Corotational or Jaumann derivative	Eq. 5.5–2
d_i	Normal-stress displacement functions	Eqs. 4.4–9 and 11
e	Base for natural logarithms	—
$\boldsymbol{F}$	Force	Eq. 1.2–50
$\boldsymbol{F}$	Time-average force in bead-rod chain	Eq. 2.3–38
$\boldsymbol{F}_k$	Kinetic force	Eq. 1.2–50
$\boldsymbol{F}_b$	Buoyancy force	Eq. 1.2–50
$\mathscr{F}$	Total thrust	Eq. 4.5–7
f	Distribution function	Eq. 2.3–3
f_1, f_2	Functions in Reiner-Rivlin equation	Eq. 8.5–4
$G(t - t')$	Relaxation modulus in linear viscolastic fluid, in Goddard-Miller equation, in single-integral codeformational model	Eq. 6.1–16 Eq. 7.5–1 Table 9.4–2
$G(t - t_0;\gamma_0)$	Relaxation modulus (following a sudden shearing displacement)	Eq. 4.4–24

$G_I, G_{II}, \ldots$	Kernel functions in corotational memory integral expansion	Eq. 6.4–1 Eq. 8.3–1
$G_1, G_2, \ldots G^1, G^2, \ldots$	Kernel functions in codeformational memory integral expansions	Eqs. 9.5–1 and 2
$\boldsymbol{g}$	Gravitational acceleration	Table 1.1–1
$\boldsymbol{g}_i, \boldsymbol{g}^i$	Base vectors in nonorthogonal coordinates	Eqs. A.8–1 and 2
$\hat{\boldsymbol{g}}_i, \hat{\boldsymbol{g}}^i$	Codeforming base vectors	Fig. 9.1–1 Table 9.1–3
g_{ij}, g^{ij}	Components of the metric tensor	Eqs. A.8–5 and 6
$\hat{g}_{ij}, \hat{g}^{ij}$	Components of the metric tensor in codeforming coordinate system	Table 9.1–2 Eq. 9.3–7
$\hat{H}$	Enthalpy per unit mass	Eq. 1B.15–2
$\mathscr{H}$	Hamiltonian	Eq. 2.3–7
H	Hookean spring constant	Eq. 2.5–1
h	Height	Fig. 1.2–6
h	Heat-transfer coefficient	Eq. 5.4–31
h_α	Scale factors in orthogonal curvilinear coordinates	Eq. 3.4–5 Eq. A.7–15
$\mathscr{I}m\{w\}$	Imaginary part of the complex number w	—
i	$\sqrt{-1}$	Eq. 1D.2–2
J	Variational function	Eq. 5.3–1
J	Creep compliance	Eq. 4.4–25
$J_e{}^0$	Steady-state creep compliance	Eq. 4.4–26 Eq. 6.2–42
k	Boltzmann's constant	Eq. 2.3–7
k	Thermal conductivity	Eq. 1.2–3
$\boldsymbol{k}_i$	Unit vector in direction of ith connector in chain	Eq. 2.3–12
L	Length (of tube, slit, etc.)	Eq. 1.2–34
L	Length of rod in bead-rod kinetic theory models	Eq. 2.3–16
$\mathscr{L}, \mathscr{L}^{-1}$	Langevin function, inverse Langevin function	Eq. 2.3–41
$\mathscr{L}, \mathscr{L}^{-1}$	Laplace transform, inverse Laplace transform	Eq. 6.3–4
M	Molecular weight	Eq. 2.2–1
$\bar{M}_n$	Number-average molecular weight	Eq. 2.2–1
$\bar{M}_w$	Weight-average molecular weight	Eq. 2.2–3

$\bar{M}_v$	Viscosity-average molecular weight	Eq. 2.4–5
$M(t - t')$	Memory function in linear viscoelasticity, and in codeformational single-integral model	Eq. 6.1–17 and Table 9.4–2
m	Parameter in power-law viscosity model	Eq. 5.1–9
m'	Parameter in power-law primary normal stress coefficient	Eq. 5.2–53
m, m_1, m_2	Elasticity dimensionless parameters	Eq. 8.4–48 Eq. 8.4–64
N	Number of beads in a chainlike model	Eq. 2.3–2
$\tilde{N}$	Avogadro's number	—
N_i	Number of moles of molecular weight M_i	Eq. 2.2–1
Nu	Nusselt number	Eq. 5.4–33
n	Parameter in power-law viscosity model	Eq. 5.1–9
n'	Parameter in power-law normal stress coefficient	Eq. 5.2–53
n	Parameter in Carreau viscosity function	Eq. 5.1–13
$\boldsymbol{n}$	Unit normal vector	Fig. 1.1–1
$O(\epsilon)$	Of the order of magnitude of ϵ	Eq. 1.2–53
P	Set of generalized momenta	Eq. 2.3–3
$P_{N,L}(r)$	Distribution function for end-to-end vector of freely jointed chain	Eq. 2.3–23
$\mathscr{P}$	Modified pressure	Table 1.2–1
p	Pressure	Eq. 1.2–1
p	Extent of reaction	Eq. 2.2–12
p	Laplace transform variable	Eq. 6.3–2
p_a	Atmospheric pressure	Eq. 1.2–66
p_H	Hole pressure error	Fig. 3.4–1
Q	Set of generalized coordinates	Eq. 2.3–3
Q	Volume rate of flow	Eq. 1.2–8
$\boldsymbol{q}$	Heat flux vector	Table 1.1–1
R	Radius (of tube, sphere, disk, cylinder, etc.)	Fig. 1.2–2 Fig. 1.2–3 Fig. 1.2–6
$\boldsymbol{R}$	Dumbbell orientation vector	Eq. 2.5–1
R_1, R_2	Radii of cylinders in bipolar coordinates	Fig. A.7–1
$\boldsymbol{R}_\nu$	Position of the νth bead with respect to center of mass	Eq. 2.3–1

$\mathscr{Re}\{w\}$	Real part of the complex number w	—
r	Radial coordinate (in spherical and in cylindrical coordinates)	Fig. A.6–1
r	End-to-end distance of a polymer molecule	Eq. 2.3–1
$\boldsymbol{r}$	Position vector	Eq. 1.1–10 Fig. A.6–1 Eq. A.2–24
S	Surface area	Fig. 1.1–1
s	Radius of gyration of a macromolecule	Eq. 2.3–2
s	Reciprocal of power-law model n	Eq. 5.2–22
T	Temperature	Eq. 1.2–3
$\mathscr{T}$	Torque	Eq. 1.2–57
t	Time	Table 1.1–1
t'	Past time	Fig. 4.1–2
$\hat{U}$	Internal energy per unit mass	Table 1.1–1
V	Volume	Fig. 1.1–1
$\hat{V}$	Volume per unit mass (same as $1/\rho$)	Eq. 1B.15–1
$\boldsymbol{v}$	Velocity vector	Table 1.1–1 Eq. 9.2–3
$\boldsymbol{v}_\infty$	Approach velocity of fluid moving toward a submerged object	Fig. 1.2–3
W	Width (of slit, film, etc.)	Eq. 1.2–8
W	Angular velocity	Fig. 1.2–5
$\boldsymbol{w}$	Fluid angular velocity (one half the "vorticity")	Problem 1C.7 Eq. 7.1–4
w_i	Mass of the ith fraction	Eq. 2.2–2
X	$\cosh \xi + \cos \theta$ in bipolar coordinates	Eq. 5.3–9 Ex. 5 after §A.7
(x,t)	Label of fluid particle at position x_1, x_2, x_3 at time t	§7.1
x,y,z	Cartesian coordinates	—
$x_i' = x_i'(x,t,t')$	Displacement functions	Eq. 9.1–3
Z	Classical partition function	Eq. 2.3–8

Greek Symbols

α	Molecular expansion factor	Eq. 2.3–43
α	Thermal diffusivity	Eq. 5.4–17

α	Dimensionless parameter in Ellis model	Eq. 5.1–14
α	Dimensionless parameter used in time-constant expression	Eq. 6.1–15
α_1,α_2	Dimensionless constants in the Bird-Carreau model	Eq. 9.4–3
$\alpha_1,\alpha_2,\alpha_{12},\ldots$	Constants in the corotational retarded motion expansion	Eq. 8.4–3
$\beta_1,\beta_2,\beta_{12},\ldots$	Constants in a codeformational retarded motion expansion	Eq. 9.5–4
$\beta^1,\beta^2,\beta^{12},\ldots$	Constants in a codeformational retarded motion expansion	Eq. 9.5–5
$\Gamma(x)$	Gamma function with argument x ($= \int_0^\infty t^{x-1}e^{-t}\,dt$)	—
$\dot{\mathbf{\Gamma}}$	Corotating rate-of-strain tensor	Eq. 6.4–2 Table 7.1–2
$\ddot{\mathbf{\Gamma}}$	Time derivative of corotating rate-of-strain tensor	Eq. 7.2–40 Table 7.1–2
γ	Shear strain	Eq. 4.4–24 Fig. 4.4–21
$\dot{\gamma}$	Velocity gradient (in steady simple shear flow), shear rate (in curvilinear steady shear flow), magnitude of the $\dot{\boldsymbol{\gamma}}$ tensor	Eq. 4.1–1 Eq. 4.1–3 Eq. 4.1–4 Eq. 5.1–8
$\dot{\boldsymbol{\gamma}}$	Rate-of-strain tensor	Eq. 1.2–2
γ_∞	Ultimate recoil	Eq. 6.2–31
$\boldsymbol{\gamma}^{[0]},\boldsymbol{\gamma}_{[0]}$	Strain tensors	Eqs. 9.2–11, 12 Table 9.1–2
$\boldsymbol{\gamma}^{(n)},\boldsymbol{\gamma}_{(n)},\boldsymbol{\gamma}^{[n]},\boldsymbol{\gamma}_{[n]}$	Codeformational kinematic tensors	Table 9.1–2
$\mathbf{\Delta}$	Deformation tensor	Table 9.1–1 Eq. 9.1–5
δ	Displacement of cylinders in bipolar coordinates	Fig. A.7–1
δ	Thickness of film	Eq. 1B.14–1 Eq. 5.2–15
$\boldsymbol{\delta}$	Unit tensor with components δ_{ij}	Eq. 1.2–1 Eq. A.3–11
δ_{ij}	Kronecker delta	Eq. 2.3–15 Eqs. A.2–1,2
$\delta(x)$	Dirac delta function (*Note definition in Eq. 6.1–9a!!*)	Eq. 2.3–22
$\boldsymbol{\delta}_i$	Unit vectors in space-fixed frame	Eq. A.2–9

$\check{\boldsymbol{\delta}}_i$	Unit vectors in corotating frame	Eq. 7.1–10 Fig. 7.1–1
$\breve{\boldsymbol{\delta}}_i$	Unit vectors associated with shear axes	Eq. 4.1–5 Fig. 4.1–2
$\mathring{\boldsymbol{\delta}}_i$	Unit vectors associated with curvilinear coordinates	Eq. 7.6–2 Fig. 7.6–1
$\boldsymbol{\delta}_r$	Unit vector in radial direction	Fig. 1.2–3
$\boldsymbol{\delta}_i\boldsymbol{\delta}_j$	Unit dyads	Fig. A.3–1
$\mathbf{E}$	Deformation tensor	Table 9.1–1 Eq. 9.1–6
ϵ	Slit width divided by outer radius in an annulus ($= 1 - \kappa$)	Eq. 5.2–42
$\dot{\epsilon}$	Elongation rate	Eq. 1B.11–1
ϵ_{ijk}	Permutation symbol	Eqs. A.2–3,4, and 5
ζ	Bead friction coefficient	§2.5
$\zeta(n)$	Riemann zeta function of n	Eq. 6.2–11a
η	Shear-rate dependent viscosity	Eq. 2.4–1
η^*	Complex viscosity	Eq. 4.4–1 Eq. 6.2–6
η',η''	In-phase and out-of-phase components of η^*	Eq. 4.4–3 Eqs. 6.2–4 and 5
η^+	Stress-growth viscosity, or shear stress growth function	Eq. 4.4–18 Eq. 6.2–23
η^-	Stress-relaxation viscosity, or shear stress relaxation function	Eq. 4.4–21 Eq. 6.2–19
$\eta^\#$	Complex viscometric function	Table 7.5–1
$\bar{\eta}$	Elongational viscosity	Eq. 1B.11–3 Eq. 4.6–6
$\bar{\eta}^+$	Elongational stress growth function	Eq. 4.6–7
$\bar{\eta}^-$	Elongational stress relaxation function	Eq. 4.6–8
η_r	Relative viscosity	Eq. 2.4–1
η_s	Solvent viscosity	Eq. 2.4–1
η_0,η_∞	Zero-shear-rate viscosity, infinite-shear-rate viscosity	§4.3
η_k	Viscosity constants in generalized Maxwell model, generalized ZFD model	Eq. 6.1–10 Eq. 7.4–2
$[\eta]$	Intrinsic viscosity	Eq. 2.4–3
$[\eta'],[\eta'']$	Components of the intrinsic complex viscosity	Eqs. 4.4–16 and 17

Θ	Dimensionless temperature difference	§5.4
θ	Angle measured downwards from the z-axis (spherical coordinates); angle measured around the z-axis (cylindrical coordinates)	Fig. A.6–1
θ	Coordinate in bipolar coordinates	Eq. 5.3–8
κ	Dilatational viscosity	Eq. 1.2–1
κ	Ratio of inner to outer radius in annulus	Eq. 1B.2–1 Eq. 5.2–35
$\boldsymbol{\kappa}$	Tensor used in describing homogeneous flows	Eq. 7.1–8
λ	Time constant for rigid dumbbells	Table 2.5–1
λ	Time constant in Carreau viscosity function	Eq. 5.1–13
λ	Time constant formed from m, n, m', n'	Eq. 5.2–54
λ	Time constant in empirical relation	Eq. 6.1–15
λ	Time constant in Spriggs-Bird model	Problem 8C.2
λ_0	Time constant in Maxwell model, and in ZFD model	Eq. 6.1–4, §7.3, and footnote 4
λ_1,λ_2	Time constants in Jeffreys model, corotational Jeffreys model, Oldroyd models, Bird-Carreau model	Eq. 6.1–7 Eq. 7.3–1 Eq. 8.1–2 Eq. 9.4–3
λ_k	Time constants in generalized Maxwell model, generalized ZFD model, Rouse model	Eq. 6.1–10 Eq. 7.4–2 Table 2.5–1
λ_H	Time constant for Hookean dumbbells	Table 2.5–1
μ	Viscosity of Newtonian fluid	Eq. 1.2–1
μ_0	Parameter in Bingham model	Eq. 5.1–16
μ_0,μ_1,μ_2	Time constants in Oldroyd model	Eq. 8.1–2
ν	Kinematic viscosity	Eq. 1D.2–1
ν_1,ν_2	Time constants in Oldroyd model	Eq. 8.1–2
ξ	Coordinate in bipolar coordinates	Eq. 5.3–8
ξ	Dimensionless radial coordinate	—
$\prod$	Product symbol	—
π	3.14159...	—
$\boldsymbol{\pi}$	Total stress tensor ($p\boldsymbol{\delta} + \boldsymbol{\tau}$)	Table 1.1–1
$\boldsymbol{\pi}_n$	Stress on surface with orientation given by unit normal vector $\boldsymbol{n}$	Fig. 1.1–2 Eq. 7.2–16

P_{ij}	Rotation matrix elements	Eq. 7.2–1 Eq. 9.2–37
ρ	Density	Table 1.1–1
$\sum$	Summation symbol	—
σ_1, σ_2	Abbreviations used for special combinations of Oldroyd model parameters	Problem 8A.4 Eq. 8C.1–1
$\mathbf{T}$	Corotational stress tensor	Table 7.1–1
$\dot{\mathbf{T}}$	Time derivative of $\mathbf{T}$	Table 7.1–1
$\boldsymbol{\tau}$	Part of $\boldsymbol{\pi}$ that vanishes at equilibrium	Eq. 1.2–1
τ	Magnitude of $\boldsymbol{\tau}$	Eq. 5.2–31
τ_R	Shear stress at the tube wall (at $r = R$)	Eq. 5.2–4 Table 5.2–1
τ_B	Shear stress at slit wall (at $x = B$)	Table 5.2–1
$\tau_{1/2}$	Parameter in Ellis model	Eq. 5.1–14
τ_0	Parameter in Bingham model	Eq. 5.1–16
$\boldsymbol{\tau}^{(n)}, \boldsymbol{\tau}_{(n)}, \boldsymbol{\tau}^{[n]}, \boldsymbol{\tau}_{[n]}$	Codeformational stress tensors and their time derivatives	Table 9.1–1
$\tau_{11} - \tau_{22}$	Primary normal stress difference in viscometric flows	§3.2
$\tau_{22} - \tau_{33}$	Secondary normal stress difference in viscometric flows	§3.2
ϕ	Angle measured around the z-axis (spherical coordinates)	Fig. A.6–1
Ψ_1, Ψ_2	Primary and secondary normal stress coefficients	Eq. 4.3–2 and 3 Eq. 5.5–1
$\Psi_{1,0}, \Psi_{2,0}$	Zero-shear-rate normal stress coefficients	§4.3
Ψ_1^*, Ψ_2^*	Complex normal stress coefficients	Eqs. 4.4–8 and 10
$\Psi_1{}^d, \Psi_2{}^d$	Normal stress displacement coefficients	Eqs. 4.4–9 and 11
$\Psi'_1, \Psi''_1, \Psi'_2, \Psi''_2$	In-phase and out-of-phase components of Ψ_1^*, Ψ_2^*	Eqs. 4.4–12 and 13
Ψ_1^+, Ψ_2^+	Stress-growth normal stress coefficients	Eqs. 4.4–19 and 20
Ψ_1^-, Ψ_2^-	Stress-relaxation normal stress coefficients	Eqs. 4.4–22 and 23
$\psi(Q)$	Configuration-space distribution function	Eq. 2.3–4
$\psi(\boldsymbol{R},t)$	Dumbbell orientational distribution function	Eq. 2.5–1
$\boldsymbol{\Omega}$	The Oseen tensor	Eq. 1.2–53
$\boldsymbol{\Omega}$	Rotation tensor	Eq. 6.4–2 Eq. 7.1–10

ω	Frequency	Eq. 1D.2–2
$\boldsymbol{\omega}$	Vorticity tensor	Eq. 5.5–4 Eq. 7.1–3

Mathematical Operations

$\boldsymbol{\nabla}$	"Del" operator	Table 1.1–1 Eq. A.4–1 Eq. A.7–5, 9 Eq. A.8–20
D/Dt	Substantial or material derivative	Table 1.1–1
$\mathscr{D}/\mathscr{D}t$	Corotational or Jaumann derivative	Eq. 5.5–1 Eq. 7.1–9
$\mathfrak{d}/\mathfrak{d}t$	Convected derivative	Tables 9.1–1 and 2
$\Gamma(n)$	$\int_0^\infty t^{n-1} \exp(-t)\, dt$ or $(2/n)\int_0^\infty s \exp(-s^{2/n})\, ds$	
,	Comma notation for covariant differentiation	Eqs. A.8–21 through 24

Parentheses, Brackets, and Braces

( ),[ ],{ }	Scalar, vector and tensor products	Appendix A
$\langle\ \rangle$	Average over a flow cross section	Eq. 1.2–8
$\langle\ \rangle$	Average over phase space	Eq. 2.3–5
$\left\{ {i \atop jk} \right\}$	Christoffel symbols	Eq. A.8–17
$\mathscr{R}e\{w\}$	Real part of w	—
$\mathscr{I}m\{w\}$	Imaginary part of w	—
$\mathscr{L}\{f\}$	Laplace transform of f	—

Roman Numerals

$I_{\dot{\gamma}}, II_{\dot{\gamma}}, III_{\dot{\gamma}}$	Scalar invariants of $\dot{\boldsymbol{\gamma}}$	Eqs. 5.1–5 to 7; Eqs. A.3–20 to 22
$\overline{II}_{\dot{\gamma}}, \overline{III}_{\dot{\gamma}}$	Scalar invariants of $\dot{\boldsymbol{\gamma}}$	Eqs. A.3–23 and 24

Superscripts

†	Transpose of a second-order tensor or matrix	Eq. 1.1–10 Eq. A.3–9
$(n),[n]$	Index giving order of various kinematic tensors	Tables 9.1–1 and 2

Subscripts

$(n),[n]$	Index giving order of various kinematic tensors	Tables 9.1–1 and 2

s	Quantity associated with solvent	
0	Zero-shear-rate quantity	

Above Symbols

$\hat{}$	Per unit mass	Eq. 1.1–11
$\hat{}$	Quantities in codeforming coordinate system	Tables 9.1–1 and 2 §9.3
$\check{}$	Quantities in corotating frame	Tables 7.1–1 and 2
$\bar{}$	Laplace transform	Eq. 6.3–10
$\dot{}$	Time derivative	Eq. 1.2–62

AUTHOR INDEX

for Volume 1

SUBJECT INDEX

for Volumes 1 and 2